Lecture Notes in Computer Science

Lecture Notes in Artificial Intelligence   14481

Founding Editor

Jörg Siekmann

Series Editors

Randy Goebel, *University of Alberta, Edmonton, Canada*
Wolfgang Wahlster, *DFKI, Berlin, Germany*
Zhi-Hua Zhou, *Nanjing University, Nanjing, China*

The series Lecture Notes in Artificial Intelligence (LNAI) was established in 1988 as a topical subseries of LNCS devoted to artificial intelligence.

The series publishes state-of-the-art research results at a high level. As with the LNCS mother series, the mission of the series is to serve the international R & D community by providing an invaluable service, mainly focused on the publication of conference and workshop proceedings and postproceedings.

Andrea Campagner · Oliver Urs Lenz ·
Shuyin Xia · Dominik Ślęzak ·
Jarosław Wąs · JingTao Yao
Editors

Rough Sets

International Joint Conference, IJCRS 2023
Krakow, Poland, October 5–8, 2023
Proceedings

 Springer

Editors
Andrea Campagner
IRCCS Istituto Ortopedico Galeazzi
Milano, Italy

Oliver Urs Lenz
Ghent University
Ghent, Belgium

Shuyin Xia
Chongqing University of Posts
and Telecommunications
Chongqing, China

Dominik Ślęzak
University of Warsaw
Warsaw, Poland

JingTao Yao
University of Regina
Regina, SK, Canada

Jarosław Wąs
AGH University of Science
and Technology
Kraków, Poland

ISSN 0302-9743   ISSN 1611-3349   (electronic)
Lecture Notes in Artificial Intelligence
ISBN 978-3-031-50958-2   ISBN 978-3-031-50959-9   (eBook)
https://doi.org/10.1007/978-3-031-50959-9

LNCS Sublibrary: SL7 – Artificial Intelligence

This Springer imprint is published by the registered company Springer Nature Switzerland AG
The registered company address is: Gewerbestrasse 11, 6330 Cham, Switzerland

Paper in this product is recyclable.

Preface

This volume contains the papers selected for presentation at IJCRS 2023, the 2023 International Joint Conference on Rough Sets, held at AGH University of Kraków on October 5–8, 2023, in Kraków, Poland. Conferences in the IJCRS series, resulting from the merger of four separate conferences tying rough sets to various paradigms (RSCTC, data analysis; RSFDGrC, granular computing; RSKT, knowledge technology; and RSEISP, intelligent systems), are held annually: the first Joint Rough Set Symposium was held in Toronto, Canada, in 2007; followed by Symposiums in Chengdu, China in 2012; Halifax, Canada, 2013; Granada and Madrid, Spain, 2014; Tianjin, China, 2015, where the acronym IJCRS was proposed; continuing with the IJCRS 2016 conference in Santiago, Chile, IJCRS 2017 in Olsztyn, Poland, IJCRS 2018 in Quy Nhon, Vietnam, IJCRS 2019 in Debrecen, Hungary, IJCRS 2020 in La Habana, Cuba (held online), IJCRS 2021 in Bratislava, Slovakia (hybrid), and IJCRS 2022 in Suzhou, China (hybrid).

Following the success of the previous conferences, IJCRS 2023 continued the tradition of a very rigorous reviewing process. The 43 papers included in these proceedings were selected from 83 submissions. Every submission was reviewed by at least two Program Committee Members and domain experts. Additional expert reviews were sought when necessary. On average, each submission received three reviews. As a result, only top-quality papers were chosen for presentation at the conference. Final camera-ready submissions were further reviewed by Program Comittee Chairs and Conference Chairs. Some authors were requested to make additional revisions. We would like to thank all the authors for contributing their papers. Without their contribution, this conference would not have been possible.

The IJCRS 2023 program was further enriched by eight Keynote Speeches, among them the one presented by Tsau Young Lin, the Founding President of the International Rough Set Society (IRSS), and the Anniversary Talk by Andrzej Skowron, IRSS Fellow and former President, who celebrated his 80th birthday during the conference. We are grateful to our Keynote Speakers, Weronika Adrian, Joel Holland, Andrzej Janusz, Tianrui Li, Tsau Young Lin, Pradipta Maji, Sheela Ramanna, and Andrzej Skowron. The IJCRS 2023 program also hosted Special Sessions on "Innovative Foundational Models for Rough Sets, Approximate Reasoning, and Granular Computing" and "Data Analytics in Cybersecurity and IoT Applications", as well as a Panel on "Intelligent Informatics". We are grateful to the Special Session Organizers Stefania Boffa, A. Mani, Marcin Michalak, and Piotr Synak, to the Panelists Jimmy Huang, Duoqian Miao, and Hung Son Nguyen, as well as to the Panel Moderator Pawan Lingras.

We are also grateful to Springer for sponsoring the Best Student Paper Award: the IJCRS 2023 Program Committee Chairs and Conference Chairs assigned the Award to the article titled "Multi-granularity Feature Fusion for Transformer-Based Single Object Tracking", authored by Ziye Wang and Duoqian Miao. The Award was assigned based on a competitive process, taking into account both scientific excellence and clarity of presentation, and was awarded during the Open Meeting of the International Rough Set Society.

IJCRS 2023 would not have been successful without the support of many people and organizations. We are indebted to the Program Committee Members and external reviewers for their effort and engagement in providing a rich and rigorous scientific program. We greatly appreciate the co-operation, support, and sponsorship of various institutions, companies, and organizations, including the AGH University of Kraków, the Strategic Partners QED Software and DeepSeas, Honorary Patronage of the Polish Ministry of Science and Higher Education and of the Mayor of Kraków, as well as the International Rough Set Society. We acknowledge the use of the Springer EquinOCS conference system for paper submission and review. We are also grateful to Springer for publishing the proceedings as a volume of LNCS/LNAI.

Last but not least, we would like to thank Anna Smyk, Tomasz Hachaj, and the whole technical organization team at the AGH University of Kraków, for their great support and endless hours spent on the conference preparations.

November 2023

Andrea Campagner
Oliver Urs Lenz
Shuyin Xia
Dominik Ślęzak
Jarosław Wąs
JingTao Yao

Organization

Honorary Chairs

Andrzej Skowron   Systems Research Institute of Polish Academy of
     Sciences, Poland
Tomasz Szmuc   AGH University of Kraków, Poland
Yiyu Yao   University of Regina, Canada

Conference Chairs

Dominik Ślęzak   University of Warsaw, Poland
Jarosław Wąs   AGH University of Kraków, Poland
JingTao Yao   University of Regina, Canada

Program Committee Chairs

Andrea Campagner   IRCCS Istituto Ortopedico Galeazzi, Italy
Oliver Urs Lenz   Ghent University, Belgium
Shuyin Xia   Chongqing University of Posts and
     Telecommunications, China

Program Committee Members

Qiusheng An   Shanxi Normal University, China
Piotr Artiemjew   University of Warmia and Mazury in Olsztyn, Poland
Nouman Azam   National University of Computer and Emerging
     Sciences, Pakistan
Mohua Banerjee   Indian Institute of Technology, India
Jan Bazan   University of Rzeszów, Poland
Urszula Bentkowska   University of Rzeszów, Poland
Stefania Boffa   University of Milano-Bicocca, Italy
Henri Bollaert   Ghent University, Belgium
Joaquín Borrego-Díaz   Universidad de Sevilla, Spain
Andrea Campagner   IRCCS Istituto Ortopedico Galeazzi, Italy
Yuming Chen   Xiamen University of Technology, China
Hongmei Chen   Southwest Jiaotong University, China
Zehua Chen   Taiyuan University of Technology, China
Davide Ciucci   University of Milano-Bicocca, Italy
Chris Cornelis   Ghent University, Belgium
Jianhua Dai   Hunan Normal University, China
Tingquan Deng   Harbin Engineering University, China
Thierry Denoeux   Université de Technologie de Compiègne, France

Murat Diker	Hacettepe University, Turkey
Shifei Ding	China University of Mining and Technology, China
Barbara Dunin-Kęplicz	University of Warsaw, Poland
Soma Dutta	University of Warmia and Mazury in Olsztyn, Poland
Hamido Fujita	Iwate Prefectural University, Japan
Can Gao	Shenzhen University, China
Yang Gao	Nanjing University, China
Anna Gomolinska	University of Białystok, Poland
Salvatore Greco	University of Catania, Italia
Jerzy Grzymala-Busse	University of Kansas, USA
Shen-Ming Gu	Zhejiang Ocean University, China
Christopher Henry	University of Winnipeg, Canada
Mengjun Hu	University of Regina, Canada
Qinghua Hu	Tianjin University, China
Xuegang Hu	Hefei University of Technology, China
Bing Huang	Nanjing Audit University, China
Amir Hussain	Edinburgh Napier University, UK
Masahiro Inuiguchi	Osaka University, Japan
Ryszard Janicki	McMaster University, Canada
Andrzej Janusz	University of Warsaw, Poland
Richard Jensen	Aberystwyth University, UK
Xiuyi Jia	Nanjing University of Science and Technology, China
Chunmao Jiang	Fujian University of Technology, China
Bin Jie	Hebei Normal University, China
Olha Kaminska	Ghent University, Belgium
Aquil Khan	Indian Institute of Technology, Indore, India
Marzena Kryszkiewicz	Warsaw University of Technology, Poland
Mihir Kumar Chakarborty	Jadavpur University, India
Guangming Lang	Changsha University of Science and Technology, China
Oliver Urs Lenz	Ghent University, Belgium
Mikołaj Leszczuk	AGH University of Kraków, Poland
Deyu Li	Shanxi University, China
Fanchang Li	Soochow University, China
Huaxiong Li	Nanjing University, China
Jinhai Li	Kunming University of Science and Technology, China
Jinjin Li	Minnan Normal University, China
Kewen Li	China University of Petroleum, China
Lei-Jun Li	Hebei Normal University, China
Tianrui Li	Southwest Jiaotong University, China
Tong-Jun Li	Zhejiang Ocean University, China
Yuefeng Li	Queensland University of Technology, Australia
Jiuzhen Liang	Changzhou University, China
Shujiao Liao	Minnan Normal University, China
Guoping Lin	Minnan Normal University, China
Yaojin Lin	Hefei University of Technology, China

Pawan Lingras	St. Mary's University, Canada
Baoxiang Liu	North China University of Science and Technology, China
Caihui Liu	Gannan Normal University, China
Dun Liu	Southwest Jiaotong University, China
Guilong Liu	Beijing Language and Culture University, China
Wenqi Liu	Kunming University of Science and Technology, China
Pradipta Maji	Indian Statistical Institute, India
A. Mani	Indian Statistical Institute, India
Jesús Medina	University of Cádiz, Spain
Jusheng Mi	Hebei Normal University, China
Duoqian Miao	Tongji University, China
Marcin Michalak	Łukasiewicz Research Network - Institute of Innovative Technologies EMAG, Poland
Fan Min	Southwest Petroleum University, China
Mikhail Moshkov	King Abdullah University of Science and Technology, Saudi Arabia
Hung Son Nguyen	University of Warsaw, Poland
Sinh Hoa Nguye	Polish-Japanese Academy of Information Technology, Poland
Agnieszka Nowak-Brzezińska	University of Silesia in Katowice, Poland
Krzysztof Pancerz	John Paul II Catholic University of Lublin, Poland
Witold Pedrycz	University of Alberta, Canada
Shenglei Pei	Qinghai Nationalities University, China
Daniel Peralta	Ghent University, Belgium
Georg Peters	Munich University of Applied Sciences, Germany
James Peters	University of Manitoba, Canada
Lech Polkowski	Polish-Japanese Institute of Information Technology, Poland
Jianjun Qi	Xidian University, China
Jin Qian	Jiangsu University of Technology, China
Yuhua Qian	Shanxi University, China
Taorong Qiu	Nanchang University, China
Sheela Ramanna	University of Winnipeg, Canada
Zbigniew Ras	University of North Carolina at Charlotte, USA
Sergio Ribeiro	Pontificia Universidade Católica do Paraná, Brazil
Marek Reformat	University of Alberta, Canada
Henryk Rybiński	Warsaw University of Technology, Poland
Hiroshi Sakai	Kyushu Institute of Technology, Japan
Lin Shang	Nanjing University, China
Ming-Wen Shao	Chinese University of Petroleum, China
Yanhong She	Xian Shiyou University, China
Marek Sikora	Silesian University of Technology, Poland
Andrzej Skowron	Systems Research Institute of Polish Academy of Sciences, Poland

Dominik Ślęzak	University of Warsaw, Poland
Roman Słowinski	Poznań University of Technology, Poland
Jingjing Song	Macau University of Science and Technology, China
Łukasz Sosnowski	Systems Research Institute of Polish Academy of Sciences, Poland
Urszula Stańczyk	Silesian University of Technology, Poland
Jaroslaw Stepaniuk	Białystok University of Technology, Poland
Lin Sun	Henan Normal University, China
Zbigniew Suraj	University of Rzeszów, Poland
Piotr Synak	DeepSeas, Switzerland
Marcin Szczuka	University of Warsaw, Poland
Marcin Szeląg	Poznań University of Technology, Poland
Anhui Tan	Zhejiang Ocean University, China
Adnan Theerens	Ghent University, Belgium
Shusaku Tsumoto	Shimane University, Japan
Alicja Wakulicz-Deja	University of Silesia in Katowice, Poland
Renxia Wan	North Minzu University, China
Baoli Wang	Yuncheng University, China
Changzhong Wang	Bohai University, China
Guoyin Wang	Chongqing University of Posts and Telecommunications, China
Piotr Wasilewski	Systems Research Institute of Polish Academy of Sciences, Poland
Lai Wei	Tongji University, China
Ling Wei	Northwest University, China
Wei Wei	Shanxi University, China
Zhihua Wei	Tongji University, China
Marcin Wolski	Maria Curie-Skłodowska University, Poland
Wei-Zhi Wu	Zhejiang Ocean University, China
Shuyin Xia	Chongqing University of Posts and Telecommunications, China
Bin Xie	Hebei Normal University, China
Jun Xie	Taiyuan University of Technology, China
Jianfeng Xu	Nanchang University, China
Jiucheng Xu	Henan Normal University, China
Weihua Xu	Southwest University, China
Zhan-Ao Xue	Henan Normal University, China
Lin Xun	Henan Normal University, China
Hailong Yang	Shaanxi Normal University, China
Jilin Yang	Sichuan Normal University, China
Tian Yang	Hunan Normal University, China
Xibei Yang	Jiangsu University of Science and Technology, China
Xin Yang	Southwestern University of Finance and Economics, China
JingTao Yao	University of Regina, Canada
Yiyu Yao	University of Regina, Canada

Dongyi Ye	Fuzhou University, China
Hong Yu	Chongqing University of Posts and Telecommunications, China
Ying Yu	East China Jiaotong University, China
Xiaodong Yue	Shanghai University, China
Zhang Zehua	Taiyuan University of Technology, China
Yanhui Zhai	Shanxi University, China
Jianming Zhan	Hubei University for Nationalities, China
Hongying Zhang	Xi'an Jiaotong University, China
Hongyun Zhang	Tongji University, China
Li Zhang	Soochow University, China
Nan Zhang	Yantai University, China
Qinghua Zhang	Chongqing University of Posts and Telecommunications, China
Xianyong Zhang	Sichuan Normal University, China
Xiaohong Zhang	Shaanxi University of Science and Technology, China
Yan Zhang	Cal State University, San Bernardino, USA
Yanping Zhang	Anhui University, China
Zehua Zhang	Taiyuan University of Technology, China
Shu Zhao	Anhui University, China
Huilai Zhi	Henan Polytechnic University, China
Caiming Zhong	Ningbo University, China
Bing Zhou	Sam Houston State University, USA
Jie Zhou	Shenzhen University, China
Xianzhong Zhou	Nanjing University, China
Beata Zielosko	University of Silesia in Katowice, Poland
Li Zou	Liaoning Normal University, China

External Reviewers

Błażej Adamczyk	EFIGO, Poland
Seiki Akama	C-Republic, Japan
Tareq Alshami	Sana'a University, Yemen
Janusz Borkowski	DeepSeas, Poland
Duarte Folgado	Fraunhofer AICOS, Portugal
Brunella Gerla	University of Insubria, Italy
Arun Kumar	Indian Institute of Technology Kanpur, India
Nicolás Madrid	University of Malaga, Spain
Krishna Balajirao Manoorkar	Vrije Universiteit Amsterdam, The Netherlands
Petra Murinová	University of Ostrava, Czech Republic
Yotaro Nakayama	BIPROGY, Japan
Agnieszka Nowak-Brzezińska	University of Silesia, Poland
Eloísa Ramírez Poussa	Universidad de Cádiz, Spain
Binbin Sang	Southwest Jiaotong University, China

Anniversary Talk

Rough Sets in Interactive Granular Computing: Toward Foundations for Intelligent Systems Interacting with Human Experts and Complex Phenomena

Andrzej Skowron[1,2] and Dominik Ślęzak[2,3,4]

[1] Systems Research Institute, Polish Academy of Sciences, ul. Newelska 6, 01-447, Warsaw, Poland
skowron@mimuw.edu.pl
[2] QED Software Sp. z o.o., ul. Miedziana 3A m. 18, 00-814, Warsaw, Poland
[3] Institute of Informatics, University of Warsaw, ul. Banacha 2, 02-097, Warsaw, Poland
[4] DeepSeas Inc., 12121 Scripps Summit Drive Suite #320, San Diego, CA, 92131, USA

Abstract. We present the current research on the Interactive Granular Computing (IGrC) model and its relationships with the human-computer interaction processes. The existing rough set approaches to approximation of concepts grounded on (partial) containment of sets are extended, for the purposes of Intelligent Systems (IS's) interacting with human experts and complex phenomena, to approximations based on compound reasoning aiming to generate the right decisions about perceived situations in the real world. This paper is a step toward developing the foundations of such IS's. The decisions of IS's are constructed along the reasoning performed by complex granules (c-granules) which are responsible for creating interfaces between informational layers and physical layers of IS's, often synchronized or learnt from the reasoning performed by humans. Depending on applications, the decisions may take different forms, *e.g.*: compound decisions represented by the collections of decisions made in a given period of time, specifications of compound structural physical objects satisfying the wanted properties, (parameterized) learning algorithms generating high quality classifiers from samples of objects, pipelines of computations preserving some given constraints, etc. Both the construction of compound decisions and reasoning are performed over information perceived by means of c-granules used by IS's as interfaces for interactions with the physical world. Such interactions of c-granules are realized by the control layer (control, in short) of these granules. The discussed approximation of complex concepts in the context of IGrC is of fundamental importance for developing foundations of IS's aiming to solve complex problems.

Keywords. Interactions · (Interactive) Granular computing · Perception · Reasoning (Judgment) · Complex granule · Informational layer · Physical layer ·

Research co-funded by Polish National Centre for Research and Development (NCBiR) grant no. POIR.01.01.01-1070/21-00.

Informational granule · Network of c-granules · Control of c-granule · Rule module · Implementational module · (Inference) Reasoning module · Decomposition module · Rough sets in IGrC

1 Introduction

The IGrC model originated as an attempt in searching for the relevant computing model for different domains such as Cyber Physical Systems, Internet of Things, Wisdom Web, Society 5.0, Modeling of Complex Adaptive Systems, Natural Computing, Multiscale Modeling or Self-Organization. In this paper, we present some new results on the research reported in [42] and related to IGrC (see also [9, 10, 18, 41, 43], the papers on IGrC[1], as well as keynote and plenary talks at: IPMU 2022 (title: *Perceptual Rough Set Approach in Interactive Granular Computing*), FedCSIS 2022 (title: *Rough Sets Turn 40: From Information Systems to Intelligent Systems*), IEEE IS 2022 (title: *Rough sets turn 40: What next?*), and OLAB 2022 (title: *Decision Support in Problem Solving by Intelligent Systems Based on Interactive Granular Computing*)).

One of fundamental aspects of IGrC is the structure of the control layer (control, in short) of complex granules (c-granules). We will discuss its components such as the implementational module and the (inference) reasoning module. We will also discuss the role of reasoning performed by the control over computations composed out of networks of c-granules and generation of compound decisions along the reasoning pipelines. Finally, we will outline the rough set approach within IGrC and emphasize the role of reasoning while constructing approximations of concepts by c-granules.

One more motivation for introducing the IGrC model was to establish a new computing framework for Intelligent Systems (IS's) that deal with complex phenomena in many different areas such as multi-agent systems, robotics, cognitive science, machine learning, computational intelligence, swarm intelligence, and complex (adaptive) systems [3, 12, 13, 15, 16, 23, 24, 38, 39, 49]. Accordingly, we present some ideas for developing the IGrC model as the basis for IS's that would be able to interact with complex phenomena and human experts. In our discussions, we refer to some relevant opinions of the Turing Award winners – Frederick Brooks [5] and Leslie Valiant[2]:

> *A fundamental question for artificial intelligence is to characterize the computational building blocks* [complex granules] *that are necessary for cognition.*

One should also note the current discussion about necessity of modifying the Turing test to put in sync the issues of language and reasoning, as well as perception and action [28, 46]. Accordingly, the computing model we are searching for cannot be based on abstract granules only as it is done in Granular Computing (GrC) (see, *e.g.*, [21, 36, 37]) because dealing with perception requires interactions with physical objects too. Hence, in IGrC, so-called informational granules (ic-granules) are used to link abstract

---

[1] see, *e.g.*, https://dblp.org/pid/s/AndrzejSkowron.html.

[2] http://people.seas.harvard.edu/ ~ valiant/researchinterests.htm.

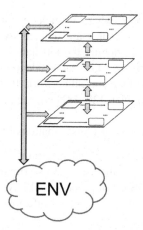

Fig. 1. Hierarchical construction of complex granules (c-granules) in continuous interaction with the physical world/environment.

and physical objects. We discuss how networks of such ic-granules are generated and changed by the control of c-granules during perceiving real world situations.

Moreover, the drawbacks of classical mathematical modeling in the case of complex phenomena (pointed out *e.g.*, by Brooks) impose the requirements on the control to be in continuous interaction with the physical world to be ready for perceiving the important changes in this world and adapt the behavior of c-granule to the observed changes (see Fig. 1). The physical world includes also human users of IS's and human experts who attempt to help IS's, often being a kind of "interface" between the system and the complex physical phenomena. One of the aspects of such cooperation between humans and IS's is in the area of data governance and data quality, *e.g.*, following the ideas of interactive data labeling and active learning [19].

The meta-equation presented in Eq. 1 characterizes the IGrC model. Let us emphasize that this model is not closed within the abstract space, contrary to the Granular Computing (GrC) model.

$$
\begin{aligned}
\text{IGrC} \; = \; \text{GrC} \; + \; & \text{INTERACTION OF PHYSICAL OBJECTS} \\
+ \; & \text{PERCEPTION} \\
+ \; & \text{REASONING (JUDGMENT)}
\end{aligned}
\tag{1}
$$

For the IGrC model networks of ic-granules [42] are of special importance. They are used by the control of c-granules for perceiving the properties of physical objects and their interactions. For a given moment of local time of c-granule, the network of ic-granules represents, in its informational layer, information about the currently perceived situation. The information is distributed between different currently used by c-granule specifications of spatio-temporal windows (addresses). Using these specifications, the control of c-granule is creating links (physical pointers) to the physical objects (parts of the physical space) corresponding to these specifications in the physical space. Any such specification w of spatio-temporal window describes a region

$\|w\|$ in $\Re^3$ of the physical space where is included the object o_w pointed out by
w. Changes of networks of ic-granules are defined by rules of the control. If the
currently perceived situation is matching the left hand side of a given rule then it is
pointed out the decision from the right hand side of the rule that has a form of
specification of transformation of the current network. It should be noted that reasoning
making it possible to resolve conflicts [29, 48] between different rules matching a given
situation should be used.

We distinguish two main kinds of transformations of networks of ic-granules:

(i) transformations aiming at enriching understanding of the currently perceived situation in the physical world with the use of transformations which in this case can be
treated as a kind of measurement of properties of physical objects or inference rules
as well as aggregation of already perceived information through some other windows and

(ii) transformations generating new decisions on the basis of the perceived history of
networks, *e.g.*, by aggregating already made decisions (*e.g.*, being constructions of
objects) into more compound ones toward receiving the trajectory of granular
computation over networks (of ic-granules) satisfying the given requirements.

In this way, along the reasoning process, there is realized the construction process
of compound decisions. They may be simple lists of decisions already performed in a
given period of time (plans [39]) or they can be in the form of compound structural
objects such as learning algorithms or specifications of compound physical objects such
as a new medicine, compound sensors or actuators. One may assume that the control
has at its disposal knowledge data bases for storing reasoning tools and tools for
generation of new decisions. It is worth mentioning that usually the control is making
decisions concerning selection of the right transformations of networks on the basis of
compressed information representing the relevant information in the current networks
rather than on the basis of the whole information represented in the network. This is
due to the high computational complexity of reasoning based on the whole networks.

An important problem refers to structural modeling of networks of ic-granules.
These networks may have hierarchical and nested structures and are compound
dynamical objects. In particular, the scope of such networks (understood as a specification of an area in the physical space on the basis of which their behavior is modeled
or discovered) should be robust (up to satisfactory degree) to often unexpected interactions with the environment and the control should be aware when unexpected
changes are becoming much higher than the expected ones (see Fig. 2).

Finally, we outline the rough set approach [31, 33, 34] grounded in the IGrC model.
In particular, approximations of concepts based on reasoning over networks of ic-
granules are far more general than the traditionally used approximations based on set
containment only [30, 31, 33, 34]. From the point of view of rough sets, the novelty
of the proposed approach is in application of the IGrC model as the basis for the design
of interfaces between the abstract informational world and the physical perceived
world. This is realized in IGrC by c-granules. These c-granules are making it possible
to continuously update information about the currently perceived situation in the
physical world what leads to better understanding the perceived situation in the
physical world. The new perceived information is granulated and aggregated with the
information stored so far in informational layers leading to discovery of the relevant

ic-granules and their networks which can be treated as computational building blocks for cognition; in the case of the rough set approach these blocks are interpreted as the relevant patterns for approximation of (complex and vague) concepts. An immediate consequence from the above considerations is that rough sets should turn into adaptive rough sets and should be supported by methods of reasoning related to data governance supporting answering queries concerning *When, How, What, Where* new data should be perceived. In the paper we also demonstrate that new reasoning methods controlling computations over networks of ic-granules are crucial in searching for approximations of complex vague concepts related to complex phenomena in the physical world.

The rest of the paper is organized as follows. In Sect. 2 some motivations for developing new computing model are discussed. Issues related to structures of networks of ic-granules are discussed in Sect. 3. Different issues of reasoning over networks of ic-granules and generation along the reasoning processes of decisions are included in Sect. 4. The outline of the rough set based approach to approximation of concepts in the framework of IGrC is discussed in Sect. 5. Finally, we present conclusions and some remarks about the future research road map.

2   Motivations – Some Comments

In the cited before references about IGrC there are widely discussed issues related to motivation for developing the IGrC model. Here, we would like to recall some of them especially important for our considerations. These are mentioned in Sect. 1 opinions which we now present in more detail.

The first one is the opinion of Frederick Brooks [5], the Turing award winner who on the basis of his experience with large projects in software engineering has written:

> *Mathematics and the physical sciences made great strides for three centuries by constructing simplified models of complex phenomena, deriving, properties from the models, and verifying those properties experimentally. This worked because the complexities ignored in the models were not the essential properties of the phenomena. It does not work when the complexities are the essence.*

This opinion is very important for IS's interacting with complex phenomena. It suggests that we do not have modeling tools for modeling IS's systems based on classical mathematical modeling only. Models developed on the current information about the perceived situations cannot be closed in the abstract space only. IS's interacting with complex phenomena should be in continuous interaction with the real world to be ready for adaptation of the currently used models. This is illustrated in Fig. 1 where hierarchical learning is illustrated with interactions of each hierarchical level with the physical environment. On each level, the discovery process of computation building blocks for cognition (granules) is running concerning the relevant relational structures as well as the relevant language of features as well as the satisfiability relation (contrary to mathematical logic where these entities are assumed to be given!).

The next opinion important for our consideration concerns the current discussion on Turing test [28]:

> *The Turing test, as originally conceived, focused on language and reasoning; problems of perception and action were conspicuously absent. The proposed tests will provide an opportunity to bring four important areas of AI research (language, reasoning, perception, and action) back into sync after each has regrettably diverged into a fairly independent area of research.*

From this opinion it follows that the right computing model for IS's cannot be closed in the abstract (mathematical) space only because for dealing with perception it is not possible to avoid issues related to dealing with physical objects, in particular interaction with these physical objects. Hence, ic-granules in IGrC were introduced linking abstract and physical objects. One should also note that the issues related to understanding the concept of interaction is central for IS's dealing with complex phenomena [27]:

> *[...] interaction is a critical issue in the understanding of complex systems of any sorts: as such, it has emerged in several well-established scientific areas other than computer science, like biology, physics, social and organizational sciences.*

3 Networks of Ic-Granules

Let us start from the basic postulates for IGrC. We assume:

- Physical objects exist in the physical space and are embedded into its parts. This is related to [14].
- Physical objects are interacting in the physical space, and thus some collections of physical objects may create dynamical systems in the physical space.
- Some properties of physical objects or their configurations as well as their interactions can be perceived by c-granules.

There are several issues related to these postulates. These issues are related to such queries as (see Fig. 2):

- How to perceive properties of these objects and their interactions ? (Where, how, when to do this).
- How to generate the relevant configurations of physical objects and modify interactions between them?
- How to ensure that the perceived properties should be robust (to a high degree) with respect to unpredictable interactions from the environment? (see [16])

We also use some postulates related to c-granules and ic-granules such as:

- the control of c-granule can
  - access some physical objects directly (these objects are directly accessible);
  - encode some information into some directly accessible physical objects;
  - decode some properties of directly accessible physical objects into its informational layer $inf\_layer(g)$;

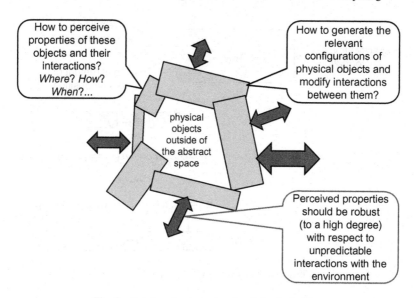

How to perceive properties of these objects and their interactions? *Where? How? When?...*

How to generate the relevant configurations of physical objects and modify interactions between them?

physical objects outside of the abstract space

Perceived properties should be robust (to a high degree) with respect to unpredictable interactions with the environment

Fig. 2. Queries about a network of physical objects.

- any ic-granule g has its informational layer $inf\_layer(g)$ and physical layer $ph\_layer(g)$:
  - informational layer $inf\_layer(g)$ consists of specifications of spatio-temporal windows w labeled by specifications tr of transformations assigned to them by the control, as well as already perceived information inf related to realization of these transformations;
  - physical layer $ph\_layer(g)$ of ic-granule g consists of physical objects pointed out by specifications of spatio-temporal windows from informational layer;
- any ic-granule in the current network of ic-granules has a distinguished *scope* being a specific spatio-temporal window; the scope of c-granule (at a given moment of time) is a union of scopes of all ic-granules of the network (at this moment of time);
- all physical objects pointed out by specifications of transformations labeling spatio-temporal windows included in the scope of a given ic-granule create three collections of physical objects: *soft_suit* consist of objects which are directly accessible by the control, *hard_suit* consists of target objects for the considered ic-granule and *link_suit* with objects responsible for transmission of interactions between *soft_suit* and *hard_suit*;
- the scopes of ic-granules in the current network of c-granule are defined by the control in such a way that the behavior of network is robust to interactions with the environment (up to satisfactory degree), *i.e.*, behavioral invariants of the network required by the control are preserved under changes caused by these interactions, if only they are below a given threshold;
- the control of c-granule can use its domain knowledge, physical laws (represented by some information granules) and already perceived information represented in informational layer to infer some properties of physical objects from the scope of

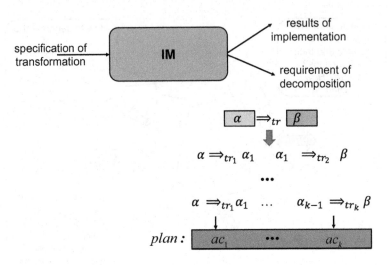

Fig. 3. Implementational module (IM).

this c-granule; hence, computations realized by c-granules are dependent on physical laws and domain knowledge as it is currently required for unconventional computing models (see, *e.g.*, [7]).

Below we include some additional comments which can help the reader to understand the role of the control of c-granule and networks of ic-granules.

The control of c-granule is able to control construction of the relevant configurations of physical objects and perceive some properties of these configurations creating dynamical systems.

The control of a given c-granule defines a specification of a spatio-temporal window, called the scope of transformation, describing a part of the physical space where the transformation should be implemented. Any such specification concerns changes which should be performed in the current network of ic-granules. In particular, this specification of transformation is expressing that some ic-granules should be eliminated, some of them should be suspended, some new ic-granules should be created, and the remaining are continuing gathering information perceiving by interaction with the physical world. As a part of the transformation specification is provided a specification of expected results of its implementation in the physical world such as the range of the expected results or a period in which it is expected that the results will be generated.

The specifications are sent by the control to its implementational module (IM) (see Fig. 3). If it is possible for IM to directly implement the specification in the physical world then ML is realizing this, otherwise ML requires from the control to decompose the specification up to the level when it will be possible to realize the specification directly in the physical space. In this way, ML is responsible for realization of so called physical semantics. In Fig. 3 the result of decomposition is illustrated as a linear plan. However, in general it can have a much more compound structure, *e.g.*, by taking into account issues of aggregation of partial results during decomposition. The reader can find more details on decomposition in [42].

Decomposition is producing a family of specifications of spatio-temporal windows and specifications of transformations required in parts of the physical space corresponding to these specifications. One should note that information labeling a window may be obtained by aggregation of information labeling other windows included in this window. Information labeling windows in a given network is updated by interaction with the physical space or as a result of reasoning with the use of domain knowledge bases or physical laws being at the disposal by the control.

The control may fix some initial conditions for interactions of perceived physical objects by encoding the relevant information into directly accessible physical objects.

Information perceived or inferred by the control related to a given spatio-temporal window is labeling this window. As it was mentioned, together with specifications are also given expected results of their realization. The expected results are compared by the control with the real ones obtained during implementation in the relevant periods of time. The results of such comparisons are used by reasoning module of the control of c-granule in the process of selection of the right transformation for realization by the control preceded by adaptation of the current set of rules, if necessary.

The control of c-granule transforms the current network of ic-granules into the new one by selecting the relevant rules from a given set of rules stored in the module of inference. Any rule is of the form $\alpha \Longrightarrow tr : \beta$ where α is a condition defined over information encoded in the informational layer of the network, tr is a specification of network transformation, and β describes expected results after realization of tr in the physical world. If α is satisfied by information from the current network then tr is a candidate for realization by IM. The module IM is resolving conflicts between such candidates to select the network transformation to be realized in the current situation.

It should be noted that in general ic-granules in the networks may have their own control. For simplicity of reasoning, in this paper we consider networks in which ic-granules (generated by the control of a given c-granule) have an empty control. The basic concepts related to c-granules and ic-granules are summarized in Table 1 and Table 2, and illustrated in Fig. 4 and Fig. 5. In general, the structure of the control may be much compound and contain many other modules (see Fig. 7).

Network of ic-granules (or ic-granule) contains a family of physical objects in parts of the physical space to which pointers are described by specifications of spatio-temporal windows (addresses) represented in informational layer of the network (of ic-granule). During activity of ic-granule, its specifications of spatio-temporal windows are also labeled by information perceived on physical objects corresponding to these windows or inferred from already perceived information.

Networks of ic-granules are changing with time. Hence, they are dynamical objects with states represented by networks of ic-granules. One should note that the networks of ic-granules are not purely mathematical (abstract) objects because they contain physical objects too. This allow us to deal in the IGrC model with issues of perception and action consistent with the opinion presented in the book [26]:

> [...] *The main idea of this book is that perceiving is a way of acting. It is something we do. Think of a blind person tap-tapping his or her way around a cluttered space, perceiving that space by touch, not all at once, but through time, by skillful probing and movement. This is or ought to be, our paradigm of what perceiving is.*

C-GRANULE: INTUITION

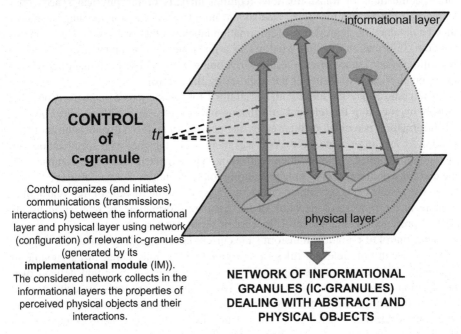

Fig. 4. A c-granule and a network of informational granules (ic-granules).

It should be noted that networks of ic-granules and c-granules can be treated as a higher order ic-granules. One can also consider societies of c-granules as higher order ic-granules. The details of higher order granules will be discussed in our next paper.

4 Reasoning over Computations on Networks of Ic-Granules and Generation of Compound Decisions Along the Reasoning Paths

The control of c-granule should be equipped with advanced reasoning tools aiming to discover the right specifications of transformations at each step of computation, for recognizing the necessity of deeper understanding the perceived situation or for aggregation the results of previously realized transformations. In reasoning different logics or ensembles of logics can be involved aiming to extend the current information about the perceived situation. Below we present an exemplary list of tasks for reasoning:

- construction from vague specification of compound objects (*e.g.*, classifiers, learning algorithms, new medicines, compound molecules) of the high quality;
- estimation of risk in realization of the selected decisions;

IC-GRANULE: INTUITION

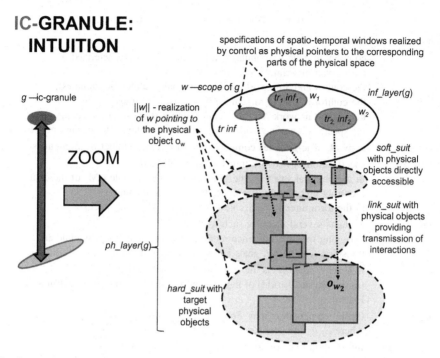

Fig. 5. Details of one of ic-granules that compose the network of ic-granules illustrated in Fig. 4.

- selection/sampling form large data sets representative and informative objects and representative and informative samples of data objects which are satisfactory for inducing classifiers with as high as possible quality of classification;
- labeling processes of the selected objects in interaction with experts or teams of experts [19];
- modification of interactions in perception of the currently perceived situation; data governance (see, *e.g.* [6]); searching for new sources of data; changing of behavior, *e.g.*, by adaptation of rules of transformation of configurations;
- implementation of configuration specifications and their changes in the physical space, estimation of expected results and perceiving real results of implementation; measuring of differences between expected and real results of implementation;
- adaptation of rules of transformation of configurations of objects and learning rules of transformation of configurations of objects;
- the control of interactions (*e.g.*, initiation, termination, suspension, creating new);
- interaction with humans (*e.g.*, evaluation of expert's competence, reliability, approximation of expertise domain of experts, negotiations and resolving conflicts between experts, planning interactions by taking into account costs of dialogue with experts or sensor measurements, checking availability of experts);
- communication/dialogue with domain experts in the case of compound labeling functions (*e.g.*, explanation of compound structures of objects and/or labels).

Table 1. Notation used in this article

Name	Interpretation
g_c	complex granule (c-granule) is a dynamic object characterized (at a given moment of local time of g_c) by the control and a network of informational granules (ic-granules); the control is aiming to achieve the goals of g_c transforming the current network N of ic granules into a new one by selecting the specification of network transformation tr (relevant to the current network)
N	network of ic-granules composed out of a finite number of ic-granules
tr	specification of network transformation realized in the physical space by the implementational module (IM) of the control (possibly preceded by decomposition of specification to the form directly realizable by IM in the physical space)
$control$	contains several modules such as reasoning module, inference module or implementational module and transition relation rel, goals and specification of family of networks Fam_net
rel	if N rel N' then N' is the real result of realization (in the physical world) of the selected transformation by the control at N (see Fig. 6)
$comp$	(finite) computation over Fam_net: N_1 rel $N_2 \dots$ rel N_k, where $N_i \in Fam\_net$ for $i = 1, \dots, k$
$trace\ (comp)$	information trace of $comp$: $inf\_l(N_1), \dots, inf\_l(N_k)$, where $inf\_l(N_i)$ is the information layer of N_i for $i = 1, \dots, k$
$goal$	goal of g_c interpreted by the control as a quality (utility) function over computations with values in $[0, 1]$
w	specification of a spatio-temporal window (in a given language)
$\|w\|$	subset of $\Re^3$, where $\Re$ is the set of reals, defined by w
o_w	physical object, $i.e.$, part of the physical space corresponding to $\|w\|$
g	informational granule (ic-granule) composed out of informational layer $inf\_layer(g)$ and physical layer $ph\_layer(g)$

These reasoning tools are supporting construction of compound objects satisfying a given specification (often expressed using complex vague concepts) like classifiers, clusters, compound sensors or robots, compound molecules or new medicine or supporting discovery of strategies for keeping the required constraints.

These tasks create new challenges for rough sets too. Examples of them are discussed in the next section.

5   Rough Sets Based on Reasoning over Networks of Ic-Granules in Foundations of IS's

In this section, we suggest that for IS's dealing with complex phenomena it is necessary to consider an extension of the rough set approach based on a new computing model making it possible to deal with perception of real situations in the physical world. Our proposal is to base such an approach on IGrC.

Table 2. Notation used in this article (cd.)

Name	Interpretation
$inf\_layer(g)$	informational layer of g consists of family of tuples (w, tr, inf), where inf is the current information updated by information perceived on o_w by g during realization by IM of the transformation specification tr
$scope(g)$	a distinguished w from $inf\_layer(g)$ where $\|w\|$ is the largest among all $\|w'\|$ from $inf\_layer(g)$
$\alpha \Longrightarrow tr : \beta$	rule, where α is a condition triggering rule defined over (relevant part of) $inf\_layer(g)$, tr is a specification of network transformation and β in the expected property of the resulting network after tr implementation
$Rule\_set$	set of rules (complex game) referring to the physical world
$ph\_layer(g)$	physical layer of g consists of parts: $soft\_suit, link\_suit, hard\_suit$ creating a dynamical physical system perceived by ic-granule g; perceived information is recorded in $inf\_layer(g)$
$soft\_suit$	consists of o_w, where w is from $inf\_layer(g)$ and o_w is directly accessible for measurement by g e.g., IM can directly realize in the physical space the specifications $enc(inf, w)$, $dec(w)$ of encoding inf in o_w and decoding information from the object after realization of $enc(inf, w)$ such that the realization of enc and next dec (i.e., $dec(enc(inf, w))$) by IM results in inf and after that $enc(dec(o_w), w)$ results in o_w provided that the environment is not disturbing these implementation processes
$hard\_suit$	consists the target objects o_w not necessarily directly accessible; information about such objects recorded in information layer is obtained by the control on the basis of reasoning about objects in $scope$ using the current information in this layer including domain knowledge or physical laws
$link\_suit$	consists objects used for transmission of interactions between $soft\_suit$ and $hard\_suit$, information about such objects recorded in information layer in case they are not directly accessible is obtained by the control on the basis of reasoning about objects from $scope$ using the current information from informational layer including domain knowledge or physical laws

From our discussion in the previous sections it follows that c-granule generates computations over networks of ic-granules which are labeled by specifications of transformations selected by the control for realization in the physical world. Hence, they can be treated as time series (not closed in the abstract space only!).

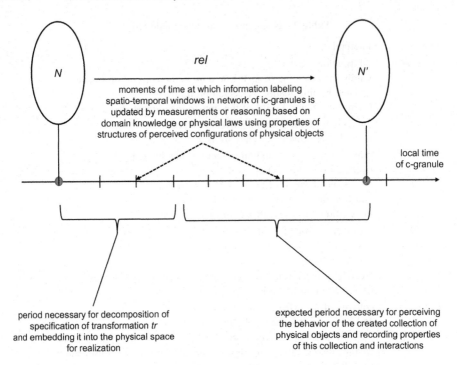

Fig. 6. Transition relation *rel*.

Let us denote by N the current network of ic-granules. The control of c-granule g is deciding on changes of N. A new network N' is obtained after performing implementation of the selected by the control at N specification of transformation tr and interaction of the new network with the environment for the required period of time (see Fig. 6). According to the selected transformation tr it may be continued perception of the situation in the physical world without changing the structure of N or perceiving will be continued after making some changes specified in tr in the structure of N.

One should note that the c-granule may have a representation of knowledge about other c-granules in the environment and cooperate or compete with them to achieve its goals. Here, one can see analogy to multiagent systems (see, *e.g.*, [39].

From generated time series of networks the control of c-granule may construct different (hierarchical) information systems (decision systems) [32, 44], *e.g.*, with objects being single networks or windows of networks. They can be represented as new ic-granules (or information granules in GrC). Such systems can be used for inducing different kinds of classifiers using hierarchical learning methods. However, it is necessary to provide for this learning the right data. Hence, one should take into considerations issues related to data governance [11] and the need of developing the relevant reasoning methods making it possible to support solving problems related to data governance.

Among the tasks to be solved by the control of c-granule the most important is the task related to the control problem. The aim of the control problem is to discover a

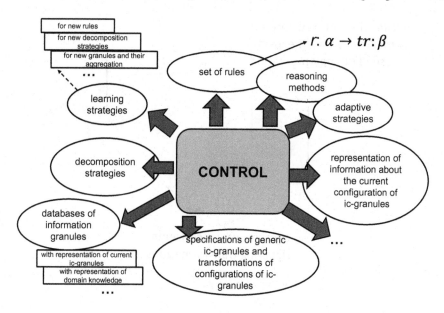

Fig. 7. The control of c-granule.

complex game (see Fig. 8) consisting of (often complex and vague) concepts labeled by specifications of transformations (of networks of ic-granules) in computations which can be treated as triggers for activating these transformations. Any such complex game consists a set of rules *Rule_set* of the form $\alpha \Longrightarrow tr : \beta$ with predecessors α describing conditions under which the rules can be initiated and successors being a specifications of transformation of netoworks tr of ic-granules with the expected results described by β. These complex games are aiming to generate finite computations with the quality higher than a given threshold (relative to the given quality or utility measure). In this problem we assume that specifications of transformations as well as complex vague concepts are selected from some given sets. Another important problem concerns discovery of such sets on the basis of which the relevant complex games may be constructed. One should note that the discovery of complex games depends on the target tasks to be realized by c-granule. In medical applications such tasks are expressed by complex vague concepts to be satisfied or preserved by the generated computation over networks of ic-granules from the given initial one. For example, this computation should belong to the lower approximation of concept *safe therapy leading to improving the health condition of the patient to the highest possible degree*. The control is aiming to achieve this goal by discovery of complex game and use it in the computation for producing the relevant plan on the basis of the discovered game (see Fig. 9). Certainly, one can relate this problem to reinforcement learning [47]. However, developing reinforcement learning methods for real world problems still is the great challenge. It seems that dialogues with experts and relevant interactions with domain knowledge bases should be involved in this process too [25, 40].

specifications of transformations
with description of expected
results of implementation

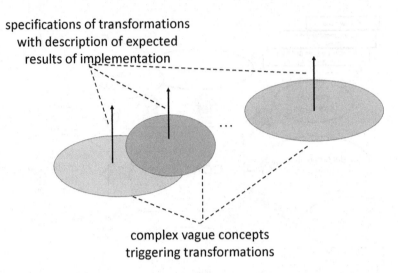

complex vague concepts
triggering transformations

Fig. 8. Complex game.

In this way we generalise the current approaches based on rough sets to the case of learning strategies for controlling computations over networks of ic-granules aiming to satisfy a given specification expressing properties of these computations.

From the above discussion it follows that the control should be equipped with advanced reasoning tools allowing among other to

- select the relevant specifications of transformations from a given set of them;
- label the selected specifications of transformations by the relevant vague concepts;
- construct approximations of these vague concepts;
- adapt the current complex games according to perceived changes.

An interesting class of challenges for the rough set based on reasoning can be expressed as follows. From a given (vague) specification construct compound objects (such as learning algorithm, cluster, classifier, compound sensor or robot, molecules, networks of servers and another hardware equipment, trajectory of computation) belonging to the lower (upper) approximation of concept *objects of high quality*. Such constructions are created along the relevant reasoning schemes using different advanced reasoning tools based on domain knowledge, dialogues with experts and/or relevant interactions with the physical objects.

An example of such a challenge, very important for IS's in supporting solution of problems, is related to approximation of functions transforming a given (often vague specification) of a problem from a given class of problems to the (semi-)optimal solution of the problem. In reality such functions can be vaguely and partially specified only. Advanced reasoning tools are necessary for making it possible to construct, along generated by these tools reasoning schemes, the approximate solution. For example, such reasoning tools can refer to decomposition of vague, specifications with the use of

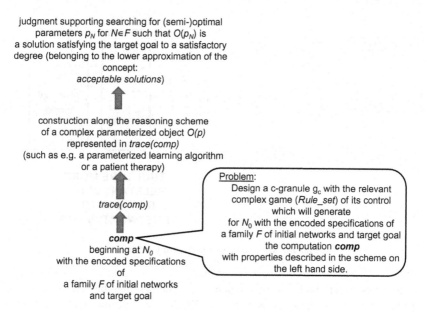

Fig. 9. Control problem.

human skills related to divide-and-conquer strategies for vague concepts (such as *e.g.*, very compound vague concept of trustworthiness[3] [1, 2, 17, 20, 22] currently investigated by many researchers) possessed by humans but not yet 'comprehended' by IS's. One may refer here to the paradigm Computing with Words due to Lotfi Zadeh [50, 51].

Developing reasoning strategies supporting designing of architectures of AI trustworthy systems is another challenge. For example, in Fig. 10 (on its right side) there is illustrated a structure relevant for AI systems preserving trustworthiness. Here, human experts in dialogue with the system are responsible for accepting tasks which are ready to be given "into hands" of such systems. The task of designing trustworthy architectures of AI systems can be understood as the task of constructing complex objects (satisfying some initial properties) from the lower approximation of the vague concept "to be trustworthy" in the vaguely specified space of architectures.

Another example is related to inducing learning algorithms generating classifiers from training samples. Here, the intuitive reasoning expressed in natural language concerning construction of learning algorithms should be embedded into the mathematical machinery [35].

Certainly, our discussion also refers to the long lasting problem of transferring intuition about proofs into the formal proofs.

From the above discussion follow several new research directions for rough sets. Among them are the following ones:

---

[3] see, *e.g.*, https://ec.europa.eu/futurium/en/ai-alliance-consultation/guidelines/1.html.

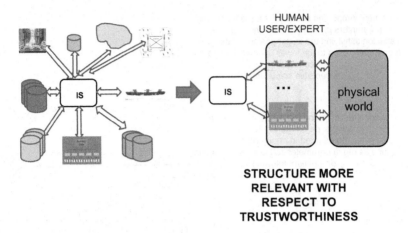

STRUCTURE MORE
RELEVANT WITH
RESPECT TO
TRUSTWORTHINESS

Fig. 10. Basic scheme for Human-Centered AI.

– Rough sets should be based on a much wider context than the context restricted to information (decision) systems only [31, 33, 34]. IS's dealing with complex phenomena should have tools for perceiving the situations in the physical world that is realized by interaction with the physical world in which classified situations are embedded. One should note that in the case of IS's dealing with complex phenomena it is not possible to base, *e.g.*, inducing classifiers on one set of attributes characterizing the perceived situation (see *e.g.*, the cited in this paper opinion by Brooks). Such models may be treated as temporary only. The system should be aware that always only a partial information about the currently perceived situation is known and it should be equipped with reasoning tools supporting searching for new relevant data and attributes (features). IS's are perceiving the current situation in the real world not at once but over a period of time during which they use different tools to perceive situation to a degree allowing the system to select and realize the right decisions. In this period of time ic-granules of networks generated by the control of c-granules of IS's are measuring some features and performing actions. Through this the generated computations over networks of ic-granules are 'shaped' to make them acceptable from the point of view of target goals, *i.e.*, information traces of computations are satisfying the required specifications at least up to satisfactory degrees (what in the rough set approach means that they are belonging to lower approximations of target concepts). Considering only abstract information granules, as is happens in GrC, is not satisfactory for IS's dealing with complex phenomena in the physical world. IGrC provides the relevant tools to build foundations for the rough set approach supporting IS's dealing with complex phenomena.

– Information (decision) system isolated from the real physical world are not satisfactory as generic granules for considerations of complex situations in the physical world. The new approach based on IGrC is providing networks of ic-granules and computations over such networks (which can be treated as a generalisation of time series because they contain both abstract and physical objects!) labeled by realized

Continuous interactions with the physical world during perceiving of the current situation aiming to understand this situation to a degree satisfactory for making the right decisions

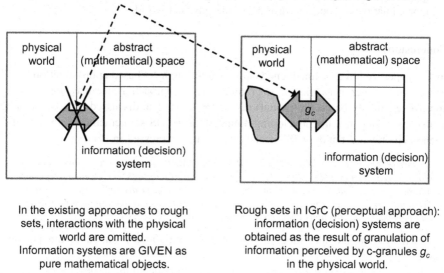

In the existing approaches to rough sets, interactions with the physical world are omitted. Information systems are GIVEN as pure mathematical objects.

Rough sets in IGrC (perceptual approach): information (decision) systems are obtained as the result of granulation of information perceived by c-granules g_c in the physical world.

Fig. 11. Rough sets in IGrC: perceptual approach.

(by the control of c-granule) transformations of networks of informational granules. Over these generalised time series information (decision) systems of different complexity (*e.g.*, hierarchical) are constructed in searching for the relevant computational building blocks for cognition (see Fig. 11).

- The approximation process of concepts is based on advanced reasoning tools (also in cooperation with human experts) carried out along computations over networks of ic-granules rather than on (partial) containment of sets only. For new applications of IS's dealing with complex phenomena, advanced reasoning tools (*e.g.*, based on dialogues with experts) should be developed which can support the existing ones, *e.g.*, in Machine Learning.
- The outlined approach to approximation of concepts seems to be important for building foundations of IS's dealing with approximations of complex vague concepts over compound objects of different kinds (such as learning algorithms, compound sensors or robots, new materials or medicines, schemes of reasoning supporting searching for solutions of problems, complex games to name a few). The discussed in the paper control problems related to construction of computations belonging to the specified regions (such as *e.g.*, lower approximations) of complex vague concepts, under assumption that there is given an initial information (such as symptoms of patient in the case of medical applications [8]) encoded in the initial networks of computations, are becoming important in building the mentioned foundations based on the rough set approach grounded in IGrC.

– The current rough set approach should be substituted by the adaptive rough set approach making it possible to adapt control strategies of c-granules used for approximation of concepts drifting in the physical world.

Conclusions

In this paper we have outlined a new approach to rough set based approximation based on reasoning over networks of ic-granules. This approach generalises the existing approaches mainly based on (partial) set containment. The discussed approach can be treated as a step toward developing foundations for IS's interacting with complex phenomena toward building IS's satisfying dreams expressed in [4]:

Tomorrow, I believe, we will use INTELLIGENT SYSTEMS to support our decisions in defining our research strategy and specific aims, in managing our experiments, in collecting our results, interpreting our data, in incorporating the findings of others, in disseminating our observations, in extending (generalizing) our experimental observations through exploratory discovery and modeling - in directions completely unanticipated.

Such systems are like modern laboratories continuously linked to the physical space.

In our further research we plan to elaborate in more detail the issues outlined in this paper and to use the discussed approach in real world projects (see, *e.g.*, [45]).

References

1. Agio, A., Omicini, A.: Measuring trustworthiness in neuro-symbolic integration. In: Ganzha, M., Maciaszek, L.A., Paprzycki, M., Ślęzak, D. (eds.) Proceedings of the 18th Conference on Computer Science and Intelligence Systems, FedCSIS 2023, Warsaw, Poland, 17–20 September, 2023. Annals of Computer Science and Information Systems, vol. 35, pp. 1–10. IEEE, PTI (2023). https://doi.org/10.15439/978-83-967447-8-4
2. Ammanath, B. (ed.): Trustworthy AI: A Business Guide for Navigating Trust and Ethics in AI. Wiley & Sons, Hobooken (2022). https://doi.org/10.1007/978-3-642-27473-2
3. Bar-Yam, Y.: Dynamics of Complex Systems. Studies in Nonlinearity. Addison-Wesley, Reading (1997)
4. Bower, J.M., Bolouri, H. (eds.): Computational Modeling of Genetic and Biochemical Networks. MIT Press, Cambridge (2001). https://doi.org/10.7551/mitpress/2018.001.0001
5. Brooks, Jr., F.P.: The Mythical Man-Month: Essays on Software Engineering, Anniversary Edition. Addison-Wesley, Boston (1995)
6. Compagnucci, M.C., Wilson, M.L., Mark Fenwick, N.F., Bärnighausen, T. (eds.): AI IN eHEALTH Human Autonomy, Data Governance and Privacy in Healthcare. Cambridge University Press, Cambridge (2022). https://doi.org/10.1017/9781108921923
7. Deutsch, D., Ekert, E., Lupacchini, R.: Machines, logic and quantum physics. Neural Comput. **6**, 265–283 (2000). https://doi.org/10.2307/421056
8. Dey, L., Jana, S., Dasgupta, T., Gupta, T.: Deciphering clinical narratives - augmented intelligence for decision making in healthcare. In: Ganzha, M., Maciaszek, L.A., Paprzycki, M., Ślęzak, D. (eds.) Proceedings of the 18th Conference on Computer Science and Intelligence Systems, FedCSIS 2023, Warsaw, Poland, 17–20 September 2023. Annals of Computer Science and Information Systems, vol. 35, pp. 11–24. IEEE, PTI (2023). https://doi.org/10.15439/978-83-967447-8-4

9. Dutta, S., Skowron, A.: Interactive Granular Computing Model for Intelligent Systems. In: Shi, Z., Chakraborty, M., Kar, S. (eds) Intelligence Science III. ICIS 2021. IFIP Advances in Information and Communication Technology, vol 623. Springer, Cham (2021). https://doi.org/10.1007/978-3-030-74826-5_4

10. Dutta, S., Skowron, A.: Toward a computing model dealing with complex phenomena: interactive granular computing. In: Nguyen, N.T., Iliadis, L., Maglogiannis, I., Trawiński, B. (eds.) ICCCI 2021. LNCS, vol. 12876, pp. 199–214. Springer, Cham (2021). https://doi.org/10.1007/978-3-030-88081-1_15

11. Eryurek, E., Gilad, U., Lakshmanan, V., Kibunguchy-Grant, A., Ashdown, J.: Data Governance: The Definitive Guide: People, Processes, and Tools to Operationalize Data Trustworthiness. O'Reilly Media, Sebastopol, CA (2021)

12. Gell-Mann, M.: The Quark and the Jaguar - Adventures in the Simple and the Complex. Brown & Co., London (1994)

13. Gershenson, C. (ed.): Design and Control of Self-organizing Systems. CopIt ArXives, Mexico City, Boston (2007)

14. Heller, M.: The Ontology of Physical Objects. Four Dimensional Hunks of Matter. Cambridge Studies in Philosophy, Cambridge University Press, Cambridge (1990)

15. Kroc, J., Sloot, P.M.A., Hoekstra, A.G. (eds.): Simulating Complex Systems by Cellular Automata. UCS, Springer, Heidelberg (2010). https://doi.org/10.1007/978-3-642-12203-3

16. Holland, J.H.: Signals and Boundaries. Building Blocks for Complex Adaptive Systems. The MIT Press, Cambridge (2014)

17. Huang, X., Kwiatkowska, M., Olejnik, M.: Reasoning about cognitive trust in stochastic multiagent systems. ACM Trans. Comput. Logic 20(4), 21:1–21:64 (2019). https://doi.org/10.1145/3329123

18. Jankowski, A.: Interactive Granular Computations in Networks and Systems Engineering: A Practical Perspective. LNNS, vol. 17. Springer, Cham (2017). https://doi.org/10.1007/978-3-319-57627-5

19. Kałuża, D., Janusz, A., Ślęzak, D.: Robust assignment of labels for active learning with sparse and noisy annotations. In: Gal, K., Nowé, A., Nalepa, G.J., Fairstein, R., Radulescu, R. (eds.) Proceedings of the 26th European Conference on Artificial Intelligence, ECAI 2023, Kraków, Poland, September 30 - October 4, 2023. Frontiers in Artificial Intelligence and Applications, vol. 372, pp. 1207–1214 (2023). https://doi.org/10.3233/FAIA230397

20. Kwiatkowska, M., Zhang, X.: When to trust AI: advances and challenges for certification of neural networks. In: Ganzha, M., Maciaszek, L.A., Paprzycki, M., Ślęzak, D. (eds.) Proceedings of the 18th Conference on Computer Science and Intelligence Systems, FedCSIS 2023, Warsaw, Poland, 17–20 September 2023. Annals of Computer Science and Information Systems, vol. 35, pp. 25–38. IEEE, PTI (2023). https://doi.org/10.15439/978-83-967447-8-4

21. Lin, T.Y., Liau, C.J., Kacprzyk, J. (eds.): Granular, Fuzzy, and Soft Computing. A Volume in the Encyclopedia of Complexity and Systems Science (Second Edition). Springer, New York (2023). https://doi.org/10.1007/978-1-0716-2628-3

22. Mariani, R., et al.: Trustworthy AI-Part 1. Computer 56, 14–18 (2023). https://doi.org/10.1109/MC.2022.3227683

23. Marr, D.: Vision. The MIT Press, Cambridge (2010)

24. Miłkowski, M.: Explaining the Computational Mind. The MIT Press, Cambridge (2013)

25. Monarch, R.M.: Human-in-the-Loop Machine Learning. Active Learning and Annotation for Human-Centered AI. Manning, Shelter Island (2021)

26. Nöe, A.: Action in Perception. The MIT Press, Cambridge (2004)

27. Omicini, A., Ricci, A., Viroli, M.: The multidisciplinary patterns of interaction from sciences to computer science. In: Goldin, D., Smolka, S., Wegner, P. (eds.) Interactive Computation: The New Paradigme, pp. 395–414. Springer, Heidelberg (2006). https://doi.org/10.1007/3-540-34874-3_15

28. Ortitz, C.L., Jr.: Why we need a physically embodied Turing test and what it might look like. AI Mag. **37**, 55–62 (2016). https://doi.org/10.1609/aimag.v37i1.2645

29. Pawlak, Z.: An inquiry into anatomy of conflicts. J. Inform. Sci. **109**, 65–78 (1998). https://doi.org/10.1016/S0020-0255(97)10072-X

30. Pawlak, Z., Skowron, A.: Rough sets: some extensions. Inf. Sci. **177**(28–40), 1 (2007). https://doi.org/10.1016/j.ins.2006.06.006

31. Pawlak, Z., Skowron, A.: Rudiments of rough sets. Inf. Sci. **177**(1), 3–27 (2007). https://doi.org/10.1016/j.ins.2006.06.003

32. Pawlak, Z.: Information systems - theoretical foundations. Inf. Syst. **6**, 205–218 (1981). https://doi.org/10.1016/0306-4379(81)90023-5

33. Pawlak, Z.: Rough sets. Int. J. Comput. Inform. Sci. **11**, 341–356 (1982). https://doi.org/10.1007/BF01001956

34. Pawlak, Z.: Rough sets: theoretical aspects of reasoning about data, system theory, knowledge engineering and problem solving, vol. 9. Kluwer Academic Publishers, Dordrecht (1991). https://doi.org/10.1007/978-94-011-3534-4

35. Pearl, J.: Causal inference in statistics: an overview. Stat. Surv. **3**, 96–146 (2009). https://doi.org/10.1214/09-SS057

36. Pedrycz, W., Skowron, S., Kreinovich, V. (eds.): Handbook of Granular Computing. Wiley, Hoboken (2008). https://doi.org/10.1002/9780470724163

37. Pedrycz, W.: Granular Computing. Analysis and Design of Intelligent Systems. CRC Press, Taylor & Francis (2013). https://doi.org/10.1201/9781315216737

38. Pylyshyn, Z.W.: Computation and Cognition. Toward a Foundation for Cognitive Science. The MIT Press, Cambridge (1984)

39. Russell, S.J., Norvig, P.: Artificial Intelligence A Modern Approach. 4th edn. Pearson Education Inc, Upper Saddle River (2021)

40. Shneiderman, B.: Human - Centered AI. Oxford University Press, Oxford (2022). https://doi.org/10.1093/oso/9780192845290.001.0001

41. Skowron, A., Jankowski, A., Dutta, S.: Interactive granular computing. Granular Comput. **1**, 95–113 (2016). https://doi.org/10.1007/s41066-015-0002-1

42. Skowron, A., Dutta, S.: Rough sets and fuzzy sets in interactive granular computing. In: Yao, J., Fujita, H., Yue, X., Miao, D., Grzymala-Busse, J., Li, F. (eds.) IJCRS 2022. LNCS, vol. 13633, pp. 19–29. Springer, Cham (2022). https://doi.org/10.1007/978-3-031-21244-4_2

43. Skowron, A., Jankowski, A.: Rough sets and interactive granular computing. Fund. Inform. **147**(2–3), 371–385 (2016). https://doi.org/10.3233/FI-2016-1413

44. Skowron, A., Ślęzak, D.: Rough sets turn 40: from information systems to intelligent systems. In: Ganzha, M., Maciaszek, L.A., Paprzycki, M., Ślęzak, D. (eds.) Proceedings of the 17th Conference on Computer Science and Intelligence Systems, FedCSIS 2022, Sofia, Bulgaria, 4–7 September 2022. Annals of Computer Science and Information Systems, vol. 30, pp. 23–34 (2022). https://doi.org/10.15439/2022F310

45. Sosnowski, Ł., Żuławińska, J., Dutta, S., Szymusik, I., Zyguła, A., Bambul-Mazurek, E.: Artificial intelligence in personalized healthcare analysis for womens' menstrual health disorders. In: Ganzha, M., Maciaszek, L.A., Paprzycki, M., Ślęzak, D. (eds.) Proceedings of the 17th Conference on Computer Science and Intelligence Systems, FedCSIS 2022, Sofia, Bulgaria, 4–7 September 2022. Annals of Computer Science and Information Systems, vol. 30, pp. 751–760 (2022). https://doi.org/10.15439/2022F59

46. Stilgoe, J.: We need a Weizenbaum test for AI. Science **381**, eadk0176 (2023). https://doi.org/10.1126/science.adk0176
47. Sutton, R.S., Barto, A.G.: Reinforcement Learning. An Introduction. 2nd edn. MIT Press, Cambridge (2018)
48. Wakulicz-Deja, A., Przybyła-Kasperek, M.: Pawlak's conflict model: directions of development. In: Ganzha, M., Maciaszek, L.A., Paprzycki, M. (eds.) Proceedings of the 2016 Federated Conference on Computer Science and Information Systems, FedCSIS 2016, Gdańsk, Poland, 11–14 September 2016. Annals of Computer Science and Information Systems, vol. 8, pp. 191–197. IEEE (2016). https://doi.org/10.15439/2016F003
49. Wąs, J.: Modeling and Simulation of Complex Collective Systems. CRC Press, Boca Raton (2023). https://doi.org/10.1201/b23388
50. Zadeh, L.A.: From computing with numbers to computing with words - from manipulation of measurements to manipulation of perceptions. IEEE Trans. Circ. Syst. **45**, 105–119 (1999). https://doi.org/10.1109/81.739259
51. Zadeh, L.A. (ed.): Computing with Words: Principal Concepts and Ideas, Studies in Fuzziness and Soft Computing, vol. 277. Springer, Heidelberg (2012). https://doi.org/10.1007/978-3-642-27473-2

Keynote Lectures

Rough-Calculus and Numerical Analysis – A Mathematical Foundation

Tsau Young (T. Y.) Lin

Data Science Institute, Granular Computing Society,
San Jose (The Capital of Silicon Valley), CA 95120, USA
prof.tylin@gmail.com

Rough set theory and college calculus are two disjoint pieces of mathematics on discrete and continuous worlds. Somehow Pawlak observed their commonality and create a subject called Rough Calculus. Numerical analysis and scientific modeling are another such a pair, in fact, a better pair: A real line can be visualized as a generalized meterstick. Markings for centimeters are labelled by integers, markings for millimeters are labelled by one decimal place ...... are labelled by n decimal places. Such sequence of labellings supports the granulation topology $\mathcal{G}$ (recall the concepts of granular computing and mathematics) that contains the old friend, the usual topology $\mathcal{U}$, as a subtopology. Such observations allow us to build a mathematical model for Approximate Arithmetic, which has been missing for centuries. An extended abstract, that needs some updating, has been announced in the Encyclopedia of Complexity and System Science in March, 2023.

Can AI and Big Data Methods Really Help in Cyber Security?

Joel Holland

DeepSeas, USA

The terminology of "Big Data" has been used in many areas. The Cyber Security problem is a great example of a Big Data problem. The amount of data that is necessary to find malicious actors within mixed corporate environments continues to expand. Finding the "dumb" or "lazy" attackers that use known attack vectors is fairly straight forward. But how do you find the more advanced technically savvy bad guys? AI is starting to play a bigger role in detection methods within cyber solutions. But how do you mix the two? It is exceeding hard to accomplish AI over truly large ever changing data sets. Training AI to find something it is seen before is a more traditional approach. However, in the cyber problem you are looking for an attacker using a new approach you don't have in your data to get around controls. In this talk we will discuss the Cyber Security detection problem and the issues around speed and size of the data combined with the ever changing attack vectors.

Can AI and Big Data Methods Really Help in Cyber Security?

Joe Holland

Applications of Tolerance-based Granular Methods (Extended Abstract)

Sheela Ramanna ⓘ

Applied Computer Science, University of Winnipeg, Canada
s.ramanna@uwinnipeg.ca

The proliferation of large-scale web corpora and social media data as well as advances in machine learning and deep learning have led to important applications in diverse Natural Language Processing (NLP) areas such as information extraction, named entity recognition (NER), text summarization and sentiment analysis to name a few. NER seeks to discover and categorize specific linguistic entities in unstructured text and relies heavily on semi-supervised learning methods due to the fact that the number of training examples are very limited. Tolerance relations provide the most general tool for studying similarity and are ideal for NLP applications.

In this talk, we investigate machine learning methods based on a tolerance form of rough [6, 10], near [7] as well fuzzy rough sets [2, 3] in two NLP areas: NER and Sentiment Analysis. We present novel semi-supervised learning algorithms (TPL and FRL) [1, 4, 8, 9] based on tolerance rough and fuzzy rough sets for NER tasks. Specifically, we address two problems: i) the challenge of labelling relational facts from large web corpora, and ii) *concept drift*. The performance of the presented algorithms is discussed in terms of bench-marked datasets and algorithms.

Another natural appearance of the tolerance relation occurs in near sets and descriptive proximity space theory. We present our recent work on tolerance near sets-based supervised learning algorithm (TSC) [5] to perform coarse-grained sentiment categorization from text by leveraging high-dimensional embedding vectors from pre-trained transformer-based models. We discuss the impact of using different embeddings on the performance of TSC on well-researched text classification datasets. We make the case that approximation structures viewed through the prism of tolerance display fluidity and integrate conceptual structures at different levels of granularity that are appropriate for NLP.

References

1. Bharadwaj, A., Ramanna, S.: Categorizing relational facts from the web with fuzzy rough sets. Knowl. Inf. Syst. **61**(3), 1695–1713 (2019). https://doi.org/10.1007/s10115-018-1250-6
2. De Cock, M., Cornelis, C.: Fuzzy rough set based web query expansion. In: Proceedings of Rough Sets and Soft Computing in Intelligent Agent and Web Technology, pp. 9–16 (2005)

3. Dubois, D., Prade, H.: Rough fuzzy sets and fuzzy rough sets*. Int. J. General Syst. **17**(2–3), 191–209 (1990)

4. Moghaddam, H., Ramanna, S.: Harvesting patterns from textual web sources with tolerance rough sets. Cell Press, Elsevier **1**(4), 100053 (2020)

5. Patel, V., Ramanna, S., Kotecha, K., Walambe, R.: Short text classification with tolerance-based soft computing method. Algorithms, MDPI **15**(8) (2022), https://www.mdpi.com/1999-4893/15/8/267

6. Pawlak, Z.: Rough sets. Int. J. Comput. Inf.Sci. **11**(5), 341–356 (1982)

7. Peters, J.: Near sets. Special theory about nearness of objects. Fundam. Informaticae **75**(1–4), 407–433 (2007)

8. Ramanna, S.: Tolerance-based granular methods: Foundations and applications in natural language processing. Intell. Decis. Technol. **17**(1), 139–1588 (2023)

9. Sengoz, C., Ramanna, S.: Learning relational facts from the web: A tolerance rough set approach. Pattern Recog. Lett. **67**(P2), 130–137 (2015)

10. Skowron, A., Stepaniuk, J.: Tolerance approximation spaces. Fundam. Inf. **27**(2–3), 245–253 (1996)

Role of Color in Histological Image Analysis: Rough-Fuzzy Computing to Deep Learning

Pradipta Maji ⓘ

Machine Intelligence Unit, Indian Statistical Institute, Kolkata, India
pmaji@isical.ac.in

In histology, microscopic images of tissue sections are examined to study the manifestation of diseases under consideration. The most important property of histological images is the enormous density of data, more cellular details, compared to other imaging modalities, which makes computer-aided diagnosis more accurate than other modalities. To facilitate pathologists' examination, tissue samples are stained with multiple contrasting histochemical reagents, which in turn highlight different tissue structures and cellular features. Hence, color in pathology plays a pivotal role as a good indicator of histological components.

One of the most common and primary problems of histological tissue analysis is the inadmissible inter and intra-specimen variation in stained tissue color. Consequently, numerical features extracted from histological images may lead to difficulty in image interpretation by automated systems, trained on a specific stain color appearance. Hence, the foremost and challenging task in stained histological image analysis is to reduce color variation present among images.

In this talk, two recently introduced approaches for stain color normalization will be discussed. While the first approach is based on rough-fuzzy computing [2], the second one is developed around generative adversarial network [1]. The rough-fuzzy circular clustering algorithm [2] is developed based on rough-fuzzy computing for stain color normalization. It judiciously integrates the merits of both fuzzy and rough sets. While the theory of rough sets deals with uncertainty, vagueness, and incompleteness in stain class definition, fuzzy set handles the overlapping nature of histochemical stains. The proposed circular clustering algorithm works on a weighted hue histogram, which considers both saturation and local neighborhood information of the given image. A new dissimilarity measure is introduced to deal with the circular nature of hue values.

On the other hand, the TredMil [1], which is based on generative adversarial network, assumes that the latent color appearance information, extracted through a color appearance encoder, and stain bound information, extracted via stain density encoder, are independent of each other. A generative module and a reconstructive module are designed accordingly to capture disentangled color appearance and stain density information. To deal with the overlapping nature of histochemical reagents, the model assumes that the latent color appearance code, extracted through the color appearance encoder, is sampled from a mixture of truncated normal distributions.

The performance of these two models, along with a comparison with state-of-the-art approaches, has been demonstrated on several data sets containing H&E stained histological images. The merits and demerits of these two approaches will also be covered in the talk.

References

1. Mahapatra, S., Maji, P.: Truncated normal mixture prior based deep latent model for color normalization of histology images. IEEE Trans. Med. Imaging **42**(6), 1746–1757 (2023). https://doi.org/10.1109/TMI.2023.3238425
2. Maji, P., Mahapatra, S.: Circular clustering in fuzzy approximation spaces for color normalization of histological images. IEEE Trans. Med. Imaging **39**(5), 1735–1745 (2020). https://doi.org/10.1109/TMI.2019.2956944

Big Data Intelligence: Challenges and Our Solutions

Tianrui Li

School of Computing and Artificial Intelligence, Southwest Jiaotong University, Chengdu 611756, China

Abstract. Big Data Intelligence is driving revolutionary changes in various industries, from business, medical treatment to government. It has also become a hot research topic in the area of artificial intelligence. This paper aims to outline some main challenges on Big Data Intelligence, e.g., data with few labels, user privacy, high dimensionality, open-world dynamics, multi-source heterogeneity. Then our solutions for Big Data Intelligence are provided, which cover the following aspects: 1) A micro-supervised disturbance learning paradigm is developed by introducing the small-perturbation ideology based on the representation probability distribution, which enables to reduce the reliance of deep representation on labels [1]; 2) A federated deep reinforcement learning framework is presented to address the problem of daily schedule recommendation, which can guarantee and protect big data privacy [2]; 3) A distributed operating-based feature selection algorithm is devised with a rough hypercuboid approach, which deals with high-dimensional big data [3]; 4) A three-way decision-based incremental learning method considering temporal-spatial multi-granularity structure is proposed, which allows to represent and learn uncertain knowledge in fuzzy open-world big data [4]; 5) Several deep learning-based models are illustrated to fuse multi-source heterogeneous big data as well as their applications in smart cities (e.g., ambulance deployment, metro operation, and air quality prediction) [5–10].

References

1. Chu, J.L., Liu, J., Wang, H.J., Meng, H., Gong, Z.G., Li, T.R.: Micro-supervised disturbance learning: a perspective of representation probability distribution. IEEE Trans. Pattern Anal. Mach. Intell. **45**(6), 7542–7558 (2023)
2. Huang, W., Liu, J., Li, T.R., Huang, T.Q., Ji, S.G., Wan, J.H.: FedDSR: daily schedule recommendation in a federated deep reinforcement learning framework. IEEE Trans. Knowl. Data Eng. **35**(4), 3912–3924 (2023)
3. Luo, C., Wang, S.Z., Li, T.R., Chen, H.M., Lv, J.C., Zhang, Y.: Spark rough hypercuboid approach for scalable feature selection. IEEE Trans. Knowl. Data Eng. **35**(3), 3130–3144 (2023)
4. Yang, X., Li, Y.J., Liu, D., Li, T.R.: Hierarchical fuzzy rough approximations with three-way multigranularity learning. IEEE Trans. Fuzzy Syst. **30**(9), 3486–3500 (2022)

5. Liu, J., Li, T.R., Ji, S.G., Xie, P., Du, S.D., Teng, F., Zhang, J.B.: Urban flow pattern mining based on multi-source heterogeneous data fusion and knowledge graph embedding. IEEE Trans. Knowl. Data Eng. **35**(2), 2133–2146 (2023)
6. Wang, Z.Y., Pan, Z.Y., Chen, S., Ji, S.G., Yi, X.W., Zhang, J.B., Wang, J.Y., Gong, Z.G., Li, T.R., Zheng, Y.: Shortening passengers' travel time: a dynamic metro train scheduling approach using deep reinforcement learning. IEEE Trans. Knowl. Data Eng. **35**(5), 5282–5295 (2023)
7. Ji, S.G., Zheng, Y., Wang, W.J., Li, T.R.: Real-time ambulance redeployment: a data-driven approach. IEEE Trans. Knowl. Data Eng. **32**(11), 2213–2226 (2020)
8. Ji, S.G., Zheng, Y., Wang, Z.Y., Li, T.R.: Alleviating users' pain of waiting: Effective task grouping for online-to-offline food delivery services. In: World Wide Web Conference (WWW 2019), pp. 773–783. (2019)
9. Du, S.D., Li, T.R., Yang, Y., Horng, S.-J.: Deep air quality forecasting using hybrid deep learning framework. IEEE Trans. Knowl. Data Eng. **33**(6), 2412–2424 (2021)
10. Yi, X.W., Zhang, J.B., Wang, Z.Y., Li, T.R., Zheng, Y.: Deep distributed fusion network for air quality prediction. In: ACM SIGKDD International Conference on Knowledge Discovery & Data Mining (KDD 2018), pp. 965–973 (2018)

Knowledge Engineering in Food Computing — Selected Problems and Applications (Extended Abstract)

Weronika T. Adrian(iD)

AGH University of Krakow, Poland
wta@agh.edu.pl

The emerging field of food computing tackles, among others, problems of knowledge acquisition, engineering, and processing in the domain of food. Food is a common human experience, yet working with recipes requires some knowledge about the process, combinations of ingredients, properties of the single constituents, and the resulting dish. This knowledge is often implicit, contextual, or culture-dependent. Making at least parts of it explicit, with some sort of formalization, opens up possibilities to develop intelligent knowledge-based solutions to assist humans in preparing and optimizing food.

One of the interesting and relevant problems is searching for substitutions in food recipes. This task may be motivated by different constraints and objectives of a person, including allergies, diets, etc. What ingredient to substitute with what and how will it influence the resulting dish are just some of the questions that require the knowledge of a dietician, or a food technologist (and sometimes: both). While machine learning-based solutions may produce proposals of ingredients that should be replaced with others based on their occurrence in similar contexts, it is not always understandable, why certain ingredients are appropriate or not, and what features of the proposed substitutes satisfy the person's goals. Thus, structured and logic-based solutions may be developed to provide transparent and explainable answers to the questions outlined above.

In this presentation, we will discuss recent research threads in the field of knowledge engineering in food computing, including methods of knowledge acquisition, modeling, and reasoning over the integrated knowledge. The talk will cover topics such as modeling food-related knowledge in the form of ontologies, ongoing efforts and international initiatives in the area, an ontology design pattern for substitution, building a knowledge graph for substitution, and logic-based solutions for selecting target ingredients to substitute and pruning "wrong" substitutes recommendations.

Model-agnostic Explanations of Black-box Prediction Models using Rough Sets – The Case of Post-Competition Analytics at KnowledgePit.ai (Keynote Abstract)

Andrzej Janusz[1,2]

[1] Institute of Informatics, University of Warsaw ul. Banacha 2, 02-097, Warsaw, Poland
[2] QED Software ul. Miedziana 3A, 00-814 Warsaw, Poland
a.janusz@mimuw.edu.pl

In my talk, I will discuss the problem of model-agnostic explainability and analytics of predictions obtained using black-box machine learning models [1]. As an example of an application, I will use a data science competition platform where researchers from around the world compete to solve real-life problems [2, 3]. Firstly, I will recall a few successful competitions and explain their scope, as well as their most notable outcomes [4, 5]. Then, I will briefly talk about our approach to the explainability of prediction errors, and its usefulness to end-users [6]. My aim will be to demonstrate how notions known from the theory of rough sets, such as the decision reducts, can be used to efficiently construct an approximation of an arbitrary set of predictions, such as a typical solution to a data science competition. I will also explain how such an approximation can be used to extract useful insights about the ML model used to generate the predictions, and about the corresponding data set. Finally, I will explain how it is all related to the problem of post-competition analytics and diagnostics of solutions submitted by the most successful teams in data science competitions [7]. I will also give examples of other types of analysis that we perform to gain a better understanding of data science problems considered in the competition.

References

1. Janusz, A., Zalewska, A., Wawrowski, Ł., Biczyk, P., Ludziejewski, J., Sikora, M., Ślęzak, D.: BrightBox – a rough set based technology for diagnosing mistakes of machine learning models. Appl. Soft Comput. 110285 (2023)
2. Grzegorowski, M., Stawicki, S.: Window-based feature extraction framework for multi-sensor data: a posture recognition case study. In: Ganzha, M., Maciaszek, L. A., Paprzycki, M., eds.: 2015 Federated Conference on Computer Science and Information Systems, FedCSIS 2015, Lódz, Poland, September 13-16, 2015. Volume 5 of Annals of Computer Science and Information Systems, pp. 397–405, IEEE (2015)

3. Vu, Q.H., Cen, L., Ruta, D., Liu, M.: Key factors to consider when predicting the costs of forwarding contracts. In: Ganzha, M., Maciaszek, L.A., Paprzycki, M., Slezak, D., eds.: Proceedings of the 17th Conference on Computer Science and Intelligence Systems, FedCSIS 2022, Sofia, Bulgaria, September 4–7, 2022. Volume 30of Annals of Computer Science and Information Systems, pp. 447–450 (2022)
4. Janusz, A., Grzegorowski, M., Michałak, M., Wróbel, L., Sikora, M., Ślęzak, D.: Predicting seismic events in coal mines based on underground sensor measurements. Eng. Appl. AI **64**, 83–94 (2017)
5. Janusz, A., Kałuża, D., Chądzyńska-Krasowska, A., Konarski, B., Holland, J., Ślęzak, D.: IEEE bigdata 2019 cup: Suspicious network event recognition. In: Baru, C.K., Huan, J., Khan, L., Hu, X., Ak, R., Tian, Y., Barga, R.S., Zaniolo, C., Lee, K., Ye, Y.F., eds.: 2019 IEEE International Conference on Big Data (IEEE BigData), Los Angeles, CA, USA, December 9–12, pp. 5881–5887, IEEE (2019)
6. Colin, J., FEL, T., Cadene, R., Serre, T.: What i cannot predict, i do not understand: A human-centered evaluation framework for explainability methods. In: Koyejo, S., Mohamed, S., Agarwal, A., Belgrave, D., Cho, K., Oh, A., eds.: Advances in Neural Information Processing Systems. Volume 35., Curran Associates, Inc., pp. 2832–2845 (2022)
7. Janusz, A., Ślęzak, D.: KnowledgePit meets BrightBox: A step toward insightful investigation of the results of data science competitions. In: Ganzha, M., Maciaszek, L.A., Paprzycki, M., Ślęzak, D., eds.: Proceedings of the 17th Conference on Computer Science and Intelligence Systems, FedCSIS 2022, Sofia, Bulgaria, September 4–7, 2022. Volume 30 of Annals of Computer Science and Information Systems, pp. 393–398 (2022)

Contents

Distances and Similarities

Hybrid Approaches

Applications

Cybersecurity and IoT

Rough Set Models

Selected Approaches to Conflict Analysis Inspired by the Pawlak Model – Case Study

Małgorzata Przybyła-Kasperek[1]([envelope]) [ORCID], Rafał Deja[2] [ORCID],
and Alicja Wakulicz-Deja[1] [ORCID]

[1] Institute of Computer Science, University of Silesia, ul. Bedzinska 39, Sosnowiec,
Poland
{malgorzata.przybyla-kasperek,alicja.wakulicz-deja}@us.edu.pl
[2] Department of Computer Science, WSB University, ul. Cieplaka 1c,
Dąbrowa Górnicza, Poland
rdeja@wsb.edu.pl

Abstract. The paper presents a comparison of selected methods of conflict analysis inspired by the Pawlak model. It examines a real-world case study, the 2020 presidential election in Poland. The study explores five distinct approaches to conflict analysis, drawing insights from this example. It highlights crucial distinctions among the models considered and provides recommendations within practical application contexts.

Keywords: Pawlak conflict model · Coalitions · Negotiations stage · Consensus

1 Introduction

Conflict is a ubiquitous aspect of everyday life. Initially, conflicts were predominantly explored in human interactions, leading to extensive research conducted primarily within the social sciences [11]. However, the game theory broadened the scope of conflict analysis and negotiation, extending its influence into domains beyond the social sciences [12,33]. The first works on conflicts from the perspective of Artificial Intelligence (AI) concerns the area of decision support systems that support people to better understand conflicts – finding the most conflicting issues or possible coalitions. Some tools for analyzing available information and suggesting the possible solutions were also proposed [7]. In recent times, conflict scenarios frequently involve intricate multi-agent systems, where we have the large number of interacting parties. Managing the numerous dependencies required to extract potential solutions manually is both impractical and non-scalable. Consequently, AI-powered conflict analysis and automated negotiation systems have become indispensable for large-scale systems. Despite the fact that these topics have been studied for more than a decade, there is still a wide range of problems that have not been proposed or discussed in the literature.

A. Campagner et al. (Eds.): IJCRS 2023, LNAI 14481, pp. 3–17, 2023.
https://doi.org/10.1007/978-3-031-50959-9_1

As a starting point of conflict analysis, the Pawlak conflict model presented in the papers [18,22] is considered. The simple model based on rough set theory [19,23] also gives great insight and understanding of any conflict. Anyway, the conclusions can be provided on the outermost level. In this paper, we describe the possible enhancements of the Pawlak model using a real political conviction conflict. The proposed enhancements allows to analyze possible coalitions and looks for consensus in negotiation process. The main goal of this paper is to review and compare selected methods of conflict analysis that are an extension of the Pawlak model and were inspired by this model. The areas of application of the discussed approaches and the key differences between the approaches are also identified. The paper is organized as follows. In the second section, an overview of the literature is presented. The third section introduces a real-world conflict scenario, along with the theoretical foundations of the analysis models and their application to the example. The fourth section compares results, discusses findings, and offers guidance on applying the analysis models. Finally, a summary is presented in the conclusion section.

2   Literature Review

The Pawlak conflict analysis model has inspired numerous researchers, leading to various extensions and approaches. For instance, Andrzej Skowron and Soma Dutta have focused on extensions designed for multi-agent systems [4,5]. Another significant approach inspired by the Pawlak model is the three-way decisions theory [16,31]. Because of restricting (in the Pawlak model) the agents values set to three (against, neutrality and favorable), the natural divisions of agents or issues into three parts can be introduced. This fits into the three-way decision theory proposed by Yao [32]. Following this approach, many researchers have studied conflict via trisecting agents, issues, and pairs of agents [6,13,14,17] using different evaluation functions [6]. As an instance, in [14] the authors proposed the use of a pair of thresholds to define the relation of coalition, neutrality and conflict. The other way of agents and issues three-section has been proposed by Sun et al. in [30] – they explored the rough set upper and lower approximation concept for this purpose. This approach has been developed in [29] by proposing a conflict analysis decision model based on rough set theory over two universes.

Furthermore, interesting applications that consider hierarchies and constraints applicable to conflict situations were demonstrated by Jarosław Stepaniuk and Andrzej Skowron [28]. Additionally, there exist approaches grounded in rough set theory for multiple criteria decision analysis [2,9]. Given the inherent high uncertainty and incomplete information associated with conflicts, rough set theory proves to be an excellent approach for such cases [10,27]. Finally, these models have found practical implementations in real-world conflict analysis scenarios, including applications in the Chinese Wall Security Policy context [15] and water resources allocation decisions [8].

3   Conflict Analysis Models – Case Study

Professor Zdzisław Pawlak proposed the conflict analysis model in the eighties of the twentieth century [18, 20, 21]. The main idea of the model is to express the views of the agents involved in a conflict on certain conflicting issues using only three values, and to store this information using an information system. Information system is a pair $S = (U, A)$, where U is the universe – the set of agents, A is a set of issues, and the set of values of $a \in A$ is equal $V^a = \{-1, 0, 1\}$. Opinion of agent $x \in U$ about issue $a \in A$ is the value $a(x)$. The meaning of this value is as follows: $a(x) = 1$ means the agent x is in favor of the issue a; $a(x) = 0$ means the agent x is neutral to the issue a; $a(x) = -1$ means the agent x is against to the issue a. In order to calculate the intensity of conflict between agents, two functions were proposed [18]. A function of distance between agents $\rho_B^* : U \times U \to [0, 1]$ for the set of issues $B \subseteq A$ is defined as follows: $\rho_B^*(x_1, x_2) = \frac{\sum_{a \in B} \phi_a^*(x_1, x_2)}{card\{B\}}$, where

$$\phi_a^*(x_1, x_2) = \begin{cases} 0 & \text{if } a(x_1)a(x_2) = 1 \text{ or } x_1 = x_2, \\ 0.5 & \text{if } a(x_1)a(x_2) = 0 \text{ and } x_1 \neq x_2, \\ 1 & \text{if } a(x_1)a(x_2) = -1. \end{cases} \tag{1}$$

A conflict function $\rho_B : U \times U \to [0, 1]$ for the set of issues $B \subseteq A$ is defined as follows:

$$\rho_B(x_1, x_2) = \frac{card\{\delta_B(x_1, x_2)\}}{card\{B\}}, \tag{2}$$

where $\delta_B(x_1, x_2) = \{a \in B : a(x_1) \neq a(x_2)\}$. When the attribute set for calculating either of the two functions matches the full attribute set $(B = A)$, we abbreviate it as ρ^* or ρ. These functions differ in how they handle agent neutrality. The distance function is more precise; if one agent in a pair is neutral, the difference between the agents equals 0.5. In contrast, the conflict function, regardless of neutrality, increments the counter to 1 for any differences in assigned values on a conflict issue between a pair of agents.

In the Pawlak model, a pair of agents $x_1, x_2 \in U$ is said to be allied $R^+(x_1, x_2)$, if $\rho^*(x_1, x_2) < 0.5$ (or $\rho(x_1, x_2) < 0.5$), in conflict $R^-(x_1, x_2)$, if $\rho^*(x_1, x_2) > 0.5$ (or $\rho(x_1, x_2) > 0.5$), neutral $R^0(x_1, x_2)$, if $\rho^*(x_1, x_2) = 0.5$ (or $\rho(x_1, x_2) = 0.5$). Set $X \subseteq U$ is a coalition if for every $x_1, x_2 \in X$, $R^+(x_1, x_2)$. The resulting coalitions are not necessarily disjoint sets, reflecting the possibility for an agent to participate in multiple coalitions due to moderate views compatible with multiple fractions.

Conflicts and decision-making are ubiquitous in nearly every aspect of our lives. This paper explores a conflict scenario within the realm of politics, specifically drawing on a real-life case from the 2020 presidential election in Poland. Consider the following example, named the "political conviction conflict". This example is derived from the Voting Lighthouse application, a product of the Center for Civic Education developed under Project No. POWR.03.01.00-00-T065/18, titled "Social and Civic Activation of Young People in the Development of Key Competencies" [1]. In this example, we have nine agents represented as $U = \{x_1, \ldots, x_9\}$ – these agents are the candidates in the presidential election: x_1 – Krzysztof Bosak; x_2 – Marek Jakubiak; x_3 – Mirosław Piotrowski; x_4 – Paweł Tanajno; x_5 – Robert Biedroń; x_6 – Stanisław Żółtek; x_7 – Szymon Hołownia; x_8 – Waldemar Witkowski; x_9 – Władysław Kosiniak-Kamysz; (two candidates in this election are missing, as in the Voting Lighthouse application there are no opinions for them) and twenty five issues $A = \{a_1, \ldots, a_{25}\}$: a_1 – Declare an emergency state in coronavirus-like situations; a_2 – Grant educational institutions more curriculum autonomy; a_3 – Prioritize elevating national identity in cultural policies; a_4 – Fund public media from the state budget; a_5 – Broaden abortion legality; a_6 – Reserve marriage for heterosexual couples; a_7 – Reduce church hierarchy influence in public affairs; a_8 – Consider easier firearm access; a_9 – Transition from coal by 2035; a_{10} – Pursue a nuclear plant in Poland; a_{11} – Allow raising animals for fur; a_{12} – Tax digital giants targeting Polish users; a_{13} – Raise taxes for high-income earners; a_{14} – Enable Swiss franc loan conversion at original cost; a_{15} – Expand President's defense policy authority; a_{16} – Strengthen judiciary independence; a_{17} – Broaden local government jurisdiction; a_{18} – Fund public housing instead of private rental subsidies; a_{19} – Establish early retirement for experienced workers; a_{20} – Significantly raise the minimum wage; a_{21} – Allow higher standard medical services payments; a_{22} – Reduce EU influence on Polish domestic policies; a_{23} – Prioritize the United States as Poland's foreign partner; a_{24} – Increase defense spending; a_{25} – Accept more labor migrants from other nations.

The views of each agent to a specific issue is presented in Table 1, where, according to the Pawlak model of conflict analysis, 1 means agree, 0 have no opinion, -1 disagree.

Table 1. Information system for the political conviction conflict.

	a_1	a_2	a_3	a_4	a_5	a_6	a_7	a_8	a_9	a_{10}	a_{11}	a_{12}	a_{13}	a_{14}	a_{15}	a_{16}	a_{17}	a_{18}	a_{19}	a_{20}	a_{21}	a_{22}	a_{23}	a_{24}	a_{25}
x_1	1	1	1	0	-1	1	-1	1	-1	1	1	0	-1	1	-1	1	1	0	1	-1	1	1	1	1	-1
x_2	1	-1	1	1	-1	1	1	1	-1	1	1	-1	-1	1	1	-1	-1	-1	1	-1	1	1	1	-1	-1
x_3	1	1	1	1	-1	1	-1	-1	-1	0	1	0	-1	1	1	1	1	1	1	-1	-1	1	0	1	-1
x_4	-1	1	0	-1	0	0	1	1	-1	1	0	1	-1	1	1	1	1	-1	-1	-1	1	1	1	-1	1
x_5	1	1	-1	1	1	-1	1	-1	1	0	-1	1	0	1	-1	1	1	1	1	1	-1	-1	1	-1	1
x_6	1	1	1	-1	-1	1	-1	1	-1	1	1	-1	-1	-1	1	1	1	-1	1	-1	1	1	1	1	1
x_7	1	1	-1	1	-1	1	1	-1	1	0	-1	1	1	-1	-1	1	1	-1	1	-1	-1	-1	1	1	1
x_8	1	1	-1	0	1	1	1	-1	-1	1	-1	1	1	-1	1	1	1	0	0	1	0	-1	1	-1	1
x_9	1	1	1	-1	-1	1	1	-1	-1	1	0	1	-1	1	1	1	1	1	1	1	1	1	1	-1	0

3.1 Conflict Analysis Using the Pawlak Model and the Distance Function

In the initial approach, we apply the Pawlak model with a distance function for conflict analysis, as discussed in [22]. We compute the distance function between agent pairs, resulting in a symmetrical matrix presented in Table 2, with zeros on the diagonal.

Table 2. Values of the distance function between agents for the political conviction conflict, ρ^*.

	x_1	x_2	x_3	x_4	x_5	x_6	x_7	x_8	x_9
x_1									
x_2	0.3								
x_3	0.2	0.38							
x_4	0.38	0.4	0.5						
x_5	0.58	0.68	0.5	0.52					
x_6	0.18	0.32	0.3	0.32	0.68				
x_7	0.48	0.62	0.44	0.54	0.26	0.46			
x_8	0.54	0.6	0.54	0.4	0.28	0.48	0.3		
x_9	0.3	0.36	0.3	0.28	0.4	0.32	0.5	0.32	

Figure 1 shows a graphical representation of the conflict situation. Agents are represented by circles in the figure. When agents are allied ($\rho^*(x, y) < 0.5$), the circles representing the agents are linked. In order to find coalitions, all cliques should be identified in the graph. So the subset of vertices such that every two vertices are linked is determined. There are seven coalitions in the example $\{x_1, x_2, x_3, x_6, x_9\}$, $\{x_1, x_2, x_4, x_6, x_9\}$, $\{x_1, x_3, x_6, x_7\}$, $\{x_4, x_6, x_8, x_9\}$, $\{x_6, x_7, x_8\}$, $\{x_5, x_7, x_8\}$ and $\{x_5, x_8, x_9\}$. As can be seen, coalitions are non-disjoint sets. Some agents show alliance with almost all other agents. As can be seen candidates Władysław Kosiniak-Kamysz and Stanisław Żółtek are in an alliance relation with virtually all other candidates for president.

As was mentioned earlier, the conflict function assigns smaller values for a pair of agents if one of the agents is neutral. In some real cases, this generates too many coalitions – based on this, no clear division can be defined.

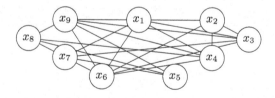

Fig. 1. A graphical representation of the political conviction conflict, the Pawlak conflict analysis model and the distance function

3.2 Conflict Analysis Using the Pawlak Model and the Conflict Function

Another way to analyze conflicts using the Pawlak approach is to use the conflict function which is described in paper [18] and defined by Formula 2. The value of the conflict function between agents is calculated for each pair of agents and are given in Table 3.

Table 3. Values of the conflict function between agents for the political conviction conflict, ρ.

	x_1	x_2	x_3	x_4	x_5	x_6	x_7	x_8	x_9
x_1									
x_2	0.36								
x_3	0.28	0.44							
x_4	0.52	0.48	0.64						
x_5	0.68	0.72	0.56	0.64					
x_6	0.24	0.32	0.36	0.4	0.72				
x_7	0.56	0.64	0.48	0.64	0.28	0.48			
x_8	0.6	0.68	0.68	0.56	0.4	0.56	0.4		
x_9	0.4	0.4	0.4	0.36	0.48	0.36	0.56	0.44	

Figure 2 shows a graphical representation of the conflict situation. The way of preparing the graph and determining the coalition is the same as before. There are five coalitions in the example $\{x_1, x_2, x_3, x_6, x_9\}$, $\{x_2, x_4, x_6, x_9\}$, $\{x_3, x_6, x_7\}$, $\{x_5, x_7, x_8\}$, $\{x_5, x_8, x_9\}$. As before, coalitions are non-disjoint sets, but here, there are fewer of them. The conflict function treats neutrality as equivalent to differing opinions among agents. Compared to the previous analysis, some agents now belong to a smaller number of coalitions. This results from a more restrictive treatment of their neutrality, potentially seen as a penalty, as it is considered as different opinion from both proponents and opponents of an issue.

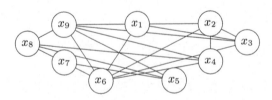

Fig. 2. A graphical representation of the political conviction conflict, the Pawlak conflict analysis model and the conflict function

3.3 Conflict Analysis Using Hierarchical Clustering for Determining Disjoint Clusters

Another approach was presented in paper [25] and consists in combination of the Pawlak approach with an agglomerative hierarchical clustering algorithm.

The major difference now is that coalitions are disjoint sets, and the method to generate coalitions relies on iteratively combining the agents with the smallest distance into groups. The agglomerative hierarchical algorithm is implemented based of the conflict function value matrix. Initially, each agent is treated as a separate cluster. Coalitions are generated iteratively as follows:

1. One pair of different clusters is selected for which the conflict function reaches a minimum value. If the selected value is less than 0.5, then agents from the selected pair of clusters are combined into one new cluster. Otherwise, the clustering process is terminated.

2. After defining a new cluster, the value of the distance between the clusters is recalculated. The following method for recalculating the value of the distance is used. Let $\hat{\rho} : 2^U \times 2^U \to [0, 1]$, where $\hat{\rho}(\{x_1\}, \{x_2\}) = \rho(x_1, x_2)$ for each $x_1, x_2 \in U$ and let C_i be a cluster formed from the merger of two clusters $C_i = C_{i,1} \cup C_{i,2}$ and let it be given a cluster C_j then

$$
\hat{\rho}(C_i, C_j) = \begin{cases} \frac{\hat{\rho}(C_{i,1}, C_j) + \hat{\rho}(C_{i,2}, C_j)}{2} & \text{if } \hat{\rho}(C_{i,1}, C_j) < 0.5 \\ & \text{and } \hat{\rho}(C_{i,2}, C_j) < 0.5 \\ \max\{\hat{\rho}(C_{i,1}, C_j), \hat{\rho}(C_{i,2}, C_j)\} & \text{if } \hat{\rho}(C_{i,1}, C_j) \geq 0.5 \\ & \text{or } \hat{\rho}(C_{i,2}, C_j) \geq 0.5 \end{cases} \tag{3}
$$

In Formula 3, the second equation ensures that agents in conflict relations are excluded from a single coalition. Relying solely on the first equation might lead to a scenario where for two values – one exceeding 0.5 and the other falling below 0.5 – the average could be less than 0.5. Subsequently, this could result that agents in conflict relations being included in one coalition in subsequent steps.

Table 3 shows the values of the conflict function between agents in the considered political conviction conflict. In the first step, we select a pair of agents for which the conflict function takes the smallest value. This will be $\rho(x_6, x_1) = 0.24$. Then we combine these agents into a cluster and recalculate the distances between a new cluster and other agents according to Formula 3. Distance function values recalculated according to the proposed method are proposed in Table 4. For example, the value of the conflict function for the pair $\{x_1, x_6\}$ and $\{x_2\}$ was calculated as follows
$\hat{\rho}(\{x_1, x_6\}, \{x_2\}) = \frac{\hat{\rho}(\{x_1\}, \{x_2\}) + \hat{\rho}(\{x_6\}, \{x_2\})}{2} = \frac{0.36 + 0.32}{2} = 0.34$
We use this formula because both values of the conflict function $\hat{\rho}(\{x_1\}, \{x_2\})$ and $\hat{\rho}(\{x_6\}, \{x_2\})$ are less than 0.5.

The remaining steps of the agglomerative hierarchical clustering follow a similar approach. The realization of the whole process is presented as a dendrogram in Fig. 3. We end the clustering process when all $\hat{\rho}$ function values are greater or equal to 0.5. This results in three distinct coalitions: $\{x_1, x_2, x_3, x_6\}$, $\{x_5, x_7, x_8\}$, $\{x_4, x_9\}$. Consequently, each presidential candidate now belongs to a single coalition, eliminating one-element isolated points among the candidates.

Table 4. Values of conflict function, stage 1 of agglomerative hierarchical clustering algorithm.

	$\{x_1, x_6\}$	$\{x_2\}$	$\{x_3\}$	$\{x_4\}$	$\{x_5\}$	$\{x_7\}$	$\{x_8\}$	$\{x_9\}$
$\{x_1, x_6\}$								
$\{x_2\}$	0.34							
$\{x_3\}$	0.32	0.44						
$\{x_4\}$	0.52	0.48	0.64					
$\{x_5\}$	0.72	0.72	0.56	0.64				
$\{x_7\}$	0.56	0.64	0.48	0.64	0.28			
$\{x_8\}$	0.6	0.68	0.68	0.56	0.4	0.4		
$\{x_9\}$	0.38	0.4	0.4	0.36	0.48	0.56	0.44	

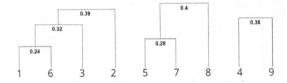

Fig. 3. Dendrogram – the implementation of agglomerative hierarchical clustering algorithm

3.4   Conflict Analysis Using Negotiations Stage

The approach discussed in this section was proposed in paper [24]. It consists of two stages, defining initial coalitions and considering neutral agents. The first stage is very similar to the Pawlak model, only the way of defining the relations between the agents changes. Let p be a real number that belongs to the interval $(0, 0.5)$. We say that agents $x_1, x_2 \in U$ are allied $R^+(x_1, x_2)$, if and only if $\rho(x_1, x_2) < 0.5 - p$. Agents $x_1, x_2 \in U$ are in a conflict $R^-(x_1, x_2)$, if and only if $\rho(x_1, x_2) > 0.5 + p$. Agents $x_1, x_2 \in U$ are neutral $R^0(x_1, x_2)$, if and only if $0.5 - p \leq \rho(x_1, x_2) \leq 0.5 + p$.

In the second stage of this approach, in addition to agents' opinions on the conflict issues, it is vital to identify their top-priority conflict issues. Let us assume that the most important issues for the agents are as follows: a_5, a_7, a_9 for agent x_1; a_3, a_5, a_9 for x_2; a_5, a_7, a_8 for x_3; a_1, a_4, a_9 for x_4; a_3, a_6, a_8 for x_5; a_4, a_5, a_7 for x_6; a_3, a_5, a_8 for x_7; a_3, a_8, a_9 for x_8; a_4, a_5, a_8 for x_9. This data is not available in the Voting Lighthouse application [1], it was assigned by the authors of the article. Let us also set p at 0.1. Consequently, any pair of agents with conflict function values belongs to the $[0.4, 0.6]$ range is considered neutral. Therefore, only agent pairs with conflict function values below 0.4 form alliances. Figure 4, generated using the data in Table 3, illustrates these alliances with connecting lines.

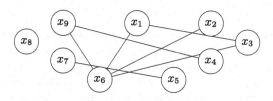

Fig. 4. A graphical representation of the political conviction conflict, conflict analysis with negotiations stage

The initial coalitions are all complete subgraphs depicted in Fig. 4: $\{x_1, x_3, x_6\}$, $\{x_2, x_6\}$, $\{x_4, x_9\}$, $\{x_6, x_9\}$, $\{x_5, x_7\}$. In the next step of the algorithm, only agents that have not been included in any initial coalition and those that are not in conflict are considered (not in conflict means $\rho(x_i, x_j) \leq 0.6$ for $x_i, x_j \in U$). For these agents, we determine the values of a generalized distance function ϕ_G. It is a function $\phi_G : U \times U \to [0, \infty)$ where

$$\phi_G(x_i, x_j) = \frac{\sum_{a \in Sign_{i,j}} |a(x_i) - a(x_j)|}{card\{Sign_{i,j}\}}$$

where $x_i, x_j \in U$ and $Sign_{i,j} \subseteq A$ is the set of significant conflicting issues for the pair of agents x_i, x_j. In the set $Sign_{i,j}$ there are issues, which are the most significant for agents x_i or x_j. Thus, for the pair of agents the average module of difference of opinion on issues that are significant for these agents is calculated. These values are shown in Table 5. In the considered example, only agent x_8 does not belong to any initial coalition. If a conflict occurs between the two agents, then the corresponding cell in Table 5 contains the sign X.

Table 5. Values of the generalized distance function between agents, ϕ_G.

	x_1	x_2	x_3	x_4	x_5	x_6	x_7	x_9
x_8	1.60	X	X	1.20	1.00	1.50	1.00	1.00

During the second stage of the cluster creating process, the negotiation process and the intensity of the conflict between the two groups of agents is determined by using the generalized distance. We define the generalized distance between agents $\rho_G : 2^U \times 2^U \to [0, \infty)$

$$\rho_G(X, Y) = \begin{cases} 0 & \text{if } card\{X \cup Y\} \leq 1 \\ \dfrac{\sum_{x_i, x_j \in X \cup Y} \phi_G(x_i, x_j)}{card\{X \cup Y\} \cdot (card\{X \cup Y\} - 1)} & \text{else} \end{cases} \quad (4)$$

where $X, Y \subseteq U$. The value of the generalized conflict function between the initial clusters and agent x_8 are calculated as follows

$$\rho_G^x(\{x_4, x_9\}, \{x_8\}) = \frac{\phi_G^x(x_4, x_8) + \phi_G^x(x_9, x_8) + \phi_G^x(x_4, x_9)}{3} = 1.0(6)$$
$$\rho_G^x(\{x_6, x_9\}, \{x_8\}) = \frac{\phi_G^x(x_6, x_8) + \phi_G^x(x_9, x_8) + \phi_G^x(x_6, x_9)}{3} = 1.1(6)$$

$$\rho_G^x(\{x_5, x_7\}, \{x_8\}) = \frac{\phi_G^x(x_5, x_8) + \phi_G^x(x_7, x_8) + \phi_G^x(x_5, x_7)}{3} = 1$$

We combine agents whose generalized distance doesn't exceed a user-defined threshold, typically set at 2 as in [24]. Agent x_8 is combined with all previously mentioned initial coalitions, resulting in these final sets: $\{x_1, x_3, x_6\}$, $\{x_2, x_6\}$, $\{x_4, x_8, x_9\}$, $\{x_6, x_8, x_9\}$ and $\{x_5, x_7, x_8\}$. This approach yields larger coalitions compared to agglomerative hierarchical clustering but smaller than the classic Pawlak model. To implement this method, we require additional information: each agent's critical issues.

3.5 Consensus Model

The final goal of analysis of any conflict is to propose the solution i.e. the consensus. Consensus is the situation which is acceptable by all the agents taking part in the conflict. Analyzing the Table 2 it can be easily noticed that such situation in the discussed political conviction conflict does not exist for an acceptable value of the distance function (here the threshold is set to 0.5).

On the other hand, there are many real examples where the consensus is found, usually within the negotiation process. In the papers [3, 26] the enhancements of the Pawlak model have been proposed to embrace the background knowledge of the conflict and allows to search for solution (acceptable situation) not visible in information table describing the conflict.

Local States. We assume [3, 26] that each agent has its information table with local states defining its current view as well as preferences i.e. $I_x = (U_x, A_x)$, where $a : U_x \rightarrow V_a$ for any $a \in A_x$ and V_a is the value set of attribute a, A_x is the set of attributes and U_x is the set of local states of the agent x. Any local state $s \in U_x$ is fully described by the information vector $Inf_{A_x}(s) = (a, a(s)) : s \in U_x$. We assume that sets $\{A_x\}$ are pairwise disjoint, i.e. $\{A_x \cap A_y = \emptyset\}$ where x and y denotes different agents. The user preferences are expressed by assigning the subjective evaluation to each state. Let $e_x : U_x \rightarrow R[0, 1]$ is the target function, then the states with greater value of e_x are assumed to be more preferred.

In discussed political conviction conflict the information about agents preferences is missing. However, it is quite common that after elections parties are creating coalitions and agreeing their views (there are issues they care about more than others). For illustrative purposes, we generated the local states for each agent by changing the view for two attributes randomly. To each of the local states, we assign the value of subjective evaluation. The exemplar local state decision table for agent x_1 is present in Table 6.

The best way to approximate the agent preferences is to infer the rules from the local state decision table. We are generating the minimal rules in the number of attributes from the left side. The set of rules forms the formula describing acceptable states of the given agent f_{s_x}.

The exemplar rules for agent x_1 preferred states are as follow: $(a_4 = 0 \wedge a_3 = 1)$. The threshold for the target function is set to $e_x = \frac{1}{3}$.

Table 6. Local states with evaluation for agent x_1 in the political conviction conflict.

a_1	a_2	a_3	a_4	a_5	a_6	a_7	a_8	a_9	a_{10}	a_{11}	a_{12}	a_{13}	a_{14}	a_{15}	a_{16}	a_{17}	a_{18}	a_{19}	a_{20}	a_{21}	a_{22}	a_{23}	a_{24}	a_{25}	e_{x_1}
1	1	1	0	-1	1	-1	1	-1	1	1	0	-1	1	-1	1	1	0	1	-1	1	1	1	1	-1	$\frac{2}{3}$
1	-1	1	1	-1	1	1	1	-1	1	1	-1	-1	1	1	-1	-1	-1	1	-1	1	1	1	-1	-1	0
1	1	1	1	-1	1	-1	-1	-1	0	1	0	-1	1	1	1	1	1	1	-1	-1	1	0	1	-1	0
-1	1	0	-1	0	0	1	1	-1	1	0	1	-1	1	1	1	1	-1	-1	-1	1	1	1	-1	1	0
1	1	-1	1	1	-1	1	-1	1	0	-1	1	0	1	-1	1	1	1	1	1	-1	-1	1	-1	1	0
1	1	1	-1	-1	1	-1	1	-1	1	1	-1	-1	-1	1	1	1	-1	1	-1	1	1	1	1	1	0
1	1	-1	1	-1	1	1	-1	1	0	-1	1	1	-1	-1	1	1	-1	1	-1	-1	-1	1	1	1	0
1	1	-1	0	1	1	1	-1	-1	1	-1	1	1	-1	1	1	1	0	0	1	0	-1	1	-1	1	0
1	1	1	-1	-1	1	1	-1	-1	1	0	1	-1	1	1	1	1	1	1	1	1	1	1	-1	0	0
1	1	1	0	-1	1	-1	1	-1	1	1	0	-1	1	-1	1	0	0	1	-1	1	1	1	1	-1	$\frac{2}{3}$
1	1	1	0	-1	1	-1	1	-1	1	1	0	-1	1	-1	1	1	-1	1	-1	1	1	1	1	-1	$\frac{2}{3}$
1	1	1	0	-1	1	-1	1	-1	1	1	0	-1	1	-1	1	0	-1	1	-1	1	1	1	1	-1	$\frac{1}{3}$

Constraints. Additionally, to understand the root cause of the conflict and to find a consensus easier, some constraints can be specified. Constraints describe the dependencies among the local states of agents. They can come from the resources limitation or just specify something crucial from the agent's perspective. We assume that finally the constrains are delivered in the form of propositional formula here denoted by f_C. The example for agent x_5 could be as follow: $(a_9 = 1 \land a_{10} = 1)$, which is interpreted that for x_5 it is vital to build the nuclear plant to move away from coal.

Situations. The situation [3,26] in the conflict is any element of the cartesian product $S(U) = \prod_{i=1}^{n} INF(x_i)$, where $n = card(U)$, is the number of agents taking part in the conflict and $INF(x)$ is the set of all possible information vectors of agent x. The situation corresponding to the global state $\bar{s} = (s_1, ..., s_n) \in U_{x_1} \times ... \times U_{x_n}$ is defined by $(Inf_{x_1}(s_1), ..., Inf_{x_n}(s_n))$.

Similar to local states, we assume the situation can be evaluated too. The main idea lies in belief that the stable solution is obtained when the common good is considered. The situation evaluation $q(S)$ can be given by the expert like the negotiator or independent organization. Based on the given threshold $q(S) \geq q_t$ we define the set of situations by prepositional formula f_S obtained by inferring the rules from the decision table corresponding to the information table of situations $S(U)$. Another way to evaluate the situations is by applying the agents local state evaluation in the calculation, i.e. the global state evaluation can be defined by $p(\bar{s}) = F(e_{x_1}, ..., e_{x_n})$, where F is a suitable function e.g. $F(r_1, ..., r_m) = \sum_{i=1}^{m} r_i$.

Consensus. The described model above with user preferences, situations evaluation and constraints forms the basis to efficiently search for consensus. Consensus is the set of situations that satisfies the boolean formula $f = \bigwedge_{x \in U} f_{s_x} \land f_C \land f_S$.

The information gathered in political conviction conflict is not enough to fully present the concepts within the presented model. We augment the information with local states and their evaluation as described in Subsect. 3.5. Also, the assumption that agents are locally using different set of attributes seems to be difficult to achieve. In reality the conflicting parties are often using different wording even when the overall meaning is similar. To match our model and available information in political conviction conflict we add the following constraints: $\forall_{x,y \in U} \forall_{a_x \in A_x, a_y \in A_y} a_x = a_y$. That means the set of agents attributes is the same. When assuming the situation is evaluated based on agents local states evaluation, we can simplify our consensus problem, for political conviction example, to the formula $f = \bigwedge_{x \in U} f_{s_x}$. Also we are not taking into account any constraints.

Continuing the experiment with political conviction conflict, we obtained the descriptions of preferable states for exemplar as presented in Table 7.

Table 7. Agents preferable states description

agent	f_{s_x}
x_1	$a_3 = 1 \land a_4 = 0$
x_2	$a_2 = -1$
x_3	$a_{23} = 0$
x_4	$a_1 = -1$
x_5	$a_6 = -1$
x_6	$a_{12} = -1 \land a_2 = 1$
x_7	$a_6 = 1 \land a_9 = 1$
x_8	$a_{19} = 0$
x_9	$a_1 = 1 \land a_{11} = 0$

Computing the formula f, it can be noticed that acceptable situation cannot be found. However, we can propose such situations for limited number of agents. For example when excluding agents x_6, x_7, and x_9 the acceptable situations are described by the formula: $(a_1 = -1) \land (a_2 = -1) \land (a_3 = 1) \land (a_6 = -1) \land (a_{19} = 0) \land (a_{23} = 0)$. Note that the solution (consensus) can be found among the situations not considered in the conflict description.

4   Comparison and Application Areas

We compare the discussed approaches and highlight their application domains. Notably, the outcomes from these models differ. The target of the approaches from Sects. 3.1, 3.2, 3.3 and 3.4 is assessing agents' alignment and forming consistent coalitions, while the model described in Sect. 3.5 seeks consensus within set constraints and preferences. Table 8 summarizes key model characteristics, necessary input data, suggested application domains and evidence for these conclusions.

Table 8. The comparison of conflict models

Model	Characteristics	Require-ments	Application areas	Evidence
Pawlak model with distance function	The same set of issues for agents. Neutrality is treated more softly. The largest number of coalitions with a greater number of agents are generated. Coalitions are non-disjoint	Agents' views on conflict situations	Any area where our goal is to determine large groups of agents/units that can cooperate with each other because they have compatible views/goals. Designation on a global scale sets of companies that can cooperate with each other	According to Formula 1, an agent who holds a neutral stance on an issue will receive a distance value of 0.5, regardless of the opinions of other agents. This ensures a lower value for the distance function and increases the likelihood of forming alliance relations between agents
Pawlak model and conflict function	The same set of issues for agents. Neutrality is treated the same as all other views. Smaller coalitions than in the previous approach are generated. Coalitions are non-disjoint	Agents' views on conflict situations	Situations where rather agents are not neutral to the conflicting issues. For example collaborative agents, where each one has two states: action, non-action toward the issue. We want to designate areas of collaboration for these agents	According to Formula 2, when an agent adopts a neutral stance on an issue, it is considered distinct from agents taking a position either in favor of or against the issue
Model with agglomerative hierarchical clustering	The same set of issues for agents. Coalitions are disjoint. Neutrality is treated the same as all other views. The smallest number and separated coalitions are generated	Agents' views on conflict situations	Applications in situations where we want to generate antagonistic groups: conflicts that exclude cooperation in many groups at the same time	According to Formula 3 and the stop condition outlined in Sect. 3.3, it is evident that the clusters are pairwise disjoint
Model with negotiations stage	The same set of issues for agents. Two stages of coalitions generation. Ability to tune the parameter that controls agents' alignment. Ability to focus on relevant issues for agents. Identification of neutral agents who can be seen as peacemakers between coalitions	Agents' views on conflict situations. The most significant issues for agents	Situations in which we are interested in recognizing smaller groups and identifying agents that are a bridge between strongly compatible groups (this strength can be controlled by a parameter)	The neutrality condition has been expanded to a broader range, significantly increasing the likelihood of an agent being in neutral relation. Conversely, the alliance threshold has been elevated to $(0.5 - p$ where $p > 0)$. As a result, fewer agents maintain in alliance relation compared to the previous approaches
Consensus model	Different set of attributes used by each agent to describe the local states. The situation is composed from local states. Looking for consensus is the main goal of the model (the coalitions are not investigated)	Back-ground knowledge including local states evaluation, situation evaluation and constraints	Any conflict can be analysed, however the background knowledge should be available. Ready to use in negotiations, where is allowed to find the solution not only within the situations considered by agents	We can individually characterize each agent's views and evaluation of reality using a set of local states U_x and a target function e_x. Furthermore, constraints are introduced in the system, which can reflect limited resources and is a significant extension compared to previous proposals

5   Conclusions

The paper presented an overview of selected conflict analysis methods based on the Pawlak model. A real-life example concerning the presidential election in 2020 in Poland was discussed. Coalitions of presidential candidates using different approaches were generated. The study conducted a comparative evaluation of the conflict analysis models in terms of their features and prerequisites. The paper illuminates noteworthy distinctions among the discussed conflict analysis methods and highlights their respective areas of application.

References

1. (CCE) The Center for Citizenship Education: Voting Lighthouse application. https://latarnikwyborczy.pl/
2. Cinelli, M., Kadziński, M., Gonzalez, M., Słowiński, R.: How to support the application of multiple criteria decision analysis? Let us start with a comprehensive taxonomy. Omega **96**, 102261 (2020)
3. Deja, R.: Conflict analysis. Int. J. Intell. Syst. **17**(2), 235–253 (2002)
4. Dutta, S., Skowron, A.: Interactive granular computing model for intelligent systems. In: Shi, Z., Chakraborty, M., Kar, S. (eds.) ICIS 2021. IAICT, vol. 623, pp. 37–48. Springer, Cham (2021). https://doi.org/10.1007/978-3-030-74826-5_4
5. Dutta, S., Skowron, A.: Toward a computing model dealing with complex phenomena: interactive granular computing. In: Nguyen, N.T., Iliadis, L., Maglogiannis, I., Trawiński, B. (eds.) ICCCI 2021. LNCS (LNAI), vol. 12876, pp. 199–214. Springer, Cham (2021). https://doi.org/10.1007/978-3-030-88081-1_15
6. Fan, Y., Qi, J., Wei, L.: A conflict analysis model based on three-way decisions. In: Nguyen, H.S., Ha, Q.-T., Li, T., Przybyła-Kasperek, M. (eds.) IJCRS 2018. LNCS (LNAI), vol. 11103, pp. 522–532. Springer, Cham (2018). https://doi.org/10.1007/978-3-319-99368-3_41
7. Giordano, R., Passarella, G., Uricchio, V., Vurro, M.: Integrating conflict analysis and consensus reaching in a decision support system for water resource management. J. Environ. Manage. **84**(2), 213–228 (2007)
8. Gong, Z.T., Sun, B.Z., Xu, Z.M., Zhang, Z.Q.: The rough set analysis approach to water resources allocation decision in the inland river basin of arid regions (ii): The conflict analysis of satisfactions of the decision. In: 2006 International Conference on Machine Learning and Cybernetics, pp. 1358–1361. IEEE (2006)
9. Greco, S., Matarazzo, B., Slowinski, R.: Rough sets theory for multicriteria decision analysis. Eur. J. Oper. Res. **129**(1), 1–47 (2001)
10. Han, X., Dleu, T., Nguyen, L., Xu, H.: Conflict analysis based on rough set in e-commerce. Int. J. Adv. Manag. Sci. **2**(1), 1–8 (2013)
11. Helmold, M., Dathe, T., Hummel, F., et al.: Successful Negotiations. Springer, Heidelberg (2022). https://doi.org/10.1007/978-3-658-35701-6
12. Hipel, K.W., Fang, L., Kilgour, D.M.: The graph model for conflict resolution: reflections on three decades of development. Group Decis. Negot. **29**, 11–60 (2020)
13. Lang, G., Luo, J., Yao, Y.: Three-way conflict analysis: a unification of models based on rough sets and formal concept analysis. Knowl. Based. Syst. **194**, 105556 (2020)
14. Lang, G., Miao, D., Cai, M.: Three-way decision approaches to conflict analysis using decision-theoretic rough set theory. Inf. Sci. **406**, 185–207 (2017)

15. Lin, T.Y.: Granular computing on binary relations analysis of conflict and Chinese wall security policy. In: Alpigini, J.J., Peters, J.F., Skowron, A., Zhong, N. (eds.) RSCTC 2002. LNCS (LNAI), vol. 2475, pp. 296–299. Springer, Heidelberg (2002). https://doi.org/10.1007/3-540-45813-1_38
16. Liu, D., Liang, D., Wang, C.: A novel three-way decision model based on incomplete information system. Knowl. Based. Syst. **91**, 32–45 (2016)
17. Luo, J., Hu, M., Lang, G., Yang, X., Qin, K.: Three-way conflict analysis based on alliance and conflict functions. Inf. Sci. **594**, 322–359 (2022)
18. Pawlak, Z.: An inquiry into anatomy of conflicts. Inf. Sci. **109**(1–4), 65–78 (1998)
19. Pawlak, Z.: Rough sets. Int. J. Comput. Inf. Sci. **11**, 341–356 (1982)
20. Pawlak, Z.: On conflicts. Int. J. Man-Mach. Stud. **21**(2), 127–134 (1984)
21. Pawlak, Z.: Anatomy of conflics. Bull. EATCS **50**, 234–246 (1993)
22. Pawlak, Z.: Some remarks on conflict analysis. Eur. J. Oper. Res. **166**(3), 649–654 (2005)
23. Pawlak, Z., Skowron, A.: Rudiments of rough sets. Inf. Sci. **177**(1), 3–27 (2007)
24. Przybyła-Kasperek, M., Wakulicz-Deja, A.: A dispersed decision-making system- the use of negotiations during the dynamic generation of a system's structure. Inf. Sci. **288**, 194–219 (2014)
25. Przybyła-Kasperek, M., Wakulicz-Deja, A.: Global decision-making system with dynamically generated clusters. Inf. Sci. **270**, 172–191 (2014)
26. Skowron, A., Deja, R.: On some conflict models and conflict resolutions. Roman. J. Inf. Sci. Technol. **3**(1–2), 69–82 (2002)
27. Skowron, A., Ramanna, S., Peters, J.F.: Conflict analysis and information systems: a rough set approach. In: Wang, G.-Y., Peters, J.F., Skowron, A., Yao, Y. (eds.) RSKT 2006. LNCS (LNAI), vol. 4062, pp. 233–240. Springer, Heidelberg (2006). https://doi.org/10.1007/11795131_34
28. Stepaniuk, J., Skowron, A.: Three-way approximation of decision granules based on the rough set approach. Int. J. Approximate Reasoning **155**, 1–16 (2023)
29. Sun, B., Chen, X., Zhang, L., Ma, W.: Three-way decision making approach to conflict analysis and resolution using probabilistic rough set over two universes. Inf. Sci. **507**, 809–822 (2020)
30. Sun, B., Ma, W., Zhao, H.: Rough set-based conflict analysis model and method over two universes. Inf. Sci. **372**, 111–125 (2016)
31. Yao, J., Medina, J., Zhang, Y., Ślęzak, D.: Formal concept analysis, rough sets, and three-way decisions (2022)
32. Yao, Y.: Three-way decisions with probabilistic rough sets. Inf. Sci. **180**(3), 341–353 (2010)
33. Zeng, Y., Li, J., Cai, Y., Tan, Q., Dai, C.: A hybrid game theory and mathematical programming model for solving trans-boundary water conflicts. J. Hydrol. **570**, 666–681 (2019)

Multi-heuristic Induction of Decision Rules

Beata Zielosko$^{(\boxtimes)}$ ⓘ, Evans Teiko Tetteh ⓘ, and Diana Hunchak

Institute of Computer Science, University of Silesia in Katowice, Będzińska 39,
41-200 Sosnowiec, Poland
{beata.zielosko,evans.tetteh}@us.edu.pl

Abstract. The main objectives of data mining tasks involve extracting knowledge from data, which can be presented in the form of distributed local sources or centralized one. Inducing decision rules from one local data source is relatively straightforward. Nevertheless, obtaining a global model of rules based on different rule-based models is a more complicated task. In the paper, a new method for inducing decision rules from different sets of rules considered as data sources that are spread out is proposed. Each data source is characterized by a set of rules that are derived from the decision table using three different heuristics. To achieve a comprehensive model that represents the knowledge found within these different models, methods for global optimization relative to length and support are proposed. Experiments were performed on datasets from UCI Machine Learning Repository taking into account the characteristics of induced rule sets, i.e., their number, length and support, and classification accuracy. Constructed global rule-based models, taking into account average values, are comparable to the best results related to local rule-based models.

Keywords: Decision rules · Heuristics · Length · Support

1 Introduction

Rule-based expert systems play an important role in the development of artificial intelligence domain. An important element of these systems is decision rules. They are taking the form of $IF\dots THEN$ sentences and allow the intuitive presentation of expert knowledge, which they are used for modelling human reasoning processes and solving problems in different domains. Decision rules are considered as one of the most popular forms of knowledge representation used in various fields of data mining and machine learning.

There exist a variety of approaches and algorithms for the induction of decision rules. Most of the approaches, with the exception of brute-force, Apriori algorithm [1], extensions of dynamic programming [3], and Boolean reasoning [14] cannot guarantee the construction of optimal rules (i.e., rules with minimum length or maximum support). Therefore, algorithms based on sequential

A. Campagner et al. (Eds.): IJCRS 2023, LNAI 14481, pp. 18–30, 2023.
https://doi.org/10.1007/978-3-031-50959-9_2

covering procedure [19], greedy algorithms [23], and different biologically inspired methods such as genetic algorithms [22], and many others [6,21], are used for rule induction. In addition, approaches for constructing rules directly from data or by using other objects such as decision trees or reducts should be mentioned.

It is known that constructing rules with minimum length or maximum support is an NP-hard problem [4,10,13]. In the paper [2], heuristics *RM*, *Poly*, and *Log*, for decision rule induction, were proposed and it was shown experimentally that they allow constructing short rules with relatively high support. Moreover, such heuristics were also applied in the feature selection area [20].

Measures such as length and support are important factors in determining the quality of decision rules, especially from the point of view of knowledge representation. Short rules are easy to understand and interpret. The choice of this measure also coincides with the Minimum Description Length principle [17] "the best hypothesis for a given set of data is the one that leads to the largest compression of data". Support plays an important role from the point of view of knowledge discovery, as it allows to map important patterns present in the data. In the case of classification, shorter rules can decrease the likelihood of model overfitting and enhance the model's ability to generalize.

Technological progress necessitates the handling of ever-expanding volumes of data, sourced from diverse data origins. This implies that this data can differ in terms of where it comes from and how it's structured. Sets of rules induced by different approaches and algorithms can be considered as distributed data sources. The field of distributed data analysis has been steadily growing in recent years [5] with a variety of assumptions. In the framework of decision rules induction, a parallel computation approach for learning a single model from a set of disjoint data sets which are distributed across a set of computers was proposed in [7]. Paper [11] describes an approach for data analysis using decision trees, rules, reducts and association rules, which can be useful in the case of distributed data mining tasks. In [12], authors presented an algorithm and heuristic for learning decision rules from a set of decision trees where trees are considered as local, distributed data sources.

In this paper, a new approach to the induction of decision rules is proposed. Data sources are represented by sets of rules induced from the decision table by three different heuristics *RM*, *Poly*, and *Log*. In order to obtain a global model representing knowledge occurring in these different rule sets, methods for optimization of rules with respect to support and length were proposed.

The presented methods of global optimization have many practical applications, especially nowadays when there is a need to process large amounts of scattered data also related to the same field. As an example, a medical case can be referenced, when, on the basis of data related to the same specialization, e.g. paediatric, located in different places in the country, there is a need to extract knowledge mapping general patterns in these data, or deviations that are also important in such application. The construction of a global model allows obtaining patterns that reflect the knowledge that is true for most of the distributed sources and for the paediatric as a domain.

In the framework of performed experiments, the global model of rules was compared with rule-based models induced by the considered heuristics applied to the decision table, and with decision rules induced from the decision tree constructed by ID3 algorithm [16]. The motivation for choosing the latter approach was that decision trees are considered as good predictors and their form of knowledge representation is compatible with decision rules. The experiments were conducted on datasets from the UCI Machine Learning Repository [8] and involved a comparison from the point of view of (i) knowledge representation, i.e. the characteristics of rule sets, their number, length and support, and (ii) classification by applying the 10-fold cross-validation method. Obtained results show that global optimization relative to support produces rules with greater support on average compared to each of the three heuristics. Global optimization of length allows us to obtain short rules, on average, with values smaller than in the case of *Poly* and *Log* heuristics and comparable with *RM* heuristic. Taking into account classification accuracy proposed methods allow to construction global model of rules with performance comparable to distributed, local rule-based classifiers and greater than the rule-based model obtained by the *ID3* algorithm.

The structure of the paper is organised as follows. Section 2 consists of information about decision rules and the main approaches for their induction, including applied selected heuristics *RM*, *Poly*, and *Log*. Section 3 describes the proposed methods for the global optimization of rules relative to length and support. Experimental results are presented in Sect. 4. Section 5 contains conclusions and future research plans.

2   Decision Rules

In this section, two main approaches for the induction of decision rules are described: one that is applied directly to the data table and the second which allows induction of decision rules based on a decision tree.

2.1   Heuristics for Induction of Decision Rules

First, we will provide the main notions, followed by the representation of knowledge from local data sources using rule-based models induced by the selected heuristics described below.

Main Notions. The main structure for tabular data representation is the decision table $T = (U, C \cup \{d\})$ [15], where U is a nonempty finite set of objects, $C = \{f_1, \ldots, f_n\}$ is a nonempty finite set of condition attributes, i.e., $f_i : U \to V_f$, where V_f is the set of values of attribute f_i, and d, $d \notin C$, is a distinguished attribute called a decision or class label, with values $V_d = \{d_1, \ldots, d_{|V_d|}\}$.

The expression

$$(f_{i_1} = a_1) \wedge \ldots \wedge (f_{i_m} = a_m) \to d = a \tag{1}$$

is called a *decision rule over* T if $f_{i_1}, \ldots, f_{i_m} \in \{f_1, \ldots, f_n\}$, $a_1, \ldots, a_m$ are values of corresponding attributes, and a is a class label.

Let T be a decision table, $N(T)$ denotes the number of rows in the table T. By $N(T, a)$ the number of rows r from T with a value of a decision attribute equal a is denoted, and $M(T, a) = N(T) - N(T, a)$. The *most common decision for* T, is denoted by $mcd(T)$, it is a decision a, such that $N(T, a)$ has a maximum value and a has a minimum index. $E(T)$ denotes the set of not constant condition attributes from T.

Removing some rows from the decision table T allows the creation of a *subtable* of T. It is denoted as $T(f_{i_1}, a_1), \ldots, (f_{i_m}, a_m)$ and it consists of rows, which at the intersection with columns $f_{i_1}, \ldots, f_{i_m}$ have values $a_1, \ldots, a_m$. A decision rule over T (1) corresponds to the subtable $T' = T(f_{i_1}, a_1), \ldots, (f_{i_m}, a_m)$ of T.

The rule (1) is called *realizable for a row* r if a row r belongs to T'. The rule (1) is called *true* for T if each row of T', for which the rule (1) is realizable, has the decision a attached to it. If the considered rule is true for T and realizable for r, then it is a *rule for* T *and* r.

The length of the rule (1) is the number of descriptors from the left-hand side of the rule and is denoted as m. The support of the rule (1) is the number of rows in T', which are labelled with the decision a. If a rule is true for T, then its support equals $N(T')$.

Description of Selected Heuristics. In the paper [2], heuristics RM, $Poly$, and Log for decision rule induction were proposed. They were selected in this paper because they allow the construction of short rules with relatively good support.

Algorithm 1 presents the pseudo-code of the work of heuristic H which is as one of the heuristics RM, $Poly$, Log, where the following notation is used:

$T^{(j+1)} = T^{(j)}(f_i, b_i)$, where j denotes an index of the subsequently obtained subtable during the work of heuristic H,

$RM(f_i, r, a) = (N(T^{(j+1)}) - N(T^{(j+1)}, a))/N(T^{(j+1)})$,

$\alpha(f_i, r, a) = N(T^{(j)}, a) - N(T^{(j+1)}, a)$,

$\beta(f_i, r, a) = M(T^{(j)}, a) - M(T^{(j+1)}, a)$, and

$M(f_i, r, a) = M(T^{(j+1)}, a) = N(T^{(j+1)}) - N(T^{(j+1)}, a)$.

The heuristic H constructs a decision rule for the table T and a row r with assigned decision a. It starts with a decision rule in which the left-hand side is empty, $\rightarrow d = a$. During the work of the algorithm, in each iteration an attribute $f_i \in \{f_1, \ldots, f_n\}$ is selected, such it fulfils heuristic H and has the minimum index. In particular,

- for RM it is a minimization of the value of $RM(f_i, r, a)$,
- for $Poly$ it is a maximization of the value of $\frac{\beta(f_i, r, a)}{\alpha(f_i, r, a) + 1}$,
- for Log it is a maximization of the value of $\frac{\beta(f_i, r, a)}{\log_2(\alpha(f_i, r, a) + 2)}$.

The heuristic H is applied sequentially, for each row r of T, so at the end of the work, the number of induced rules equals $|U|$.

The following Example 1 demonstrates calculations executed by all heuristics.

Algorithm 1. Greedy heuristic H for construction of a decision rule for T and r, H is one of the heuristics RM, $Poly$, Log.

Require: Decision table T with condition attributes $f_1,\ldots,f_n$, row $r=(b_1,\ldots,b_n)$
Ensure: Decision rule for T and r
  begin
  $Q \leftarrow \emptyset$;
  $j \leftarrow 0$;
  $T^{(j)} \leftarrow T$;
  while all rows in $T^{(j)}$ are not assigned the same decision a **do**
    select $f_i \in \{f_1,\ldots,f_n\}$ with the minimum index fulfilling the heuristic H:

    – RM: minimization of the value of $RM(f_i,r,a)$,
    – $Poly$: maximization of the value of $\frac{\beta(f_i,r,a)}{\alpha(f_i,r,a)+1}$,
    – Log: maximization of the value of $\frac{\beta(f_i,r,a)}{\log_2(\alpha(f_i,r,a)+2)}$.

    $T^{(j+1)} \leftarrow T^{(j)}(f_i,b_i)$;
    $Q \leftarrow Q \cup \{f_i\}$;
    $j = j+1$;
  end while
  $\bigwedge_{f_i \in Q}(f_i = b_i) \rightarrow d = a$, where a is a decision value.
  end

Example 1. It will present how heuristic H constructs a decision rule for the decision table T_0, row r_1 with the assigned decision A. The decision table T_0 has three condition attributes, so there are considered three subtables: $T_1^{(1)} = T_0^{(0)}(f_1,0), T_2^{(1)} = T_0^{(0)}(f_2,0)$ *and* $T_3^{(1)} = T_0^{(0)}(f_3,1)$.

$$T_0 = \begin{array}{c|c|c|c|c|} & f_1 & f_2 & f_3 & d \\ \hline r_1 & 0 & 0 & 1 & A \\ r_2 & 2 & 1 & 1 & B \\ r_3 & 2 & 0 & 1 & A \\ r_4 & 2 & 1 & 0 & B \end{array} \quad T_1^{(1)} = \begin{array}{|c|c|c|c|} f_1 & f_2 & f_3 & d \\ \hline 0 & 0 & 1 & A \end{array}$$

$$T_2^{(1)} = \begin{array}{c|c|c|c|c|} & f_1 & f_2 & f_3 & d \\ \hline r_1 & 0 & 0 & 1 & A \\ r_3 & 2 & 0 & 1 & A \end{array} \quad T_3^{(1)} = \begin{array}{c|c|c|c|c|} & f_1 & f_2 & f_3 & d \\ \hline r_1 & 0 & 0 & 1 & A \\ r_2 & 2 & 1 & 1 & B \\ r_3 & 2 & 0 & 1 & A \end{array}$$

Heuristic *Poly*: $\frac{\beta(f_1,r_1,A)}{\alpha(f_1,r_1,A)+1} = \frac{2}{2}$, $\frac{\beta(f_2,r_1,A)}{\alpha(f_2,r_1,A)+1} = \frac{2}{1}$, $\frac{\beta(f_3,r_1,A)}{\alpha(f_3,r_1,A)+1} = \frac{1}{1}$, so the rule $f_2 = 0 \rightarrow d = A$ is obtained.

Heuristic *Log*: $\frac{\beta(f_1,r_1,A)}{\log_2(\alpha(f_1,r_1,A)+2)} = \frac{2}{\log_2 3}$, $\frac{\beta(f_2,r_1,A)}{\log_2(\alpha(f_2,r_1,A)+2)} = \frac{2}{\log_2 2}$, $\frac{\beta(f_3,r_1,A)}{\log_2(\alpha(f3,r1,A)+2)} = \frac{1}{\log_2 2}$, so the rule $f2 = 0 \rightarrow d = A$ is obtained.

Heuristic *RM*: $RM(f_1,r_1,A) = 0, RM(f_2,r_1,A) = 0, RM(f_3,r_1,A) = \frac{1}{3}$, so the rule $f1 = 0 \rightarrow d = A$ is obtained.

An example of global optimization relative to length of induced decision rules is included in Sect. 3.

2.2   Decision Rules Derived from Decision Trees

A decision tree is a popular machine learning algorithm used for both classification and regression tasks [9]. It models decisions and their possible consequences
in a tree-like structure. Each internal node represents a decision based on a feature, each branch represents an outcome of that decision, and each leaf node
represents a predicted class (for classification) or a predicted value (for regression). The induction of rules based on decision trees involves translating the
hierarchical structure of a decision tree into a set of human-readable rules that
outline the conditions for making predictions. To create these rules, we traverse
the decision tree from the root to each leaf/terminal node, noting the attribute
conditions encountered along the way. At each internal node, the condition associated with the chosen branch is added to the rule. This process continues until
a leaf node is reached, at which point the predicted class (for classification) or
the predicted value (for regression) is assigned to the rule.

There are various decision tree algorithms for various use cases [9], however,
in this paper, the Iterative Dichotomiser 3 ($ID3$) by Ross Quinlan [16] was used.
The ID3 algorithm initializes with the original set of objects as the root node. In
each iteration, it assesses unused attributes in the current set, computing their
entropy (2). The attribute with the lowest entropy or highest information gain is
chosen. The set is then divided by this attribute, creating subsets. The process
recurs on each subset, only considering unselected attributes. Recursion stops if:

1. All subset elements have the same class or decision value.
2. No attributes remain, but subset elements differ in class. The leaf node is
   assigned the most common class.
3. No examples exist in the subset due to unmatched attribute values. A leaf
   node is made, labeled with the parent node's most common class.

The algorithm constructs a decision tree, where internal nodes signify attributes
for data division, and leaf nodes denote the final subset's class label.

Entropy, denoted as $H(S)$, quantifies the level of uncertainty present within
a given S.

$$H(S) = \sum_{x \in X} -p(x) \log_2 p(x) \tag{2}$$

where,
S - the current dataset with only unselected attributes,
X - the set of values of decision attribute in S
$p(x)$ - the ratio of the count of elements in class x to the count of elements in
set S.

3   Global Optimization

The aim of the global optimization of decision rules is the construction of a rule-
based model which can be considered as general knowledge of different sets of
rules induced from the decision table. In the paper, global optimization relative
to length and global optimization relative to support are proposed.

It is assumed that decision rules induced by different heuristics are assigned to rows of considered decision table T. By $rule^{H,r}$ is denoted a decision rule induced by the given heuristic H for a row r of T. Such a rule is indicated by two measures: length $rule_{len}^{H,r}$ and support $rule_{supp}^{H,r}$.

Global Optimization Relative to Length. By $Rul_{len}(T,r)$ is denoted a set of rules assigned to a given row after global optimization relative to length. During this process among rules induced by different heuristics and assigned to a given row r, only those are selected which have a minimum value of length $Opt_{len}(T,r)$.

$$Opt_{len}(T,r) = \min\{rule_{len}^{H,r} : H = RM, Poly, Log\},$$

and among all rules corresponding to a given row and heuristics RM, $Poly$, and Log respectively, only these are selected where

$$rule_{len}^{H,r} = Opt_{len}(T,r).$$

As a result of the global optimization relative to the length, each row r of T has assigned set of decision rules $Rul_{len}(T,r)$.

Example 2. For decision table T_0 and row r_1 presented at Example 1, each heuristic induced rule with length equal to 1, so as a result of global optimization relative to length the set of decision rules $Rul_{len}(T_0, r_1) = \{f_2 = 0 \rightarrow d = A, f1 = 0 \rightarrow d = A\}$ assigned to row r_1 is obtained.

Global Optimization Relative to Support. By $Rul_{supp}(T,r)$ is denoted a set of rules assigned to a given row after global optimization relative to support. During this process among rules induced by different heuristics and assigned to a given row r, only those rules are selected which have a maximum value of support $Opt_{supp}(T,r)$.

$$Opt_{supp}(T,r) = \max\{rule_{supp}^{H,r} : H = RM, Poly, Log\},$$

and among all rules corresponding to a given row and heuristics RM, $Poly$, and Log respectively, only these are selected where

$$rule_{supp}^{H,r} = Opt_{supp}(T,r).$$

As a result of the global optimization relative to the support, each row r of T has assigned set of decision rules $Rul_{supp}(T,r)$.

4   Experimental Results

Experiments were performed on ten datasets from UCI ML Repository [8]. For those which contains missing values, each such value was replaced with the most common value of the corresponding attribute.

The aim of the experiments was to analyze experimentally and compare the proposed methods of global optimization relative to length and support, from the point of view of knowledge representation and classification. Obtained global models of rules were compared with local rule-based models induced by the heuristics *RM*, *Poly*, and *Log*. Additionally, they were compared with the model of decision rules derived from decision tree induced by *ID3* algorithm. Such an algorithm was selected as it allows work with categorical data, besides it is known that decision trees are good predictors and their form of knowledge representation is compatible with decision rules.

The experiments were performed using the Python language and ChefBoost framework [18].

Table 1 presents a number of unique rules obtained for global optimization relative to length, support and for each heuristic *RM*, *Poly* and *Log*. For the comparisons results obtained by the *ID3* algorithm are also included. Based on the presented results, it is possible to see that method based on global optimization relative to support produces a smaller number of rules compared to global optimization relative to length. In the case of rules induced by individual heuristics, the number of rules is smaller than in the case of global optimization with respect to length as well as support. This situation is due to the fact that in the case of global optimization several decision rules can be assigned to one row of the decision table. The smallest number of unique rules occurs in the case of the ID3 algorithm.

Table 2 presents the minimum, average (relative to the number of unique rules) and maximum length of decision rules obtained for each heuristic *RM*, *Poly* and *Log* respectively. It is shown that heuristics induce short decision rules. Obtained values can be compared relative to the number of attributes. In this case, for heuristic *Log* and dataset chess-kr-vs-kp with 36 attributes, the maximum length of rules is 11. On average, heuristic *RM* induces the shortest decision rules.

Table 3 presents the minimum, average (relative to the number of unique rules) and maximum support of decision rules obtained for each heuristic *RM*, *Poly* and *Log* respectively. Taking into account the maximum support of decision rules, for three datasets cars, chess-kr-vs-kp, and nursery and each heuristic *RM*, *Poly* and *Log*, the values are relatively big compared to the number of rows in the dataset. On average, the biggest values were obtained by heuristic *Poly*.

Table 4 presents minimum, average (relative to the number of unique rules) and maximum length and support obtained after global optimization respectively relative to length and support. Results related to the *ID3* algorithm are also included. It is possible to see that the global optimization relative to length induces short decision rules. On average, the values are comparable to results obtained by heuristic *RM* (Table 2). In the case of global optimization relative to support, on average, the results increase the values obtained by each of the heuristic *RM*, *Poly* and *Log* presented in Table 3. It is also apparent that the proposed global optimization methods produce better results than rules induced from the decision tree and the *ID3* algorithm.

Table 1. Number of unique decision rules obtained by global optimization relative to length, support, heuristics *RM*, *Poly* and *Log*, and *ID*3 algorithm.

| Dataset | Attr | Rows | Global optimization | | *RM* | *Poly* | *Log* | *ID*3 |
			length	support				
balance-scale	4	625	303	303	303	303	303	401
breast-cancer	9	266	250	222	175	169	143	141
cars	6	1728	465	465	291	301	292	296
chess-kr-vs-kp	36	3196	271	150	220	358	177	49
flags	26	194	143	112	123	98	87	88
house-votes	16	279	71	49	55	50	41	27
lymphography	18	148	79	77	66	55	53	54
nursery	8	12960	1596	1576	855	973	872	839
soybean-large	35	292	158	153	141	81	82	109
tic-tac-toe	9	958	355	251	268	204	356	218
Average			369.1	335.8	249.7	259.2	240.6	222.2

Table 2. Minimum, average and maximum length of decision rules induced by heuristics *RM*, *Poly* and *Log*.

| Dataset | Attr | *RM* | | | *Poly* | | | *Log* | | |
		Min	Avg	Max	Min	Avg	Max	Min	Avg	Max
balance-scale	4	3	3.41	4	3	3.41	4	3	3.41	4
breast-cancer	9	1	3.52	8	3	5.05	8	1	3.29	6
cars	6	1	5.44	6	1	5.44	6	1	5.45	6
chess-kr-vs-kp	36	1	4.68	23	3	13.66	22	1	5.33	11
flags	26	1	2.23	9	3	6.88	13	1	3.26	6
house-votes	16	2	3.29	5	2	4.38	7	2	3.56	7
lymphography	18	1	2.56	5	1	4.85	8	1	2.85	5
nursery	8	1	5.87	8	1	5.93	8	1	5.79	8
soybean-large	35	1	3.27	35	1	6.79	13	1	4.59	8
tic-tac-toe	9	3	4.32	7	3	4.70	7	3	4.20	6
Average		3.86			6.11			4.17		

Table 5 presents the accuracy of classification and standard deviation, for global optimization relative to length, global optimization relative to support and heuristics *RM*, *Poly* and *Log*, and *ID*3 algorithm. The classifiers were constructed using 10 fold cross-validation method and standard voting. Accuracy of classification is the number of correctly recognized instances relative to the number of rows from test table. Obtained results show that on average, classification accuracies for global optimization relative to length and support are very

Table 3. Minimum, average and maximum support of decision rules induced by heuristics *RM*, *Poly* and *Log*.

Dataset	Rows	RM			Poly			Log		
		Min	Avg	Max	Min	Avg	Max	Min	Avg	Max
balance-scale	625	1	3.38	5	1	3.38	5	1	3.38	5
breast-cancer	266	1	3.25	24	1	4.43	23	1	4.10	25
cars	1728	1	8.14	576	1	7.97	576	1	8.11	576
chess-kr-vs-kp	3196	1	76.40	743	1	119.16	743	1	75.11	743
flags	194	1	2.59	18	2	7.09	22	1	5.68	22
house-votes	279	1	32.22	95	2	43.02	95	2	40.02	95
lymphography	148	1	7.70	32	2	13.15	27	1	10.83	32
nursery	12960	1	31.06	4320	1	29.11	4320	1	30.50	4320
soybean-large	292	1	3.59	23	1	5.63	38	1	5.96	38
tic-tac-toe	958	1	13.21	90	1	15.04	90	2	13.04	90
Average		18.15			24.80			19.67		

Table 4. Minimum, average and maximum length and support of decision rules obtained by global optimization relative to length and support and *ID3* algorithm.

Dataset	Global - length			Global - support			ID3 - length			ID3 - support		
	Min	Avg	Max	Min	Avg	Max	Min	Avg	Max	Min	Avg	Max
balance-scale	3	3.41	4	1	3.38	5	3	3.86	4	1	1.56	5
breast-cancer	1	3.22	6	1	4.64	25	2	3.82	7	1	1.89	12
cars	1	5.49	6	1	6.11	576	1	5.52	6	1	5.84	576
chess-kr-vs-kp	1	4.51	11	1	181.17	743	1	9.24	16	1	65.22	743
flags	1	2.22	4	2	7.38	22	1	3.30	7	1	2.20	10
house-votes	2	3.28	5	2	41.00	95	2	5.63	8	1	10.33	95
lymphography	1	2.61	4	2	14.12	32	2	3.39	6	1	2.74	25
nursery	1	5.74	8	1	28.72	4320	1	6.66	8	1	15.45	4320
soybean-large	1	3.20	7	1	4.81	38	2	4.50	8	1	2.68	24
tic-tac-toe	3	4.14	6	2	18.00	90	3	5.78	7	1	4.39	90
Average	3.78			30.93			5.17			11.23		

close. The results are also comparable taking into account classification accuracies obtained by each heuristic *RM*, *Poly* and *Log* separately. Both the proposed optimization methods and each of the heuristics provide slightly higher classification accuracy than the *ID3* algorithm. It should be also noted that on average, standard deviation values are almost the same for all rule-based classifiers.

Table 5. Accuracy and standard deviation of rule-based classifiers.

Dataset	Global-length		Global-support		RM		Poly		Log		ID3	
	Acc	Std	Acc	Std	Acc	Std	Acc	Std	Acc	Std	Acc	Std
balance-scale	0.786	0.04	0.773	0.04	0.774	0.05	0.773	0.05	0.766	0.05	0.622	0.07
breast-cancer	0.692	0.07	0.673	0.07	0.665	0.10	0.737	0.04	0.699	0.08	0.646	0.11
cars	0.905	0.03	0.887	0.03	0.918	0.04	0.931	0.04	0.925	0.02	0.922	0.03
chess-kr-vs-kp	0.994	0.00	0.993	0.01	0.991	0.01	0.986	0.01	0.992	0.01	0.997	0.00
flags	0.656	0.16	0.701	0.10	0.646	0.15	0.666	0.12	0.669	0.11	0.634	0.13
house-votes	0.943	0.04	0.943	0.03	0.953	0.04	0.932	0.05	0.943	0.04	0.935	0.05
lymphography	0.837	0.08	0.810	0.09	0.810	0.09	0.783	0.13	0.845	0.08	0.758	0.11
nursery	0.989	0.01	0.988	0.00	0.986	0.01	0.980	0.01	0.987	0.00	0.985	0.00
soybean-large	0.787	0.09	0.804	0.07	0.774	0.08	0.811	0.06	0.798	0.09	0.712	0.06
tic-tac-toe	0.927	0.03	0.938	0.02	0.947	0.02	0.938	0.02	0.939	0.02	0.843	0.05
Average	0.852	0.05	0.851	0.05	0.846	0.06	0.854	0.05	0.856	0.05	0.805	0.06

5 Conclusions

In the paper, methods for obtaining a single global model based on rule sets induced by different heuristics which can be considered as distributed data sources, were proposed. Quality of induced rules was assessed by their length and support. These factors are important from the point of view of knowledge representation because short rules are easier for human beings to understand, and support allows us to discover major patterns in the data. The proposed methods of global model induction are based on the selection of decision rules characterized by small length - in the case of global optimization relative to length, and high support - in the case of global optimization relative to support.

Creating a global rule model on the basis of different sets of rules is important from the point of view of practical applications, e.g. when a company has many branches scattered in different locations and there is a need to obtain knowledge about the patterns that are true for most branches, or even for the whole company.

Obtained results show that the proposed methods for constructing global models include short rules with relatively good support. On average, in the case of length, the values are comparable with the results of the RM heuristic, and in the case of support, the values are not far from the results of the Poly heuristic. Taking into account classification accuracy both global optimization relative to length and support yield results comparable to those obtained by the local data models induced by heuristic RM, Poly and Log. In addition, the results are better than in the case of the ID3 algorithm.

In future works, other approaches for induction of rule sets will be considered and other approaches for constructing a global model, including decision trees, will be studied.

References

1. Agrawal, R., Srikant, R.: Fast algorithms for mining association rules in large databases. In: Bocca, J.B., Jarke, M., Zaniolo, C. (eds.) VLDB, pp. 487–499. Morgan Kaufmann (1994)
2. Alsolami, F., Amin, T., Moshkov, M., Zielosko, B., Zabinski, K.: Comparison of heuristics for optimization of association rules. Fund. Inform. **166**(1), 1–14 (2019). https://doi.org/10.3233/FI-2019-1791
3. Amin, T., Chikalov, I., Moshkov, M., Zielosko, B.: Dynamic programming approach for partial decision rule optimization. Fund. Inform. **119**(3–4), 233–248 (2012). https://doi.org/10.3233/FI-2012-735
4. Bonates, T., Hammer, P.L., Kogan, A.: Maximum patterns in datasets. Discret. Appl. Math. **156**(6), 846–861 (2008)
5. Fu, Y.: Distributed data mining: an overview. Newsl. IEEE Tech. Committ. Distrib. Process. **4**(3), 5–9 (2001)
6. Grzegorowski, M., Ślęzak, D.: On resilient feature selection: computational foundations of R-C-reducts. Inf. Sci. **499**, 25–44 (2019). https://doi.org/10.1016/j.ins.2019.05.041
7. Hall, L.O., Chawla, N., Bowyer, K.W., Kegelmeyer, W.P.: Learning rules from distributed data. In: Zaki, M.J., Ho, C.-T. (eds.) LSPDM 1999. LNCS (LNAI), vol. 1759, pp. 211–220. Springer, Heidelberg (2000). https://doi.org/10.1007/3-540-46502-2_11
8. Kelly, M., Longjohn, R., Nottingham, K.: The UCI machine learning repository. https://archive.ics.uci.edu. Accessed June 2023
9. Kotsiantis, S.B.: Decision trees: a recent overview. Artif. Intell. Rev. **39**, 261–283 (2013)
10. Moshkov, M., Piliszczuk, M., Zielosko, B.: Greedy algorithm for construction of partial association rules. Fund. Inform. **92**(3), 259–277 (2009)
11. Moshkov, M., Zielosko, B., Tetteh, E.T.: Selected data mining tools for data analysis in distributed environment. Entropy **24**(10) (2022). https://doi.org/10.3390/e24101401
12. Moshkov, M., Zielosko, B., Tetteh, E.T., Glid, A.: Learning decision rules from sets of decision trees. In: Buchmann, R.A., et al. (eds.) Information Systems Development: Artificial Intelligence for Information Systems Development and Operations (ISD2022 Proceedings), Cluj-Napoca, Romania, 31 August–2 September 2022. Risoprint/Association for Information Systems (2022)
13. Nguyen, H.S., Ślęzak, D.: Approximate reducts and association rules. In: Zhong, N., Skowron, A., Ohsuga, S. (eds.) RSFDGrC 1999. LNCS (LNAI), vol. 1711, pp. 137–145. Springer, Heidelberg (1999). https://doi.org/10.1007/978-3-540-48061-7_18
14. Pawlak, Z., Skowron, A.: Rough sets and Boolean reasoning. Inf. Sci. **177**(1), 41–73 (2007)
15. Pawlak, Z., Skowron, A.: Rudiments of rough sets. Inf. Sci. **177**(1), 3–27 (2007)
16. Quinlan, J.R.: Induction of decision trees. Mach. Learn. **1**, 81–106 (1986)
17. Rissanen, J.: Modeling by shortest data description. Automatica **14**(5), 465–471 (1978)
18. Serengil, S.I.: ChefBoost: a lightweight boosted decision tree framework (2021). https://doi.org/10.5281/zenodo.5576203
19. Sikora, M., Matyszok, P., Wróbel, L.: SCARI: separate and conquer algorithm for action rules and recommendations induction. Inf. Sci. **607**, 849–868 (2022). https://doi.org/10.1016/j.ins.2022.06.026

20. Stańczyk, U., Zielosko, B.: Heuristic-based feature selection for rough set approach. Int. J. Approximate Reasoning **125**, 187–202 (2020). https://doi.org/10.1016/j.ijar.2020.07.005
21. Stawicki, S., Ślęzak, D., Janusz, A., Widz, S.: Decision bireducts and decision reducts - a comparison. Int. J. Approximate Reasoning **84**, 75–109 (2017). https://doi.org/10.1016/j.ijar.2017.02.007
22. Wróblewski, J.: Theoretical foundations of order-based genetic algorithms. Fund. Inform. **28**(3–4), 423–430 (1996). https://doi.org/10.3233/FI-1996-283414
23. Zielosko, B., Piliszczuk, M.: Greedy algorithm for attribute reduction. Fund. Inform. **85**(1–4), 549–561 (2008)

Algebraic Formulations and Geometric Interpretations of Decision-Theoretic Rough Sets

Jianfeng Xu[1,4]($\boxtimes$), Duoqian Miao[2], Li Zhang[3,4], and Yiyu Yao[4]

[1] School of Software, Nanchang University, Nanchang 330047, Jiangxi, China
jianfeng_x@ncu.edu.cn
[2] Department of Computer Science and Technology, Tongji University,
Shanghai 201804, China
dqmiao@tongji.edu.cn
[3] School of Science, Beijing University of Posts and Telecommunications,
Beijing 100876, China
[4] Department of Computer Science, University of Regina, Regina, SK S4S 0A2,
Canada
Yiyu.Yao@uregina.ca

Abstract. Decision-theoretic rough sets (DTRS) are a probabilistic generalization of rough sets based on Bayesian decision theory. Existing studies on DTRS mainly focus on algebraic approaches. They investigate the formal properties of cost functions of three actions (i.e., assigning an object to the positive, boundary, or negative regions) and the procedure for determining a pair of thresholds by minimizing the overall cost of a rough-set based three-way classification. The objective of this paper is to propose a new direction of research towards the visualization of DTRS. As a complement and an alternative to algebraic approaches, we examine geometric interpretations of DTRS. The geometric approaches are intuitively appealing, easy-to-grasp, and easy-to-use. By looking at visual representations of the various costs, the thresholds, and the geometric relationships between the costs and thresholds, we gain new insights into, and a deeper understanding of, DTRS. Geometric approaches can help practitioners use and apply quickly and effectively DTRS. Combining algebraic approaches and geometric approaches is instrumental in pursuing future research on DTRS.

Keywords: Decision-theoretic rough set (DTRS) · Two-way Decision · Three-way decision · Geometric interpretation

1 Introduction

A fundamental idea of Pawlak rough sets [12,13] is three-way classification under incomplete information or knowledge. For a given subset of objects representing the set of instances of a concept, rough sets use three pairwise disjoint positive, negative, and boundary regions to approximate the concept. There are many extensions of rough sets [2,3,7,17]. Decision-theoretic rough sets (DTRS) are a probabilistic generalization of Pawlak rough sets based on Bayesian decision theory [22,23,28]. Since its introduction, DTRS model has received attentions from many researchers and has been widely applied into various fields [1,8,9,14, 20,31,32].

Existing studies on rough sets and DTRS mainly focus on algebraic formulations, which has produced fundamental theoretical results and practical applications. Recently, Xu et al. [18] gave some geometric interpretations of DTRS. Their preliminary results motivate this paper on a combination of algebraic and geometric approaches to study decision-theoretic rough sets. Our main objective is to complement the algebraic methods by geometric visual methods and tools.

Visual decision-making utilizes an interactive visual interface to implement physical interpretations and facilitate decision-making [10,11,15]. It combines data representation, model analysis, human-machine interaction, and visualization technology. It is widely applied in big data analytics, knowledge graphs, healthcare, biomedicine, industrial mining, and other fields [5,6,16]. Visual decision technology not only enables an intuitive display of huge volume of data and a visual interpretation of an abstract theory, but also facilitates effective communication among stakeholders and users. It is expected that, when used wisely and appropriately, visual decision making technology may play an important role in future studies on rough sets.

As a matter of fact, several studies have already touched upon, either explicitly or implicitly, geometric approaches to rough sets and three-way decision [2,19,26,27]. However, a systematic study still does not exist. As a first step towards visual approaches to rough sets, this paper presents both algebraic and geometric analysis and interpretation of DTRS. The results suggest a new research direction called visual three-way decision, which goes beyond three-way decision with DTRS.

2   Algebraic Formulations

In this section, we review the Bayesian decision procedure [4] for algebraic formulation of models of probabilistic two-way and three-way classifications with respect to a Pawlak approximation space. The latter is commonly known as decision-theoretic rough sets (DTRS) model [28,29].

2.1   Bayesian Decision Procedure

In concept learning, formation, and classification, one typically represents a concept by a pair of a set of attributes, called the intension of the concept, and a set of objects to which the concept applies, called the extension of the concept. Pawlak [12,13] theory of rough sets concerns concept approximations under situations where a finite number of attributes are used to describe objects. Due to the expressive power of a limited set of attributes, some objects have the same description and, therefore, cannot be differentiated. The indistinguishability of objects is formally described by an equivalence relation on a universe of objects, which induces a partition of the universe of objects. Equivalence classes are the basic building blocks for formulating two-way and three-way classifications.

Definition 1. *Suppose that U is a finite universe of objects and E is an equivalence relation on U, that is, E is reflexive, symmetric, and transitive. The equivalence class containing an object $x \in U$ is given by:*

$$[x]_E = [x] = \{y \in U \mid xEy\}. \tag{1}$$

The family of all equivalence classes defines a partition, or quotient space, of U:

$$U/E = \{[x] \mid x \in U\}. \tag{2}$$

By following the notional system of Yang and Yao [20,30], we call the pair apr = $(U, U/E)$ an approximation space. Moreover, $[x] \in U/E$ is called a granular object in the quotient space U/E.

Definition 2. *For a subset of objects $C \subseteq U$ representing the extension of a concept or a class, its complement is given by $\overline{C} = U - C$. The conditional probability that an object $y \in U$ is an instance of C given that $y \in [x]$ is defined by:*

$$\Pr(C|[x]) = \frac{|C \cap [x]|}{|[x]|}, \tag{3}$$

where $|\cdot|$ denotes the cardinality of a set. By the law of probability, we have $\Pr(\overline{C}|[x]) = 1 - \Pr(C|[x])$.

A Bayesian decision procedure for classification within an approximation space is briefly described as follows. For a concept with a subset of objects $C \subseteq U$ as its extension, we can formulate a set of two states $W = \{P : C, N : \overline{C}\}$, where P and N denote that an object is in C and not in C, respectively. For building an n-way classification scheme, we have a finite set of n possible actions $A = \{a_1, \cdots, a_n\}$. Let $\Pr(C|[x])$ and $\Pr(\overline{C}|[x])$ denote, respectively, the conditional probabilities of an object in C and $\overline{C}$ given that the object is in $[x]$. Suppose that $\lambda(a|C)$ and $\lambda(a|\overline{C})$ are, respectively, the loss, cost, or risk for taking action $a \in A$ for an object, if the object is in C and in $\overline{C}$. If an action $a \in A$ is taken for all objects in $[x]$, the expected loss for the equivalence class $[x]$ can be calculated as follows:

$$R(a|[x]) = \lambda(a|C)\Pr(C|[x]) + \lambda(a|\overline{C})\Pr(\overline{C}|[x]). \tag{4}$$

Let $\tau : U/E \longrightarrow A$ denote a decision function for equivalence classes in U/E, where $\tau([x])$ denotes the action taken for all objects in the equivalence class $[x]$. The overall risk of τ is the following summation:

$$R(\tau) = \sum_{[x] \in U/E} R(\tau([x])|[x]). \tag{5}$$

The minimum risk Bayesian decision procedure suggests that one should choose a decision function τ that minimizes the overall risk in Eq. (5).

By the fact that the expected losses of different equivalence classes are independent of each other, we can in fact minimize Eq. (5) by choosing an action that minimizes the expected loss of each equivalence class. More specifically, for an equivalence class $[x] \in U/E$, we have the following optimization problem:

$$\arg_{a \in A} R(a|[x]) = \arg_{a \in A}(\lambda(a|C)\Pr(C|[x]) + \lambda(a|\overline{C})\Pr(\overline{C}|[x])), \qquad (6)$$

where arg denotes the choice of an argument that minimizing a function. With this general formulation, we can examine in particular two-way and three-way classification models.

2.2   Two-Way Classification

In a two-way classification model, with respect to a given concept $C \subseteq U$, we divide U into two regions, namely, a positive region $\text{POS}(C)$ and a negative region $\text{NEG}(C)$. Accordingly, the set of actions $A = \{a_P, a_N\}$ consists of two actions:

$$a_P: \text{accept } x \in C, \text{ i.e., decide } x \in \text{POS}(C),$$
$$a_N: \text{reject } x \in C, \text{ i.e., decide } x \in \text{NEG}(C).$$

That is, for a two-way classification, we either accept or reject an object to be an instance of the concept. Assume that the loss function is given by the following 2×2 matrix,

Action	State	
	$P: C$	$N: \overline{C}$
a_P: accept	$\lambda_{PP} = \lambda(a_P|C)$	$\lambda_{PN} = \lambda(a_P|\overline{C})$
a_N: reject	$\lambda_{NP} = \lambda(a_N|C)$	$\lambda_{NN} = \lambda(a_N|\overline{C})$

The expected costs of the two actions are given by:

$$R(a_P|[x]) = \lambda_{PP}\Pr(C|[x]) + \lambda_{PN}\Pr(\overline{C}|[x]),$$
$$R(a_N|[x]) = \lambda_{NP}\Pr(C|[x]) + \lambda_{NN}\Pr(\overline{C}|[x]). \qquad (7)$$

By the Bayesian decision procedure, we can easily have the following rules for defining an optimal decision function: for $[x] \in U/E$,

If $R(a_P|[x]) \leq R(a_N|[x])$, take action a_P, i.e., decide $x \in \text{POS}(C)$;

If $R(a_N|[x]) < R(a_P|[x])$, take action a_N, i.e., decide $x \in \text{NEG}(C)$.

When the two actions have the same cost, a tie-breaking criterion may be used by selecting any action. In specifying the two rules, we take the action a_P when a_P and a_N has the same cost.

To simply the two rules, we make the following assumption on the loss function:

$$(c1) \qquad \lambda_{PP} < \lambda_{NP}, \quad \lambda_{NN} < \lambda_{PN}.$$

Assumption (c1) states that the cost of a correction decision is less than the cost of an incorrect decision, which seems to be very reasonable. Under the Assumption (c1), by using the equality $\Pr(\overline{C}|[x]) = 1 - \Pr(C|[x])$, we can easily obtain the following simpler form of the two rules:

If $\Pr(C|[x]) \geq \gamma$, take action a_P, i.e., decide $x \in \mathrm{POS}(C)$;

If $\Pr(C|[x]) < \gamma$, take action a_N, i.e., decide $x \in \mathrm{NEG}(C)$,

where

$$\gamma = \frac{\lambda_{PN} - \lambda_{NN}}{(\lambda_{PN} - \lambda_{NN}) + (\lambda_{NP} - \lambda_{PP})} = \left(1 + \frac{\lambda_{NP} - \lambda_{PP}}{\lambda_{PN} - \lambda_{NN}}\right)^{-1}, \qquad (8)$$

and $0 < \gamma < 1$. That is, the value of γ is in fact determined by the ratio $(\lambda_{NP} - \lambda_{PP})/(\lambda_{PN} - \lambda_{NN})$. Different loss functions may therefore produce the same threshold as long as the ratio remains to the same. Based on the analysis, we obtain a two-way classification model.

Definition 3. *In an approximation space apr $= (U, U/E)$, given a threshold $0 \leq \gamma \leq 1$ and a target concept $C \subseteq U$, we can make a two-way classification to approximate C as follows:*

$$\mathrm{POS}_\gamma(C) = \{x \in U \mid \Pr(C|[x]) \geq \gamma\},$$
$$\mathrm{NEG}_\gamma(C) = \{x \in U \mid \Pr(C|[x]) < \gamma\}. \qquad (9)$$

Moreover, an optimal threshold γ can be determined by a loss function.

In formulating a two-way classification, we may view the conditional probability $\Pr(C|[x])$ as a measure of confidence in accepting or rejecting an object to be an instance of C given that the object is in $[x]$. We accept an object to be an instance of the concept, if the confidence is above a certain level; otherwise, we reject the object. The Bayesian decision procedure establishs a solid foundation for us to determine the required level of confidence with the associated cost.

2.3   Three-Way Classification

In a three-way classification model, namely, the DTRS model [23, 28, 29], with respect to a given concept $C \subseteq U$, we divide U into three regions, namely, a positive region $\mathrm{POS}(C)$, a negative region $\mathrm{NEG}(C)$, and a boundary region $\mathrm{BND}(C)$. Accordingly, the set of actions $A = \{a_P, a_N, a_B\}$ consists of three actions:

a_P: accept $x \in C$, i.e., decide $x \in \mathrm{POS}(C)$,

a_N: reject $x \in C$, i.e., decide $x \in \mathrm{NEG}(C)$,

a_B: neither accept nor reject $x \in C$, i.e., decide $x \in \mathrm{BND}(C)$.

That is, for a three-way classification, we either accept, reject, or non-commitment an object to be an instance of the concept. Assume that the loss function is given by the following 3×2 matrix,

Action	State	
	$P\colon C$	$N\colon \overline{C}$
a_P: accept	$\lambda_{PP} = \lambda(a_P|C)$	$\lambda_{PN} = \lambda(a_P|\overline{C})$
a_N: reject	$\lambda_{NP} = \lambda(a_N|C)$	$\lambda_{NN} = \lambda(a_N|\overline{C})$
a_B: neither accept nor reject	$\lambda_{BP} = \lambda(a_B|C)$	$\lambda_{BN} = \lambda(a_B|\overline{C})$

The expected costs of the three actions are given by:

$$R(a_P|[x]) = \lambda_{PP}\Pr(C|[x]) + \lambda_{PN}\Pr(\overline{C}|[x]),$$
$$R(a_N|[x]) = \lambda_{NP}\Pr(C|[x]) + \lambda_{NN}\Pr(\overline{C}|[x]),$$
$$R(a_B|[x]) = \lambda_{BP}\Pr(C|[x]) + \lambda_{BN}\Pr(\overline{C}|[x]). \tag{10}$$

By the Bayesian decision procedure, we can easily have the following rules for defining an optimal decision function: for $[x] \in U/E$,

If $R(a_P|[x]) < R(a_N|[x])$ and $R(a_P|[x]) \le R(a_B|[x])$,
take action a_P, i.e., decide $x \in \text{POS}(C)$;
If $R(a_N|[x]) < R(a_P|[x])$ and $R(a_N|[x]) \le R(a_B|[x])$,
take action a_N, i.e., decide $x \in \text{NEG}(C)$;
If $R(a_B|[x]) < R(a_P|[x])$ and $R(a_B|[x]) < R(a_N|[x])$,
take action a_B, i.e., decide $x \in \text{BND}(C)$.

In the three rules, when two or three actions have the same cost, we break ties according to the ordering of three actions a_P, a_N, a_B.

To simply the three rules, we make the following assumptions on the loss function:

$$\text{(C1)} \quad \lambda_{PP} \le \lambda_{BP} < \lambda_{NP}, \quad \lambda_{NN} \le \lambda_{BN} < \lambda_{PN}.$$

Assumption (C1) is an extended version of assumption (c1). It states that the cost of a correction decision is less than the cost of an incorrect decision, and the cost of a_B lies between a correct and an incorrect decision and may be the same as a correct decision. Since a non-commitment decision a_B in fact falls between a_P and a_N, Assumption (C1) seems to be reasonable.

Under the Assumption (C1), by using the equality $\Pr(\overline{C}|[x]) = 1 - \Pr(C|[x])$, we can easily obtain the following simpler form of the three rules:

If $\Pr(C|[x]) \ge \alpha$ and $\Pr(C|[x]) > \gamma$, take action a_P, i.e., decide $x \in \text{POS}(C)$;
If $\Pr(C|[x]) \le \beta$ and $\Pr(C|[x]) < \gamma$, take action a_N, i.e., decide $x \in \text{NEG}(C)$;
If $\Pr(C|[x]) > \beta$ and $\Pr(C|[x]) < \alpha$, take action a_B, i.e., decide $x \in \text{BND}(C)$,

where

$$\alpha = \frac{\lambda_{PN} - \lambda_{BN}}{(\lambda_{PN} - \lambda_{BN}) + (\lambda_{BP} - \lambda_{PP})} = \left(1 + \frac{\lambda_{BP} - \lambda_{PP}}{\lambda_{PN} - \lambda_{BN}}\right)^{-1},$$

$$\beta = \frac{\lambda_{BN} - \lambda_{NN}}{(\lambda_{BN} - \lambda_{NN}) + (\lambda_{NP} - \lambda_{BP})} = \left(1 + \frac{\lambda_{NP} - \lambda_{BP}}{\lambda_{BN} - \lambda_{NN}}\right)^{-1},$$

$$\gamma = \frac{\lambda_{PN} - \lambda_{NN}}{(\lambda_{PN} - \lambda_{NN}) + (\lambda_{NP} - \lambda_{PP})} = \left(1 + \frac{\lambda_{NP} - \lambda_{PP}}{\lambda_{PN} - \lambda_{NN}}\right)^{-1}, \quad (11)$$

and $0 < \alpha \leq 1$, $0 \leq \beta < 1$, and $0 < \gamma < 1$. The three regions give a three-way approximation of C, which is called a probabilistic rough set induced by C.

The threshold α is obtained by comparing the costs of actions a_P and a_B, the threshold β by comparing the costs of actions a_N and a_B, and γ by comparing the costs of actions a_P and a_N. Under Assumption (C1), the first form of β is well-defined. However, the second form of β is well-defined only if $\lambda_{BN} < \lambda_{NN}$. When $\lambda_{BN} = \lambda_{PN}$, we use the first form that gives $\beta = 0$.

To further simply the three rules, we make another assumption:

$$(C2) \quad \frac{\lambda_{BP} - \lambda_{PP}}{\lambda_{PN} - \lambda_{BN}} < \frac{\lambda_{NP} - \lambda_{BP}}{\lambda_{BN} - \lambda_{NN}}.$$

The Assumptions (C1) and (C2) imply that $0 \leq \beta < \gamma < \alpha \leq 1$. It follows that only a pair of thresholds (α, β) is needed and the three rules can be further simplified:

If $\Pr(C|[x]) \geq \alpha$, take action a_P, i.e., decide $x \in \mathrm{POS}(C)$;

If $\Pr(C|[x]) \leq \beta$, take action a_N, i.e., decide $x \in \mathrm{NEG}(C)$;

If $\beta < \Pr(C|[x]) < \alpha$, take action a_B, i.e., decide $x \in \mathrm{BND}(C)$.

Based on the analysis, we obtain a three-way classification model.

Definition 4. *In an approximation space apr $= (U, U/E)$, given a pair thresholds $0 \leq \beta < \alpha \leq 1$ and a target concept $C \subseteq U$, we can make a three-way classification to approximate C as follows:*

$$\mathrm{POS}_{(\alpha,\beta)}(C) = \{x \in U \mid \Pr(C|[x]) \geq \alpha\},$$
$$\mathrm{NEG}_{(\alpha,\beta)}(C) = \{x \in U \mid \Pr(C|[x]) \leq \beta\},$$
$$\mathrm{BND}_{(\alpha,\beta)}(C) = \{x \in U \mid \beta < \Pr(C|[x]) < \alpha\}. \quad (12)$$

Moreover, a pair of optimal thresholds α and β can be determined by a loss function.

Compared with a two-way decision model, we only accept or reject an object if our confidence levels are sufficiently high, namely, at or above α for acceptance, and at or below β for rejection. Otherwise, we neither accept nor reject the object. A decision of non-commitment in fact has a lower cost than either acceptance or rejection. This is a main advantage of three-way classification model.

3   Geometric Interpretations

We now examine geometric interpretations of two-way and three-way classifications derived from the algebraic formulations.

3.1   Geometric Interpretation of the Bayesian Classification

By the law of probability $\Pr(\overline{C}|[x]) = 1 - \Pr(C|[x])$, the cost of action $a \in A$ defined by Eq. (4) in the Bayesian decision procedure can be expressed as a linear function of $\Pr(C|[x])$ as follows:

$$\begin{aligned} R(a|[x]) &= \lambda(a|C)\Pr(C|[x]) + \lambda(a|\overline{C})\Pr(\overline{C}|[x]) \\ &= (\lambda(a|C) - \lambda(a|\overline{C}))\Pr(C|[x]) + \lambda(a|\overline{C}). \end{aligned} \tag{13}$$

We can draw a line in a two-dimensional space to represent the relation between the cost of an action and the conditional probability. Suppose that the horizontal-axis represents the value of conditional probability $p = \Pr(C|[x])$ and the vertical-axis represents the value cost $r = R(a|[x])$. The relationship between p and r, i.e., $r = (\lambda(a|C) - \lambda(a|\overline{C}))p + \lambda(a|\overline{C})$, can be described by a line R_a in the probability-risk, or p-r space, as shown in Fig. 1. The value of $R(a|[x])$ is $\lambda(a|\overline{C})$ when $\Pr(C|[x]) = 0$ and is $\lambda(a|C)$ when $\Pr(C|[x]) = 1$. The slope of the line is given by $\lambda(a|C) - \lambda(a|\overline{C})$. When $\lambda(a|C) > \lambda(a|\overline{C})$, $R(a|[x])$ is monotonically increasing with respect to $\Pr(C|[x])$; when $\lambda(a|C) < \lambda(a|\overline{C})$, $R(a|[x])$ is monotonically decreasing with respect to $\Pr(C|[x])$.

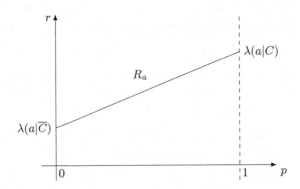

Fig. 1. Cost of an action $a \in A$ versus probability

For a set of n actions $A = \{a_1, \dots, a_n\}$, we can draw n lines in the p-r space. For a given conditional probability p_0, the n lines have n possibly different intersections with the line $p = p_0$, we can simply choose an action that has the lowest r value from the n intersections.

3.2 Geometric Interpretation of Two-Way Classification

In two-way classification, for $\Pr(C|[x]) + \Pr(\overline{C}|[x]) = 1$, Eq. (7) of the cost of actions a_P and a_N in the $p\text{-}r$ space can be expressed as two linear functions of $\Pr(C|[x])$ as follows:

$$
\begin{aligned}
R(a_P|[x]) &= \lambda_{PP}\Pr(C|[x]) + \lambda_{PN}\Pr(\overline{C}|[x]) \\
&= (\lambda_{PP} - \lambda_{PN})\Pr(C|[x]) + \lambda_{PN}, \\
R(a_N|[x]) &= \lambda_{NP}\Pr(C|[x]) + \lambda_{NN}\Pr(\overline{C}|[x]) \\
&= (\lambda_{NP} - \lambda_{NN})\Pr(C|[x]) + \lambda_{NN}.
\end{aligned}
\tag{14}
$$

In the probability-risk space, the two lines R_P and R_N in Fig. 2 denote, respectively, the relationships between the expected costs and probability of the two actions a_P and a_N. The value of $R(a_P|[x])$ is λ_{PN} when $\Pr(C|[x]) = 0$ and is λ_{PP} when $\Pr(C|[x]) = 1$. The slope of line R_P is given by $\lambda_{PP} - \lambda_{PN}$. When $\lambda_{PP} > \lambda_{PN}$, R_P is monotonically increasing with respect to $\Pr(C|[x])$; when $\lambda_{PP} < \lambda_{PN}$, R_P is monotonically decreasing with respect to $\Pr(C|[x])$. Similarly, the value of $R(a_N|[x])$ is λ_{NN} when $\Pr(C|[x]) = 0$ and is λ_{NP} when $\Pr(C|[x]) = 1$. The slope of line R_N is given by $\lambda_{NP} - \lambda_{NN}$. When $\lambda_{NP} > \lambda_{NN}$, R_P is monotonically increasing with respect to $\Pr(C|[x])$; when $\lambda_{NP} < \lambda_{NN}$, R_P is monotonically decreasing with respect to $\Pr(C|[x])$.

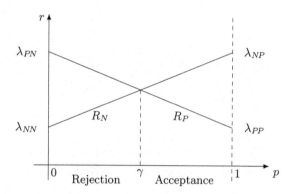

Fig. 2. Two-way classification

Assumption (c1), i.e., $\lambda_{NP} > \lambda_{PP}$ and $\lambda_{PN} > \lambda_{NN}$, is represented by putting λ_{NP} above λ_{PP} on line $p = 1$, and putting λ_{PN} above λ_{NN} on line $p = 0$. It follows that the two lines R_P and R_N must have an intersection at $p = \gamma$, where γ is between 0 and 1 and given by:

$$
\gamma = \left(1 + \frac{\lambda_{NP} - \lambda_{PP}}{\lambda_{PN} - \lambda_{NN}}\right)^{-1}.
\tag{15}
$$

From Fig. 2, we can see that

$$\text{if } \Pr(C|[x]) = \gamma, R(a_P|[x]) = R(a_N|[x]);$$
$$\text{if } \Pr(C|[x]) > \gamma, R(a_P|[x]) < R(a_N|[x]);$$
$$\text{if } \Pr(C|[x]) < \gamma, R(a_P|[x]) > R(a_N|[x]).$$

That is, we make an acceptance decision when the probability falls within the interval $[\gamma, 1]$ and a rejection decision when the probability falls within the interval $[0, \gamma)$.

According to Eq. (15), the ratio $(\lambda_{NP} - \lambda_{PP})/(\lambda_{PN} - \lambda_{NN})$ determines the value of γ. Different loss functions may produce the same γ. For example, if we increase the cost λ_{NP} and, at the same time, decrease the cost of λ_{NN}, we would have the same γ value, as long as we keep the ratio unchanged. This is equivalent to rotate line R_N in Fig. 2 counter-clockwise by fixing the point $(\gamma, (\lambda_{PP} - \lambda_{PN})\gamma + \lambda_{PN})$. Likewise, we may also rotate line R_P clockwise and rotate line R_N either clockwise or counter-clockwise. Figure 3 shows the result of rotating both R_P and R_N clockwise, resulting lines R'_P and R'_N, respectively. It follows that $(\lambda_{NP} - \lambda_{PP})/(\lambda_{PN} - \lambda_{NN}) = (\lambda'_{NP} - \lambda'_{PP})/(\lambda'_{PN} - \lambda'_{NN})$ and $\gamma = \gamma'$.

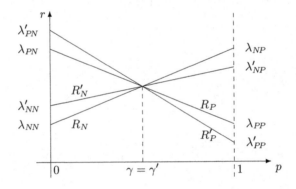

Fig. 3. Two-way classification: Different loss functions

For a given loss function, λ_{PP}, λ_{PN}, λ_{NP}, and λ_{NN}, we may produce a standardized loss function by setting:

$$\lambda^s_{PP} = \lambda_{PP} - \lambda_{PP} = 0,$$
$$\lambda^s_{PN} = \lambda_{PN} - \lambda_{NN},$$
$$\lambda^s_{NP} = \lambda_{NP} - \lambda_{PP},$$
$$\lambda^s_{NN} = \lambda_{NN} - \lambda_{NN} = 0. \tag{16}$$

As shown by Fig. 4, the standardized loss function is obtained from the original loss function by reducing the corresponding costs of actions a_P and a_N by λ_{PP}

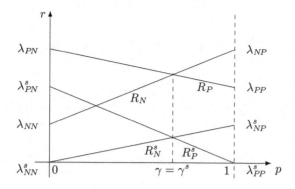

Fig. 4. Two-way classification: a loss function and its standardized loss function

and λ_{NN} for objects in C and in $\overline{C}$, respectively. Based on the analysis, we have $(\lambda_{NP} - \lambda_{PP})/(\lambda_{PN} - \lambda_{NN}) = (\lambda^s_{NP} - 0)/(\lambda^s_{PN} - 0) = \lambda^s_{NP}/\lambda^s_{PN}$ and $\gamma = \gamma^s$. Like the ratio form, we only need to consider the costs of incorrect decisions in a standardized loss function.

3.3   Geometric Interpretation of Three-Way Classification

In three-way classification, for $\Pr(C|[x]) + \Pr(\overline{C}|[x]) = 1$, Eq. (10) of the cost of actions a_P, a_N, and a_B in the p-r space can be expressed as three linear functions of $\Pr(C|[x])$ as follows:

$$\begin{aligned}
R(a_P|[x]) &= \lambda_{PP} \Pr(C|[x]) + \lambda_{PN} \Pr(\overline{C}|[x]) \\
&= (\lambda_{PP} - \lambda_{PN}) \Pr(C|[x]) + \lambda_{PN}, \\
R(a_N|[x]) &= \lambda_{NP} \Pr(C|[x]) + \lambda_{NN} \Pr(\overline{C}|[x]) \\
&= (\lambda_{NP} - \lambda_{NN}) \Pr(C|[x]) + \lambda_{NN}, \\
R(a_B|[x]) &= \lambda_{BP} \Pr(C|[x]) + \lambda_{BN} \Pr(\overline{C}|[x]) \\
&= (\lambda_{BP} - \lambda_{BN}) \Pr(C|[x]) + \lambda_{BN}.
\end{aligned} \tag{17}$$

They are represented by three lines R_P, R_N, and R_B in the probability-risk space, as shown in Fig. 5. The two lines R_P and R_N are the same as the case of the two-way classification. For the third line R_B, the value of $R(a_B|[x])$ is λ_{BN} when $\Pr(C|[x]) = 0$ and is λ_{BP} when $\Pr(C|[x]) = 1$. The slope of line R_B is given by $\lambda_{BP} - \lambda_{BN}$. When $\lambda_{BP} > \lambda_{BN}$, R_B is monotonically increasing with respect to $\Pr(C|[x])$; when $\lambda_{BP} < \lambda_{BN}$, R_B is monotonically decreasing with respect to $\Pr(C|[x])$.

For Assumption (C1), i.e., $\lambda_{PP} \leq \lambda_{BP} < \lambda_{NP}$ and $\lambda_{NN} \leq \lambda_{BN} < \lambda_{PN}$, similar to the case of two-way classification, we put λ_{NP} above λ_{PP} on line $p = 1$ and λ_{PN} above λ_{NN} on line $p = 0$. At the same time, we put λ_{BP} somewhere in the middle of λ_{NP} and λ_{PP} on line $p = 1$ and put λ_{BN} somewhere in the middle of λ_{NN} and λ_{PN} on line $p = 1$. The two lines R_P and R_B must have an intersection

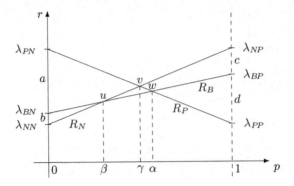

Fig. 5. Three-way classification

at $p = \alpha$ between 0 and 1; the two lines R_N and R_B must have an intersection at $p = \beta$ between 0 and 1; the two lines R_P and R_N must have an intersection at $p = \gamma$ between 0 and 1. The values of α, β, and γ are given by:

$$\alpha = \left(1 + \frac{\lambda_{BP} - \lambda_{PP}}{\lambda_{PN} - \lambda_{BN}}\right)^{-1},$$

$$\beta = \left(1 + \frac{\lambda_{NP} - \lambda_{BP}}{\lambda_{BN} - \lambda_{NN}}\right)^{-1},$$

$$\gamma = \left(1 + \frac{\lambda_{NP} - \lambda_{PP}}{\lambda_{PN} - \lambda_{NN}}\right)^{-1}, \tag{18}$$

where γ is the threshold of a two-way classification defined by Eq. (15).

Let

$$a = \lambda_{PN} - \lambda_{BN}, \qquad b = \lambda_{BN} - \lambda_{NN};$$
$$c = \lambda_{NP} - \lambda_{BP}, \qquad d = \lambda_{BP} - \lambda_{PP}. \tag{19}$$

From Fig. 5, we can see that the three thresholds α, β, and γ can be expressed as:

$$\alpha = \left(1 + \frac{d}{a}\right)^{-1},$$

$$\beta = \left(1 + \frac{c}{b}\right)^{-1},$$

$$\gamma = \left(1 + \frac{d+c}{a+b}\right)^{-1}. \tag{20}$$

Assumption (C2) can be expressed as follows:

$$\frac{\lambda_{BP} - \lambda_{PP}}{\lambda_{PN} - \lambda_{BN}} < \frac{\lambda_{NP} - \lambda_{BP}}{\lambda_{BN} - \lambda_{NN}} \iff \frac{d}{a} < \frac{c}{b}, \tag{21}$$

which implies that $\beta < \alpha$. It can be verified that

$$\frac{d}{a} < \frac{c}{b} \implies \frac{d}{a} < \frac{d+c}{a+b} < \frac{c}{b}. \tag{22}$$

Therefore, we have $\beta < \gamma < \alpha$.

The geometric interpretation of Assumption (C2) and the relationship between three thresholds can be explained as follows. In Fig. 5, the intersections of R_N and R_B, R_N and R_P, and R_B and R_P are denoted by u, v, and w, respectively. Triangles $(\lambda_{PN}, w, \lambda_{BN})$ and $(w, \lambda_{BP}, \lambda_{PP})$ are similar triangles. Their heights are α and $1-\alpha$, respectively. Similarly, Triangles $(\lambda_{BN}, u, \lambda_{NN})$ and $(u, \lambda_{NP}, \lambda_{BP})$ are similar triangles. Their heights are β and $1 - \beta$, respectively. Therefore, by the properties of similar triangles, we can re-express Assumption (C2) as:

$$\frac{1-\alpha}{\alpha} = \frac{d}{a} < \frac{c}{b} = \frac{1-\beta}{\beta}, \tag{23}$$

which immediately leads to $\beta < \alpha$. By using a similar argument and Eq. (22), we can also show that $\beta < \gamma$ and $\gamma < \alpha$.

As seem from Fig. 5, under the condition $\beta < \gamma < \alpha$, a_P has the minimum value when probability falls within the interval $[\alpha, 1]$, we make an acceptance decision; a_N has the minimum value when probability falls within the interval $[0, \beta]$, we make a rejection decision; a_B has the minimum value when probability falls within the interval (β, α), we neither accept nor reject and make a non-commitment decision.

4 Conclusions

In this paper, we have examined both algebraic formulations and geometric interpretations of two-way and three-way classifications. While the algebraic formulation are slightly different formulations of the decision-theoretic rough sets model [22,23,28,29], the geometric interpretations are new and extend the preliminary results of Xu et al. [18]. The geometric studies provide intuitive and visual understanding of the decision-theoretic rough sets models. In particular, geometric interpretations help us understand better the loss functions, the relationship between cost and probability, and the meaning of thresholds. They are useful in guiding us to derive and interpret different loss functions and their consequences for classification. A combination of algebraic and geometric approaches provides a fuller and richer figure of DTRS. This new direction of research deserves a more thorough study.

Three-way classification with DTRS represents a narrow sense of three-way decision [21–23,25]. A wide sense of three-way decision concerns thinking, problem-solving, and computing in threes [24,26,27]. One may expect that the similar geometric approaches can be used to study three-way decision in general. In other words, the investigation of this paper points at a new research avenue for three-way decision, which may be called visual three-way decision. Visual three-way decision is thinking, problem solving, and computing visually in threes. The main ideas are to use visual methods and tools to facilitate three-way decision in practical applications. Considering the importance of visual thinking in human cognition and problem solving, visual three-way decision is a worthy research topic.

Acknowledgements. This work was supported in part by a Discovery Grant from NSERC, Canada, and National Natural Science Foundations of China (Grant No. 62266032), China Scholarship Council (Grant No. 202006825060), Jiangxi Natural Science Foundations (Grant No. 20202BAB202018), Jiangxi Training Program for Academic and Technical Leaders in Major Disciplines-Leading Talents Project (Grant No. 20225BCI22016).

References

1. Abdullah, S., Al-Shomrani, M.M., Liu, P., Ahmad, S.: A new approach to three-way decisions making based on fractional fuzzy decision theoretical rough set. Int. J. Intell. Syst. **37**, 2428–2457 (2022)
2. Deng, J., Zhan, J., Herrera-Viedma, E., Herrera, F.: Regret theory-based three-way decision method on incomplete multi-scale decision information systems with interval fuzzy-numbers. IEEE Trans. Fuzzy Syst. **31**, 982–996 (2022)
3. Duan, J., Wang, G., Hu, X., Xia, D., Wu, D.: Mining multigranularity decision rules of concept cognition for knowledge graphs based on three-way decision. Inf. Process. Manage. **60**, 103365 (2023)
4. Duda, R.O., Hart, R.O.: Pattern Classification and Scene Analysis. Wiley, New York (1973)
5. Gold, J.I., Stocker, A.A.: Visual decision-making in an uncertain and dynamic world. Annu. Rev. Vision Sci. **3**, 227–250 (2017)
6. Kril, D.N., Brodhead, M.T., Moorehouse, A.G.: Further evaluation of a decision-making algorithm supporting visual analysis of time-series data. Behav. Interv. **37**, 1030–1042 (2022)
7. Li, W., Yang, B.: Three-way decisions with fuzzy probabilistic covering-based rough sets and their applications in credit evaluation. Appl. Soft Comput. **136**, 110–144 (2023)
8. Liang, D., Pedrycz, W., Liu, D.: Determining three-way decisions with decision-theoretic rough sets using a relative value approach. IEEE Trans. Syst. Man Cybern.: Syst. **47**, 1785–1799 (2016)
9. Liu, F., Liu, Y., Abdullah, S.: Three-way decisions with decision-theoretic rough sets based on covering-based Q-rung orthopair fuzzy rough set model. J. Intell. Fuzzy Syst. **40**, 9765–9785 (2021)
10. Muntwiler, C., Eppler, M.J.: Improving decision making through visual knowledge calibration. Manag. Decis. **61**, 2374–2390 (2023)
11. Padilla, L.M., Creem-Regehr, S.H., Hegarty, M., Stefanucci, J.K.: Decision making with visualizations: a cognitive framework across disciplines. Cogn. Res.: Principles Implicat. **3**, 1–25 (2018)
12. Pawlak, Z.: Rough sets. Int. J. Comput. Inf. Sci. **11**, 341–356 (1982)
13. Pawlak, Z.: Rough Sets: Theoretical Aspects of Reasoning About Data. Kluwer Academic Publishers, Dordrecht (1991)
14. Qian, J., Liu, C., Yue, X.: Multigranulation sequential three-way decisions based on multiple thresholds. Int. J. Approximate Reasoning **105**, 396–416 (2019)

15. Smith, J., Legg, P., Matovic, M., Kinsey, K.: Predicting user confidence during visual decision making. ACM Trans. Interact. Intell. Syst. (TiiS) **8**, 1–30 (2018)
16. Sridhar, V.H., et al.: The geometry of decision-making in individuals and collectives. Proc. Nat. Acad. Sci. **118**, e2102157118 (2021)
17. Sun, T., Mei, X., Sun, X., Cai, Y., Ji, Y., Fan, Z.: Real-time monitoring and control of the breakthrough stage in ultrafast laser drilling based on sequential three-way decision. IEEE Trans. Industr. Inf. **19**, 5422–5432 (2022)
18. Xu, J., Miao, D., Zhang, Y.: Piece-wise delay cost-sensitive three-way decisions (in Chinese). J. Softw. **33**, 3754–3775 (2022)
19. Xu, J., Zhang, Y., Miao, D.: Three-way confusion matrix for classification: a measure driven view. Inf. Sci. **507**, 772–794 (2020)
20. Yang, J.L., Yao, Y.Y.: Granular rough sets and granular shadowed sets: three-way approximations in Pawlak approximation spaces. Int. J. Approximate Reasoning **142**, 231–247 (2022)
21. Yao, Y.: Three-way decision: an interpretation of rules in rough set theory. In: Wen, P., Li, Y., Polkowski, L., Yao, Y., Tsumoto, S., Wang, G. (eds.) RSKT 2009. LNCS (LNAI), vol. 5589, pp. 642–649. Springer, Heidelberg (2009). https://doi.org/10.1007/978-3-642-02962-2_81
22. Yao, Y.Y.: Three-way decisions with probabilistic rough sets. Inf. Sci. **180**, 341–353 (2010)
23. Yao, Y.Y.: The superiority of three-way decisions in probabilistic rough set models. Inf. Sci. **181**, 1080–1096 (2011)
24. Yao, Y.Y.: Three-way decision and granular computing. Int. J. Approximate Reasoning **103**, 107–123 (2018)
25. Yao, Y.Y.: Three-way granular computing, rough sets, and formal concept analysis. Int. J. Approximate Reasoning **116**, 106–125 (2020)
26. Yao, Y.Y.: The geometry of three-way decision. Appl. Intell. **51**, 6298–6325 (2021)
27. Yao, Y.Y.: Symbols-meaning-value (SMV) space as a basis for a conceptual model of data science. Int. J. Approximate Reasoning **144**, 113–128 (2022)
28. Yao, Y.Y., Wong, S.K.M.: A decision theoretic framework for approximating concepts. Int. J. Man Mach. Stud. **37**, 793–809 (1992)
29. Yao, Y.Y., Wong, S.K.M., Lingras, P.: A decision-theoretic rough set model. In: Ras, Z.W., Zemankova, M., Emrich, M.L. (eds.) Methodologies for Intelligent Systems, vol. 5, pp. 17–24. North-Holland, New York (1990)
30. Yao, Y.Y., Yang, J.L.: Granular fuzzy sets and three-way approximations of fuzzy sets. Int. J. Approximate Reasoning **161**, 109003 (2023)
31. Zhang, Q., Xie, Q., Wang, G.: A novel three-way decision model with decision-theoretic rough sets using utility theory. Knowl.-Based Syst. **159**, 321–335 (2018)
32. Zhao, X.R., Hu, B.Q.: Three-way decisions with decision-theoretic rough sets in multiset-valued information tables. Inf. Sci. **507**, 684–699 (2020)

Reduction of Binary Attributes: Rough Set Theory Versus Formal Concept Analysis

Piotr Wasilewski[1,2]($\boxtimes$) (ID), Janusz Kacprzyk[1,3] (ID), and Sławomir Zadrożny[1,3] (ID)

[1] Systems Research Institute, Polish Academy of Sciences,
ul. Newelska 6, 01-447 Warsaw, Poland
[2] Faculty of Computer Science, Dalhousie University, 6050 University Avenue,
Halifax, NS, Canada
pwasilew@ibspan.waw.pl
[3] WIT – Warsaw School of Information Technology,
Newelska 6, 01-447 Warsaw, Poland

Abstract. The paper compares the concepts of reduction of binary attributes in rough set theory (RST) and the reduction of unary attributes or dychotomic attributes in formal concept analysis (FCA). We present some basics of both theories together with a brief presentation of elements of the theory of set spaces used in the paper as a platform for mentioned comparison. Then we deliver some results on binary attribute reduction in RST and attribute reduction in FCA. We characterize independence of sets of binary attributes in RST by complete algebras of sets completely generated by completely irredundant families of sets. Then by means of complete algebras of sets and indiscernibility relations with respect to families of sets we investigate some families of FCA-attributes. And finally we present some formal context for which we prove that RST-binary attribute reduction and FCA-unary attribute reduction give the same results.

Keywords: formal concept analysis · concept lattices · social choice theory · social choice function · voting procedure · voting criteria

1 Introduction

Concepts modelling is a key issue in knowledge representation. In the most general view, concepts modelling may be seen as a subject of the study within the framework of the *granular computing* [1] which somehow combines very many specific approaches such as fuzzy logic, rough set theory (RST) or formal concept analysis (FCA), to name just a few. The two latter theories share a starting point to concepts modelling, i.e., both assume a set of objects characterized by a set of attributes. Then, they employ their own methods of concepts modelling and employing to knowledge representation emerging from the original description of objects.

The RST is apparently a richer theory but the FCA may be a more suitable tool in some scenarios. In particular, the latter theory seems to be very suitable

A. Campagner et al. (Eds.): IJCRS 2023, LNAI 14481, pp. 46–61, 2023.
https://doi.org/10.1007/978-3-031-50959-9_4

to analyse small sets of objects characterized by attributes which apply or do not apply to particular objects. That is, an object can either *have an attribute* or not have it but such a *unary* attribute should not be interpreted as a classical binary attribute with the values, e.g., 0 and 1 with the former and the latter representing, respectively, the absence of the attribute and its possession by an object. For example, if particular attributes represent possession by an object of some *properties* or meeting some *criteria*, then the unary interpretation seems to be more appropriate as it may help to avoid the temptation to misinterpret the objects as binary vectors what in some cases may be not well founded.

In both the RST and the FCA the operation of *attributes reduction* plays an important role in concepts modelling and knowledge representation. This is meant, generally, as identifying/filtering out these attributes which do not contribute to the conceptual structure exposed by the data (objects) under consideration. Both, RST and FCA use their own approach to attributes reduction what is in line with their different interpretation of the attributes. However, our goal is to get a more in depth view of the similarities and differences between the concepts of attributes reduction in both theories. Hence, we provide such a formal comparison of both concepts, based on the previous work of the first author [31]. In particular, we employ the theory of *set spaces* [31] which may be seen, to some extent, as a generalization of both RST and FCA, providing a convenient platform to carry out the aforementioned comparison.

The rest of the paper is organized as follows. In Sect. 2 we present basics of rough sets and formal concept analysis. Section 3 is devoted to presentation of set spaces serving as a platform for comparison of rough set theory and formal concept analysis. We conclude Sect. 3 recalling the main result of the paper [31] concerning the case when both theories "agree" on conceptual structures which they produce, i.e. a characterization of the situation where extents of concepts from conceptual lattice from FCA are exactly definable sets in RST. Section 4 delivers results of the paper on comparison of reduction of binary attributes in rough set theory with attribute reduction in formal concept analysis. We characterize independence of sets of binary attributes in RST by complete algebras of sets completely generated by completely irredundant families of sets. Then, by means of complete algebras of sets and indiscernibility relations with respect to families of sets we investigate some families of FCA-attributes. And finally we present some formal context for which we prove that RST-binary attribute reduction and FCA-unary attribute reduction give the same results. Section 4 is followed by Conclusions section.

2   Rudiments of Rough Sets and Formal Concept Analysis

In this section we are going to present basic concepts of rough set theory (RST) and formal concept analysis (FCA). Both RST and FCA provide some mathematical approaches to concepts modelling, however both theories differ with respect to the very idea of concept and with respect to supported operations on concepts returning also concepts. RST based approach to concepts modelling represents ideas of George Boole who proposed extensional formulation of concepts, identifying concepts with their extensions (denotations). In such an approach unions of concepts (extensions) as well as intersections of concepts (extensions) are also concepts. Moreover George Boole proposed the new operation on concepts, namely a complementation of a given concept. On the other hand, FCA offers an approach to concepts modelling rooted in traditional logic where a concept is modelled as a pair of elements: the *extension* (the extent) of a concept which is a set of objects which are examples of this concept and the *intension* (the intent) of a concept which is a set of attributes (or properties) shared by all objects examples of this concept. This represents an intensional point of view where concepts are determined by both objects and attributes. In FCA based approach unions of concepts are not just unions of sets of objects (their extensions).

2.1   Elements of Rough Set Theory

Zdzisław Pawlak introduced rough sets [23,24] for the study of information systems [15,21,22]. Information systems are structures representing information by means of objects and their attributes. In rough set theory *attributes* are understood as functions of the form (we adopt here a slightly more general definition where an attribute takes a set of elements as its value, what represents uncertainty about its actual value):

$$a : U \longrightarrow \wp(Val_a)$$

where U is a set of objects, Val_a a set of values of attribute a and $\wp(Val_a)$ is the power set of Val_a. An *information system* is a triple of the form $\langle U, At, \{Val_a\}_{a \in At} \rangle$ where U is a set of objects, At is a set of attributes, and $\{Val_a\}_{a \in At}$ is the family of value domains of attributes from the set At. Information system $\langle U, At, \{Val_a\}_{a \in At} \rangle$ is *total* (or *complete*) iff $a(x) \neq \emptyset$ for every $x \in U$ and $a \in At$. Information system $\langle U, At, \{Val_a\}_{a \in At} \rangle$ is *deterministic* iff $card(a(x)) = 1$ for every $x \in U$ and $a \in At$, otherwise $\langle U, At, \{Val_a\}_{a \in At} \rangle$ is *indeterministic*.

In rough set theory information represented in information systems is analysed by means of indiscernibility relations. Let $\langle U, At, \{Val_a\}_{a \in At} \rangle$ be an information system and let $B \subseteq At$. An *indiscernibility relation* $\sim_B$ is defined as follows

$$(x, y) \in \sim_B \Leftrightarrow a(x) = a(y) \tag{1}$$

for any $a \in B$ and $x, y \in U$. One can note that $\sim_B$ is an equivalence relation on U.

Let $\langle U, At, \{Val_a\}_{a \in At} \rangle$ be an information system and let $B \subseteq At$. An attribute $a \in B$ is *indispensable* in B iff $\sim_B \neq \sim_{B \setminus \{a\}}$. Set $B \subseteq At$ is *independent* iff every attribute in B is indispensable, otherwise B is *dependent*. A set $C \subseteq B$ is *reduct* iff C is independent and $\sim_C = \sim_B$. According to Pawlak, indiscernibility relations represent knowledge derived from information systems. Keeping this interpretation in mind and looking at the Eq. (1) one can note that replacing a set of attributes with its reduct does not reduce the knowledge.

Indiscernibility relations are also useful for handling functional dependencies between families of attributes: let $\langle U, At, \{Val_a\}_{a \in At}\rangle$ be an information system and $A, B \subseteq At$, family B *depends on* family A iff $\sim_A \subseteq \sim_B$. One of the main tools for calculating indiscernibility relations and reducts are discernibility matrices [26]

In rough set theory, knowledge is founded on the ability to discern between objects and, in an algebraic approach to RST, is represented by abstract structures, referred to as *approximation spaces*. The approximation space is an ordered pair (U, R), where U is a set of objects and R is an equivalence relation on U. *Definable sets in approximation space* (U, R) are subsets of U which are unions of equivalence classes of relation R. If it does not lead to a confusion we call them shortly *definable sets*. Subsets of U which are not definable are called *rough sets*. The family of definable sets of the space (U, R) is denoted by $\mathsf{Def}_R(U)$. In [24] Pawlak called the equivalence classes of a relation R as *atoms*. There are two reasons for this: first, the family $\mathsf{Def}_R(U)$ is closed for complements, arbitrary unions and intersections i.e. the algebra $(\mathsf{Def}_R(U); \cup, \cap, ^c, \emptyset, U)$ is a complete field of sets, where $X^c := U \setminus X$, and so algebra $(\mathsf{Def}_R(U); \cup, \cap, ^c, \emptyset, U)$ is a complete and atomic Boolean algebra and atoms of this algebra are exactly equivalence classes of the relation R. The second reason is that using this term Palwak underlined the intuition that equivalence classes are basic, indivisible, building blocks of knowledge which is achieved by abstraction from particular information about objects. One can note that every information system $\langle U, At, \{Val_a\}_{a \in At}\rangle$ determines an approximation space $(U, \sim_B)$, for any $B \subseteq At$. Hence, indiscernibility relations can be viewed as bridges from a "concrete" level of information to an abstract level of knowledge. Following George Boole's extensional view of *concepts* Pawlak called subsets of U as *concepts*.

For any approximation space (U, R) and for $X \subseteq U$ two operators are defined as follows:

$$R_*(X) = \bigcup \{Y \in U_{/R} : Y \subseteq X\} \qquad R^*(X) = \bigcup \{Y \in U_{/R} : Y \cap X \neq \emptyset\}.$$

Sets $R_*(X)$ and $R^*(X)$ are called *lower* and *upper approximations* of a set $X \subseteq U$ respectively. One can note that if set $X \subseteq U$ is definable, then $R_*(X) = R^*(X)$, otherwise $R_*(X) \subsetneq R^*(X)$.

On the basis of approximation spaces Pawlak introduced structures representing knowledge named *knowledge bases*. A *knowledge base* is a pair $(U, \mathcal{R})$ such that $\mathcal{R} \subseteq Eq(U)$ where $Eq(U)$ is the family of all equivalence relations on U. A relation $R \in \mathcal{R}$ is *dispensable in* $\mathcal{R}$ if $\bigcap(\mathcal{R}) = \bigcap(\mathcal{R} \setminus \{R\})$, otherwise R is *indispensable*. If each $R \in \mathcal{R}$ is indispensable, then the family $\mathcal{R}$ is *independent*, otherwise $\mathcal{R}$ is *dependent*.

2.2   Elements of Formal Concept Analysis

The *formal context* is a triple (G, M, I), where G and M are sets of *objects* and *attributes*, respectively, while $I \subseteq G \times M$ is a binary relation [5]. Sets G and M are called *extent* and *intent* of formal context (G, M, I) respectively. When

$(g, m) \in I$ then we say that the *object g has the attribute m*, or that *the attribute m is possessed by the object g*. Sometimes we write also *g I m*. The relation *I* is also called *incidence relation*.

For a formal context (G, M, I) we define two operators $i : \mathcal{P}(G) \longrightarrow \mathcal{P}(M)$ and $e : \mathcal{P}(M) \longrightarrow \mathcal{P}(G)$, as follows:

$$X \mapsto X^i = \{m \in M : g \ I \ m, \ \forall \ g \in X\},$$

$$Y \mapsto Y^e = \{g \in G : g \ I \ m, \ \forall \ m \in Y\},$$

for each $X \subseteq G$, $Y \subseteq M$. Operators ^{i} and ^{e} are called *the intension operator* and *the extension operator*, respectively. Operators ^{i} and ^{e} are dual in the sense of order theory.

In the formal concept analysis literature there is commonly used practice to denote operators ^{i} and ^{e} by the same symbol $\cdot'$ (the prime symbol) [5]. Formal properties of the intension and extension operators are presented in Table 1 using prime notation. This practice simplifying notation and makes calculations easier while it should not cause any confusion: if it is clear that $X \subseteq G$ or $Y \subseteq M$ then this determines the meaning of X' and Y'. For example, in Table 1, since $X \subseteq G$, then formula (3a) $X' = X'''$ can be rewritten as $X^i = X^{iei}$. Looking at Table 1 one can note also that operators ^{i} and ^{e} are dual in the sense of order theory and moreover they form *Galois connection*. In addition, operators ^{i} and ^{e} combined together create closure operators: ^{ie} is a closure operator on the set G (properties (1a) - (3a) in Table 1) and ^{ei} is a closure operator on the set M (properties (1b) - (3b) in Table 1).

Table 1. Basic properties of the intension and extension operators for sets of objects $X, X_1, X_2 \subseteq G$, and sets of attributes $Y, Y_1, Y_2 \subseteq M$ of the context $\langle G, M, I \rangle$

1a. $X_1 \subseteq X_2 \Rightarrow X_2' \subseteq X_1'$	1b. $Y_1 \subseteq Y_2 \Rightarrow Y_2' \subseteq Y_1'$
2a. $X \subseteq X''$	2b. $Y \subseteq Y''$
3a. $X' = X'''$	3b. $Y' = Y'''$
4a $X'''' = X''$	4b. $Y'''' = Y''$
5a $(X_1 \cup X_2)' = X_1' \cap X_2'$	5b $(Y_1 \cup Y_2)' = Y_1' \cap Y_2'$
6. $X \subseteq Y' \Leftrightarrow Y \subseteq X' \Leftrightarrow X \times Y \subseteq I.$	

A set $\{m\}^e$ or simply m^e, can be interpreted as an extent of the attribute m. Instead of m^e we will also write $/m/$. Following that notation, $/\Gamma/ := \{ /m/ : m \in \Gamma\}$ for $\Gamma \subseteq M$, i.e. $/\Gamma/$ denotes the family of extents of attributes from the family Γ. For any formal context (G, M, I) we can define the *indiscernibility relation* $\simeq_M$ on G with respect to the set of attributes M in the following way:

$$(x, y) \in \simeq_M :\Leftrightarrow (x, m) \in I \Leftrightarrow (y, m) \in I \ for \ each \ m \in M$$

for $x, y \in G$.

A formal concept of the context $\langle G, M, I \rangle$ is a pair (A, B) where $A \subseteq G$, $B \subseteq M$, $A^i = B$ and $B^e = A$. A is called the *extent* and B is called the *intent* of the concept (A, B). The family of all formal concepts of context (G, M, I) is denoted by $\mathfrak{B}(G, M, I)$. The following relation $\preccurlyeq$ may be defined on the family $\mathfrak{B}(G, M, I)$:

$$(A_1, B_1) \preccurlyeq (A_2, B_2) \Leftrightarrow A_1 \subseteq A_2 \quad \text{or, equivalently:} \quad B_2 \subseteq B_1.$$

where $(A_1, B_1), (A_2, B_2) \in \mathfrak{B}(G, M, I)$. If $(A_1, B_1) \preccurlyeq (A_2, B_2)$, then we say that (A_1, B_1) is a *subconcept* of (A_2, B_2) and (A_2, B_2) is called a *superconcept* of (A_1, B_1). The relation $\preccurlyeq$ is a partial order on the family of concepts $\mathfrak{B}(G, M, I)$ and it is called *the hierarchical order* (or simply *order*). Moreover, one can show that $\langle \mathfrak{B}(G, M, I), \leq \rangle$ is a complete lattice, called the *concept lattice* of the context (G, M, I). We denote that lattice by $\underline{\mathfrak{B}}(G, M, I)$. In addition, we denote the family of extents of all formal concepts of the context (G, M, I) by $\mathfrak{B}_{ext}(G, M, I)$. Let us note, that $\langle \mathfrak{B}_{ext}(G, M, I), \subseteq \rangle$ is a lattice isomorphic to the concept lattice $\underline{\mathfrak{B}}(G, M, I)$. The algebraic description of concept lattices is given by the Basic Theorem on Concept Lattices [5,33] which states that the concept lattice $\underline{\mathfrak{B}}(G, M, I)$ is a complete lattice in which infimum and supremum are given by:

$$\bigwedge_{i \in I}(A_i, B_i) = (\bigcap_{i \in I} A_i (\bigcup_{i \in I} B_i)''), \quad \bigvee_{i \in I}(A_i, B_i) = ((\bigcup_{i \in I} A_i)'', \bigcap_{i \in I} B_i).$$

3 Set Spaces - A Platform for Comparison of RST and FCA

This section is devoted to the presentation of *set spaces* [31]. Let U be any set and let $\mathbb{C} \subseteq \wp(U)$. We call the pair $(U, \mathbb{C})$ *a set space*. For any set space $(U, \mathbb{C})$, by $Sg(\mathbb{C})$ we denote the least algebra (field) of sets on U containing family $\mathbb{C}$, by $\sigma(\mathbb{C})$ the least σ-algebra (σ-field) of sets on U containing family $\mathbb{C}$, and by $Sg^\star(\mathbb{C})$ the least complete algebra (complete field) of sets on U containing family $\mathbb{C}$. By $At(Sg^\star(\mathbb{C}))$ we denote the set of atoms of the algebra $Sg^\star(\mathbb{C})$. By $Sg^\cap(\mathbb{C})$ we denote the least closure system on U containing $\mathbb{C}$ i.e. the family $\mathbb{A} \subseteq \wp(U)$ closed for arbitrary intersections and containing family $\mathbb{C}$. Dually, by $Sg^\cup(\mathbb{C})$ we denote the least family of sets closed for arbitrary unions of sets and containing the family $\mathbb{C}$.

Let $(U, \mathbb{C})$ be a set space such that $\mathbb{C} = \{C_i\}_{i \in I}$. By a *generalized component over the family* $\mathbb{C}$ (shortly: g-component) we call any intersection of the following form:

$$\bigcap_{i \in I} e_i C_i \tag{2}$$

where $e_i \in \{0, 1\}$ and $1 C_i := C_i$, $0 C_i := C_i^c$ for each $i \in I$, and $C_i^c := U \setminus C_i$. We denote the family of all generalized components over the family $\mathbb{C}$ by $\prod_\mathbb{C}$ and the family of all nonempty g-components over the family $\mathbb{C}$ by $\prod_\mathbb{C}^+$.

For any set space $(U, \mathbb{C})$ we define two operators on U:

$$\mathbb{C}_*(X) := \bigcup \{A \in Sg^\star(\mathbb{C}) : A \subseteq X\} \qquad \mathbb{C}^*(X) := \bigcap \{A \in Sg^\star(\mathbb{C}) : X \subseteq A\}.$$

We call operators $\mathbb{C}_*$, $\mathbb{C}^*$ the *lower* and *the upper approximation*, respectively. We call the elements of the algebra $Sg^\star(\mathbb{C})$ the *sets definable in set space* $(U, \mathbb{C})$ or shortly *definable sets* when it does not lead to a confusion.

Let $(U, \mathbb{C})$ be a set space, then the family $\mathbb{C}$ is *completely irredundant* if for all sets $B \in \mathbb{C}$, $B \notin Sg^\star(\mathbb{C} \setminus \{B\})$, otherwise the family $\mathbb{C}$ is *completely redundant*.

Let U be a nonempty set. For any set $C \in \wp(U)$ we define the *indiscernibility relation on U with respect to C* denoted as $\approx_C$, in the following way

$$x \approx_C y \iff_{def} x \in C \Leftrightarrow y \in C.$$

For any set space $(U, \mathbb{C})$ we define *an indiscernibility relation on U with respect to the family $\mathbb{C}$* denoted as $\approx_\mathbb{C}$, in the following way:

$$(x, y) \in \approx_\mathbb{C} \iff_{def} (x, y) \in \bigcap_{C \in \mathbb{C}} \approx_C . \tag{3}$$

Thus, the following implication holds $\forall \mathbb{A}, \mathbb{B} \subseteq \wp(U)$:

$$\mathbb{A} \subseteq \mathbb{B} \Rightarrow \approx_\mathbb{B} \subseteq \approx_\mathbb{A} .$$

One can note that the reverse implication does not hold in general. However, in the next section we will show sufficient and necessary condition for this reverse implication to hold.

Let $(U, \mathbb{C})$ be a set space , then by Proposition 4.14 in [31]:

$$\approx_\mathbb{C} = \approx_{Sg(\mathbb{C})} = \approx_{\sigma(\mathbb{C})} = \approx_{Sg^\star(\mathbb{C})} .$$

If $U \neq \emptyset$ and $\mathbb{C}, \mathbb{D} \subseteq \mathcal{P}(U)$, then by Theorem 4.16 from [31] partially recalled in Sect. 3 we get:

$$\approx_\mathbb{C} \subseteq \approx_\mathbb{D} \iff Sg^\star(\mathbb{D}) \subseteq Sg^\star(\mathbb{C}).$$

One can strengthen inclusions to identities in the above formula.

For every approximation space (U, R) there is a set space $(U, \mathbb{C})$ such that $R = \approx_\mathbb{C}$. Moreover, for any approximation space (U, R) and for any set space $(U, \mathbb{C})$ the following conditions are equivalent:

(1) $\forall X \subseteq U : R_*(X) = \mathbb{C}_*(X)$ and $R^*(X) = \mathbb{C}^*(X)$,
(2) $R = \approx_\mathbb{C}$,
(3) $\mathrm{Def}_R(U) = Sg^\star(\mathbb{C})$.

Now, we will show the links between the characteristic concepts of RST and FCA. One can note that for any formal context $\langle G, M, I \rangle$ the following equation holds:

$$\simeq_M = \approx_{/M/} .$$

Let us recall definition of the *context exactness* [31]: a formal context $\langle G, M, I \rangle$ is *exact* iff the following conditions are satisfied:

(i) there exists a non-empty subset of attributes $M_0 \subseteq M$ such that $M_0 := \{m \in M : \exists A \in G_{/\simeq_M}, /m/ = A^c\}$,
(ii) $/M/ \subseteq Sg^{\cap}(/M_0/)$.

In FCA if a given concept have some subconcepts, then usually its extent is not a union of its subconcepts extents. So *exact contexts* are contexts where the extent of the every concept is a union of its subconcepts extents and moreover the concept extents in the exact context are exactly unions of equivalence classes of the indiscernibility relation with respect to all attributes from the exact context intent.

Let us adopt the following notation [31]: for any family of sets $\mathbb{C}$ let $[\mathbb{C}]^c := \{C^c : C \in \mathbb{C}\}$, where $C^c := U \setminus C$. By $\not\simeq_M$ we denote the opposite relation to the relation $\simeq_M$, i.e. $\not\simeq_M := (G \times G) \setminus \simeq_M$. Now we can recall main results from the last section of [31]. We start with Lemma 8.2 in [31]: the context $\langle G, M, I \rangle$ is exact if and only if

$$[G_{/\simeq_M}]^c \subseteq /M/ \subseteq Sg^{\cap}([G_{/\simeq_M}]^c) = Sg^{\star}(/M/) = Sg^{\cup}(G_{/\simeq_M}).$$

Now we present characterization of context exactness by Theorem 8.3 from [31]: let $\langle G, M, I \rangle$ be a formal context. then the following conditions are equivalent:

(1) the context $\langle G, M, I \rangle$ is exact,
(2) $\mathfrak{B}_{ext}(G, M, I) = \mathrm{Def}_{\simeq_M}(G)$,
(3) $Sg^{\cap}(/M/) = Sg^{\star}(/M/)$,
(4) $\mathfrak{B}_{ext}(G, M, I) = \mathfrak{B}_{ext}(G, G, \not\simeq_M)$.

4 Reduction of Binary Attributes in RST and Reduction of Attributes in FCA - A Comparison

Before proceeding to the comparison announced in the title of this section we need to remind a few more properties of the set spaces which are a platform for the said comparison. We start this section with showing that complete irredundancy of families of sets is hereditary on subsets of their subsets, i.e. the following holds:

Lemma 1. *Let $(U, \mathbb{C})$ be a set space. If $\mathbb{C}$ is completely irredundant, then $\forall \mathbb{A} \subseteq \mathbb{C} : \mathbb{A}$ is completely irredundant.*

Proof: Let $(U, \mathbb{C})$ be a set space such that $\mathbb{C}$ is completely irredundant and let $\mathbb{A} \subseteq \mathbb{C}$. Let us assume that $\mathbb{A}$ is not completely irredundant. Then there exists $D \in \mathbb{A}$ such that $D \in Sg^{\star}(\mathbb{A} \setminus \{D\})$. Since $\mathbb{A} \subseteq \mathbb{C}$, then $\mathbb{A} \setminus \{D\} \subseteq \mathbb{C} \setminus \{D\}$. Thus $Sg^{\star}(\mathbb{A} \setminus \{D\}) \subseteq Sg^{\star}(\mathbb{C} \setminus \{D\})$. Therefore $D \in Sg^{\star}(\mathbb{C} \setminus \{D\})$ what contradicts the assumption that $\mathbb{C}$ is completely irredundant. Thus we have shown that $\mathbb{A} \subseteq \mathbb{C}$ is completely irredundant.   □

Proposition 1. *Let $(U, \mathbb{C})$ be a set space. Then the following conditions are equivalent:*

(1) $\forall A, \mathbb{B} \subseteq \mathbb{C}: \approx_{\mathbb{B}} \subseteq \approx_A \Rightarrow A \subseteq \mathbb{B}$,
(2) *the family $\mathbb{C}$ is completely irredundant.*

Proof:
($\Rightarrow$) Let us assume that the family $\mathbb{C}$ is not completely irredundant and thus there exists $B \in \mathbb{C}$ such that $B \in Sg^*(\mathbb{C} \setminus \{B\})$. Let us note that $\mathbb{C} \setminus \{B\} \subseteq Sg^*(\mathbb{C} \setminus \{B\})$. Since $B \in Sg^*(\mathbb{C} \setminus \{B\})$ thus $(\mathbb{C} \setminus \{B\}) \cup \{B\} \subseteq Sg^*(\mathbb{C} \setminus \{B\})$ so $\mathbb{C} \subseteq Sg^*(\mathbb{C} \setminus \{B\})$. Therefore $Sg^*(\mathbb{C}) \subseteq Sg^*(Sg^*(\mathbb{C} \setminus \{B\})) = Sg^*(\mathbb{C} \setminus \{B\})$ and so $Sg^*(\mathbb{C}) \subseteq Sg^*(\mathbb{C} \setminus \{B\})$. Because of that $\mathbb{C} \setminus \{B\} \subseteq \mathbb{C}$ we know that $Sg^*(\mathbb{C} \setminus \{B\}) \subseteq Sg^*(\mathbb{C})$. Thus we showed that $Sg^*(\mathbb{C}) = Sg^*(\mathbb{C} \setminus \{B\})$, this implies that $\approx_{Sg^*(\mathbb{C})} = \approx_{Sg^*(\mathbb{C}\setminus\{B\})}$. From this and by Propositon 4.14 from [31] we get that

$$\approx_{\mathbb{C}} = \approx_{Sg^*(\mathbb{C})} = \approx_{Sg^*(\mathbb{C}\setminus\{B\})} = \approx_{\mathbb{C}\setminus\{B\}}.$$

Thus $\approx_{\mathbb{C}} = \approx_{\mathbb{C}\setminus\{B\}}$ and in particular $\approx_{\mathbb{C}\setminus\{B\}} \subseteq \approx_{\mathbb{C}}$ but $\mathbb{C} \not\subseteq \mathbb{C} \setminus \{B\}$ and thus (1) does not hold. Hence, we have proved that negated (2) implies negated (1) and thus the implication (1) $\Rightarrow$ (2) holds.

($\Leftarrow$) Let us assume that $\mathbb{C}$ is completely irredundant, $A, \mathbb{B} \subseteq \mathbb{C}$, $\approx_{\mathbb{B}} \subseteq \approx_A$ and assume also that $A \not\subseteq \mathbb{B}$. Thus there is $D \in A \setminus \mathbb{B}$. Since $D \notin \mathbb{B}$, then $\mathbb{B} = \mathbb{B}\setminus\{D\}$. Thus $Sg^*(\mathbb{B}) = Sg^*(\mathbb{B} \setminus \{D\})$. Since $\approx_{\mathbb{B}} \subseteq \approx_A$, then $Sg^*(A) \subseteq Sg^*(\mathbb{B})$. Since $\mathbb{B} \subseteq \mathbb{C}$, then $\mathbb{B} \setminus \{D\} \subseteq \mathbb{C} \setminus \{D\}$ and so $Sg^*(\mathbb{B} \setminus \{D\}) \subseteq Sg^*(\mathbb{C} \setminus \{D\})$. Thus

$$D \in A \setminus \mathbb{B} \subseteq Sg^*(A) \subseteq Sg^*(\mathbb{B}) = Sg^*(\mathbb{B} \setminus \{D\}) \subseteq Sg^*(\mathbb{C} \setminus \{D\}).$$

Thus $D \in Sg^*(\mathbb{C} \setminus \{D\})$. By assumptions $D \in A \setminus \mathbb{B}$ and $A \subseteq \mathbb{C}$, we get that $D \in \mathbb{C}$. Therefore we have shown that $D \in \mathbb{C}$ and $D \in Sg^*(\mathbb{C} \setminus \{D\})$ what contradicts the assumption that family $\mathbb{C}$ is completely irredundant. $\square$

Theorem 1. *Let $\langle U, At, \{Val_a\}_{a\in At}\rangle$ be a binary information system i.e. $\forall a \in At : Val_a = \{0,1\}$. Let $B \subseteq At$ and $\mathbb{C}(B) := \{C_b\}_{b\in B}$, where $C_b = \{x \in U : b(x) = 1\}$. Then the following conditions are equivalent:*

(1) *a set $B \subseteq At$ is independent,*
(2) *the family $\mathbb{C}(B)$ is completely irredundant.*

Proof: In order to prove this proposition we prove the following equivalence: set $B \subseteq At$ is dependent $\Leftrightarrow$ family $\mathbb{C}(B)$ is completely redundant.

($\Rightarrow$) Let us assume that $B \subseteq At$ is dependent, then there is $b \in B$ such that $\sim_B = \sim_{B\setminus\{b\}}$. Let us note that $\approx_{\mathbb{C}(D)} = \sim_D$ (cf., (1) and (3), respectively) for any $D \subseteq At$ therefore $\approx_{\mathbb{C}(B)} = \sim_B = \sim_{B\setminus\{b\}} = \approx_{\mathbb{C}(B)\setminus\{C_b\}}$. Thus $\approx_{\mathbb{C}(B)} = \approx_{\mathbb{C}(B)\setminus\{C_b\}}$, what implies $C_b \in Sg^*(\mathbb{C}(B))$. Equation $\approx_{\mathbb{C}(B)} = \approx_{\mathbb{C}(B)\setminus\{C_b\}}$ by Theorem 4.16 from [31] partially recalled in Sect. 3 is equivalent to $Sg^*(\mathbb{C}(B)) = Sg^*(\mathbb{C}(B)\setminus\{C_b\})$, so $C_b \in Sg^*(\mathbb{C}(B)\setminus\{C_b\})$. Therefore family $\mathbb{C}(B)$ is completely redundant.

($\Leftarrow$) Let family $\mathbb{C}(B)$ is completely redundant. Thus there exists $b \in B$ such that $C_b \in Sg^*(\mathbb{C}(B) \setminus \{C_b\})$ and then $(\mathbb{C}(B) \setminus \{C_b\}) \cup \{C_b\} \subseteq Sg^*(\mathbb{C} \setminus \{C_b\})$ so

$\mathbb{C}(B) \subseteq Sg^*(\mathbb{C}(B) \setminus \{C_b\})$. Therefore $Sg^*(\mathbb{C}(B)) \subseteq Sg^*(Sg^*(\mathbb{C}(B) \setminus \{C_b\})) = Sg^*(\mathbb{C}(B) \setminus \{C_b\})$ since Sg^* is a closure operator on set $\wp(\wp(U))$. Thus $Sg^*(\mathbb{C}(B)) \subseteq Sg^*(\mathbb{C}(B) \setminus \{C_b\})$. Because of $(\mathbb{C}(B) \setminus \{C_b\}) \subseteq \mathbb{C}(B)$ we get $Sg^*(\mathbb{C}(B) \setminus \{C_b\}) \subseteq Sg^*(\mathbb{C}(B))$ since operator Sg^* is monotonic. Thus we have shown that $Sg^*(\mathbb{C}(B)) \subseteq Sg^*(\mathbb{C}(B) \setminus \{C_b\}) \subseteq Sg^*(\mathbb{C}(B))$ so $Sg^*(\mathbb{C}(B)) = Sg^*(\mathbb{C}(B) \setminus \{C_b\})$. This is equivalent to $\approx_{\mathbb{C}(B)} = \approx_{(\mathbb{C}(B) \setminus \{C_b\})}$. Thus because $\sim_B = \approx_{\mathbb{C}(B)}$ and $\sim_{(B \setminus \{b\})} = \approx_{(\mathbb{C}(B) \setminus \{C_b\})}$ then $\sim_B = \approx_{\mathbb{C}(B)} = \approx_{\mathbb{C}(B) \setminus \{C_b\}} = \sim_{B \setminus \{b\}}$ and so $\sim_B = \sim_{B \setminus \{b\}}$. Therefore set $B \subseteq At$ is dependent. $\square$

Now we are going to compare the concept of reducibility of attributes in rough set theory and in formal concept analysis. In fact in FCA the concept of attribute reducibility have the same name like in RST but in the former case its meaning is more narrow. Therefore referring to the concept of reduction of attributes in formal concept analysis we will use the name *FCA-reduction of attributes*. Recall also that or any formal context $\langle G, M, I \rangle$ and an attribute $m \in M$ an *attribute concept* is the formal concept of the following form: (m^e, m^{ei}). Thus, let $\langle G, M, I \rangle$ be a formal context. We say that an attribute $m \in M$ is *FCA-reducible* if and only if its attribute concept is reducible in formal concept analysis, i.e.

$$(m^e, m^{ei}) = \bigwedge (n^e, n^{ei})$$

or all $n \in M$ such that $(m^e, m^{ei}) \preccurlyeq (n^e, n^{ei})$ and $(m^e, m^{ei}) \neq (n^e, n^{ei})$. We write $(m^e, m^{ei}) \prec (n^e, n^{ei})$ to denote that $(m^e, m^{ei}) \preccurlyeq (n^e, n^{ei})$ and $(m^e, m^{ei}) \neq (n^e, n^{ei})$.

Proposition 2. *Let $\langle G, M, I \rangle$ be a formal context and $m \in M$. Then the following conditions are equivalent:*

(1) *attribute m is* FCA-reducible,
(2) $/m/ \in Sg^{\cap}(/M \setminus \{m\}/)$.

Proof: Let $\langle G, M, I \rangle$ be a formal context and let $m_0 \in M$. Then the following conditions are equivalent:

$m_0 \in M$ is FCA-reducible,

$/m_0/ = \bigcap \{/n/ \in /M/ : (m_0^e, m_0^{ei}) \prec (n^e, n^{ei})\}$,

$/m_0/ = \bigcap \{/n/ \in /M/ : m_0^e \subseteq n^e \ \& \ m_0 \neq n)\}$,

$/m_0/ = \bigcap \{/n/ \in /M/ : /m_0/ \subseteq /n/ \ \& \ /m_0/ \neq /n/)\}$,

$/m_0/ \in Sg^{\cap}(\bigcap \{/n/ \in /M/ : /m_0/ \subseteq /n/ \ \& \ /m_0/ \neq /n/)\})$,

$/m_0/ \in Sg^{\cap}(/M \setminus \{m_0\}/)$ and $m_0 \in M$,

$\square$

Since for any family $\mathbb{C} \subseteq \wp(U)$, $Sg^{\cap}(\mathbb{C}) \subseteq Sg^*(\mathbb{C})$, then we get that $Sg^{\cap}(/M/) \subseteq Sg^*(/M/)$. In view of this and of the Proposition 2 one can see that the idea of reduction of attributes in FCA is narrower than in RST. The main reason for this is that in FCA while checking reducibility we are taking into account only intersections of attribute extensions while in the light of Theorem 1 in RST the reduction of attributes is based on all Boolean set-operations instead of intersections only.

However, in FCA one can mimic the way in which reducibility is treated in RST by means of manipulating formal contexts. Thus the rest of this section is devoted to methods of information reduction in FCA aimed at proposing such formal contexts for which inclusion of the form $Sg^{\cap}(/M/) \subseteq Sg^*(/M/)$ will be strengthened to identity $Sg^{\cap}(/M/) = Sg^*(/M/)$.

Definition 1. *Let $\langle G, M, I \rangle$ be a formal context. Then with $M^{\complement}$ and $I^{\complement}$ we will denote the context $\langle G, M^{\complement}, I^{\complement} \rangle$ where:*

$$M^{\complement} = \{m^{\complement} : m \in M\}, \tag{4}$$

$$I^{\complement} = \{(g, m^{\complement}) : (g, m) \notin I\}. \tag{5}$$

Thus, the attributes of $M^{\complement}$ are in a sense dual/complementary to the original attributes of M: an attribute $m^{\complement}$ holds for an object g if the original attribute m does not hold for this object. We will denote as $M_{\cup}$ the sum of the set of attributes M of the original context $\langle G, M, I \rangle$ and the set of attributes $M^{\complement}$ of the new context $\langle G, M^{\complement}, I^{\complement} \rangle$, i.e., $M_{\cup} = M \cup M^{\complement}$.

Theorem 2. *Let $(U, \mathbb{C})$ be a set space and let $\mathbb{D} := \{C^c : C \in \mathbb{C}\}$. Then the following holds:*

(1) $Sg^*(\mathbb{C}) = Sg^*(\mathbb{D}) = Sg^*(\mathbb{C} \cup \mathbb{D})$,
(2) $\approx_{\mathbb{C}} = \approx_{\mathbb{D}} = \approx_{\mathbb{C} \cup \mathbb{D}}$.

Proof: Let $(U, \mathbb{C})$ be a set space and let $\mathbb{D} := \{C^c : C \in \mathbb{C}\}$.

(1) In order to show $Sg^*(\mathbb{C} \cup \mathbb{D}) \subseteq Sg^*(\mathbb{C})$ we will show that $\mathbb{D} \subseteq Sg^*(\mathbb{C})$. For any set $C \in \mathbb{C}$, $C^c \in Sg^*(\mathbb{C})$ since $Sg^*(\mathbb{C})$ is a complete field of sets completely generated by family $\mathbb{C}$ (i.e. the least complete field of sets containing family $\mathbb{C}$). Since $\forall C \in \mathbb{C} : C^c \in Sg^*(\mathbb{C})$ then $\mathbb{D} \subseteq Sg^*(\mathbb{C})$. By definition of $Sg^*(\mathbb{C})$ we get also $\mathbb{C} \subseteq Sg^*(\mathbb{C})$. Since $\mathbb{D} \subseteq Sg^*(\mathbb{C})$ and $\mathbb{C} \subseteq Sg^*(\mathbb{C})$, then $\mathbb{C} \cup \mathbb{D} \subseteq Sg^*(\mathbb{C})$. Thus $Sg^*(\mathbb{C} \cup \mathbb{D}) \subseteq Sg^*(Sg^*(\mathbb{C})) = Sg^*(\mathbb{C})$ since Sg^* is closure operator on power set $\wp(U)$. Therefore $Sg^*(\mathbb{C} \cup \mathbb{D}) \subseteq Sg^*(\mathbb{C})$.

Since $\mathbb{C} \subseteq \mathbb{C} \cup \mathbb{D}$, then by monotonicity of operator Sg^* it follows that $Sg^*(\mathbb{C}) \subseteq Sg^*(\mathbb{C} \cup \mathbb{D})$. Therefore $Sg^*(\mathbb{C}) = Sg^*(\mathbb{C} \cup \mathbb{D})$, where $\mathbb{D} := \{C^c : C \in \mathbb{C}\}$.

Now in order to show $Sg^*(\mathbb{D}) = Sg^*(\mathbb{C} \cup \mathbb{D})$ we will show that $\mathbb{C} \subseteq Sg^*(\mathbb{D})$. Since $\mathbb{D} := \{C^c : C \in \mathbb{C}\} \subseteq Sg^*(\mathbb{D})$ and since $Sg^*(\mathbb{D})$ as a complete algebra of sets is closed under complementation of sets thus $[\mathbb{D}]^c \subseteq Sg^*(\mathbb{D})$, and as $[\mathbb{D}]^c = \mathbb{C}$, thus $\mathbb{C} \subseteq Sg^*(\mathbb{D})$. The further proof proceeds analogously like in the case of proving $Sg^*(\mathbb{C}) = Sg^*(\mathbb{C} \cup \mathbb{D})$.

Since $Sg^*(\mathbb{C}) = Sg^*(\mathbb{C} \cup \mathbb{D})$ and $Sg^*(\mathbb{D}) = Sg^*(\mathbb{C} \cup \mathbb{D})$, then this implies that

$$Sg^*(\mathbb{C}) = Sg^*(\mathbb{D}) = Sg^*(\mathbb{C} \cup \mathbb{D})$$

(2) This is equivalent to (1) by Proposition 4.15 from [31] recalled also in this paper in Sect. 3.   □

From Theorem 2 we get the following corollary for families of attributes in formal concept analysis:

Corollary 1. *Let $\langle G, M, I \rangle$ be a formal context. Then the following holds:*

(1) $Sg^*(/M/) = Sg^*(/M^{\complement}/) = Sg^*(/M \cup M^{\complement}/),$
(2) $\simeq_M = \simeq_{M^{\complement}} = \simeq_{M \cup M^{\complement}}.$

i.e. sets of attributes M, $M^{\complement}$ and $M \cup M^{\complement}$ determine the same indiscernibility relation whereas extensions of attributes from M, $M^{\complement}$ and $M \cup M^{\complement}$ determine the same complete atomic Boolean algebra of sets.

Proof: Let $\langle G, M, I \rangle$ be a formal context. In order to prove (1) it is enough to consider $\mathbb{C} := /M/$, thus $\mathbb{D} = /M^{\complement}/$ and $\mathbb{C} \cup \mathbb{D} = /M \cup M^{\complement}/$ and then use Theorem 2. In order to prove (2) one can note that the following facts hold:

$$\approx_{/M/} = \simeq_M, \quad \approx_{/M^{\complement}/} = \simeq_{M^{\complement}}, \quad \approx_{/M \cup M^{\complement}/} = \simeq_{M \cup M^{\complement}}$$

Then by Theorem 2 of this paper, condition (1) of this corollary on the basis of the above equations is equivalent to

$$\approx_{/M/} = \approx_{/M^{\complement}/} = \approx_{/M \cup M^{\complement}/},$$

thus

$$\simeq_M = \simeq_{M^{\complement}} = \simeq_{M \cup M^{\complement}}.$$

                                          □

From Propositions 4.3 and 4.7 from [31] we have the following equalities:

(1) $At(Sg^*(\mathbb{C})) = \prod_{\mathbb{C}}^{+},$
(2) $\prod_{\mathbb{C}}^{+} = U_{/\approx_{\mathbb{C}}}.$

one can get the following corollary:

Corollary 2. *Let $\langle G, M, I \rangle$ be a formal context. Then the following equations hold:*

$$At(Sg^*(/M/)) = \prod_{/M/}^{+} = G_{/\simeq_M}.$$

where $\prod_{/M/}^{+}$ denotes the family of all nonempty g-components over the family $/M/$; cf. (2).
Proof: Let $\langle G, M, I \rangle$ be a formal context. According to Propositions 4.3 and 4.7 for any family $\mathbb{C} \subseteq \wp(G)$ the following equations hold respectively: $At(Sg^*(\mathbb{C})) = \prod_{\mathbb{C}}^{+}$ and $\prod_{\mathbb{C}}^{+} = G_{/\approx_{/M/}}.$ Thus for $/M/ \subseteq \wp(G)$ we get $At(Sg^*(/M/)) = \prod_{/M/}^{+}$ and $\prod_{/M/}^{+} = G_{/\approx_{/M/}}.$ Now one can note that

$\approx_{/M/} = \simeq_M$ thus, since $\approx_{/M/}$ and $\simeq_M$ are equivalence relations, by the Abstraction Principle we get $G_{/\approx_{/M/}} = G_{/\simeq_M}$.   □

For any set space $(U, \mathbb{C})$ let us recall that $[\mathbb{C}]^c := \{C^c : C \in \mathbb{C}\}$, where $C^c := U \setminus C$. From Corollary 2 we infer the following proposition about the structure of $Sg^\star(/M/)$ as the complete and atomic Boolean algebra:

Proposition 3. *Let $\langle G, M, I \rangle$ be a formal context. By $CoAt(Sg^\star(/M/))$ we denote the family of coatoms of the complete algebra of sets $Sg^\star(/M/)$ i.e. the family of maximal sets in $Sg^\star(/M/)$ which are different from G. Then the following conditions hold:*

(1) $CoAt(Sg^\star(/M/)) = [G_{/\simeq_M}]^c$,

(2) $[G_{/\simeq_M}]^c = \{\not\simeq_M (x)\}_{x \in G}$.

Proof: Let $\langle G, M, I \rangle$ be a formal context.

(1) Since $At(Sg^\star(/M/)) = G_{/\simeq_M}$, then $G_{/\simeq_M} \subseteq Sg^\star(/M/)$. Since $Sg^\star(/M/)$ is complete field of sets, so $Sg^\star(/M/)$ is closed for complements of sets, thus $[G_{/\simeq_M}]^c \subseteq Sg^\star(/M/)$. Since $\langle Sg^\star(/M/); \cup, \cap, ^c, \emptyset, G \rangle$ is a complete and atomic Boolean algebra, then complements of atoms are precisely coatoms of algebra $\langle Sg^\star(/M/); \cup, \cap, ^c, \emptyset, G \rangle$, thus $CoAt(Sg^\star(/M/)) = [G_{/\simeq_M}]^c$.

(2) Let $x \in G$. We will show that $(x_{/\simeq_M})^c = \not\simeq_M (x)$. Let $z \in G$, then the following conditions are equivalent:

$$z \in (x_{/\simeq_M})^c$$
$$z \notin x_{/\simeq_M}$$
$$\forall y \in x_{/\simeq_M} : (z, y) \notin \simeq_M$$
$$\forall y \in x_{/\simeq_M} : z \not\simeq y$$
$$z \in \not\simeq_M (x).$$

Since $z \in G$ was chosen arbitrarily, then $(x_{/\simeq_M})^c = \not\simeq_M (x)$. Because of $x \in G$ was chosen arbitrarily as well, therefore we have already shown that

$$[G_{/\simeq_M}]^c = \{\not\simeq_M (x)\}_{x \in G}.$$

                                                                                    □

Definition 2. *Let $\langle G, M, I \rangle$ be a formal context. Then*

$$M^* := \{\not\simeq_M (x) : x \in G\}.$$

One can note that relation $\in$ can also be an incidence relation in those formal contexts, in which attributes are sets, i.e. $(g, m) \in I \Leftrightarrow g \in m$.

Theorem 3. *Let $\langle G, M, I \rangle$ be a formal context. Then for formal context $\langle G, M^*, \in \rangle$ the following equations hold:*

$$Sg^\cap(M^*) = Sg^\star(/M/) = Sg^*(M^*).$$

By Propositions 3.1 and 3.2 we get $CoAt(Sg^\star(/M/)) = \{\not\simeq_M (x)\}_{x \in G}$. Therefore $CoAt(Sg^\star(/M/)) = M^*$. Since $Sg^\star(/M/)$ is complete and atomic Boolean algebra, it is also coatomic i.e. for every $A \in Sg^\star(/M/)$ and every $B \in CoAt(Sg^\star(/M/))$ either $A \subseteq B$ or $A \cap B = \emptyset$. Let $M^*(A) := \{B \in CoAt(Sg^\star(/M/)) : A \subseteq B\}$. Then one can show that $A = \bigcap M^*(A)$. This means by definition that $A \in Sg^\cap(M^*)$. Since $A \in Sg^\star(/M/)$ was chosen arbitrarily, then we showed that $Sg^\star(/M/) \subseteq Sg^\cap(M^*)$. Let us note that $M^* \subseteq Sg^\star(/M/)$, thus $Sg^\cap(M^*) \subseteq Sg^\star(/M/)$, since $Sg^\star$ is a closure operator on set G. Therefore $Sg^\cap(M^*) = Sg^\star(/M/)$.

Let us note that $M^* \subseteq Sg^\star(/M/)$ and $/M/ \subseteq Sg^\star(M^*)$. Therefore $Sg^\star(M^*) \subseteq Sg^\star(Sg^\star(/M/)) = Sg^\star(/M/)$ and $Sg^\star(/M/) \subseteq Sg^\star(Sg^\star(M^*)) = Sg^\star(M^*)$ since operator $Sg^\star$ is a closure operator on G and so it is also monotonic. Thus $Sg^\star(M^*) \subseteq Sg^\star(/M/)$ and $Sg^\star(/M/) \subseteq Sg^\star(M^*)$ and so $Sg^\star(/M/) = Sg^\star(M^*)$ what concludes the proof of the second equation.   □

We end this paper presenting the following corollary of above Theorem 3 and Theorem 3.8 from the paper [31]:

Corollary 3. *Let $\langle G, M, I \rangle$ be a formal context. Then for formal context $\langle G, M^*, \in \rangle$ the following conditions hold:*

(1) *the context $\langle G, M, I \rangle$ is exact,*
(2) $\mathfrak{B}_{ext}(G, M, I) = Def_{\simeq_M}(G)$,
(3) $\mathfrak{B}_{ext}(G, M, I) = \mathfrak{B}_{ext}(G, G, \not\simeq_M)$.

5   Conclusions

Rough sets theory and formal concept analysis are two prominent examples of approaches to concepts modelling. Information systems and formal contexts in the former and in the latter case, respectively, serve for characterizing sets of objects under consideration with the sets of attributes. Despite the different nature of the attributes used in these two approaches, an important concept shared by them is the operation of reduction of the attributes. In the paper we compare these operations employing the theory of the set spaces as a common platform and we draw interesting conclusions regarding these operation, on an abstract level of algebraic considerations. In particular, we show that the idea of reduction of attributes in FCA is narrower than in RST. However the full attributes reduction in RST can be reconstructed in FCA by means of manipulating formal contexts.

It should be, however, emphasized that our interest is not purely theoretical. In our previous works [8,10,32] in the area of group decision making, we have used both theories to analyse popular voting procedures with respect to their possession or not of some desired properties proposed in the literature. Thus, we are going to exploit the results reported in the current paper to continue and deepen mentioned comparison of voting procedures. The insights collected during the work reported in this paper are encouraging and promising.

References

1. Bargiela, A., Pedrycz, W.: Granular Computing: An Introduction. Kluwer Academic Publishers, Amsterdam (2003)
2. Fedrizzi, M., Kacprzyk, J., Nurmi, H.: How different are social choice functions: a rough sets approach. Qual. Quant. Int. J. Methodol. **30**(1), 87–99 (1996)
3. Fishburn, P.C.: The Theory of Social Choice functions. Princeton University Press, Princeton (1973)
4. Fishburn, P.C.: Social choice functions. Soc. Ind. Appl. Math. Rev. **16**(1), 63–90 (1974)
5. Ganter, B., Wille, R.: Formal Concept Analysis: Mathematical Foundation. Springer, Heidelberg (1999). https://doi.org/10.1007/978-3-642-59830-2
6. Kacprzyk, J.: Group decision making with a fuzzy majority. Fuzzy Sets Syst. **18**, 105–118 (1986)
7. Kacprzyk, J., Fedrizzi, M., Nurmi, H.: Group decision making and consensus under fuzzy preferences and fuzzy majority. Fuzzy Sets Syst. **49**, 21–31 (1992)
8. Kacprzyk, J., Merigó, J.M., Nurmi, H., Zadrożny, S.: Multi-agent systems and voting: how similar are voting procedures. In: Lesot, M.-J., et al. (eds.) IPMU 2020. CCIS, vol. 1237, pp. 172–184. Springer, Cham (2020). https://doi.org/10.1007/978-3-030-50146-4_14
9. Kacprzyk, J., Nurmi, H., Zadrożny, S.: Reason vs. rationality: from rankings to tournaments in individual choice. In: Mercik, J. (ed.) Transactions on Computational Collective Intelligence XXVII. LNCS, vol. 10480, pp. 28–39. Springer, Cham (2017). https://doi.org/10.1007/978-3-319-70647-4_2
10. Kacprzyk, J., Nurmi, H., Zadrozny, S.: Towards a comprehensive similarity analysis of voting procedures using rough sets and similarity measures. In: Skowron, A., Suraj, Z. (eds.) Rough Sets and Intelligent Systems - Professor Zdzisław Pawlak in Memoriam. Intelligent Systems Reference Library, vol. 42, pp. 359–380. Springer, Heidelberg (2013). https://doi.org/10.1007/978-3-642-30344-9_13
11. Kacprzyk, J., Zadrozny, S.: Towards a general and unified characterization of individual and collective choice functions under fuzzy and nonfuzzy preferences and majority via the ordered weighted average operators. Int. J. Intell. Syst. **24**, 4–26 (2009)
12. Kacprzyk, J., Zadrozny, S.: Towards human consistent data driven decision support systems using verbalization of data mining results via linguistic data summaries. Bull. Polish Acad. Sci. Techn. Sci. **58**(3), 359–370 (2010)
13. Kelly, J.S.: Social Choice Theory. Springer, Heidelberg (1988). https://doi.org/10.1007/978-3-662-09925-4
14. Lin, T.Y., Liau, C.J., Kacprzyk, J. (eds.): Granular, Fuzzy, and Soft Computing: A Volume in the Encyclopedia of Complexity and Systems Science Series. 1st edn. Springer, Cham (2023)
15. Lipski, W.: Informational systems with incomplete information. In: 3rd International Symposium on Automata, Languages and Programming, Edinburgh, Scotland, pp. 120–130 (1976)
16. Nurmi, H.: Comparing Voting Systems. D. Reidel, Dordrecht (1987)
17. Nurmi, H.: Voting Paradoxes and How to Deal With Them. Springer, Heidelberg (1999)
18. Nurmi, H.: The choice of voting rules based on preferences over criteria. In: Kamiński, B., Kersten, G.E., Szapiro, T. (eds.) GDN 2015. LNBIP, vol. 218, pp. 241–252. Springer, Cham (2015). https://doi.org/10.1007/978-3-319-19515-5_19

19. Nurmi, H., Kacprzyk, J.: On fuzzy tournaments and their solution concepts in group decision making. Eur. J. Oper. Res. **51**(2), 223–232 (1991)
20. Nurmi, H., Kacprzyk, J., Zadrożny, S.: Voting systems in theory and practice. In: Szapiro, T., Kacprzyk, J. (eds.) Collective Decisions: Theory, Algorithms And Decision Support Systems. SSDC, vol. 392, pp. 3–16. Springer, Cham (2022). https://doi.org/10.1007/978-3-030-84997-9_1
21. Orłowska, E., Pawlak, Z.: Representation of nondeterministic information. Theoret. Comput. Sci. **29**, 27–39 (1984)
22. Pawlak, Z.: Information systems - theoretical foundations. Inf. Syst. **6**, 205–218 (1981)
23. Pawlak, Z.: Rough sets. Int. J. Comput. Inf. Sci. **18**, 341–356 (1982)
24. Pawlak, Z.: Rough Sets. Theoretical Aspects of Reasoning About Data. Kluwer Academic Publishers, Dordrecht (1991)
25. Pedrycz, W., Skowron, A., Kreinovich, V. (eds.): Handbook on Granular Computing. Wiley, New York (2009)
26. Rauszer, C., Skowron, A.: The discernibility matrices and functions in information systems. In: R. Słowiński, (Ed.) Intelligent Decision Support. Handbook of Applications and Advances in the Rough Set Theory, pp. 331–362. Kluwer (1991)
27. Stumme, G.: Conceptual knowledge discovery and data mining with formal concept analysis. Tutorial slides at the European Conference on Machine Learning and Principles and Practice of Knowledge Discovery in Databases ECML/PKDD'2002
28. Wasilewski, P.: Dependency and supervenience. In: L. Czaja (ed.) Proceedings of the Concurrence, Specification and Programming (CS&P'2003), vol. 2, pp. 550–560. University of Warsaw Press (2003)
29. Wasilewski, P.: On selected similarity relations and their applications into cognitive science (in Polish). Unpublished doctoral dissertation, Jagiellonian University: Department of Logic, Krakow, Poland (2004)
30. Wasilewski, P.: Concept lattices vs. approximation spaces. In: Ślęzak, D., Wang, G., Szczuka, M., Düntsch, I., Yao, Y. (eds.) RSFDGrC 2005. LNCS (LNAI), vol. 3641, pp. 114–123. Springer, Heidelberg (2005). https://doi.org/10.1007/11548669_12
31. Wasilewski, P.: Algebras of definable sets vs. concept lattices. Fundamenta Informaticae **167**(3), 235–256 (2019)
32. Wasilewski, P. Kacprzyk, J., Zadrozny, S.: On some concept lattice of social choice functions. In: M. Paprzycki (ed.) Proceedings of 18th Conference on Computer Sciences and Intelligent Systems FedCSIS 2023 (2023)
33. Wille, R.: Restructuring lattice theory: an approach based on hierarchies of concepts. In: Rival, I. (ed.) Ordered Sets. NATO Advanced Study Institutes Series, vol. 83, pp. 445–470. Reidel, Dordrecht (1982)

An Acceleration Method for Attribute Reduction Based on Attribute Synthesis

Chengzhi Shi[1], Taihua Xu[1,2(✉)], Fuhao Cheng[3], Xibei Yang[1],
and Jianjun Chen[1]

[1] School of Computer, Jiangsu University of Science and Technology, Zhenjiang,
Jiangsu, China
18852645655@163.com
[2] Key Laboratory of Oceanographic Big Data Mining and Application of Zhejiang
Province, Zhejiang Ocean University, Zhoushan, Zhejiang, China
[3] Shanghai Waigaoqiao Shipbuilding Co., Ltd., Shanghai, China

Abstract. Attribute reduction plays a crucial role in eliminating redundant attributes of data. As an effective means to deal with numerical data, neighborhood rough set model has been widely used in attribute reduction. In this model, the determination of sample neighborhood relies on the calculation of distance between samples, which need to traverse all attributes of samples. This way will result in huge time consumption for high-dimensional data because the time consumption of solving reduction is closely related to the computational efficiency of sample neighborhood. In view of this, an attribute synthesis method based on the attribute similarity is put forward for solving above drawbacks. In this paper, firstly, all attributes are divided by K-means clustering into multiple attribute clusters. Secondly, the attributes in the same cluster are synthesized into a new pseudo-attribute. Then a new decision system can be formed by all the pseudo-attributes. Thirdly, the pseudo-attribute with the greatest importance is selected through the forward greedy search strategy in the new decision system. Finally, let the original attributes corresponding to the selected pseudo-attribute be an attribute subset, then we determine whether the subset satisfies the constraint condition of reduction. If not, the pseudo-attribute with the second greatest importance is considered to conduct above step, until the attribute subset satisfying the constraint condition is calculated and output as the reduction. In order to verify the effectiveness of the proposed method, the comparative experiments are conducted on 4 UCI standard datasets and 4 face datasets. The experimental results show that the proposed method can not only significantly reduce the time consumption of attribute reduction, but also relatively improve the classification performance of the reduction. Moreover, the more attributes the sample own, the more significant improvement the method has.

Keywords: Attribute reduction · Neighborhood rough set · Attribute synthesis · Pseudo-attribute · Attributes cluster

A. Campagner et al. (Eds.): IJCRS 2023, LNAI 14481, pp. 62–74, 2023.
https://doi.org/10.1007/978-3-031-50959-9_5

1 Introduction

Rough set theory [1] is an effective tool to deal with imprecise, inconsistent and incomplete information [2]. Currently, rough set theory has been successfully applied to artificial intelligence, data mining, pattern recognition, and other areas [2–8]. As a core concept of rough set theory, attribute reduction [12–18] has attracted many scholars' attention because it can remove redundant and irrelevant attributes in the data. The attribute reduction algorithms commonly used in rough set theory can be roughly classified into 2 categories: the exhaustive algorithm [9,10] and the heuristic algorithm [24]. For example, both the discernibility matrix method [9] and the backtrack method [10,11] belong to the exhaustive algorithm. Although the exhaustive algorithm can obtain all the reductions, it is time-consuming for calculating the attribute reduction of large-scale complex data. On the contrary, the heuristic algorithm is preferred due to its high efficiency.

Neighborhood rough set model [11,20] was proposed to break the limitation of classical rough set on numerical data. In this model, the determination of sample neighborhood relies on the calculation of distance between samples, which need to traverse all attributes of samples. This way will result in huge time consumption for high-dimensional data because the time consumption of solving reduction is closely related to the computational efficiency of sample neighborhood. For this reason, it is necessary to design a more efficient method for calculating the reduction.

Given that the time complexity of calculating the sample neighborhood is positively correlated to the number of attributes, we speculate that reducing the number of attributes is a feasible approach to improve computational efficiency of attribute reduction. In this paper, we try to utilize attribute synthesis strategy to reduce the number of attributes. As a result, an acceleration method for attribute reduction based on attribute synthesis is proposed. Firstly, all attributes are divided by K-means clustering into several attribute clusters. Secondly, the attributes in the same cluster are synthesized into a new pseudo-attribute. Then a new decision system can be formed by all the pseudo-attributes. Thirdly, the pseudo-attribute with the greatest importance is selected through the forward greedy search strategy [19] in the new decision system. Once the most important attribute is picked out, the synthesized original attributes need to be focus on. Finally, let the focused original attributes be an attribute subset, which need to be determined whether it satisfies the constraint condition of reduction. If not, the pseudo-attribute with the second greatest importance is considered to conduct above step, until the attribute subset satisfying the constraint condition is calculated and output as the reduction.

In brief, the major contributions have two aspects.

1. This paper first proposes a method of attribute synthesis to improve the reduction complexity. It can successfully reduce the dimension of high-dimensional data and effectively compress the search space of attributes. Thus the purpose of reducing the time consumption of attribute reduction can be achieved.

2. The proposed attribute synthesis strategy can improve the classification accuracy. At the same time, the more attributes the sample own, the greater the classification accuracy is.

The rest of this paper is arranged as bellow. In Sect. 2, it mainly introduces the basic concepts of neighborhood rough set, the measurement criterion and importance function of attribute reduction. In addition, a forward greedy search strategy is also described. In Sect. 3, a new forward greedy search strategy based on attribute synthesis is proposed. In Sect. 4, experimental results and related analysis are displayed. Section 5 summarizes the full text.

2   Preliminaries

2.1   Neighborhood Rough Set

A decision system can be defined as a two-tuple $DS = <U, A \cup \{d\}>$, where $U = \{x_1, x_2, \ldots, x_m\}$ is a non-empty finite set composed of m samples, called the universe of discourse. A is the collection of all conditional attributes, and d is the decision attribute. Let $IND(\{d\})$ be the equivalence relation with respect to the decision attribute d, thus there is $IND(\{d\}) = \{(x, y) \in U \times U : d(x) = d(y)\}$, where $d(x)$ is the decision value of the sample x in U. Given a decision system DS, then $U/IND(\{d\}) = \{X_1, X_2, \ldots, X_s\}$ denotes the partition of universe U, which is induced by decision attribute d. $\forall X_k \in U/IND(\{d\})$, it represents the kth decision class. $[x_i]_k$ represents the set of samples that belong to the kth decision class with sample x_i.

Definition 1 [24]. *Let $DS = <U, A \cup \{d\}>$ be a decision system. $\forall B \subseteq A$, the neighborhood relationship is defined as follows:*

$$N_B = \{(x_i, x_j) \in U \times U : r(x_i, x_j) \leq \delta\}, \tag{1}$$

in which, $\forall x_i, x_j \in U$, $r(x_i, x_j)$ represents the distance between x_i and x_j over B, δ ($\delta \geq 0$) is the neighborhood radius.

Then from Eq. 1, for any $x_i \in U$, the neighborhood of x_i with respect to B is defined as follows:

$$\delta_B(x_i) = \{x_j \mid x_j \in U, r(x_i, x_j) \leq \delta\}. \tag{2}$$

Definition 2 [24]. *Let $DS = <U, A \cup \{d\}>$ be a decision system. For $\forall B \subseteq A$, the upper and lower approximations of d with respect to B are defined as follows:*

$$\underline{N_B^\delta}(d) = \bigcup_{k=1}^{s} \underline{N_B^\delta}(X_k), \tag{3}$$

$$\overline{N_B^\delta}(d) = \bigcup_{k=1}^{s} \overline{N_B^\delta}(X_k), \tag{4}$$

in which, $\forall X_k \in U/IND(\{d\})$, the upper and lower approximations of X_k are defined as follows:

$$\underline{N}_B^{\delta}(X_k) = \{x_i \mid x_i \in U, \delta_B(x_i) \subseteq X_k\}, \tag{5}$$

$$\overline{N_B^{\delta}}(X_k) = \{x_i \mid x_i \in U, \delta_B(x_i) \cap X_k \neq \varnothing\}. \tag{6}$$

2.2 Attribute Reduction

The essence of attribute reduction is to perform feature selection under different measure criteria. Depending on the different application requirements, numerous scholars have defined different measure criteria, such as approximate quality [24], conditional entropy [21–23], decision error rate [24], etc. Approximate quality [24] is a commonly used measure criterion in attribute reduction. It reflects the proportion of the samples whose neighborhood belongs to the same decision class to all the samples in the domain under certain conditional attributes.

Definition 3 [24]. *Let $DS = <U, A \cup \{d\}>$ be a decision system. For $\forall B \subseteq A$, the approximate quality of d with respect to B is defined as follows:*

$$\gamma_B^{\delta}(d) = \frac{\left|N_B^{\delta}(d)\right|}{|U|}, \tag{7}$$

in which, $|\bullet|$ represents the cardinality of the set $\bullet$. Obviously, from Eq. 7, $0 \leq \gamma_B^{\delta}(d) \leq 1$.

As another measure criteria commonly used in attribute reduction, conditional entropy [21–23] reflects the ability of conditional attributes to identify decision attributes. There are many definitions of conditional entropy, one of which is shown in Definition 4.

Definition 4 [21]. *For $\forall B \subseteq A$, the conditional entropy of d with respect to B is defined as follows:*

$$ENT_B^{\delta}(d) = -\frac{1}{U} \sum_{x_i \in U} \log \frac{|\delta_B(x_i) \cap [x_i]_k|}{|\delta_B(x_i)|}. \tag{8}$$

Obviously, from Eq. 8, the smaller the value of conditional entropy, the greater the ability of conditional attributes to identify decision attributes.

At present, the commonly used attribute reduction strategies include exhaustive search, optimization, forward greedy search, etc. Among them, the forward greedy search strategy using iterative method is widely used due to its high efficiency. The fitness function is a crucial concept in the forward greedy search strategy, as it enables the evaluation of the significance of conditional attributes for classification. Based on the measure criteria described in Definitions 3 and 4, the following importance function is designed to evaluate candidate attributes.

Definition 5 [21]. *Let* $DS = <U, A \cup \{d\}>$ *be a decision system. For* $\forall B \subseteq A$, $\forall a \in A - B$, *the importance of attribute* a *to attribute set* B *is defined as follows:*

$$\text{SIG}_\gamma(a, B, d) = \gamma^\delta_{B \cup \{a\}}(d) - \gamma^\delta_B(d), \qquad (9)$$

$$\text{SIG}_{ENT}(a, B, d) = ENT^\delta_B(d) - ENT^\delta_{B \cup \{a\}}(d). \qquad (10)$$

Equation 9 takes the approximate quality as the measure criterion. Obviously, the more improvement the attribute a achieves on the approximate quality, the greater the importance attribute a has. Equation 10 takes the conditional entropy as the measure criterion. Obviously, the more deterioration the attribute a achieves on the conditional entropy, the greater the importance attribute a has. Attribute importance reflects the change of approximate quality or conditional entropy after attribute a is added into attribute subset B. If the attribute importance of a is not greater than zero, then attribute a is redundant and cannot be added into the reduction set.

Definition 6 [24]. *Let* $DS = <U, A \cup \{d\}>$ *be a decision system.* $\forall R \subseteq A$, *define* R *be a reduction with respect to the approximate quality* γ *iff:*

1. $\gamma^\delta_R(d) \geq \gamma^\delta_A(d)$.
2. $\forall B \subset R, \gamma^\delta_B(d) < \gamma^\delta_R(d)$.

When the above two conditions in Definition 6 are satisfied, the value of the approximate quality based on the reduction set R is not lower than that based on condition attribute set A, and R is the smallest reduction set.

Definition 7 [24]. *Let* $DS = <U, A \cup \{d\}>$ *be a decision system.* $\forall R \subseteq A$, *define* R *be a reduction with respect to the conditional entropy* ENT *iff:*

1. $ENT^\delta_R(d) \leq ENT^\delta_A(d)$.
2. $\forall B \subset R, ENT^\delta_B(d) > ENT^\delta_R(d)$.

When the above two conditions in Definition 7 are satisfied, the value of the conditional entropy based on the reduction set R is not higher than that based on the condition attribute set A, and R is the smallest reduction set.

For a decision system, the traditional heuristic attribute reduction algorithm is introduced in Algorithm 1, where φ represents the measure criteria which can be approximate quality in Definition 6 and conditional entropy in Definition 7.

Algorithm 1. Forward Greedy Search Algorithm For Solving Reduction.

Require: A decision system $DS = <U, A \cup \{d\}>$ the radius is δ.
Ensure: A reduction on φ: R.
  1: Calculate $\varphi_A^\delta(d)$;
  2: $R = \varnothing$;
  3: $\forall a \in A - R$, calculate the importance $\text{SIG}_\varphi(a, R, d)$ of attribute a;
  4: Select the most important attribute b, let $R = R \cup \{b\}$;
  5: Calculate $\varphi_R^\delta(d)$;
  6: If $\varphi_R^\delta(d)$ satisfies the constraint condition, go to 7; otherwise, repeat 3-5;
  7: **return** R.

The traditional forward greedy search strategy (Algorithm 1) takes the empty set as the starting point. In iterative way, it adds the attribute with the largest attribute importance into the reduction set each time. Until that the obtained reduction set meet the reduction constraints. The time complexity of Algorithm 1 is $O\left(|U|^2 \cdot |A|^2\right)$ where $|U|$ is the number of samples in the domain, and $|A|$ is the number of conditional attributes.

3 Reduction Based on Attribute Synthesis

In this section, a forward greedy search strategy based on attribute synthesis method is proposed to improve the time consumption of Algorithm 1 on attribute reduction.

In the proposed method, a choice can be made between a distance-based measure and a correlation-based measure for clustering all attributes. However, it should be noted that correlation-based measures typically assume linearity in data relationships and may exhibit suboptimal performance when such linearity is absent. Additionally, the computation of correlation becomes more intricate in high-dimensional datasets. To circumvent these aforementioned issues, the author opt for employing a distance-based measure.

The new method can reduce the data dimension by considering the similarities between attributes. Firstly, all attributes are divided into different attribute clusters. Since the attributes in same attribute cluster have high similarity, they can be synthesized into a new pseudo-attribute. So that a new decision system can be formed based on above multiple pseudo-attributes. In the process of attribute reduction, it is only necessary to use the forward greedy search strategy to handle the candidate pseudo-attributes of the new decision system. The pseudo-attribute with the largest importance will be picked up. Finally, we use constraint conditions of reduction to test the set of original attributes that synthesized into the picked pseudo-attribute. If the constraint is not satisfied, the pseudo-attribute with the second largest importance is picked up and restored to original attributes which are judged by constraint. This process repeats until that the constraint conditions of reduction are satisfied. The specific implementation is shown in Algorithm 2.

Algorithm 2. Forward Greedy Search Strategy Based On Attribute Synthesis.

Require: A decision system $DS = <U, A \cup \{d\}>$ the radius is δ.
Ensure: A reduction on φ: R.
1: Calculate $\varphi_A^\delta(d)$, R $= \varnothing$;
2: Use K-means clustering algorithm to get multiple attribute clusters;
3: Average the values of several attributes in multiple attribute clusters, and synthesize them into multiple pseudo-attributes;
4: Form a new decision system $DS' = <U, A' \cup \{d\}>$ for pseudo-attributes of all samples;
5: $R' = \varnothing$;
6: $\forall a \in A' - R'$, compute the importance $\text{SIG}_\varphi(a, R', d)$ of pseudo-attribute a;
7: Select the most important attribute b, let $R' = R' \cup \{b\}$;
8: Restore the pseudo-attributes in R' to multiple original attributes and put them into R;
9: Calculate $\varphi_R^\delta(d)$;
10: If $\varphi_R^\delta(d)$ satisfies the constraint condition, go to 11; otherwise, repeat 6-9;
11: **return** R.

The author employs the K-means clustering in Algorithm 2 to group attributes, taking into account the following factors: 1) The first reason is interpretability. The cluster center point generated by K-means clustering represents the average of the actual data points, aligning with the concept of pseudo-attribute proposed in this paper; 2) The second reason lies in its high computational efficiency and fewer parameters. K-means clustering demonstrates feasibility for clustering high-dimensional data; 3) Lastly, K-means clustering exhibits generalization capability as it does not assume any specific data distribution, making it applicable to diverse types of data.

The time complexity of Algorithm 2 is $O\left(|U| \cdot |A| \cdot |A'|\right) + O\left(|U|^2 \cdot |A'|^2\right)$ where $|U|$ is the number of samples in the domain, $|A'|$ is the number of attribute clusters obtained by using the K-means clustering algorithm, i.e., the value of K. At the same time, $|A'|$ is also the number of pseudo-attributes that are synthesized. In this paper, we set the value of K as $\lceil |A|/2 \rceil$ through sufficient experiments. As a result, $|A'| = K = \lceil |A|/2 \rceil \approx |A|/2$. Then the time complexity of Algorithm 2 is about $O\left(|U| \cdot |A|^2/2\right) + O\left(|U|^2 \cdot |A|^2/4\right)$. However, the time complexity of Algorithm 1 is $O\left(|U|^2 \cdot |A|^2\right)$. Obviously, Algorithm 2 can reduce the time consumption of reduction.

4 Experimental Analysis

In order to verify the effectiveness of Algorithm 2, 4 groups of UCI datasets and 4 groups of face datasets are used to test Algorithms 1 and 2. It is noted that the Algorithm 1 is comparative algorithm. The basic descriptions of the data is shown in Table 1.

It should be noted that the model used in this paper is neighborhood rough set. Under different neighbor radii, different reduction results will be obtained. And then the meaning or generalization performance is not the same. Therefore, in order to observe the change trend of reduction performance and find the reduction results with better generalization performance, it is necessary to perform attribute reduction under multiple radii. In this paper, 10 different radii δ with values of 0.025, 0.05, 0.075, ... , 0.25 were selected for the comprehensive experimental results.

In order to verify the effectiveness of the proposed algorithm, the 5-fold cross validation method is used. The specific process is: the samples in the data are divided into five same size sample subgroups, that is, $U_1, U_2, \ldots, U_5$. Any 4 subgroups are integrated into a integrated subgroup which is taken as training set for deriving reduction results. The remaining 1 subgroup can then be taken as testing set for classification. In this way, above calculation process will be repeated 5 rounds.

Table 1. Data sets description

ID	Datasets	Number of samples	Number of attributes	Number of decision classes
1	Sonar_Nor	208	60	2
2	Libras Movement	360	90	15
3	Musk (Version 1)	476	166	2
4	LSVT Voice Rehabilitation	126	256	2
5	Yale_32 × 32	165	1024	15
6	ORL_32 × 32	400	1024	40
7	Yale_64 × 64	165	4096	15
8	ORL_64 × 64	400	4096	40

In this experiment, approximation quality and conditional entropy are selected as the measure criteria of reduction. Then, the following 4 procedures can be obtained: 'TS-A', 'TS-E', 'NS-A' and 'NS-E'. Where, 'TS-A' and 'TS-E' represent the Algorithm 1 using approximate quality and conditional entropy, respectively. 'NS-A' and 'NS-E' represent the Algorithm 2 using approximate quality and conditional entropy, respectively. The obtained attribute reduction results, derived by above 4 procedures, are trained by KNN and SVM classifiers for predicting the decision labels of testing samples.

4.1 Time Consumption Comparison

In this subsection, 4 procedures ('TS-A', 'TS-E', 'NS-A' and 'NS-E') are used to compute the reduction results of 8 datasets. The time consumption of them are recorded and compared, as shown in Fig. 1. Observing Fig. 1, it is found that the performance of the proposed 'NS-A' and 'NS-E' are significantly better than 'TS-A' and 'TS-E' in terms of time consumption over 8 datasets.

The following conclusions can be drawn: 1) Based on the two measure criteria of approximate quality and conditional entropy, Algorithm 2 can significantly reduce the time consumption of computing reduction compared with Algorithm 1; 2) Compared with the approximate quality, two attribute reduction procedures based on conditional entropy ('TS-E' and 'NS-E') have lower time consumption.

4.2   Classification Accuracy Comparison

In this subsection, KNN and SVM classifiers are used to train the attribute reduction results of 4 procedures, for predicting the decision labels of testing samples. Then the testing samples are classified to verify the classification accuracy. Tables 2 and 3 show the average and maximum values of the classification accuracy with respect to KNN classifier on 10 different radii. Tables 4 and 5 show the average and maximum values of the classification accuracy with respect to SVM classifier on 10 different radii.

Table 2. Average classification accuracies based on KNN classifier

ID	TS-A	NS-A	TS-E	NS-E
1	0.8025 ± 0.0128	$\mathbf{0.8260 \pm 0.0235}$	0.7932 ± 0.0211	$\mathbf{0.8413 \pm 0.0233}$
2	0.7725 ± 0.0154	$\mathbf{0.7733 \pm 0.0147}$	0.7753 ± 0.0118	$\mathbf{0.7789 \pm 0.0158}$
3	0.7662 ± 0.0240	$\mathbf{0.7834 \pm 0.0267}$	0.7509 ± 0.0261	$\mathbf{0.7826 \pm 0.0421}$
4	0.7122 ± 0.0195	$\mathbf{0.7179 \pm 0.0537}$	0.7101 ± 0.0385	$\mathbf{0.7783 \pm 0.0325}$
5	0.4570 ± 0.0644	$\mathbf{0.5109 \pm 0.0412}$	0.5133 ± 0.0256	$\mathbf{0.5479 \pm 0.0259}$
6	0.7998 ± 0.0225	$\mathbf{0.8425 \pm 0.0184}$	0.7950 ± 0.0369	$\mathbf{0.8318 \pm 0.0279}$
7	0.5600 ± 0.0418	$\mathbf{0.6218 \pm 0.0244}$	0.6067 ± 0.0296	$\mathbf{0.6164 \pm 0.0345}$
8	0.7997 ± 0.0170	$\mathbf{0.8300 \pm 0.0190}$	0.7747 ± 0.0296	$\mathbf{0.8128 \pm 0.0255}$

Table 3. Maximum classification accuracies based on KNN classifier

ID	TS-A	NS-A	TS-E	NS-E
1	0.8224	**0.8655**	0.8220	**0.8747**
2	**0.7944**	**0.7944**	0.7944	**0.7972**
3	0.7962	**0.8151**	0.7898	**0.8427**
4	0.7458	**0.8086**	0.7628	**0.8249**
5	0.5576	**0.5697**	0.5576	**0.5939**
6	0.8450	**0.8675**	0.8375	**0.8750**
7	0.6061	**0.6667**	0.6424	**0.6788**
8	0.8275	**0.8500**	0.8050	**0.8500**

Table 4. Average classification accuracies based on SVM classifier

ID	TS-A	NS-A	TS-E	NS-E
1	0.7302 ± 0.0158	$\mathbf{0.7643 \pm 0.0174}$	00.7302 ± 0.0221	$\mathbf{0.7593 \pm 0.0220}$
2	0.5231 ± 0.0214	$\mathbf{0.6053 \pm 0.0207}$	0.5286 ± 0.0148	$\mathbf{0.6003 \pm 0.0313}$
3	0.6597 ± 0.0378	$\mathbf{0.6969 \pm 0.0443}$	0.6891 ± 0.0209	$\mathbf{0.7308 \pm 0.0152}$
4	0.6801 ± 0.0132	$\mathbf{0.7208 \pm 0.0446}$	0.6889 ± 0.0241	$\mathbf{0.7570 \pm 0.0342}$
5	0.2939 ± 0.0718	$\mathbf{0.4648 \pm 0.0451}$	0.3024 ± 0.0289	$\mathbf{0.4758 \pm 0.0336}$
6	0.7090 ± 0.0341	$\mathbf{0.8288 \pm 0.0186}$	0.7085 ± 0.0482	$\mathbf{0.8048 \pm 0.0410}$
7	0.3945 ± 0.0651	$\mathbf{0.5885 \pm 0.0349}$	0.3661 ± 0.0388	$\mathbf{0.5442 \pm 0.0500}$
8	0.6560 ± 0.0438	$\mathbf{0.7955 \pm 0.0162}$	0.6380 ± 0.0330	$\mathbf{0.7898 \pm 0.0331}$

Table 5. Maximum classification accuracies based on SVM classifier

ID	TS-A	NS-A	TS-E	NS-E
1	0.7508	**0.8031**	0.7646	**0.8031**
2	0.5611	**0.6361**	0.5444	**0.6472**
3	0.7142	**0.7585**	0.7250	**0.7522**
4	0.7148	**0.8086**	0.7375	**0.8086**
5	0.3455	**0.5152**	0.3333	**0.5455**
6	0.7450	**0.8500**	0.7650	**0.8675**
7	0.4909	**0.6303**	0.4182	**0.6303**
8	0.7275	**0.8200**	0.6900	**0.8450**

According to above Tables 2, 3, 4 and 5, the following conclusions can be drawn: 1) For the two measure criteria of approximate quality and conditional entropy, Algorithm 2 can improve the classification accuracy compared with Algorithm 1, and the effect obtained by conditional entropy is more obvious; 2) For KNN and SVM classifiers, Algorithm 2 also performs better on classification accuracies compared with Algorithm 1, and the classification effect in SVM classifier is more obvious; 3) For face datasets with higher data dimensions, Algorithm 2 obtains better classification accuracies than Algorithm 1, indicating that Algorithm 2 is more suitable for processing high-dimensional data; 4) The maximum classification accuracy of Algorithm 2 surpasses that of Algorithm 1 for both KNN and SVM classifiers, with a particularly pronounced improvement observed in that Algorithm 2 exhibits superior classification potential and feasibility compared to Algorithm 1; 5) The maximum classification accuracy of Algorithm 2 in the KNN classifier is higher compared that of the SVM classifier, indicating a superior upper limit of classification performance for Algorithm 2 in the KNN classifier.

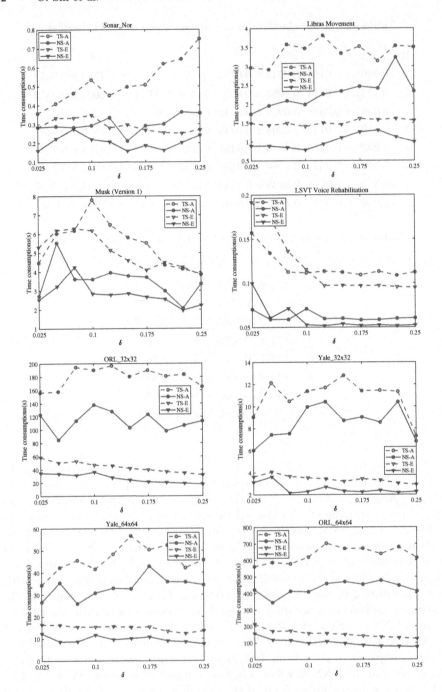

Fig. 1. Time consumption comparison for computing reduction

5   Conclusions and Future Perspectives

In the process of attribute reduction based on neighborhood rough set model, all attributes of samples need to be traversed for computing sample neighborhood. This way will result in huge time consumption for high-dimensional data. In order to solve this problem, we proposed an attribute synthesis method. Firstly, all attributes are divided by K-means clustering into multiple attribute clusters. Secondly, the attributes in the same cluster are synthesized into a new pseudo-attribute. Then a new decision system can be formed by all the pseudo-attributes. Thirdly, the pseudo-attribute with the greatest importance is selected through the forward greedy search strategy in the new decision system. Finally, let the original attributes corresponding to the selected pseudo-attribute be an attribute subset, then we determine whether the subset satisfies the constraint condition of reduction. If not, the pseudo-attribute with the second greatest importance is considered to conduct above step, until the attribute subset satisfying the constraint condition is calculated and output as the reduction. The proposed attribute reduction algorithm based on attribute synthesis was experimentally tested on 8 datasets, revealing its ability to not only reduce the time consumption of attribute reduction but also enhance the classification accuracy of reduction.

On the basis of this work, the following issues will be further explored:

1. This paper only used the approximate quality and conditional entropy as the measure criteria of attribute reduction. Other measure methods can be further considered, such as neighborhood discrimination index, decision error rate and so on.
2. The reduction is accelerated only from the attribute aspect. Actually, it can be considered from the perspective of the number of samples.

Acknowledgements. This work is supported by the National Natural Science Foundation of China (Grant Nos. 62006099, 62076111), and the Key Laboratory of Oceanographic Big Data Mining & Application of Zhejiang Province (No. OBDMA202104).

References

1. Pawlak, Z.: Rough Sets: Theoretical Aspects of Reasoning About Data. Kluwer Academic Publishers (1992)
2. Wang, J.B., Wu, W.Z., Tan, A.H.: Multi-granulation-based knowledge discovery in incomplete generalized multi-scale decision systems. Int. J. Mach. Learn. Cybernet. **13**(12), 3963–3979 (2022)
3. Hu, Q.H., An, S., Yu, D.R.: Soft fuzzy rough sets for robust feature evaluation and selection. Inf. Sci. **180**(22), 4384–4400 (2010)
4. Qian, Y.H., Liang, X.Y., Wang, Q., et al.: Local rough set: a solution to rough data analysis in big data. Int. J. Approximate Reasoning **97**(1), 38–63 (2018)
5. Ju, H.R., Li, H.X., Yang, X.B., et al.: Cost-sensitive rough set: a multi-granulation approach. Knowl.-Based Syst. **123**(1), 137–153 (2017)

6. Yao, Y.Y., Zhang, X.Y.: Class-specific attribute reducts in rough set theory. Inf. Sci. **418–419**, 601–618 (2017)
7. Slowinski, R., Vanderpooten, D.: A generalized definition of rough approximations based on similarity. IEEE Trans. Knowl. Data Eng. **12**(2), 331–336 (2000)
8. Xu, T.H., Wang, G.Y., Yang, J.: Finding strongly connected components of simple digraphs based on granulation strategy. Int. J. Approximate Reasoning **118**, 64–78 (2020)
9. Dubois, D., Prade, H.: Rough fuzzy sets and fuzzy rough sets. Int. J. Gener. Syst. **17**, 191–209 (1990)
10. Lin, T.Y.: Granular Computing on binary relations I: data mining and neighborhood systems. In: Skoworn, A., Polkowshi, L. (eds.) Rough Sets in Knowledge Discovery, pp. 107–121. Physica-Verlag (1998)
11. Hu, Q.H., Yu, D.R., Xie, Z.X.: Neighborhood classifiers. Expert Syst. Appl. **34**, 866–876 (2008)
12. Gong, Z.C., Liu, Y.X., Xu, T.H., et al.: Unsupervised attribute reduction: improving effectiveness and efficiency. Int. J. Mach. Learn. Cybernet. **13**(11), 3645–3662 (2022)
13. Fang, Y., Cao, X.M., Wang, X., et al.: Three-way sampling for rapid attribute reduction. Inf. Sci. **609**, 26–45 (2022)
14. Wu, Z.J., Mei, Q.Y., Zhang, Y., et al.: A distributed attribute reduction algorithm for high-dimensional data under the spark framework. Int. J. Comput. Intell. Syst. **15**(1), 1–14 (2022)
15. Yang, T.L., Li, Z.W., Li, J.J.: Attribute reduction for set-valued data based on prediction label. Int. J. Gener. Syst. 1–31 (2023)
16. Gao, C., Zhou, J., Xing, J., et al.: Parameterized maximum-entropy-based three-way approximate attribute reduction. Int. J. Approximate Reasoning **151**, 85–100 (2022)
17. Chen, Q., Xu, T.H., Chen, J.J.: Attribute reduction based on lift and random sampling. Symmetry **14**(9), 1828 (2022)
18. Yan, W.W., Ba, J., Xu, T.H., Yu, H.L., Shi, J.L., Han, B.: Beam-influenced attribute selector for producing stable reduct. Mathematics **10**(4), 553 (2022)
19. Jiang, Z.H., Yang, X.B., Yu, H.L., et al.: Accelerator for multi-granularity attribute reduction. Knowl.-Based Syst. **177**, 145–158 (2019)
20. Gao, Y., Chen, X.J., Yang, X.B., et al.: Neighborhood attribute reduction: a multicriterion strategy based on sample selection. Information **9**, 282–302 (2018)
21. Hu, Q.H., Che, X.J., Zhang, L., et al.: Rank entropy based decision trees for monotonic classification. IEEE Trans. Knowl. Data Eng. **24**, 2052–2064 (2012)
22. Liu, K.Y., Yang, X.B., Yu, H.L., et al.: Rough set based semi-supervised feature selection via ensemble selector. Knowl.-Based Syst. **165**, 282–296 (2019)
23. Gao, C., Lai, Z.H., Zhou, J., et al.: Granular maximum decision entropy-based monotonic uncertainty measure for attribute reduction. Int. J. Approximate Reasoning **104**, 9–24 (2019)
24. Yang, X.B., Liang, S.C., Yu, H.L., et al.: Pseudo-label neighborhood rough set: measures and attribute reductions. Int. J. Approximate Reasoning **105**, 115–129 (2019)

Attribute Reduction Based on the Multi-annulus Model

Yan Liu[1(✉)], Jingjing Song[1], Taihua Xu[1,2], and Jianjun Chen[1]

[1] School of Computer, Jiangsu University of Science and Technology,
Zhenjiang 212100, Jiangsu, China
859247755@qq.com
[2] Key Laboratory of Oceanographic Big Data Mining and Application of Zhejiang
Province, Zhejiang Ocean University, Zhoushan 316022, Zhejiang, China

Abstract. Attribute reduction is one of the important methods in data mining preprocessing. In order to reduce the time consumption of attribute reduction, from the perspective of data distribution, the multi-annulus model is established for samples under different labels. The multi-annulus model can adaptively generate multiple annuluses with different radii according to the different data densities, and the intersection between annuluses is regarded as the boundary domain of annuluses, which introduces a new angle to solve the attribute reduction problem. The key to solving the attribute reduction of the multi-annulus model is to take the quality in the annulus as the metric criterion. Additionally, the radius ratio of the annulus and the boundary region of the annulus is used as the weight of the quality in the annulus. To obtain the change of the quality in the annulus caused by each candidate attribute added to the attribute reduction pool, the forward greedy strategy is employed. Its essence is to divide the unevenly distributed data into corresponding annuluses and reduce the preprocessing process of the data itself, so as to accelerate the process of attribute reduction. Compared with the other three mainstream attribute reduction algorithms, the final experimental results show that the proposed algorithm can greatly shorten the time of attribute reduction by introducing the multi-annulus model.

Keywords: Attribute reduction · Annulus model · Multi-annulus model · Hierarchical processing · Feature selection

1 Introduction

Attribute reduction [1–5], as one of the important feature selection methods, aims to eliminate redundant and less important attributes, obtain effective reduced attributes, and maintain the overall classification performance of the system unchanged. This allows the reduced attributes to replace the original high-dimensional data for analysis and decision-making, ultimately reducing the complexity of decision-making. With the advent of the big data era, the scale of

A. Campagner et al. (Eds.): IJCRS 2023, LNAI 14481, pp. 75–86, 2023.
https://doi.org/10.1007/978-3-031-50959-9_6

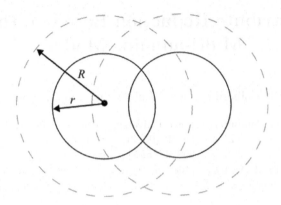

Fig. 1. Illustration of Multiple Annulues Model

data generated in numerous application fields has increased rapidly. Correspondingly, the number of redundant attributes has also increased, which not only interferes with the structural information of similar samples but also destroys the structural information between different samples. Therefore, it is urgent to reduce the dimensionality of attributes before data analysis and decision-making, and attribute reduction becomes an effective and important solution. However, finding the minimal attribute reduction has been proven to be an NP-hard problem [6], and seeking a feasible and efficient attribute reduction algorithm is still a challenging research task.

It is worth noting that some existing methods require data encapsulation before attribute reduction, such as neighborhood rough sets [7], which require setting a neighborhood radius for each sample, calculating whether each sample belongs to the same neighborhood with other samples, and then performing attribute reduction based on this neighborhood relationship. In addition, the neighborhood radius suitable for problem-solving still relies on a searching strategy to determine. This searching strategy will have a significant impact on the time efficiency of the algorithm. To address this issue, a multi-annulus model is established based on samples with different labels while fully considering the distribution of samples between different labels. As shown in Fig. 1, the inner annulus radii r and outer annulus radii R of the k-th decision are established.

The specific steps to establish a multi-annulus model are as follows: First, samples in each label are stratified according to different degrees of discreteness, forming multi-annulus, which essentially transforms the encapsulation and judgment of relatively complex data into a logical division of data. Second, the intersection areas generated between the annuluses are considered as the annulus boundary domain, and the degree of discreteness of the samples jointly determines the weight of calculating the quality of the annulus. Finally, the inner annulus quality is used as a measurement criterion, and the inner annulus quality of each attribute set is calculated using the forward greedy algorithm. When the inner annulus quality of the attribute set is greater than the inner annulus quality of all attributes, the final attribute reduction result is obtained. It should be further noted that the size of the inner annulus quality, as the standard for measuring the importance of attributes, changes with the addition of different attribute groups. The larger the inner annulus quality, the more compact the sample distribution under the current attribute group, which intuitively indicates that this group of attributes can highlight the key features of the samples. Correspondingly, the smaller the inner annulus quality, the less constraining the current attribute group is on the samples, making it difficult to preserve the original features of the samples.

The rest of this paper is organized as follows. In Sect. 2, the multi-annulus model is constructed through existing research and the corresponding definition is given. In Sect. 3, The attribute reduction based on the multi-annulus model is established. The comparative experimental results and analyses are shown in Sect. 4. Finally, the conclusions are summarized in Sect. 4.5.

2 Construction of Multi-annulus Model

It is worth noting that some existing methods require encapsulation processing of data before attribute reduction [8–10]. For instance, the three-way decision analysis [11,12] has analyzed the three regions of rough set theory and integrated them with decision theory to establish three-way decision rules, which consistent with human cognition. In order to avoid the risk of making incorrect decisions, delayed decision-making is employed. Formal concept analysis [13,14] proposes the concept of concept information granules and rules for concept approximation, emphasizing the extraction of potential rules from data analysis. In addition, the appropriate neighborhood radius for problem solving still needs to be determined by searching strategies, which will greatly affect the time efficiency of the algorithm. To address this issue, considering the distribution of samples among different labels, a multi-annulus model is established based on samples with different labels.

Let $BS = \langle U, AT \cup \{D\} \rangle$ be a knowledge system, where AT is the set of all conditional attributes, $U = \{x_1, x_2, \ldots, x_n\}$ is a non-empty finite domain in the real number space of n samples, and D is the decision attribute set with L decision classes. The cluster of all indistinguishable relationships related to AT in BS is denoted as $IND(BS)$. If $P \subseteq AT$ is the attribute subset that satisfies $IND(BS) = IND(P)$, then P is called the attribute reduction of AT. In this section, we will mainly introduce the establishment of the multi-annulus model on the dataset and apply the model to attribute reduction.

When classifying samples with different decision attributes, the features extracted from the same class samples are usually taken as the basis for judging whether they belong to a certain decision class. For example, in the "Iris" dataset, the Virginia iris is significantly larger than other types of irises in terms of sepal length, petal length, and petal width, while irises with wider sepals and shorter petal lengths are unique features of the mountain irises.

Therefore, in order to facilitate the acquisition of sample features under a certain decision attribute, the samples are first divided into L classes according to the differences in decision attributes, that is, the partition of all samples in a knowledge system BS with respect to the decision attribute D: $U/\text{IND}(D) = \{X_1, X_2, \ldots, X_L\}$.

Definition 1: Given a knowledge system BS, and an equivalent partition by decision attribute D, the sample center of the k-th decision class is defined as:

$$C_k = \text{mean}(X_k), \tag{1}$$

where mean is used to calculate the average of the feature vectors.

When generating different circular annuluses based on the sample center of each decision class, it is expected that the inner radius of each annulus can be adaptively generated according to the data density. It is not difficult to understand that by specifying the maximum amount of information allowed within the annulus, the length of the annulus radius can be inversely proportional to the data density.

Definition 2: Let ϵ be a given ratio parameter $(0 < \epsilon < 1)$, and define the maximum amount of information contained in the inner and outer annuluses as:

$$\alpha = \epsilon \cdot |\, X_k \,|, \tag{2}$$

where $0.5 \leq \epsilon < 1$, and $|\, X_k \,|$ denotes the number of samples in the k-th decision class.

To divide the samples into the corresponding information capacity annuluses, there are several steps involved. First, the distance between each sample and the sample center of the corresponding decision class is calculated. Next, the distance vector after natural sorting is obtained. Then, the samples with α step length in the distance vector as the inner annulus samples are calculated, while the remaining samples are considered as the outer annulus samples.

Definition 3: The inner annulus samples of the k-th decision class will be defined as:

$$G_k = \underbrace{[x_u^s, x_{u+1_s}^s, \ldots, x_{v-1_s}^s, x_{v_s}^s]}_{|u-v|+1=\alpha} \in S_\Delta(x_i, C_k), \tag{3}$$

where $S_\Delta(x_i, C_k)$ returns the ascending order of the Euclidean distances from each instance x_i to the sample center C_k, and x_s is the sorted sample. In addition, the annulus model roughly reflects the degree of dispersion of the data distribution to a certain extent, that is, the smaller the ratio of the inner and outer annulus radius, the more concentrated the data, and the effective features of the data are better preserved in the inner annulus. Conversely, it indicates that the data distribution is relatively dispersed, and it is difficult to capture the characteristics of the data.

Definition 4: The inner and outer annulus radii of the k-th decision class will be defined as:

$$r_k = \max(G_k),$$
$$R_k = \max(S_\Delta(x_i, C_k)). \tag{4}$$

After determining the inner annulus radius of the annulus model, we can obtain the intersection area of the annuluses belonging to two different decision classes, which we call the annulus boundary domain. Intuitively, samples of the two classes in the annulus boundary domain have fewer distinguishable features. Similarly, as the complement of the annulus boundary domain samples in the inner annulus samples, the samples in the non-annulus boundary domain often have strong distinguishing features.

Definition 5: The samples in the p-th decision class and the q-th decision class that are in the inner annulus intersection, i.e., in the annulus boundary domain will be defined as:

$$BR_p^q = R_p \cap R_q = \{x_i \mid x_i \in G_p, x_i \in G_q\}. \tag{5}$$

Definition 6: The samples in the non-annulus boundary domain of the p-th decision class and the q-th decision class will be defined as:

$$\begin{aligned} IE_p &= \{x_i \mid x_i \in G_p, x_i \notin BR_p^q\}, \\ IE_q &= \{x_i \mid x_i \in G_q, x_i \notin BR_p^q\}. \end{aligned} \tag{6}$$

When establishing other multiple annuluses, each inner annulus will intersect with multiple other annuluses to generate annulus boundary domains, so it is necessary to record the intersections generated by different annuluses.

Definition 7: Let the decision class $d_p \in D$, define the set of other decision classes intersecting with the inner annulus of d_p as:

$$ID(p) = \{q \mid r_p < d(C_p, C_q), d_q \in D\}, \tag{7}$$

where $d(\cdot, \cdot)$ returns the Euclidean distance between two vectors.

3   Attribute Reduction Based on the Multi-annulus Model

Definition 8: Let A, B be two different decision classes, define the degree to which A contains B as:

$$IN(A, B) = \frac{Card(BR_A^B)}{Card(G_A)}, \tag{8}$$

where BR_A^B is the samples in the annulus boundary domain of decision classes A and B, and G_A is the set of samples in decision class A. At the same time, the degree of inclusion $IN(A, B)$ to some extent reflects the separability of the classification problem in the given attribute space. The larger the degree of inclusion, the larger the overlap area of sets A and B, i.e., the annulus boundary domain, and the smaller the separability of the two classes of samples.

The annulus boundary domain and non-boundary domain objectively reflect the state of the sample distribution and also reflect the constraint ability of the attribute group on the sample distribution. Therefore, it can be an important part of evaluating the dependence of decision attribute D on the subset of conditional attributes B.

Definition 9: Given the knowledge system BS, $\forall B \subseteq AT$, define the degree of dependence of decision attribute D on the entire subset of conditional attributes B as:

$$\gamma_B(D) = \sum_{p=1}^{L} \omega_1 \mid BR_p^q \mid + \omega_2 \mid IE_p \mid, \tag{9}$$

where $\omega_1 = \frac{R_p + R_q}{R_p r_q + R_q r_p}$, $\omega_2 = \frac{r_p}{R_q}$, $q \in ID(p)$ and $p \neq q$.

The dependence degree is a function obtained by calculating the weights of the annulus boundary domain and non-boundary domain under the premise of considering the data distribution, and can be used as an evaluation index of the importance of attribute sets.

Definition 10 [5]: Given the knowledge system BS, let $\forall B \subseteq AT$, $\forall a \in A - B$, define the importance of a relative to B as:

$$SIG(a, B, D) = \gamma_{a \cup B}(D) - \gamma_B(D). \tag{10}$$

The attribute importance includes three elements: the attribute itself, the attribute subset, and the decision variable. Based on the attribute importance index, we can construct a greedy attribute reduction algorithm. Each time, the candidate attribute with the highest importance is selected and added to the attribute reduction pool. When the importance of all remaining attributes is 0, i.e., at this time, adding any remaining attribute will not cause any change in the dependence function, thereby eliminating a large number of redundant and poorly distinguishable features and obtaining the final attribute reduction result. Therefore, the output attribute reduction result still maintains the original data's distinguishing features as a whole. The description of the attribute reduction algorithm based on the multi-annulus model can be found in Algorithm 1.

Algorithm 1. Attribute reduction algorithm based on the multi-annulus model

1. Initialization: $\emptyset \rightarrow red$, $\gamma_B(D) = 0$;
2. For any $a_k \in A - red$, calculate: $SIG(a_k, B, D) = \gamma_{red \cup a_k}(D) - \gamma_{red}(D)$;
3. While $SIG(a_k, red, D) > 0$:
   a. $\forall a \in AT - red$;
   b. Establish a multi-annulus model;
   c. Find the attribute a_k that satisfies:
      $$SIG(a_k, red, D) = \max(SIG(a, red, D));$$
   c. $red = red \cup \{a_k\}$;
4. End
5. Return Red.

This algorithm provides a way to derive the attribute reduction based on the multi-annulus model. It iteratively selects the attribute with the highest importance and includes it in the attribute reduction result until no more significant attributes are available.

In Algorithm 1, Step 1 initializes an attribute reduction set red and the dependence degree of attribute D on the entire subset of conditional attributes B, denoted as γ_B. Step 2 iterates through each complement a_k of red, and calculates the attribute importance of a_k relative to B, denoted as $SIG(a_k, B, D)$, using formulas (9) and (10). If $SIG(a_k, B, D) > 0$, Step 3 is executed continuously to find the attribute with the highest importance a_k and add it to the attribute reduction set until the attribute importance no longer increases or $A - red = \emptyset$. Finally, the attribute reduction result red is obtained.

4 Experiments

4.1 Datasets

In order to verify the performance of the proposed algorithm, we selected 10 groups of data from the UCI dataset for algorithm comparison experiments. Table 1 shows the basic information of the selected datasets, where m, $attrs$, and l represent the number of samples, the number of attributes, and the number of decision classes in the dataset, respectively. Next, we compare the multi-annulus attribute reduction algorithm (referred to as the multi-annulus method) with the forward greedy algorithm [7] of neighborhood rough sets (referred to as the neighborhood method), the fast attribute reduction algorithm [15] based on symmetry and decision filtering (referred to as the symmetry method), and the attribute partitioning method [16] of granule rough set attribute reduction (referred to as the granular method) on 10 datasets. According to Hu et al.'s research, the neighborhood radius of the neighborhood method and symmetry

Table 1. Basic information of experimental datasets

ID	datasets	m	$attrs$	l
1	Banknote Authentication	1372	4	2
2	Iris	150	4	3
3	Page-blocks	5473	10	5
4	Solar Flare	1389	9	2
5	Statlog	2310	18	7
6	Synthetic Control Chart Time Series	600	60	6
7	Wdbc	569	30	2
8	Wilt	4839	5	2
9	Wine	178	13	3
10	Wireless Indoor Localization	2000	7	4

method is set to 0.125. In addition, after experimental verification, the unique parameter ϵ of the multi-annulus method proposed in this paper is set to 0.5. Moreover, each algorithm is tested using a five-fold cross-validation method, and the best-performing data is displayed in bold.

4.2   Time Consumption Comparison

In this section, we compare the time consumption of the neighborhood method (Algorithm 1), symmetry method (Algorithm 2), multi-annulus method (Algorithm 3), and the proposed method (Algorithm 4) for attribute reduction. The specific time consumption is shown in Table 2. The following conclusions can be drawn from the analysis of Table 2:

Table 2. Time consumption of different attribute reductions algorithms

ID	Algorithm1	Algorithm 2	Algorithm 3	Algorithm 4
1	0.6326	0.0808	**0.0589**	0.0742
2	0.0171	0.0076	0.0063	**0.0036**
3	12.1128	7.9039	7.8931	**7.5667**
4	0.7855	0.0680	0.0592	**0.0519**
5	88.9080	11.6724	2.2325	**1.1211**
6	17.0822	7.4489	4.0532	**1.4137**
7	5.0398	0.3223	0.1353	**0.0731**
8	3.0654	0.6732	**0.2164**	0.4268
9	0.1705	0.0245	0.0325	**0.0174**
10	7.1051	0.7723	**0.0365**	0.2053
MEAN	13.4919	2.2974	1.4724	**1.0954**

(1) In general, the time spent on deriving attribute reduction using the multi-annulus method is much less than that using the neighborhood method, and compared with the symmetry method, the time efficiency is also significantly improved. In terms of the average time consumed for deriving attribute reduction, the multi-annulus method is still better than the other two methods, and the average time acceleration ratios of the multi-annulus method compared with the neighborhood method and symmetry method are 12.3 and 2.1, respectively, further verifying the performance of the multi-annulus method in deriving attribute reduction.

(2) From the perspective of datasets with a larger number of decision-making classes, such as the Statlog dataset (ID = 5), attribute reduction is carried out under the premise of 7 decision-making classes. The time consumption of the neighborhood method and the symmetry method has significantly

increased, to 88.9080 s and 11.6724 s respectively, while the time consumption of the multi-annulus method is only 1.1211 s. The speedup ratios compared with the other two algorithms are 79.3 and 10.4 respectively. This is because the neighborhood method needs to calculate whether each object is a positive domain sample under different decision classes. Although the symmetry method has already accelerated this process through bucketing strategy and neighborhood symmetry, it still cannot avoid a large amount of computation generated when judging positive domain samples. The multi-annulus model established based on sample distribution, using the intersections produced between different annuluses as a method to calculate importance, can greatly enhance the efficiency of reducts. Hence, it can be seen that introducing the multi-annulus model can indeed accelerate the process of simplification solution after sample segmentation.

Table 3. Comparison of time consumption of different reducts algorithms

ID	CART classification accuracy				SVM classification accuracy			
	Algorithm 1	Algorithm 2	Algorithm 3	Algorithm 4	Algorithm 1	Algorithm 2	Algorithm 3	Algorithm 4
1	0.9357	**0.9449**	0.9412	0.9357	**0.9759**	0.9744	0.8723	0.8768
2	0.9316	**0.9550**	0.9334	0.9450	0.9116	0.8316	0.9097	**0.9183**
3	0.9554	**0.9584**	0.9254	0.9360	**0.9192**	0.9191	0.8931	0.9024
4	0.8084	0.8066	0.8064	**0.8165**	0.8228	0.8228	0.8133	**0.8243**
5	0.9128	**0.9306**	0.9098	0.9234	**0.9040**	0.8979	0.8901	0.9010
6	0.7429	**0.7866**	0.7865	0.7700	0.7233	0.7441	0.7728	**0.8729**
7	0.9051	**0.9209**	0.9123	0.8602	0.9516	0.9595	**0.9605**	0.8637
8	0.9711	0.9697	0.9421	**0.9740**	0.9460	0.9460	0.9432	**0.9461**
9	0.8608	0.8412	0.8389	**0.8834**	0.8804	**0.9312**	0.8492	0.8005
10	0.9566	0.9592	0.9503	**0.9600**	0.9715	**0.9738**	0.8968	0.9640
Mean	0.8994	**0.9073**	0.8946	0.9004	**0.9006**	0.9001	0.8801	0.8870

4.3 Classification Accuracy

In this section of the experiment, two classifiers, Classification and Regression Tree (CART) [17] and Support Vector Machine (SVM) [18], are used to compare the neighborhood method, symmetry method, and multi-annulus method. Table 3 shows the classification accuracy obtained by different algorithms and the mean accuracy of each algorithm on different datasets.

From the analysis of Table 3, we can conclude that, overall, the multi-annulus method does not differ significantly in classification performance compared to the other two methods, whether under CART or SVM classifiers. This indicates that the multi-annulus model reducts algorithm can produce satisfactory reducts. The multi-annulus method is an algorithm based on data distribution. For some datasets, such as those with ID = 3 and ID = 7, the classification accuracy may be affected due to the high degree of data dispersion. Combined with Table 2, it is not difficult to conclude that the reducts based on the multi-annulus model can not only shorten the time efficiency of obtaining reducts but also maintain considerable classification performance.

4.4 Parameter Sensitivity

In the introduction of Definition 2 in the second section of this paper, the maximum information content of the inner and outer annuluses is determined by setting the annulus parameter ϵ, i.e., the size of ϵ indirectly determines the inner and outer radii of each annulus model, thereby affecting the result of different annulus intersections and the final dependency calculation. To further explore the sensitivity of the multi-annulus method to the unique parameter ϵ, we select four datasets with relatively poor classification accuracy in Table 3 and set the multi-annulus method parameter ϵ to 0.5–1. The classification accuracy with the change of ϵ on CART and SVM classifiers is shown in Figure 2.

From Fig. 2, it can be seen that with the increase of the information content of the inner annulus, i.e., the increase of ϵ, the classification accuracy obtained by the multi-annulus method in CART and SVM classifiers generally shows an upward trend. This is because the more information the inner annulus contains, the easier it is to intersect with other annuluses, thus obtaining better distinguishing key features. In addition, the obtained classification accuracy gradually stabilizes after the parameter is set to 0.5. Increasing the parameter setting at this time will further improve the classification accuracy, but it will also increase the time consumption of processing intersecting samples. Therefore, this experiment sets the parameter to 0.5, which ensures the accuracy of classification and also improves the time efficiency.

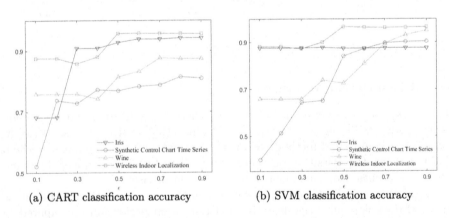

(a) CART classification accuracy   (b) SVM classification accuracy

Fig. 2. Variation of SVM and CART classification accuracy with parameters

4.5   Conclusion

The multi-annulus model proposed in this paper is a new spatial structure suitable for single-label reducts. Its purpose is to divide samples according to data distribution, deal with samples that intersect with other annuluses at different levels separately, and thus obtain effective distinguishing features. Comparative experiments with two classic reducts algorithms on 10 UCI datasets show that the algorithm achieves significant acceleration effects while obtaining reducts with comparable classification performance. The following issues are worth further study:

(1) The reducts based on the multi-annulus model is divided from the perspective of data distribution, and we can try to apply the multi-annulus model to other classification problems, such as dealing with multi-label dataset classification problems, etc.
(2) We can attempt to combine our framework with other feature selection techniques.

Acknowledgements. This work was supported by the National Natural Science Foundation of China (Nos. 62006099, 62076111) and the Key Laboratory of Oceanographic Big Data Mining & Application of Zhejiang Province (No. OBDMA202104).

References

1. Chen, Y., Liu, K., Song, J., Fujita, H., Yang, X., Qian, Y.: Attribute group for attribute reduction. Inf. Sci. **535**, 64–80 (2020)
2. Li, J., Yang, X., Song, X., Li, J., Wang, P., Yu, D.: Neighborhood attribute reduction: a multi-criterion approach. Int. J. Mach. Learn. Cybern. **10**, 731–742 (2019)
3. Khan, M.T., Azam, N., Khalid, S., Yao, J.T.: A three-way approach for learning rules in automatic knowledge-based topic models. Int. J. Approximate Reasoning **82**, 210–226 (2017)
4. Afridi, M.K., Azam, N., Yao, J.T.: Variance based three-way clustering approaches for handling overlapping clustering. Int. J. Approximate Reasoning **118**, 47–63 (2020)
5. Zhang, X., Yao, Y.: Tri-level attribute reduction in rough set theory. Expert Syst. Appl. **190**, 116187 (2022)
6. Azam, N., Zhang, Y., Yao, J.T.: Evaluation functions and decision conditions of three-way decisions with game-theoretic rough sets. Eur. J. Oper. Res. **261**, 704–714 (2017)
7. Hu, Q., Yu, D., Liu, J., Wu, C.: Neighborhood rough set based heterogeneous feature subset selection. Inf. Sci. **178**, 3577–3594 (2008)
8. Wang, P., Shi, H., Yang, X., Mi, J.: Three-way k-means: integrating k-means and three-way decision. Int. J. Mach. Learn. Cybernet. **10**, 2767–2777 (2019)
9. Zhai, J., Wang, X., Zhang, S., Hou, S.: Tolerance rough fuzzy decision tree. Inf. Sci. **465**, 425–438 (2018)
10. Liu, Y., et al.: Discernibility matrix based incremental feature selection on fused decision tables. Int. J. Approximate Reasoning **118**, 1–26 (2020)
11. Bao Qing Hu: Three-way decisions space and three-way decisions. Inf. Sci. **281**, 21–52 (2014)

12. Zhang, Y., Yao, J.T.: Gini objective functions for three-way classifications. Int. J. Approximate Reasoning **81**, 103–114 (2017)
13. Li, M., Wang, G.: Approximate concept construction with three-way decisions and attribute reduction in incomplete contexts. Knowl.-Based Syst. **91**, 165–178 (2016)
14. Zhao, S., et al.: An accelerator for rule induction in fuzzy rough theory. IEEE Trans. Fuzzy Syst. **29**, 3635–3649 (2021)
15. Wang, N., Peng, Z., Cui, L.: EasiFFRA:a fast feature reduction algorithm based on neighborhood rough set. J. Comput. Res. Dev. **56**, 2578–2588 (2019). (in Chinese)
16. Ba, J., Chen, Y., Yang, X.: Attribute partition strategy for quick searching reducts based on granular ball rough set. J. Nanjing Univ. Sci. Technol. **45**, 394–400 (2021). (in Chinese)
17. Rutkowski, L., Jaworski, M., Pietruczuk, L., Duda, P.: The cart decision tree for mining data streams. Inf. Sci. **266**, 1–15 (2014)
18. Chauhan, V.K., Dahiya, K., Sharma, A.: Problem formulations and solvers in linear SVM: a review. Artif. Intell. Rev. **52**, 803–855 (2019)

Foundations

Deterministic and Nondeterministic Decision Trees for Decision Tables with Many-Valued Decisions from Closed Classes

Azimkhon Ostonov$^{(\boxtimes)}$ and Mikhail Moshkov

Computer, Electrical and Mathematical Sciences and Engineering Division and Computational Bioscience Research Center, King Abdullah University of Science and Technology (KAUST), Thuwal 23955-6900, Saudi Arabia
{azimkhon.ostonov,mikhail.moshkov}@kaust.edu.sa

Abstract. Decision rules and decision trees are studied intensively in rough set theory. The following questions seem to be important for this theory: relations between decision trees and decision rule systems, and dependence of the complexity of decision trees and decision rule systems on the complexity of the set of attributes attached to columns of the decision table. In this paper, instead of decision rule systems we study nondeterministic decision trees that can be considered as representations of decision rule systems. We consider classes of decision tables with many-valued decisions (multi-label decision tables) closed relative to removal of attributes (columns) and changing sets of decisions assigned to rows. For tables from an arbitrary closed class, we study functions that characterize the dependence in the worst case of the minimum complexity of deterministic and nondeterministic decision trees on the complexity of the set of attributes attached to columns. We enumerate all types of behavior of these functions. We also study the dependence in the worst case of the minimum complexity of deterministic decision trees on the minimum complexity of nondeterministic decision trees. This study leads to understanding of the nontrivial relationships between deterministic decision trees and systems of decision rules represented by nondeterministic decision trees.

Keywords: Decision tables with many-valued decisions · Closed classes of decision tables · Deterministic decision trees · Nondeterministic decision trees

1 Introduction

Decision rules and decision trees are studied intensively in rough set theory [22–26, 31]. The following questions seem to be important for this theory: relations between decision trees and decision rule systems, and dependence of the complexity of decision trees and decision rule systems on the complexity of the set of attributes attached to columns of the decision table.

© The Author(s), under exclusive license to Springer Nature Switzerland AG 2023
A. Campagner et al. (Eds.): IJCRS 2023, LNAI 14481, pp. 89–104, 2023.
https://doi.org/10.1007/978-3-031-50959-9_7

In this paper, instead of decision rule systems we study nondeterministic decision trees that can be considered as representations of decision rule systems. We consider classes of decision tables with many-valued decisions (multi-label decision tables) closed relative to removal of attributes (columns) and changing sets of decisions assigned to rows. For tables from an arbitrary closed class, we study functions that characterize the dependence in the worst case of the minimum complexity of deterministic and nondeterministic decision trees on the complexity of the set of attributes attached to columns. We enumerate all types of behavior of these functions.

We also study the dependence in the worst case of the minimum complexity of deterministic decision trees on the minimum complexity of nondeterministic decision trees. This study leads to understanding of the nontrivial relationships between deterministic decision trees and systems of decision rules represented by nondeterministic decision trees.

A decision table with many-valued decisions is a rectangular table in which columns are labeled with attributes, rows are pairwise different and each row is labeled with a nonempty finite set of decisions. Rows are interpreted as tuples of values of the attributes. For a given row, it is required to find a decision from the set of decisions attached to the row. To this end, we can use the following queries: we can choose an attribute and ask what is the value of this attribute in the considered row. We study two types of algorithms based on these queries: deterministic and nondeterministic decision trees. One can interpret nondeterministic decision trees for a decision table as a way to represent an arbitrary system of true decision rules for this table that cover all rows. We consider so-called bounded complexity measures that characterize the time complexity of decision trees, for example, the depth of decision trees.

Decision tables with many-valued decisions often appear in data analysis. Our approach is closer to multi-label learning [6,33,34] in which each decision from the set attached to a row is considered correct, than to superset learning in which the set attached to a row is the set of possible decisions containing the correct one [8]. Moreover, decision tables with many-valued decisions are common in such areas as combinatorial optimization, computational geometry, and fault diagnosis, where they are used to represent and explore problems [2,18].

Decision trees [1,2,7,15,16,28,30] and decision rule systems [4,5,9,10,17,18, 24,26] are widely used as classifiers, as a means for knowledge representation, and as algorithms for solving various problems of combinatorial optimization, fault diagnosis, etc. Decision trees and rules are among the most interpretable models in data analysis [12].

The depth of deterministic and nondeterministic decision trees for computation Boolean functions (variables of a function are considered as attributes) was studied quite intensively [3,11,14,32].

We study classes of decision tables with many-valued decisions closed under removal of columns (attributes) and changing the decisions (really, sets of decisions). The most natural examples of such classes are closed classes of decision tables generated by information systems introduced by Pawlak [21]. An information system consists of a set of objects (universe) and a set of attributes (functions) defined on the universe and with values from a finite set. A problem over an information system is specified by a finite number of attributes that divide the universe into nonempty domains in which these attributes have fixed values. A nonempty finite set of decisions is attached to each domain. For a given object from the universe, it is required to find a decision from the set attached to the domain containing this object.

A decision table with many-valued decisions corresponds to this problem in a natural way: columns of this table are labeled with the considered attributes, rows correspond to domains and are labeled with sets of decisions attached to domains. The set of decision tables corresponding to problems over an information system forms a closed class generated by this system. Note that the family of all closed classes is essentially wider than the family of closed classes generated by information systems. In particular, the union of two closed classes generated by two information systems is a closed class. However, generally, there is no an information system that generates this class.

Various classes of objects that are closed under different operations are intensively studied. Among them, in particular, are classes of Boolean functions closed under the operation of superposition [27] and minor-closed classes of graphs [29]. Decision tables represent an interesting mathematical object deserving mathematical research, in particular, the study of closed classes of decision tables.

This paper continues the study of closed classes of decision tables that began with work [13] and continued with works [19,20]. In [13], we studied the dependence of the minimum depth of deterministic decision trees and the depth of deterministic decision trees constructed by a greedy algorithm on the number of attributes (columns) for conventional decision tables from classes closed under operations of removal of columns and changing of decisions.

In [19], we considered classes of decision tables with many-valued decisions closed under operations of removal of columns, changing of decisions, permutation of columns, and duplication of columns. We studied relationships among three parameters of these tables: the complexity of a decision table (if we consider the depth of decision trees, then the complexity of a decision table is the number of columns in it), the minimum complexity of a deterministic decision tree, and the minimum complexity of a nondeterministic decision tree. We considered rough classification of functions characterizing relationships and enumerated all possible seven types of the relationships.

In [20], we considered classes of decision tables with 0–1-decisions (each row is labeled with the decision 0 or the decision 1assigned to rows. For tables from an arbitrary closed class, we studied the dependence of the minimum complexity of deterministic decision trees on various parameters of the tables: the minimum complexity of a test, the complexity of the set of attributes attached to columns, and the minimum complexity of a strongly nondeterministic decision tree. We also studied the dependence of the minimum complexity of strongly nondeterministic decision trees on the complexity of the set of attributes attached to columns. Note that a strongly nondeterministic decision tree can be interpreted as a set of true decision rules that cover all rows labeled with the decision 1.

Let A be a class of decision tables with many-valued decisions closed under removal of columns and changing of decisions, and ψ be a bounded complexity measure. In this paper, we study three functions: $\mathcal{F}_{\psi,A}(n)$, $\mathcal{G}_{\psi,A}(n)$, and $\mathcal{H}_{\psi,A}(n)$.

The function $\mathcal{F}_{\psi,A}(n)$ characterizes the growth in the worst case of the minimum complexity of a deterministic decision tree for a decision table from A with the growth of the complexity of the set of attributes attached to columns of the table. We prove that the function $\mathcal{F}_{\psi,A}(n)$ is either bounded from above by a constant, or grows as a logarithm of n, or grows almost linearly depending on n (it is bounded from above by n and is equal to n for infinitely many n). These results are generalizations of results obtained in [20] for closed classes of decision tables with 0–1-decisions.

The function $\mathcal{G}_{\psi,A}(n)$ characterizes the growth in the worst case of the minimum complexity of a nondeterministic decision tree for a decision table from A with the growth of the complexity of the set of attributes attached to columns of the table. We prove that the function $\mathcal{G}_{\psi,A}(n)$ is either bounded from above by a constant or grows almost linearly depending on n (it is bounded from above by n and is equal to n for infinitely many n).

The function $\mathcal{H}_{\psi,A}(n)$ characterizes the growth in the worst case of the minimum complexity of a deterministic decision tree for a decision table from A with the growth of the minimum complexity of a nondeterministic decision tree for the table. We indicated the condition for the function $\mathcal{H}_{\psi,A}(n)$ to be defined everywhere. Let $\mathcal{H}_{\psi,A}(n)$ be everywhere defined. We proved that this function is either bounded from above by a constant, or is greater than or equal to n for infinitely many n. In particular, for any nondecreasing function φ such that $\varphi(n) \geq n$ and $\varphi(0) = 0$, the function $\mathcal{H}_{\psi,A}(n)$ can grow between $\varphi(n)$ and $\varphi(n) + n$. We indicated also conditions for the function $\mathcal{H}_{\psi,A}(n)$ to be bounded from above by a polynomial on n.

There is a similarity between some results obtained in this paper for closed classes of decision tables and results from the book [16] obtained for problems over information systems. However, the results of the present paper are more general.

Proofs of the considered statements are too long for the conference paper. They will be presented in its journal extension.

The rest of the paper is organized as follows: Sect. 2 contains main definitions and notation, Sect. 3 contains main results, and Sect. 4 – short conclusions.

2   Main Definitions and Notation

Denote $\omega = \{0, 1, 2, \ldots\}$, $\mathcal{S}(\omega)$ the set of nonempty finite subsets of the set ω and, for any $k \in \omega \backslash \{0, 1\}$, denote $E_k = \{0, 1, \ldots, k-1\}$. Let $P = \{f_i : i \in \omega\}$ be the set of *attributes* (really, names of attributes). Two attributes $f_i, f_j \in P$ are considered *different* if $i \neq j$.

f_2 f_4 f_3	
1 1 1	$\{1\}$
0 1 1	$\{0,1,2\}$
1 1 0	$\{1,3\}$
0 0 1	$\{2\}$
1 0 0	$\{3\}$
0 0 0	$\{2,3\}$

Fig. 1. Decision table from $\mathcal{M}_2^\infty$

2.1 Decision Tables

First, we define the notion of a decision table.

Definition 1. *Let $k \in \omega\backslash\{0,1\}$. Denote by $\mathcal{M}_k^\infty$ the set of rectangular tables filled with numbers from E_k in each of which rows are pairwise different, each row is labeled with a set from $\mathcal{S}(\omega)$ (set of decisions), and columns are labeled with pairwise different attributes from P. Rows are interpreted as tuples of values of these attributes. Empty tables without rows belong also to the set $\mathcal{M}_k^\infty$. We will use the same notation Λ for these tables. Tables from $\mathcal{M}_k^\infty$ will be called decision tables with many-valued decisions (decision tables).*

Two tables from $\mathcal{M}_k^\infty$ are *equal* if one can be obtained from another by permutation of rows with attached to them sets of decisions. For a table $T \in \mathcal{M}_k^\infty$, we denote by $\Delta(T)$ the set of rows of the table T and by $\Pi(T)$ we denote the intersection of sets of decisions attached to rows of T. Decisions from $\Pi(T)$ are called *common decisions* for the table T.

Example 1. Figure 1 shows a decision table from $\mathcal{M}_2^\infty$.

Denote by $\mathcal{M}_k^\infty\mathcal{C}$ the set of tables from $\mathcal{M}_k^\infty$ in each of which there exists a common decision. Let $\Lambda \in \mathcal{M}_k^\infty\mathcal{C}$.

Let T be a nonempty table from $\mathcal{M}_k^\infty$. Denote by $P(T)$ the set of attributes attached to columns of the table T. We denote by $\Omega_k(T)$ the set of finite words over the alphabet $\{(f_i,\delta) : f_i \in P(T), \delta \in E_k\}$ including the empty word λ. For any $\alpha \in \Omega_k(T)$, we now define a subtable $T\alpha$ of the table T. If $\alpha = \lambda$, then $T\alpha = T$. If $\alpha \neq \lambda$ and $\alpha = (f_{i_1},\delta_1)\cdots(f_{i_m},\delta_m)$, then $T\alpha$ is the table obtained from T by removal of all rows that do not satisfy the following condition: in columns labeled with attributes $f_{i_1},\ldots,f_{i_m}$, the row has numbers $\delta_1,\ldots,\delta_m$, respectively.

We now define two operations on decision tables: removal of columns and changing of decisions. Let $T \in \mathcal{M}_k^\infty$.

Definition 2. *Removal of columns. Let $D \subseteq P(T)$. We remove from T all columns labeled with the attributes from the set D. In each group of rows equal on the remaining columns, we keep the first one. Denote the obtained table by $I(D,T)$. In particular, $I(\emptyset,T) = T$ and $I(P(T),T) = \Lambda$. It is obvious that $I(D,T) \in \mathcal{M}_k^\infty$.*

f_2 f_3		
1	1	$\{1\}$
0	1	$\{0,1\}$
1	0	$\{0,1\}$
0	0	$\{0\}$

Fig. 2. Decision table obtained from the decision table shown in Fig. 1 by removal of a column and changing of decisions

Definition 3. *Changing of decisions. Let $\nu : E_k^{|P(T)|} \to \mathcal{S}(\omega)$ (by definition, $E_k^0 = \emptyset$). For each row $\bar{\delta}$ of the table T, we replace the set of decisions attached to this row with $\nu(\bar{\delta})$. We denote the obtained table by $J(\nu, T)$. It is obvious that $J(\nu, T) \in \mathcal{M}_k^\infty$.*

Definition 4. *Denote $[T] = \{J(\nu, I(D,T)) : D \subseteq P(T), \nu : E_k^{|P(T) \setminus D|} \to \mathcal{S}(\omega)\}$. The set $[T]$ is the closure of the table T under the operations of removal of columns and changing of decisions.*

Example 2. Figure 2 shows the table $J(\nu, I(D, T_0))$, where T_0 is the table shown in Fig. 1, $D = \{f_4\}$ and $\nu(x_1, x_2) = \{\min(x_1, x_2), \max(x_1, x_2)\}$.

Definition 5. *Let $A \subseteq \mathcal{M}_k^\infty$ and $A \neq \emptyset$. Denote $[A] = \bigcup_{T \in A} [T]$. The set $[A]$ is the closure of the set A under the considered two operations. The class (the set) of decision tables A will be called a closed class if $[A] = A$.*

A closed class of decision tables will be called *nontrivial* if it contains nonempty decision tables.

Let A_1 and A_2 be closed classes of decision tables from $\mathcal{M}_k^\infty$. Then $A_1 \cup A_2$ is a closed class of decision tables from $\mathcal{M}_k^\infty$.

2.2   Deterministic and Nondeterministic Decision Trees

A *finite tree with root* is a finite directed tree in which exactly one node called the *root* has no entering edges. The nodes without leaving edges are called *terminal* nodes.

Definition 6. *A k-decision tree is a finite tree with root, which has at least two nodes and in which*

- *The root and edges leaving the root are not labeled.*
- *Each terminal node is labeled with a decision from the set ω.*
- *Each node, which is neither the root nor a terminal node, is labeled with an attribute from the set P. Each edge leaving such node is labeled with a number from the set E_k.*

Example 3. Figures 3 and 4 show 2-decision trees.

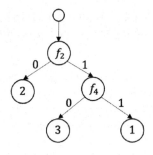

Fig. 3. A deterministic decision tree for the decision table shown in Fig. 1

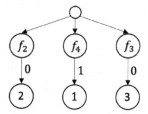

Fig. 4. A nondeterministic decision tree for the decision table shown in Fig. 1

We denote by $\mathcal{T}_k$ the set of all k-decision trees. Let $\Gamma \in \mathcal{T}_k$. We denote by $P(\Gamma)$ the set of attributes attached to nodes of Γ that are neither the root nor terminal nodes. A *complete path* of Γ is a sequence $\tau = v_1, d_1, \ldots, v_m, d_m, v_{m+1}$ of nodes and edges of Γ in which v_1 is the root of Γ, v_{m+1} is a terminal node of Γ and, for $j = 1, \ldots, m$, the edge d_j leaves the node v_j and enters the node v_{j+1}. Let $T \in \mathcal{M}_k^\infty$. If $P(\Gamma) \subseteq P(T)$, then we correspond to the table T and the complete path τ a word $\pi(\tau) \in \Omega_k(T)$. If $m = 1$, then $\pi(\tau) = \lambda$. If $m > 1$ and, for $j = 2, \ldots, m$, the node v_j is labeled with the attribute f_{i_j} and the edge d_j is labeled with the number δ_j, then $\pi(\tau) = (f_{i_2}, \delta_2) \cdots (f_{i_m}, \delta_m)$. Denote $T(\tau) = T\pi(\lambda)$.

Definition 7. *Let* $T \in \mathcal{M}_k^\infty \backslash \{\Lambda\}$. *A deterministic decision tree for the table* T *is a k-decision tree* Γ *satisfying the following conditions:*

- *Only one edge leaves the root of* Γ.
- *For any node, which is neither the root nor a terminal node, edges leaving this node are labeled with pairwise different numbers.*
- $P(\Gamma) \subseteq P(T)$.
- *For any row of* T, *there exists a complete path* τ *of* Γ *such that the considered row belongs to the table* $T(\tau)$.
- *For any complete path* τ *of* Γ, *either* $T(\tau) = \Lambda$ *or the decision attached to the terminal node of* τ *is a common decision for the table* $T(\tau)$.

Example 4. The 2-decision tree shown in Fig. 3 is a deterministic decision tree for the decision table shown in Fig. 1.

Definition 8. *Let $T \in \mathcal{M}_k^\infty \backslash \{\Lambda\}$. A nondeterministic decision tree for the table T is a k-decision tree Γ satisfying the following conditions:*

- $P(\Gamma) \subseteq P(T)$.
- *For any row of T, there exists a complete path τ of Γ such that the considered row belongs to the table $T(\tau)$.*
- *For any complete path τ of Γ, either $T(\tau) = \Lambda$ or the decision attached to the terminal node of τ is a common decision for the table $T(\tau)$.*

Example 5. The 2-decision tree shown in Fig. 4 is a nondeterministic decision tree for the decision table shown in Fig. 1.

2.3   Complexity Measures

Denote by B the set of all finite words over the alphabet $P = \{f_i : i \in \omega\}$, which contains the empty word λ and on which the word concatenation operation is defined.

Definition 9. *A complexity measure is an arbitrary function $\psi : B \to \omega$ that has the following properties: for any words $\alpha_1, \alpha_2 \in B$,*

- *$\psi(\alpha_1) = 0$ if and only if $\alpha_1 = \lambda$ – positivity property.*
- *$\psi(\alpha_1) = \psi(\alpha_1')$ for any word α_1' obtained from α_1 by permutation of letters – commutativity property.*
- *$\psi(\alpha_1) \leq \psi(\alpha_1\alpha_2)$ – nondecreasing property.*
- *$\psi(\alpha_1\alpha_2) \leq \psi(\alpha_1) + \psi(\alpha_2)$ – boundedness from above property.*

The following functions are complexity measures:

- Function h for which, for any word $\alpha \in B$, $h(\alpha) = |\alpha|$, where $|\alpha|$ is the length of the word α. This function is called the *depth*.
- An arbitrary function $\varphi : B \to \omega$ such that $\varphi(\lambda) = 0$, for any $f_i \in B$, $\varphi(f_i) > 0$ and, for any nonempty word $f_{i_1} \cdots f_{i_m} \in B$,

$$\varphi(f_{i_1} \cdots f_{i_m}) = \sum_{j=1}^{m} \varphi(f_{i_j}). \tag{1}$$

This function is called the *weighted depth*.
- An arbitrary function $\rho : B \to \omega$ such that $\rho(\lambda) = 0$, for any $f_i \in B$, $\rho(f_i) > 0$, and, for any nonempty word $f_{i_1} \cdots f_{i_m} \in B$, $\rho(f_{i_1} \cdots f_{i_m}) = \max\{\rho(f_{i_j}) : j = 1, \ldots, m\}$.

Definition 10. *A bounded complexity measure is a complexity measure ψ, which has the boundedness from below property: for any word $\alpha \in B$, $\psi(\alpha) \geq |\alpha|$.*

Any complexity measure satisfying the equality (1), in particular the function h, is a bounded complexity measure. One can show that if functions ψ_1 and ψ_2 are complexity measures, then the functions ψ_3 and ψ_4 are complexity measures, where for any $\alpha \in B$, $\psi_3(\alpha) = \psi_1(\alpha) + \psi_2(\alpha)$ and $\psi_4(\alpha) = \max(\psi_1(\alpha), \psi_2(\alpha))$. If the function ψ_1 is a bounded complexity measure, then the functions ψ_3 and ψ_4 are bounded complexity measures.

Definition 11. *Let ψ be a complexity measure. We extend it to the set of all finite subsets of the set P. Let D be a finite subset of the set P. If $D = \emptyset$, then $\psi(D) = 0$. Let $D = \{f_{i_1}, \ldots, f_{i_m}\}$ and $m \geq 1$. Then $\psi(D) = \psi(f_{i_1} \cdots f_{i_m})$.*

Definition 12. *Let ψ be a complexity measure. We extend it to the set of finite words Ω in the alphabet $\{(f_i, \delta) : f_i \in P, \delta \in \omega\}$ including the empty word λ. Let $\alpha \in \Omega$. If $\alpha = \lambda$, then $\psi(\alpha) = 0$. Let $\alpha = (f_{i_1}, \delta_1) \cdots (f_{i_m}, \delta_m)$ and $m \geq 1$. Then $\psi(\alpha) = \psi(f_{i_1} \cdots f_{i_m})$.*

2.4 Parameters of Decision Trees and Tables

Definition 13. *Let ψ be a complexity measure. We extend the function ψ to the set $\mathcal{T}_k$. Let $\Gamma \in \mathcal{T}_k$. Then $\psi(\Gamma) = \max\{\psi(\pi(\tau))\}$, where the maximum is taken over all complete paths τ of the decision tree Γ. For a given complexity measure ψ, the value $\psi(\Gamma)$ will be called the complexity of the decision tree Γ. The value $h(\Gamma)$ will be called the depth of the decision tree Γ.*

Let ψ be a complexity measure. We now describe the functions ψ^d, ψ^a, m_ψ, W_ψ, S_ψ, and N defined on the set $\mathcal{M}_k^\infty$ and functions Z, G defined on the set $\mathcal{M}_2^\infty$ and taking values from the set ω. By definition, the value of each of these functions for Λ is equal 0. We also describe the function l_ψ defined on the set $\mathcal{M}_k^\infty \times \omega$. By definition, the value of this function for tuple (Λ, n), $n \in \omega$, is equal 0. Let $T \in \mathcal{M}_k^\infty \setminus \{\Lambda\}$ and $n \in \omega$.

- $\psi^d(T) = \min\{\psi(\Gamma)\}$, where the minimum is taken over all deterministic decision trees Γ for the table T.
- $\psi^a(T) = \min\{\psi(\Gamma)\}$, where the minimum is taken over all nondeterministic decision trees Γ for the table T.
- $m_\psi(T) = \max\{\psi(f_i) : f_i \in P(T)\}$.
- $W_\psi(T) = \psi(P(T))$.
- Let $\bar{\delta}$ be a row of the table T. Denote $S_\psi(T, \bar{\delta}) = \min\{\psi(D)\}$, where the minimum is taken over all subsets D of the set $P(T)$ such that in the set of columns of T labeled with attributes from D the row $\bar{\delta}$ is different from all other rows of the table T. Then $S_\psi(T) = \max\{S_\psi(T, \bar{\delta})\}$, where the maximum is taken over all rows $\bar{\delta}$ of the table T.
- $N(T)$ is the number of rows in the table T.
- Let $T \in \mathcal{M}_2^\infty$. A decision table $Q \in \mathcal{M}_2^\infty$ will be called *complete* if $N(Q) = 2^{|P(Q)|}$. Then $Z(T)$ is the maximum number of columns in a complete table from $[T]$ if such tables exist and 0 otherwise.

- Let $T \in \mathcal{M}_2^\infty$. A word $\alpha \in \Omega_2(T)$ will be called *annihilating* word for the table T if $T\alpha = \Lambda$ and α does not contain letters (f_i, δ) and (f_i, σ) such that $\delta \neq \sigma$. An annihilating word α for the table T will be called *irreducible* if any subword of α obtained from α by removal of some letters and different from α is not annihilating. Then $G(T)$ is the maximum length of an irreducible annihilating word for the table T if such words exist and 0 otherwise.
- Denote $\Omega_k^n(T) = \{\alpha : \alpha \in \Omega_k(T), \psi(\alpha) \leq n\}$. A finite set $U \subseteq \Omega_k^n(T)$ will be called (ψ, n)-*cover* of the table T if $\bigcup_{\alpha \in U} \Delta(T\alpha) = \Delta(T)$. The (ψ, n)-cover U will be called *irreducible* if each proper subset of U is not a (ψ, n)-cover of the table T. Then $l_\psi(T, n)$ is the maximum cardinality of an irreducible (ψ, n)-cover of the table T. It is clear that $\{\lambda\}$ is an irreducible (ψ, n)-cover of the table T. Therefore $l_\psi(T, n) \geq 1$.

Example 6. We denote by T_0 the decision table shown in Fig. 1. One can show that $h^d(T_0) = 2$, $h^a(T_0) = 1$, $m_h(T_0) = 1$, $W_h(T_0) = 3$, $S_h(T_0) = 2$, $N(T_0) = 6$, $Z(T_0) = 2$, $G(T_0) = 3$, $l_h(T_0, 0) = 1$, $l_h(T_0, 1) = 3$ and $l_h(T_0, n) = 6$ for any $n \in \omega \backslash \{0, 1\}$.

3   Main Results

In this section, we consider results obtained for the functions $\mathcal{F}_{\psi,A}$, $\mathcal{G}_{\psi,A}$, and $\mathcal{H}_{\psi,A}$ and discuss closed classes of decision tables generated by information systems.

3.1   Function $\mathcal{F}_{\psi,A}$

Let ψ be a bounded complexity measure and A be a nontrivial closed class of decision tables from $\mathcal{M}_k^\infty$. We now define a function $\mathcal{F}_{\psi,A} : \omega \to \omega$. Let $n \in \omega$. Then $\mathcal{F}_{\psi,A}(n) = \max\{\psi^d(T) : T \in A, W_\psi(T) \leq n\}$.

The function $\mathcal{F}_{\psi,A}$ characterizes the growth in the worst case of the minimum complexity of deterministic decision trees for decision tables from A with the growth of the complexity of the sets of attributes attached to columns of these tables.

Let $D = \{n_i : i \in \omega\}$ be an infinite subset of the set ω in which, for any $i \in \omega$, $n_i < n_{i+1}$. We now define a function $H_D : \omega \to \omega$. Let $n \in \omega$. If $n < n_0$, then $H_D(n) = 0$. If, for some $i \in \omega$, $n_i \leq n < n_{i+1}$, then $H_D(n) = n_i$.

Theorem 1. *Let ψ be a bounded complexity measure and A be a nontrivial closed class of decision tables from $\mathcal{M}_k^\infty$. Then $\mathcal{F}_{\psi,A}$ is an everywhere defined nondecreasing function such that $\mathcal{F}_{\psi,A}(n) \leq n$ for any $n \in \omega$ and $\mathcal{F}_{\psi,A}(0) = 0$. For this function, one of the following statements holds:*

(a) If the functions S_ψ and N are bounded from above on the class A, then there exists a positive constant c_0 such that $\mathcal{F}_{\psi,A}(n) \leq c_0$ for any $n \in \omega$.

(b) *If the function S_ψ is bounded from above on the class A and the function N is not bounded from above on the class A, then there exist positive constants c_1, c_2, c_3, c_4 such that $c_1 \log_2 n - c_2 \leq \mathcal{F}_{\psi,A}(n) \leq c_3 \log_2 n + c_4$ for any $n \in \omega \backslash \{0\}$.*

(c) *If the function S_ψ is not bounded from above on the class A, then there exists an infinite subset D of the set ω such that $H_D(n) \leq \mathcal{F}_{\psi,A}(n)$ for any $n \in \omega$.*

3.2   Function $\mathcal{G}_{\psi,A}$

Let ψ be a bounded complexity measure and A be a nontrivial closed class of decision tables from $\mathcal{M}_k^\infty$. We now define a function $\mathcal{G}_{\psi,A}$. Let $n \in \omega$. Then $\mathcal{G}_{\psi,A}(n) = \max\{\psi^a(T) : T \in A, W_\psi(T) \leq n\}$.

The function $\mathcal{G}_{\psi,A}$ characterizes the growth in the worst case of the minimum complexity of nondeterministic decision trees for decision tables from A with the growth of the complexity of the sets of attributes attached to columns of these table.

Theorem 2. *Let ψ be a bounded complexity measure and A be a nontrivial closed class of decision tables from $\mathcal{M}_k^\infty$. Then $\mathcal{G}_{\psi,A}$ is an everywhere defined nondecreasing function such that $\mathcal{G}_{\psi,A}(n) \leq n$ for any $n \in \omega$ and $\mathcal{G}_{\psi,A}(0) = 0$. For this function, one of the following statements holds:*

(a) *If the function S_ψ is bounded from above on the class A, then there exists a positive constant c such that $\mathcal{G}_{\psi,A}(n) \leq c$ for any $n \in \omega$.*

(b) *If the function S_ψ is not bounded from above on the class A, then there exists an infinite subset D of the set ω such that $H_D(n) \leq \mathcal{G}_{\psi,A}(n)$ for any $n \in \omega$.*

3.3   Function $\mathcal{H}_{\psi,A}$

Let ψ be a bounded complexity measure and A be a nontrivial closed class of decision tables from $\mathcal{M}_k^\infty$. We now define possibly partial function $\mathcal{H}_{\psi,A} : \omega \to \omega$. Let $n \in \omega$. If the set $\{\psi^d(T) : T \in A, \psi^a(T) \leq n\}$ is infinite, then the value $\mathcal{H}_{\psi,A}(n)$ is undefined. Otherwise, $\mathcal{H}_{\psi,A}(n) = \max\{\psi^d(T) : T \in A, \psi^a(T) \leq n\}$.

The function $\mathcal{H}_{\psi,A}$ characterizes the growth in the worst case of the minimum complexity of deterministic decision trees for decision tables from A with the growth of the minimum complexity of nondeterministic decision trees for these tables.

We now define possibly partial function $L_{\psi,A} : \omega \to \omega$. Let $n \in \omega$. If the set $\{l_\psi(T,n) : T \in A\}$ is infinite, then the value $L_{\psi,A}(n)$ is not defined. Otherwise, $L_{\psi,A}(n) = \max\{l_\psi(T,n) : T \in A\}$. One can show that in this case $L_{\psi,A}(n) \geq 1$.

The following statement describes the criterion for the function $\mathcal{H}_{\psi,A}$ to be everywhere defined.

Theorem 3. *Let ψ be a bounded complexity measure and A be a nontrivial closed class of decision tables from $\mathcal{M}_k^\infty$. The function $\mathcal{H}_{\psi,A}$ is everywhere defined if and only if the function $L_{\psi,A}$ is everywhere defined.*

We now describe two possible types of behavior for everywhere defined function $\mathcal{H}_{\psi,A}$.

Theorem 4. *Let ψ be a bounded complexity measure, A be a nontrivial closed class of decision tables from $\mathcal{M}_k^\infty$, and the function $\mathcal{H}_{\psi,A}$ be everywhere defined. Then $\mathcal{H}_{\psi,A}$ is a nondecreasing function and $\mathcal{H}_{\psi,A}(0) = 0$. For this function, one of the following statements holds:*

(a) *If the function ψ^d is bounded from above on the class A, then there is a nonnegative constant c such that $\mathcal{H}_{\psi,A}(n) \le c$ for any $n \in \omega$.*
(b) *If the function ψ^d is not bounded from above on the class A, then there exists an infinite subset D of the set ω such that $\mathcal{H}_{\psi,A}(n) \ge H_D(n)$ for any $n \in \omega$.*

Remark 1. From Theorem 1 it follows that the function ψ^d is bounded from above on the class A if and only if the functions S_ψ and N are bounded from above on the class A.

The following statement shows a wide spectrum of behavior of the everywhere defined function $\mathcal{H}_{\psi,A}$ in the case when the function ψ^d is not bounded from above on the class A.

Theorem 5. *Let $\varphi : \omega \to \omega$ be a nondecreasing function such that $\varphi(n) \ge n$ for any $n \in \omega$ and $\varphi(0) = 0$. Then there exist a closed class A of decision tables from $\mathcal{M}_k^\infty$ and a bounded complexity measure ψ such that the function $\mathcal{H}_{\psi,A}$ is everywhere defined and $\varphi(n) \le \mathcal{H}_{\psi,A}(n) \le \varphi(n) + n$ for any $n \in \omega$.*

Let A be a nontrivial closed class of decision tables from $\mathcal{M}_2^\infty$ and ψ be a bounded complexity measure. Let $n \in \omega$. Denote $A_\psi(n) = \{T : T \in A, m_\psi(T) \le n\}$. Since $\Lambda \in A$, both sets $A_\psi(n)$ and $\{Z(T) : T \in A_\psi(n)\}$ are nonempty sets. We now define probably partial function $Z_{\psi,A} : \omega \to \omega$. If $\{Z(T) : T \in A_\psi(n)\}$ is an infinite set, then the value $Z_{\psi,A}(n)$ is not defined. Otherwise, $Z_{\psi,A}(n) = \max\{Z(T) : T \in A_\psi(n)\}$.

Let $n \in \omega$. Since $\Lambda \in A$, the set $\{G(T) : T \in A_\psi(n)\}$ is nonempty. We now define probably partial function $G_{\psi,A} : \omega \to \omega$. If $\{G(T) : T \in A_\psi(n)\}$ is an infinite set, then the value $G_{\psi,A}(n)$ is not defined. Otherwise, $G_{\psi,A}(n) = \max\{G(T) : T \in A_\psi(n)\}$.

The following statement describes the criterion for the everywhere defined function $\mathcal{H}_{\psi,A}$ to be bounded from above by a polynomial. For simplicity, we consider here only decision tables from the set $\mathcal{M}_2^\infty$.

Theorem 6. *Let A be a nontrivial closed class of decision tables from $\mathcal{M}_2^\infty$, ψ be a bounded complexity measure and the function $\mathcal{H}_{\psi,A}$ be everywhere defined. Then a polynomial p_0 such that $\mathcal{H}_{\psi,A}(n) \le p_0(n)$ for any $n \in \omega$ exists if and only if there exist polynomials p_1, p_2, and p_3 such that $Z_{\psi,A}(n) \le p_1(n)$, $G_{\psi,A}(n) \le p_2(n)$ and $L_{\psi,A}(n) \le 2^{p_3(n)}$ for any $n \in \omega$.*

3.4   Family of Closed Classes of Decision Tables

Let U be a set and $\Phi = \{f_0, f_1, \ldots\}$ be a finite or countable set of functions (attributes) defined on U and taking values from E_k. The pair (U, Φ) is called a k-*information system*. A *problem* over (U, Φ) is an arbitrary tuple $z = (U, \nu, f_{i_1}, \ldots, f_{i_n})$, where $n \in \omega \backslash \{0\}$, $\nu : E_k^n \to S(\omega)$ and $f_{i_1}, \ldots, f_{i_n}$ are functions from Φ with pairwise different indices $i_1, \ldots, i_n$. The problem z is to determine a value from the set $\nu(f_{i_1}(u), \ldots, f_{i_n}(u))$ for a given $u \in U$. Various examples of k-information systems can be found in [15].

We denote by $T(z)$ a decision table from $\mathcal{M}_k^\infty$ with n columns labeled with attributes $f_{i_1}, \ldots, f_{i_n}$. A row $(\delta_1, \ldots, \delta_n) \in E_k^n$ belongs to the table $T(z)$ if and only if the system of equations $\{f_{i_1}(x) = \delta_1, \ldots, f_{i_n}(x) = \delta_n\}$ has a solution from the set U. This row is labeled with the set of decisions $\nu(\delta_1, \ldots, \delta_n)$.

Let the algorithms for the problem z solving be algorithms in which each elementary operation consists in calculating the value of some attribute from the set $\{f_{i_1}, \ldots, f_{i_n}\}$ on a given element $u \in U$. Then, as a model of the problem z, we can use the decision table $T(z)$, and as models of algorithms for the problem z solving – deterministic and nondeterministic decision trees for the table $T(z)$.

Denote by $\mathcal{Z}^\infty(U, \Phi)$ the set of problems over (U, Φ) and $\mathcal{A}^\infty(U, \Phi) = \{T(z) : z \in \mathcal{Z}^\infty(U, \Phi)\}$. One can show that $\mathcal{A}^\infty(U, \Phi) = [\mathcal{A}^\infty(U, \Phi)]$, i.e., $\mathcal{A}^\infty(U, \Phi)$ is a closed class of decision tables from $\mathcal{M}_k^\infty$ *generated* by the information system (U, Φ).

Closed classes of decision tables generated by k-information systems are the most natural examples of closed classes. However, the notion of a closed class is essentially wider. In particular, the union $\mathcal{A}^\infty(U_1, \Phi_1) \cup \mathcal{A}^\infty(U_2, \Phi_2)$, where (U_1, Φ_1) and (U_2, Φ_2) are k-information systems, is a closed class, but generally, we cannot find an information system (U, Φ) such that $\mathcal{A}^\infty(U, \Phi) = \mathcal{A}^\infty(U_1, \Phi_1) \cup \mathcal{A}^\infty(U_2, \Phi_2)$.

3.5   Example of Information System

Let $\mathbb{R}$ be the set of real numbers and $F = \{f_i : i \in \omega\}$ be the set of functions defined on $\mathbb{R}$ and taking values from the set E_2 such that, for any $i \in \omega$ and $a \in \mathbb{R}$,

$$f_i(a) = \begin{cases} 0, & a < i, \\ 1, & a \geq i. \end{cases}$$

Let ψ be a bounded complexity measure and $A = \mathcal{A}^\infty(\mathbb{R}, F)$. One can prove the following statements:

- The function N is not bounded from above on the set A.
- The function S_ψ is bounded from above on the set A if and only if there exists a constant $c_0 > 0$ such that $\psi(f_i) \leq c_0$ for any $i \in \omega$.
- The function $L_{\psi,A}$ is everywhere defined if and only if, for any $n \in \omega$, the set $\{i : i \in \omega, \psi(f_i) \leq n\}$ is finite.
- A polynomial p such that $L_{\psi,A}(n) \leq 2^{p(n)}$ for any $n \in \omega$ exists if and only if there exists a polynomial q such that $|\{i : i \in \omega, \psi(f_i) \leq n\}| \leq 2^{q(n)}$ for any $n \in \omega$.

- For any $n \in \omega$, $Z_{\psi,A}(n) \le 1$.
- For any $n \in \omega$, $G_{\psi,A}(n) \le 2$.

4 Conclusions

In this paper, we studied relationships among three parameters of tables from closed classes of decision tables with many-valued decisions: the minimum complexity of a deterministic decision tree, the minimum complexity of a nondeterministic decision tree, and the complexity of the set of attributes attached to columns. Future research will be devoted to the study of relationships among time and space complexity of deterministic and nondeterministic decision trees for decision tables from closed classes.

Acknowledgements. Research reported in this publication was supported by King Abdullah University of Science and Technology (KAUST). The authors are grateful to the anonymous reviewers for useful comments.

References

1. AbouEisha, H., Amin, T., Chikalov, I., Hussain, S., Moshkov, M.: Extensions of Dynamic Programming for Combinatorial Optimization and Data Mining. Intelligent Systems Reference Library, vol. 146. Springer, Heidelberg (2019). https://doi.org/10.1007/978-3-319-91839-6
2. Alsolami, F., Azad, M., Chikalov, I., Moshkov, M.: Decision and Inhibitory Trees and Rules for Decision Tables with Many-valued Decisions. Intelligent Systems Reference Library, vol. 156. Springer, Heidelberg (2020). https://doi.org/10.1007/978-3-030-12854-8
3. Blum, M., Impagliazzo, R.: Generic oracles and oracle classes (extended abstract). In: 28th Annual Symposium on Foundations of Computer Science, Los Angeles, California, USA, 27–29 October 1987, pp. 118–126. IEEE Computer Society (1987)
4. Boros, E., Hammer, P.L., Ibaraki, T., Kogan, A.: Logical analysis of numerical data. Math. Program. **79**, 163–190 (1997)
5. Boros, E., Hammer, P.L., Ibaraki, T., Kogan, A., Mayoraz, E., Muchnik, I.B.: An implementation of logical analysis of data. IEEE Trans. Knowl. Data Eng. **12**(2), 292–306 (2000)
6. Boutell, M.R., Luo, J., Shen, X., Brown, C.M.: Learning multi-label scene classification. Pattern Recognit. **37**(9), 1757–1771 (2004)
7. Breiman, L., Friedman, J.H., Olshen, R.A., Stone, C.J.: Classification and Regression Trees. Wadsworth and Brooks (1984)
8. Campagner, A., Ciucci, D., Hüllermeier, E.: Rough set-based feature selection for weakly labeled data. Int. J. Approx. Reason. **136**, 150–167 (2021)
9. Chikalov, I., et al.: Three Approaches to Data Analysis - Test Theory, Rough Sets and Logical Analysis of Data. Intelligent Systems Reference Library, vol. 41. Springer, Heidelberg (2013). https://doi.org/10.1007/978-3-642-28667-4
10. Fürnkranz, J., Gamberger, D., Lavrac, N.: Foundations of Rule Learning. Cognitive Technologies. Springer, Heidelberg (2012). https://doi.org/10.1007/978-3-540-75197-7

11. Hartmanis, J., Hemachandra, L.A.: One-way functions, robustness, and the non-isomorphism of NP-complete sets. In: Proceedings of the Second Annual Conference on Structure in Complexity Theory, Cornell University, Ithaca, New York, USA, 16–19 June 1987. IEEE Computer Society (1987)
12. Molnar, C.: Interpretable Machine Learning. A Guide for Making Black Box Models Explainable. 2nd edn. (2022). christophm.github.io/interpretable-ml-book/
13. Moshkov, M.: On depth of conditional tests for tables from closed classes. In: Markov, A.A. (ed.) Combinatorial-Algebraic and Probabilistic Methods of Discrete Analysis (in Russian), pp. 78–86. Gorky University Press, Gorky (1989)
14. Moshkov, M.: About the depth of decision trees computing Boolean functions. Fundam. Inform. **22**(3), 203–215 (1995)
15. Moshkov, M.: Time complexity of decision trees. Trans. Rough Sets **3**, 244–459 (2005)
16. Moshkov, M.: Comparative Analysis of Deterministic and Nondeterministic Decision Trees. Intelligent Systems Reference Library, vol. 179. Springer, Heidelberg (2020). https://doi.org/10.1007/978-3-030-41728-4
17. Moshkov, M., Piliszczuk, M., Zielosko, B.: Partial Covers, Reducts and Decision Rules in Rough Sets - Theory and Applications. Studies in Computational Intelligence, vol. 145. Springer, Heidelberg (2008). https://doi.org/10.1007/978-3-540-69029-0
18. Moshkov, M., Zielosko, B.: Combinatorial Machine Learning - A Rough Set Approach. Studies in Computational Intelligence, vol. 360. Springer, Heidelberg (2011). https://doi.org/10.1007/978-3-642-20995-6
19. Ostonov, A., Moshkov, M.: Comparative analysis of deterministic and nondeterministic decision trees for decision tables from closed classes. arXiv:2304.10594 [cs.CC] (2023). https://doi.org/10.48550/arXiv.2304.10594
20. Ostonov, A., Moshkov, M.: Deterministic and strongly nondeterministic decision trees for decision tables from closed classes. arXiv:2305.06093 [cs.CC] (2023). https://doi.org/10.48550/arXiv.2305.06093
21. Pawlak, Z.: Information systems theoretical foundations. Inf. Syst. **6**(3), 205–218 (1981)
22. Pawlak, Z.: Rough sets. Int. J. Parallel Program **11**(5), 341–356 (1982)
23. Pawlak, Z.: Rough classification. Int. J. Man Mach. Stud. **20**(5), 469–483 (1984)
24. Pawlak, Z.: Rough Sets - Theoretical Aspects of Reasoning About Data. Theory and Decision Library: Series D, vol. 9. Kluwer (1991)
25. Pawlak, Z., Polkowski, L., Skowron, A.: Rough set theory. In: Wah, B.W. (ed.) Wiley Encyclopedia of Computer Science and Engineering. Wiley (2008). https://doi.org/10.1002/9780470050118.ecse466
26. Pawlak, Z., Skowron, A.: Rudiments of rough sets. Inf. Sci. **177**(1), 3–27 (2007)
27. Post, E.: Two-Valued Iterative Systems of Mathematical Logic. Annals of Mathematics Studies, vol. 5. Princeton University Press, Princeton-London (1941)
28. Quinlan, J.R.: C4.5: Programs for Machine Learning. Morgan Kaufmann (1993)
29. Robertson, N., Seymour, P.D.: Graph minors. XX. Wagner's conjecture. J. Comb. Theory Ser. B **92**(2), 325–357 (2004)
30. Rokach, L., Maimon, O.: Data Mining with Decision Trees - Theory and Applications. Series in Machine Perception and Artificial Intelligence, vol. 69. World Scientific (2007)

31. Skowron, A., Rauszer, C.: The discernibility matrices and functions in information systems. In: Slowinski, R. (ed.) Intelligent Decision Support - Handbook of Applications and Advances of the Rough Sets Theory. Theory and Decision Library, vol. 11, pp. 331–362. Springer, Dordrecht (1992). https://doi.org/10.1007/978-94-015-7975-9_21
32. Tardos, G.: Query complexity, or why is it difficult to separate $NP^A \cap coNP^A$ from P^A by random oracles A? Comb. **9**(4), 385–392 (1989)
33. Vens, C., Struyf, J., Schietgat, L., Dzeroski, S., Blockeel, H.: Decision trees for hierarchical multi-label classification. Mach. Learn. **73**(2), 185–214 (2008)
34. Zhou, Z., Zhang, M., Huang, S., Li, Y.: Multi-instance multi-label learning. Artif. Intell. **176**(1), 2291–2320 (2012)

Paraconsistent Logics: A Survey Focussing on the Rough Set Approach

Bidhan Saha[1]($\boxtimes$), Mohua Banerjee[1], and Soma Dutta[2]

[1] Department of Mathematics and Statistics,
Indian Institute of Technology Kanpur, Kanpur, India
{bsaha,mohua}@iitk.ac.in
[2] Department of Mathematics and Computer Science,
University of Warmia and Mazury, Olsztyn, Poland
soma.dutta@matman.uwm.edu.pl

Abstract. A survey of approaches yielding paraconsistent logics is made and is summarised through a diagram. The rough set theoretic approach is included in the survey, and it is the focus in the second part of the work. Several new paraconsistent systems are presented, that are obtained by weakening existing rough modus ponens rules.

Keywords: Paraconsistent logics · Ex Contradiction Quodlibet (ECQ) · Rough set approach · Modal logic $S5$

1 Introduction

In common sense reasoning, one often derives meaningful conclusions from contradictory information, that is, information which contains both a statement and its negation. Classical logic does not allow such inferences. Many researchers have tried to deal with the issue by formulating "inconsistency-tolerant systems" [26]. Such systems violate either of two classical properties: the principle of explosion or the law of non-contradiction. Violation of the principle of explosion forms the basis of *paraconsistent* logics. The paper presents a survey of various techniques which give rise to paraconsistent logics. There are several existing surveys of paraconsistent systems, for instance in [1,24,26,31,38,40]. However, this survey is an attempt to compile almost all the major techniques by which paraconsistent systems are arrived at. Furthermore, we focus on paraconsistent logics arising out of rough set theory – this is not part of any of the other surveys. Utilising the notion of rough truth [36] and various forms of rough modus ponens [6,17], paraconsistent systems have been obtained. Furthering this direction of work, we get new paraconsistent systems by weakening existing rough modus ponens rules.

Let us give the formal versions of the classical principles mentioned above. Define a logic $\mathcal{L}$ as a tuple $(FOR, \vdash)$, where FOR represents the set of formulas,

and $\vdash$ is the consequence relation[1] ($\vdash \subseteq \mathcal{P}(FOR) \times FOR$). FOR is based on an enumerable language with the connective negation ($\neg$). Let $\Gamma \cup \{\alpha, \beta\} \subseteq FOR$.
(I) The *Principle of Explosion or Ex Contradictione Quodlibet* (ECQ):
$\forall \Gamma \forall \alpha \forall \beta (\Gamma \cup \{\alpha, \neg \alpha\} \vdash \beta)$.
(II) The *Law of Non-contradiction* (LNC): $\vdash \neg(\alpha \wedge \neg \alpha)$.
A logic $\mathcal{L}$ is said to be *explosive* if ECQ holds.

Łukasiewicz's 3-valued logic [14] and Kleene's 3-valued logic [30] are examples of inconsistency-tolerant systems that violate LNC. However, ECQ holds in both – so these are not referred to as paraconsistent systems. Note that in classical logic, ECQ and LNC are equivalent. The above two logics show that the principles may not be comparable in the context of inconsistency-tolerant logics. In fact, there are also logics, such as the Logic of Paradox [38], where ECQ fails but LNC holds.

The survey of approaches giving paraconsistent logics (i.e. where ECQ fails) is summarised by the diagram given in Fig. 1. In Sect. 2, we briefly outline the motivations behind each approach (including the rough set theoretic one), and cite some logics that are obtained by following the approach. In Sect. 3, new paraconsistent systems arising from the rough set theoretic approach are presented. Section 4 concludes the article.

2   Different Approaches to Paraconsistency

Let us now present the different approaches, mentioning how ECQ is violated in each case yielding paraconsistent logics. Figure 1 gives a brief representation of the complete survey. The rectangular nodes, other than the topmost node, represent different techniques of obtaining paraconsistent logics violating ECQ.

2.1   The 3-Valued Approach

Let the three truth-values be denoted as t, m, f, where t, f correspond to the classical truth values *true* and *false* respectively, and m represents the *middle-value*. The question of whether the third value should be regarded as "close" to the notion of truth or not is then addressed – in the former case, it is termed as a *designated* value. Paraconsistency may result, when m and t are both designated values. Let us see how. We assume that the logical connective of negation ($\neg$) in the language behaves classically on $\{t, f\}$, i.e., $\neg t = f$, $\neg f = t$, while $\neg m = m$. So for any propositional variable p in the language, if p takes the value m, $\neg p$ also evaluates to m. The semantic consequence relation $\models$ is defined such that whenever the premises take designated values under any valuation, the conclusion must also get a designated value under that valuation. It is then clear

---

[1] The consequence relation $\vdash$ is used as a meta-linguistic symbol. In the context of a particular logic this can be a relation obtained semantically or syntactically; moreover sometimes the same notion of consequence is represented by an operator from $\mathcal{P}(FOR)$ to $\mathcal{P}(FOR)$. The use of notation would be clear from the contexts.

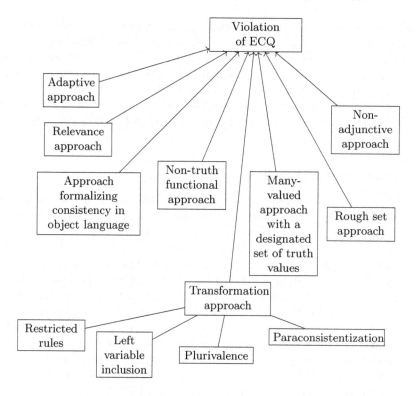

Fig. 1. Paraconsistent logics: Different approaches

that for any distinct propositional variables p, q and a valuation that gives p the value m and q the value f, $\{p, \neg p\} \nvDash q$. An example for this approach is provided by the Logic of Paradox (LP) mentioned earlier in Sect. 1. This 3-valued logic was proposed to deal with logical paradoxes; in this logic the truth-value m represents *both true and false*.

2.2 Non-adjunctive Approach

A non-adjunctive system is one that does not validate the *law of adjunction*, i.e. there exist formulas α, β such that $\{\alpha, \beta\} \nvdash \alpha \wedge \beta$, where $\wedge$ is the logical connective of *conjunction*. In particular, if $\beta := \neg \alpha$ then $\{\alpha, \neg \alpha\} \nvdash \alpha \wedge \neg \alpha$, which ensures that such a system violates ECQ. The first known non-adjunctive paraconsistent logic is the Discussive (or Discursive) logic (J) proposed by Jaśkowski [23,29]. The motivation of not adopting the law of adjunction in J, arose from the perspective of a discussion where different participants may offer some contradictory information, idea or opinion. According to an individual participant, a statement in a discussion may be true and consistent, but may be discordant with the opinion of other participants. Let us imagine that each participant in the discussion is represented by a single world and the opinion of that participant is considered to be true in that very world; i.e. each such set of sentences is true in at least one world. This sense of world based semantics might have been the reason for which Jaśkowski considered the normal modal logic S5 to model the semantics for J. The language of J is that of S5 (see [23]). The semantic consequence relation ($\vDash_J$) is defined for J as follows: $\Gamma \vDash_J \alpha$ if and only if $\Diamond\Gamma \vDash_{S5} \Diamond\alpha$, where "$\Diamond$" is the possibility modal operator and $\Diamond\Gamma := \{\Diamond\gamma : \gamma \in \Gamma\}$. For any propositional variable p, it is easy to show in S5 that $\{\Diamond p, \Diamond\neg p\} \nvDash_{S5} \Diamond(p \wedge \neg p)$ and thus $\{p, \neg p\} \nvDash_J p \wedge \neg p$.

2.3   Non-truth Functional Approach

Classically, the truth value of a compound formula, is wholly determined by the truth value(s) of the propositional variables that occur in the formula, and the logical connectives are referred to as being *truth functional*. In particular, if negation is not truth functional, the possibility of violation of ECQ arises – as we observe in the following examples. Consider Da-costa's C_n-systems [22] and Diderik Batens' system PI [8]. In both these logics, negation is taken to be weaker than that in classical logic. One direction of classical negation holds: if a statement is false then its negation is true. However, it is possible that a statement and its negation are both true. But then $\{p, \neg p\} \nvDash q$ for distinct propositional variables p, q. A non-truth functional semantics for negation is also considered by Tuziak [45] in case of the paraconsistent extensions of the positive fragment CPL^+ of CPL, the classical propositional logic. Other than one extension that is explosive, one can argue that ECQ is violated in the rest, in a manner similar to the cases of C_n and PI.

2.4   Transformation Approach

The transformation approach gives us different ways to obtain a paraconsistent logic from an explosive logic. The basic principle is to define a new logic by restricting the inferences of the mother logic in such a manner that the new logic becomes non-explosive. We place under this approach, the methods defined in [7,15,25,39]. For other methods of variable inclusion readers are referred to [16].

In [25] the authors have presented a method called *paraconsistentization*, where the consequence relation of the transformed system is based on those derivations of the parent consequence relation that are obtained from a *consistent* subset of the premise set. Let us describe this formally.

Given a logic $\mathcal{L} := (FOR, \vdash)$ and $\Gamma \subseteq FOR$, the set $C(\Gamma) := \{\alpha : \Gamma \vdash \alpha\}$, is said to be the set of all consequences of Γ. Γ is called C - *consistent* if $C(\Gamma) \neq FOR$, otherwise it is called C - inconsistent. The *paraconsistentizaion consequence relation* $\vdash^{\mathbb{P}} \subseteq \mathcal{P}(FOR) \times FOR$ is defined as $\Gamma \vdash^{\mathbb{P}} \alpha$, if and only if there is some C-consistent subset Γ' of Γ such that $\Gamma' \vdash \alpha$. In [25], a set of conditions on C is imposed so that the corresponding $(FOR, \vdash^{\mathbb{P}})$ becomes paraconsistent.

As examples, one can show that the consequence relations of Łukasiewicz's 3-valued logic (L_3) [14], Gödel's 3-valued logic (G_3) [41] and Kleene's 3-valued logic (K_3) [30] (in all of which ECQ holds) satisfy the said conditions. Hence the paraconsistentization of these logics are indeed paraconsistent.

Two other methods for converting a system satisfying ECQ into one that violates the principle, have been presented in [7, 15]. These are termed as the *left variable inclusion* and *restricted rules* methods.

A new logic $\mathcal{L}^l := (FOR, \vdash^l)$, based on $\mathcal{L}$, is defined as follows: $\Gamma \vdash^l \alpha$ if and only if there is a $\Gamma' \subseteq \Gamma$ such that $Var(\Gamma') \subseteq Var(\alpha)$ and $\Gamma' \vdash \alpha$, where $Var(\gamma)$ is the set of all propositional variable(s) contained in γ, and $Var(\Gamma) := \cup\{Var(\gamma) : \gamma \in \Gamma\}$. The new logic $\mathcal{L}^l$ is known as the *left variable inclusion companion* of $\mathcal{L}$.

In the restricted rules method, one Hilbert-style logic is obtained from another by imposing restrictions on logical rules. Suppose $\mathcal{L}$ is a Hilbert-style logic with $A(\subseteq FOR)$ as the set of axioms and $R_{\mathcal{L}}(\subseteq \mathcal{P}(FOR) \times FOR)$ as the set of rules of inference. Based on $\mathcal{L}$, a new Hilbert-style logic $\mathcal{L}^{re} := (FOR, \vdash^{re})$ is defined with the same set A of axioms and set of rules $R_{\mathcal{L}^{re}} := \{\frac{\Gamma}{\alpha} \in R_{\mathcal{L}} \mid Var(\Gamma) \subseteq Var(\alpha)\}$. $\mathcal{L}^{re}$ is known as the *restricted rules companion* of $\mathcal{L}$.

In [7], it is shown that if $\mathcal{L}$ is a Hilbert-style logic then $\mathcal{L}^l$ is stronger than[2] $\mathcal{L}^{re}$ in the sense that for all Γ and α, $\Gamma \vdash^{re} \alpha$ implies $\Gamma \vdash^l \alpha$. Moreover, the authors showed that if the mother logic contains a formula α such that α is not a theorem and there is some propositional variable p which is not included in α, then both the logics, obtained by applying left variable inclusion and restricted rules, are paraconsistent. The same work also presents a sufficient condition specifying when $\mathcal{L}^l$ and $\mathcal{L}^{re}$, obtained from the same mother logic, would be identical. CPL serves as an example for which the left variable inclusion and restricted rules companions are identical, and paraconsistent.

Plurivalence [39] is another way of transforming an explosive logic to a non-explosive one. In this method a mother logic (two-valued or many-valued) with a univalent semantic consequence relation $\models_u$ is converted to a logic by generating a plurivalent semantic consequence relation $\models_p$. However, in [39], the author mainly focused on converting a many-valued logic with a univalent consequence relation to its "plurivalent" counterpart. A univalent interpretation $(M, \mathcal{V})$ is considered based on a structure $M := (V, D, \delta)$, consisting of a non-empty set V of truth values, a designated set $D (\subseteq V)$ of truth values, a set δ of truth functions for the logical connectives, and a valuation $\mathcal{V}$, assigning a unique truth value to each propositional variable. $\mathcal{V}$ is extended over the set of all formulas in the usual recursive manner, and $\models_u^M$ is also defined in the usual manner: $\Gamma \models_u^M \alpha$ if and only if for all interpretations $(M, \mathcal{V})$, if $\mathcal{V}(\gamma) \in D$ for all $\gamma \in \Gamma$ then $\mathcal{V}(\alpha) \in D$. Now given $(M, \mathcal{V})$, the respective plurivalent interpretation $(M, \mathcal{R})$ is obtained by replacing the valuation $\mathcal{V}$ by a relation $\mathcal{R}$ which relates every propositional variable to a subset of V, and the relation is suitably extended [39] over the whole set of formulas. Further, it is defined that $\mathcal{R}$ *designates* α if and only if there is some $v \in D$ such that $\alpha \mathcal{R} v$ holds. The plurivalent semantic consequence $\models_p^M$ is defined as follows: $\Gamma \models_p^M \alpha$ if and only if for all interpretations $(M, \mathcal{R})$, if $\mathcal{R}$ designates every member of Γ then $\mathcal{R}$ designates α. It is observed that for

---

[2] This notion is more formally introduced in Sect. 3.1.

any plurivalent semantics which contains the truth-values t (true) and f (false) such that negation ($\neg$) behaves classically for t, f, ECQ is violated by $\models_p^M$. As an example, the plurivalent counterpart of CPL is in fact, LP (ref. Sect. 2.1).

2.5 Logics of Formal Inconsistency

The key feature of the logics of formal inconsistency (LFI) is to bring in a unary 'consistency' operator $\circ$ in the object language so that some formulas can be segregated from those which are tolerant to inconsistency. The first reported such family of logics was proposed by Carnielli and Marcos in [18,20]. A finite set of compound formula(s) involving only α, denoted as $\bigcirc(\alpha)$, is introduced. With respect to this set a contradictory premise may behave classically. If $\bigcirc(\alpha)$ is a singleton, it is simply denoted as $\circ\alpha$.

Further, a notion of *gentle explosion* is defined. A set Γ of formulas is said to be *gently explosive* if (i) there is α such that $\bigcirc(\alpha) \cup \{\alpha\}$ and $\bigcirc(\alpha) \cup \{\neg\alpha\}$ are both non-trivial[3], and (ii) for all α, β, $\Gamma \cup \bigcirc(\alpha) \cup \{\alpha, \neg\alpha\} \models \beta$. A logic $\mathcal{L}$ is said to satisfy *Gentle Principle of Explosion* (gPPS) if all $\Gamma (\subseteq FOR)$ are gently explosive. Then an LFI is defined as follows.

Definition 1. *A logic $\mathcal{L}$ is said to be LFI if ECQ does not hold but (gPPS) holds.*

Thus, the definition of LFI allows explosion in a controlled way; it allows $\{\alpha, \neg\alpha\}$ to be explosive in the presence of $\bigcirc\alpha$, where neither α nor $\neg\alpha$ alone leads to triviality in the presence of $\bigcirc\alpha$. Let us note that Da Costa's hierarchy of C_n-systems, discussed in Sect. 2.3, are all examples of LFI. As mentioned in [18], Discussive logic (J) (ref. Sect. 2.2) is also an LFI, in which $\bigcirc(\alpha) = \circ\alpha :=$ $\neg(\alpha \rightarrow_d \neg(\alpha \vee \neg\alpha)) \rightarrow_d (\neg\alpha \rightarrow_d \neg(\alpha \vee \neg\alpha))$.

A wide variety of paraconsistent logics fall under the general definition of LFIs. For instance, a fundamental LFI logic (mbC) is proposed as an extension of CPL^+ by adding $\circ$ in the language and axioms $\circ\alpha \rightarrow (\alpha \rightarrow (\neg\alpha \rightarrow \beta))$ and $\alpha \vee \neg\alpha$, where the former one presents a form of *gentle explosion*.

Moreover, mbC also shares properties of the non-truth functional approach discussed in Sect. 2.3. A valuation $\mathcal{V}$ for mbC is a function that assigns the values t and f to all formulas where the semantics for negation ($\neg$) and consistency operator ($\circ$) is given below.

(1) $\mathcal{V}(\neg\alpha) = f$ implies $\mathcal{V}(\alpha) = t$.
(2) $\mathcal{V}(\circ\alpha) = t$ implies $\mathcal{V}(\alpha) = f$ or $\mathcal{V}(\neg\alpha) = f$.

Clearly, both $\neg$ and $\circ$ are non-truth functional. Thus, based on the usual definition of semantic consequence it can be shown that $\{\circ p, p\} \nvDash_{mbC} q$ and $\{\circ p, \neg p\} \nvDash_{mbC} q$ for some distinct p, q; however, as $\circ p, p, \neg p$ cannot be true together, $\{\circ p, p, \neg p\} \models_{mbC} \beta$ for all β. So mbC satisfies (gPPS). Moreover, from

---

[3] A subset $\Gamma \subseteq FOR$ is said to be *trivial* if $\Gamma \vdash \alpha$ for every formula α, otherwise it is called *non-trivial*.

Sect. 2.3 it is straightforward that mbC violates ECQ. Hence mbC is an LFI. In a step by step manner adding some further axiom schemes and imposing some further conditions on ∘, Carnielli and Marcos have developed a series of interesting LFIs. For more details readers are referred to [19,20].

2.6 Adaptive Approach

The adaptive approach aims to identify the appearance of inconsistency within a derivation and develop strategies to adapt changes in the derivation so that it violates ECQ. Diderik Batens is the forerunner in developing different strategies for adaptive reasoning [9], among which the strategies dealing with inconsistencies lead towards paraconsistent systems. An adaptive logic consists of three components, namely (i) a lower limit logic (LLL) (ii) a set of abnormalities, and (iii) an adaptive strategy. LLL consists of a number of inferential rules that are accepted as a basis for a particular adaptive system. The set of abnormalities consists of some formulas that help to decide when an already derived formula need not be a conclusion. An adaptive strategy describes how to handle the applications of inference rules based on the set of abnormalities. The LLL and the collection of abnormalities jointly define the strategy to be adapted in the derivation chain, and on the other hand the set of abnormalities and the strategies jointly define which formulas in the derivation chain are to be marked for throwing out of consideration. The upper limit logic (ULL) can be obtained by extending LLL with the condition that no abnormality is logically derivable. ULL includes the inferential rules (and/or axioms) of LLL as well as some *supplementary rules* (and/or axioms) that can be used during the reasoning process in the absence of abnormalities. Thus, based on the set of inference rules and axioms added to LLL, there can be different adaptive logics lying between LLL and ULL.

Let us consider an example of an adaptive logic $\mathbf{CLuN}^m$ which is obtained from the lower limit logic $\mathbf{CLuN}^4$ [11,12] where abnormal formulas are of the form $\alpha \wedge \neg\alpha$ and classical logic CPL is used as an ULL.

Let $\Gamma := \{p, q, \neg p, \neg p \vee r, \neg q \vee s\}$ be a premise set, where p, q, r, s are all propositional variables. In the logic $\mathbf{CLuN}$, r cannot be deduced from Γ (see [10]). Below, let us present how the adaptive notion of derivation accommodates something additional in the logic $\mathbf{CLuN}^m$.

---

[4] The logic $\mathbf{CLuN}$, developed by Diderik Batens, is a predicative paraconsistent logic. The propositional part of $\mathbf{CLuN}$ is obtained by adding the axiom-schema $(\alpha \rightarrow \neg\alpha) \rightarrow \neg\alpha$ to CPL^+. The adaptive logic $\mathbf{CLuN}^m$ is obtained from $\mathbf{CLuN}$ based on a strategy, called *minimal abnormality.*.

1   p premise
2   q premise
3    $\neg p$ premise
4    $\neg p \vee r$ premise
5    $\neg q \vee s$ premise
6   r from 1 and 4 and the consistent behaviour of p $\sqrt{}$
7   s from 2 and 5 and the consistent behaviour of q
8    $p \wedge \neg p$ from 1 and 3

The first five steps of the derivation chain are obtained directly from Γ. At step 6, r is derived from steps 1 and 4 under the condition p *behaves consistently* as the formula $p \wedge \neg p$ has not been derived till now in the derivation chain. Similarly at step 7, s is derived under the condition q *behaves consistently*. However at stage 8, $p \wedge \neg p$ is derived from the steps 1 and 3. That is, the inconsistent behaviour of p becomes prominent at stage 8 and thus r can no longer be guaranteed as a deduction under $\mathbf{CLuN}^m$. So, after the inconsistent behaviour of p becomes apparent at stage 8, the line 6 is marked with ($\sqrt{}$) to indicate that a previously derived formula is no longer possible to derive. Hence ECQ fails to hold in $\mathbf{CLuN}^m$. In general, we have $C_{\mathbf{CLuN}}(\Gamma) \subset C_{\mathbf{CLuN}^m}(\Gamma) \subset C_{CPL}(\Gamma)$, where C denotes a consequence operator.

2.7 Relevance Approach

T.J. Smiley, in 1959, attempted to impose a notion of relevance between the premises and conclusion of an inference relation by defining the logical entailment [43] as follows: $\{\alpha_1, \alpha_2, ..., \alpha_n\} \vdash \beta$ if and only if $(\alpha_1 \wedge \alpha_2 \wedge ... \wedge \alpha_n) \rightarrow \beta$ is a classical tautology and neither $\neg(\alpha_1 \wedge \alpha_2 \wedge ... \wedge \alpha_n)$ nor β is a tautology. Thus, $\{\alpha, \neg\alpha\} \vdash \beta$ fails to hold as $\neg(\alpha \wedge \neg\alpha)$ is a classical tautology. Contrary to classical context here the semantic consequence is not defined based on the material implication which allows anything to follow from a false formula. The mentioned notion of entailment is not reflexive as clearly $\{\alpha \wedge \neg\alpha\} \nvdash \alpha \wedge \neg\alpha$.

The relevant logics of Anderson and Belnap (1975), known as *first-degree entailment* (*FDE*), contains formulas of the form $\alpha \rightarrow \beta$ called first-degree entailments, where α, β do not contain any occurrence of the connective $\rightarrow$. Belnap and Dunn around 1977 [13] proposed a four-valued semantics for *FDE* containing the truth values $N = \emptyset$, $T = \{t\}$, $F = \{f\}$, $B = \{t, f\}$ where T, F behave classically and $\neg B = B$, $\neg N = N$. Considering $D := \{T, B\}$ as the designated set of truth values, the semantic consequence for FDE is defined as: $\Gamma \vDash_{fde} \alpha$ if and only if for all valuation $\mathcal{V}$, if $\mathcal{V}(\alpha) \in D$ for all $\alpha \in \Gamma$ then $\mathcal{V}(\beta) \in D$ [33,42]. Clearly $\{p, \neg p\} \nvDash_{fde} q$, where p, q are distinct propositional variables.

2.8 Rough Set Approach

Rough sets were introduced by Pawlak in 1982 [35]. In literature, one finds different approaches linking rough sets and paraconsistency. For instance, in

[47,48], a notion of "paraconsistent set" is developed. Contrary to a set over a universe U, a paraconsistent set incorporates a four-valued membership function which also can be represented as a set over $U \cup \neg U$ where $\neg U = \{\neg x : x \in U\}$ and for any $X \subseteq U \cup \neg U$, $\neg x \in X$ denotes that there is an evidence that x is not in X. Then, a notion of paraconsistent rough set is defined by bringing in rough approximation space in the context of paraconsistent set. Consequently, "whether an object is an instance of a certain concept" is approximated. Such approximations are often relevant in abstracting rules from a database.

Similar attempts can be found in the area of paraconsistent rough description logics [46], where typically the aim has been to develop a language which can represent assertions like "an object is an instance of a concept", "two objects are related by a relation" etc., and design a reasoning strategy based on such assertions. As databases may contain inconsistent information and the rough set based approximations are defined based on a database, interpreting or evaluating the assertions generated from a database using the tools of paraconsistent set and paraconsistent rough set come naturally. However, in none of the above mentioned works the intention has been to develop a proof theory for deductive reasoning based on such a language.

In [32,44,49], some logics are developed based on the notion of paraconsistent rough sets. However, the consequence relation of most of them are explosive. Among the logics presented in [44], DDT is a first-order extension of Belnap's logic involving two negations (weak and strong) and it can be shown that the consequence relation is non-explosive only with respect to the weak negation.

In contrast to the above-mentioned rough set approaches to paraconsistency, in [32] the authors developed a proof theory for the decision logic of rough sets based on a four-valued tableau calculi. The original work of decision logic of rough sets was by Pawlak [37], addressing consistent decision tables. This idea is extended in [32] by introducing a variable precision rough set model and respectively allowing four values, namely *true, false, uncertain, inconsistent*, to describe whether an object is an instance of a concept. In the tableau calculi, for four-valued logic two negations are introduced. With respect to the strong negation ($\sim$) the rule for explosion, denoted by $\{Tp, T \sim p\}$, is dropped and with respect to the weak negation ($\neg$) violation of LNC is shown. Although, in the work the authors presented a logic for decision systems which is paraconsistent in nature, the language and its interpretation are developed keeping decision systems in mind, and the logical inference is not defined for deriving a formula from a non-empty set of formulae – as is the focus of the present paper.

Relations between 3-valued systems (algebras, logics) and rough sets have been discussed in several papers, e.g. [3,4,21,28,34]. The 3-valued approach adopted in [3] in particular, yields a few deductive systems of reasoning including a paraconsistent logic $\mathcal{L}^I$ which we briefly present here. The language of $\mathcal{L}^I$ has a countable set of propositional variables $P := \{p, q, r, ...\}$ and connectives $\neg, \rightarrow$. The formulas are defined using the scheme: $p \mid \neg \alpha \mid \alpha \rightarrow \beta$. Let FOR^I denote the set of all formulas. A sequent calculus for $\mathcal{L}^I$ is formulated that is shown to be sound and complete with respect to a semantics based on a non-deterministic matrix (Nmatrix). Let us describe the semantics. The Nmatrix corresponding to $\mathcal{L}^I$ is defined as $\mathcal{M}^I := (\mathcal{T}^I, \mathcal{D}^I, \mathcal{O}^I)$ where,

- $\mathcal{T}^I := \{t(true), f(false), u(unknown)\}$ is a set of truth values,
- $\mathcal{D}^I := \{t, u\}$ is the set of designated values,
- $\mathcal{O}^I := \{\neg^{\mathcal{M}}, \rightarrow^{\mathcal{M}}\}$, with $\neg^{\mathcal{M}}$ and $\rightarrow^{\mathcal{M}}$ as interpretations of $\neg$ and $\rightarrow$ respectively given by the following tables:

$\neg^{\mathcal{M}}$	
f	t
u	u
t	f

$\rightarrow^{\mathcal{M}}$	f	u	t
f	t	t	t
u	u	$\{u,t\}$	t
t	f	u	t

A valuation v in an Nmatrix is a function $v : FOR^I \rightarrow \mathcal{T}^I$ that satisfies the following condition: $v(*(\alpha_1, ..., \alpha_n)) \in *^{\mathcal{M}}(v(\alpha_1), ..., v(\alpha_n))$ for any n-ary connective $*$ in $\mathcal{L}^I$.

- A formula α is satisfied by a valuation v, in symbols $v \vDash \alpha$, if $v(\alpha) \in \mathcal{D}^I$.
- A sequent $\Gamma \Rightarrow \Delta$ is satisfied by a valuation v, written as $v \vDash \Gamma \Rightarrow \Delta$, if and only if either v does not satisfy some formula in Γ or v satisfies some formula in Δ.
- A sequent $\Gamma \Rightarrow \Delta$ is valid ($\vDash \Gamma \Rightarrow \Delta$), if it is satisfied by all valuations v.
- The consequence relation ($\vdash_{\mathcal{M}}$) on FOR^I is defined as: for any $T, S \subseteq FOR^I$, $T \vdash_{\mathcal{M}} S$ if and only if there exist finite sets $\Gamma \subseteq T$ and $\Delta \subseteq S$ such that the sequent $\Gamma \Rightarrow \Delta$ is valid.

Soundness and completeness of $\mathcal{L}^I$ with respect to the consequence relation $\vdash_{\mathcal{M}}$ is established. Now it is clear from the definition of $\vdash_{\mathcal{M}}$ that $\mathcal{L}^I$ violates ECQ. Indeed, if we choose two distinct propositional variables p, q and a valuation v such that $v(p) := u$ and $v(q) := f$ then $\{p, \neg p\} \nvdash_{\mathcal{M}} q$.

It should also be mentioned here that in [3], two kinds of *determinizations* of $\mathcal{L}^I$ have been derived: "Łukasiewicz determinization" and "Kleene determinization". It is shown that the Łukasiewicz determinization is equivalent to the paraconsistent logic J_3, while the Kleene determinization is equivalent to the paraconsistent logic Pac [2]. Therefore $\mathcal{L}^I$ may be looked upon as a "common denominator" of J_3 and Pac.

It has been well-established over the years that rough set theory has many intricacies which cannot be captured by any single approach. We now turn to the notions of *rough truth* proposed in [36] and *rough consequence* introduced in [6], and paraconsistent deductive systems that have been obtained through work based on these notions.

A rough set may be viewed as a triple (X, R, A), where $A \subseteq X$ and (X, R) is a Pawlakian approximation space, i.e. X is a non-empty set and R an equivalence relation on it. It was noticed early on that an $S5$ formula α may be interpreted in an $S5$ model (X, R, v) as a rough set (X, R, A), with A being the set of possible worlds where α is true under the valuation v, i.e. $A = v(\alpha)$. Moreover, $v(L\alpha) = \underline{v(\alpha)}$, the lower approximation of $v(\alpha)$, and $v(M\alpha) = \overline{v(\alpha)}$, the upper approximation of $v(\alpha)$, where L, M denote the necessity and possibility operators respectively. One of the basic notions in rough set theory is that of rough equality: sets A and B are *roughly equal* if they have the same lower and upper approximations. In $S5$, one can see that rough equality of α, β would be represented by the formula $\alpha \approx \beta := (L\alpha \leftrightarrow L\beta) \wedge (M\alpha \leftrightarrow M\beta)$. This is because in any $S5$ model (X, R, v) with $A = v(\alpha)$ and $B = v(\beta)$, $\alpha \approx \beta$ would be true under v if and only if A, B are roughly equal. *Rough consequence* was based on the idea of *rough modus ponens*, which, loosely put, stipulates that for any $S5$ formulas α, α', β, if α, $\alpha \approx \alpha'$ and $\alpha' \to \beta$ are all "derivable" from a set Γ of $S5$ formulas, then β should also follow from Γ. Formally, the rough consequence relation $\mathrel{|\!\sim}$ was defined in the backdrop of $S5$:

$\Gamma \mathrel{|\!\sim} \alpha$, in case α is a member of Γ, or is an $S5$ theorem, or is derived through the rule(RMP): $\dfrac{\Gamma \mathrel{|\!\sim} \alpha \quad \Gamma \mathrel{|\!\sim} \beta \to \gamma \quad \Gamma \vdash \alpha \approx \beta}{\Gamma \mathrel{|\!\sim} \gamma}$,

where $\vdash$ denotes the $S5$ consequence relation. One may notice that whenever $\alpha = \beta$, the rule (RMP) becomes standard modus ponens (MP). A study was made in [5,6,17] by weakening RMP in different ways (again in the backdrop of $S5$), and considering logics based on the resulting rules. This included a rule based on the formula $\alpha \rightsquigarrow \beta := (L\alpha \to L\beta) \wedge (M\alpha \to M\beta)$, representing *rough inclusion*. Following is a summary of rules considered in these works:

$(MP_\approx)$: $\dfrac{\Gamma \mathrel{|\!\sim} \alpha \quad \Gamma \mathrel{|\!\sim} \beta \to \gamma \quad \vdash \alpha \approx \beta}{\Gamma \mathrel{|\!\sim} \gamma}$ and $(MP_\rightsquigarrow)$: $\dfrac{\Gamma \mathrel{|\!\sim} \alpha \quad \Gamma \mathrel{|\!\sim} \beta \to \gamma \quad \vdash \alpha \rightsquigarrow \beta}{\Gamma \mathrel{|\!\sim} \gamma}$

(RMP_1): $\dfrac{\Gamma \mathrel{|\!\sim} \alpha \quad \vdash \beta \to \gamma \quad \vdash M\alpha \to M\beta}{\Gamma \mathrel{|\!\sim} \gamma}$ and (RMP_2): $\dfrac{\vdash \alpha \quad \Gamma \mathrel{|\!\sim} \beta \to \gamma \quad \vdash L\alpha \to L\beta}{\Gamma \mathrel{|\!\sim} \gamma}$

(RMP_1) and (RMP_2) were used to define the logic Lr. It was seen later that (RMP_2), in fact, follows from (RMP_1), and (RMP_1) is equivalent to the rule:

(R_1): $\dfrac{\Gamma \mathrel{|\!\sim} \alpha \quad \vdash M\alpha \to M\beta}{\Gamma \mathrel{|\!\sim} \beta}$

The logic Lr was extended in [5] by adding another rule:

(R_2): $\dfrac{\Gamma \mathrel{|\!\sim} M\alpha \quad \Gamma \mathrel{|\!\sim} M\beta}{\Gamma \mathrel{|\!\sim} M\alpha \wedge M\beta}$

This new logic is denoted by $\mathcal{L}_\mathcal{R}$. In [5] it was shown that Jaśkowski's discussive logic J (discussed in Sect. 2.2) is equivalent to $\mathcal{L}_\mathcal{R}$.

The following are from [17]:

MP_1: $\dfrac{\Gamma \mathrel{|\!\sim} \alpha \quad \Gamma \mathrel{|\!\sim} \beta \to \gamma \quad \vdash M\alpha \to M\beta}{\Gamma \mathrel{|\!\sim} \gamma}$ $\qquad MP_2$: $\dfrac{\Gamma \mathrel{|\!\sim} \alpha \quad \Gamma \mathrel{|\!\sim} \beta \to \gamma \quad \vdash L\alpha \to L\beta}{\Gamma \mathrel{|\!\sim} \gamma}$

MP_3: $\dfrac{\Gamma \mathrel{|\!\sim} \alpha \quad \Gamma \mathrel{|\!\sim} \beta \to \gamma \quad \vdash \alpha \to \beta}{\Gamma \mathrel{|\!\sim} \gamma}$ $\qquad MP_4$: $\dfrac{\Gamma \mathrel{|\!\sim} \alpha \quad \Gamma \mathrel{|\!\sim} \beta \to \gamma \quad \vdash M\alpha \to \beta}{\Gamma \mathrel{|\!\sim} \gamma}$

MP_5: $\dfrac{\Gamma \mathrel{|\!\sim} \alpha \quad \Gamma \mathrel{|\!\sim} \beta \to \gamma \quad \vdash L\alpha \to M\beta}{\Gamma \mathrel{|\!\sim} \gamma}$ $\qquad MP_0$: $\dfrac{\Gamma \mathrel{|\!\sim} \alpha \quad \Gamma \mathrel{|\!\sim} \alpha \to \gamma}{\Gamma \mathrel{|\!\sim} \gamma}$

For each of the rules MP_i, $i = 0, ..., 5, \rightsquigarrow, \approx$, a rough consequence relation $\mathrel{|\!\sim}_i$ is defined as mentioned earlier. The corresponding logics are denoted as Lr_i. In [5], it was shown that ECQ fails to hold in both Lr and $\mathcal{L}_\mathcal{R}$. ECQ also fails for the system Lr_4, as observed in [17]. In all the other Lr_i systems, ECQ holds. However, we shall see in the next section that one can define several other rough consequence relations that yield paraconsistent logics.

3   Some New Paraconsistent Logics

Let us consider a set $\Gamma \cup \{\alpha, \beta, \gamma\}$ of $S5$ formulas. We define new rules of inference by weakening MP_i, $i = 1, ..., 5, \rightsquigarrow, \approx$, as follows:

$$\frac{\Gamma \vdash\!\!\!\!\!/\; \alpha \quad \vdash \beta \rightarrow \gamma \quad \vdash \alpha \rightarrow_i \beta}{\Gamma \vdash\!\!\!\!\!/\; \gamma},$$

where $\alpha \rightarrow_i \beta$ denotes the $S5$-implication in MP_i (e.g. $\alpha \rightarrow_1 \beta := M\alpha \rightarrow M\beta$).

Proposition 1. *The new rule defined above is equivalent to* $\dfrac{\Gamma \vdash\!\!\!\!\!/\; \alpha \quad \vdash \alpha \rightarrow_i \gamma}{\Gamma \vdash\!\!\!\!\!/\; \gamma}$, *for* $i = 1, ..., 5, \rightsquigarrow$.

Proof. We give the proof for $i = 1$. The other cases have similar proofs. Suppose the rule $\dfrac{\Gamma \vdash\!\!\!\!\!/\; \alpha \quad \vdash \beta \rightarrow \gamma \quad \vdash M\alpha \rightarrow M\beta}{\Gamma \vdash\!\!\!\!\!/\; \gamma}$ holds. Let $\Gamma \vdash\!\!\!\!\!/\; \alpha$, $\vdash M\alpha \rightarrow M\gamma$. Since $\vdash \gamma \rightarrow \gamma$, $\Gamma \vdash\!\!\!\!\!/\; \gamma$ holds as well.

Conversely, let $\dfrac{\Gamma \vdash\!\!\!\!\!/\; \alpha \quad \vdash M\alpha \rightarrow M\gamma}{\Gamma \vdash\!\!\!\!\!/\; \gamma}$ hold, and $\Gamma \vdash\!\!\!\!\!/\; \alpha$, $\vdash \beta \rightarrow \gamma$, $\vdash M\alpha \rightarrow M\beta$. Now $\vdash \beta \rightarrow \gamma$ implies $\vdash M\beta \rightarrow M\gamma$. Therefore $\vdash M\alpha \rightarrow M\gamma$ holds as well. Then by assumption, $\Gamma \vdash\!\!\!\!\!/\; \gamma$. □

Notation 1. Henceforth, $\overline{MP_i}$ will denote the rule $\dfrac{\Gamma \vdash\!\!\!\!\!/\; \alpha \quad \vdash \alpha \rightarrow_i \gamma}{\Gamma \vdash\!\!\!\!\!/\; \gamma}$, for $i = 1, ..., 5, \rightsquigarrow, \approx$. In other words,

$\overline{MP_1}$: $\dfrac{\Gamma \vdash\!\!\!\!\!/\; \alpha \quad \vdash M\alpha \rightarrow M\gamma}{\Gamma \vdash\!\!\!\!\!/\; \gamma}$ $\overline{MP_2}$: $\dfrac{\Gamma \vdash\!\!\!\!\!/\; \alpha \quad \vdash L\alpha \rightarrow L\gamma}{\Gamma \vdash\!\!\!\!\!/\; \gamma}$

$\overline{MP_3}$: $\dfrac{\Gamma \vdash\!\!\!\!\!/\; \alpha \quad \vdash \alpha \rightarrow \gamma}{\Gamma \vdash\!\!\!\!\!/\; \gamma}$ $\overline{MP_4}$: $\dfrac{\Gamma \vdash\!\!\!\!\!/\; \alpha \quad \vdash M\alpha \rightarrow \gamma}{\Gamma \vdash\!\!\!\!\!/\; \gamma}$

$\overline{MP_5}$: $\dfrac{\Gamma \vdash\!\!\!\!\!/\; \alpha \quad \vdash L\alpha \rightarrow M\gamma}{\Gamma \vdash\!\!\!\!\!/\; \gamma}$ $\overline{MP_\rightsquigarrow}$: $\dfrac{\Gamma \vdash\!\!\!\!\!/\; \alpha \quad \vdash \alpha \rightsquigarrow \gamma}{\Gamma \vdash\!\!\!\!\!/\; \gamma}$

$\overline{MP_\approx}$: $\dfrac{\Gamma \vdash\!\!\!\!\!/\; \alpha \quad \vdash \alpha \approx \gamma}{\Gamma \vdash\!\!\!\!\!/\; \gamma}$.

For $i = \approx$, we have one direction of Proposition 1.

Observation 1. $\dfrac{\Gamma \vdash\!\!\!\!\!/\; \alpha \quad \vdash \beta \rightarrow \gamma \quad \vdash \alpha \approx \beta}{\Gamma \vdash\!\!\!\!\!/\; \gamma}$ *implies* $\overline{MP_\approx}$.

3.1   The Systems $\overline{Lr_i}$

A logic is defined for each rule $\overline{MP_i}$, $i = 1, ..., 5, \rightsquigarrow, \approx$, as mentioned in Sect. 2.8. Formally, the consequence relation $\vdash\!\!\!\!\!/_{\overline{i}}$ for $\overline{Lr_i}$ is given as follows.

Definition 2. $\Gamma \vdash\!\!\!\!\!/_{\overline{i}} \alpha$ *if and only if there is a sequence* $\alpha_1, \alpha_2, ..., \alpha_n (= \alpha)$ *such that for each* $\alpha_j (j = 1, ..., n)$, *one of the following holds.*
(i) $\alpha_j \in \Gamma$ ... (ov).
(ii) α_j *is an $S5$ theorem*   ... (S5).
(iii) α_j *is derived from some of* $\alpha_1, ..., \alpha_{i-1}$ *by* $\overline{MP_i}$.

Let us consider the $S5$-implications in $\overline{MP_i}$, $i = 1, 2, 3$, namely $M\alpha \rightarrow M\beta, L\alpha \rightarrow L\beta, \alpha \rightarrow \beta$. One may abbreviate these implications as $\Delta_i \alpha \rightarrow \Delta_i \beta$, where Δ_i is respectively M, L and "1" for $i = 1, 2, 3$. We then have the following.

Theorem 1. *For $i = 1, 2, 3$, $\Gamma \vdash_{\overline{i}} \alpha$, if and only if $\vdash \Delta_i \alpha$, or there is a β in Γ with $\vdash \Delta_i \beta \to \Delta_i \alpha$.*

Proof. By induction on the number n of steps of derivation of α from Γ.
Basis $n = 0$: $\vdash \alpha$, or $\alpha \in \Gamma$. If $\vdash \alpha$, for case $i = 3$, we are done. For the cases $i = 1, 2$, observe that $\vdash \alpha$ implies $\vdash \Delta_i \alpha$. So we are done in these cases as well. Now suppose $\alpha \in \Gamma$. Since $\vdash \Delta_i \alpha \to \Delta_i \alpha$, the result obtains.
Induction step: We sketch the proof. Suppose α is derived by $\overline{MP_i}$ from $\Gamma \vdash_{\overline{i}} \gamma$ and $\vdash \Delta_i \gamma \to \Delta_i \alpha$ for some γ. Then by induction, either $\vdash \Delta_i \gamma$ or $\vdash \Delta_i \beta \to \Delta_i \gamma$ for some $\beta \in \Gamma$. If $\vdash \Delta_i \gamma$ holds, then from $\vdash \Delta_i \gamma \to \Delta_i \alpha$, $\vdash \Delta_i \alpha$ holds as well. Otherwise $\vdash \Delta_i \beta \to \Delta_i \gamma$ for some $\beta \in \Gamma$, and then $\vdash \Delta_i \beta \to \Delta_i \alpha$ holds.

Conversely, suppose $\vdash \Delta_i \alpha$. Then $\Gamma \vdash_{\overline{i}} \Delta_i \alpha$, by (S5). So we are done for $i = 3$. If $i = 1, 2$, since $\vdash \Delta_i \Delta_i \alpha \to \Delta_i \alpha$, $\Gamma \vdash_{\overline{i}} \alpha$ holds by $\overline{MP_i}$. For the other case, suppose $\vdash \Delta_i \beta \to \Delta_i \alpha$ for some $\beta \in \Gamma$. Now $\Gamma \vdash_{\overline{i}} \beta$ holds by (ov) and then $\Gamma \vdash_{\overline{i}} \alpha$ holds by $\overline{MP_i}$. □

Theorem 2. $\overline{Lr_1}, \overline{Lr_2}, \overline{Lr_3}$ *are paraconsistent logics.*

Proof. Consider two distinct propositional variables p, q. We use Theorem 1.

(i) Since $\nvdash Mq$, $\nvdash Mp \to Mq$ and $\nvdash M(\neg p) \to Mq$, therefore $\{p, \neg p\} \nvdash_{\overline{1}} q$ by Theorem 1. Hence $\overline{Lr_1}$ is paraconsistent.

(ii) $\nvdash Lq$, $\nvdash Lp \to Lq$ and $\nvdash L(\neg p) \to Lq$ give $\{p, \neg p\} \nvdash_{\overline{2}} q$.

(iii) $\nvdash q$, $\nvdash p \to q$ and $\nvdash \neg p \to q$ give $\{p, \neg p\} \nvdash_{\overline{3}} q$. □

Notation 2. For logics L_1 and L_2 with the same language, $L_1 \preceq L_2$ denotes that L_2 is stronger than L_1, i.e. if α is derivable from Γ in L_1 then α is derivable from Γ in L_2, for all α, Γ.

Lemma 1. $\overline{Lr_4} \preceq \overline{Lr_3}$.

Proof. It suffices to show that $\overline{MP_4}$ is derivable in $\overline{Lr_3}$. Let $\Gamma \vdash \alpha$ and $\vdash M\alpha \to \gamma$. Since $\vdash M\alpha \to \gamma$ implies $\vdash \alpha \to \gamma$, then by $\overline{MP_3}$, $\Gamma \vdash \gamma$. □

Theorem 3. $\overline{Lr_4}$ *is paraconsistent.*

Proof. Follows from Lemma 1 and Theorem 2. □

Theorem 4.

(i) $\Gamma \vdash_{\overline{\sim}} \alpha$, *if and only if either $\vdash \alpha$ or there is a β in Γ with $\vdash (M\beta \to M\alpha) \wedge (L\beta \to L\alpha)$.*

(ii) $\Gamma \vdash_{\overline{\approx}} \alpha$, *if and only if either $\vdash \alpha$ or there is a β in Γ with $\vdash (M\beta \leftrightarrow M\alpha) \wedge (L\beta \leftrightarrow L\alpha)$.*

Proof. We give the proof of (i), which is by induction on the number n of steps of derivation of α from Γ. (ii) can be proved in a similar manner.
Basis: $n = 0$. Either $\vdash \alpha$ or $\alpha \in \Gamma$. If $\vdash \alpha$, there is nothing to show. In the other case, $\vdash (M\alpha \leftrightarrow M\alpha) \wedge (L\alpha \leftrightarrow L\alpha)$ gives the result.

In the induction step, we consider the possibility that α is derived by $\overline{MP_{\leadsto}}$ from $\Gamma \mathrel{\vdash_{\leadsto}} \gamma$ and $\vdash (M\gamma \to M\alpha) \wedge (L\gamma \to L\alpha)$, for some γ. Then by induction hypothesis, either $\vdash \gamma$ or $\vdash (M\beta \to M\gamma) \wedge (L\beta \to L\gamma)$, for some $\beta \in \Gamma$. If $\vdash \gamma$ holds then $\vdash L\gamma$ holds as well. Now $\vdash L\gamma \to L\alpha$ holds as $\vdash (M\gamma \to M\alpha) \wedge (L\gamma \to L\alpha)$ holds. Therefore $\vdash L\alpha$ as well as $\vdash \alpha$ holds. Otherwise $\vdash (M\beta \to M\gamma) \wedge (L\beta \to L\gamma)$, for some $\beta \in \Gamma$. Then $\vdash (M\beta \to M\alpha) \wedge (L\beta \to L\alpha)$ holds as, $\vdash (M\gamma \to M\alpha) \wedge (L\gamma \to L\alpha)$ holds.

Conversely, suppose $\vdash \alpha$ holds, then $\Gamma \mathrel{\vdash_{\leadsto}} \alpha$, by (S5). For the other case, suppose $\vdash (M\beta \to M\alpha) \wedge (L\beta \to L\alpha)$ holds for some $\beta \in \Gamma$. Now $\Gamma \mathrel{\vdash_{\leadsto}} \beta$ holds by (ov) and then $\Gamma \mathrel{\vdash_{\leadsto}} \alpha$ holds by $\overline{MP_{\leadsto}}$. $\square$

Theorem 5. $\overline{Lr_{\leadsto}}$ *and* $\overline{Lr_{\approx}}$ *are paraconsistent logics.*

Proof. Take two distinct propositional variables p, q. Since $\nvdash q$, $\nvdash (Mp \to Mq) \wedge (Lp \to Lq)$ and $\nvdash (M(\neg p)) \to Mq) \wedge (L(\neg p)) \to Lq)$, therefore, $\{p, \neg p\} \nvdash_{\leadsto} q$ as well as $\{p, \neg p\} \nvdash_{\approx} q$ by Theorem 4. Hence $\overline{Lr_{\leadsto}}$ and $\overline{Lr_{\approx}}$ are both paraconsistent logics. $\square$

3.2   Paraconsistency of $\overline{Lr_5}$

The paraconsistency of $\overline{Lr_5}$ is established using the fact that $\overline{Lr_3}$ is paraconsistent (Theorem 2).

Let $\mathcal{L}_{S5}$ and $\mathcal{L}_{CPL}$ denote the languages of $S5$ and CPL respectively. We define a translation $* : \mathcal{L}_{S5} \to \mathcal{L}_{CPL}$ as follows:

- $*(p) := p$, where p is a propositional variable.
- $*(\alpha \vee \beta) := *(\alpha) \vee *(\beta)$.
- $*(\alpha \wedge \beta) := *(\alpha) \wedge *(\beta)$.
- $*(\alpha \to \beta) := *(\alpha) \to *(\beta)$.
- $*(L\alpha) := *(\alpha)$.
- $*(M\alpha) := *(\alpha)$.

The image of a formula α under the above translation, $*(\alpha)$, is referred to as the *PC-transform* of α in [27]. A modal logic $Triv$ is defined in [27] such that $S5 \preceq Triv$. Moreover $Triv$ collapses into CPL, in the sense that every formula α in $Triv$ is equivalent to its *PC*-transform $*(\alpha)$. Then for all α, $\vdash_{Triv} \alpha$ if and only if $\vdash_{CPL} *(\alpha)$. Thus one obtains

Lemma 2. *For any S5 formula* α, $\vdash \alpha$ *implies* $\vdash_{CPL} *(\alpha)$.

We are then able to establish

Lemma 3. $\Gamma \mathrel{\vdash_5} \alpha$ *implies* $*(\Gamma) \mathrel{\vdash_3} *(\alpha)$, *where* $*(\Gamma) = \{*(\gamma) : \gamma \in \Gamma\}$.

Proof. By induction on the number n of steps of derivation of α from Γ.
Basis: $n = 0$. Either $\vdash \alpha$ or $\alpha \in \Gamma$. If $\vdash \alpha$ then $\vdash_{CPL} *(\alpha)$ by Lemma 2, and so $\vdash *(\alpha)$ holds. Hence $*(\Gamma) \mathrel{\vdash_3} *(\alpha)$, by ($S5$). Otherwise $\alpha \in \Gamma$, then $*(\alpha) \in *(\Gamma)$, therefore $*(\Gamma) \mathrel{\vdash_3} *(\alpha)$, by (ov).

For the induction step, suppose α is derived by $\overline{MP_5}$ from $\Gamma \vdash_{\overline{5}} \gamma$ and $\vdash L\gamma \rightarrow M\alpha$, for some γ. Then by the induction hypothesis, $*(\Gamma)\vdash_{\overline{3}} *(\gamma)$. Using Lemma 2, $\vdash_{CPL} *(\gamma) \rightarrow *(\alpha)$ holds, and therefore $\vdash *(\gamma) \rightarrow *(\alpha)$ holds as well. So by $\overline{MP_3}$, $*(\Gamma)\vdash_{\overline{3}} *(\alpha)$. □

Theorem 6. $\overline{Lr_5}$ *is a paraconsistent logic.*

Proof. Let p, q be two distinct propositional variables. Then $\{p, \neg p\} \not\vdash_{\overline{3}} q$, as shown in the proof of Theorem 2. Since $*(\{p, \neg p\}) = \{p, \neg p\}$ and $*(q) = q$, by Lemma 3, $\{p, \neg p\} \not\vdash_{\overline{5}} q$. □

4   Conclusions

In this paper, we have presented a survey on the major approaches that yield paraconsistent logics. One of the salient features of this survey is the inclusion of the *transformation approach* and *rough set approach*. We focus on how new paraconsistent logics may be obtained, based on the *rough set approach*. Several new logics are proposed by considering a weakened version of rough modus ponens rules.

In [17], there is a study of the relationship between the logics that are based on the different kinds of rough modus ponens rules. An immediate task would be to investigate relations of the logics obtained in this work with the ones studied in [17]. Appropriate semantics for the new logics also need to be explored, to give a complete idea of these paraconsistent systems.

References

1. Arruda, A.I.: A survey of paraconsistent logic. In: Arruda, A., Chuaqui, R., Da Costa, N. (eds.) Mathematical Logic in Latin America. Studies in Logic and the Foundations of Mathematics, vol. 99. pp. 1–41. Elsevier (1980)
2. Avron, A.: Natural 3-valued logics-characterization and proof theory. J. Symbolic Logic **56**(1), 276–294 (1991)
3. Avron, A., Konikowska, B.: Rough sets and 3-valued logics. Stud. Logica. **90**(1), 69–92 (2008)
4. Banerjee, M.: Rough sets and 3-valued Łukasiewicz logic. Fund. Inform. **31**(3–4), 213–220 (1997)
5. Banerjee, M.: Logic for rough truth. Fund. Inform. **71**(2–3), 139–151 (2006)
6. Banerjee, M., Chakraborty, M.K.: Rough consequence. Bull. Polish Acad. Sci. (Math.) **41**(4), 299–304 (1993)
7. Basu, S.S., Chakraborty, M.K.: Restricted rules of inference and paraconsistency. Log. J. IGPL **30**(3), 534–560 (2022)
8. Batens, D.: Paraconsistent extensional propositional logics. Logique Anal. **23**(90–91), 195–234 (1980)
9. Batens, D.: A universal logic approach to adaptive logics. Log. Univers. **1**(1), 221–242 (2007)

10. Batens, D.: Tutorial on inconsistency-adaptive logics. In: Beziau, J.-Y., Chakraborty, M., Dutta, S. (eds.) New Directions in Paraconsistent Logic. SPMS, vol. 152, pp. 3–38. Springer, New Delhi (2015). https://doi.org/10.1007/978-81-322-2719-9_1

11. Batens, D., De Clercq, K.: A rich paraconsistent extension of full positive logic. Logique Anal. **47**(185/188), 227–257 (2004)

12. Batens, D., De Clercq, K., Kurtonina, N.: Embedding and interpolation for some paralogics. The propositional case. Rep. Math. Log. **33**, 29–44 (1999)

13. Belnap Jr, N.D.: A useful four-valued logic. In: Dunn, J.M., Epstein, G. (eds.), Modern Uses of Multiple-valued Logic. Episteme, vol. 2, pp. 5–37. Springer, Dordrecht (1977). https://doi.org/10.1007/978-94-010-1161-7_2

14. Boicescu, V., Filipoiu, A., Georgescu, G., Rudeanu, S.: Łukasiewicz-Moisil Algebras. North Holland (1991)

15. Bonzio, S., Moraschini, T., Pra Baldi, M.: Logics of left variable inclusion and Płonka sums of matrices. Arch. Math. Logic **60**(1–2), 49–76 (2021)

16. Bonzio, S., Paoli, F., Pra Baldi, M.: Logics of variable inclusion. Springer, Cham (2022). https://doi.org/10.1007/978-3-031-04297-3

17. Bunder, M.W., Banerjee, M., Chakraborty, M.K.: Some rough consequence logics and their interrelations. In: Peters, J.F., Skowron, A. (eds.) Transactions on Rough Sets VIII. LNCS, vol. 5084, pp. 1–20. Springer, Heidelberg (2008). https://doi.org/10.1007/978-3-540-85064-9_1

18. Carnielli, W., Coniglio, M.E., Marcos, J.: Logics of formal inconsistency. In: Gabbay, D., Guenthner, F. (eds.) Handbook of Philosophical Logic, vol. 14, pp. 1–93. Springer, Dordrecht (2007). https://doi.org/10.1007/978-1-4020-6324-4_1

19. Carnielli, W., Coniglio, M.E.: Paraconsistent Logic: Consistency, Contradiction and Negation. LEUS, vol. 40. Springer, Cham (2016). https://doi.org/10.1007/978-3-319-33205-5

20. Carnielli, W.A., Marcos, J.: A taxonomy of C-systems. In: Carnielli, W.A., Marcelo, C., D'ottaviano, I.M.L. (eds.) Paraconsistency, vol. 228, pp. 1–94. Dekker, New York (2002)

21. Ciucci, D., Dubois, D.: Three-valued logics, uncertainty management and rough sets. In: Peters, J.F., Skowron, A. (eds.) Transactions on Rough Sets XVII. LNCS, vol. 8375, pp. 1–32. Springer, Heidelberg (2014). https://doi.org/10.1007/978-3-642-54756-0_1

22. da Costa, N.C.A., Alves, E.H.: A semantical analysis of the calculi C_n. Notre Dame J. Formal Logic **18**(4), 621–630 (1977)

23. da Costa, N.C.A., Doria, F.A.: On Jaśkowski's discussive logics. Stud. Logica. **54**(1), 33–60 (1995)

24. da Costa, N.C.A., Krause, D., Bueno, O.: Paraconsistent logics and paraconsistency. In: Jacquette, D. (ed.) Philosophy of Logic. Handbook of the Philosophy of Science, pp. 791–911. North-Holland, Amsterdam (2007)

25. de Souza, E.G., Costa-Leite, A., Dias, D.H.B.: On a paraconsistentization functor in the category of consequence structures. J. Appl. Non-Class. Log. **26**(3), 240–250 (2016)

26. Dutta, S., Chakraborty, M.K.: Consequence–inconsistency interrelation: in the framework of paraconsistent logics. In: Beziau, J.-Y., Chakraborty, M., Dutta, S. (eds.) New Directions in Paraconsistent Logic. SPMS, vol. 152, pp. 269–283. Springer, New Delhi (2015). https://doi.org/10.1007/978-81-322-2719-9_12

27. Hughes, G.E., Cresswell, M.J.: A New Introduction to Modal Logic. Routledge, Milton Park (1996)

28. Iturrioz, L.: Rough sets and three-valued structures. In: Orłowska, E. (ed.) Logic at Work: Essays Dedicated to the Memory of Helena Rasiowa. Studies in Fuzziness and Soft Computing, vol. 24, pp. 596–603. Physica, Heidelberg (1999)
29. Kotas, J.: The axiomatization of S. Jaśkowski's discussive system. Stud. Logica **33**, 195–200 (1974)
30. Malinowski, G.: Kleene logic and inference. Bull. Sect. Logic **43**(1/2), 43–52 (2014)
31. Middelburg, C.A.: A survey of paraconsistent logics. arXiv preprint arXiv:1103.4324 (2011)
32. Nakayama, Y., Akama, S., Murai, T.: Four-valued tableau calculi for decision logic of rough set. Procedia Comput. Sci. **126**, 383–392 (2018)
33. Omori, H., Wansing, H.: 40 years of FDE: an introductory overview. Stud. Logica. **105**(6), 1021–1049 (2017)
34. Pagliani, P.: Rough set theory and logic-algebraic structures. In: Orłowska, E. (ed.) Incomplete Information: Rough Set Analysis. Studies in Fuzziness and Soft Computing, vol. 13, pp. 109–190. Physica, Heidelberg (1998). https://doi.org/10.1007/978-3-7908-1888-8_6
35. Pawlak, Z.: Rough sets. Internat. J. Comput. Inform. Sci. **11**(5), 341–356 (1982)
36. Pawlak, Z.: Rough logic. Bull. Polish Acad. Sci. (Tech. Sci.) **35**(5–6), 253–258 (1987)
37. Pawlak, Z.: Rough Sets: Theoretical Aspects of Reasoning about Data. Kluwer Academic Publishers, Amsterdam (1991)
38. Priest, G.: Paraconsistent logic. In: Gabbay, D.M., Guenthner, F. (eds.) Handbook of Philosophical Logic, vol. 6, pp. 287–393. Springer, Netherlands, Dordrecht (2002). https://doi.org/10.1007/978-94-017-0460-1_4
39. Priest, G.: Plurivalent logics. Australas. J. Log. **11**(1), 2–13 (2014)
40. Priest, G., Routley, R.: Introduction: paraconsistent logics. Stud. Logica. **43**(1–2), 3–16 (1984)
41. Robles, G.: A Routley-Meyer semantics for Gödel 3-valued logic and its paraconsistent counterpart. Log. Univers. **7**, 507–532 (2013)
42. Shramko, Y., Zaitsev, D., Belikov, A.: First-degree entailment and its relatives. Stud. Logica. **105**(6), 1291–1317 (2017)
43. Smiley, T.J.: Entailment and deducibility. Proc. Aristot. Soc. **59**, 233–254 (1959)
44. Tsoukiàs, A.: A first order, four-valued, weakly paraconsistent logic and its relation with rough sets semantics. Found. Comput. Decision Sci. **27**(2), 77–96 (2002)
45. Tuziak, R.: Paraconsistent extensions of positive logic. Bull. Sect. Logic **25**(1), 15–20 (1996)
46. Viana, H., Alcântara, J.A., Martins, A.T.: Searching contexts in paraconsistent rough description logic. J. Braz. Comput. Soc. **21**(1), Art. 7, 13 (2015)
47. Vitória, A., Małuszyński, J., Szałas, A.: Modeling and reasoning with paraconsistent rough sets. Fund. Inform. **97**(4), 405–438 (2009)
48. Vitória, A., Szałas, A., Małuszyński, J.: Four-valued extension of rough sets. In: Wang, G., Li, T., Grzymala-Busse, J.W., Miao, D., Skowron, A., Yao, Y. (eds.) RSKT 2008. LNCS (LNAI), vol. 5009, pp. 106–114. Springer, Heidelberg (2008). https://doi.org/10.1007/978-3-540-79721-0_19
49. Wu, H.: Paraconsistent rough set algebras. In: Chen, Y., Zhang, S. (eds.) AILA 2022. Communications in Computer and Information Science, vol. 1657, pp. 166–179. Springer, Singapore (2022). https://doi.org/10.1007/978-981-19-7510-3_13

Hexagons of Opposition in Linguistic Three-Way Decisions

Stefania Boffa$^{(\boxtimes)}$ and Davide Ciucci

University of Milano-Bicocca, DISCo, Viale Sarca 336, 20126 Milan, Italy
{stefania.boffa,davide.ciucci}@unimib.it

Abstract. In three-way decision theory, three disjoint sets covering a given universe, are determined: the positive, negative, and boundary regions. They correspond to three types of decisions on their objects: acceptance, rejection, and abstention or non-commitment. A linguistic approach for identifying the three regions relies on specific evaluative linguistic expressions, such as "very big", "roughly small", "not small", "medium", and so forth.

In this article, we construct hexagons of opposition using the regions generated by different evaluative linguistic expressions. Then, we explore the logical relations between the vertices of different hexagons.

Keywords: Hexagon of opposition · Aristotle Square · Evaluative Linguistic Expressions · Three-way Decisions

1 Introduction

In three-way decision theory, three disjoint sets covering a given universe, are determined. They are called *positive*, *negative*, and *boundary regions* and correspond to three types of decisions on their objects: acceptance, rejection, and abstention or non-commitment. Three-way decisions are naturally interpreted in Rough Set Theory, as well as in other frameworks [18,19,21]. A linguistic approach to finding the three regions was recently introduced in [3] and deals with a particular class of *evaluative linguistic expressions*. These are expressions of the human language like "very big", "roughly small", "not small", "medium", and so on, and the related theory is constructed in a formal system of higher-order fuzzy logic (fuzzy type theory) [12,14]. Such a model called *linguistic three-way decision*, gives a new interpretation to the acceptance, rejection, and non-commitment regions, which is more understandable by users that do not have mathematical knowledge. This is an advantage from the standpoint of *Explainable Artificial Intelligence (XAI)* the approach to AI focusing on the ability of machines to give sound motivations about their decisions and behaviour [10]. Three-way decision with evaluative linguistic expressions is strictly connected to a novel and linguistic generalization of the classical notion of rough sets [3].

In this article, the linguistic regions are organized to form a hexagon of opposition, which is an evolution of an important tool of Aristotle's logic: the square of opposition. The *square of opposition* (also called *Aristotle square*) is a mathematical chart exhibiting the relationship between four logical propositions [15]. The possible links represented by the square of opposition are called *relations of contradictory, contrary, sub-contrary, and subalternation* and have the following meaning: two propositions are

- *contradictories* if and only if they cannot be true together and cannot be false together;
- *contraries* if and only if they can be false together but cannot be true together;
- *sub-contraries* if and only if they can be true together but cannot be false together;
- *subalterns* (or *superalterns*) if and only if one implies the other.

The hexagon of opposition can be obtained by adding two new logical propositions to those of Aristotle square together with the related relations or by overlapping three Aristotle squares [2,11]. This diagram is considered a powerful instrument to schematize the connections between concepts in various situations. Therefore, it is repurposed in different fields, for instance, in Rough set Theory, Formal Concept Analysis, Probability Theory, and Possibility Theory [7,8,16,20]. Lately, hexagons have been used to solve concrete problems: detecting influential news in online communities [1] and evaluating emotional dynamics in social media conversations [9]. Another type of hexagon of opposition (different to the classical one in its structure) is proposed in fuzzy formal concept analysis [5].

After recalling some preliminary notions in Sect. 2, we present the results of this article in Sect. 3:

- In Subsect. 3.1, we show that the regions POS, NEG, and BND generated by a given evaluative linguistic expression and their unions $POS \cup NEG$, $POS \cup BND$, and $NEG \cup BND$, can be placed into the vertices of a hexagon of opposition. Furthermore, we discuss how the logical relations between the linguistic regions reflect the three-way decision philosophy. For example, the relation of contrary between the positive and negative regions means that an object cannot be both accepted and rejected.
- In Subsect. 3.2, we analyze the logical relations between the linguistic regions deriving from two different evaluative linguistic expressions Ev_1 and Ev_2. In particular, we discover the relations of contrary, sub-contrary, and subalternation existing between the vertices of the hexagons corresponding to Ev_1 and Ev_2. This aspect is fundamental to deepen the understanding of how the choice of the initial evaluative linguistic expression affects the final decision on the objects.

Let us recall that hexagons of opposition were already introduced in the context of three-way decision theory, where the three regions (called *probabilistic regions*) are generated by using a pair of thresholds and the concept of probabilistic rough sets [1]. Moreover, the logical relations between the probabilistic regions deriving by different pairs of thresholds were studied in [4].

2 Preliminaries

In this section, we give the basic notions of evaluative linguistic expressions, three-way decisions and the hexagon of opposition.

2.1 Three-Way Decisions with Evaluative Linguistic Expressions

Evaluative Linguistic Expressions. Let us recall the notions of the theory of evaluative linguistic expressions that are essential to explain our results.

Evaluative linguistic expressions (evaluative expressions for short) are expressions that commonly appear in the human language when people judge, evaluate, give opinions, and so on. The *pure evaluative expressions* are the simplest ones and are composed of an adjective that could be preceded by an adverb. Examples are *very tall*, *extremely boring*, and *expensive*. Other pure evaluative expressions are fuzzy numbers like about twenty-five. The rest of the evaluative expressions can be composed of pure evaluative expressions and the connective *not, and,* and *or*. In this article, we confine to the evaluative expressions involving the adjectives *small, medium,* and *big* because they are employed to evaluate the size of sets. Examples are *not small, very big, medium,* and *extremely big*.

Evaluative expressions are characterized by the notions of intention, context, and extension. The meaning of an evaluative linguistic expression is modelled by its intention, which is a function assigning to each context another mapping called extension (see [12,14] for more details). In this article, we usually deal with the standard context $\langle 0, 0.5, 1 \rangle$, and so, extensions of evaluative expressions are maps from $[0,1]$ to $[0,1]$. Thus, we can say that a special fuzzy set $Ev :$ $[0,1] \rightarrow [0,1]$ represents a given evaluative expression in the context $\langle 0, 0.5, 1 \rangle$. We denote the collection of the extensions of all evaluative expressions in the context $\langle 0, 0.5, 1 \rangle$ with the symbol $\mathcal{E}$. An example of extension of the evaluative expression *not small* is the function $\neg Sm : [0,1] \rightarrow [0,1]$ defined by

$$\neg Sm(x) = \begin{cases} 1 & \text{if } x \in [0.275, 1], \\ 1 - \dfrac{(0.275 - x)^2}{0.02305} & \text{if } x \in (0.16, 0.275), \\ \dfrac{(x - 0.0745)^2}{0.01714} & \text{if } x \in (0.0745, 0.16], \\ 0 & \text{if } x \in [0, 0.0745]. \end{cases} \tag{1}$$

In [13], $\neg Sm$ is also used to construct the formula of the fuzzy quantifier *many*.

See [12,14] for more details about the theory of evaluative linguistic expressions.

Three-Way Decisions with Evaluative Linguistic Expressions. From now on, we consider a finite universe U, a subset X of U, a pair of thresholds (α, β) so that $0 \leq \beta < \alpha \leq 1$, and an equivalence relation $\mathcal{R}$ on U (i.e., $\mathcal{R}$ is reflexive, symmetric and transitive). Moreover, we use the symbol $[x]_{\mathcal{R}}$ to denote the equivalence class of $x \in U$.

In order to define the linguistic regions determined by $Ev \in \mathcal{E}$, we need to consider for each $x \in U$ the value $Ev\left(\dfrac{|[x]_{\mathcal{R}} \cap X|}{|[x]_{\mathcal{R}}|}\right)$, which is the evaluation of the size of $X \cap [x]_{\mathcal{R}}$ w.r.t. the size of $[x]_{\mathcal{R}}$ by using Ev[1]. For example, if $Ev = \neg Sm$, then $\neg Sm\left(\dfrac{|[x]_{\mathcal{R}} \cap X|}{|[x]_{\mathcal{R}}|}\right)$ measures "*how much the size of $X \cap [x]_{\mathcal{R}}$ is not small w.r.t. the size of $[x]_{\mathcal{R}}$*". In other words, we are saying that "*the size of the set of elements of $[x]_{\mathcal{R}}$ that also belong to X is not small with the truth degree $\neg Sm\left(\dfrac{|[x]_{\mathcal{R}} \cap X|}{|[x]_{\mathcal{R}}|}\right)$*".

Remark 1. By [6], we know that $\neg Sm\left(\dfrac{|[x]_{\mathcal{R}} \cap X|}{|[x]_{\mathcal{R}}|}\right)$ coincides with the formula of the quantifier *many*[2]. This means that $\neg Sm\left(\dfrac{|[x]_{\mathcal{R}} \cap X|}{|[x]_{\mathcal{R}}|}\right)$ is the truth degree to which "*many objects of $[x]_{\mathcal{R}}$ are in X*".

For each evaluative expression, a triple of subsets of U is determined as follows [3].

Definition 1. *Let $Ev \in \mathcal{E}$, the (α, β)-linguistic positive, negative, and boundary regions induced by Ev are respectively the following:*

(i) $POS^{Ev}_{(\alpha,\beta)}(X) = \left\{x \in U \mid Ev\left(\dfrac{|[x]_{\mathcal{R}} \cap X|}{|[x]_{\mathcal{R}}|}\right) \geq \alpha\right\}$;

(ii) $NEG^{Ev}_{(\alpha,\beta)}(X) = \left\{x \in U \mid Ev\left(\dfrac{|[x]_{\mathcal{R}} \cap X|}{|[x]_{\mathcal{R}}|}\right) \leq \beta\right\}$;

(iii) $BND^{Ev}_{(\alpha,\beta)}(X) = \left\{x \in U \mid \beta < Ev\left(\dfrac{|[x]_{\mathcal{R}} \cap X|}{|[x]_{\mathcal{R}}|}\right) < \alpha\right\}$.

Thus, an object x belongs to the (α, β)-linguistic positive region when the size of $[x]_{\mathcal{R}} \cap X$ (w.r.t. the size of $[x]_{\mathcal{R}}$) evaluated by Ev is at least α. Analogously, x belongs to the (α, β)-linguistic negative region when the size of $[x]_{\mathcal{R}} \cap X$ (w.r.t. the size of $[x]_{\mathcal{R}}$) evaluated by Ev is at most β. Finally, the remaining elements of U form the (α, β)-linguistic boundary region.

Definition 1 also leads to the notion of linguistic rough sets, which are generalizations of Pawlak rough sets:

---

[1] The function such that $X \mapsto \dfrac{|X|}{|U|}$ for each $X \subseteq U$ is understood as a *normalized fuzzy measure* [17].

[2] This is a fuzzy quantifier $\mathcal{S}_{many}$ assigning a value of $[0,1]$ to each pair of fuzzy sets. In [6], it has been proven that $\mathcal{S}_{many}(A, B) = \neg Sm\left(\dfrac{|A \cap B|}{|A|}\right)$, when A and B are classical set of the given universe.

Definition 2. *Let $Ev \in \mathcal{E}$, the (α, β)-linguistic rough set of X determined by $\mathcal{R}$ and Ev is the pair $(\mathcal{L}_{(\alpha,\beta)}^{Ev}(X), \mathcal{U}_{(\alpha,\beta)}^{Ev}(X))$, where*

$$\mathcal{L}_{(\alpha,\beta)}^{Ev}(X) = POS_{(\alpha,\beta)}^{Ev}(X) \text{ and } \mathcal{U}_{(\alpha,\beta)}^{Ev}(X) = POS_{(\alpha,\beta)}^{Ev}(X) \cup BND_{(\alpha,\beta)}^{Ev}(X).$$

$\mathcal{L}_{(\alpha,\beta)}^{Ev}(X)$ and $\mathcal{U}_{(\alpha,\beta)}^{Ev}(X)$ are respectively called (α, β)-linguistic lower and upper approximations of X determined by $\mathcal{R}$ and Ev.

In the sequel, we need the following theorem.

Theorem 1. *The (α, β)-linguistic positive, negative, and boundary regions induced by Ev form a tri-partition of U^3.*

2.2   Hexagons of Opposition

Definition 3. *Let $A \subseteq U$, P_A denotes a property such that "$x \in U$ satisfies P_A if and only if $x \in A$". Equivalently, we can say that "P_A is the property of belonging to A".*

Definition 4. *Let $A, B \subseteq U$. Then,*

(i) P_A and P_B are contraries if and only if $A \cap B = \emptyset$;
(ii) P_A and P_B are sub-contraries if and only if $A \cup B = U$;
(iii) P_B is subaltern of P_A if and only if $A \subseteq B$;
(iv) P_A and P_B are contradictories if and only if $A = U \setminus B$.

Remark 2. Of course, if P_A and P_B are contradictories, then $A \cap B = \emptyset$ and $A \cup B = U$.

For convenience, we place the sets instead of their properties into the vertices of the hexagon. Moreover, in the sequel, we could say that a logical relation holds between A and B to indicate that it holds between P_A and P_B.

Definition 5. *Let $A, B, C, D, E, F \subseteq U$. Then, $A, B, C, D, E,$ and F form an hexagon of opposition if and only if*

(i) P_A and P_B are contraries, as well as P_A and P_C, and P_B and P_C;
(ii) P_A and P_E are contradictories, as well as P_B and P_D, and P_C and P_F;
(iii) P_D and P_E are sub-contraries, as well as P_D and P_F, and P_E and P_F;
(iv) P_D is sub-altern of P_A and P_C, P_E is sub-altern of P_B and P_C, and P_F is sub-altern of P_A and P_B.

The hexagon of opposition having $A, B, C, D, E,$ and F as vertices is represented by Fig. 1, where we use the lines ——, - - -, ⟶, and ═══ to denote the relations of contrary, contradictory, subalternation, and sub-contrary, respectively.

A hexagon of opposition arises whenever a tri-partition of a universe is given.

Theorem 2 [7]. *Let $\{A, B, C\}$ be a partition of U. Then, $A, B, C, A \cup C, B \cup C,$ and $A \cup B$ form a hexagon of opposition as in Fig. 2.*

---

[3] By a tri-partition of U we mean a collection of three mutually disjoint subsets covering U. So, notice that the limit cases where one or two sets are empty are included.

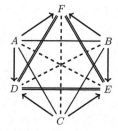

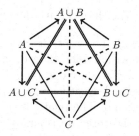

Fig. 1. Hexagon of opposition with $A, B, C, D, E, F \subseteq U$.

Fig. 2. Hexagon of opposition related to the tri- partition $\{A, B, C\}$ of U.

3   Hexagons of Opposition in Linguistic Three-Way Decisions

This section firstly presents a hexagon of opposition generated by $Ev \in \mathcal{E}$. After that, it investigates the logical relations involving the vertices of the hexagons of opposition generated by $Ev_1, Ev_2 \in \mathcal{E}$.

3.1   Hexagons of Opposition with Evaluative Linguistic Expressions

In this subsection, we arrange the (α, β)-linguistic regions and their unions to form a hexagon of opposition.

Theorem 3. *Let $Ev \in \mathcal{E}$. Then, $POS^{Ev}_{(\alpha,\beta)}$, $NEG^{Ev}_{(\alpha,\beta)}$, $BND^{Ev}_{(\alpha,\beta)}$, $POS^{Ev}_{(\alpha,\beta)} \cup BND^{Ev}_{(\alpha,\beta)}$, $NEG^{Ev}_{(\alpha,\beta)} \cup BND^{Ev}_{(\alpha,\beta)}$, and $POS^{Ev}_{(\alpha,\beta)} \cup NEG^{Ev}_{(\alpha,\beta)}$ form a hexagon of opposition as in Fig. 3.*

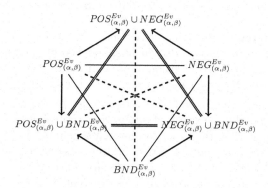

Fig. 3. Hexagon of opposition related to $POS^{Ev}_{(\alpha,\beta)}$, $NEG^{Ev}_{(\alpha,\beta)}$, and $BND^{Ev}_{(\alpha,\beta)}$.

Proof. The thesis clearly follows from Theorems 1 and 2.

Remark 3. The relations represented by the hexagon reflect the basic behaviour of the decision rules. Indeed, the contrary relations indicate that an object should be subject to only one type of decision. For example, since the (α, β)-linguistic positive and negative regions are contraries, an object cannot be both accepted and rejected. The relations of sub-contrary correspond to the fact that at least one of the three decisions must be made on each object. The meaning of the relations of subalternation is trivial, for instance, being accepted implies being accepted or not-committed. The following is the meaning of the relations of contradictory: when we know that one of the three decisions is not made on an object, then one of the other two decisions must be made. For example, the contradictory between $POS_{(\alpha,\beta)}^{Ev}$ and $NEG_{(\alpha,\beta)}^{Ev} \cup BND_{(\alpha,\beta)}^{Ev}$ is equivalent to that a decision of non-acceptance on an object x of U necessarily implies a decision of rejection or non-commitment on x.

Example 1. Let us focus on an example already presented in [3], where

- $U = \{u_1, \ldots, u_{32}\}$ is a set of users of online communities and
- the equivalence relation $\mathcal{R}$ on U is defined as follows: let $u_i, u_j \in U$,

$$u_i \mathcal{R} u_j \text{ if and only if } u_i \text{ and } u_j \text{ belong to the same community.}$$

Then, $\mathcal{R}$ partitions the universe of the users into six equivalence classes: $C_1 = \{u_1, \ldots, u_5\}$, $C_2 = \{u_6, \ldots, u_{10}\}$, $C_3 = \{u_{11}, \ldots, u_{15}\}$, $C_4 = \{u_{16}, \ldots, u_{20}\}$, $C_5 = \{u_{21}, \ldots, u_{25}\}$, and $C_6 = \{u_{26}, \ldots, u_{32}\}$.

Also, we deal with the set

$$X_{Sport} = \{u_{10}, u_{11}, u_{12}, u_{18}, u_{19}, u_{20}, u_{21}, u_{22}, u_{23}, u_{24}, u_{26}\}$$

made of all users of U interested in the topic Sport. Using linguistic three-way decisions, we can select the most appropriate communities among $C_1, \ldots, C_6$ to which propose sports news.

In order to determine the linguistic regions, we choose $(\alpha, \beta) = (0.8, 0.2)$ and the evaluative expression *not small*, which is modelled by the function $\neg Sm \in \mathcal{E}$ defined by (1). By (1) and Definition 1, we can easily compute the $(0.8, 0.2)$-linguistic regions: $POS_{(0.8, 0.2)}^{\neg Sm}(X_{Sport}) = C_3 \cup C_4 \cup C_5$, $NEG_{(0.8, 0.2)}^{\neg Sm}(X_{Sport}) = C_1$, and $BND_{(0.8, 0.2)}^{\neg Sm}(X_{Sport}) = C_2 \cup C_6$ (see [3] for more details). Then, Fig. 4 depicts the corresponding hexagon of opposition, according to Theorem 3.

In line with Remark 3, the hexagon represents a simple way to view the relationship between the three different types of decisions on the users. For instance, considering that C_1 and $C_2 \cup C_3 \cup C_4 \cup C_5 \cup C_6$ are contradictories, we can graphically see that the users of the communities different from C_1 could receive the sports news.

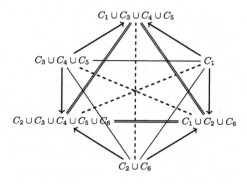

Fig. 4. Hexagon of opposition with $POS^{\neg Sm}_{(0.8,0.2)}$, $NEG^{\neg Sm}_{(0.8,0.2)}$, and $BND^{\neg Sm}_{(0.8,0.2)}$.

3.2   Comparing Hexagon of Oppositions

This subsection analyzes the logical relations between the vertices of the hexagons of opposition arising from the same pair of thresholds (α, β), but different evaluative expressions $Ev_1, Ev_2 \in \mathcal{E}$. Therefore, the meaning of such relations is discussed focusing on how the final decision on the objects of U changes when we choose different evaluative expressions.

In the sequel, we consider this order on $\mathcal{E}$: let $Ev_1, Ev_2 \in \mathcal{E}$,

$$Ev_1 \preceq Ev_2 \text{ if and only if } Ev_1(a) \leq Ev_2(a) \text{ for each } a \in [0,1].$$

For simplicity, given $Ev_i \in \mathcal{E}$, we use the symbols POS_i, NEG_i, and BND_i instead of $POS^{Ev_i}_{(\alpha,\beta)}$, $NEG^{Ev_i}_{(\alpha,\beta)}$, and $BND^{Ev_i}_{(\alpha,\beta)}$, respectively. Then, we want to discover the logical relations concerning the vertices of the hexagons depicted in Fig. 5.

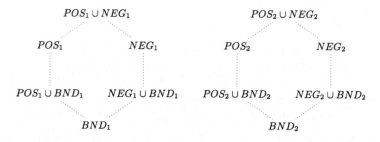

Fig. 5. Hexagons based on Ev_1 and Ev_2, respectively.

Relations of Subalternation

Theorem 4. *Let $Ev_1, Ev_2 \in \mathcal{E}$ such that $Ev_1 \preceq Ev_2$. Then,*

(a) P_{POS_2} is subaltern of P_{POS_1},
(b) P_{NEG_1} is subaltern of P_{NEG_2}.

Proof.(a) Let $x \in POS_1$, then $Ev_1 \left(\frac{|X \cap [x]_R|}{|[x]_R|} \right) \geq \alpha$ from Definition 1(i). Since
$Ev_2 \left(\frac{|X \cap [x]_R|}{|[x]_R|} \right) \geq Ev_1 \left(\frac{|X \cap [x]_R|}{|[x]_R|} \right)$, we get $x \in POS_2$. Consequently, $POS_1 \subseteq POS_2$. Finally, by Definition 4 (iii), P_{POS_2} is subaltern of P_{POS_1}.

(b) Let $x \in NEG_2$, then $Ev_2 \left(\frac{|X \cap [x]_R|}{|[x]_R|} \right) \leq \beta$ from Definition 1 (ii). Since
$Ev_2 \left(\frac{|X \cap [x]_R|}{|[x]_R|} \right) \geq Ev_1 \left(\frac{|X \cap [x]_R|}{|[x]_R|} \right)$, we get $x \in NEG_1$. Then, $NEG_2 \subseteq NEG_1$. By Definition 4 (iii), we can conclude that P_{NEG_1} is subaltern of P_{NEG_2}.

Figure 6 shows the logical relations listed by Theorem 4.

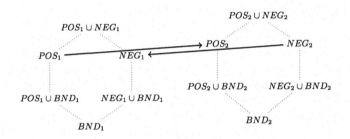

Fig. 6. Relations of subalternation of Theorem 4.

Theorem 5. *Let* $Ev_1, Ev_2 \in \mathcal{E}$ *such that* $Ev_1 \preceq Ev_2$. *Then,*

(a) $P_{POS_2 \cup BND_2}$ is subaltern of P_{POS_1},
(b) $P_{POS_2 \cup NEG_2}$ is subaltern of P_{POS_1},
(c) $P_{POS_2 \cup BND_2}$ is subaltern of P_{BND_1},
(d) $P_{POS_2 \cup BND_2}$ is subaltern of $P_{POS_1 \cup BND_1}$.

Proof.(a) By Theorem 4 (a), $POS_1 \subseteq POS_2$. Hence, $POS_1 \subseteq POS_2 \cup BND_2$. Then, by Definition 4 (iii), $P_{POS_2 \cup BND_2}$ is subaltern of P_{POS_1}.

(b) By Theorem 4(a), $POS_1 \subseteq POS_2$. Hence, $POS_1 \subseteq POS_2 \cup NEG_2$. By Definition 4 (iii), $P_{POS_2 \cup NEG_2}$ is subaltern of P_{POS_1}.

(c) Let $x \in BND_1$, then $\beta < Ev_1 \left(\frac{|X \cap [x]_\mathcal{R}|}{|[x]_\mathcal{R}|} \right) < \alpha$ from Definition 1 (iii). Since $Ev_2 \left(\frac{|X \cap [x]_\mathcal{R}|}{|[x]_\mathcal{R}|} \right) \geq Ev_1 \left(\frac{|X \cap [x]_\mathcal{R}|}{|[x]_\mathcal{R}|} \right)$, we have $Ev_2 \left(\frac{|X \cap [x]_\mathcal{R}|}{|[x]_\mathcal{R}|} \right) > \beta$, which means that $x \in BND_2 \cup POS_2$. Hence, $BND_1 \subseteq BND_2 \cup POS_2$. By Definition 4 (iii), $P_{POS_2 \cup BND_2}$ is subaltern of P_{BND_1}.

(d) By Theorem 4(b), $NEG_2 \subseteq NEG_1$. Since NEG_1 and NEG_2 are respectively the complements of $POS_1 \cup BND_1$ and $POS_2 \cup BND_2$, we consequently get $POS_1 \cup BND_1 \subseteq POS_2 \cup BND_2$. Then, by Definition 4 (iii), $P_{POS_2 \cup BND_2}$ is subaltern of $P_{POS_1 \cup BND_1}$.

Figure 7 shows the logical relations listed by Theorem 5.

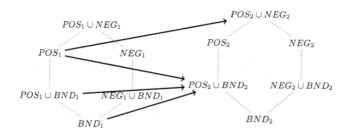

Fig. 7. Relations of subalternation of Theorem 5.

Theorem 6. *Let $Ev_1, Ev_2 \in \mathcal{E}$ such that $Ev_1 \preceq Ev_2$. Then,*

(a) $P_{NEG_1 \cup BND_1}$ is subaltern of P_{BND_2},
(b) $P_{NEG_1 \cup BND_1}$ is subaltern of P_{NEG_2},
(c) $P_{NEG_1 \cup BND_1}$ is subaltern of $P_{NEG_2 \cup BND_2}$,
(d) $P_{POS_1 \cup NEG_1}$ is subaltern of P_{NEG_2}.

Proof.(a) By Theorem 5(b), $POS_1 \subseteq POS_2 \cup NEG_2$. Since $NEG_1 \cup BND_1$ and BND_2 are respectively the complements of POS_1 and $POS_2 \cup NEG_2$, we get $BND_2 \subseteq NEG_1 \cup BND_1$. Then, by Definition 4 (iii), $P_{NEG_1 \cup BND_1}$ is subaltern of P_{BND_2}.
(b) By Theorem 4(b), $NEG_2 \subseteq NEG_1$. So, we can immediately conclude that $NEG_2 \subseteq NEG_1 \cup BND_1$. By Definition 4 (iii), $P_{NEG_1 \cup BND_1}$ is subaltern of P_{NEG_2}.
(c) By Theorem 4(a), $POS_1 \subseteq POS_2$. Since $NEG_1 \cup BND_1$ and $NEG_2 \cup BND_2$ are respectively the complements of POS_1 and POS_2, we lastly get $NEG_2 \cup BND_2 \subseteq NEG_1 \cup BND_1$. Thus, by Definition 4 (iii), $P_{NEG_1 \cup BND_1}$ is subaltern of $P_{NEG_2 \cup BND_2}$.
(d) By Theorem 4(b), $NEG_2 \subseteq NEG_1$. So, we can immediately conclude that $NEG_2 \subseteq POS_1 \cup NEG_1$. By Definition 4 (iii), $P_{POS_1 \cup NEG_1}$ is subaltern of P_{NEG_2}.

Figure 8 shows the logical relations listed by Theorem 6.

Relations of Contrary

Theorem 7. *Let $Ev_1, Ev_2 \in \mathcal{E}$ such that $Ev_1 \preceq Ev_2$. Then,*

(a) P_{POS_1} and P_{NEG_2} are contraries,
(b) P_{POS_1} and P_{BND_2} are contraries,
(c) P_{POS_1} and $P_{NEG_2 \cup BND_2}$ are contraries,
(d) $P_{POS_1 \cup BND_1}$ and P_{NEG_2} are contraries,
(e) P_{BND_1} and P_{NEG_2} are contraries.

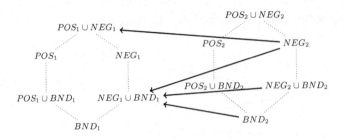

Fig. 8. Relations of subalternation of Theorem 6.

Proof.(a) By Theorem 4(b), $NEG_2 \subseteq NEG_1$. Moreover, $POS_1 \cap NEG_1 = \emptyset$ from Theorem 3 (recall that P_{POS_1} and P_{NEG_1} are contraries). Then, $POS_1 \cap NEG_2 = \emptyset$. Hence, by Definition 4 (i), P_{POS_1} and P_{NEG_2} are contraries.

(b) By Theorem 4(a), $POS_1 \subseteq POS_2$. Moreover, $POS_2 \cap BND_2 = \emptyset$ from Theorem 3 (recall that P_{POS_2} and P_{BND_2} are contraries). Then, $POS_1 \cap BND_2 = \emptyset$. By Definition 4 (i), P_{POS_1} and P_{BND_2} are contraries.

(c) Using items (a) and (b) of this theorem, we get $POS_1 \cap NEG_2 = \emptyset$ and $POS_1 \cap BND_2 = \emptyset$. Consequently, $POS_1 \cap (NEG_2 \cup BND_2) = \emptyset$. By Definition 4 (i), P_{POS_1} and $P_{NEG_2 \cup BND_2}$ are contraries.

(d) By Theorem 4(b), $NEG_2 \subseteq NEG_1$. Moreover, $NEG_1 \cap (POS_1 \cup BND_1) = \emptyset$ from Theorem 3 (P_{NEG_1} and P_{POS_1} are contraries as well as P_{NEG_1} and P_{BND_1}). Then, $NEG_2 \cap (POS_1 \cup BND_1) = \emptyset$. Thus, by Definition 4 (i), $P_{POS_1 \cup BND_1}$ and P_{NEG_2} are contrary.

(e) By Theorem 4(b), $NEG_2 \subseteq NEG_1$. Furthermore, $NEG_1 \cap BND_1 = \emptyset$ from Theorem 3 (P_{NEG_1} and P_{BND_1} are contrary). Therefore, $NEG_2 \cap BND_1 = \emptyset$. By Definition 4 (i), P_{BND_1} and P_{NEG_2} are contraries.

Figure 9 shows the relations of opposition listed by Theorem 7.

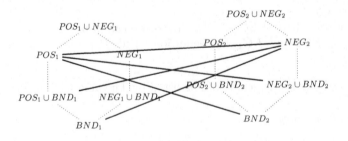

Fig. 9. Relations of contrary of Theorem 7.

Relations of Sub-contrary

Theorem 8. *Let $Ev_1, Ev_2 \in \mathcal{E}$ such that $Ev_1 \preceq Ev_2$. Then,*

(a) $P_{NEG_1 \cup BND_1}$ and P_{POS_2} are sub-contraries,

(b) $P_{NEG_1 \cup BND_1}$ and $P_{POS_2 \cup BND_2}$ are sub-contraries,

(c) $P_{NEG_1 \cup BND_1}$ and $P_{POS_2 \cup NEG_2}$ are sub-contraries,

(d) P_{NEG_1} and $P_{POS_2 \cup BND_2}$ are sub-contraries,

(e) $P_{POS_1 \cup NEG_1}$ and $P_{POS_2 \cup BND_2}$ are sub-contraries.

Proof.(a) By Theorem 4(a), $POS_1 \subseteq POS_2$. Also, by Theorem 3, P_{POS_1} and $P_{NEG_1 \cup BND_1}$ are contradictories. So, by Remark 2, $POS_1 \cup (NEG_1 \cup BND_1) = U$. Thus, $POS_2 \cup (NEG_1 \cup BND_1) = U$. That is, $P_{NEG_1 \cup BND_1}$ and P_{POS_2} are sub-contraries from Definition 4 (ii).

(b) $POS_2 \cup (NEG_1 \cup BND_1) = U$ from item (a) of this theorem. This implies that $(POS_2 \cup BND_2) \cup (NEG_1 \cup BND_1) = U$. Then, by Definition 4 (ii), $P_{NEG_1 \cup BND_1}$ and $P_{POS_2 \cup BND_2}$ are sub-contraries.

(c) $POS_2 \cup (NEG_1 \cup BND_1) = U$ from item (a) of this theorem. This implies that $(POS_2 \cup NEG_2) \cup (NEG_1 \cup BND_1) = U$. By Definition 4 (ii), $P_{NEG_1 \cup BND_1}$ and $P_{POS_2 \cup NEG_2}$ are sub-contraries.

(d) By Theorem 4(b), $NEG_2 \subseteq NEG_1$. Furthermore, by Theorem 3, P_{NEG_2} and $P_{POS_2 \cup BND_2}$ are contradictories. Then, by Remark 2, $NEG_2 \cup (POS_2 \cup BND_2) = U$. As a consequence, $NEG_1 \cup (POS_2 \cup BND_2) = U$. By Definition 4 (ii), P_{NEG_1} and $P_{POS_2 \cup BND_2}$ are sub-contraries.

(e) By Theorem 6(d), $NEG_2 \subseteq NEG_1 \cup POS_1$. Additionally, as in the proof of item (d), we get $NEG_2 \cup (POS_2 \cup BND_2) = U$. Then, $(NEG_1 \cup POS_1) \cup (POS_2 \cup BND_2) = U$. Ultimately, by Definition 4 (ii), $P_{POS_1 \cup NEG_1}$ and $P_{POS_2 \cup BND_2}$ are sub-contraries.

Figure 10 shows the logical relations listed by Theorem 8.

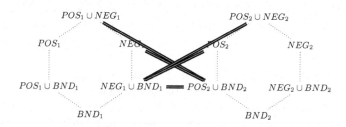

Fig. 10. Relations of sub-contrary of Theorem 8.

3.3 Discussion on the Logical Relations Between Hexagons

First of all, we can notice that relations of contradiction between hexagons generally do not exist. Exceptions occur in some trivial cases, for instance when one of the following equality holds: $POS_1 = POS_2$, $NEG_1 = NEG_2$, or $BND_1 = BND_2$. Indeed, if $POS_1 = POS_2$, then $NEG_1 \cup BND_1 = NEG_2 \cup BND_2$.

Thus, since P_{POS_1} and $P_{NEG_1 \cup BND_1}$ are contradictories from Theorem 3, it is trivial that P_{POS_1} and $P_{NEG_2 \cup BND_2}$ are contradictories too.

Then, we notice that we get a symmetrical diagram by putting together Figs. 6, 7, and 8, which are all the diagrams representing the relations of subalternation (see Fig. 11).

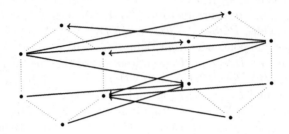

Fig. 11. All the relations of subalternation between the hexagons related to Ev_1 and Ev_2.

In addition, the hexagons that are connected by the relations of contrary in Fig. 9, form a symmetrical chart too. The same holds for the diagram depicted by Fig. 10 and arising from the relations of sub-contrary.

Such symmetry can be analytically translated and described as follows. Consider the function f assigning to each vertex of the hexagon of Ev_1 a vertex of the hexagon of Ev_2 such that

$$f(POS_1) = NEG_2, f(NEG_1) = POS_2, f(POS_1 \cup BND_1) = NEG_2 \cup BND_2,$$
$$f(POS_1 \cup NEG_1) = POS_2 \cup NEG_2, f(NEG_1 \cup BND_1) = POS_2 \cup BND_2,$$
$$\text{and } f(BND_1) = BND_2.$$

Observe that f transforms each vertex of the first hexagon into a vertex of the second hexagon by exchanging the positive and negative regions (i.e. $P_1 \mapsto N_2$ and $N_1 \mapsto P_2$) and by leaving unchanged the boundary region ($B_1 \mapsto B_2$).

So, let A_1 and A_2 be two vertices of the hexagons related to Ev_1 and Ev_2, a logical relation holds between A_1 and A_2 if and only if it holds between $f(A_1)$ and $f^{-1}(A_2)$. For example, by Theorem 7 (c), POS_1 and $NEG_2 \cup BND_2$ are contraries; also, by Theorem 7 (d), $f(POS_1) = NEG_2$ and $f^{-1}(NEG_2 \cup BND_2) = POS_1 \cup BND_1$ are contraries too.

Finally, let us analyze the meaning of the logical relations between different hexagons in terms of decisions. So, the hexagons of Ev_1 and Ev_2 lead to different decision procedures denoted with 1 and 2, which are connected as follows. Let $x \in U$,

- if x is accepted by 1, then x is also accepted by 2 (Theorem 4(a)),
- if x is rejected by 2, then x is also rejected by 1 (Theorem 4(b)),

- if x is non-committed by 1, then x cannot be rejected by 2 (Theorem 5(c)),
- if x is not rejected by 1, then x is also not rejected by 2 (Theorem 5(d)),
- if x is non-committed by 2, then x cannot be accepted by 1 (Theorem 6(a)),
- if x is not accepted by 2, then x is not accepted by 1 (Theorem 6(c)),
- x can not be both accepted by 1 and rejected by 2 (Theorem 7 (a)),
- x can not be both accepted by 1 and non-committed by 2 (Theorem 7(b)),
- x can not be both accepted by 1 and non-accepted by 2 (Theorem 7(c)),
- x can not be both rejected by 2 and non-rejected by 1 (Theorem 7(d)),
- x can not be both rejected by 2 and non-committed by 1 (Theorem 7(e)),
- x must be either non-accepted by 1 or accepted by 2 (Theorem 8(a)),
- x must be either non-accepted by 1 or non-rejected by 2 (Theorem 8(b)),
- x must be either non-accepted by 1 or committed by 2 (Theorem 8(c)),
- x must be either rejected by 1 or non-rejected by 2 (Theorem 8(d)),
- x must be either committed (accepted or rejected) by 1 or non-rejected by 2 (Theorem 8(e)).

Notice that items (a) and (b) of Theorem 5 are not significant for explaining how 1 and 2 are connected, considering that they do not add knowledge w.r.t. Theorem 4(a). The same is for the items (b) and (d) of Theorem 6, which derive from Theorem 4 (b).

4   Conclusions

The contribution of this work is twofold. Firstly, it represents a step forward in the study of the extensions of Aristotle's square. Secondly, it provides new tools for possible applications in the context of three-way decision theory. We plan to extend this article by finding the logical relations between hexagons generated by the same evaluative expression, but different pairs of thresholds (α, β) and (α', β'). Also, the logical relations between hexagons could be investigated in a more complex situation, when the hexagons are generated by different evaluative expressions and different pairs of thresholds.

References

1. Abbruzzese, R., Gaeta, A., Loia, V., Lomasto, L., Orciuoli, F.: Detecting influential news in online communities: an approach based on hexagons of opposition generated by three-way decisions and probabilistic rough sets. Inf. Sci. **578**, 364–377 (2021)
2. Béziau, J.Y.: The power of the hexagon. Log. Univers. **6**(1–2), 1–43 (2012)
3. Boffa, S., Ciucci, D.: Three-way decisions with evaluative linguistic expressions. arXiv preprint arXiv:2307.11766 (2023)
4. Boffa, S., Ciucci, D., Murinová, P.: Comparing hexagons of opposition in probabilistic rough set theory. In: Ciucci, D., et al. (eds.) Information Processing and Management of Uncertainty in Knowledge-Based Systems. IPMU 2022. CCIS, vol. 1601, pp. 622–633. Springer, Cham (2022). https://doi.org/10.1007/978-3-031-08971-8_51

5. Boffa, S., Murinová, P., Novák, V.: Graded polygons of opposition in fuzzy formal concept analysis. Int. J. Approx. Reason. **132**, 128–153 (2021)
6. Boffa, S., Murinová, P., Novák, V.: A proposal to extend relational concept analysis with fuzzy scaling quantifiers. Knowl.-Based Syst. **231**, 107452 (2021)
7. Ciucci, D., Dubois, D., Prade, H.: Oppositions in rough set theory. In: Li, T., et al. (eds.) RSKT 2012. LNCS (LNAI), vol. 7414, pp. 504–513. Springer, Heidelberg (2012). https://doi.org/10.1007/978-3-642-31900-6_62
8. Dubois, D., Prade, H.: From blanché's hexagonal organization of concepts to formal concept analysis and possibility theory. Log. Univers. **6**, 149–169 (2012)
9. Gaeta, A.: Evaluation of emotional dynamics in social media conversations: an approach based on structures of opposition and set-theoretic measures. Soft Comput. **27**(15), 10893–10903 (2023)
10. Moral, A., Castiello, C., Magdalena, L., Mencar, C.: Explainable Fuzzy Systems. Springer, Cham (2021). https://doi.org/10.1007/978-3-030-71098-9
11. Moretti, A.: Why the logical hexagon? Log. Univers. **6**(1), 69–107 (2012)
12. Novák, V.: A comprehensive theory of trichotomous evaluative linguistic expressions. Fuzzy Sets Syst. **159**(22), 2939–2969 (2008)
13. Novák, V.: A formal theory of intermediate quantifiers. Fuzzy Sets Syst. **159**(10), 1229–1246 (2008)
14. Novák, V., Perfilieva, I., Dvorak, A.: Insight into Fuzzy Modeling. John Wiley & Sons, Hoboken (2016)
15. Parsons, T.: The traditional square of opposition (1997)
16. Pfeifer, N., Sanfilippo, G.: Probabilistic squares and hexagons of opposition under coherence. Int. J. Approx. Reason. **88**, 282–294 (2017)
17. Sugeno, M.: Theory of fuzzy integrals and its applications. Doctoral Thesis, Tokyo Institute of Technology (1974)
18. Yao, Y.: Three-way decision: an interpretation of rules in rough set theory. In: Wen, P., Li, Y., Polkowski, L., Yao, Y., Tsumoto, S., Wang, G. (eds.) RSKT 2009. LNCS (LNAI), vol. 5589, pp. 642–649. Springer, Heidelberg (2009). https://doi.org/10.1007/978-3-642-02962-2_81
19. Yao, Y.: Three-way decisions with probabilistic rough sets. Inf. Sci. **180**(3), 341–353 (2010)
20. Yao, Y.: Three-way granular computing, rough sets, and formal concept analysis. Int. J. Approx. Reason. **116**, 106–125 (2020)
21. Yao, Y., et al.: An outline of a theory of three-way decisions. In: RSCTC, vol. 7413, pp. 1–17 (2012)

Algebraic Models for Qualified Aggregation in General Rough Sets, and Reasoning Bias Discovery

A. Mani(✉)

Machine Intelligence Unit, Indian Statistical Institute, Kolkata,
203, B. T. Road, Kolkata 700108, India
a.mani.cms@gmail.com, amani.rough@isical.ac.in
https://www.logicamani.in

Abstract. In the context of general rough sets, the act of combining two things to form another is not straightforward. The situation is similar for other theories that concern uncertainty and vagueness. Such acts can be endowed with additional meaning that go beyond structural conjunction and disjunction as in the theory of *-norms and associated implications over L-fuzzy sets. In the present research, algebraic models of acts of combining things in generalized rough sets over lattices with approximation operators (called rough convenience lattices) is invented. The investigation is strongly motivated by the desire to model skeptical or pessimistic, and optimistic or possibilistic aggregation in human reasoning, and the choice of operations is constrained by the perspective. Fundamental results on the weak negations and implications afforded by the minimal models are proved. In addition, the model is suitable for the study of discriminatory/toxic behavior in human reasoning, and of ML algorithms learning such behavior.

Keywords: Abstract Approximations · Rough Implications · Algebraic Semantics · Skeptical Reasoning · Overly Optimistic Reasoning · L-Fuzzy Implications · Granular Operator Spaces · Rough Convenience Lattices · Algorithmic Bias Discovery

1 Introduction

In any context, the act of combining two things involves meta-level or semantic assumptions. These are more involved in the context of general rough sets because of the increased complexity of associated domains of discourse. Any generalized conjunction-like operation is referred to as an *aggregation*, while a generalized disjunction-like operation as a *co-aggregation* (valuations are not

---

This research is supported by Woman Scientist Grant No. WOS-A/PM-22/2019 of the Department of Science and Technology.

assumed). The situation is similar for other mature theories that concern uncertainty, vagueness or imprecision. For example, the theory of *-norms and associated implications are extensively investigated over L-fuzzy sets [1,2]. The purpose of the present research is to invent (or construct), and investigate models of somewhat related acts of combining things over lattices with approximation operators, and without explicit negations. At the application level, this research is strongly motivated by the desire to model skeptical or pessimistic or cautious, and overly optimistic reasoning over concepts (in human reasoning), and the choice of operations is constrained by the perspective.

Fundamental semantic results that formally address the following constraint on domains of discourse: When things are implied and negated in some perspectives then they are being approximated and vice versa are proposed. More specifically, concrete algebraic models (that involve a surprisingly weak set of axioms) in which the principle is valid are shown to exist in this research.

Several algebraic models of the operations of combining objects or rough objects (in several senses) in the context of classical, general and granular rough sets are known [12,21,25,27]. However, not many impose a meaning constraint that amount to combining in skeptical or biased or bigoted or overly optimistic ways. These concepts can possibly be attained relatively through partial orders on approximations. For example, if the lower approximation l_1 approximates better than another lower approximation l_2, then the latter is relatively more skeptical than the former. Consequently, aggregations of the l_2-approximations of objects must be more skeptical than that of aggregations of l_1-approximations. In the present research, the relative aspect is hidden because in most cases, the collections of rough objects (in various senses) form a lattice.

In the context of Pawlak/classical rough sets, it is proved by the present author [19] that aggregations f (interpreted as *rough dependence*), defined by the equation $f(a,b) = a^l \cap b^l$ can be used to define algebraic models that make no reference to approximations. The intent in the paper was to establish the differences between rough sets and probability theory from a *dependence* perspective. For a fuller discussion, the reader is referred to Sect. 5. However, the status of this operation and related ones in the abstract rough set literature is not known. This fundamental problem is solved in this research in suitably minimalist frameworks without negation operations. The framework is far weaker than the general approximation algebras considered in the paper [8], and is a specific version of a high general granular operator space [21,23] without the granularity requirement. Specific set-theoretic subclasses of high general granular operator spaces are also covered.

The organization of this paper is as follows. Necessary background is outlined in the next section. The model(s) are invented in the third section. Illustrative concrete and abstract instances of the models are explored in the section on examples. Connections between qualified aggregation and rough dependence are clarified in the fifth section.

2   Background

Definition 1. *A L-Fuzzy set* [11] *is a map* $\varphi : X \longmapsto L$, *where* X *is a set and* $L = \langle \underline{L}, \leqslant \rangle$ *is a quasi-ordered set. The set of all L-fuzzy sets will be denoted by* $\mathbb{F}(X, L)$.

2.1   T-Norms, S-Norms, Uninorms, and Implications

While T-norms, S-norms and uninorms are primarily viewed as operations on lattices or partially ordered sets for the algebraic and logical models of L-fuzzy sets, they can be utilized in the algebraic models of entirely different phenomena. The mentioned T/S/uni-norms are well-known algebraic operations in the algebra literature because the topological constraints of the unit interval context are not imposed. The conditions that make related considerations stand out from those on corresponding order-compatible algebras are boundedness and the role of additional operations such as those of generalized implications and negations. Some essential concepts (for more details, see for example [2,3,7,27,33]) are mentioned here.

Definition 2. *Let* $P = \langle \underline{P}, \leqslant, e \rangle$ *be a partially ordered set (poset) on the set* $\underline{P}$ *with distinguished element (or 0-ary operation)* e, *then any order-compatible binary associative operation* · *on it with identity element* · *is referred to as a* pseudo uni-norm. *A commutative pseudo uni-norm is a* uni-norm. *If* e *is the greatest (respectively least) element of* P *then* · *is a* pseudo t-norm *(respectively* pseudo s-norm*).*

The set of all pseudo uninorms on the poset P is denoted by $\mathcal{U}(P, e)$. It forms a poset in the induced point-wise order. Both t-norms and t-conorms are uninorms.

Definition 3. *If* L *is a bounded lattice with bottom* $\bot$ *and top* $\top$, *then a t-norm* $\odot$ *is a commutative, associative order compatible monoidal operation with* $\top$ *being the identity. A* s-norm *(or* t-conorm*) is a commutative, associative order compatible monoidal operation with* $\bot$ *being the identity.*

Consider the conditions possibly satisfied by a map $n : L \longmapsto L$:

$$n(\bot) = \top \ \& \ n(\top) = \bot \tag{N1}$$
$$(\forall a, b)(a \leqslant b \longrightarrow n(b) \leqslant n(a)) \tag{N2}$$
$$(\forall a)n(n(a)) = a \tag{N3}$$
$$n(a) \in \{\bot, \top\} \text{ if and only if } a = \bot \text{ or } a = \top \tag{N4}$$

n is a *negation* if and only if it satisfies N1 and N2, while n is a *strong negation* if and only if it satisfies all the four conditions.

2.2 Implication Operations

Implications satisfy a wide array of properties as they depend on the other permitted operations. Here some relevant ones are mentioned.

A function $\beth : L^2 \longmapsto L$ is an *implication* [2] if it satisfies (for any $a, b, c \in L$) the following:

$$\text{If } a \leqslant b \text{ then } \beth bc \leqslant \beth ac \qquad \text{(First Place Antitonicity FPA)}$$
$$\text{If } b \leqslant c \text{ then } \beth ab \leqslant \beth ac \qquad \text{(Second Place Monotonicity SPM)}$$
$$\beth \bot \bot = \top \qquad \text{(Boundary Condition 1: BC1)}$$
$$\beth \top \top = \top \qquad \text{(Boundary Condition 2: BC2)}$$
$$\beth \top \bot = \bot \qquad \text{(Boundary Condition 3: BC3)}$$

Infix notation is preferred for algebraic reasons. The set of all implications on the lattice L will be denoted by $\mathcal{I}(L)$. It can be endowed with a bounded lattice structure under the induced order

$$\beth_1 \preceq \beth_2 \text{ if and only if } (\forall a, b \in L)\beth_1 ab \leqslant \beth_2 ab.$$

The top $\beth_\top$ and bottom $\beth_\bot$ implications are defined as follows:

- If $a = \top$ & $b = \bot$ then $\beth_\top ab = \bot$, otherwise $\beth_\top ab = \top$.
- If $a = \bot$ & $b = \top$ then $\beth_\bot ab = \top$, otherwise $\beth_\bot ab = \bot$.

Some other properties of interest in this paper are

$$\beth \top x = x \qquad \text{(LNP)}$$
$$\beth a(\beth bc) = \beth(b\beth(ac)) \qquad \text{(Exchange Principle EP)}$$
$$\beth ab = 1 \text{ if and only if } a \leqslant b \qquad \text{(Ordering Property, OP)}$$
$$\beth a(\beth ab) = \beth ab \qquad \text{(Iterative Boolean Law, IBL)}$$
$$b \leqslant \beth ab \qquad \text{(Consequent Boundary, CB)}$$

Further, note that Tarski algebras are the same thing as implication algebras [22,29]. A few full dualities relating to classes of such algebras are known. One of this is a duality for finite Tarski sets [5,6] or covering approximation spaces.

Definition 4. *A Tarski algebra (or an implication algebra) is an algebra of the form* $S = \langle \underline{S}, \beth, \top \rangle$ *of type* $2,0$ *that satisfies (in the following, [29])*

$$\beth \top a = a \qquad \text{(Left Neutrality, LNP)}$$
$$\beth aa = \top \qquad \text{(Identity Principle, IP)}$$
$$\beth a(\beth bc) = \beth(\beth ab)(\beth ac) \qquad \text{(T3)}$$
$$\beth(\beth ab)b = \beth(\beth ba)a \qquad \text{(T4)}$$

A join-semilattice order $\leqslant$ is definable in a IA as below:

$$(\forall a, b)\, a \leqslant b \leftrightarrow \beth ab = \top;\, \text{the join is}\, a \vee b = \beth(\beth ab)b$$

Filters or deductive systems of an IA S are subsets $K \subseteq S$ that satisfy

$$1 \in K \,\&\, (\forall a, b)(a, \beth ab \in K \longrightarrow b \in K)$$

The set of all filters $\mathcal{F}(S)$ is an algebraic, distributive lattice whose compact elements are all those filters generated by finite subsets of S.

3   Model of Rough Skeptic and Pessimistic Reasoning

For concreteness, a minimal base model over which the theory will be invented is defined next. Generalizations to weaker order, antichains, and partial approximation operations are considered in a separate paper.

Definition 5. *An algebra of the form* $B = \langle \underline{B}, l, u, \vee, \wedge, \bot, \top \rangle$ *with* $(\underline{B}, \vee, \wedge, \bot, \top)$ *being a bounded lattice will be said to be a* rough convenience lattice *(RCL) if the following conditions are additionally satisfied ($\leqslant$ is the associated lattice order, and the operations* l *and* u *are generalized lower and upper approximation operators respectively):*

$$(\forall x) x^{ll} = x^l \leqslant x \leqslant x^u \leqslant x^{uu} \tag{lu1}$$

$$(\forall a, b)(a \leqslant b \longrightarrow a^l \leqslant b^l) \tag{l-mo}$$

$$(\forall a, b)(a \leqslant b \longrightarrow a^u \leqslant b^u) \tag{u-mo}$$

$$(\forall a, b)\, a^l \vee b^l \leqslant (a \vee b)^l \,\&\, a^u \vee b^u = (a \vee b)^u \tag{lu2}$$

$$(\forall a, b)(a \wedge b)^l = a^l \wedge b^l \,\&\, (a \wedge b)^u \leqslant a^u \wedge b^u \tag{lu3}$$

$$\top^u = \top \,\&\, \bot^l = \bot = \bot^u \tag{topbot}$$

Proposition 1. *In Definition 5, lu2 and lu3 follow from lu1, l-mo and u-mo.*

The concept is weaker than that of a general abstract approximation space. Note that by default no relation between the lower and upper approximations are assumed. Further, nothing is assumed about negations or complementation. A special case of a rough convenience lattice is a set-HGOS under additional conditions. However, note that no granularity-related restrictions are imposed on a rough convenience lattice.

An element $a \in \mathcal{B}$ will be said to be *lower definite* (resp. *upper definite*) if and only if $a^l = a$ (resp. $a^u = a$) and *definite*, when it is both lower and upper definite. For possible concepts of rough objects the reader is referred to the paper [21].

Definition 6. *By a* roughly consistent object *(RCO) (respectively* lower RCO, upper RCO*) will be meant a set of elements of the RCL* $\underline{B}$ *of the form* $H = \{x; (\forall b \in H) x^l = b^l \ \& \ x^u = b^u\}$ *(respectively* $H_l = \{x; (\forall b \in H_l) x^l = b^l\}$ *and* $H_u = \{x; (\forall b \in H_u) x^u = b^u\}$ *). The set of all roughly consistent objects is partially ordered by the set inclusion relation. Relative to this order,* maximal *roughly consistent objects will be referred to as* rough objects. *Analogously,* lower *and* upper rough objects *will be spoken of. The collection of such objects will respectively be denoted by* $R(B)$, $R_l(B)$ *and* $R_u(B)$.

Proposition 2. *In a RCL B, every maximal roughly consistent object is an interval of the form* (x^l, x^u) *for some* $x \in Bs$. *The converse holds as well.*

Proof. The result follows from the monotonicity of l and u, and lul. □

Definition 7. *By the* rough order $\Subset$ *on* $R(B)$, *will be meant the relation* $\Subset$ *defined by* $(x^l, x^u) \Subset (a^l, a^u)$ *if and only if* $(x^l \leqslant a^l$ *and* $x^u \leqslant a^u)$.

It can be shown that $\Subset$ is a bounded partial lattice order on $R(B)$. The least element of $R(B)$ is $(\bot, \bot)$, and its greatest element is the interval $(\top, \top)$. The meet and join operations of B induce partial lattice operations on $R(B)$ – these are investigated separately.

Definition 8. *In a RCL B, let for any* $a, b \in B$

$$a \cdot b := a^l \wedge b^l \text{ and } a \otimes b = a^u \vee b^u.$$

The operations $\cdot$ *and* $\otimes$ *will respectively be referred to as* cautious co-aggregation *CCA, and* optimistic aggregation *OA respectively.*

The operation $\cdot$ can as well be interpreted as a pessimistic co-aggregation. The appropriateness of the competing interpretations is dependent on the relation with the **OA** or on other context-specific features.

Theorem 1. *The CCA operation defined above satisfies all the following:*

$$(\forall a, b) \, a \cdot b = b \cdot a \qquad \text{(Ccomm)}$$
$$(\forall a, b, e) \, a \cdot (b \cdot e) = (a \cdot b) \cdot e \qquad \text{(Casso)}$$
$$(\forall a, b)(a \leqslant b \longrightarrow a \cdot e \leqslant b \cdot e) \qquad \text{(Cm)}$$
$$(\forall a) \, a \cdot \bot = \bot \qquad \text{(Cb)}$$

Proof. For any $a, b \in B$, $a \cdot b = a^l \wedge b^l = b^l \wedge a^l = b \cdot a$. This proves Ccomm. Associativity can be proved as follows. For any $a, b, e \in B$,

$$a \cdot (b \cdot e) = a^l \wedge (b \cdot e)^l = \qquad \text{(by definition)}$$
$$= a^l \wedge (b^l \wedge e^l)^l = a^l \wedge b^{ll} \wedge e^{ll} \qquad \text{(by lu3)}$$
$$= (a^l \wedge b^l) \wedge e^l = (a \cdot b) \cdot e \qquad \text{(by lu1.)}$$

Monotonicity follows from the monotonicity of l.
Finally, $a \cdot \bot = a^l \wedge \bot^l = a^l \wedge \bot = \bot$. □

Theorem 2. *Omitting the initial universal quantifiers,*

$$a \otimes b = b \otimes a \quad \text{(Acomm)}$$
$$(a^{uu} = a^u \ \& \ b^{uu} = b^u \ \& \ e^{uu} = e^u \longrightarrow a \otimes (b \otimes e) = (a \otimes b) \otimes e) \quad \text{(wAasso1)}$$
$$a \otimes ((b \vee e) \otimes a) = ((a \vee b) \otimes c) \otimes c \quad \text{(wAsso2)}$$
$$(a \leqslant b \longrightarrow a \otimes e \leqslant b \otimes e) \quad \text{(Am)}$$
$$a \otimes \top = \top \quad \text{(Ab)}$$

Proof. Acomm follows from the commutativity of $\vee$.
  For any $a, b, e \in B$,

$$a \otimes (b \otimes e) = a^u \vee (b \otimes e)^u = \qquad \text{(by definition)}$$
$$= a^u \vee (b^u \vee e^u)^u = a^u \vee b^{uu} \vee e^{uu}, \qquad \text{(by lu2)}$$
$$\text{However, } (a \otimes b) \otimes e = a^{uu} \vee b^{uu} \vee e^u \qquad \text{(by lu2)}$$
$$\text{So the premise of wAsso1 ensures it.}$$

Using the definition of $\otimes$ on the LHS and RHS of wAsso2, it can be seen that

$$a \otimes ((b \vee e) \otimes a) = a^u \vee ((b \vee e)^u \vee a^u)^u = \qquad \text{(by definition)}$$
$$a^{uu} \vee b^{uu} \vee e^{uu}$$
$$\text{(by Acomm, lu2, u-mo, lu1)}$$
$$\text{Similarly, } ((a \vee b) \otimes c) \otimes c = a^{uu} \vee b^{uu} \vee c^{uu} \vee c^u = \qquad \text{(by definition, lu2)}$$
$$a^{uu} \vee b^{uu} \vee e^{uu} \qquad \text{(by u-mo)}$$
$$\text{This proves wAsso2.} \qquad \text{(by =)}$$

  Monotonicity of $\otimes$ can be proved from the monotonicity of $\vee$ and that of u.
□

Definition 9. *Two generalized negations are definable on a RCL, B as follows:*

$$\neg a = \inf\{z : z \in B \ \& \ a \otimes z = \top\} \quad \text{(addneg)}$$
$$\sim a = \sup\{z : z \in B \ \& \ a \cdot z = \bot\} \quad \text{(mulneg)}$$

These satisfy the properties specified in the next two theorems.

Theorem 3.

$$(\forall a)\neg\neg a \leqslant a^u \qquad (\text{WN3-N})$$
$$(\forall a,b)(a \leqslant b \longrightarrow \neg b \leqslant (\neg b)^u \leqslant (\neg a)^u) \qquad (\text{WN2-N})$$
$$\neg\bot \leqslant \top \ \& \ \neq \top = \bot \qquad (\text{WN1-N})$$

Proof. $\neg\neg a \otimes \neg a = \top$. Therefore,

$$(\neg\neg a)^u \vee (\neg a)^u = \top = (\neg a)^u \vee a^u$$

By the definition of $\neg$, it follows that $\neg\neg a \leqslant (\neg\neg a)^u \leqslant a^u$ as $(\neg a)^u$ is in both the equalities.

To see this suppose $\neg\neg a > a$, then $\neg\neg a \vee a = \neg\neg a$ This implies $(\neg\neg a)^u \vee a^u = (\neg\neg a)^u$, and $a^u \vee (\neg a)^u = \top$ contradicts the definition of $\neg$. This proves WN3-N.

Clearly, $(\neg b)^u \vee b^u = \top = (\neg a)^u \vee a^u$. If $a \leqslant b$ then $a^u \leqslant b^u$. So, $(\neg a)^u \vee a^u \vee b^u = \top$. This yields $(\neg a)^u \vee b^u = \top$. By the definition of $\neg$ and monotonicity of $\vee$, it is necessary that $(\neg b)^u \leqslant \vee(\neg a)^u$. This proves WN2-N.

If $a^u \vee \top^u = \top$ for some a, then a can be any element of the universe because $\top^u = \top$. Of these the smallest is $\bot$. Therefore, $\neq \top = \bot$ holds. However, $\neg\bot$ is the infimum of the elements whose upper approximation is $\top$, and so $\neg\bot \leqslant \top$. WN1-N is thus proved. $\qquad\square$

Theorem 4.

$$(\forall a,b)(a \leqslant b \longrightarrow (\sim b)^l \leqslant (\sim a)^l \leqslant \sim a) \qquad (\text{WN2-S})$$
$$\bot \leqslant \sim \top \ \& \ \top = \sim \bot \qquad (\text{WN3-S})$$

Proof. Clearly, $(\sim b)^l \wedge b^l = \bot = (\sim a)^l \wedge a^l$. If $a \leqslant b$ then $a^l \leqslant b^l$. So, $(\sim b)^l \wedge a^l = \bot$. By the definition of $\sim$, this means $(\sim b)^l \leqslant (\sim a)^l \leqslant \sim a$, and proves WN2-S.

If $a^l \wedge \top^l = \bot$ for some a, then a must necessarily be an element of the universe satisfying $a^l = \bot$. By definition, it is clear that in general, a need not coincide with $\bot$.

If $a^l \wedge \bot^l = \bot$, then a can be any element of the universe. The largest of these is $\top$. Therefore, $\sim \bot = \top$. This proves WN3-S. $\qquad\square$

The above means that $\sim$ is a weak negation. The properties of the negation improve when the RCL satisfies a weak complementation c that satisfies the conditions

$$(\forall x)x^{cc} \leqslant x \qquad (\text{c1})$$
$$(\forall x)x^c \wedge x = \bot \qquad (\text{c2})$$

The above two conditions ensure that for any a, $\sim a \leqslant a^{lc}$.

3.1   Implications

Given the definitions of negation, and $*$-norms, a natural candidate for a definition of implication is given by the following equation

$$\beth_\neg ab = (\neg a) \otimes b \qquad \text{(Negimplication)}$$

Theorem 5. *The operation* $\beth_\neg$ *satisfies the properties: FPA, IP, SPM, BC1, BC2, and BC3. So it is an implication operation.*

Proof. If $a \leqslant c$, then $\neg c \leqslant (\neg c)^u \leqslant (\neg a)^u$. Therefore, $(\neg c)^u \vee b^u \leqslant (\neg a)^u \vee b^u$ From this it follows that $\beth_\neg cb \leqslant \beth_\neg ab$. So FPA holds.

Suppose $b \leqslant c$ then for any $a \in B$ $\beth_\neg ab = (\neg a)^u \vee b^u$ and $\beth_\neg ac = (\neg a)^u \vee c^u$ (by definition). Monotonicity of u ensures that $(\neg a)^u \vee b^u \leqslant (\neg a)^u \vee c^u$. Therefore, SPM must hold.

$\beth_\neg \bot\bot = (\neg\bot)^u \vee \bot^u = \top$ (by definition). So BC1 holds.
$\beth_\neg \top\top = (\neg\top)^u \vee (\top)^u = \top$ (by definition). So BC2 holds.
$\beth_\neg \top\bot = (\neg\top)^u \vee (\bot)^u = (\bot)^u = \bot$. So BC3 holds.
For any a, $\beth_\neg aa = (\neg a)^u \vee (a)^u = \top$. So IP holds. $\qquad\square$

Other possibilities are

$$\beth_o ab = (\neg a) \vee b \qquad \text{(negvee)}$$
$$\beth_\backsim ab = (\backsim a) \cdot b \qquad \text{(simplication)}$$
$$\beth_s ab = (\backsim a) \wedge b \qquad \text{(simwed)}$$

Of these $\beth_\backsim$ is most interesting, and has the following properties:

Theorem 6. *In a RCL B,* $\beth_\backsim$ *satisfies FPA, SPM, BC3, IBL and converse of CB.*

Proof. FPA: Suppose $a \leqslant b$ for any $a, b \in B$. $\beth_\backsim bc = (\backsim b)^l \wedge c^l$, and $\beth_\backsim ac = (\backsim a)^l \wedge c^l$, By WN2-S, it follows that $(\backsim b)^l \wedge c^l \leqslant (\backsim a)^l \wedge c^l$. This ensures FPA.

SPM: Suppose $a \leqslant b$ for any $a, b \in B$. $\beth_\backsim cb = (\backsim c)^l \wedge b^l$, and $\beth_\backsim ca = (\backsim c)^l \wedge a^l$. Under the assumption $(\backsim c)^l \wedge a^l \leqslant (\backsim c)^l \wedge b^l$. SPM follows from this.

BC3: $\beth_\backsim \top\top = (\backsim \top)^l \wedge \top^l = \bot$. So BC3 holds.

IBL: For any $a, b \in B$, $ia(\beth ab) = (\backsim a)^l \wedge ((\backsim a)^l \wedge b^l)^l = (\backsim a)^l \wedge (\backsim a)^{ll} \wedge b^{ll} = (\backsim a)^l \wedge (\backsim a)^l \wedge b^l = (\backsim a)^l \wedge b^l = \beth_\backsim ab$. This proves IBL.

The converse of CB holds because $(\backsim a)^l \wedge b^l \leqslant b$ is satisfied for all possible values of a. $\qquad\square$

It can be verified that BC1, BC2, and IP do not hold in general for $\sqsupset_\smile$.

The operation $\cdot$ can be naturally interpreted as a pessimistic or skeptical aggregation because it essentially selects a common part of two lower approximations (that are not restricted in their badness). Two related operations are $\odot$ and $\times$ can be defined by (for any $a, b \in B$) $a \odot b := a^l \vee b^l$ and $a \times b := a^u \wedge b^u$. The operation $\otimes$ on the other hand is optimistic at every stage of the reasoning process. First, the possibilistic upper approximation operators are used, and then the one that certainly contains the upper approximations is constructed. $\odot$ and $\times$ are essentially intermediate operations.

3.2 Concrete and Abstract Algebraic Models

It is shown that a rough convenience lattice and closely related abstract algebraic systems have a far richer structure than is assumed in the literature. In concrete terms, every RCL can be naturally enhanced to the following algebraic system.

Definition 10. *By a* Concrete RCL Aggregation Algebra *(CRCLAA) will be meant an algebra of the form*

$$B = \langle \underline{B}, \otimes, \cdot, \vee, \wedge, l, u, \neg, \sim, \bot, \top \rangle$$

with $\langle \underline{B}, \vee, \wedge, l, u, \bot, \top \rangle$ *being a* RCL, *and the operations* $\otimes, \cdots, \otimes, \cdot, \neg,$ *and* $\sim$ *are as defined in the previous subsections.*

While the operations $\cdot$ and $\otimes$ are terms derived in the signature of the RCL, the other operations are defined by imposing a perspective on them. This suggests that an abstract property-based definition of an algebra of the same type may not be always equivalent to a CRCLAA. Additionally, it makes sense to retain the implications and omit negations.

Definition 11. *By an* Abstract RCL Aggregation Negation Algebra *(CRCLANA) will be meant an algebra of the form*

$$B = \langle \underline{B}, \otimes, \cdot, \vee, \wedge, l, u, \neg, \sim, \bot, \top \rangle$$

that satisfies the following conditions:

$$\langle \underline{B}, \vee, \wedge, l, u, \bot, \top \rangle \text{ is a RCL.} \tag{rcl}$$
$$\cdot \text{ satisfies Ccomm, Casso, Cm, and Cb.} \tag{cdotc}$$
$$\otimes \text{ satisfies Acomm, wAsso1, wAsso2, Am, and Ab.} \tag{otimc}$$
$$\neg \text{ satisfies WN1-N, WN2-N, and WN3-N.} \tag{negc}$$
$$\sim \text{ satisfies WN2-S, and WN3-S.} \tag{simc}$$

Definition 12. *An Abstract RCL Aggregation Implication Algebra (CRCLAIA) shall be an algebra of the form* $B = \langle \underline{B}, \otimes, \cdot, \vee, \wedge, l, u, \sqsupset_\neg, \sqsupset_\neg, \perp, \top \rangle$ *that satisfies:*

$$\langle \underline{B}, \vee, \wedge, l, u, \perp, \top \rangle \ \text{is a RCL.} \tag{rcl}$$

$$\cdot \ \text{satisfies Ccomm, Casso, Cm, and Cb.} \tag{cdotc}$$

$$\otimes \ \text{satisfies Acomm, wAsso1, wAsso2, Am, and Ab.} \tag{otimc}$$

$$\sqsupset_\neg \ \text{satisfies FPA, SPM, BC3, and IBL.} \tag{imsc}$$

$$\sqsupset_\neg \ \text{satisfies FPA, IP, SPM, BC1, BC2, and BC3.} \tag{inegc}$$

The above allows the following interesting problems. It may be noted that the associated contexts in logic are not known because the defining conditions are not strong enough.

Problem 1. Under what additional conditions are CRCLANA and CRCLAIA representable as concrete RCL aggregation algebras?

4   Illustrative Examples

In academic learning contexts, all stakeholders approximate concepts within their own frameworks, and perspectives [24]. However, the learning context admits of common languages of discourse – it is very important that this be *large and expressive* enough. In practical terms, this means that the admitted basic predicates or functions should be many in number, and be endowed with minimalist properties relative to what may be possible in associated contexts. Below an abstract and a concrete example are constructed.

4.1   Abstract Example

Let $B = \{\perp, \top, a, b, c, e, f\}$ be endowed with the lattice order depicted in Figure 1. Suppose the lower and upper approximations are respectively $\{(\perp, \perp), (\top, e), (a, c), (b, b), (c, c), (e, c), (f, \perp)\}$ and $\{(\perp, \perp), (\top, \top), (a, a), (b, \top), (c, e), (e, e), (f, b)\}$ respectively. The operations $\otimes, \cdot, \neg,$ and $\sim$ are then computable as in the three tables, while the implications follow (Fig. 1 and Tables 1, 2 and 3).

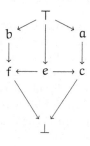

Fig. 1. Bounded Lattice

Table 1. ⊗-Table

⊗	⊥	T	a	b	c	e	f
⊥	⊥	T	a	T	e	e	b
T	T	T	T	T	T	T	T
a	a	T	a	T	T	T	T
b	T	T	T	T	T	T	T
c	e	T	T	T	e	e	T
e	e	T	T	T	e	e	T
f	b	T	T	T	T	T	b

Table 2. ·-Table

⊗	⊥	T	a	b	c	e	f
⊥	⊥	T	a	T	e	e	b
T	T	T	T	T	T	T	T
a	a	T	a	T	T	T	T
b	T	T	T	T	T	T	T
c	e	T	T	T	e	e	T
e	e	T	T	T	e	e	T
f	b	T	T	T	T	T	b

Table 3. Negations

Neg	⊥	T	a	b	c	e	f
¬	b	⊥	⊥	⊥	⊥	⊥	⊥
~	T	f	b	c	f	f	T

4.2   Detection of Reasoning and Algorithm Bias

Skeptical aggregation is often a feature of negative bias in human reasoning. Suppose a toxic person with decision-making powers is constrained by their environment from explicitly discriminating against specific groups of people. Then they are likely to discriminate by adopting additional distracting strategies and empty agendas. The effect of such practices can be analyzed through the aggregation strategies adopted. In fact, serious political analysts frequently try to do precisely that.

The behavior of biased or defective algorithms is typically reflected in the data used, and produced by it (because the results produced at different stages are again a form of data). Analysis of empirical bias can possibly be deduced from the associated data sets. If it can be shown that bias is due to the algorithm learning from biased data, then it means that algorithm is not safe for the purpose. Such generalities can be analyzed with the proposed methodology.

Many types of models are possible for information tables that are the result of systemic bias in the data collection process or due to external factors. These may be partly reflected in the data, and in such circumstances the methods invented in this research may be applicable. External factors may be taken into account through the approximation operators (about which the assumptions are left to the practitioner). The essence of the procedure is outlined below:

- Let $C_1, \ldots C_k$ be subsets of B that potentially correspond to specific objects that are discriminated against.
- Let $E_1, \ldots E_k$ be subsets of B that potentially correspond to specific objects that are unduly favored.
- Let $F_1, \ldots F_k$ be subsets of B that potentially correspond to specific objects that help in bias determination
- It is assumed that a similar pattern of bias is maintained by the process.
- Compute $C_i \cdot F_i$, $C_i \otimes F_i$, $E_i \cdot F_i$, and $E_i \otimes F_i$
- Let the principal lattice filters generated by these be respectively $\exists(C_i \cdot F_i)$, $\exists(C_i \otimes F_i)$, $\exists(E_i \cdot F_i)$, and $\exists(E_i \otimes F_i)$.
- If B is finite, compute the cardinalities of the principal lattice filters generated by the four.

- A simple measure of bias $\flat(C, E)$ is defined in Equation biase. It will be referred to as the *flat bias measure*.

$$\flat(C, E) = 1 - \frac{1}{k} \sum \frac{\text{Card}(\dashv(C_i \cdot F_i))}{\text{Card}(\dashv(E_i \cdot F_i))} \qquad \text{(biase)}$$

If it is certain that the co-aggregations are justified, then they can be used for the *sharp bias measure* defined in Equation sharpe

$$\eth(C, E) = 1 - \frac{1}{k} \sum \frac{\text{Card}(\dashv(C_i \otimes F_i)) - \text{Card}(\dashv(C_i \cdot F_i))}{\text{Card}(\dashv(E_i \otimes F_i)) - \text{Card}(\dashv(E_i \cdot F_i))} \qquad \text{(sharpe)}$$

5   Skeptical Aggregation and Rough Dependence

A theory of rough dependence, and associated measures for subclasses of granular rough sets (in the axiomatic sense) is invented by the present author in earlier papers [17–19]. It concerns the extent to which an object depends on another expressed in terms of rough objects of different types. The representation is used in the context of contrasting it with that of probabilistic dependence. In fact, it is proved by her [19] that the models of dependence based probability and models of rough dependence do not share too many axioms. Subsequent research led to the invention of a theory of non-stochastic rough randomness and large-minded reasoners [26].

Definition 13. *Let* $B = \langle \underline{B}, \mathcal{G}, l, u, \vee, \wedge, \perp, \top \rangle$ *be a structure with* $\underline{B}$ *being a subset of a powerset* $\wp(S)$, $\langle \underline{B}, l, u, \vee, \wedge, \perp, \top \rangle$ *being a RCL,* $\wedge = \cap$, $\vee = \cup$, $\top = S$, *and* $\perp = \emptyset$, *and* $\mathcal{G} \subseteq B$ *is a granulation on* B *in the axiomatic sense* [15,21,23] *(t being a term function in the algebraic language of the RCL):*

$$(\forall x \exists a_1, \dots a_r \in \mathcal{G})\, t(a_1, y_2, \dots a_r) = x^l \qquad \text{(Weak RA, WRA)}$$
$$\text{and } (\forall x)\, (\exists a_1, \dots a_r \in \mathcal{G})\, t(a_1, a_2, \dots a_r) = x^u,$$

$$(\forall a \in \mathcal{G})(\forall x \in \wp(\underline{S}))\, (a \subseteq x \longrightarrow a \subseteq x^l), \qquad \text{(Lower Stability, LS)}$$

$$(\forall x, a \in \mathcal{G})(\exists z \in \wp(\underline{S}))\, x \subset z, \,\& \, a \subset z \, \& \, z^l = z^u = z, \qquad \text{(Full Underlap, FU)}$$

B *will then be referred to as a set granular RCL (sGRCL).*

It is easy to see that all sGRCLs are set HGOS as well.

In a sGRCL B, if $\nu(B)$ is the collection of definite objects in some sense,

Definition 14. *The $\mathcal{G}v$-infimal degree of dependence $\beta_{i\tau v}$ of x on z is defined by*

$$\beta_{i\mathcal{G}v}(x, z) = \inf_{v(B)} \bigcup \{g : g \in \mathcal{G} \ \& \ g \subseteq x \ \& \ g \subseteq z\}. \tag{1}$$

The infimum is over the $v(B)$ elements contained in the union.
   The $\mathcal{G}v$-supremal degree of dependence $\beta_{s\mathcal{G}v}$ of x on z is defined by

$$\beta_{s\mathcal{G}v}(x, z) = \sup_{v(B)} \bigcup \{g : g \in \mathcal{G} \ \& \ g \subseteq x \ \& \ g \subseteq z\}. \tag{2}$$

The supremum is over the $v(S)$ element containing the union.

If unions of granules are always definite, then the two concepts coincide.

Theorem 7. *In classical rough sets with $\mathcal{G}$ being the set of equivalence classes and $v(B) = \delta_l(B)$ - the set of lower definite elements, then*

$$\beta_i xz = x^l \cap z^l = \beta_s xz$$

The converse of $(x \odot y = 0 \longrightarrow \beta_i xy = 0)$ is not true in general.

The following proposition can be deduced

Proposition 3. *In the context of classical rough sets, the degrees of rough dependence (relative to $v(B)$ being the set of lower definite elements) coincides with the skeptical aggregation operation.*

However, in slightly more general set-theoretical contexts, rough dependence does not coincide with skeptical aggregation. This follows from the above definition (additionally, readers may refer to Sect. 7 of the paper [19]).

6   Directions

It should be stressed that there is much scope for reducing the axioms assumed in studies on rough sets over residuated lattices or ortho-lattices [4]. This research contributes to this broad project in the spirit of reverse mathematics that seeks a minimum of axioms for a result. Dualities, somewhat related to recent results [9], for CRCLANA and CRCLAIA are of interest.
   The terms *pessimistic,* and *optimistic* are used in different senses in the rough set and AIML literature. In the so-called multi-granulation studies [28] that concern contexts with multiple rough approximations (or multiple relations or granulations) on the same universe, it is used as an adjective for specific derived approximations. However, these are studied under other names in many older papers [13,16,30]. Algebraic aspects are explored by the present author [16], and others [10]. Three-way decision strategies are additionally studied *to seek common ground while eliminating differences* [32]. Modal logic of the point-wise approximations in some of these contexts are explored in more recent work [14].

The present study is about models of aggregation and co-aggregation from a rough set view, and therefore it is not directly related to these as the idea necessarily involves systems of approximations from different sources (and is relative to at least two such sources). For example, by regulating the nature of the lower approximation, even extremely biased or bigoted views can be expressed by the aggregation f mentioned earlier. As this is not really part of a multi source scenario, possible connections are research topics. Implication operations from a rough set perspective are studied in many related models such as quasi-boolean algebras [31]. The results proved here show that many of the assumptions are not essential. A detailed study will appear separately.

In forthcoming papers, the semantics is extended to antichains of mutually distinct objects, building on earlier work of the present author [20]. Further applications to concept modeling in education research and teaching contexts are areas of her ongoing research [24].

References

1. Baczynski, M., Jayaram, B.: Fuzzy Implications. Springer, Heidelburg (2008). https://doi.org/10.1007/978-3-540-69082-5
2. Bedregal, B., Beliakov, G., Bustince, H., Fernandez, J., Pradera, A., Reiser, R.: (S, N)-implications on bounded lattices. In: Baczyński, M., Beliakov, G., Sola, H.B., Pradera, A. (eds.) Advances in Fuzzy Implication Functions. Studies in Fuzziness and Soft Computing, vol. 300, pp. 105–124. Springer, Heidelberg (2013). https://doi.org/10.1007/978-3-642-35677-3_5
3. Bedregal, B., Santiago, R., Madeira, A., Martins, M.: Relating Kleene algebras with pseudo uninorms. In: Areces, C., Costa, D. (eds.) DaLi 2022. LNCS, vol. 13780, pp. 37–55. Springer, Cham (2022). https://doi.org/10.1007/978-3-031-26622-5_3
4. Cattaneo, G., Ciucci, D.: Algebraic methods for orthopairs and induced rough approximation spaces. In: Mani, A., Cattaneo, G., Düntsch, I. (eds.) Algebraic Methods in General Rough Sets. TM, pp. 553–640. Springer, Cham (2018). https://doi.org/10.1007/978-3-030-01162-8_7
5. Celani, S.: Modal Tarski algebras. Rep. Math. Logic **39**, 113–126 (2005)
6. Celani, S., Cabrer, L.: Topological duality for Tarski algebras. Algebra Univers. **58**(1), 73–94 (2008)
7. Chakraborty, M., Dutta, S.: Theory of Graded Consequence. Logic in Asia. Springer, Singapore (2019). https://doi.org/10.1007/978-981-13-8896-5
8. Ciucci, D.: Approximation algebra and framework. Fund. Inform. **94**, 147–161 (2009)
9. Düntsch, I., Orlowska, E.: Discrete dualities for groupoids. Rend. Istit. Mat. Univ. Trieste **53**, 1–19 (2021). https://doi.org/10.13137/2464-8728/33304
10. Gegeny, D., Kovacs, L., Radeleczki, S.: Lattices defined by multigranular rough sets. Int. J. Approximate Reasoning **151**, 413–429 (2022)
11. Goguen, J.A.: L-fuzzy sets. J. Math. Anal. Appl. **18**, 145–174 (1967)
12. Järvinen, J.: Lattice theory for rough sets. In: Peters, J.F., Skowron, A., Düntsch, I., Grzymała-Busse, J., Orłowska, E., Polkowski, L. (eds.) Transactions on Rough Sets VI. LNCS, vol. 4374, pp. 400–498. Springer, Heidelberg (2007). https://doi.org/10.1007/978-3-540-71200-8_22

13. Khan, M.A.: Multiple-source approximation systems, evolving information systems and corresponding logics. Trans. Rough Sets **20**, 146–320 (2016)
14. Khan, M.A., Patel, V.S.: A simple modal logic for reasoning in multi granulation rough models. ACM Trans. Comput. Log. **19**(4), 1–23 (2018)
15. Mani, A.: Dialectics of counting and the mathematics of vagueness. In: Peters, J.F., Skowron, A. (eds.) Transactions on Rough Sets XV. LNCS, vol. 7255, pp. 122–180. Springer, Heidelberg (2012). https://doi.org/10.1007/978-3-642-31903-7_4
16. Mani, A.: Towards logics of some rough perspectives of knowledge. In: Suraj, Z., Skowron, A. (eds.) Rough Sets and Intelligent Systems - Professor Zdzisław Pawlak in Memoriam. Intelligent Systems Reference Library, vol. 43, pp. 419–444. Springer, Heidelberg (2013). https://doi.org/10.1007/978-3-642-30341-8_22
17. Mani, A.: Ontology, rough Y-systems and dependence. Internat. J. Comp. Sci. and Appl. **11**(2), 114–136 (2014). special Issue of IJCSA on Computational Intelligence
18. Mani, A.: Algebraic semantics of proto-transitive rough sets. Trans. Rough Sets **XX**(LNCS 10020), 51–108 (2016)
19. Mani, A.: Probabilities, dependence and rough membership functions. Int. J. Comput. Appl. **39**(1), 17–35 (2016). https://doi.org/10.1080/1206212X.2016.1259800
20. Mani, A.: Knowledge and consequence in AC Semantics for general rough sets. In: Wang, G., Skowron, A., Yao, Y., Ślęzak, D., Polkowski, L. (eds.) Thriving Rough Sets. SCI, vol. 708, pp. 237–268. Springer, Cham (2017). https://doi.org/10.1007/978-3-319-54966-8_12
21. Mani, A.: Algebraic methods for granular rough sets. In: Mani, A., Cattaneo, G., Düntsch, I. (eds.) Algebraic Methods in General Rough Sets. TM, pp. 157–335. Springer, Cham (2018). https://doi.org/10.1007/978-3-030-01162-8_3
22. Mani, A.: Algebraic representation, dualities and beyond. In: Mani, A., Cattaneo, G., Düntsch, I. (eds.) Algebraic Methods in General Rough Sets. TM, pp. 459–552. Springer, Cham (2018). https://doi.org/10.1007/978-3-030-01162-8_6
23. Mani, A.: Comparative approaches to granularity in general rough sets. In: Bello, R., Miao, D., Falcon, R., Nakata, M., Rosete, A., Ciucci, D. (eds.) IJCRS 2020. LNCS (LNAI), vol. 12179, pp. 500–517. Springer, Cham (2020). https://doi.org/10.1007/978-3-030-52705-1_37
24. Mani, A.: Mereology for STEAM and Education Research. In: Chari, D., Gupta, A. (eds.) EpiSTEMe 9, vol. 9, pp. 122–129. TIFR, Mumbai (2022). https://www.researchgate.net/publication/359773579
25. Mani, A., Düntsch, I., Cattaneo, G. (eds.): Algebraic Methods in General Rough Sets. Trends in Mathematics, Birkhauser Basel (2018). https://doi.org/10.1007/978-3-030-01162-8
26. Mani, A., Mitra, S.: Large minded reasoners for soft and hard cluster validation –some directions, pp. 1–16. Annals of Computer and Information Sciences, PTI (2023)
27. Pagliani, P., Chakraborty, M.: A Geometry of Approximation: Rough Set Theory: Logic Algebra and Topology of Conceptual Patterns. Springer, Berlin (2008). https://doi.org/10.1007/978-1-4020-8622-9
28. Qian, Y., Liang, J.Y., Yao, Y.Y., Dang, C.Y.: MGRS: a multi granulation rough set. Inf. Sci. **180**, 949–970 (2010)
29. Rasiowa, H.: An Algebraic Approach to Nonclassical Logics, Studies in Logic, vol. 78. North Holland, Warsaw (1974)
30. Rauszer, C.: Rough logic for multi-agent systems. In: Masuch, M., Polos, L. (eds.) Logic at Work'92, LNCS, vol. 808, pp. 151–181. Dodrecht (1991)

31. Saha, A., Sen, J., Chakraborty, M.K.: Algebraic structures in the vicinity of pre-rough algebra and their logics II. Inf. Sci. **333**, 44–60 (2016). https://doi.org/10.1016/j.ins.2015.11.018
32. Xue, Z., Zhao, L., Sun, L., Zhang, M., Xue, T.: Three-way decision models based on multigranulation support intuitionistic fuzzy rough sets. Int. J. Approximate Reasoning **124**, 147–172 (2020)
33. Yager, R.: On some new class of implication operators and their role in approximate reasoning. Inf. Sci. **167**, 193–216 (2004)

Two-Sorted Modal Logic for Formal and Rough Concepts

Prosenjit Howlader[✉] and Churn-Jung Liau

Institute of Information Science, Academia Sinica, Taipei 115, Taiwan
prosen@mail.iis.sinica.edu.tw, liaucj@iis.sinica.edu.tw

Abstract. In this paper, we propose two-sorted modal logics for the representation and reasoning of concepts arising from rough set theory (RST) and formal concept analysis (FCA). These logics are interpreted in two-sorted bidirectional frames, which are essentially formal contexts with converse relations. The logic **KB** contains ordinary necessity and possibility modalities and can represent rough set-based concepts. On the other hand, the logic **KF** has window modality that can represent formal concepts. We study the relationship between **KB** and **KF** by proving a correspondence theorem. It is then shown that, using the formulae with modal operators in **KB** and **KF**, we can capture formal concepts based on RST and FCA and their lattice structures.

Keywords: Modal logic · Formal concept analysis · Rough set theory

1 Introduction

Rough set theory (RST) [14] and formal concept analysis (FCA) [17] are both well-established areas of study with a variety of applications in fields like knowledge representation and data analysis. There has been a great deal of research on the intersections of RST and FCA over the years, including those by Kent [11], Saquer et al. [15], Hu et al. [10], Düntsch and Gediga [4], Yao [20], Yao et al. [21], Meschke [13], Ganter et al. [5] and Conradie et al. [3].

Central notions in FCA are formal contexts and their associated concept lattices. A formal context (or simply context) is a triple $\mathbb{K} := (G, M, I)$ where $I \subseteq G \times M$. A given context induces two maps $+ : (\mathcal{P}(G), \subseteq) \to (\mathcal{P}(M), \supseteq)$ and $- : (\mathcal{P}(M), \supseteq) \to (\mathcal{P}(G), \subseteq)$, where for all $A \in \mathcal{P}(G)$ and $B \in \mathcal{P}(M)$:

$$A^+ = \{m \in M \mid \text{ for all } g \in A \ \ gIm\},$$

$$B^- = \{g \in G \mid \text{ for all } m \in B \ \ gIm\}.$$

A pair of set (A, B) is called a *formal concept* (or simply concept) if $A^+ = B$ and $A = B^-$. The set $\mathcal{FC}$ of all concepts forms a complete lattice and is called a *concept lattice*.

This work is partially supported by National Science and Technology Council (NSTC) of Taiwan under Grant No. 110-2221-E-001-022-MY3.

On the other hand, the basic construct of the original RST is the *Pawlakian approximation space* (W, E), where W is the universe and E is an equivalence relation on W. Then, by applying notions of modal logic to RST, Yao et al [22] proposed generalised approximation space (W, E) with E being any binary relation on W. In addition, they also suggested to use a binary relation between two universes of discourse, containing objects and properties respectively, as another generalised formulation of approximation spaces. The rough set model over two universes is thus a formal context in FCA. Düntsch et al. [4] defined sufficiency, dual sufficiency, possibility and necessity operators based on a rough set model over two universes, where necessity and possibility operators are, in fact, rough set approximation operators. Based on these operators, Düntsch et al. [4] and Yao [20] introduced property oriented concepts and object oriented concepts respectively.

For a context $\mathbb{K} := (G, M, I)$, $I(x) := \{y \in M : xIy\}$ and $I^{-1}(y) := \{x \in G : xIy\}$ are the I-neighborhood and I^{-1}-neighbourhood of x and y respectively. For $A \subseteq G$, and $B \subseteq M$, the pairs of dual approximation operators are defined as:

$$B_I^{\diamond^{-1}} := \{x \in G : I(x) \cap B \neq \emptyset\}, \qquad B_I^{\square^{-1}} := \{x \in G : I(x) \subseteq B\}.$$
$$A_{I^{-1}}^{\diamond} := \{y \in M : I^{-1}(y) \cap A \neq \emptyset\}, \qquad A_{I^{-1}}^{\square} := \{y \in M : I^{-1}(y) \subseteq A\}.$$

If there is no confusion about the relation involved, we shall omit the subscript and denote $B_I^{\diamond^{-1}}$ by $B^{\diamond^{-1}}$, $B_I^{\square^{-1}}$ by $B^{\square^{-1}}$ and similarly for the case of A. A pair (A, B) is a *property oriented concept* of $\mathbb{K}$ iff $A^{\diamond} = B$ and $B^{\square^{-1}} = A$; and it is an *object oriented concept* of $\mathbb{K}$ iff $A^{\square} = B$ and $B^{\diamond^{-1}} = A$. As in the case of FCA, the set $\mathcal{OC}$ of all object oriented concepts and the set $\mathcal{PC}$ of all property oriented concepts form complete lattices, which are called *object oriented concept lattice* and *property oriented concept lattice* respectively.

For any concept (A, B), the set A is called its *extent* and B is called its *intent*. For concept lattices $\mathcal{X} = \mathcal{FC}, \mathcal{PC}, \mathcal{OC}$, the set of all extents and intents of $\mathcal{X}$ are denoted by $\mathcal{X}_{ext}$ and $\mathcal{X}_{int}$, respectively.

Proposition 1. *For a context $\mathbb{K} := (G, M, I)$, the following holds.*

(a) $\mathcal{FC}_{ext} = \{A \subseteq G \mid A^{+-} = A\}$ *and* $\mathcal{FC}_{int} = \{B \subseteq M \mid B^{-+} = B\}$.

(b) $\mathcal{PC}_{ext} = \{A \subseteq G \mid A^{\diamond\square^{-1}} = A\}$ *and* $\mathcal{PC}_{int} = \{B \subseteq M \mid B^{\square^{-1}\diamond} = B\}$.

(c) $\mathcal{OC}_{ext} = \{A \subseteq G \mid A^{\square\diamond^{-1}} = A\}$ *and* $\mathcal{OC}_{int} = \{B \subseteq M \mid B^{\diamond^{-1}\square} = B\}$.

It can be shown that the sets $\mathcal{FC}_{ext}, \mathcal{PC}_{ext}$ and $\mathcal{OC}_{ext}$ form complete lattices and are isomorphic to the corresponding concept lattices. Analogously, the sets $\mathcal{FC}_{int}, \mathcal{PC}_{int}$ and $\mathcal{OC}_{int}$ form complete lattices and are dually isomorphic to the corresponding concept lattices. Therefore, a concept can be identify with its extent or intent. The relationship between these two kinds of rough concept lattices and concept lattices of FCA are investigated in [19]. In particular, the following theorem is proved.

Theorem 1 [19]. *For a context $\mathbb{K} = (G, M, I)$ and the complemented context $\mathbb{K}^c = (G, M, I^c)$, the following holds.*

(a) The concept lattice of $\mathbb{K}$ is isomorphic to the property oriented concept lattice of $\mathbb{K}^c$.
(b) The property oriented concept lattice of $\mathbb{K}$ is dually isomorphic to the object oriented concept lattice of $\mathbb{K}$.
(c) The concept lattice of $\mathbb{K}$ is dually isomorphic to the object oriented concept lattice of $\mathbb{K}^c$.

In addition, to deal with the negation of concept, the notions of *semiconcepts* and *protoconcepts* are introduced in [18]. Algebraic studies of these notions led to the definition of double Boolean algebras and pure double Boolean algebras [18]. These structures have been investigated by many authors [1,9,16,18]. There is also study of logic corresponding to these algebraic structures [7,8].

The operators used in formal and rough concepts correspond to modalities used in modal logic [2,6]. In particular, the operator used in FCA is the window modality (sufficiency operator) [6] and those used in RST are box (necessity operator) and diamond (possibility operator) [2]. Furthermore, a context is a two-sorted structure consisting of a set of objects and a set of properties. Considering these facts, our goal in this work is to formulate two-sorted modal logics that are sound and complete with respect to the class of all contexts and can represent all the three kinds of concepts and their lattices.

To achieve the goal, we first introduce the notion of *two-sorted bidirectional frame*, which is simply a formal context extended with the converse of the binary relation. Then, we propose two-sorted modal logics **KB** and **KF** as representation formalism for rough and formal concepts respectively, and two-sorted bidirectional frames serve as semantic models of the logics. We also prove the soundness and completeness of the proposed logics with respect to the semantic models.

Next, we will review basic definitions and main results of general many-sorted polyadic modal logic. Then, in Sect. 2.1, we define the logic **KB** and characterize the pairs of formula that represent property and object oriented concepts of context. The logic **KF** and formal concept are discussed in Sect. 2.2. We revisit the three concept lattices and their relations in terms of logic in Sect. 3. Finally, we summarize the paper and indicate directions of future work in Sect. 4.

1.1 Many-Sorted Polyadic Modal Logic

The many-sorted polyadic modal logic is introduced in [12]. The alphabet of the logic consists of a many-sorted signature (S, Σ), where S is the collection of sorts and Σ is the set of modalities, and an S-*indexed* family $P := \{P_s\}_{s \in S}$ of propositional variables, where $P_s \neq \emptyset$ and $P_s \cap P_t = \emptyset$ for distinct $s, t \in S$. Each modality $\sigma \in \Sigma$ is associated with an arity $s_1 s_2 \ldots s_n \to s$. For any $n \in \mathbb{N}$, we denote $\Sigma_{s_1 s_2 \ldots s_n s} = \{\sigma \in \Sigma \mid \sigma : s_1 s_2 \ldots s_n \to s\}$

For an (S, Σ)-modal language $\mathcal{ML}_S$, the set of formulas is an S-index family $Fm_S := \{Fm_s \mid s \in S\}$, defined inductively for each $s \in S$ by

$$\phi_s ::= p_s \mid \neg \phi_s \mid \phi_s \wedge \phi_s \mid \sigma(\phi_{s_1} \ldots \phi_{s_n}) \mid \sigma^\square(\phi_{s_1} \ldots \phi_{s_n}),$$

where $p_s \in P_s$ and $\sigma \in \Sigma_{s_1 s_2 \ldots s_n s}$.

A *many-sorted relational frame* is a pair $\mathfrak{F} := (\{W_s\}_{s \in S}, \{R_\sigma\}_{\sigma \in \Sigma})$ where $W_s \neq \emptyset$, $W_{s_i} \cap W_{s_j} = \emptyset$ for $s, s_i \neq s_j \in S$ and $R_\sigma \subseteq W_s \times W_{s_1} \ldots \times W_{s_n}$ if $\sigma \in \Sigma_{s_1 s_2 \ldots s}$. The class of all many-sorted relational frames is denoted as $\mathbb{SRF}$. A *valuation* v is an S-indexed family of maps $\{v_s\}_{s \in S}$, where $v_s : P_s \to \mathcal{P}(W_s)$. A many-sorted model $\mathfrak{M} := (\mathfrak{F}, v)$ consists of a many-sorted frame $\mathfrak{F}$ and a valuation v. The satisfaction of a formula in a model $\mathfrak{M}$ is defined inductively as follows.

Definition 1. Let $\mathfrak{M} := (\{W_s\}_{s \in S}, \{R_\sigma\}_{\sigma \in \Sigma}, v)$ be a many-sorted model, $w \in W_s$ and $\phi \in Fm_s$ for $s \in S$. We define $\mathfrak{M}, w \models_s \phi$ by induction over ϕ as follows:

1. $\mathfrak{M}, w \models_s p$ iff $w \in v_s(p)$
2. $\mathfrak{M}, w \models_s \neg\phi$ iff $\mathfrak{M}, w \not\models_s \phi$
3. $\mathfrak{M}, w \models_s \phi_1 \wedge \phi_2$ iff $\mathfrak{M}, w \models_s \phi_1$ and $\mathfrak{M}, w \models_s \phi_2$
4. If $\sigma \in \Sigma_{s_1 s_2 \ldots s}$, then $\mathfrak{M}, w \models_s \sigma(\phi_1, \phi_2 \ldots \phi_n)$ iff there is $(w_1, w_2 \ldots w_n) \in W_{s_1} \times W_{s_2} \ldots W_{s_n}$ such that $(w, w_1, w_2 \ldots w_n) \in R_\sigma$ and $\mathfrak{M}, w_i \models_{s_i} \phi_i$ for every $i \in \{1, 2 \ldots n\}$
5. If $\sigma^\square \in \Sigma_{s_1 s_2 \ldots s_n s}$, then $\mathfrak{M}, w \models_s \sigma^\square(\phi_1, \phi_2 \ldots \phi_n)$ iff for all $(w_1, w_2 \ldots w_n) \in W_{s_1} \times W_{s_2} \ldots W_{s_n}$ such that $(w, w_1, w_2 \ldots w_n) \in R_\sigma$ implies that $\mathfrak{M}, w_i \models_{s_i} \phi_i$ for some $i \in \{1, 2 \ldots n\}$

Definition 2 [12]. Let $\mathfrak{M}$ be an (S, Σ)-model. Then, for a set Φ_s of formula, $\mathfrak{M}, w \models_s \Phi_s$ if $\mathfrak{M}, w \models_s \phi$ for all $\phi \in \Phi_s$.

Let $\mathcal{C}$ be a class of models. Then, for a set $\Phi_s \cup \{\phi\} \subseteq Fm_s$, ϕ is a local semantic consequence of Φ_s over $\mathcal{C}$ and denoted as $\Phi_s \models_s^{\mathcal{C}} \phi$ if $\mathfrak{M}, w \models_s \Phi_s$ implies $\mathfrak{M}, w \models_s \phi$ for all models $\mathfrak{M} \in \mathcal{C}$. If $\mathcal{C}$ is the class of all models, we omit the superscript and denote it as $\Phi_s \models_s \phi$.

If Φ_s is empty, we say ϕ is valid in $\mathcal{C}$ and denoted it as $\mathcal{C} \models_s \phi$. When $\mathcal{C}$ is the class of all models based on a given frame $\mathfrak{F}$, we also denote it by $\mathfrak{F} \models_s \phi$.

To characterize the local semantic consequence, the modal system $\mathbf{K}_{(S,\Sigma)} := \{\mathbf{K}_s\}_{s \in S}$ is proposed in [12], where $\mathbf{K}_s$ is the axiomatic system in Fig. 1 in which $\sigma \in \Sigma_{s_1 \ldots s_n, s}$:

When the signature is clear from the context, the subscripts may be omitted and we simply write the system as $\mathbf{K}$.

Definition 3 [12]. Let $\Lambda \subseteq Fm_S$ be an S-sorted set of formulas. The normal modal logic defined by Λ is $\mathbf{K}\Lambda := \{\mathbf{K}\Lambda_s\}_{s \in S}$ where $\mathbf{K}\Lambda_s := \mathbf{K}_s \cup \{\lambda' \in Fm_s \mid \lambda'$ is obtained by uniform substitution applied to a formula $\lambda \in \Lambda_s\}$.

Definition 4 [12]. A sequence of formulas $\phi_1, \phi_2, \ldots \phi_n$ is called a $\mathbf{K}\Lambda$-proof for the formula ϕ if $\phi_n = \phi$ and ϕ_i is in $\mathbf{K}\Lambda_{s_i}$ or inferred from $\phi_1, \ldots, \phi_{i-1}$ using modus pones and universal generalization. If ϕ has a proof in $\mathbf{K}\Lambda$, we say that ϕ is a theorem and write $\vdash_s^{\mathbf{K}\Lambda} \phi$. Let $\Phi \cup \{\phi\} \subseteq Fm_s$ be a set of formulas. Then, we say that ϕ is provable form Φ, denoted by $\Phi \vdash_s^{\mathbf{K}\Lambda} \phi$, if there exist $\phi_1, \ldots, \phi_n \in \Phi$ such that $\vdash_s^{\mathbf{K}\Lambda} (\phi_1 \wedge \ldots \wedge \phi_n) \to \phi$. In addition, the set Φ is $\mathbf{K}\Lambda$-inconsistent if $\bot$ is provable from it, otherwise it is $\mathbf{K}\Lambda$-consistent.

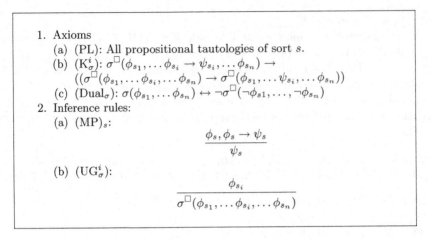

1. Axioms
   (a) (PL): All propositional tautologies of sort s.
   (b) (K_σ^i): $\sigma^\square(\phi_{s_1}, \ldots \phi_{s_i} \to \psi_{s_i}, \ldots \phi_{s_n}) \to$
       $((\sigma^\square(\phi_{s_1}, \ldots \phi_{s_i}, \ldots \phi_{s_n}) \to \sigma^\square(\phi_{s_1}, \ldots \psi_{s_i}, \ldots \phi_{s_n}))$
   (c) $(Dual_\sigma)$: $\sigma(\phi_{s_1}, \ldots \phi_{s_n}) \leftrightarrow \neg\sigma^\square(\neg\phi_{s1}, \ldots, \neg\phi_{s_n})$
2. Inference rules:
   (a) $(MP)_s$:
   $$\frac{\phi_s, \phi_s \to \psi_s}{\psi_s}$$
   (b) (UG_σ^i):
   $$\frac{\phi_{s_i}}{\sigma^\square(\phi_{s_1}, \ldots \phi_{s_i}, \ldots \phi_{s_n})}$$

Fig. 1. The axiomatic system $\mathbf{K}_s$

Proposition 2 [12]. $\mathbf{K}\Lambda$ is strongly complete with respect to a class of models $\mathcal{C}$ if and only if any consistent set Γ of formulas is satisfied in some model from $\mathcal{C}$.

Definition 5 [12]. The canonical model is

$$\mathfrak{M}^{\mathbf{K}\Lambda} := (\{W_s^{\mathbf{K}\Lambda}\}_{s\in S}, \{R_\sigma^{\mathbf{K}\Lambda}\}_{\sigma\in\Sigma}, V^{\mathbf{K}\Lambda})$$

where

(a) for any $s \in S$, $W_s^{\mathbf{K}\Lambda} = \{\Phi \subseteq Fm_s \mid \Phi$ is maximally $\mathbf{K}\Lambda$-consistent$\}$,
(b) for any $\sigma \in \Sigma_{s_1\ldots s_n, s}$, $w \in W_s^{\mathbf{K}\Lambda}, u_1 \in W_{s_1}^{\mathbf{K}\Lambda}, \ldots u_n \in W_{s_n}^{\mathbf{K}\Lambda}$, $R_\sigma^{\mathbf{K}\Lambda} w u_1 \ldots u_n$
    if and only if $(\psi_1, \ldots, \psi_n) \in u_1 \times u_2 \times \ldots \times u_n$ implies that $\sigma(\psi_1, \ldots, \psi_n) \in w$.
(c) $V^{\mathbf{K}\Lambda} = \{V_s^{\mathbf{K}\Lambda}\}$ is the valuation defined by $V_s^{\mathbf{K}\Lambda}(p) = \{w \in W_s^{\mathbf{K}\Lambda} \mid p \in w\}$
    for any $s \in S$ and $p \in P_s$.

Lemma 1 [12]. If $s \in S$, $\phi \in Fm_s$, $\sigma \in \Sigma_{s_1\ldots s_n, s}$ and $w \in W_s^{\mathbf{K}\Lambda}$ then the following hold:

(a) $R_\sigma^{\mathbf{K}\Lambda} w u_1 \ldots u_n$ if and only if for any formulas $\psi_1, \ldots, \psi_n, \sigma^\square(\psi_1, \ldots, \psi_n) \in w$
    implies $\psi_i \in u_i$ for some $i \in \{1, 2, \ldots, n\}$.
(b) If $\sigma(\psi_1, \ldots, \psi_n) \in w$ then for any $i \in \{1, 2 \ldots, n\}$ there is $u_i \in W_{s_i}^{\mathbf{K}\Lambda}$ such
    that $\psi_1 \in u_1, \ldots, \psi_n \in u_n$ and $R_\sigma^{\mathbf{K}\Lambda} w u_1 \ldots u_n$.
(c) $\mathfrak{M}^{\mathbf{K}\Lambda}, w \models_s \phi$ if and only if $\phi \in w$.

Proposition 3 [12]. If Φ_s is a $\mathbf{K}\Lambda$-consistent set of formulas then it is satisfied in the canonical model.

These results implies the soundness and completeness of $\mathbf{K}$ directly.

Theorem 2. $\mathbf{K}$ is sound and strongly complete with respect to the class of all (S, Σ)-models, that is, for any $s \in S$, $\phi \in Fm_s$ and $\Phi_s \subseteq Fm_s$, $\Phi_s \vdash_s^{\mathbf{K}} \phi$ if and only if $\Phi_s \models_s \phi$.

2   Two-Sorted Modal Logic and Concept Lattices

In this section, we present the logics **KB** and **KF** and discuss their relationship with rough and formal concepts.

2.1   Two-Sorted Modal Logic and Concept Lattices in Rough Set Theory

Let us consider a special kind of two-sorted signature $(\{s_1, s_2\}, \Sigma)$ where $\Sigma = \Sigma_1 \uplus \Sigma_2$ is the direct sum of two sets of unary modalities such that $\Sigma_1 = \Sigma_{s_1 s_2}$ and $\Sigma_2 = \Sigma_{s_2 s_1} = \Sigma_1^{-1} := \{\sigma^{-1} : \sigma \in \Sigma_1\}$. We say that the signature is bidirectional. Modal languages built over bidirectional signatures are interpreted in bidirectional frames.

Definition 6. For the signature above, a two-sorted bidirectional frame is a quadruple :

$$\mathfrak{F}_2 := (W_1, W_2, \{R_\sigma\}_{\sigma \in \Sigma_1}, \{R_{\sigma^{-1}}\}_{\sigma \in \Sigma_1})$$

where W_1, W_2 are non-empty disjoint sets and $R_\sigma \subseteq W_2 \times W_1$, $R_{\sigma^{-1}}$ is the converse of R_σ. The class of all two-sorted bidirectional frame is denoted as $\mathbb{BSFR}_2$.

The logic system **KB** for two-sorted bidirectional frames is define as **K**Λ where Λ consists of the following axioms:

(B) $p \to (\sigma^{-1})^\square \sigma p$ and $q \to \sigma^\square \sigma^{-1} q$ where $p \in P_{s_1}$ and $q \in P_{s_2}$.

Theorem 3. **KB** is sound with respect to class $\mathbb{BSFR}_2$ of all two-sorted bidirectional frame.

Proof. The proof is straightforward. Here we give the proof for the axiom $p \to (\sigma^{-1})^\square \sigma p$. Let $\mathfrak{M}$ be a model based on the frame $\mathfrak{F}_2$ defined above and $\mathfrak{M}, w_1 \models_{s_1} p$ for some $w_1 \in W_1$. Now, for any $w_2 \in W_2$ such that $R_{\sigma^{-1}} w_1 w_2$, we have $\mathfrak{M}, w_2 \models_{s_2} \sigma p$ because $R_\sigma w_2 w_1$ follows from the converse of relation. This leads to $\mathfrak{M}, w_1 \models_{s_1} (\sigma^{-1})^\square \sigma p$ immediately.

The completeness theorem is proved using the canonical model of **KB**, which is an instance of that constructed in Definition 5. Hence,

$$\mathfrak{M}^{\mathbf{KB}} := (\{W_{s_1}^{\mathbf{KB}}, W_{s_2}^{\mathbf{KB}}\}, \{R_\sigma^{\mathbf{KB}}, R_{\sigma^{-1}}^{\mathbf{KB}}\}_{\sigma \in \Sigma}, V^{\mathbf{KB}})$$

It is easy to see that the model satisfies the following properties for $x \in W_{s_1}^{\mathbf{KB}}$ and $y \in W_{s_2}^{\mathbf{KB}}$:

(a) $R_\sigma^{\mathbf{KB}} yx$ iff $\phi \in x$ implies that $\sigma\phi \in y$ for any $\phi \in Fm_{s_1}$.
(b) $R_{\sigma^{-1}}^{\mathbf{KB}} xy$ iff $\phi \in y$ implies that $\sigma^{-1}\phi \in x$ for any $\phi \in Fm_{s_2}$.

Theorem 4. KB is strongly complete with respect to class of all two-sorted bidirectional models, that is for any $s \in \{s_1, s_2\}$, $\phi \in Fm_s$ and $\Phi_s \subseteq Fm_s$, $\Phi_s \models_s^{\mathrm{BSFR}_2} \phi$ implies that $\Phi_s \vdash_s^{\mathbf{KB}} \phi$.

Proof. It is sufficient to show that the canonical model is a bidirectional frame. Then, the result follows from Propositions 2 and 3. Let $x \in W_{s_1}^{\mathbf{KB}}$ and $y \in W_{s_2}^{\mathbf{KB}}$ and assume $(y, x) \in R_\sigma^{\mathbf{KB}}$. Then, for any $\phi \in y$, we have $\sigma^\square \sigma^{-1} \phi \in y$ by axiom (B), which in turns implies $\sigma^{-1} \phi \in x$ by Lemma 1. Hence, $(x, y) \in R_{\sigma^{-1}}^{\mathbf{KB}}$ by property (b) of the canonical model. Analogously, we can show that $(x, y) \in R_{\sigma^{-1}}^{\mathbf{KB}}$ implies $(y, x) \in R_\sigma^{\mathbf{KB}}$. That is, $R_{\sigma^{-1}}^{\mathbf{KB}}$ is indeed the converse of $R_\sigma^{\mathbf{KB}}$.

To represent rough concepts, we consider a particular bidirectional signature $(\{s_1, s_2\}, \{\Diamond, \Diamond^{-1}\})$ (i.e. the signature that Σ_1 is a singleton containing the modality $\Diamond$). As usual, we denote the dual modalities of $\Diamond$ and $\Diamond^{-1}$ by $\square$ and $\square^{-1}$ respectively. Let $\mathcal{SF}_2$ denote the class of all bidirectional frames over the signature and let $\mathcal{K}$ be the set of all contexts. Then, there is a bijective correspondence between $\mathcal{K}$ and $\mathcal{SF}_2$ given by $(G, M, I) \mapsto (G, M, I^{-1}, I)$. Note that I^{-1} and I respectively correspond to modalities $\Diamond$ and $\Diamond^{-1}$ under the mapping. We use $Fm(\mathbf{RS}) := \{Fm(\mathbf{RS})_{s_1}, Fm(\mathbf{RS})_{s_2}\}$ and $\mathbf{KB}_2$ to denote the indexed family of formulas and its logic system over the particular signature respectively. By Theorems 3 and 4, $\mathbf{KB}_2$ is sound and complete with respect to the class $\mathcal{SF}_2$ and hence $\mathcal{K}$.

Example 1. In a typical application of FCA to association rule mining, a formal context can represent transaction data, where G is the set of customers of all ages groups and M is the set of items. In such a scenario, for the bidirectional frame (G, M, I^{-1}, I), $g \in G$, and $m \in M$, gIm means that the customer g has bought the item m. Let $\mathfrak{M} = (G, M, I^{-1}, I, v)$ be a model based on the frame, $\phi \in Fm(\mathbf{RS})_{s_1}$ and $v_1(\phi)$ represent the group of customers of 30 to 50 age group, $\psi \in Fm(\mathbf{RS})_{s_2}$ and $v_2(\psi)$ represent the item set of electronic products. Then, the formulas $\square \phi$ and $\square^{-1} \psi$ in the model $\mathfrak{M}$ may be interpreted as follows.

- For $g \in G$, $\mathfrak{M}, g \models_{s_2} \square^{-1} \psi$ means that all items bought by g are in the item set $v_2(\psi)$, or the customer g only buys electronic products.
- For $m \in M$, $\mathfrak{M}, m \models_{s_2} \square \phi$ means that all customers buying m are in the 30 to 50 age group.

Let us denote the truth set of a formula $\phi \in Fm(\mathbf{RS})_{s_i} (i = 1, 2)$ in a model $\mathfrak{M}$ by $[[\phi]]_{\mathfrak{M}} := \{w \in W_i \mid \mathfrak{M}, w \models_{s_i} \phi\}$. We usually omit the subscript and simply write $[[\phi]]$.

Proposition 4. Let $\mathbb{K} := (G, M, I)$ be a context and $\mathfrak{M} := (G, M, I^{-1}, I, v)$ be a model based on its corresponding frame. Then, the relationship between approximation operators and modal formulas is as follows:

(i) $[[\phi]]^\Diamond = [[\Diamond \phi]]$ and $[[\phi]]^\square = [[\square \phi]]$ for $\phi \in Fm(\mathbf{RS})_{s_1}$.
(ii) $[[\phi]]^{\Diamond^{-1}} = [[\Diamond^{-1} \phi]]$ and $[[\phi]]^{\square^{-1}} = [[\square^{-1} \phi]]$ for $\phi \in Fm(\mathbf{RS})_{s_2}$.

Definition 7. Let $\mathfrak{C} := \{(G, M, I^{-1}, I)\}$ be a frame based on the context $\mathbb{K} = (G, M, I)$. Then, we define

(a) $Fm_{PC_{ext}} := \{\phi \in Fm(\mathbf{RS})_{s_1} \mid \models^{\mathfrak{C}}_{s_1} \Box^{-1}\Diamond\phi \leftrightarrow \phi\}$ and $Fm_{PC_{int}} := \{\phi \in Fm(\mathbf{RS})_{s_2} \mid \models^{\mathfrak{C}}_{s_2} \Diamond\Box^{-1}\phi \leftrightarrow \phi\}$

(b) $Fm_{OC_{ext}} := \{\phi \in Fm(\mathbf{RS})_{s_1} \mid \models^{\mathfrak{C}}_{s_1} \Diamond^{-1}\Box\phi \leftrightarrow \phi\}$ and $Fm_{OC_{int}} := \{\phi \in Fm(\mathbf{RS})_{s_2} \mid \models^{\mathfrak{C}}_{s_2} \Box\Diamond^{-1}\phi \leftrightarrow \phi\}$

(c) $Fm_{PC} := \{(\phi, \psi) \mid \phi \in Fm_{PC_{ext}}, \psi \in Fm_{PC_{int}}, \models^{\mathfrak{C}}_{s_1} \phi \leftrightarrow \Box^{-1}\psi, \models^{\mathfrak{C}}_{s_2} \Diamond\phi \leftrightarrow \psi\}$

(d) $Fm_{OC} := \{(\phi, \psi) \mid \phi \in Fm_{OC_{ext}}, \psi \in Fm_{OC_{int}}, \models^{\mathfrak{C}}_{s_1} \phi \leftrightarrow \Diamond^{-1}\psi, \models^{\mathfrak{C}}_{s_2} \Box\phi \leftrightarrow \psi\}$

Obviously, when $(\phi, \psi) \in Fm_{PC}$, $([[\phi]], [[\psi]]) \in \mathcal{PC}$ for any models based on $\mathfrak{C}$. Hence, Fm_{PC} consists of pairs of formulas representing property oriented concepts. Analogously, Fm_{OC} provides the representation of object oriented concepts. Note that these sets are implicitly parameterized by the underlying context and should be indexed with $\mathbb{K}$. However, for simplicity, we usually omit the index.

2.2 Two Sorted Modal Logic and Concept Lattice in Formal Concept Analysis

To represent formal concepts, we consider another two-sorted bidirectional signature $(\{s_1, s_2\}, \{\boxminus, \boxminus^{-1}\})$, where $\Sigma_{s_1 s_2} = \{\boxminus\}$ and $\Sigma_{s_2 s_1} = \{\boxminus^{-1}\}$, and the logic **KF** based on it. Syntactically, the signature is the same as that for $\mathbf{KB}_2$ except we use different symbols to denote the modalities. Hence, formation rules of formulas remain unchanged and we denote the indexed family of formulas by $Fm(\mathbf{KF}) = \{Fm(\mathbf{KF})_{s_1}, Fm(\mathbf{KF})_{s_2}\}$. In addition, while both **KF** and $\mathbf{KB}_2$ are interpreted in bidirectional models, the main difference between them is on the way of their modalities being interpreted.

Definition 8. Let $\mathfrak{M} := (W_1, W_2, R, R^{-1}, v)$. Then,

(a) For $\phi \in Fm(\mathbf{KF})_{s_1}$ and $w \in W_2$, $\mathfrak{M}, w \models_{s_2} \boxminus\phi$ iff for any $w' \in W_1$, $\mathfrak{M}, w' \models_{s_1} \phi$ implies $R(w, w')$

(b) For $\phi \in Fm(\mathbf{KF})_{s_2}$ and $w \in W_1$, $\mathfrak{M}, w \models_{s_1} \boxminus^{-1}\phi$ iff for any $w' \in W_2$, $\mathfrak{M}, w' \models_{s_2} \phi$ implies $R^{-1}(w, w')$

Example 2. Continuing with Example 1, let $\phi \in Fm(\mathbf{KF})_{s_1}$, $\psi \in Fm(\mathbf{KF})_{s_2}$, $v_1(\phi)$, and $v_2(\psi)$ remain unchanged. Then, the intuitive meanings of the formulas $\boxminus\phi$ and $\boxminus^{-1}\psi$ are as follows:

– For $g \in G$, $\mathfrak{M}, g \models_{s_1} \boxminus^{-1}\psi$, g buys all electronic products.
– For $m \in M$, $\mathfrak{M}, m \models_{s_2} \boxminus\phi$, all customers in the 30 to 50 age group $v_1(\phi)$ buy m.

The logic system $\mathbf{KF} := \{\mathbf{KF}_{s_1}, \mathbf{KF}_{s_2}\}$ is shown in Fig. 2.

We define a translation $\rho : Fm(\mathbf{KF}) \rightarrow Fm(\mathbf{RS})$ where $\rho = \{\rho_1, \rho_2\}$ is defined as follows:

1. Axioms:
   (PL) Propositional tautologies of sort s_i for $i = 1, 2$.
   $(K_{\boxminus}^1)$ $\boxminus(\phi_1 \wedge \neg\phi_2) \rightarrow (\boxminus\neg\phi_1 \rightarrow \boxminus\neg\phi_2)$ for $\phi_1, \phi_2 \in Fm(\mathbf{KF})_{s_1}$
   (B^1) $\phi \rightarrow \boxminus^{-1}\boxminus\phi$ for $\phi \in Fm(\mathbf{KF})_{s_1}$
   $(K_{\boxminus^{-1}}^2)$ $\boxminus^{-1}(\psi_1 \wedge \neg\psi_2) \rightarrow (\boxminus^{-1}\neg\psi_1 \rightarrow \boxminus^{-1}\neg\psi_2)$ for $\psi_1, \psi_2 \in Fm(\mathbf{KF})_{s_2}$
   (B^2) $\psi \rightarrow \boxminus\boxminus^{-1}\psi$ for $\psi \in Fm(\mathbf{KF})_{s_2}$
2. Inference rules:
   - $(MP)_s$: for $s \in \{s_1, s_2\}$ and $\phi, \psi \in Fm_s$
   $$\frac{\phi, \phi \rightarrow \psi}{\psi}$$
   - $(UG_{\boxminus}^1)$: for $\phi \in Fm(\mathbf{KF})_{s_1}$,
   $$\frac{\neg\phi}{\boxminus\phi}$$
   - $(UG_{\boxminus^{-1}}^2)$: for $\psi \in Fm(\mathbf{KF})_{s_2}$,
   $$\frac{\neg\psi}{\boxminus^{-1}\psi}$$

Fig. 2. The axiomatic system **KF**

1. $\rho_i(p) := p$ for all $p \in P_{s_i}$ for $i = 1, 2$.
2. $\rho_i(\phi \wedge \psi) := \rho_i(\phi) \wedge \rho_i(\psi)$ for $\phi, \psi \in Fm(\mathbf{KF})_{s_i}$, $i = 1, 2$.
3. $\rho_i(\neg\phi) := \neg\rho_i(\phi)$ for $\phi \in Fm(\mathbf{KF})_{s_i}$, $i = 1, 2$.
4. $\rho_1(\boxminus\phi) := \Box\neg\rho_1(\phi)$ for $\phi \in Fm_{\mathbf{KF}_{s_1}}$.
5. $\rho_2(\boxminus^{-1}\phi) := \Box^{-1}\neg\rho_2(\phi)$ for $\phi \in Fm_{\mathbf{KF}_{s_2}}$.

Theorem 5. For any formula $\phi \in Fm_{\mathbf{KF}_{s_i}}(i = 1, 2)$ the following hold.

(a) $\Phi \vdash^{\mathbf{KF}} \phi$ if and only if $\rho(\Phi) \vdash^{\mathbf{KB}_2} \rho(\phi)$ for any $\Phi \subseteq Fm_{\mathbf{KF}_{s_i}}$.
(b) Let $\mathfrak{M} := (W_1, W_2, I, I^{-1}, v)$ be a model and $\mathfrak{M}^c := (W_1, W_2, I^c, (I^{-1})^c, v)$ be the corresponding complemented model, $w \in W_i$, $\mathfrak{M}, w \models_{s_i} \phi$ if and only if $\mathfrak{M}^c, w \models_{s_i} \rho(\phi)$ for all $i = 1, 2$.
(c) ϕ is valid in the class $\mathcal{SF}_2$ if and only if $\rho(\phi)$ is valid in $\mathcal{SF}_2$.

Proof. (a). We can prove it by showing that ϕ is an axiom in **KF** if and only if $\rho(\phi)$ is an axiom in $\mathbf{KB}_2$, and for each rule in **KF**, there is a translation of it in $\mathbf{KB}_2$ and vice verse.
(b). By induction on the complexity of formulas, as usual, the proof of basis and Boolean cases are straightforward. For $\phi = \boxminus\psi$, let us assume any $w \in W_2$. Then, by Definition 8, $\mathfrak{M}, w \models_{s_2} \boxminus\phi$ iff for all $w' \in W_1$, $I^c ww'$ implies that $\mathfrak{M}, w' \models_{s_1} \neg\psi$. By induction hypothesis, this means that for all $w' \in W_1$, $I^c ww'$ implies that $\mathfrak{M}^c, w' \models_{s_1} \neg\rho(\psi)$. That is, $\mathfrak{M}^c, w \models_{s_2} \Box\neg\rho(\phi)$. By

definition of ρ, this is exactly $\mathfrak{M}^c, w \models_{s_2} \rho(\phi)$. The case of $\phi = \boxminus^{-1}\psi$ is proved analogously.

(c). This follows immediately from (b).

Proposition 5. (a). For $\phi_1, \phi_2 \in Fm(\mathbf{KF})_{s_1}$, $\dfrac{\phi_1 \rightarrow \phi_2}{\boxminus\phi_2 \rightarrow \boxminus\phi_1}$

(b). For $\phi_1, \phi_2 \in Fm(\mathbf{KF})_{s_2}$, $\dfrac{\phi_1 \rightarrow \phi_2}{\boxminus^{-1}\phi_2 \rightarrow \boxminus^{-1}\phi_1}$

Proof. We only prove (a) and the proof of (b) is similar.

$$\vdash^{\mathbf{KF}} \phi_1 \rightarrow \phi_2$$
$$\vdash^{\mathbf{KB}_2} \rho(\phi_1) \rightarrow \rho(\phi_2)(\text{Theorem 5 (a)})$$
$$\vdash^{\mathbf{KB}_2} (\rho(\phi_1) \rightarrow \rho(\phi_2)) \rightarrow (\neg\rho(\phi_2) \rightarrow \neg\rho(\phi_1))(\text{PL})$$
$$\vdash^{\mathbf{KB}_2} \neg\rho(\phi_2) \rightarrow \neg\rho(\phi_1)(\text{MP})$$
$$\vdash^{\mathbf{KB}_2} \Box(\neg\rho(\phi_2) \rightarrow \neg\rho(\phi_1))(\text{UG})$$
$$\vdash^{\mathbf{KB}_2} \Box(\neg\rho(\phi_2) \rightarrow \neg\rho(\phi_1)) \rightarrow (\Box\neg\rho(\phi_2) \rightarrow \Box\neg\rho(\phi_1))(\text{K})$$
$$\vdash^{\mathbf{KB}_2} \Box\neg\rho(\phi_2) \rightarrow \Box\neg\rho(\phi_1)(\text{MP})$$
$$\vdash^{\mathbf{KF}} \boxminus\phi_2 \rightarrow \boxminus\phi_1(\text{Theorem 5(a)})$$

Theorem 6. $\mathbf{KF}$ is sound and strongly complete with respect to the class $\mathcal{SF}_2$.

Proof. This follows from Theorem 5 and the fact that $\mathbf{KB}_2$ is sound and strongly complete with respect to $\mathcal{SF}_2$.

Proposition 6. Recalling the definition of truth set, we have

(i) $[[\boxminus\phi]] = [[\phi]]^+$ for $\phi \in Fm(\mathbf{KF})_{s_1}$
(ii) $[[\boxminus^{-1}\phi]] = [[\phi]]^-$ for $\phi \in Fm(\mathbf{KF})_{s_2}$.

Definition 9. Let $\mathfrak{C} := \{(G, M, I^{-1}, I)\}$ be a frame based on the context (G, M, I). Then, we define

(a) $Fm_{FC_{ext}} := \{\phi \in Fm(\mathbf{KF})_{s_1} \mid \models_{s_1}^{\mathfrak{C}} \boxminus^{-1}\boxminus\phi \leftrightarrow \phi\}$ and $Fm_{FC_{int}} := \{\phi \in Fm(\mathbf{KF})_{s_2} \mid \models_{s_2}^{\mathfrak{C}} \boxminus\boxminus^{-1}\phi \leftrightarrow \phi\}$

(b) $Fm_{FC} := \{(\phi, \psi) \mid \phi \in Fm_{PC_{ext}}, \psi \in Fm_{PC_{int}}, \models_{s_1}^{\mathfrak{C}} \phi \leftrightarrow \boxminus^{-1}\psi, \models_{s_2}^{\mathfrak{C}} \boxminus\phi \leftrightarrow \psi\}$

In other words, the set Fm_{FC} represents formal concepts induced from the context (G, M, I).

3   Logical Representation of Three Concept Lattices

We have seen that a certain pairs of formulas in the logic **KF** and **KB$_2$** can represent concepts in FCA and RST respectively. The observation suggests the definition below.

Definition 10. Let $\phi \in Fm(\mathbf{RS})_{s_1}$, $\psi \in Fm(\mathbf{RS})_{s_2}$, $\eta \in Fm(\mathbf{KF})_{s_1}$, and $\gamma \in Fm(\mathbf{KF})_{s_2}$. Then, for a context (G, M, I), we say that

(a) (ϕ, ψ) is a *(logical) property oriented concept* of $\mathbb{K}$ if $(\phi, \psi) \in Fm_{PC}$.
(b) (ϕ, ψ) is a *(logical) object oriented concept* of $\mathbb{K}$ if $(\phi, \psi) \in Fm_{OC}$.
(c) (η, γ) is a *(logical) formal concept* of $\mathbb{K}$ if $(\eta, \gamma) \in Fm_{FC}$.

We now explore the relationships between the three notions and their properties. In what follows, for a context $\mathbb{K} = (G, M, I)$, we usually use $\mathfrak{C}_0 := \{(G, M, I^{-1}, I)\}$ and $\mathcal{C}_1 := \{(G, M, (I^c)^{-1}), I^c\}$ to denote frames corresponding to $\mathbb{K}$ and $\mathbb{K}^c$ respectively.

Proposition 7. Let $\mathbb{K} := (G, M, I)$ be a context. Then,

(a) (ϕ, ψ) is a property oriented concept of $\mathbb{K}$ iff $(\neg\phi, \neg\psi)$ is an object oriented concept of $\mathbb{K}^c$ for $\phi \in Fm(\mathbf{RS})_{s_1}$ and $\psi \in Fm(\mathbf{RS})_{s_2}$.
(b) (ϕ, ψ) is a formal concept of $\mathbb{K}$ iff $(\rho(\phi), \neg\rho(\psi))$ is a property oriented concept of $\mathbb{K}^c$ for $\phi \in Fm(\mathbf{KF})_{s_1}$ and $\psi \in Fm(\mathbf{KF})_{s_2}$.
(c) (ϕ, ψ) is a formal concept of $\mathbb{K}$ iff $(\neg\rho(\phi), \rho(\psi))$ is an object oriented concept of $\mathbb{K}^c$ for $\phi \in Fm(\mathbf{KF})_{s_1}$ and $\psi \in Fm(\mathbf{KF})_{s_2}$..

Proof. (a) Suppose that (ϕ, ψ) is a property oriented concept of $\mathbb{K}$, then by definition, $\models^{\mathfrak{C}_0}_{s_1} \Box^{-1}\Diamond\phi \leftrightarrow \phi$, $\models^{\mathfrak{C}_0}_{s_2} \Diamond\Box^{-1}\psi \leftrightarrow \psi$, $\models^{\mathfrak{C}_0}_{s_1} \phi \leftrightarrow \Box^{-1}\psi$, and $\models^{\mathfrak{C}_0}_{s_2} \Diamond\phi \leftrightarrow \psi$. Hence, we have the following derivation,

$$\models^{\mathfrak{C}_0}_{s_1} \Box^{-1}\Diamond\phi \leftrightarrow \phi$$
$$\models^{\mathfrak{C}_0}_{s_1} (\Box^{-1}\Diamond\phi \leftrightarrow \phi) \leftrightarrow (\neg\phi \leftrightarrow \neg\Box^{-1}\Diamond\phi)$$
$$\models^{\mathfrak{C}_0}_{s_1} \neg\phi \leftrightarrow \neg\Box^{-1}\Diamond\phi$$
$$\models^{\mathfrak{C}_0}_{s_1} \neg\phi \leftrightarrow \Diamond^{-1}\Box\neg\phi$$

Therefore, $\neg\phi \in Fm_{PC_{ext}}$. Similarly, by $\models^{\mathfrak{C}_0}_{s_2} \Diamond\Box^{-1}\psi \leftrightarrow \psi$, contraposition and modus ponens, we can show that $\neg\psi \in Fm_{PC_{int}}$.
Using $\models^{\mathfrak{C}_0}_{s_1} \phi \leftrightarrow \Box^{-1}\psi$, $\models^{\mathfrak{C}_0}_{s_2} \Diamond\phi \leftrightarrow \psi$, contraposition and modus ponens , we can show that $(\neg\phi, \neg\psi) \in Fm_{OC}$.
We can also prove the converse direction by replacing $\phi, \psi, \Diamond, \Box$ with $\neg\phi, \neg\psi, \Box, \Diamond$ respectively.

(b) Because (ϕ, ψ) is a formal concept, we have $\models^{\mathfrak{C}_0}_{s_1} \boxminus^{-1}\boxminus\phi \leftrightarrow \phi$, $\models^{\mathfrak{C}_0}_{s_2} \boxminus\boxminus^{-1} \psi \leftrightarrow \psi$, $\models^{\mathfrak{C}_0}_{s_1} \phi \leftrightarrow \boxminus^{-1}\psi$, and $\models^{\mathfrak{C}_0}_{s_2} \boxminus\phi \leftrightarrow \psi$. By $\models^{\mathfrak{C}_0}_{s_1} \boxminus^{-1}\boxminus\phi \leftrightarrow \phi$ and Theorem 5 (a), we have $\models^{\mathcal{C}_1}_{s_1} \rho(\boxminus^{-1}\boxminus\phi) \leftrightarrow \rho(\phi)$ which implies that $\models^{\mathcal{C}_1}_{s_1} \Box^{-1}\Diamond\rho(\phi) \leftrightarrow \rho(\phi)$. By $\models^{\mathfrak{C}_0}_{s_2} \boxminus\boxminus^{-1} \psi \leftrightarrow \psi$ and Theorem 5 (a), we have $\models^{\mathcal{C}_1}_{s_2} \Box\neg\Box^{-1}\neg\rho(\psi) \leftrightarrow \rho(\psi)$, which implies that $\models^{\mathcal{C}_1}_{s_2} \Diamond\Box^{-1}\neg\rho(\psi) \leftrightarrow \neg\rho(\psi)$.

Similarly, we can show that $\models^{\mathcal{C}_1}_{s_1} \rho(\phi) \leftrightarrow \Box^{-1}\neg\rho(\psi)$, $\models^{\mathcal{C}_1}_{s_2} \Diamond\rho(\phi) \leftrightarrow \neg\rho(\psi)$. Therefore, $(\rho(\phi), \rho(\psi))$ is a property oriented concept for (G, M, I^c). The proof for the converse direction is similar.

(c) It follows from (a) and (b) immediately.

Now, we can define a relation $\equiv_1$ on the set Fm_{PC} as follows: For $(\phi, \psi), (\phi', \psi') \in Fm_{PC}$, $(\phi, \psi) \equiv_1 (\phi', \psi')$ if and only if $\models^{\mathcal{C}_0} \phi \leftrightarrow \phi'$.

Analogously, we can define $\equiv_2$ and $\equiv_3$ on the set Fm_{OC} and Fm_{FC}, respectively. Obviously, $\equiv_1, \equiv_2$ and $\equiv_3$ are all equivalence relations. Let $Fm_{PC}/\equiv_1$, $Fm_{OC}/\equiv_2$, and $Fm_{FC}/\equiv_3$ be the sets of equivalence classes.

Proposition 8. For $(\phi, \psi), (\phi', \psi') \in Fm_X$, $(\phi, \psi) \equiv_i (\phi', \psi')$ iff $\models^{\mathcal{C}_0} \psi \leftrightarrow \psi'$, where $i \in \{1, 2, 3\}$ for $X \in \{PC, OC, FC\}$ respectively.

Proof. Let us prove the case of FC as an example. Suppose $(\phi, \psi), (\phi', \psi') \in Fm_{FC}$ and $(\phi, \psi) \equiv_3 (\phi', \psi')$. Then, $\models^{\mathcal{C}_0}_{s_1} \phi \leftrightarrow \phi'$, which implies $\models^{\mathcal{C}_0}_{s_2} \Box\phi \leftrightarrow \Box\phi'$ according to the semantics of **KF**. In addition, by definition of Fm_{FC}, $\models^{\mathcal{C}_0}_{s_2} \Box\phi \leftrightarrow \psi$, and $\models^{\mathcal{C}_0}_{s_2} \Box\phi' \leftrightarrow \psi'$. Hence, $\models^{\mathcal{C}_0}_{s_2} \psi \leftrightarrow \psi'$.

Proofs for other two cases are similar.

Proposition 9. Let $X \in \{PC_{ext}, OC_{ext}, FC_{ext}\}$ and $Y \in \{PC_{int}, OC_{int}, FC_{int}\}$. Then,

(a) Fm_X and Fm_Y are closed under conjunction.
(b) If $\phi \in Fm_X$ and $\psi \in Fm_Y$, then $\circ\phi \in Fm_Y$ and $\circ^{-1}\psi \in Fm_X$, where $\circ \in \{\Box, \Box, \Diamond\}$ depending on X and Y according to their respective definitions.

Proof. (a). We prove the case of $Fm_{FC_{ext}}$ as an example and other cases can be proved in a similar way. Let $\phi, \phi' \in Fm_{FC_{ext}}$. Then, $\models^{\mathcal{C}_0}_{s_1} \Box^{-1}\Box\phi \leftrightarrow \phi$ and $\models^{\mathcal{C}_0}_{s_1} \Box^{-1}\Box\phi' \leftrightarrow \phi'$. By using the translation ρ and Theorem 5, we have both $\models^{\mathcal{C}_0}_{s_1} \Box^{-1}\Box(\phi \wedge \phi') \to \Box^{-1}\Box\phi$ and $\models^{\mathcal{C}_0}_{s_1} \Box^{-1}\Box(\phi \wedge \phi') \to \Box^{-1}\Box\phi'$. Hence, we can derive $\models^{\mathcal{C}_0}_{s_1} \Box^{-1}\Box(\phi \wedge \phi') \to (\phi \wedge \phi')$. Also, with the translation, we have $\models^{\mathcal{C}_0}_{s_1} (\phi \wedge \phi') \to \Box^{-1}\Box(\phi \wedge \phi')$ because the formula is mapped to an instance of axiom (B). Hence, $\phi \wedge \phi' \in Fm_{FC_{ext}}$.

(b). Let us prove the case of $\phi \in Fm_{FC_{ext}}$ as an example. Assume that $\phi \in Fm_{FC_{ext}}$ and $\circ = \Box$. Then, according to the semantics of $\Box$, $\models^{\mathcal{C}_0}_{s_1} \Box^{-1}\Box \phi \leftrightarrow \phi$ implies $\models^{\mathcal{C}_0}_{s_1} \Box\Box^{-1}\Box\phi \leftrightarrow \Box\phi$. Hence $\Box\phi \in Fm_{FC_{int}}$. Similarly, if $\psi \in Fm_{FC_{int}}$, then $\Box^{-1}\psi \in Fm_{FC_{ext}}$.

From the proposition, we can derive the following corollary immediately.

Corollary 1. (a) $(\phi_1, \psi_1), (\phi_2, \psi_2) \in Fm_{PC}$ implies that $(\phi_1 \wedge \phi_2, \Diamond(\phi_1 \wedge \phi_2))$ and $(\Box^{-1}(\psi_1 \wedge \psi_2), \psi_1 \wedge \psi_2) \in Fm_{PC}$.
(b) $(\phi_1, \psi_1), (\phi_2, \psi_2) \in Fm_{OC}$ implies that $(\phi_1 \wedge \phi_2, \Box(\phi_1 \wedge \phi_2))$ and $(\Diamond^{-1}(\psi_1 \wedge \psi_2), \psi_1 \wedge \psi_2) \in Fm_{OC}$.
(c) $(\phi_1, \psi_1), (\phi_2, \psi_2) \in Fm_{FC}$ implies that $(\phi_1 \wedge \phi_2, \Box(\phi_1 \wedge \phi_2))$ and $(\Box^{-1}(\psi_1 \wedge \psi_2), \psi_1 \wedge \psi_2) \in Fm_{FC}$.

Now we can define the following structures:

$(Fm_{PC}/\equiv_1, \vee_1, \wedge_1)$, where $[(\phi, \psi)], [(\phi', \psi')] \in Fm_{PC}/\equiv_1$,

$$[(\phi, \psi)] \wedge_1 [(\phi', \psi')] := [(\phi \wedge \phi', \Diamond(\phi \wedge \phi'))]$$

$$[(\phi, \psi)] \vee_1 [(\phi', \psi')] := [(\Box^{-1}(\psi \wedge \psi'), (\psi \wedge \psi'))]$$

$(Fm_{OC}/\equiv_2, \vee_2, \wedge_2)$, where $[(\phi, \psi)], [(\phi', \psi')] \in Fm_{OC}/\equiv_2$,

$$[(\phi, \psi)] \wedge_2 [(\phi', \psi')] := [(\phi \wedge \phi', \Box(\phi \wedge \phi'))]$$

$$[(\phi, \psi)] \vee_2 [(\phi', \psi')] := [(\Diamond^{-1}(\psi \wedge \psi'), (\psi \wedge \psi'))]$$

$(Fm_{FC}/\equiv_3, \vee_3, \wedge_3)$, where $[(\phi, \psi)], [(\phi', \psi')] \in Fm_{FC}/\equiv_3$,

$$[(\phi, \psi)] \wedge_3 [(\phi', \psi')] := [(\phi \wedge \phi', \boxminus(\phi \wedge \phi'))]$$

$$[(\phi, \psi)] \vee_3 [(\phi', \psi')] := [(\boxminus^{-1}(\psi \wedge \psi'), (\psi \wedge \psi'))]$$

Theorem 7. For a context $\mathbb{K}$, $(Fm_{PC}/\equiv_1, \vee_1, \wedge_1)$, $(Fm_{OC}/\equiv_2, \vee_2, \wedge_2)$ and $(Fm_{FC}/\equiv_3, \vee_3, \wedge_3)$, are lattices.

Proof. We give proof for the structure $(Fm_{FC}/\equiv_3, \vee_3, \wedge_3)$ and the proofs of other cases are similar. Let $(\phi, \psi), (\phi_1, \psi_1), (\phi', \psi'), (\phi'_1, \psi'_1) \in Fm_{FC}$ such that $(\phi, \psi) \equiv_3 (\phi_1, \psi_1)$ and $(\phi', \psi') \equiv_3 (\phi'_1, \psi'_1)$. By Corollary 1, $(\phi \wedge \phi', \boxminus(\phi \wedge \phi'))$, $(\boxminus^{-1}(\psi \wedge \psi'), \psi \wedge \psi'), (\phi_1 \wedge \phi'_1, \boxminus(\phi_1 \wedge \phi'_1))$ and $(\boxminus^{-1}(\psi_1 \wedge \psi'_1), \psi_1 \wedge \psi'_1) \in Fm_{FC}$. Now $(\phi, \psi) \equiv_3 (\phi_1, \psi_1)$ and $(\phi', \psi') \equiv_3 (\phi'_1, \psi'_1)$ implies that $\models^{\mathfrak{C}_0} \phi \leftrightarrow \phi_1$ and $\models^{\mathfrak{C}_0} \phi' \leftrightarrow \phi'_1$. By Proposition 8, $\models^{\mathfrak{C}_0} \psi \leftrightarrow \psi_1$ and $\models^{\mathfrak{C}_0} \psi' \leftrightarrow \psi'_1$. $\models^{\mathfrak{C}_0} \phi \wedge \phi' \leftrightarrow \phi_1 \wedge \phi'_1$ and $\models^{\mathfrak{C}_0} \psi \wedge \psi' \leftrightarrow \psi_1 \wedge \psi'_1$ which implies that $(\phi \wedge \phi', \boxminus(\phi \wedge \phi')) \equiv_3 (\phi_1 \wedge \phi'_1, \boxminus(\phi_1 \wedge \phi'_1))$ and $(\boxminus^{-1}(\psi \wedge \psi'), \psi \wedge \psi') \equiv_3 (\boxminus^{-1}(\psi_1 \wedge \psi'_1), \psi_1 \wedge \psi'_1)$. Hence, $\wedge_3$ and $\vee_3$ are well-defined operations. Their commutativity and associativity follow from the fact that $\vdash^{\mathbf{KF}} \phi \wedge \psi \leftrightarrow \psi \wedge \phi$ and $\vdash^{\mathbf{KF}} (\phi \wedge \psi) \wedge \gamma \leftrightarrow \phi \wedge (\psi \wedge \gamma)$. Now we will show that for all $[(\phi_1, \psi_1)], [(\phi_2, \psi_2)] \in Fm_{FC}/\equiv_3$, $[(\phi_1, \psi_1)] \wedge ([(\phi_1, \psi_1] \vee [(\phi_2, \psi_2)]) = [(\phi_1, \psi_1)]$ which is equivalent to $[(\phi_1 \wedge \boxminus^{-1}(\psi_1 \wedge \psi_2), \boxminus(\phi_1 \wedge \boxminus^{-1}(\psi_1 \wedge \psi_1)))] = [(\phi_1, \psi_1)]$. We know that $\models^{\mathfrak{C}_0} \phi_1 \wedge \boxminus^{-1}(\psi_1 \wedge \psi_2) \to \phi_1$. In addition,

$$\models^{\mathfrak{C}_0}_{s_1} \phi_1 \leftrightarrow \boxminus^{-1}\psi_1 \text{ as } (\phi_1, \psi_1) \in Fm_{FC}$$

$$\models^{\mathfrak{C}_0}_{s_2} \psi_1 \wedge \psi_2 \to \psi_1$$

$$\models^{\mathfrak{C}_0}_{s_1} \boxminus^{-1}\psi_1 \to \boxminus^{-1}(\psi_1 \wedge \psi_2) \text{ by Proposition 5}$$

$$\models^{\mathfrak{C}_0}_{s_1} \phi_1 \to \boxminus^{-1}(\psi_1 \wedge \psi_2)$$

$$\models^{\mathfrak{C}_0}_{s_1} \phi_1 \to \phi_1 \wedge \boxminus^{-1}(\psi_1 \wedge \psi_2)$$

So $\models^{\mathfrak{C}_0}_{s_1} \phi_1 \leftrightarrow \boxminus^{-1}(\psi_1 \wedge \psi_2)$ which implies that $[(\phi_1, \psi_1)] \wedge ([(\phi_1, \psi_1] \vee [(\phi_2, \psi_2)]) = [(\phi_1, \psi_1)]$. Analogously, we can show that $[(\phi_1, \psi_1)] \vee ([(\phi_1, \psi_1] \wedge [(\phi_2, \psi_2)]) = [(\phi_1, \psi_1)]$. Hence $(Fm_{FC}/\equiv_3, \vee_3, \wedge_3)$ is a lattice.

Theorem 8. Let $\mathbb{K}$ be a context and let $\mathbb{K}^c$ be its corresponding complemented context. Let Fm_{FC} be the set of logical formal concepts of $\mathbb{K}$ and let Fm_{PC} and Fm_{OC} be the sets of logical property oriented concepts and logical object oriented concepts of $\mathbb{K}^c$ respectively. Then,

(a) $(Fm_{FC}/\equiv_3, \vee_3, \wedge_3)$ and $(Fm_{PC}/\equiv_1, \vee_1, \wedge_1)$ are isomorphic.
(b) $(Fm_{PC}/\equiv_1, \vee_1, \wedge_1)$ and $(Fm_{OC}/\equiv_2, \vee_2, \wedge_2)$ are dually isomorphic.
(c) $(Fm_{FC}/\equiv_3, \vee_3, \wedge_3)$ and $(Fm_{OC}/\equiv_2, \vee_2, \wedge_2)$ are dually isomorphic.

Proof. (a) By Proposition 7, the mapping $h : Fm_{FC}/\equiv_3 \to Fm_{PC}/\equiv_1$ defined by $h([(\phi, \psi)]) := [(\rho(\phi), \rho(\neg\psi))]$ is well-defined and surjective. Now $h([(\phi_1, \psi_1)]) = h([(\phi_2, \psi_2)])$ implies $[\rho((\phi_1), \rho(\neg\psi_1))] = [(\rho(\phi_2), \rho(\neg\psi_2))]$, which in turn implies $\models^{\mathfrak{C}_1} \rho(\phi_1) \leftrightarrow \rho(\phi_2)$, and by Theorem 5, $\models^{\mathfrak{C}_0} \phi_1 \leftrightarrow \phi_2$. This means that $[(\phi_1, \psi_1)] = [(\phi_2, \psi_2)]$. Thus, h is injcetive, and as a result, h is a bijection. In addition,

$$\begin{aligned} h([(\phi_1, \psi_1)] \wedge_3 [(\phi_2, \psi_2)]) &= h([(\phi_1 \wedge \phi_2, \boxminus(\phi_1 \wedge \phi_2))]) \\ &= ([\rho(\phi_1 \wedge \phi_2), \rho(\neg \boxminus (\phi_1 \wedge \phi_2))]) \\ &= ([\rho(\phi_1 \wedge \phi_2), \Diamond\rho(\phi_1 \wedge \phi_2)]) \\ &= h([(\phi_1, \psi_1)]) \wedge_1 h([(\phi_2, \psi_2)]) \end{aligned}$$

Therefore, h is an isomorphism.
(b) Analogously, we can show that $f : Fm_{PC}/\equiv_1 \to Fm_{OC}/\equiv_1$ such that $f([(\phi, \psi)]) := [(\neg\phi, \neg\psi)]$ is a dual isomorphism.
(c) It follows from (a) and (b) immediately.

4 Conclusion and Future Direction

In this paper, we show that concepts based on RST and FCA can be represented in two dual instances of two-sorted modal logics **KB** and **KF**. An interesting question is how to deal with both kinds of concepts in a single framework. To address the question, we apparently need a signature including all modalities in **KB** and **KF** together. For that, the Boolean modal logic proposed in [6] may be helpful. Hence, to investigate many-sorted Boolean modal logic and its representational power for concepts based on both RST and FCA will be an important direction in our future work.

As a formal context consists of objects, properties, and a relation between them, the relationship between objects and properties can change over time. Hence, to model and analyze the dynamics of contexts is also desirable. Using two-sorted bidirectional relational frames, we can model contexts at some time. Therefore, integrating temporal logic with many-sorted modal logic will provide an approach to model dynamics of contexts. This is another possible direction for further research.

References

1. Balbiani, P.: Deciding the word problem in pure double Boolean algebras. J. Appl. Log. **10**(3), 260–273 (2012)
2. Blackburn, P., De Rijke, M., Venema, Y.: Modal Logic. Cambridge University Press, Cambridge (2001)
3. Conradie, W., et al.: Rough concepts. Inf. Sci. **561**, 371–413 (2021)
4. Düntsch, I., Gediga, G.: Modal-style operators in qualitative data analysis. In: Vipin, K., et al. (eds.) Proceedings of the 2002 IEEE International Conference on Data Mining, pp. 155–162. IEEE Computer Society (2002)
5. Ganter, B., Meschke, C.: A formal concept analysis approach to rough data tables. In: Peters, J.F., et al. (eds.) Transactions on Rough Sets XIV. LNCS, vol. 6600, pp. 37–61. Springer, Heidelberg (2011). https://doi.org/10.1007/978-3-642-21563-6_3
6. Gargov, G., Passy, S., Tinchev, T.: Modal environment for Boolean speculations. In: Skordev, D.G. (eds) Mathematical Logic and Its Applications, pp. 253–263, Springer, Boston (1987). https://doi.org/10.1007/978-1-4613-0897-3_17
7. Howlader, P., Banerjee, M.: Kripke contexts, double Boolean algebras with operators and corresponding modal systems. J. Logic Lang. Inform. **32**, 117–146 (2023)
8. Howlader, P., Banerjee, M.: A non-distributive logic for semiconcepts and its modal extension with semantics based on Kripke contexts. Int. J. Approximate Reasoning **153**, 115–143 (2023)
9. Howlader, P., Banerjee, M.: Topological representation of double Boolean algebras. Algebra Univer. **84**, Paper No. 15, 32 (2023)
10. Hu, K., Sui, Y., Lu, Y., Wang, J., Shi, C.: Concept approximation in concept lattice. In: Cheung, D., Williams, G.J., Li, Q. (eds.) PAKDD 2001. LNCS (LNAI), vol. 2035, pp. 167–173. Springer, Heidelberg (2001). https://doi.org/10.1007/3-540-45357-1_21
11. Kent, R.E.: Rough concept analysis. In: Ziarko, W.P. (ed.) Rough Sets. Fuzzy Sets and Knowledge Discovery, pp. 248–255. Springer, London (1994). https://doi.org/10.1007/978-1-4471-3238-7_30
12. Leuştean, L., Moangă, N., Şerbănuţă, T.F.: A many-sorted polyadic modal logic. Fund. Inform. **173**(2–3), 191–215 (2020)
13. Meschke, C.: Approximations in concept lattices. In: Kwuida, L., Sertkaya, B. (eds.) ICFCA 2010. LNCS (LNAI), vol. 5986, pp. 104–123. Springer, Heidelberg (2010). https://doi.org/10.1007/978-3-642-11928-6_8
14. Pawlak, Z.: Rough sets. Int. J. Comput. Inform. Sci. **11**(5), 341–356 (1982)
15. Saquer, J., Deogun, J.S.: Concept approximations based on rough sets and similarity measures. Int. J. Appl. Math. Comput. Sci. **11**(3), 655–674 (2001)
16. Vormbrock, B., Wille, R.: Semiconcept and protoconcept algebras: the basic theorems. In: Ganter, B., Stumme, G., Wille, R. (eds.) Formal Concept Analysis. LNCS (LNAI), vol. 3626, pp. 34–48. Springer, Heidelberg (2005). https://doi.org/10.1007/11528784_2
17. Wille, R.: Restructuring lattice theory: an approach based on hierarchies of concepts. In: Ferré, S., Rudolph, S. (eds.) ICFCA 2009. LNCS (LNAI), vol. 5548, pp. 314–339. Springer, Heidelberg (2009). https://doi.org/10.1007/978-3-642-01815-2_23
18. Wille, R.: Boolean concept logic. In: Ganter, B., Mineau, G.W. (eds.) ICCS-ConceptStruct 2000. LNCS (LNAI), vol. 1867, pp. 317–331. Springer, Heidelberg (2000). https://doi.org/10.1007/10722280_22

19. Yao, Y.: A comparative study of formal concept analysis and rough set theory in data analysis. In: Tsumoto, S., Słowiński, R., Komorowski, J., Grzymała-Busse, J.W. (eds.) RSCTC 2004. LNCS (LNAI), vol. 3066, pp. 59–68. Springer, Heidelberg (2004). https://doi.org/10.1007/978-3-540-25929-9_6
20. Yao, Y.Y.: Concept lattices in rough set theory. In: IEEE Annual Meeting of the Fuzzy Information Processing Society-NAFIPS, vol. 2, pp. 796–801. IEEE (2004)
21. Yao, Y.Y., Chen, Y.: Rough set approximations in formal concept analysis. In: Peters, J.F., et al. (eds.) Transactions on Rough Sets V, pp. 285–305. Springer, Berlin (2006)
22. Yao, Y.Y., Lin, T.Y.: Generalization of rough sets using modal logics. Intell. Autom. Soft Comput. **2**(2), 103–119 (1996)

Kryszkiewicz's Relation for Indiscernibility of Objects in Data Tables Containing Missing Values

Michinori Nakata[1]([⊠]), Norio Saito[1], Hiroshi Sakai[2], and Takeshi Fujiwara[3]

[1] Faculty of Management and Information Science, Josai International University,
1 Gumyo, Togane, Chiba 283-8555, Japan
nakatam@ieee.org, saitoh_norio@jiu.ac.jp
[2] Department of Mathematics and Computer Aided Sciences, Faculty of Engineering,
Kyushu Institute of Technology, Tobata, Kitakyushu 804-8550, Japan
sakai@mns.kyutech.ac.jp
[3] Faculty of Informatics, Tokyo University of Information Sciences, 4-1 Onaridai,
Wakaba-ku, Chiba 265-8501, Japan
fujiwara@rsch.tuis.ac.jp

Abstract. We check Kryszkiewicz's approach for rough sets in data tables containing missing values in terms of Lipski's approach based on possible world semantics. The relation for indiscernibility of objects used by Kryszkiewicz, which most authors use, derives a pair of lower and upper approximations as the actual approximations to a target set. This means that Kryszkiewicz's approach is accompanied by information loss because what Lipski's approach derives is the lower and upper bounds of the actual approximations. It is clarified that Kryszkiewicz's relation for indiscernibility is equal to the union of possible indiscernibility relations in Lipski's approach. As a result, the lower and the upper approximation derived from Kryszkiewicz's relation are equal to the lower bound of the actual lower approximation and the upper bound of the actual upper approximation, respectively. Bridging the gap between the two approaches, we propose another relation for indiscernibility that is equal to the intersection of possible indiscernibility relations. By using Kryszkiewicz's relation and the proposed relation, Kryszkiewicz's approach can derive the same approximations as Lipski's one. Therefore, we can keep using Kryszkiewicz's approach without information loss.

Keywords: rough sets · incomplete information · missing values · lower and upper approximations · possible world semantics

1 Introduction

The framework of rough sets, proposed by Pawlak [1], is used as an effective tool in the field of data mining and related topics. In the rough sets lower and upper approximations, which correspond to inclusion and intersection operations, are

A. Campagner et al. (Eds.): IJCRS 2023, LNAI 14481, pp. 170–184, 2023.
https://doi.org/10.1007/978-3-031-50959-9_12

derived using the indiscernibility of objects. The indiscernibility is expressed by a relation showing two objects are indiscernible if the values characterizing the objects are equal.

Pawlak's framework is constructed under data tables containing only complete information. As a matter of fact, it is well-known that real data tables usually contain incomplete information [2,3]. This leads to that Kryszkiewicz [4,5] deals with missing values.

Kryszkiewicz extended the relation of indiscernibility so that we can deal with the indiscernibility of objects in data tables containing missing values, which is called Kryszkiewicz's approach. It is very attractive to use this approach because the indiscernibility of objects characterized by missing values can be handled by a relation. So most authors use Kryszkiewicz's approach [5–15] and its extended versions are proposed [7,14,16–18].

These studies using Kryszkiewicz's approach, however, have two drawbacks. One is that information loss occurs [19,20]. This leads to poor approximations [14,19]. The other is that these have a common characteristic of deriving a pair of lower and upper approximations to a target set of objects in spite of dealing with missing values. This is incompatible with Lipski's approach based on possible world semantics. Lipski showed the limitation of obtaining information granules from data tables with incomplete information [21]. According to Lipski, we cannot obtain the actual answer but can have nothing to obtain the lower and upper bounds of the actual answer. This is true for rough sets under data tables containing missing values. In other words, what we obtain is the lower and upper bounds of the actual lower approximation and the lower and upper bounds of the actual upper approximation.

We check Kryszkiewicz's relation and lower and upper approximations derived from the relation from the viewpoint of possible world semantics. Our aim is to clarify the relationship between Kryszkiewicz's approach and Lipski's approach and to find out the way that we can keep using Kryszkiewicz's approach without information loss.

The paper is organized as follows. In Sect. 2, the traditional approach of Pawlak is briefly addressed under a complete information table. In Sect. 3, we apply Lipski's approach to rough sets under an incomplete information table. The lower and upper bounds of lower and upper approximations are shown. In Sect. 4, we check approximations derived from Kryszkiewicz's relation. It is shown that information loss occurs and poor approximations are derived. In Sect. 5, we clarify the relationship between Kryszkiewicz's approach and Lipski's approach. As a result, it is obtained that the lower approximation of Kryszkiewicz's approach is equal to the lower bound of the actual lower approximation while the upper approximation is equal to the upper bound of the actual upper approximation. Finally, it is clarified how to make Kryszkiewicz's approach compatible with Lipski's approach. In Sect. 6, we address conclusions.

2 Pawlak's Approach

A data set is a set of objects that is concretely represented as a two-dimensional information table, where each row represents properties of an object and each column denotes a property called an attribute. Pawlak dealt with a complete information table with no incomplete information. Complete information table CT consists of $(U, AT, \cup_{a \in AT} V_a)$, where U is a non-empty finite set of objects called the universe and AT is a non-empty finite set of attributes such that $\forall a \in AT : U \to V_a$, and set V_a is called the domain of attribute a:

$$CT = \{o \in U \mid \forall a \in AT \; \exists v \in V_a \; a(o) = v\}, \tag{1}$$

where $a(o)$ is the value of attribute a for o. Binary relation I_A for indiscernibility of objects in U on subset $A \subseteq AT$ of attributes is:

$$I_A = \{(o, o') \in U \times U \mid \forall a \in A \; a(o) = a(o')\}. \tag{2}$$

This relation, called an indiscernibility relation, is reflexive, symmetric, and transitive. From the indiscernibility relation, equivalence class $E(o)_A \, (= \{o' \mid (o, o') \in I_A\})$ containing object o is obtained. This is also the set of objects that is indiscernible with object o, called the indiscernible class containing o on A. Family FE_A of equivalence classes on A is derived from the indiscernibility relation:

$$FE_A = \cup_{o \in U} \{E(o)_A\}. \tag{3}$$

The objects are uniquely partitioned using the family. Lower approximation $\underline{apr}(T)_A$ and upper approximation $\overline{apr}(T)_A$ of target set T of objects by FE_A are:

$$\underline{apr}(T)_A = \{o \in U \mid E(o)_A \subseteq T\}, \tag{4}$$
$$\overline{apr}(T)_A = \{o \in U \mid E(o)_A \cap T \neq \emptyset\}. \tag{5}$$

Example 1

Let complete information table CT be obtained as follows:

$$CT$$

O	a_1	a_2	a_3
1	y	v	b
2	y	v	c
3	x	u	c
4	x	v	c
5	y	w	b

In information table CT, $U = \{o_1, o_2, o_3, o_4, o_5\}$. Domains V_{a_1}, V_{a_2}, and V_{a_3} of attributes a_1, a_2, and a_3 are $\{x, y\}$, $\{u, v, w\}$, and $\{b, c\}$, respectively. Indiscernibility relation I_{a_1} on a_1 is:

$$I_{a_1} = \{(o_1, o_1), (o_1, o_2), (o_1, o_5), (o_2, o_1), (o_2, o_2), (o_2, o_5), (o_3, o_3), (o_3, o_4), (o_4, o_3),$$
$$(o_4, o_4), (o_5, o_1), (o_5, o_2), (o_5, o_5)\}.$$

Equivalence classes containing each object on a_1 are:

$$E(o_1)_{a_1} = E(o_2)_{a_1} = E(o_5)_{a_1} = \{o_1, o_2, o_5\},$$
$$E(o_3)_{a_1} = E(o_4)_{a_1} = \{o_3, o_4\}.$$

Family FE_{a_1} of equivalence classes on a_1 is:

$$FE_{a_1} = \{\{o_1, o_2, o_5\}, \{o_3, o_4\}\}.$$

Let target set T be $\{o_2, o_3, o_4\}$ that is specified by constraint $a_3 = c$. Lower approximation $\underline{apr}(T)_{a_1}$ and upper approximation $\overline{apr}(T)_{a_1}$ by FE_{a_1} are:

$$\underline{apr}(T)_{a_1} = \{o_3, o_4\},$$
$$\overline{apr}(T)_{a_1} = \{o_1, o_2, o_3, o_4, o_5\}.$$

3 Lipski's Approach

Lipski [21] derived the set of possible tables from an information table containing missing values by using possible world semantics. He showed the lower and the upper bound of the actual result by aggregating results obtained from the possible tables. We call this series of procedures Lipski's approach.

Following in Lipski's footsteps, we first obtain the set of possible tables from an incomplete information table. Second, we apply the ways addressed in the previous section to each possible table. Third, the results from the possible tables are aggregated using intersection and union operations.

3.1 Possible Tables and Possible Indiscernibility Relations

Possible table pt_A on set A of attributes is a table in which each missing value for every attribute $a \in A$ in incomplete information table IT is replaced with value $v \in V_a$.

$$pt_A = \{o \in U \mid \forall a \in A \exists v \in V_a \; a_{pt}(o) = v \land \forall a \notin A \; a_{pt}(o) = a_{IT}(o)\}, \quad (6)$$

where $a_{pt}(o)$ and $a_{IT}(o)$ are the values of a for o in possible table pt and in incomplete information table IT, respectively. When missing values exist on set A of attributes in incomplete information table IT, set PT_A of possible tables on A is:

$$PT_A = \{pt_{A,1}, \ldots, pt_{A,n}\}, \quad (7)$$

where every possible table $pt_{A,i}$ has an equal possibility that it is actual, number n of possible tables is equal to $\Pi_{a \in A}|V_a|^{m_a}$, the number of missing values is m_a on attribute a, and $|V_a|$ is the cardinality of domain V_a.

Attribute $a \in A$ has a value in V_a in all possible tables on A. Possible indiscernibility relation $PI_{A,i}$ is derived from possible table $pt_{A,i}$.

$$PI_{A,i} = \{(o, o') \in U \times U \mid \forall a \in A \; a(o)_i = a(o')_i\}, \quad (8)$$

where $a(o)_i$ is the value of attribute a for o in $pt_{A,i}$.

Example 2

Let incomplete information table IT be obtained as follows:

$$IT$$

O	a_1	a_2	a_3
1	x	u	b
2	y	v	b
3	x	$*$	c
4	$*$	u	b
5	$*$	v	c
6	y	u	c
7	y	v	c

In information table IT, universe U is $\{o_1, o_2, o_3, o_4, o_5, o_6, o_7\}$ and domains V_{a_1}, V_{a_2}, and V_{a_3} of attributes a_1, a_2, and a_3 are the same as CT in Example 1. We obtain twelve $(= 2 \times 2 \times 3)$ possible tables $pt_{\{a_1,a_2\},1}, \cdots, pt_{\{a_1,a_2\},12}$ from IT on $\{a_1, a_2\}$ because missing value $*$ on attribute a_1 of objects o_4 and o_5 is replaced by one of domain elements x and y of attribute a_1 and missing value $*$ on attribute a_2 of object o_3 is replaced by one of domain elements u, v and w of attribute a_2.

$$pt_{\{a_1,a_2\},1}$$

O	a_1	a_2	a_3
1	x	u	b
2	y	v	b
3	x	u	c
4	x	u	b
5	x	v	c
6	y	u	c
7	y	v	c

$$pt_{\{a_1,a_2\},2}$$

O	a_1	a_2	a_3
1	x	u	b
2	y	v	b
3	x	v	c
4	x	u	b
5	x	v	c
6	y	u	c
7	y	v	c

$$pt_{\{a_1,a_2\},3}$$

O	a_1	a_2	a_3
1	x	u	b
2	y	v	b
3	x	w	c
4	x	u	b
5	x	v	c
6	y	u	c
7	y	v	c

$$pt_{\{a_1,a_2\},4}$$

O	a_1	a_2	a_3
1	x	u	b
2	y	v	b
3	x	u	c
4	x	u	b
5	y	v	c
6	y	u	c
7	y	v	c

$$pt_{\{a_1,a_2\},5}$$

O	a_1	a_2	a_3
1	x	u	b
2	y	v	b
3	x	v	c
4	x	u	b
5	y	v	c
6	y	u	c
7	y	v	c

$$pt_{\{a_1,a_2\},6}$$

O	a_1	a_2	a_3
1	x	u	b
2	y	v	b
3	x	w	c
4	x	u	b
5	y	v	c
6	y	u	c
7	y	v	c

$$pt_{\{a_1,a_2\},7}$$

O	a_1	a_2	a_3
1	x	u	b
2	y	v	b
3	x	u	c
4	y	u	b
5	x	v	c
6	y	u	c
7	y	v	c

$$pt_{\{a_1,a_2\},8}$$

O	a_1	a_2	a_3
1	x	u	b
2	y	v	b
3	x	v	c
4	y	u	b
5	x	v	c
6	y	u	c
7	y	v	c

$$pt_{\{a_1,a_2\},9}$$

O	a_1	a_2	a_3
1	x	u	b
2	y	v	b
3	x	w	c
4	y	u	b
5	x	v	c
6	y	u	c
7	y	v	c

$$pt_{\{a_1,a_2\},10}$$

O	a_1	a_2	a_3
1	x	u	b
2	y	v	b
3	x	u	c
4	y	u	b
5	y	v	c
6	y	u	c
7	y	v	c

$$pt_{\{a_1,a_2\},11}$$

O	a_1	a_2	a_3
1	x	u	b
2	y	v	b
3	x	v	c
4	y	u	b
5	y	v	c
6	y	u	c
7	y	v	c

$$pt_{\{a_1,a_2\},12}$$

O	a_1	a_2	a_3
1	x	u	b
2	y	v	b
3	x	w	c
4	y	u	b
5	y	v	c
6	y	u	c
7	y	v	c

o_5 is discernible with o_3 on $\{a_1, a_2\}$ in $pt_{\{a_1,a_2\},1}$, whereas o_5 is indiscernible with o_3 on $\{a_1, a_2\}$ in $pt_{\{a_1,a_2\},2}$. In other words, $pt_{\{a_1,a_2\},1}$ corresponds to the case where o_5 is discernible with o_3 on $\{a_1, a_2\}$, whereas $pt_{\{a_1,a_2\},2}$ does to the case where o_5 is indiscernible with object o_3 .

By applying formula (8) to each possible table, possible indiscernibility relation $PI_{\{a_1,a_2\},i}$ on $\{a_1, a_2\}$ for $i = 1, 12$ is:

$$PI_{\{a_1,a_2\},1} = \{(o_1,o_1),(o_2,o_2),(o_3,o_3),(o_4,o_4),(o_5,o_5),(o_6,o_6),(o_7,o_7),(o_1,o_3),$$
$$(o_3,o_1),(o_1,o_4),(o_4,o_1),(o_2,o_7),(o_7,o_2),(o_3,o_4),(o_4,o_3)\},$$

$$PI_{\{a_1,a_2\},2} = \{(o_1,o_1),(o_2,o_2),(o_3,o_3),(o_4,o_4),(o_5,o_5),(o_6,o_6),(o_7,o_7),(o_1,o_4),$$
$$(o_4,o_1),(o_2,o_7),(o_7,o_2),(o_3,o_5),(o_5,o_3)\},$$

$$PI_{\{a_1,a_2\},3} = \{(o_1,o_1),(o_2,o_2),(o_3,o_3),(o_4,o_4),(o_5,o_5),(o_6,o_6),(o_7,o_7),(o_1,o_4),$$
$$(o_4,o_1),(o_2,o_7),(o_7,o_2)\},$$

$$PI_{\{a_1,a_2\},4} = \{(o_1,o_1),(o_2,o_2),(o_3,o_3),(o_4,o_4),(o_5,o_5),(o_6,o_6),(o_7,o_7),(o_1,o_3),$$
$$(o_3,o_1),(o_1,o_4),(o_4,o_1),(o_2,o_7),(o_7,o_2),(o_3,o_4),(o_4,o_3),(o_2,o_5),$$
$$(o_5,o_2),(o_5,o_7),(o_7,o_5)\},$$

$$PI_{\{a_1,a_2\},5} = \{(o_1,o_1),(o_2,o_2),(o_3,o_3),(o_4,o_4),(o_5,o_5),(o_6,o_6),(o_7,o_7),(o_1,o_4),$$
$$(o_4,o_1),(o_2,o_5),(o_5,o_2),(o_5,o_7),(o_7,o_5),(o_2,o_7),(o_7,o_2)\},$$

$$PI_{\{a_1,a_2\},6} = \{(o_1,o_1),(o_2,o_2),(o_3,o_3),(o_4,o_4),(o_5,o_5),(o_6,o_6),(o_7,o_7),(o_1,o_4),$$
$$(o_4,o_1),(o_2,o_5),(o_5,o_2),(o_5,o_7),(o_7,o_5),(o_2,o_7),(o_7,o_2)\},$$

$$PI_{\{a_1,a_2\},7} = \{(o_1,o_1),(o_2,o_2),(o_3,o_3),(o_4,o_4),(o_5,o_5),(o_6,o_6),(o_7,o_7),(o_1,o_3),$$
$$(o_3,o_1),(o_2,o_7),(o_7,o_2),(o_4,o_6),(o_6,o_4)\},$$

$$PI_{\{a_1,a_2\},8} = \{(o_1,o_1),(o_2,o_2),(o_3,o_3),(o_4,o_4),(o_5,o_5),(o_6,o_6),(o_7,o_7),(o_2,o_7),$$
$$(o_7,o_2),(o_3,o_5),(o_5,o_3),(o_4,o_6),(o_6,o_4)\},$$

$$PI_{\{a_1,a_2\},9} = \{(o_1,o_1),(o_2,o_2),(o_3,o_3),(o_4,o_4),(o_5,o_5),(o_6,o_6),(o_7,o_7),(o_2,o_7),$$
$$(o_7,o_2),(o_4,o_6),(o_6,o_4)\},$$

$$PI_{\{a_1,a_2\},10} = \{(o_1,o_1),(o_2,o_2),(o_3,o_3),(o_4,o_4),(o_5,o_5),(o_6,o_6),(o_7,o_7),(o_1,o_3),$$
$$(o_3,o_1),(o_2,o_5),(o_5,o_2),(o_2,o_7),(o_7,o_2),(o_4,o_6),(o_6,o_4),(o_5,o_7),$$
$$(o_7,o_5)\},$$

$$PI_{\{a_1,a_2\},11} = \{(o_1,o_1),(o_2,o_2),(o_3,o_3),(o_4,o_4),(o_5,o_5),(o_6,o_6),(o_7,o_7),(o_2,o_5),$$
$$(o_5,o_2),(o_5,o_7),(o_7,o_5),(o_2,o_7),(o_7,o_2),(o_4,o_6),(o_6,o_4)\},$$

$$PI_{\{a_1,a_2\},12} = \{(o_1,o_1),(o_2,o_2),(o_3,o_3),(o_4,o_4),(o_5,o_5),(o_6,o_6),(o_7,o_7),(o_2,o_5),$$
$$(o_5,o_2),(o_5,o_7),(o_7,o_5),(o_2,o_7),(o_7,o_2),(o_4,o_6),(o_6,o_4)\}.$$

3.2   Possible Equivalence Classes

Possible equivalence class $PE(o)_{A,i}$ containing object o in $pt_{A,i}$ is:

$$PE(o)_{A,i} = \{o' \in U \mid (o, o') \in PI_{A,i}\}. \tag{9}$$

Minimum possible equivalence class $PE(o)_{A,min}$ and maximum possible equivalence class $PE(o)_{A,max}$ containing object o on A are:

$$PE(o)_{A,min} = \cap_i PE(o)_{A,i}, \tag{10}$$

$$PE(o)_{A,max} = \cup_i PE(o)_{A,i}. \tag{11}$$

Family $FPE(o)_A$ of possible equivalence classes containing o consists of those containing o in the possible tables.

$$FPE(o)_A = \cup_i \{PE(o)_{A,i}\}. \tag{12}$$

The family has a lattice structure with the minimum and the maximum element, which are the minimum and the maximum possible equivalence class, respectively [22]. Family $FPE_{A,i}$ of equivalence classes in possible table $pt_{A,i}$ is obtained from possible indiscernibility relation $PI_{A,i}$.

$$FPE_{A,i} = \cup_{o \in U} \{PE(o)_{A,i}\}. \tag{13}$$

3.3 Aggregation of Possible Indiscernibility Relations

We have two aggregations of possible indiscernibility relations the common and the whole indiscernibility relation. Common indiscernibility relation CPI_A is the intersection of $PI_{A,i}$:

$$CPI_A = \cap_i PI_{A,i}. \tag{14}$$

Whole indiscernibility relation WPI_A is the union of $PI_{A,i}$:

$$WPI_A = \cup_i PI_{A,i}. \tag{15}$$

Family$\{CPI_A, PI_{A,1}, \cdots, PI_{A,n}, WPI_A\}$ has a lattice structure with the minimum element CPI_A and the maximum element WPI_A [22].

Proposition 1

$$PE(o)_{A,min} = \{o' \in U \mid (o', o) \in CPI_A\},$$
$$PE(o)_{A,max} = \{o' \in U \mid (o', o) \in WPI_A\}.$$

This proposition shows that the minimum and maximum equivalence classes expressed by formulae (10) and (11) can be also derived from the common and the whole indiscernibility relation.

Example 3

By applying formulae (14) and (15) to possible indiscernibility relations of Example 2, common indiscernibility relation $CPI_{\{a_1,a_2\}}$ and whole indiscernibility relations $WPI_{\{a_1,a_2\}}$ are:

$$CPI_{\{a_1,a_2\}} = \{(o_1,o_1),(o_2,o_2),(o_3,o_3),(o_4,o_4),(o_5,o_5),(o_2,o_7),(o_7,o_2)\},$$
$$WPI_{\{a_1,a_2\}} = \{(o_1,o_1),(o_1,o_3),(o_1,o_4),(o_2,o_2),(o_2,o_5),(o_2,o_7),(o_3,o_1),$$
$$(o_3,o_3),(o_3,o_4),(o_3,o_5),(o_4,o_1),(o_4,o_3),(o_4,o_4),(o_4,o_6),(o_5,o_2),$$
$$(o_5,o_3),(o_5,o_5),(o_5,o_7),(o_6,o_4),(o_6,o_6),(o_7,o_2),(o_7,o_5),(o_7,o_7)\}.$$

By applying the formulae in Proposition 1 to $CPI_{\{a_1,a_2\}}$ and $WPI_{\{a_1,a_2\}}$, minimum possible equivalence class $PE(o_j)_{\{a_1,a_2\},min}$ and maximum possible equivalence class $PE(o_j)_{\{a_1,a_2\},max}$ containing object o_j on $\{a_1,a_2\}$ for $j = 1,7$ are:

$$PE(o_1)_{\{a_1,a_2\},min} = \{o_1\},$$
$$PE(o_1)_{\{a_1,a_2\},max} = \{o_1,o_3,o_4\},$$
$$PE(o_2)_{\{a_1,a_2\},min} = \{o_2.o_7\},$$
$$PE(o_2)_{\{a_1,a_2\},max} = \{o_2,o_5,o_7\},$$
$$PE(o_3)_{\{a_1,a_2\},min} = \{o_3\},$$
$$PE(o_3)_{\{a_1,a_2\},max} = \{o_1,o_3,o_4,o_5\},$$
$$PE(o_4)_{\{a_1,a_2\},min} = \{o_4\},$$
$$PE(o_4)_{\{a_1,a_2\},max} = \{o_1,o_3,o_4,o_6\},$$
$$PE(o_5)_{\{a_1,a_2\},min} = \{o_5\},$$
$$PE(o_5)_{\{a_1,a_2\},max} = \{o_2,o_3,o_5,o_7\},$$
$$PE(o_6)_{\{a_1,a_2\},min} = \{o_6\},$$
$$PE(o_6)_{\{a_1,a_2\},max} = \{o_4,o_6\},$$
$$PE(o_7)_{\{a_1,a_2\},min} = \{o_2.o_7\},$$
$$PE(o_7)_{\{a_1,a_2\},max} = \{o_2,o_5,o_7\},$$

3.4   Lower and Upper Bounds of the Actual Approximations

When target set T of objects is specified, lower approximation $\underline{apr}(T)_{A,i}$ and upper approximation $\overline{apr}(T)_{A,i}$ in possible table pt_i are:

$$\underline{apr}(T)_{A,i} = \{o \in U \mid PE(o)_{A,i} \subseteq T\}, \tag{16}$$
$$\overline{apr}(T)_{A,i} = \{o \in U \mid PE(o)_{A,i} \cap T \neq \emptyset\}. \tag{17}$$

Minimum lower approximation $\underline{apr}(T)_{A,min}$ that is the lower bound of the actual lower approximation, maximum lower approximation $\underline{apr}(T)_{A,max}$ that is the upper bound of the actual lower approximation, minimum upper approximation $\overline{apr}(T)_{A,min}$ that is the lower bound of the actual upper approximation, and maximum upper approximation $\overline{apr}(T)_{A,max}$ that is the upper bound of the actual upper approximation are:

$$\underline{apr}(T)_{A,min} = \cap_i \underline{apr}(T)_{A,i}, \tag{18}$$

$$\underline{apr}(T)_{A,max} = \cup_i \underline{apr}(T)_{A,i}, \tag{19}$$

$$\overline{apr}(T)_{A,min} = \cap_i \overline{apr}(T)_{A,i}, \tag{20}$$

$$\overline{apr}(T)_{A,max} = \cup_i \overline{apr}(T)_{A,i}. \tag{21}$$

Proposition 2

$$\underline{apr}(T)_{A,min} = \{o \in U \mid PE(o)_{A,max} \subseteq T\},$$
$$\underline{apr}(T)_{A,max} = \{o \in U \mid PE(o)_{A,min} \subseteq T\},$$
$$\overline{apr}(T)_{A,min} = \{o \in U \mid PE(o)_{A,min} \cap T \neq \emptyset\},$$
$$\overline{apr}(T)_{A,max} = \{o \in U \mid PE(o)_{A,max} \cap T \neq \emptyset\}.$$

This proposition shows that the lower and upper bounds of approximations can be also derived from using minimum and maximum possible equivalence classes. The actual lower and upper approximations exist between $\underline{apr}(T)_{A,min}$ and $\underline{apr}(T)_{A,max}$ and between $\overline{apr}(T)_{A,min}$ and $\overline{apr}(T)_{A,max}$, respectively. This is the results obtained from Lipski's approach.

Example 4
Let target set T be $\{o_1, o_2, o_4\}$ that is specified by constraint $a_3 = b$. The lower and upper bounds of approximations are obtained from the formulae in Proposition 2. By using the minimum and maximum possible equivalence classes obtained in Example 3, minimum lower approximation $\underline{apr}(T)_{\{a_1,a_2\},min}$, maximum lower approximation $\underline{apr}(T)_{\{a_1,a_2\},max}$, minimum upper approximation $\overline{apr}(T)_{\{a_1,a_2\},min}$, and maximum upper approximation $\overline{apr}(T)_{\{a_1,a_2\},max}$ are:

$$\underline{apr}(T)_{\{a_1,a_2\},min} = \emptyset,$$
$$\underline{apr}(T)_{\{a_1,a_2\},max} = \{o_1, o_4\},$$
$$\overline{apr}(T)_{\{a_1,a_2\},min} = \{o_1, o_2, o_4, o_7\},$$
$$\overline{apr}(T)_{\{a_1,a_2\},max} = \{o_1, o_2, o_3, o_4, o_5, o_6, o_7\}.$$

The actual lower approximation exists between $\emptyset$ and $\{o_1, o_4\}$ and the actual upper approximation exists between $\{o_1, o_2, o_4, o_7\}$ and $\{o_1, o_2, o_3, o_4, o_5, o_6, o_7\}$
.

4   Kryszkiewicz's Relation for Indiscernibility of Objects

Kryszkiewicz proposed a formula expressing a binary relation for object indistinguishability in information tables containing missing values [4,5]. The formula is expressed as follows:

$$I_A^K = \{(o, o') \in U \times U \mid \forall a \in A \ \ a(o) = a(o') \lor a(o) = * \lor a(o') = *\}. \tag{22}$$

We call this relation Kryszkiewicz's relation. The relation is reflective, and symmetric, but not transitive. Set $E_A^{\overline{K}}(o)$ of objects that are considered as equivalent to object o is derived from the relation:

$$E_A^{\overline{K}}(o) = \{o' \in U \mid (o, o') \in I_A^{\overline{K}}\}, \tag{23}$$

The set is not an equivalence class. Let T be a target set of objects specified by a constraint on a set of attributes. By using $E_A^{\overline{K}}(o)$, lower approximation $\underline{apr}_A^{\overline{K}}(T)$ and upper approximation $\overline{apr}_A^{\overline{K}}(T)$ are:

$$\underline{apr}(T)_A^{\overline{K}} = \{o \in U \mid E(o)_A^{\overline{K}} \subseteq T\}, \tag{24}$$

$$\overline{apr}(T)_A^{\overline{K}} = \{o \in U \mid E(o)_A^{\overline{K}} \cap T \neq \emptyset\}. \tag{25}$$

These approximations are not compatible with those derived from Lipski's approach because Lipski's approach shows that the actual approximations cannot be obtained and what we can obtain is the lower and the upper bounds of the actual approximations, as is described in the previous section.

In addition, these formulae have the drawbacks pointed out by [14,19], as is shown in the below example.

Example 5
We check Kryszkiewicz's relation by using incomplete information table IT in Example 2. Let target set T be $\{o_1, o_2, o_4\}$, which is the same as in Example 2. Using formula (22) in IT, we obtain the following binary relation for indiscernibility:

$$I_{\{a_1,a_2\}}^{\overline{K}} = \{(o_1, o_1), (o_1, o_3), (o_1, o_4), (o_2, o_2), (o_2, o_5), (o_2, o_7), (o_3, o_1),$$
$$(o_3, o_3), (o_3, o_4), (o_3, o_5), (o_4, o_1), (o_4, o_3), (o_4, o_4), (o_4, o_6), (o_5, o_2),$$
$$(o_5, o_3), (o_5, o_5), (o_5, o_7), (o_6, o_4), (o_6, o_6), (o_7, o_2), (o_7, o_5), (o_7, o_7)\}.$$

Class $E_{\{a_1,a_2\}}^{\overline{K}}(o_j)$ derived from the relation for $j = 1, 7$ is:

$$E_{\{a_1,a_2\}}^{\overline{K}}(o_1) = \{o_1, o_3, o_4\},$$

$$E_{\{a_1,a_2\}}^{\overline{K}}(o_2) = \{o_2, o_5, o_7\},$$

$$E_{\{a_1,a_2\}}^{\overline{K}}(o_3) = \{o_3, o_4, o_5\},$$

$$E_{\{a_1,a_2\}}^{\overline{K}}(o_4) = \{o_1, o_3, o_4, o_6\},$$

$$E_{\{a_1,a_2\}}^{\overline{K}}(o_5) = \{o_2, o_3, o_5\},$$

$$E_{\{a_1,a_2\}}^{\overline{K}}(o_6) = \{o_4, o_6\},$$

$$E_{\{a_1,a_2\}}^{\overline{K}}(o_7) = \{o_2, o_5, o_7\}.$$

Lower approximation $\underline{apr}^{\overline{K}}_{\{a_1,a_2\}}(T)$ and upper approximation $\overline{apr}^{\overline{K}}_{\{a_1,a_2\}}(T)$ from formulae (24) and (25) are:

$$\underline{apr}(T)^{\overline{K}}_{\{a_1,a_2\}} = \emptyset,$$

$$\overline{apr}(T)^{\overline{K}}_{\{a_1,a_2\}} = \{o_1, o_2, o_3, o_4, o_5, o_6, o_7\}. = U$$

Note that nothing is obtained for the lower and upper approximations.

Example 5 shows that we have poor results for approximations in the case of using Kryszkiewicz's relation because only the maximum possible classes of objects are taken into account. For example, possible classes of objects equivalent to object o_1 on $\{a_1, a_2\}$ in IT of Example 2 are $\{o_1\}$, $\{o_1, o_3\}$, $\{o_1, o_4\}$, and $\{o_1, o_3, o_4\}$, but only the maximum possible class $\{o_1, o_3, o_4\}$ is derived from $I^{\overline{K}}_{\{a_1,a_2\}}$. In other words, only the possibility that missing value $*$ may be equal to a value is considered, but the opposite possibility is neglected. Clearly, information loss occurs.

5 Relationship Between Kryszkiewicz's Approach and Lipski's Approach

Indeed, Kryszkiewicz's relation is not compatible with Lipski's approach, but we have the following proposition.

Proposition 3

$$I^{\overline{K}}_A = WPI_A.$$

This proposition shows that the binary relation for indiscernibility used by Kryszkiewicz is equal to the whole indiscernibility relation, the union of possible indiscernibility relations in possible tables.

Proposition 4

$$\underline{apr}(T)^{\overline{K}}_A = \underline{apr}(T)_{A,min},$$

$$\overline{apr}(T)^{\overline{K}}_A = \overline{apr}(T)_{A,max}.$$

The proposition shows that the lower and upper approximations derived from Kryszkiewicz's relation are equal to ones derived from using the whole indiscernibility relation in possible world semantics. In other words, the lower approximation derived from Kryszkiewicz's relation is the lower bound of the actual lower approximation, whereas the upper approximation derived from Kryszkiewicz's relation is the upper bound of the actual upper approximation.

This suggests that the incompatibility between Kryszkiewicz's approach and Lipski's approach is resolved by adding the relation for indiscernibility corresponding to the common indiscernibility relation. The expression corresponding to the common indiscernibility relation under the notation of Kryszkiewicz is:

$$I_A^K = \{(o, o') \in U \times U \mid (o = o') \vee$$
$$(\forall a \in A \ a(o) = a(o') \wedge a(o) \neq * \wedge a(o') \neq *)\}. \tag{26}$$

Proposition 5

$$I_A^K = CPI_A.$$

$E_A^K(o)$ of each object is derived from I_A^K using formula (26) similarly to $E_A^{\overline{K}}(o)$. By using $E_A^K(o)$, lower approximation $\underline{apr}_A^K(T)$ and upper approximation $\overline{apr}_A^K(T)$ are:

$$\underline{apr}(T)_A^K = \{o \in U \mid E(o)_A^K \subseteq T\}, \tag{27}$$
$$\overline{apr}(T)_A^K = \{o \in U \mid E(o)_A^K \cap T \neq \emptyset\}. \tag{28}$$

Proposition 6

$$\underline{apr}(T)_A^K = \underline{apr}(T)_{A,max},$$
$$\overline{apr}(T)_A^K = \overline{apr}(T)_{A,min}.$$

Extending Kryszkiewicz's approach by adding formula (26) to formula (22), we can resolve the incompatibility between Kryszkiewicz's approach and Lipski's approach. In other words, we can keep using Kryszkiewicz's approach in which missing values are embedded into the formula expressing indiscernibility for objects under complete information.

Example 6

Using formula (26) in IT in Example 2, we obtain the following binary relation for indiscernibility:

$$I_{\{a_1,a_2\}}^K = \{(o_1, o_1), (o_2, o_2), (o_2, o_7), (o_3, o_3), (o_4, o_4), (o_5, o_5), (o_6, o_6), (o_7, o_2),$$
$$(o_7, o_7)\}.$$

Class $E_{\{a_1,a_2\}}^K(o_j)$ derived from the relation for $j = 1, 7$ is:

$$E_{\{a_1,a_2\}}^{\overline{K}}(o_1) = \{o_1\},$$
$$E_{\{a_1,a_2\}}^{\overline{K}}(o_2) = \{o_2, o_7\},$$

$$E^{\overline{K}}_{\{a_1,a_2\}}(o_3) = \{o_3\},$$

$$E^{\overline{K}}_{\{a_1,a_2\}}(o_4) = \{o_4\},$$

$$E^{\overline{K}}_{\{a_1,a_2\}}(o_5) = \{o_5\},$$

$$E^{\overline{K}}_{\{a_1,a_2\}}(o_6) = \{o_6\},$$

$$E^{\overline{K}}_{\{a_1,a_2\}}(o_7) = \{o_2,o_7\}.$$

Let target set T be $\{o_1,o_2,o_4\}$, which is the same as in Example 2. Lower approximation $\underline{apr}^K_{\{a_1,a_2\}}(T)$ and upper approximation $\overline{apr}^K_{\{a_1,a_2\}}(T)$ from formulae (27) and (28) are:

$$\underline{apr}(T)^K_{\{a_1,a_2\}} = \{o_1,o_4\},$$

$$\overline{apr}(T)^K_{\{a_1,a_2\}} = \{o_1,o_2,o_4,o_7\}.$$

6   Conclusions

We have described rough sets in incomplete information tables with missing values whose values are unknown. Many authors use Kryszkiewicz's relation in order to express the indiscernibility between values. The relation, however, involves information loss and is not compatible with Lipski's approach based on possible world semantics. As a result, it creates poor results for approximations.

In order to resolve the above point, we have checked Kryszkiewicz's relation for the indiscernibility of objects from the viewpoint of Lipski's approach. As a result, Kryszkiewicz's relation is equal to the whole indiscernibility relation that is the union of indiscernibility possible relations derived from possible tables. This means that the lower and the upper approximation derived from Kryszkiewicz's relation are the lower bound of the actual lower approximation and the upper bound of the actual upper approximation, respectively.

The drawback of Kryszkiewicz's relation can be resolved by adding the relation corresponding to the common indiscernibility relation that is the intersection of possible indiscernibility relation. As a result, the extension of Kryszkiewicz's approach is compatible with Lipski's approach based on possible world semantics. No information loss occurs in applying rough sets to incomplete information tables with missing values. Therefore, we can keep using Kryszkiewicz's approach giving the relation of indiscernibility for objects in incomplete information tables.

The future work will be to confirm the effect of the newly added relation on previous research through several experiments.

Acknowledgment. Part of this work is supported by JSPS KAKENHI Grant Number JP20K11954.

References

1. Pawlak, Z.: Rough Sets: Theoretical Aspects of Reasoning about Data. Kluwer Academic Publishers, Dordrecht (1991). https://doi.org/10.1007/978-94-011-3534-4
2. Parsons, S.: Current approaches to handling imperfect information in data and knowledge bases. IEEE Trans. Knowl. Data Eng. **8**, 353–372 (1996)
3. Parsons, S.: Addendum to current approaches to handling imperfect information in data and knowledge bases. IEEE Trans. Knowl. Data Eng. **10**, 862 (1998)
4. Kryszkiewicz, M.: Rough set approach to incomplete information systems. Inf. Sci. **112**, 39–49 (1998)
5. Kryszkiewicz, M.: Rules in incomplete information systems. Inf. Sci. **113**, 271–292 (1999)
6. Greco, S., Matarazzo, B., Słowinski, R.: Handling missing values in rough set analysis of multi-attribute and multi-criteria decision problems. In: Zhong, N., Skowron, A., Ohsuga, S. (eds.) RSFDGrC 1999. LNCS (LNAI), vol. 1711, pp. 146–157. Springer, Heidelberg (1999). https://doi.org/10.1007/978-3-540-48061-7_19
7. Grzymala-Busse, J.W.: Data with missing attribute values: generalization of indiscernibility relation and rule induction. Trans. Rough Sets **I**, 78–95 (2004)
8. Grzymala-Busse, J.W.: Characteristic relations for incomplete data: a generalization of the indiscernibility relation. Trans. Rough Sets **IV**, 58–68 (2005)
9. Latkowski, R.: Flexible indiscernibility relations for missing values. Fund. Inf. **67**, 131–147 (2005)
10. Leung, Y., Li, D.: Maximum consistent techniques for rule acquisition in incomplete information systems. Inf. Sci. **153**, 85–106 (2003)
11. Nakata, M., Sakai, H.: Rough sets handling missing values probabilistically interpreted. In: Ślezak, D., Wang, G., Szczuka, M., Düntsch, I., Yao, Y. (eds.) RSFDGrC 2005. LNCS (LNAI), vol. 3641, pp. 325–334. Springer, Heidelberg (2005). https://doi.org/10.1007/11548669_34
12. Nakata, M., Sakai, H.: Twofold rough approximations under incomplete information. Int. J. Gen Syst **42**, 546–571 (2013). https://doi.org/10.1080/17451000.2013.798898
13. Sakai, H.: Twofold rough approximations under incomplete information. Fund. Inf. **48**, 343–362 (2001)
14. Stefanowski, J., Tsoukiàs, A.: Incomplete information tables and rough classification. Comput. Intell. **17**, 545–566 (2001)
15. Sun, L., Wanga, L., Ding, W., Qian, Y., Xu, J.: Neighborhood multi-granulation rough sets-based attribute reduction using lebesgue and entropy measures in incomplete neighborhood decision systems. Knowl.-Based Syst. **192**, 105373 (2020)
16. Wang, G.: Extension of rough set under incomplete information systems. In: 2002 IEEE World Congress on Computational Intelligence. IEEE International Conference on Fuzzy Systems. FUZZ-IEEE 2002, pp. 1098–1103 (2002) https://doi.org/10.1109/FUZZ.2002.1006657
17. Nguyen, D.V., Yamada, K., Unehara, M.: Extended tolerance relation to define a new rough set model in incomplete information systems. Adv. Fuzzy Syst. Article ID 372091, 10 (2013). https://doi.org/10.1155/2013/372091
18. Rady, E.A., Abd El-Monsef, M.M.E., Adb El-Latif, W.A.: A modified rough sets approach to incomplete information systems. J. Appl. Math. Decis. Sci. Article ID 058248, 13 (2007)
19. Nakata, M., Sakai, H.: Applying rough sets to information tables containing missing values. In: Proceedings of 39th International Symposium on Multiple-Valued Logic, pp. 286–291. IEEE Press (2009). DOI: https://doi.org/10.1109/ISMVL.2009.1

20. Yang, T., Li, Q., Zhou, B.: Related family: a new method for attribute reduction of covering information systems. Inf. Sci. **228**, 175–191 (2013)
21. Lipski, W.: On semantics issues connected with incomplete information databases. ACM Trans. Database Syst. **4**, 262–296 (1979)
22. Nakata, M., Saito, N., Sakai, H., Fujiwara, T.: Structures derived from possible tables in an incomplete information table. In: Proceedings in SCIS 2022, p. 6. IEEE Press (2022) https://doi.org/10.1109/SCISISIS55246.2022.10001919

Algebraic, Topological, and Mereological Foundations of Existential Granules

A. Mani[✉][iD]

Machine Intelligence Unit, Indian Statistical Institute, Kolkata, 203, B. T. Road,
Kolkata 700108, India
a.mani.cms@gmail.com, amani.rough@isical.ac.in
https://www.logicamani.in

Abstract. In this research, new concepts of existential granules that determine themselves are invented, and are characterized from algebraic, topological, and mereological perspectives. Existential granules are those that determine themselves initially, and interact with their environment subsequently. Examples of the concept, such as those of granular balls, though inadequately defined, algorithmically established, and insufficiently theorized in earlier works by others, are already used in applications of rough sets and soft computing. It is shown that they fit into multiple theoretical frameworks (axiomatic, adaptive, and others) of granular computing. The characterization is intended for algorithm development, application to classification problems and possible mathematical foundations of generalizations of the approach. Additionally, many open problems are posed and directions provided.

Keywords: Existential Granules · Adaptive Granules · Axiomatic Granular Computing · Topological Vector Spaces · Granular Operator Spaces · Granular Balls · Ball K-Means Algorithm · Clean Rough Randomness

1 Introduction

In the philosophy literature, existentialism is about basic questions of human existence, individuality, and interactions with the environment. In this research, the adjective *existential* is used in relation to self-determination of objects and subsequent transformations in relation to objects of similar type in their environment. Consequently, they are a form of adaptive objects, and not all adaptive objects are existential.

In the axiomatic approach to granularity [15,16,18], granules are typically specified by conditions. Though external generation procedures are not specified in the literature, they can be expected to be mostly compatible with the methodology. In the precision-based approach [11,13,34], precision-levels

---

This research is supported by Woman Scientist Grant No. WOS-A/PM-22/2019 of the Department of Science and Technology.

A. Campagner et al. (Eds.): IJCRS 2023, LNAI 14481, pp. 185–200, 2023.
https://doi.org/10.1007/978-3-031-50959-9_13

define admissible granules. However, a mere specification of a precision-level, rarely defines a granule or generates one. Even if it does, it is not that the precision level is intrinsic to the granule. Generation procedures can certainly be added to precision-based granules subject to compatibility with the conditions. Adaptive granules [28] are expected to adapt to processes, and even they may be generated initially through some procedures. The existential aspect in these are their initial self-determination, and subsequent transformation in response to interactions. These ideas can be defined in other categorical perspectives of the same theories.

Closed and open balls and spherical surfaces are pretty standard objects in studies on metrizable spaces and related topologies. However, they are not interpreted as such in some empirical machine learning practices, and if so, how are they interpreted? This is one of the problems tackled in this research.

In recent research, some improved algorithms for K-means clustering and classification issues are developed [31–33]. The algorithms are explained through ideas of *granular balls*. However, they are not sufficiently theorized in the mentioned papers, and in others that make use of the concept [5,8,25,27]. The basic assumptions, and possible generalizations of the concept are investigated by the present author here. New concepts of existential granules are proposed as a severe generalization of the concept, and are shown to be compatible with her axiomatic frameworks for granules [15,16,18]. A somewhat underspecified example of the formation of existential granules in real life contexts is the following: When law enforcers try to comb a forested area for possible illegal activity with information from drones and other sources, then they typically form multiple teams to cover distinct sub-areas. Based on the result from the combing operation completed, they are likely to redefine the sub-areas to be searched again. The sequence of set of subareas at each stage will stabilize when all relevant areas are checked. The sub-areas may be regarded as granules that transform themselves at each stage of the operation due to new information, and the state of the search operation(s) on other areas. Finally, they become stable at some later stage.

Clean rough randomness as a not-necessarily stochastic or algorithmically randomness concept is recently introduced by the present author [19,21], and is capable of modeling many algorithms such as those that operate over entire sets of tolerances. The problem of precisely formalizing the adaptive aspect of the algorithms using related functions is additionally posed.

The following sections are organized as follows. Some background is provided in the next section. Existential ball K-Means (BKM) algorithms are analyzed, related partial algebras are invented and a soft generalization where the algorithm works is specified in the third section. The granular ball methodology is formalized through a reading of related algorithms in the next. A formal approach to existential granules is invented in the fifth section and the problem of *appropriately* formalizing the BKM algorithm in the rough randomness perspective is formulated. Further directions are considered in the sixth.

2   Background

Distances are often intended to model qualitative ideas of being different in numeric terms. Therefore, it is not required to satisfy most conditions typical of a metric in a metric space. A *distance function* on a set S is a function $\sigma : S^2 \longmapsto \mathfrak{R}_+$ that satisfies

$$(\forall a)\sigma(a, a) = 0 \qquad\qquad\qquad \text{(distance)}$$

The collection $\mathcal{B} = \{B_\sigma(x, r) : x \in S \ \& \ r > 0\}$ of all r-spheres generated by σ is a weak base for the topology τ_σ defined by

$$V \in \tau_\sigma \text{ if and only if } (\forall x \in V \exists r > 0)B_\sigma(x, r) \subseteq V$$

Consider the conditions:

Identity$(\forall a, b)(\sigma(a, b) = 0 \leftrightarrow a = b)$,
Symmetry$(\forall a, b)\sigma(a, b) = \sigma(b, a)$,
Triangle$(\forall a, b, c) \ \sigma(a, b) \leqslant \sigma(a, c) + \sigma(c, b)$, and
Pseudo-Identity$(\forall a, b)(\sigma(a, b) = 0 \longrightarrow a = b)$.

σ is said to be a metric or semimetric or pseudometric respectively as it satisfies all the four or identity and symmetry or the last three conditions respectively. A quasi-metric is a distance that satisfies the triangle inequality, while a distance that satisfies the triangle inequality up to a constant $k > 0$ (k-triangle: $(\forall a, b, c) \ k\sigma(a, b) \leqslant \sigma(a, c) + \sigma(c, b)$) is called a weak quasi-metric. Given a distance function on a set, a topology does not automatically follow. This holds as well for semimetrics [6]. All generalized metric spaces will be collectively referred to as ∗-metric spaces.

If x is a point in a ∗-metric space (X, σ), and H a subset of X then the distance of x from H is given by $\underline{\sigma}(x, H) = \inf\{\sigma(x, a) : a \in H\}$. The distance between two subsets H and F can be measured with the Hausdorff distance σ_h or the infimal distance σ_I(these are not metrics):

$$\sigma_h(H, F) = \max\{\sup_{x \in H} \underline{\sigma}(x, F), \sup_{x \in F} \underline{\sigma}(H, x)\} \ \& \ \sigma_I(H, F) = \inf\{\sigma(a, b) : a \in H, b \in F\}.$$

The former is a metric on the set of compact subsets if σ is a metric.

2.1   Topological Vector Spaces

Some familiarity with topological vector spaces (TVS) [2,30] will be assumed. A ∗-metric vector space is a pair (X, σ), with X being a vector space over the real field, and σ a ∗-metric such that the operations are jointly continuous (that is if $(x_n) \to x$ and $(b_n) \to b$ in X, and $(\alpha_n) \to \alpha$ in $\mathfrak{R}$, then $(\alpha_n x_n + b_n) \to \alpha x + b$.)

Consider the following properties of a function $p : X \longmapsto \mathfrak{R}_+$ for any x, $b \in X$

$$p(0) = 0 \qquad \text{(PN1)}$$

$$p(x) \geqslant 0 \qquad \text{(PN2)}$$

$$p(-x) = p(x) \qquad \text{(PN3)}$$

$$p(x+b) \leqslant p(x) + p(b) \qquad \text{(PN4)}$$

$$\text{Continuity of scalar multiplication} \qquad \text{(PN5)}$$

$$\text{If } p(x) = 0 \text{ then } x = 0 \qquad \text{(PNT)}$$

$$p(\alpha x) = |\alpha| p(x) \qquad \text{(SN1)}$$

p is said to be a *paranorm* (respectively *seminorm*) if it satisfies PN1-PN5 (respectively PN2,PN4 and SN1). It is *total*, if it satisfies PNT. All seminorms are paranorms, and a total seminorm is a *norm*.

A paranormed space is a pair (X, p) where p is a paranorm over the vector space X. It is complete if (X, σ) is complete, where $\sigma(a, b) = p(a - b)$. Every pseudometric vector space can be endowed with a paranorm from which it is derived.

It should be noted that *-metrics that have nothing to do with any intended topology are sometimes used in ML practice.

2.2   Partial Algebraic Systems

For basics of partial algebras, the reader is referred to [4,14].

Definition 1. *A partial algebra* P *is a tuple of the form*

$$\langle \underline{P}, f_1, f_2, \ldots, f_n, (r_1, \ldots, r_n) \rangle$$

with $\underline{P}$ *being a set,* f_i's *being partial function symbols of arity* r_i. *The interpretation of* f_i *on the set* $\underline{P}$ *should be denoted by* $f_i^{\underline{P}}$, *but the superscript will be dropped in this paper as the application contexts are simple enough. If predicate symbols enter into the signature, then* P *is termed a* partial algebraic system.

In this paragraph the terms are not interpreted. For two terms s, t, $s \overset{\omega}{=} t$ shall mean, if both sides are defined then the two terms are equal (the quantification is implicit). $\overset{\omega}{=}$ is the same as the existence equality (also written as $\overset{e}{=}$) in the present paper. $s \overset{\omega^*}{=} t$ shall mean if either side is defined, then the other is and the two sides are equal (the quantification is implicit). Note that the latter equality can be defined in terms of the former as

$$(s \overset{\omega}{=} s \longrightarrow s \overset{\omega}{=} t) \,\&\, (t \overset{\omega}{=} t \longrightarrow s \overset{\omega}{=} t)$$

Various kinds of morphisms can be defined between two partial algebras or partial algebraic systems of the same or even different types. For two partial algebras of the same type

$$X = \langle \underline{X}, f_1, f_2, \ldots, f_n \rangle \text{ and } W = \langle \underline{W}, g_1, g_2, \ldots, g_n \rangle,$$

a map $\varphi : X \longmapsto W$ is said to be a

- *morphism* if for each i,

$$(\forall (x_1, \ldots x_k) \in \text{dom}(f_i)) \varphi(f_i(x_1, \ldots, x_k)) = g_i(\varphi(x_1), \ldots, \varphi(x_k))$$

- *closed morphism*, if it is a morphism and the existence of $g_i(\varphi(x_1), \ldots, \varphi(x_k))$ implies the existence of $f_i(x_1, \ldots, x_k)$.

Usually it is more convenient to work with closed morphisms.

3   Existential Granular K-Means Algorithm

The end product of a hard or soft clustering can often be interpreted as a granulation. The so-called ball granular computing [32] is not properly formalized from a mathematical perspective as the goal of the authors is to stress the performance of their algorithms. Its origin is obviously related to the ball K-means algorithm [33]. A critical analysis with some generalization of the last mentioned method is proposed first after reconsidering the basic assumptions implicit in it.

Let the dataset of points be V that is a subset of the real topological vector space X with pseudometric (or metric) σ which in turn is equivalent to a paranorm. V is not usually closed under the algebraic operations induced from X. Algebraic closure on real data is often more complex as additional layers of meaning based on bounds or types may be of interest. Some questions that can shape the semantic domain and therefore relevant algebraic models are

1. Should the value of operations beyond V be considered?
2. Are the interpretations of operations over X meaningful for the context of V? To what extent should they be permitted? Does the smallest subspace $\text{Alg}(V)$ containing V suffice?
3. Should the interpretations of the operations over X be reinterpreted (at least partly) over V. This is especially useful when the values or bounds imposed by V are alone meaningful.

Depending on the answers to these, the appropriate algebraic operations on the balls may be selected and this lead to a natural generalization of algorithm.

The basic steps of the ball k-means algorithm are

Subregion Form an arbitrary clustering $E_1, E_2, \ldots E_k$ of V.
MCT Compute the mean c_i for each subset E_i.

Radius Taking the greatest distance among points in E_i from c_i as the radius r_i, generate the ball C_i for each i.

Neighbors Define the relation $\eta C_j C_i$ (for C_j is a neighbor of C_i) if and only if $\sigma c_i c_j < 2r_i$. For C_i, let N_{C_i} be its set of neighbor balls (granules).

Stable If $N_{C_i} \neq \emptyset$, then its stable region is defined by

- $St_\sigma(C_i) = B(c_i, 0.5 \min \sigma(c_i, c) : c \in N_{C_i})$,
- and its active area by $AA(C_i) = C_i \setminus St(C_i)$.

Annular Regions Let $Card(N_C) = k$, then for $i \in [1, k]$, the ith annular region on C, A_i^C is $\{x : \sigma(c, c_i) < 2\sigma(x, c) \leqslant \sigma(c, c_{i+1})\}$ for $i < k$ and $x : \sigma(c, c_i) < 2\sigma(x, c) \leqslant r$ for $i = k$.

The BKM algorithm and related considerations can be extended to any metric TVS. However, the results cannot be guaranteed for semi-metric spaces or pseudo-metric spaces, in general.

For each i, the radius at the first iteration $r_i = Sup\{\sigma(x, c_i) : x \in E_i\}$. Given two ball clusters C_i and C_j with centers c_i and c_j respectively at a fixed iteration level, define

$$\eta C_j C_i \text{ iff } \sigma(c_i, c_j) < 2r_i.$$

η is a reflexive, non-symmetric relation in general. C_j is a *neighbor* of C_i if and only if $\eta C_i C_j$ For C_i, let N_{C_i} be its set of neighbor balls (granules). If $N_{C_i} \neq \emptyset$, then its stable region is defined by

$$St(C_i) = B(c_i, 0.5 \min \sigma(c_i, c) : c \in N_{C_i}).$$

The active area $AA(C_i) = C_i \setminus St(C_i)$.

The term *i-closest* is not defined in the paper [33]. It is simply the closest neighbor cluster(s). As it is based on distance between centers, uniqueness cannot be guaranteed. Let $Card(N_C) = k$, then for $i \in [1, k]$, the ith annular region on C, A_i^C is $\{x : \sigma(c, c_i) < 2\sigma(x, c) \leqslant \sigma(c, c_{i+1})\}$ for $i < k$ and $x : \sigma(c, c_i) < 2\sigma(x, c) \leqslant r$ for $i = k$.

The ball k-means algorithm can be reinterpreted as a granular approximation procedure of an unknown clustering that is supposed to exist. The steps in the approximation being guided by η, stable regions, and annular regions as proved in Theorem 1 ([33]). Stable regions may be read as partial lower approximations of the initial granules at that stage.

Theorem 1. 1. *If C_i is a neighbor of C, then some non-stable points of C may be moved into C_i.*

2. *If C_j is not a neighbor of C, then no points of C can be moved into C_j*

3. *For a given C with center c, and $Card(N_C) = k$, the points in the ith (i $\leqslant$ k) annular space of C can only be moved within the first i-closest neighbor clusters and itself.*

4. *If $c_i^{(t)}$ is the center of the ball C_i in the tth iteration, then if $\sigma(c_i^{(t-1)}, c_j^{(t-1)}) \geqslant 2r_i^{(t)} + \sigma(c_i^{(t)}, c_i^{(t-1)}) + \sigma(c_j^{(t)}, c_j^{(t-1)})$, then C_j cannot be a neighbor ball of C_i in the current iteration. So the computation of the center distance is avoided.*

It can additionally be proved that

Theorem 2. *If* X *is a paranormed TVS, then the BKM algorithm terminates in a soft clustering.*

3.1   Partial Algebras for BKM Variants

Let $B_r^V(c)$ denote the closed ball $\{x \,:\, x \in V, \& \, \sigma(x, c) \leqslant r\}$ with center c and radius r over V (it will also be referred to as the *cautious closed ball*), then the following algebraic partial/total operations are definable for any $a, b \in B_r^V(c)$ and any $\alpha, \beta \in \mathfrak{R}$

$$\alpha a \oslash \beta b = \begin{cases} \alpha a + \beta b, & \text{if } \alpha a, \beta b, \alpha a + \beta b \in B_r^V(c) \\ \text{undefined}, & \text{otherwise} \end{cases}$$

On the closed ball $B_r^X(c) = \{x : x \in X, \& \, \sigma(x, c) \leqslant r\}$ too, a similar operation $\oplus$ may be interpreted (relative to the operations on X).

$$\alpha a \oplus \beta b = \begin{cases} \alpha a \oplus \beta b, & \text{if } \alpha a + \beta b \in B_r^X(c) \\ \text{undefined}, & \text{otherwise} \end{cases}$$

Theorem 3. *In the above context,* $B_r^X(c)$ *satisfies*

$$a \oplus b \overset{s}{=} b \oplus a \qquad \text{(weak* comm)}$$
$$a \oplus (b \oplus c) \overset{\omega}{=} (a \oplus b) \oplus c \qquad \text{(weak assoc)}$$
$$\alpha(\beta a) \overset{\omega}{=} (\alpha\beta)a \qquad \text{(weak scal1)}$$
$$\alpha a \oplus \beta a \overset{s}{=} (\alpha + \beta)a \qquad \text{(weak* scal2)}$$
$$a \oplus 0 \overset{s}{=} 0 \oplus a \qquad \text{(weak* 0)}$$
$$(\forall a, b, c)(a \oplus b = 0 = a \oplus c \longrightarrow b = c) \qquad \text{(inverse)}$$

Proof. The weak versions of the equalities hold when both sides are defined, while the stronger version ($\overset{s}{=}$) require any one of the sides to be defined. Weak* commutativity is obvious. Weak associativity holds, and its stronger version does not because $a \oplus b$ may not be defined, even though $a \oplus (b \oplus c)$ is.

Theorem 4. *In the above context,* $B_r^V(c)$ *satisfies* $\text{dom}(\oslash) \subset \text{dom}(\oplus)$ *and*

$$a \oslash b \overset{s}{=} b \oslash a \qquad \text{(weak* comm)}$$
$$a \oslash (b \oslash c) \overset{\omega}{=} (a \oslash b) \oslash c \qquad \text{(weak assoc)}$$
$$\alpha(\beta a) \overset{\omega}{=} (\alpha\beta)a \qquad \text{(weak scal1)}$$
$$\alpha a \oslash \beta a \overset{s}{=} (\alpha + \beta)a \qquad \text{(weak* scal2)}$$
$$a \oslash 0 \overset{s}{=} 0 \oslash a \qquad \text{(weak* 0)}$$
$$(\forall a, b, c)(a \oslash b = 0 = a \oslash c \longrightarrow b = c) \qquad \text{(inverse)}$$

4   The Granular Ball Methodologies

A version of the granular ball methods for classification can be found in the preprint [31]. Readers are left wondering whether a norm is even being used (Eqn 2 in page 3), while the exact partitioning of parent balls into child balls (in Definition 1) is impossible. However, the algorithm is relatively clear, and examples for the intent of Definition 1 are provided. Earlier versions have additional mathematical issues (see the discussion at https://pubpeer.com/publications/4354287243FC39A66DD432BC41046B).

The methods for adaptive granules involves quality checks based on purity of granules (relative to proportion of labels), and heterogeneity of overlap of granules. Otherwise, the essential methods are variations of the one used in the ball K-means algorithm.

1. The data set may be partly labelled.
2. Regard whole data set or a sample V as a closed sphere with center $c = \frac{1}{n} \sum (x_i)$ and radius as $r = \frac{1}{n} \sum \sigma(x_i, c)$.
3. Split the current granular ball into k sub balls using ball K-means. It should be noted that splitting is not a partitioning operation.
4. Check quality of granular Balls through simple purity measures based on ratio of majority label.
5. Stop if the purity measure is OK. Otherwise, repeat the process on the balls derived.

The adaptive version is similar but with further stages of splitting whenever a granular ball at the current iteration has heterogeneous overlap with another granular ball, and an accelerated granular ball generation process. Child balls and parent balls are further used in the quality checks, and the ball K-means algorithm is avoided.

4.1   Fixing the Mathematics

From the intent of definition 1 (of parent and child balls), it is clear to the present author that the *real data points within a ball* are confused with the ball. A mathematical way of correcting can be through differential geometry, and at least concepts of orientation are essential. The code and algorithm are however based on distances from centers, and labels. A minimal fix that avoids the geometry is the following:

Definition 2. *Let V be a finite subset of a real normed finite dimensional TVS X, and $B_r^V(c)$ be a ball, and $\{B_{r_i}^V(c_i)\}$ a finite sequence of n number of balls, all interpreted in V (with centers in X) then $B_r^V(c)$ is a* major ball *and $\{B_{r_i}^V(c_i)\}$ are* minor balls *if and only if the following holds:*

$$\bigcup_i B_{r_i}^V(c_i) = \{x : x \in B_{r_i}^V(c_i) \text{ for any } i\} = B_r^V(c) \qquad \text{(sum)}$$

$$\bigcap_i B_{r_i}^V(c_i) = \emptyset \qquad \text{(collectionwise disjoincy)}$$

This definition makes no sense when $B_r^V(c) = B_r^X(c)$ (and actually under much weaker conditions). Further, minor/child balls should rather be pairwise disjoint (for any $i \neq j\, B_{r_i}^V(c_i) \cap B_{r_j}^V(c_j) = \emptyset$) in Definition 1 of [31].

5 Existential Granulations

To accommodate multiple nonequivalent concepts of granules, and granulations, a loose definition of *existential granule* is proposed first. Suppose $X = \langle \underline{X}, \mathcal{F}, \mathcal{G}\rangle$ is a triple with $\underline{X}$ being a set, $\mathcal{F}$ a mathematical structure on it, and $\mathcal{G} \subseteq \wp(\underline{X})$ a granulation on it. A granule $G \in \mathcal{G}$ will be said to be *existential* if and only if there exists a subset E of G and an operator $\partial : \wp(X) \longmapsto \wp(X)$ such that $G = \partial(E)$ and $(\exists n \geqslant 1)\partial^{n+1}(E) = \partial^n(E)$. A granulation is existential, if it is a collection of existential granules determined by the features encoded by its constituent points. That is, it is essentially self-determining up to a point. It is possible to argue on this concept being existential in numerous ways – it is exactly the reason for naming it *existential* as opposed to *self-determined*. The idea of non-crisp granules is known in both the axiomatic, precision-based and adaptive theories of granularity. Existential granules have a precise generation aspect motivated by the problem of reducing computational load. Formalization in the axiomatic abstract perspective requires an additional closure operator as defined below, while computational aspects require further specialization. The definitions below are relatively more convenient for abstract approaches [15, 16, 18]. *The main questions of this approach are about formalizing the known applications, representing the operation ∂, and suitability of the restrictions of admissible granules.*

Definition 3. *A high mereological approximation Space (**mash**) $\mathbb{S}$ is a partial algebraic system of the form $\mathbb{S} = \langle \underline{\mathbb{S}}, l, u, \mathbf{P}, \leqslant, \vee, \wedge, \bot, \top\rangle$ with $\underline{\mathbb{S}}$ being a set, l, u being operators $: \underline{\mathbb{S}} \longmapsto \underline{\mathbb{S}}$ satisfying the following ($\underline{\mathbb{S}}$ is replaced with $\mathbb{S}$ if clear from the context. $\vee$ and $\wedge$ are idempotent partial operations and $\mathbf{P}$ is a binary predicate.):*

$$(\forall x)\mathbf{P}xx \qquad\qquad\qquad (\text{PT1})$$

$$(\forall x, b)(\mathbf{P}xb\ \&\ \mathbf{P}bx \longrightarrow x = b) \qquad\qquad (\text{PT2})$$

$$(\forall a, b)a \vee b \overset{\omega}{=} b \vee a;\ (\forall a, b)a \wedge b \overset{\omega}{=} b \wedge a \qquad (\text{G1})$$

$$(\forall a, b)(a \vee b) \wedge a \overset{\omega}{=} a;\ (\forall a, b)(a \wedge b) \vee a \overset{\omega}{=} a \qquad (\text{G2})$$

$$(\forall a, b, c)(a \wedge b) \vee c \overset{\omega}{=} (a \vee c) \wedge (b \vee c) \qquad (\text{G3})$$

$$(\forall a, b, c)(a \vee b) \wedge c \overset{\omega}{=} (a \wedge c) \vee (b \wedge c) \tag{G4}$$

$$(\forall a, b)(a \leqslant b \leftrightarrow a \vee b = b \leftrightarrow a \wedge b = a) \tag{G5}$$

$$(\forall a \in \mathbb{S})\, \mathbf{P}a^l a \; \& \; a^{ll} = a^l \; \& \; \mathbf{P}a^u a^{uu} \tag{UL1}$$

$$(\forall a, b \in \mathbb{S})(\mathbf{P}ab \longrightarrow \mathbf{P}a^l b^l \; \& \; \mathbf{P}a^u b^u) \tag{UL2}$$

$$\perp^l = \perp \; \& \; \perp^u = \perp \; \& \; \mathbf{P}\top^l\top \; \& \; \mathbf{P}\top^u\top \tag{UL3}$$

$$(\forall a \in \mathbb{S})\, \mathbf{P}\perp a \; \& \; \mathbf{P}a\top \tag{TB}$$

In a *high general granular operator space* (GGS), defined below, aggregation and co-aggregation operations ($\vee$, $\wedge$) are conceptually separated from the binary parthood ($\mathbf{P}$), and a basic partial order relation ($\leqslant$). Parthood is assumed to be reflexive, antisymmetric, and not necessarily transitive. It may satisfy additional generalized transitivity conditions in many contexts. Real-life information processing often involves many non-evaluated instances of aggregations (fusions), commonalities (conjunctions) and implications because of laziness or supporting metadata or for other reasons – this justifies the use of partial operations. Specific versions of a GGS and granular operator spaces have been studied in the research paper [16]. Partial operations in GGS permit easier handling of adaptive granules [28] through morphisms– concrete methods need to use the frameworks of clear rough random functions. Note further that it is not assumed that $\mathbf{P}a^{uu}a^u$. The universe $\mathbb{S}$ may be a set of collections of attributes, labeled or unlabeled objects among other things. A *high general existential granular operator space* (eGGS) can be obtained from a GGS by simply restricting the predicate γ as follows:

$$\gamma x \text{ if and only if } x \in \mathcal{G} = \mathcal{I}(\partial)$$

Definition 4. *A High General Granular Operator Space (GGS)* $\mathbb{S}$ *is a partial algebraic system of the form*

$$\mathbb{S} = \langle \underline{\mathbb{S}}, \gamma, l, u, \mathbf{P}, \leqslant, \vee, \wedge, \perp, \top \rangle$$

with $\underline{\mathbb{S}} = \langle \underline{\mathbb{S}}, l, u, \mathbf{P}, \leqslant, \vee, \wedge, \perp, \top \rangle$ *being a* **mash**, *γ being a unary predicate that determines $\mathcal{G}$ (by the condition γx if and only if $x \in \mathcal{G}$) an admissible granulation(defined below) for $\mathbb{S}$. Further, γx will be replaced by $x \in \mathcal{G}$ for convenience. Let $\mathbb{P}$ stand for proper parthood, defined via $\mathbb{P}ab$ if and only if $\mathbf{P}ab$ & $\neg\mathbf{P}ba$). A granulation* is said to be admissible if there exists a term operation t formed from the weak lattice operations such that the following three conditions hold:

$$(\forall x \exists x_1, \ldots x_r \in \mathcal{G})\, t(x_1, x_2, \ldots x_r) = x^l \qquad \text{(Weak RA, WRA)}$$
$$\text{and } (\forall x \, \exists x_1, \ldots x_r \in \mathcal{G})\, t(x_1, x_2, \ldots x_r) = x^u,$$

$$(\forall a \in \mathcal{G})(\forall x \in \underline{\mathbb{S}}))\,(\mathbf{P}ax \longrightarrow \mathbf{P}ax^l), \qquad \text{(Lower Stability, LS)}$$

$$(\forall x,\, a \in \mathcal{G} \exists z \in \underline{\mathbb{S}})\, \mathbb{P}xz, \,\&\, \mathbb{P}az \,\&\, z^l = z^u = z, \qquad \text{(Full Underlap, FU)}$$

Definition 5. *A High General Existential Granular Operator Space (eGGS) $\mathbb{S}$ is a GGS in which the predicate γ is replaced by a unary operation $\eth$ that satisfies*

$$\gamma x \text{ if and only if } x \in \mathcal{G} = \mathfrak{I}(\eth) \qquad \text{(G1)}$$

$$(\forall x)(\exists n \geqslant 1)\eth^{n+1}(x) = \eth^n(x) \qquad \text{(G2)}$$

Existential granular versions of the following particular classes can be defined by analogy.

Definition 6. • *In the above definition, if the anti-symmetry condition PT2 is dropped, then the resulting system will be referred to as a Pre-GGS. If the restriction $\mathbf{P}a^l a$ is removed from UL1 of a **pre-GGS**, then it will be referred to as a Pre*-GGS.*
• *In a **GGS** (resp **Pre*-GGS**), if the parthood is defined by $\mathbf{P}ab$ if and only if $a \leqslant b$ then the **GGS** is said to be a high granular operator space **GS** (resp. **Pre*-GS**).*
• *A higher granular operator space (**HGOS**) (resp **Pre*-HGOS**) $\mathbb{S}$ is a **GS** (resp **Pre*-GS**) in which the lattice operations are total.*
• *In a higher granular operator space, if the lattice operations are set theoretic union and intersection, then the **HGOS** (resp. Pre*-HGOS) will be said to be a set HGOS (resp. set Pre*-HGOS). In this case, $\underline{\mathbb{S}}$ is a subset of a power set, and the partial algebraic system reduces to $\mathbb{S} = \langle \underline{\mathbb{S}}, \gamma, l, u, \subseteq, \cup, \cap, \perp, \top \rangle$ with $\underline{\mathbb{S}}$ being a set, γ being a unary predicate that determines $\mathcal{G}$ (by the condition γx if and only if $x \in \mathcal{G}$). Closure under complementation is not guaranteed in it.*

5.1 Clean Rough Randomness and Models of Algorithms

Some essential aspects of clean rough randomness [19,21] are repeated for convenience, and the problem of formalizing the studied algorithms is in the perspective is formulated.

Many types of randomness are known in the literature. Stochastic randomness, often referred to as randomness, is often misused without proper justification. In the paper [10], a phenomenon is defined to be *stochastically random* if it has probabilistic regularity in the absence of other types of regularity. In this definition, the concept of regularity may be understood as *mathematical regularity* in some sense. Generalizations of mathematical probability theory through hybridization with rough sets from a stochastic perspective are explained in the book [12]. This approach is not ontologically consistent with pure rough reasoning or explainable AI as its focus is on modeling the result of numeric simplifications in a measure-theoretic context.

Empirical studies show that humans cannot estimate measures of stochastic randomness and weakenings thereof in real life properly [1]. This is consistent with the observation that connections in the rough set literature between specific versions of rough sets and subjective probability theories (Bayesian or frequentist) are not good approximations. In fact, rough inferences are grounded in some non-stochastic comprehension of attributes (their relation with the approximated object in terms of number or relative quantity and quality) [20,23].

The idea of *rough randomness* is defined by the present author [19] as follows:*a phenomenon is clean roughly random (C-roughly random) if it can be modeled by general rough sets or a derived process thereof.* In concrete situations, such a concept should be realizable in terms of C-roughly random functions or predicates defined below (readers should note that any one of the concepts of rough objects in the literature [16] such as *a non crisp object* or a *pair of definite objects of the form* (a, b) *satisfying* **P**ab among others are permitted):

Definition 7. *Let* A_τ *be a collection of approximations of type* τ, *and* E *a collection of rough objects defined on the same universe* S, *then by a* C-rough random function of type-1 *(CRRF1) will be meant a partial function* $\xi : A_\tau \longmapsto E$.

Definition 8. *Let* A_τ *be a collection of approximations of type* τ, S *a subset of* $\wp(S)$, *and* $\mathfrak{R}$ *the set of reals, then by a* C-rough random function of type-2 *(CRRF2) will be meant a function* $\chi : A_\tau \times S \longmapsto \mathfrak{R}$.

Definition 9. *Let* A_τ *be a collection of approximations of type* τ, *and* F *a collection of objects defined on the same universe* S, *then by a* C-rough random function of type-3 *(CRRF3) will be meant a function* $\mu : A_\tau \longmapsto F$.

Definition 10. *Let* $\mathcal{O}_\tau$ *be a collection of approximation operators of type* τ_l *or* τ_u, *and* E *a collection of rough objects defined on the same universe* S, *then by a* C-rough random function of type-H *(CRRFH) will be meant a partial function*

$$\xi : \mathcal{O}_\tau \times \wp(S) \longmapsto E.$$

It is obvious that a CRRF1 and CRRF2 are independent concepts, while a total CRRF1 is an CRRF3, and CRRFH is distinct (though related to CRRF3). The set of all such functions will respectively be denoted by CRRF1(S, E, τ), CRRF2$(S, \mathfrak{R}, \tau)$, CRRF3(S, F, τ), and CRRFH(S, E, τ). For detailed examples, the reader is referred to the earlier papers [19,21]

Example 1. Let S be a set with a pair of lower (l) and upper (u) approximations satisfying (for any $a, b, x \subseteq S$)

$$x^{ll} \subseteq x^l \subseteq x^u \qquad \text{(l-id, int-cl)}$$

$$a \subseteq b \longrightarrow a^l \subseteq b^l \qquad \text{(l-mo)}$$

$$a \subseteq b \longrightarrow a^u \subseteq b^u \qquad \text{(u-mo)}$$

$$\emptyset^l = \emptyset \ \& \ S^u = S \qquad \text{(l-bot, u-top)}$$

The above axioms are minimalist, and most general approaches satisfy them. In addition, let

$$\mathcal{A}_\tau = \{x : (\exists a \subseteq S) \, x = a^l \text{ or } x = a^u \qquad (1)$$

$$E_1 = \{(a^l, a^u) : a \in S\} \qquad \text{(E1)}$$

$$F = \{a : a \subseteq S \ \& \ \neg \exists b \, b^l = a \lor b^u = a\} \qquad \text{(E0)}$$

$$E_2 = \{b : b^u = b \ \& \ b \subseteq S\} \qquad \text{(E2)}$$

$$\xi_1(a) = (a, b^u) \text{ for some } b \subseteq S \qquad \text{(xi1)}$$

$$\xi_2(a) = (b^l, a) \text{ for some } b \subseteq S \qquad \text{(xi2)}$$

$$\xi_3(a) = (e, f) \in E_1 \ \& \ e = a \text{ or } f = a \qquad \text{(xi3)}$$

E_1 in the above is a set of rough objects, and a number of algebraic models are associated with it [16]. A partial function $f : \mathcal{A}_\tau \longmapsto E_1$ that associates $a \in \mathcal{A}_\tau$ with a minimal element of E_1 that covers it in the inclusion order is a CRRF of type 1. For general rough sets, this CRRF can be used to define algebraic models and explore duality issues [17], and for many cases associated these are not investigated. A number of similar maps with value in understanding models [20] can be defined. Rough objects are defined and interpreted in a number of other ways including F or E_2.

Conditions xi1-xi3 may additionally involve constraints on b, e and f. For example, it can be required that there is no other lower or upper approximation included between the pair or that the second component is a minimal approximation covering the first. It is easy to see that

Theorem 5. ξ_i for $i = 1, 2, 3$ are CRRF of type-1.

Example 2. In the above example, rough inclusion functions, membership, and quality of approximation functions [7,29] can be used to define CRRF2s. An example is the function ξ_5 defined by

$$\xi_5(a, b) = \frac{Card(b \setminus a)}{Card(b)} \qquad (1)$$

5.2   Formalizing the BKM Algorithms

The ball K-means algorithm can potentially be formalized by rough random functions of type 3 in several ways. For this purpose, one can use a single RRF-3 φ and a number of classical lower and upper approximation that describe each update on the original k clusters sequentially or use a sequence of RRF-3 s with pairs of classical lower and upper approximations to describe the updates. Therefore, the real problem is of finding and formulating the most appropriate formalization. How does one restrict the choice of approximation operations?

All crisp clusterings form partitions, and therefore all such clusterings form the granulation of Pawlak rough sets over the universe in question. This is the suggested origin of the *classical and upper approximations*.

6   Further Directions

It might appear to easy to cast the ball K-means and granular ball algorithms in the interactive granular computing perspective. It is already shown that such is not essential. The proposal of interactive granules and related computing (IGrC) is formulated in relation to a certain perception of the basic semantic domain, and is primarily intended to reduce the complexity of decision-making in application contexts [24,28]. Some objects are supposed to be *non-mathematical objects* at a level of discourse, and possess some properties of granularity. The use of complex granule (c-granule) comprising abstract objects, physical objects, as well as objects linking abstract and physical objects, by itself, and their rule-based approach apparently constrains the authors to that view. States of c-granules are represented by networks of informational granules (ic-granules) linking abstract and physical objects. Such c-granules are intended for modeling perceptions of physical processes in the real world. However, the mathematical approach to such cases is through improved sequences of models, and objects, and through better choice of semantic domains. Data drives nothing, it is for us to invent models that make any driving to be possible at all.

The obvious idea of replacing the hyper-sphere with a smooth hypersurface is possible in theory, and justifiable if the geometry is relevant. However, the computational complexity may increase substantially. Actually, no hyperspheres or balls are used in the both the BKM and granular ball algorithms. It is only in the imagination of the authors. If the shapes generated by points are really of interest then other metrics and the Hausdorff-Gromov distance [22] may be used painfully. The possible mathematical generalization of the proposed method requires justification in applied problems. The geometrical shape of granules typically matter in the domain of topological data analysis [9], spatial mereology [3] and near sets [26]. Such approaches have steep requirements on the domain for easier computing. In future studies, existential granules will be explored in greater depth.

References

1. Beach, L., Braun, G.: Laboratory studies of subjective probability: a status report. In: Wright, G., Ayton, P. (eds.) Subjective Probability, pp. 107–128. John Wiley (1994)
2. Bogachev, V.I., Smolyanov, O.G.: Topological Vector Spaces and Their Applications. Springer Monographs in Mathematics. Springer. Heidelberg (2017). https://doi.org/10.1007/978-3-319-57117-1
3. Burkhardt, H., Seibt, J., Imaguire, G., Gerogiorgakis, S. (eds.): Handbook of Mereology. Philosophia Verlag, Germany (2017)
4. Burmeister, P.: A Model-Theoretic Oriented Approach to Partial Algebras. Akademie-Verlag (1986, 2002)
5. Chen, Y., Wang, P., Yang, X., Mi, J., Liu, D.: Granular ball guided selector for attribute reduction. Knowl.-Based Syst. **229**, 107326 (2021). https://doi.org/10.1016/j.ins.2023.119071
6. Gagrat, M., Naimpally, S.: Proximity approach to semi-metric and developable spaces. Pac. J. Math. **44**(1), 93–105 (1973)
7. Gomolinska, A.: Rough approximation based on weak q-RIFs. Trans. Rough Sets **X**, 117–135 (2009)
8. Ji, X., Peng, J., Zhao, P., Yao, S.: Extended rough sets model based on fuzzy granular ball and its attribute reduction. Inf. Sci. (2023). https://doi.org/10.1016/j.ins.2023.119071
9. Kim, W., Memoli, F.: Persistence over posets. Not. Am. Math. Soc. **2761**, 1214–1224 (2023). https://doi.org/10.1090/noti2761
10. Kolmogorov, A.N.: On the logical foundations of probability theory. In: Shiryayev, A.N. (ed.) Selected Works of A. N. Kolmogorov, vol. 2, chap. 53, pp. 515–519. Kluwer Academic, Nauka (1986)
11. Lin, T.Y.: Granular computing-1: the concept of granulation and its formal model. Int. J. Granular Comput. Rough Sets Int. Syst. **1**(1), 21–42 (2009)
12. Liu, B.: Uncertainty Theory, Studies in Fuzziness and Soft Computing, vol. 154. Springer, Heidelberg (2004). https://doi.org/10.1007/978-3-540-73165-8_5
13. Liu, G.: The axiomatization of the rough set upper approximation operations. Fund. Inf. **69**(23), 331–342 (2006)
14. Ljapin, E.S.: Partial Algebras and Their Applications. Academic, Kluwer (1996)
15. Mani, A.: Dialectics of counting and the mathematics of vagueness. Trans. Rough Sets **XV**(LNCS 7255), 122–180 (2012)
16. Mani, A.: Algebraic methods for granular rough sets. In: Mani, A., Düntsch, I., Cattaneo, G. (eds.) Algebraic Methods in General Rough Sets, pp. 157–336. Trends in Mathematics, Birkhauser Basel (2018)
17. Mani, A.: Representation, duality and beyond. In: Mani, A., Düntsch, I., Cattaneo, G. (eds.) Algebraic Methods in General Rough Sets, pp. 459–552. Trends in Mathematics, Birkhauser Basel (2018)
18. Mani, A.: Comparative approaches to granularity in general rough sets. In: Bello, R., Miao, D., Falcon, R., Nakata, M., Rosete, A., Ciucci, D. (eds.) IJCRS 2020. LNCS (LNAI), vol. 12179, pp. 500–517. Springer, Cham (2020). https://doi.org/10.1007/978-3-030-52705-1_37
19. Mani, A.: Rough randomness and its application. J. Calcutta Math. Soc. 1–15 (2023). https://doi.org/10.5281/zenodo.7762335. https://zenodo.org/record/7762335
20. Mani, A., Düntsch, I., Cattaneo, G. (eds.): Algebraic Methods in General Rough Sets. Trends in Mathematics, Birkhauser Basel (2018). https://doi.org/10.1007/978-3-030-01162-8

21. Mani, A., Mitra, S.: Large minded reasoners for soft and hard cluster validation - some directions, pp. 1–16. Annals of Computer and Information Sciences, PTI (2023)
22. Memoli, F.: The gromov-hausdorff distance: a brief tutorial on some of its quantitative aspects. Actes des rencontres du C.I.R.M. **3**(1), 89–96 (2013)
23. Pagliani, P., Chakraborty, M.: A Geometry of Approximation: Rough Set Theory: Logic, Algebra and Topology of Conceptual Patterns. Springer, Heidelberg (2008). https://doi.org/10.1007/978-1-4020-8622-9_15
24. Pedrycz, W., Skowron, A., Kreinovich, V.: Handbook of Granular Computing. John Wiley, Hoboken (2008)
25. Peng, X., Wang, P., Xia, S., Wang, C., Chen, W.: VPGB: a granular-ball based model for attribute reduction and classification with label noise. Inf. Sci. **611**, 504–521 (2022). https://doi.org/10.1016/j.ins.2022.08.066
26. Peters, J.F.: Topology of Digital Images. ISRL, vol. 63. Springer, Heidelberg (2014). https://doi.org/10.1007/978-3-642-53845-2
27. Qian, W., Xu, F., Huang, J., Qian, J.: A novel granular ball computing-based fuzzy rough set for feature selection in label distribution learning. Knowl.-Based Syst. **278**, 110898 (2023). https://doi.org/10.1016/j.knosys.2023.110898
28. Skowron, A., Jankowski, A., Dutta, S.: Interactive granular computing. Granular Comput. **1**(2), 95–113 (2016)
29. Stepaniuk, J.: Rough-Granular Computing in Knowledge Discovery and Data Mining. Studies in Computational Intelligence, vol. 152, Springer, Heidelberg (2009). 10.1007/978-3-540-70801-8
30. Wilansky, A.: Modern Methods in Topological Vector Spaces. McGraw-Hill, New York (1978)
31. Xia, S., Dai, X., Wang, G., Gao, X., Giem, E.: An efficient and adaptive granular-ball generation method in classification problem. Arxiv (2022)
32. Xia, S., Liu, Y., Ding, X., Wang, G., Yu, H., Luo, Y.: Granular ball computing classifiers for efficient, scalable and robust learning. Inf. Sci. **483**, 136–152 (2019)
33. Xia, S., et al.: A fast adaptive k-means algorithm. IEE Trans. Pattern Anal. Mach. Intell. 1–13 (2020). https://doi.org/10.1109/TPAMI.2020.3008694
34. Yao, Y.Y.: Information granulation and rough set approximation. Int. J. Intell. Syst. **16**, 87–104 (2001)

Aggregation Operators on Shadowed Sets Deriving from Conditional Events and Consensus Operators

Stefania Boffa[1]([✉])[ID], Andrea Campagner[2], Davide Ciucci[1][ID], and Yiyu Yao[3]

[1] Dipartimento di Informatica, Sistemistica e Comunicazione, University of
Milano-Bicocca, viale Sarca 336, 20126 Milan, Italy
stefania.boffa@unimib.it
[2] IRCCS Ospedale Galeazzi - Sant'Ambrogio, Milan, Italy
[3] Department of Computer Science, University of Regina, Regina, Canada

Abstract. We introduce particular aggregation operators on shadowed sets, which derive from the operations between conditional events and from the consensus operator. Considering that shadowed sets arise as approximations of fuzzy sets, we also present and study special classes of aggregation functions that can be approximated by the considered operations on shadowed sets.

Keywords: Shadowed sets · Aggregation functions · Aggregation of shadowed sets · Conditional events · Consensus operator

1 Introduction

Shadowed sets were initially proposed by Pedrycz for approximating fuzzy sets by using only three possible membership values: 0 (non-membership), 1 (membership), and [0,1] (intermediate membership [16] or uncertainty). Thus, given a universe X, a fuzzy set $f : X \to [0,1]$ can be transformed into a shadowed set $\mathcal{S}(f) : X \to \{0, [0,1], 1\}$ by means of a particular mapping $\mathcal{S}$ that can be defined in several ways [27]. In this article, we adopt the most abstract approach (that is a generalization of the decision-theoretic one [8,10,26]), where $\mathcal{S}$ solely depends on a pair of thresholds α and β of the real interval $[0,1]$ so that $\alpha < \beta$.

Shadowed sets have captured the interest of many scholars, who have explored both their theoretical properties as well as their applications (see [3,4,11,15] for some examples). In particular, different operators to aggregate shadowed sets have been proposed in the literature [2,6,9,17,18,22]. This article, which is a continuation of [2], contributes to this body of knowledge in two directions. First, we introduce novel aggregation operators on shadowed sets by transforming some operations already existing in literature, which are related to conditional events [23] and orthopairs [5]. In an earlier paper, Boffa et al. [2] systematically studied special aggregation operators on shadowed sets. In Sect. 3, we define new shadowed set operations and study their mathematical properties. We notice that

A. Campagner et al. (Eds.): IJCRS 2023, LNAI 14481, pp. 201–215, 2023.
https://doi.org/10.1007/978-3-031-50959-9_14

the present work is the continuation of [2] and thus the numbering of operations follows the one in that paper. Second, in Sect. 4, the attention is shifted to the analysis of the correspondence between the fuzzy and shadowed operators. In detail, for each operation $*$ on shadowed sets and for each pair of thresholds (α, β), we find an aggregation function $\otimes$ on fuzzy sets such that $\mathcal{S}(f \otimes g) = \mathcal{S}(f) * \mathcal{S}(g)$ for each $f, g \in [0, 1]^X$. In this way, $*$ can be interpreted as an approximation of $\otimes$ by means of the transformation $\mathcal{S}$.

2   Preliminaries

In this section, we provide the necessary background on shadowed sets, conditional events, orthopairs, and aggregation functions.

2.1   Shadowed Sets

A *shadowed set* is a mapping $s : X \to \{0, [0, 1], 1\}$, which is obtained from a fuzzy set $f : X \to [0, 1]$ by means of a transformation $\mathcal{S} : \mathcal{F}_X \mapsto \mathcal{S}_X$, where $\mathcal{S}_X = \{0, [0, 1], 1\}^X$ and $\mathcal{F}_X = [0, 1]^X$.

Different approaches exist to define a transformation, such as the first one proposed by Pedrycz [16] and others based on information-theoretic [3,12,21, 24] or decision-theoretic [8,10,26] notions. In this article, we adopt a general and abstract definition: a transformation $\mathcal{S}_{(\alpha,\beta)} : \mathcal{F}_X \mapsto \mathcal{S}_X$ form a fuzzy to a shadowed set is thresholding step function, which can be expressed as

$$\mathcal{S}_{(\alpha,\beta)}(f)(x) = \begin{cases} 1 & \text{if } \beta < f(x) \le 1; \\ [0,1] & \text{if } \alpha \le f(x) \le \beta; \\ 0 & \text{if } 0 \le f(x) < \alpha; \end{cases} \tag{1}$$

where $0 < \alpha < \beta < 1$. Shadowed sets can be put in a one-to-one correspondence with conditional events and orthopairs, as we are going to explain

Conditional Events. A conditional event is an ordered pair of elements of a set, which represents an algebra of subsets of the domain of a probability space [23]. Given the Boolean Algebra $\mathcal{B} = (\{0, 1\}, \wedge, \vee, \neg, 0, 1)$, the set of conditional events on $\mathcal{B}$ is defined as:

$$\mathcal{C}(\{0, 1\}) = \{(a, b) \in \{0, 1\} \times \{0, 1\} \mid a \wedge b = 0\}.$$

That is, $\mathcal{C}(\{0, 1\}) = \{(0, 1), (1, 0), (0, 0)\}$. Generally, the conditional event (a, b) represents the rule "if b then a". In the case of $\mathcal{C}(\{0, 1\})$, the pairs $(1, 0)$ and $(0, 1)$ are interpreted as true and false, respectively; while the pair $(0, 0)$ means undefined, namely it is not possible or it does not make sense to say if a conditional event is true or false by using the available knowledge.

There exists a one-to-one correspondence between $\{0, [0, 1], 1\}$ and the set of conditional events of a Boolean algebra, which is defined by the mapping $\alpha : \mathcal{C}(\{0, 1\}) \to \{0, [0, 1], 1\}$ such that

$$\alpha((0, 1)) = 0, \quad \alpha((1, 0)) = 1, \quad \text{and} \quad \alpha((0, 0)) = [0, 1]. \tag{2}$$

Orthopairs. An orthopair is a set with uncertainty, i.e., for some objects we are unable to say if they belong or not to a given set. Formally, an orthopairs on a set X is a pair $O = (P, N)$ such that $P, N \subseteq X$ and $P \cap N = \emptyset$. P is called the *positive region* and is understood as the set of all elements of X that certainly belong to the set represented by O. Analogously, N is called the *negative region* and is understood as the set of all elements of X that certainly do not belong to the set represented by O. Starting from O, a third set called the *boundary region* is defined as $Bnd = X \setminus (P \cup N)$ and is made of all elements of X such that we do not know if they belong to the set represented by O. As shown in [6], it is easy to observe that an orthopair $O = (P, N)$ on X corresponds to the shadowed set $s : X \to \{0, [0, 1], 1\}$ defined as follows: let $x \in X$,

$$s(x) = \begin{cases} 1 & \text{if } x \in P; \\ 0 & \text{if } x \in N; \\ [0, 1] & \text{if } x \in Bnd; \end{cases} \tag{3}$$

and, vice-versa, a shadowed set $s : X \to \{0, [0, 1], 1\}$ uniquely determines the orthopair $O = (\{x \in X \mid s(x) = 1\}, \{x \in X \mid s(x) = 0\})$ on X. Thus, the positive and negative regions of an orthopair O are respectively equivalent to the subsets of the initial universe mapped into 1 and 0 by a shadowed set s; while the boundary region of O is equivalent to the collection of all elements associated with the intermediate membership by s.

We notice that the correspondences given in equations (2),(3) are not the only possible ones from a mathematical standpoint, but are those that correspond to the usual interpretation of uncertainty, true/positive and false/negative in the three settings.

2.2   Aggregation Functions

An *aggregation function* is a mapping $A : [0, 1] \times [0, 1] \mapsto [0, 1]$, that is usually intended as a way to model generalizations of set-theoretic operations (such as the union or intersection) in generalized settings (e.g., in fuzzy set theory or in decision making) [22]. Aggregation functions are usually assumed to satisfy some further properties [9]:

(i) $A(0, 0) = 0$ and $A(1, 1) = 1$ (*boundary condition*).
(ii) (a) If $x \leq y$ and $z \leq w$, then $A(x, z) \leq A(y, w)$, for each $x, y, z, w \in [0, 1]$ (*monotonicity*). Also, we say that A is *increasing*.
    (b) If $x \leq y$ and $z \leq w$, then $A(x, z) \geq A(y, w)$, for each $x, y, z, w \in [0, 1]$ (*monotonicity*). Also, we say that A is *decreasing*.
(iii) $A(x, y) = A(y, x)$, for each $x, y \in [0, 1]$ (*commutativity*).
(iv) $A(x, x) = x$, for each $x \in [0, 1]$ (*idempotence*).
(v) $A(A(x, y), z) = A(x, A(y, z))$, for each $x, y, z \in [0, 1]$ (*associativity*).

Moreover, an aggregation function A has a *neutral element* $e \in [0, 1]$ if and only if $A(x, e) = x = A(e, x)$, for each $x \in [0, 1]$.

Definition 1. *Let A and $\tilde{A}$ be aggregation functions, then A is the* dual operator *of $\tilde{A}$ if and only if*

$$\tilde{A}(x,y) = 1 - (A(1-x, 1-y)). \tag{4}$$

In this article, we also deal with uninorms:

An aggregation function $A : [0,1] \times [0,1] \to [0,1]$ is a *uninorm* if and only if it is commutative, associative, non-decreasing, and there exists a neutral element $e \in [0,1]$. Additionally, A is *conjunctive* if and only if $A(1,0) = 0$; A is *disjunctive* if and only if $A(1,0) = 1$ [25].

3 How to Aggregate Shadowed Sets

In this section, we present some operations to aggregate shadowed sets. They derive from the operations on conditional events and the so-called consensus operator on orthopairs.

3.1 Using Operations on Conditional Events

Several operations arise by extending the Boolean conjunction $\wedge$ to the set of conditional events $\mathcal{C}(\{0,1\})$ and requiring that $(0,1) \wedge (0,1) = (0,1)$, $(0,1) \wedge (1,0) = (0,1)$, $(1,0) \wedge (0,1) = (0,1)$, and $(1,0) \wedge (1,0) = (1,0)$. Nine of these operations are idempotent, commutative and Boolean polynomials of their arguments [23]. In addition, among them, two coincide with the Sobociński conjunction and the Kleene conjunction [13,20]. In [2], we already studied these two operations, where they were denoted as $*_2$ and $*_5$, and defined as in Table 1.

Table 1. Sobociński and Kleene conjunctions translated on $\{0, [0,1], 1\}$

$*_2$	0	[0,1]	1
0	0	0	0
[0,1]	0	[0,1]	1
1	0	1	1

$*_5$	0	[0,1]	1
0	0	0	0
[0,1]	0	[0,1]	[0,1]
1	0	[0,1]	1

The translation on shadowed sets of the remaining seven operations is given in Table 2[1]. The operations $*_{15}, \ldots, *_{21}$ are non-monotone (i.e., neither increasing nor decreasing), and only $*_{20}$ and $*_{21}$ are associative. Moreover, $*_{20}$ is already known in literature as *weak Kleene conjunction* [1].

The corresponding disjunctions $+_2$, $+_5$, and $+_{15}, \ldots, +_{21}$ on $\{0, [0,1], 1\}$ respectively listed in Tables 3 and 4, can be analogously obtained by translating the disjunctions on conditional events provided in [23] and using the function α given by (2).

---

[1] We indicate such operations with the symbols $*_{15}, \ldots, *_{21}$ since other operations $*_1, \ldots, *_{14}$ are already defined on shadowed sets in [2].

Table 2. Conjunctions of conditional events translated on $\{0, [0,1], 1\}$. Each conjunction $*_i$ is obtained by completing the empty boxes of the first sub-table (showing the common parts of the seven operations) with the row i of the second sub-table.

$*_i$	0	[0,1]	1
0	0		0
[0,1]		[0,1]	
1	0		1

n.	$0 *_i [0,1]$	$[0,1] *_i 0$	$[0,1] *_i 1$	$1 *_i [0,1]$
15	[0,1]	[0,1]	0	0
16	[0,1]	[0,1]	1	1
17	1	1	[0,1]	[0,1]
18	1	1	1	1
19	1	1	0	0
20	[0,1]	[0,1]	[0,1]	[0,1]
21	0	0	0	0

Table 3. Sobociński and Kleene disjunctions translated on $\{0, [0,1], 1\}$

$+_2$	0	[0,1]	1
0	0	0	1
[0,1]	0	[0,1]	1
1	1	1	1

$+_5$	0	[0,1]	1
0	0	[0,1]	1
[0,1]	[0,1]	[0,1]	1
1	1	1	1

Table 4. Disjunctions of conditional events translated on $\{0, [0,1], 1\}$. Each disjunction $+_i$ is obtained by completing the empty boxes of the first sub-table (showing the common parts of the seven operations) with the row i of the second sub-table.

$+_i$	0	[0,1]	1
0	0		1
[0,1]		[0,1]	
1	1		1

n.	$0 +_i [0,1]$	$[0,1] +_i 0$	$[0,1] +_i 1$	$1 +_i [0,1]$
15	1	1	[0,1]	[0,1]
16	0	0	[0,1]	[0,1]
17	[0,1]	[0,1]	0	0
18	0	0	0	0
19	1	1	0	0
20	[0,1]	[0,1]	[0,1]	[0,1]
21	1	1	1	1

As in the case of their dual operations, all the disjunctions of Table 4 are idempotent, commutative, and non-monotone. Furthermore, $+_{20}$ and $+_{21}$ are associative, and $+_{20}$ corresponds to the so-called *Kleene weak disjunction*.

3.2 Using Orthopair-Like Operations

Some operations on orthopairs have been already translated on $\{0, 1/2, 1\}$ using the bijection between orthopairs and three-valued functions [7]. Thus, they can be equivalently defined on $\{0, [0,1], 1\}$ as well.

However, another operation $*_C$ on shadowed sets arises from the so-called *consensus operator*, which is defined in [14] as follows: let (P_i, N_i) and (P_j, N_j) be orthopairs on X,

$$(P_i, N_i) \odot_C (P_j, N_j) = ((P_i \setminus N_j) \cup (P_j \setminus N_i), \ (N_i \setminus P_j) \cup (N_j \setminus P_i)).$$

The operation $*_C$ that is reported in Table 5, is commutative, increasing, and non-associative. Clearly, the boundary condition is satisfied as well.

Table 5. Consensus operation on $\{0, [0, 1], 1\}$

$*_C$	0	$[0, 1]$	1
0	0	0	$[0, 1]$
$[0, 1]$	0	$[0, 1]$	1
1	$[0, 1]$	1	1

In addition, by interpreting $[0, 1]$ as $\frac{1}{2}$, it is easy to check that $*_C$ can be obtained also as the average of $*_2$ and $+_2$: let $x, y \in \{0, [0, 1], 1\}$,

$$x *_C y = \frac{(x *_2 y) + (x +_2 y)}{2}.$$

As in the case of $\odot_C$, the operation $*_C$ can be interpreted in terms of the consensus between two information sources: indeed, $x *_C y \in \{0, 1\}$ if and only if x and y are not in conflict, otherwise $x *_C y = [0, 1]$.

4 Aggregation Functions Generating Operations on Shadowed Sets

In this section, for each operation on shadowed sets proposed in Sect. 3, we define a class of aggregation functions so that there exists a transformation $\mathcal{S}_{(\alpha, \beta)}$ with $\alpha \neq \beta$ and $0 < \alpha \leq 1/2 \leq \beta < 1$ acting as a homomorphism between fuzzy sets and shadowed sets. So, such aggregation functions can be faithfully approximated by means of operators on shadowed sets. In symbols, given an operation $*$ on shadowed sets, an aggregation function $\otimes$ of this type satisfies the equation[2]

$$\mathcal{S}_{(\alpha, \beta)}(f \otimes g) = \mathcal{S}_{(\alpha, \beta)}(f) * \mathcal{S}_{(\alpha, \beta)}(g) \text{ for each } f, g \in [0, 1]^X. \tag{5}$$

Furthermore,

- we prove that each aggregation function assigned to $*$ preserves the algebraic properties of $*$; an exception is the idempotence holding only for some of these aggregation functions (Propositions 1 2, and 4);

---

[2] The function $f \otimes g \in [0, 1]^X$ is defined as follows: $(f \otimes g)(x) = f(x) \otimes g(x)$ for each $x \in X$. Similarly, the function $\mathcal{S}_{(\alpha, \beta)}(f) * \mathcal{S}_{(\alpha, \beta)}(g) \in \{0, [0, 1], 1\}^X$ is defined as follows: $(\mathcal{S}_{(\alpha, \beta)}(f) * \mathcal{S}_{(\alpha, \beta)}(g))(x) = \mathcal{S}_{(\alpha, \beta)}(f)(x) * \mathcal{S}_{(\alpha, \beta)}(g)(x)$ for each $x \in X$.

- let $i \in \{15, \ldots, 21\}$, we show that each aggregation function assigned to the disjunction $+_i$ is the dual operation according to Definition 1 of an aggregation function assigned to the conjunction $*_i$ (see Proposition 3).

Remark 1. We implicitly assume that the thresholds α and β used to transform a fuzzy set into a shadowed set are always the same for every fuzzy set, as in the decision-theoretic formulation. On the other hand, this is not necessarily true in Pedrycz's formulation, where a fuzzy set f is approximated using the thresholds α^* and $1 - \alpha^*$ such that α^* is the minimum point of a function constructed on f [16].

Remark 2. The goals of this section are already achieved for the operations $*_2$, $*_5$, $+_2$, and $+_5$ in [2]. Firstly, we defined the classes of aggregation functions $\{\otimes_2^\gamma \mid \gamma \in (0,1)\}$ and $\{\oplus_2^\gamma \mid \gamma \in (0,1)\}$ so that Eq. (5) holds for the pairs $(\otimes_2^\alpha, *_2)$ and $(\oplus_2^\beta, +_2)$ when $\alpha \neq \beta$ and $\alpha \leq 1/2 \leq \beta$. After that, we proved that

- for each $\alpha, \beta \in (0,1)$ such that $\alpha < \beta$, $\otimes_2^\alpha$ and $\oplus_2^\beta$ are respectively *conjunctive* and *disjunctive uninorms* having α and β as neutral elements;
- for each $\gamma \in (0,1)$, $\oplus_2^\gamma$ is the dual operator of $\otimes_2^\gamma$ according to Definition 1;
- the *minimum t-norm* $\otimes_M$ and its dual *maximum t-conorm* $\oplus_M$[3] are respectively associated to $*_5$ and $+_5$ by means of Eq. (5).

4.1 Aggregation Functions Generating Conditional-Event Conjunctions on Shadowed Sets

The following is a list of classes of aggregation functions that correspond to operations on shadowed sets deriving from the conjunctions of conditional events defined in Subsect. 3.1.

Definition 2. *Let* $\alpha, \beta, \gamma \in (0,1)$ *such that* $\alpha < \beta$, *we set*

$$x \otimes_{15}^\gamma y = \begin{cases} 0 & \text{if } x \leq \gamma < y, \text{ or } y \leq \gamma < x, \\ \max(x,y) & \text{otherwise.} \end{cases} \quad (6)$$

$$x \otimes_{16}^{\alpha\beta} y = \begin{cases} \min(x,y) & \text{if } x < \alpha, \text{ and } y > \beta, \text{ or } x > \beta \text{ and } y < \alpha, \\ \max(x,y) & \text{otherwise.} \end{cases} \quad (7)$$

$$x \otimes_{17}^{\alpha\beta} y = \begin{cases} 1 & \text{if } x < \alpha \text{ and } \alpha \leq y \leq \beta, \text{ or } \alpha \leq x \leq \beta \text{ and } y < \alpha, \\ \min(x,y) & \text{otherwise.} \end{cases} \quad (8)$$

---

[3] $\otimes_M$ and $\oplus_M$ are defined as follows: $x \otimes_M y = \min(x,y)$ and $x \oplus_M y = \max(x,y)$, for each $x, y \in [0,1]$ [19].

$$x \otimes_{18}^{\alpha\beta} y = \begin{cases} 1 & \text{if } x < \alpha \text{ and } \alpha \leq y \leq \beta, \text{ or } \alpha \leq x \leq \beta \text{ and } y < \alpha, \\ \max(x,y) & \text{if } x,y \geq \alpha \\ \min(x,y) & \text{otherwise.} \end{cases}$$

(9)

$$x \otimes_{19}^{\alpha\beta} y = \begin{cases} 1 & \text{if } x < \alpha \text{ and } \alpha \leq y \leq \beta, \text{ or } \alpha \leq x \leq \beta \text{ and } y < \alpha, \\ 0 & \text{if } \alpha \leq x \leq \beta \text{ and } y > \beta, \text{ or } x > \beta \text{ and } \alpha \leq y \leq \beta, \\ \min(x,y) & \text{otherwise.} \end{cases}$$

(10)

$$x \otimes_{20}^{\gamma} y = \begin{cases} \max(x,y) & \text{if } x,y \leq \gamma, \\ \min(x,y) & \text{otherwise.} \end{cases}$$

(11)

$$x \otimes_{21}^{\gamma} y = \begin{cases} \min(x,y) & \text{if } x,y \leq \gamma \text{ or } x,y > \gamma, \\ 0 & \text{otherwise.} \end{cases}$$

(12)

The next theorem shows that the aggregation functions of Definition 2 and the operations listed in Table 2 are connected by means of Eq. (5).

Theorem 1. *Let $\alpha, \beta \in (0,1)$ such that $\alpha \neq \beta$ and $\alpha \leq 1/2 \leq \beta$. Then, let $f, g \in [0,1]^X$,*

(a) $\mathcal{S}_{(\alpha,\beta)}(f \otimes_i^{\beta} g) = \mathcal{S}_{(\alpha,\beta)}(f) *_i \mathcal{S}_{(\alpha,\beta)}(g)$ for each $i \in \{15, 20, 21\}$;
(b) $\mathcal{S}_{(\alpha,\beta)}(f \otimes_i^{\alpha\beta} g) = \mathcal{S}_{(\alpha,\beta)}(f) *_i \mathcal{S}_{(\alpha,\beta)}(g)$ for each $i \in \{16, 17, 18, 19\}$.

Proof. Let us prove that Eq. (5) holds for $\otimes_{15}^{\beta}$ and $*_{15}$. Let $x \in X$, we firstly can show that

$$\mathcal{S}_{(\alpha,\beta)}(f \otimes_{15}^{\beta} g)(x) = 0 \quad \text{if and only if} \quad \mathcal{S}_{(\alpha,\beta)}(f)(x) *_{15} \mathcal{S}_{(\alpha,\beta)}(g)(x) = 0. \quad (13)$$

Suppose that $\mathcal{S}_{(\alpha,\beta)}(f \otimes_{15}^{\beta} g)(x) = 0$. Then, by (1), $(f \otimes_{15}^{\beta} g)(x) < \alpha$, namely $f(x) \otimes_{15}^{\beta} g(x) < \alpha$. By (6), $f(x) \otimes_{15}^{\beta} g(x) = 0$ or $f(x) \otimes_{15}^{\beta} g(x) = \max(f(x), g(x))$.

- If $f(x) \otimes_{15}^{\beta} g(x) = 0$, then it must be true that $f(x) \leq \beta < g(x)$ or $g(x) \leq \beta < f(x)$, using (6) again. Thus, by (1), "$\mathcal{S}_{(\alpha,\beta)}(f)(x) \in \{0, [0,1]\}$ and $\mathcal{S}_{(\alpha,\beta)}(g)(x) = 1$" or "$\mathcal{S}_{(\alpha,\beta)}(g)(x) \in \{0, [0,1]\}$ and $\mathcal{S}_{(\alpha,\beta)}(f)(x) = 1$". Thus, by Table 2 (see the definition of $*_{15}$), we get $\mathcal{S}_{(\alpha,\beta)}(f)(x) *_{15} \mathcal{S}_{(\alpha,\beta)}(g)(x) = a *_{15} 1$ (or $1 *_{15} a$), where $a \in \{0, [0,1]\}$.
  Therefore, by Table 2, $\mathcal{S}_{(\alpha,\beta)}(f)(x) *_{15} \mathcal{S}_{(\alpha,\beta)}(g)(x) = 0$.
- Suppose that $f(x) \otimes_{15}^{\beta} g(x) = \max(f(x), g(x))$. Since $\mathcal{S}_{(\alpha,\beta)}(f \otimes_{15}^{\beta} g)(x) = 0$, $(f \otimes_{15}^{\beta} g)(x) < \alpha$ from (1). Then, $f(x) < \alpha$ and $g(x) < \alpha$. Using (1) again, we get $\mathcal{S}_{(\alpha,\beta)}(f)(x) = \mathcal{S}_{(\alpha,\beta)}(g)(x) = 0$. Finally, by Table 2, we have $\mathcal{S}_{(\alpha,\beta)}(f)(x) *_{15} \mathcal{S}_{(\alpha,\beta)}(g)(x) = 0$.

The previous implications can be easily inverted to prove that "if $\mathcal{S}_{(\alpha,\beta)}(f)(x) *_5 \mathcal{S}_{(\alpha,\beta)}(g)(x) = 0$, then $\mathcal{S}_{(\alpha,\beta)}(f \otimes_{15}^{\beta} g)(x) = 0$". So, we can conclude that (13) holds.

Now, let us prove that let $x \in X$,

$$\mathcal{S}_{(\alpha,\beta)}(f \otimes_{15}^{\beta} g)(x) = [0,1] \quad \text{if and only if} \quad \mathcal{S}_{(\alpha,\beta)}(f)(x) *_{15} \mathcal{S}_{(\alpha,\beta)}(g)(x) = [0,1].$$
$$(14)$$

By (1), if $\mathcal{S}_{(\alpha,\beta)}(f \otimes_{15}^{\beta} g)(x) = [0,1]$, then $\alpha \leq (f \otimes_{15}^{\beta} g)(x) \leq \beta$. By (6) and by $\alpha \in (0,1)$, $(f \otimes_{15}^{\beta} g)(x) > 0$ and $(f \otimes_{15}^{\beta} g)(x) = \max(f(x), g(x))$. Consequently, "$f(x) \geq \alpha$ or $g(x) \geq \alpha$" and "$f(x) \leq \beta$ and $g(x) \leq \beta$". The last two respectively imply that "$\mathcal{S}_{(\alpha,\beta)}(f)(x) > 0$ or $\mathcal{S}_{(\alpha,\beta)}(g)(x) > 0$" and "$\mathcal{S}_{(\alpha,\beta)}(g)(x), \mathcal{S}_{(\alpha,\beta)}(f)(x) \neq 1$". Consequently,

$$\mathcal{S}_{(\alpha,\beta)}(f)(x) *_{15} \mathcal{S}_{(\alpha,\beta)}(g)(x) \in \{[0,1] *_{15} [0,1], 0 *_{15} [0,1], [0,1] *_{15} [0,1]\}.$$

Namely, by Table 2, $\mathcal{S}_{(\alpha,\beta)}(f)(x) *_{15} \mathcal{S}_{(\alpha,\beta)}(g)(x) = [0,1]$.

The previous implications can be easily inverted to prove that "if $\mathcal{S}_{(\alpha,\beta)}(f)(x) *_{15} \mathcal{S}_{(\alpha,\beta)}(g)(x) = [0,1]$, then $\mathcal{S}_{(\alpha,\beta)}(f \otimes_{15}^{\beta} g)(x) = [0,1]$".

So, we can conclude that (14) holds. By (13) and (14), we are sure that let $x \in X$, $\mathcal{S}_{(\alpha,\beta)}(f \otimes_{15}^{\beta} g)(x) = 1$ if and only if $\mathcal{S}_{(\alpha,\beta)}(f)(x) *_{15} \mathcal{S}_{(\alpha,\beta)}(g)(x) = 1$.

The other equations can be analogously proved using (1), Definition 2, and Table 2.

Example 1. Consider the fuzzy sets $f : \{x_1, \ldots, x_5\} \rightarrow [0,1]$ and $g : \{x_1, \ldots, x_5\} \rightarrow [0,1]$, which are defined by Table 6.

Table 6. Definition of the functions f and g

x	x_1	x_2	x_3	x_4	x_5		x	x_1	x_2	x_3	x_4	x_5
$f(x)$	0.1	0	0.9	0	0.8		$g(x)$	0.4	1	1	0.5	0

We focus on the operation $\otimes_{15}^{\gamma}$ and choose $\alpha = 0.2$ and $\beta = 0.7$ as thresholds, which satisfy the hypothesis of Theorem 1. Therefore, by Tables 7 and 8, we can view that $\mathcal{S}_{(0.2,0.7)}(f \otimes_{15}^{0.7} g)(x) = \mathcal{S}_{(0.2,0.7)}(f)(x) *_{15} \mathcal{S}_{(0.2,0.7)}(g)(x)$, for each $x \in \{x_1, \ldots, x_5\}$, in line with Theorem 1.

Table 7. Definition of $f \otimes_{15}^{0.7} g$ and $\mathcal{S}_{(0.2,0.7)}(f \otimes_{15}^{0.7} g)$

x	x_1	x_2	x_3	x_4	x_5
$(f \otimes_{15}^{0.7} g)(x)$	0.4	0	1	0.5	0
$\mathcal{S}_{(0.2,0.7)}(f \otimes_{15}^{0.7} g)(x)$	[0,1]	0	1	[0,1]	0

Table 8. Definition of $\mathcal{S}_{(0.2,0.7)}(f)$, $\mathcal{S}_{(0.2,0.7)}(g)$, and $\mathcal{S}_{(0.2,0.7)}f *_{15} \mathcal{S}_{(0.2,0.7)}g$

x	x_1	x_2	x_3	x_4	x_5
$\mathcal{S}_{(0.2,0.7)}(f)(x)$	0	0	1	0	1
$\mathcal{S}_{(0.2,0.7)}(g)(x)$	[0,1]	1	1	[0,1]	0
$\mathcal{S}_{(0.2,0.7)}(f)(x) *_{15} \mathcal{S}_{(0.2,0.7)}(g)(x)$	[0,1]	0	1	[0,1]	0

The next proposition determines the properties holding for the aggregation functions of Definition 2. These last satisfy the same algebraic properties of the corresponding operations on shadowed sets, except the idempotence that exclusively holds for $\otimes_{16}^{\alpha\beta}$ and $\otimes_{20}^{\gamma}$.

Proposition 1. *Let $\alpha, \beta, \gamma \in (0,1)$ such that $\alpha < \beta$. Then,*

(a) *$\otimes_i^{\gamma}$ and $\otimes_j^{\alpha\beta}$ are non-monotone and satisfy the boundary condition, for each $i \in \{15, 20, 21\}$ and for each $j \in \{16, 17, 18, 19\}$;*

(b) *$\otimes_{16}^{\alpha\beta}$ and $\otimes_{20}^{\gamma}$ are idempotent and $\otimes_{15}^{\gamma}$ $\otimes_{20}^{\gamma}$ and $\otimes_{21}^{\delta}$ are commutative;*

(c) *$\otimes_{20}^{\gamma}$ and $\otimes_{21}^{\gamma}$ are associative;*

(d) *$\otimes_{20}^{\gamma}$ has 1 as neutral element.*

Proof. (a) We can show that $\otimes_{15}^{\gamma}$ is non-monotone with a counterexample. Let $\gamma = 0.5$.

- Consider $0.2 \leq 0.3$ and $0.4 \leq 0.6$. By (6), $0.2 \otimes_{15}^{0.5} 0.4 = \max(0.2, 0.4) = 0.4$ and $0.3 \otimes_{15}^{0.5} 0.6 = 0$. Clearly, $0.2 \otimes_{15}^{0.5} 0.4 > 0.3 \otimes_{15}^{0.5} 0.6$. Then, $\otimes_{15}^{0.5}$ is not increasing.
- Consider $0.2 \leq 0.6$ and $0.4 \leq 0.7$. By (6), $0.2 \otimes_{15}^{0.5} 0.4 = \max(0.2, 0.4) = 0.4$ and $0.6 \otimes_{15}^{0.5} 0.7 = 0.7$. Clearly, $0.2 \otimes_{15}^{0.5} 0.4 < 0.6 \otimes_{15}^{0.5} 0.7$. Then, $\otimes_{15}^{0.5}$ is not decreasing.

Similarly, we can prove that $\otimes_{20}^{\gamma}, \otimes_{21}^{\gamma}, \otimes_{16}^{\alpha\beta}, \otimes_{17}^{\alpha\beta}, \otimes_{18}^{\alpha\beta}$, and $\otimes_{19}^{\alpha\beta}$ are non-monotone.

It is easy to understand that $\otimes_{15}^{\gamma}$ satisfies the boundary condition. Indeed, it is trivial that $0 \otimes_{15}^{\gamma} 0 = 0$ from (6). Furthermore, $1 \otimes_{15}^{\gamma} 1$ cannot be 0 because $\gamma < 1$. Then, $1 \otimes_{15}^{\gamma} 1 = \max(1,1) = 1$.

Analogously, we can show that the boundary condition is satisfied by all other operations of item (a).

(b) $\otimes_{16}^{\alpha\beta}$ is idempotent: $x \otimes_{16}^{\alpha\beta} x \in \{\min(x,x), \max(x,x)\}$ from (7). That is, $x \otimes_{16}^{\alpha\beta} x = x$. The same trivially holds for $\otimes_{20}^{\gamma}$. The proof that $\otimes_{15}^{\gamma}, \otimes_{20}^{\gamma}$, and $\otimes_{21}^{\gamma}$ are commutative immediately follows from their definition.

(c) In order to show that $\otimes_{20}^{\gamma}$ is associative, we have to distinguish several cases.

- Let $x, y, z \in [0,1]$ such that $x, y, z \leq \gamma$. Then, both $(x \otimes_{20}^{\gamma} y) \otimes_{20}^{\gamma} z$ and $x \otimes_{20}^{\gamma} (y \otimes_{20}^{\gamma} z)$ coincide with $\max(x, y, z)$.
- Let $x, y, z \in [0,1]$ such that $x, y \leq \gamma$ and $z > \gamma$. Then, two cases can occur: $x \leq y < z$ or $y \leq x < z$. If $x \leq y < z$, then $(x \otimes_{20}^{\gamma} y) \otimes_{20}^{\gamma} z = y \otimes_{20}^{\gamma} = y$ and $x \otimes_{20}^{\gamma} (y \otimes_{20}^{\gamma} z) = x \otimes_{20}^{\gamma} y = y$. Hence, $(x \otimes_{20}^{\gamma} y) \otimes_{20}^{\gamma} z$ is equal to $x \otimes_{20}^{\gamma} (y \otimes_{20}^{\gamma} z)$. The case $y \leq x < z$ is symmetrical to the previous one, so the related proof is omitted.
- We can easily verify that the the equality $(x \otimes_{20}^{\gamma} y) \otimes_{20}^{\gamma} z = x \otimes_{20}^{\gamma} (y \otimes_{20}^{\gamma} z)$ holds in all other cases.

The associativity of $\otimes_{21}^{\gamma}$ can be analogously verified.

(d) Since $\gamma < 1$, $x \otimes_{20}^{\gamma} 1 = \min(x, 1) = x$ for each $x \in [0,1]$ (see (11)).

4.2 Aggregation Functions Generating Conditional-Event Disjunctions on Shadowed Sets

The following is a list of classes of aggregation functions that correspond to operations on shadowed sets deriving from the disjunctions of conditional events defined in Subsect. 3.1.

Definition 3. *Let $\alpha, \beta, \gamma \in (0,1)$ such that $\alpha < \beta$. Then,*

$$x \oplus_{15}^{\gamma} y = \begin{cases} 1 & \text{if } y < \gamma \leq x \text{ or } x < \gamma \leq y, \\ \min(x,y) & \text{otherwise.} \end{cases} \quad (15)$$

$$x \oplus_{16}^{\alpha\beta} y = \begin{cases} \max(x,y) & \text{if } y < \alpha < \beta < x, \text{ or } x < \alpha < \beta < y, \\ \min(x,y) & \text{otherwise.} \end{cases} \quad (16)$$

$$x \oplus_{17}^{\alpha\beta} y = \begin{cases} 0 & \text{if } \alpha \leq y \leq \beta < x, \text{ or } \alpha \leq x \leq \beta < y, \\ \max(x,y) & \text{otherwise.} \end{cases} \quad (17)$$

$$x \oplus_{18}^{\alpha\beta} y = \begin{cases} 0 & \text{if } \alpha \leq y \leq \beta < x, \text{ or } \alpha \leq x \leq \beta < y, \\ \min(x,y) & \text{if } x,y \leq \beta, \\ \max(x,y) & \text{otherwise.} \end{cases} \quad (18)$$

$$x \oplus_{19}^{\alpha\beta} y = \begin{cases} 0 & \text{if } \alpha \leq y \leq \beta < x \text{ or } \alpha \leq x \leq \beta < y, \\ 1 & \text{if } y < \alpha \leq x \leq \beta \text{ or } x < \alpha \leq y \leq \beta, \\ \max(x,y) & \text{otherwise.} \end{cases} \quad (19)$$

$$x \oplus_{20}^{\gamma} y = \begin{cases} \min(x,y) & \text{if } x,y \geq \gamma, \\ \max(x,y) & \text{otherwise.} \end{cases} \quad (20)$$

$$x \oplus_{21}^{\gamma} y = \begin{cases} \max(x,y) & \text{if } x,y \geq \gamma \text{ or } x,y < \gamma, \\ 1 & \text{otherwise.} \end{cases} \quad (21)$$

The next Theorem shows the connection between the aggregation functions of Definition 3 and the operations listed in Table 4.

Theorem 2. *Let $\alpha, \beta \in (0,1)$ such that $\alpha \neq \beta$ and $\alpha \leq 1/2 \leq \beta$. Then, let $f, g \in [0,1]^X$,*

(a) $\mathcal{S}_{(\alpha,\beta)}(f \oplus_i^{\alpha} g) = \mathcal{S}_{(\alpha,\beta)}(f) +_i \mathcal{S}_{(\alpha,\beta)}(g)$ for each $i \in \{15, 20, 21\}$;
(b) $\mathcal{S}_{(\alpha,\beta)}(f \oplus_i^{\alpha\beta} g) = \mathcal{S}_{(\alpha,\beta)}(f) +_i \mathcal{S}_{(\alpha,\beta)}(g)$ for each $i \in \{16, 17, 18, 19\}$.

Proof. The proof is similar to that of Theorem 1. It follows from (1), Table 4, and Definition 3.

Table 9. Definition of $f \oplus_{15}^{0.2} g$ and $\mathcal{S}_{(0.2,0.7)}(f \oplus_{15}^{0.2} g)$

x	x_1	x_2	x_3	x_4	x_5
$(f \oplus_{15}^{0.2} g)(x)$	1	1	1	1	1
$\mathcal{S}_{(0.2,0.7)}(f \oplus_{15}^{0.2} g)(x)$	1	1	1	1	1

Table 10. Definition of $\mathcal{S}_{(0.2,0.7)}(f)$, $\mathcal{S}_{(0.2,0.7)}(g)$, and $\mathcal{S}_{(0.2,0.7)}f +_{15} \mathcal{S}_{(0.2,0.7)}g$

x	x_1	x_2	x_3	x_4	x_5
$\mathcal{S}_{(0.2,0.7)}(f)(x)$	0	0	1	0	1
$\mathcal{S}_{(0.2,0.7)}(g)(x)$	$[0,1]$	1	1	$[0,1]$	0
$\mathcal{S}_{(0.2,0.7)}(f)(x) +_{15} \mathcal{S}_{(0.2,0.7)}(g)(x)$	1	1	1	1	1

Example 2. Consider the functions f and g of Example 1. We focus on the operation $\oplus_{15}^{\gamma}$ and choose the thresholds $\alpha = 0.2$ and $\beta = 0.7$, which satisfy the hypothesis of Theorem 2. Therefore, by Tables 9 and 10, we can view that $\mathcal{S}_{(0.2,0.7)}(f \oplus_{15}^{0.2} g)(x) = \mathcal{S}_{(0.2,0.7)}(f)(x) +_{15} \mathcal{S}_{(0.2,0.7)}(g)(x)$, for each $x \in \{x_1, \ldots, x_5\}$, in line with Theorem 2.

We determine some properties holding for the aggregation functions of Definition 3. These last satisfy the same algebraic properties of the corresponding operations on shadowed sets, except the idempotence that exclusively holds for $\oplus_{16}^{\alpha\beta}$ and $\oplus_{20}^{\gamma}$.

Proposition 2. *Let $\alpha, \beta, \gamma \in (0,1)$ such that $\alpha < \beta$. Then,*

(a) *$\oplus_i^{\gamma}$ and $\oplus_j^{\alpha\beta}$ are non-monotone and satisfy the boundary condition, for each $i \in \{15, 20, 21\}$ and $j \in \{16, 17, 18, 19\}$;*
(b) *$\oplus_{16}^{\alpha\beta}$ and $\oplus_{20}^{\gamma}$ are idempotent;*
(c) *$\oplus_{15}^{\gamma}, \oplus_{20}^{\gamma}$ and $\oplus_{21}^{\gamma}$ are commutative;*
(d) *$\oplus_{20}^{\gamma}$ and $\oplus_{21}^{\gamma}$ are associative;*
(e) *$\oplus_{20}^{\gamma}$ has 0 as neutral element.*

Proof. The proof is similar to that of Proposition 1 and it follows from Definition 3 and the items $(i) - (v)$ of Subsect. 2.2. $\qquad\square$

Aggregation function of Definition 3 are the dual operations of those of Definition 2, namely they fulfill Eq. (4).

Proposition 3. *Let $\alpha, \beta, \gamma \in (0,1)$ such that $\alpha < \beta$, and let $\alpha' = 1 - \alpha$, $\beta' = 1 - \beta$ and $\gamma' = 1 - \gamma$. Then,*

(a) *$\oplus_i^{\gamma'}$ is the dual operation of $\otimes_i^{\gamma}$ for each $i \in \{15, 20, 21\}$;*
(b) *$\oplus_i^{\beta'\alpha'}$ is the dual operation of $\otimes_i^{\alpha\beta}$ for each $i \in \{16, 17, 18, 19\}$.*

Proof. Consider the aggregation functions $\oplus_{15}^{\gamma}$ and $\otimes_{15}^{\gamma}$. Let $x, y, z \in [0,1]$, we intend to show that $x \oplus_{15}^{\gamma} y = 1 - ((1-x) \otimes_{15}^{1-\gamma} (1-y))$. So, we have to analyze all the possible cases.

– Let $y < \gamma \leq x$. By (15), $x \oplus_{15}^{\gamma} y = 1$. Moreover, $y < \gamma \leq x$ implies that $1 - y \leq 1 - \gamma < 1 - x$. Consequently, $(1-y) \otimes_{15}^{1-\gamma} (1-y) = 0$ from (6). Then, $1 - ((1-y) \otimes_{15}^{1-\gamma} (1-y)) = 1$.

– Let $x < \gamma \leq y$. The proof is symmetrical to that of the previous case. So, it is omitted.

– Suppose that $x, y < \gamma$ or $x, y \geq \gamma$, then $1 - x, 1 - y > 1 - \gamma$ or $1 - x, 1 - y \leq \gamma$. Suppose that $y \leq x$ (the case $x \leq x$ is symmetrical). By (6) and (15), we get $x \oplus_{15}^{\gamma} y = \max(x, y) = x$ and $(1-x) \otimes_{15}^{1-\gamma} (1-y) = \min(1-x, 1-y) = 1 - x$. Hence, $1 - ((1-y) \otimes_{15}^{1-\gamma} (1-y)) = 1 - (1-x) = x$.

Analogously, we can verify the duality for the other pairs of aggregation functions.

Example 3. Let $\gamma = 0.5$, we can easily check that $0.2 \oplus_{15}^{\gamma} 0.8 = 1 - ((1-0.2) \otimes_{15}^{1-\gamma} (1-0.8))$. Indeed, $0.2 \oplus_{15}^{\gamma} 0.8 = 1$ and $1 - ((1-0.2) \otimes_{15}^{1-\gamma} (1-0.8)) = 1 - (0.8 \otimes_{15}^{1-\gamma} 0.2) = 1 - 0 = 1$.

4.3 Aggregation Functions Generating $*_C$

We now define a class of aggregation functions corresponding to the consensus operator on shadowed sets $*_C$ given in Subsect. 3.2 and preserving all properties of $*_C$.

Definition 4. *Let* $\alpha, \beta \in (0, 1)$ *such that* $\alpha < \beta$. *Then, we set*

$$x \otimes_C^{\alpha\beta} y = \begin{cases} \min(x, y) & \text{if } x, y \leq \beta, \\ \max(x, y) & \text{if } x > \beta \text{ and } y \geq \alpha, \text{ or } y > \beta \text{ and } x \geq \alpha \\ 1/2 & \text{otherwise.} \end{cases} \quad (22)$$

Theorem 3. *Let* $\alpha, \beta \in (0, 1)$ *such that* $\alpha \neq \beta$ *and* $\alpha \leq 1/2 \leq \beta$, *then for each* $f, g \in [0, 1]^X$, $\mathcal{S}_{(\alpha,\beta)}(f \otimes_C^{\alpha\beta} g) = \mathcal{S}_{(\alpha,\beta)}(f) *_C \mathcal{S}_{(\alpha,\beta)}(g)$.

Proof. The proof is similar to that of Theorem 1. It follows from (1), Table 5, and Definition 4.

Example 4. Consider the functions f and g defined in Example 1. We choose the thresholds $\alpha = 0.2$ and $\beta = 0.7$, which satisfy the hypothesis of Theorem 3.

Therefore, by Tables 11 and 12, we can view that $\mathcal{S}_{(0.2,0.7)}(f \otimes_C^{0.2\ 0.7} g)(x) = \mathcal{S}_{(0.2,0.7)}(f)(x) *_C \mathcal{S}_{(0.2,0.7)}(g)(x)$, for each $x \in \{x_1, \ldots, x_5\}$, in line with Theorem 3.

Table 11. Definition of $f \otimes_C^{0.2\ 0.7} g$ and $\mathcal{S}_{(0.2,0.7)}(f \otimes_C^{0.2\ 0.7} g)$

x	x_1	x_2	x_3	x_4	x_5
$(f \otimes_C^{0.2\ 0.7} g)(x)$	0.1	0.5	1	0	0.5
$\mathcal{S}_{(0.2,0.7)}(f \otimes_C^{0.2\ 0.7} g)(x)$	0	[0,1]	1	0	[0,1]

Table 12. Definition of $\mathcal{S}_{(0.2,0.7)}(f)$, $\mathcal{S}_{(0.2,0.7)}(g)$, and $\mathcal{S}_{(0.2,0.7)}f *_C \mathcal{S}_{(0.2,0.7)}g$

x	x_1	x_2	x_3	x_4	x_5
$\mathcal{S}_{(0.2,0.7)}(f)(x)$	0	0	1	0	1
$\mathcal{S}_{(0.2,0.7)}(g)(x)$	[0,1]	1	1	[0,1]	0
$\mathcal{S}_{(0.2,0.7)}(f)(x) *_C \mathcal{S}_{(0.2,0.7)}(g)(x)$	0	[0,1]	1	0	[0,1]

Proposition 4. *Let* $\alpha, \beta \in (0,1)$ *such that* $\alpha < \beta$. *Then,* $\otimes_C^{\alpha\beta}$ *is commutative, increasing, and non-associative. Moreover, it satisfies the boundary condition.*

Proof. The proof is similar to that of Proposition 1 and it follows from Definition 4 and the items $(i) - (iii)$, and (v) of Subsect. 2.2.

5 Conclusions and Future Works

This article defines new aggregation operators on shadowed sets and investigates their correspondence with aggregation operators on fuzzy sets. The proposed aggregation operators connect shadowed sets, conditional events, and orthpairs, who all play important roles in uncertain reasoning and decision making. In the paper, we mainly considered the applications of the latter two for shadowed sets. It may also be useful to apply results of shadowed sets for the other two, to achieve greater insight on the mutual connections among these knowledge representation formalisms. In the future, we plan to extend this study by determining aggregation functions that can be approximated by shadowed set operations by means of transformations different from equation (1). Also, as mentioned in [2], we could start from our results to find the way to approximate many-valued logics with three-valued (or more generally n-valued) logics.

References

1. Bochvar, D.A., Bergmann, M.: On a three-valued logical calculus and its application to the analysis of the paradoxes of the classical extended functional calculus. Hist. Philos. Logic **2**(1–2), 87–112 (1981)
2. Boffa, S., Campagner, A., Ciucci, D., Yao, Y.: Aggregation operators on shadowed sets. Inf. Sci. **595**, 313–333 (2022)
3. Campagner, A., Dorigatti, V., Ciucci, D.: Entropy-based shadowed set approximation of intuitionistic fuzzy sets. Int. J. Intell. Syst. **35**(12), 2117–2139 (2020)
4. Casillas, J., Cordón, O., Triguero, F.H., Magdalena, L. (eds.): Interpretability Issues in Fuzzy Modeling. Studies in Fuzziness and Soft Computing, vol. 128. Springer, Heidelberg (2013). https://doi.org/10.1007/978-3-540-37057-4
5. Ciucci, D.: Orthopairs: a simple and widely usedway to model uncertainty. Fund. Inf. **108**(3–4), 287–304 (2011)
6. Ciucci, D.: Orthopairs and granular computing. Granular Comput. **1**(3), 159–170 (2016)

7. Ciucci, D., Dubois, D., Lawry, J.: Borderline vs. unknown: comparing three-valued representations of imperfect information. Int. J. Approx. Reason. **55**(9), 1866–1889 (2014)
8. Deng, X., Yao, Y.: Mean-value-based decision-theoretic shadowed sets. In: 2013 Joint IFSA World Congress and NAFIPS Annual Meeting (IFSA/NAFIPS), pp. 1382–1387. IEEE (2013)
9. Fodor, J.: Aggregation functions in fuzzy systems. In: Aspects of Soft Computing, Intelligent Robotics and Control, pp. 25–50. Springer, Heidelberg (2009). https://doi.org/10.1007/978-3-642-03633-0_2
10. Gao, M., Zhang, Q., Zhao, F., Wang, G.: Mean-entropy-based shadowed sets: a novel three-way approximation of fuzzy sets. Int. J. Approx. Reason. **120**, 102–124 (2020)
11. He, S., Pan, X., Wang, Y.: A shadowed set-based todim method and its application to large-scale group decision making. Inf. Sci. **544**, 135–154 (2021)
12. Ibrahim, M., William-West, T., Kana, A., Singh, D.: Shadowed sets with higher approximation regions. Soft. Comput. **24**, 17009–17033 (2020)
13. Kleene, S.C., De Bruijn, N., de Groot, J., Zaanen, A.C.: Introduction to metamathematics, vol. 483. van Nostrand New York (1952)
14. Lawry, J., Dubois, D.: A bipolar framework for combining beliefs about vague propositions. In: Proceedings of the Thirteenth International Conference on Principles of Knowledge Representation and Reasoning, pp. 530–540 (2012)
15. Mitra, S., Pedrycz, W., Barman, B.: Shadowed c-means: integrating fuzzy and rough clustering. Pattern Recogn. **43**(4), 1282–1291 (2010)
16. Pedrycz, W.: Shadowed sets: representing and processing fuzzy sets. IEEE Trans. Syst. Man Cybern. - PART B: Cybern. **28**(1), 103–109 (1998)
17. Pedrycz, W.: From fuzzy sets to shadowed sets: interpretation and computing. Int. J. Intell. Syst. **24**(1), 48–61 (2009)
18. Pedrycz, W., Vukovich, G.: Granular computing with shadowed sets. Int. J. Intell. Syst. **17**(2), 173–197 (2002)
19. Schweizer, B., Sklar, A.: Espaces métriques aléatories. Comptes Rendus Hebdomadaires des Seances de l'Academie des Sciences **247**(23), 2092–2094 (1958)
20. Sobociński, B.: Axiomatization of a partial system of three-value calculus of propositions. Institute of Applied Logic (1952)
21. Tahayori, H., Sadeghian, A., Pedrycz, W.: Induction of shadowed sets based on the gradual grade of fuzziness. IEEE Trans. Fuzzy Syst. **21**(5), 937–949 (2013)
22. Torra, V., Narukawa, Y.: Modeling Decisions: Information Fusion and Aggregation Operators. Springer, Heidelberg (2007). https://doi.org/10.1007/978-3-540-68791-7
23. Walker, E.A.: Stone algebras, conditional events, and three valued logic. IEEE Trans. Syst. Man Cybern. **24**(12), 1699–1707 (1994)
24. William-West, T., Ibrahim, A., Kana, A.: Shadowed set approximation of fuzzy sets based on nearest quota of fuzziness. Ann. Fuzzy Math. Inf. **4**(1), 27–38 (2019)
25. Yager, R.R., Rybalov, A.: Uninorm aggregation operators. Fuzzy Sets Syst. **80**(1), 111–120 (1996)
26. Yao, Y., Wang, S., Deng, X.: Constructing shadowed sets and three-way approximations of fuzzy sets. Inf. Sci. **412**, 132–153 (2017)
27. Zhou, J., Pedrycz, W., Gao, C., Lai, Z., Yue, X.: Principles for constructing three-way approximations of fuzzy sets: a comparative evaluation based on unsupervised learning. Fuzzy Sets Syst. **413**, 74–98 (2020)

Pawlak, Belnap and the Magical Number Seven

Salvatore Greco[1(✉)] and Roman Słowiński[2,3]

[1] Department of Economics and Business, University of Catania, Corso Italia, 55, 95129 Catania, Italy
salgreco@unict.it
[2] Institute of Computing Science, Poznań University of Technology, 60-965 Poznań, Poland
[3] Institute for Systems Research, Polish Academy of Sciences, 01-447 Warsaw, Italy

Abstract. We are considering the algebraic structure of the Pawlak-Brouwer-Zadeh lattice to distinguish vagueness due to imprecision from ambiguity due to coarseness. We show that a general class of many-valued logics useful for reasoning about data emerges from this context. All these logics can be obtained from a very general seven-valued logic which, interestingly enough, corresponds to a reasoning system developed by Jaina philosophers four centuries BC. In particular, we show how the celebrated Belnap four-valued logic can be obtained from the very general seven-valued logic based on the Pawlak-Brouwer-Zadeh lattice.

1 Introduction

The Brouwer-Zadeh lattice [7] was introduced as an algebraic structure to handle vagueness through representation of each concept X of universe U by pair (A, B), $A, B \subseteq U, A \cap B = \emptyset$, where A is the necessity kernel (the set of objects from U belonging to X without any doubt) and B is the non-possibility kernel (the set of objects from U that for sure do not belong to X). In rough set theory [15,16], the Brouwer-Zadeh lattice can be seen as an abstract model [4–6] representing each concept X through pair of elements (A, B) where A is the lower approximation (interior) and B is the complement of the upper approximation (exterior) of X. In [10], an extension of the Brouwer-Zadeh lattice, called Pawlak-Brouwer-Zadeh lattice, has been proposed, where a new operator, called Pawlak operator, assigns the pair (C, D) to each concept X represented by pair (A, B), such that C and D represent the lower approximations of A and B, respectively. The rough set theory of Pawlak operator has been discussed in [12]. In this paper, we reconsider the Pawlak-Brouwer-Zadeh lattice from Pawlak's perspective of reasoning about data [16] and Belnap's idea of automated computer reasoning [1]. In particular, we demonstrate that within the Pawlak-Brouwer-Zadeh lattice, several many-valued logics naturally arise from a very general seven-valued logic, which is useful for modeling automated data reasoning. It is interesting to observe that this seven-valued logic corresponds to a reasoning system of argumentation

A. Campagner et al. (Eds.): IJCRS 2023, LNAI 14481, pp. 216–228, 2023.
https://doi.org/10.1007/978-3-031-50959-9_15

proposed by Jaina philosophers four centuries BC [3,18]. Furthermore, the seven-valued logic is interesting from a cognitive psychology perspective. According to the influential and highly cited article 'The Magical Number Seven, Plus or Minus Two: Some Limits on Our Capacity for Processing Information' by Miller [13], it is suggested that individuals can effectively handle approximately seven stimuli simultaneously. This limit applies to both one-dimensional absolute judgment and short-term memory.

To give an intuition of the seven-valued logic and the other logics deriving from it, let us consider the following example. Consider a database containing data about the symptoms and related diagnosed diseases for a number of patients. We can imagine that for each considered disease, there are three possible types of diagnosis:

- the patient has the disease,
- the patient does not have the disease,
- it cannot be said whether the patient has or not the disease (because, for instance, other tests have to be done).

On the other hand, for each patient x there may be a number of patients in the database with the same symptoms. In this situation, it seems reasonable to analyze the database by classifying patients according to the diagnoses received by them and all other patients with the same symptoms. Proceeding in this way, there will be seven possible states of truth of the following proposition: "A patient with the same symptoms as x has the disease". Let us name this proposition by $DISx$ and enumerate the possible cases:

- all the patients with the same symptoms as x have the disease, so that, according to the available data, proposition $DISx$ is true,
- some patients with the same symptoms as x have the disease and for the others one cannot say if there is or not the disease, so that, according to the available data, proposition $DISx$ is sometimes true,
- for all the patients with the same symptoms as x one cannot say if there is or not the disease, so that, according to the available data, proposition $DISx$ is unknown,
- some of the patients with the same symptoms as x have the disease and the others have not the disease, so that, according to the available data, proposition $DISx$ is contradictory,
- among the patients with the same symptoms as x there are some with the disease, some without the diseases and some for which one does not know if there is or not the disease, so that, according to the available data, proposition $DISx$ is fully contradictory,
- some patients with the same symptoms as x do not have the disease and for the others one cannot say if there is or not the disease, so that, according to the available data, proposition $DISx$ is sometimes false,
- all the patients with the same symptoms as x have not the disease, so that, according to the available data, proposition $DISx$ is false.

The above seven situations are, of course, very detailed, so for practical reasons it could be more convenient to aggregate some of them. For example, one could distinguish between the following two situations, the first of which would suggest some treatment, while in the second situation it would be more appropriate to wait for the possible appearance of some other symptoms:

- the case in which among the patients with the same symptoms there is some patient with the disease and there is no patient without the disease, that is the case in which proposition $DISx$ is true or sometimes true, and
- the other cases.

To support decisions about triage, a reasonable approach could be the following:

- if proposition $DISx$ is true or sometimes true, patient x should be hospitalized,
- if proposition $DISx$ is false or sometimes false, patient x should not be hospitalized,
- in all other cases, an expert doctor should be called to examine patient x to make the decision about hospitalization.

To diagnose a complex disease, the database could be used as follows:

- if proposition $DISx$ is true or sometimes true, patient x is diagnosed as having the disease,
- if proposition $DISx$ is unknown, some further medical tests are required for patient x,
- if proposition $DISx$ is contradictory or fully contradictory, an expert doctor should be called to examine patient x to make the diagnosis,
- if proposition $DISx$ is false or sometimes false, patient x is not diagnosed as having the disease.

One can see that the basic case related to the general database query defines a seven-valued logic with truth values "true", "sometimes true", "unknown", "contradictory", "fully contradictory", "sometimes false" and "false". The reasoning about the treatment decision is based on a two-valued logic - apply or not the treatment - derived from the basic seven-valued logic. The envisaged protocol for triage gives an example of a three valued-logic grounded again on the basic seven-valued logic. Finally, the diagnostic procedure proposes a possible four-valued logic that can also be obtained from the seven-valued logic.

Let us explain now how the basic seven-valued logic is related to the rough set concept and to the Pawlak-Brouwer-Zadeh lattice. First, observe that the set of all patients in the data set having the same symptoms as patient x represents the equivalence class $[x]_R$ with respect to the indiscernibility relation R defined in terms of symptoms on universe U which is the data set of patients. In other words, we consider indiscernible two patients w and z in the data set if w has the same symptoms as z. Observe that each disease is represented in the data set by pair (A, B) where A is the set of patients from U having the disease, and B is the set of patients from U not having the disease. The operators of the

Pawlak-Brouwer-Zadeh lattice permit to define other interesting sets of patients from U. For example, applying the Brouwer negation to (A, B) we get $(A, B)^\approx = (B, U - B)$, and consequently we obtain set $U - B$, i.e., the complement in U of the set of patients without the disease, or, in other words, the set of patients for whom there is the disease or it is unknown if there is the disease. In the same perspective, applying the Kleene negation $(A, B)^- = (B, A)$ and the Brouwer negation to (A, B) we obtain $(A, B)^{-\approx} = (A, U - A)$, so that we get set $U - A$, i.e., the complement in U of the set of patients with the disease, or, in other words, the set of patients for whom there is not the disease or it is unknown if there is or there is not the disease. Other interesting sets that can be obtained using the operators of the Pawlak-Brouwer-Zadeh lattice are $A \cup B$, i.e., the set of all patients for whom there is the disease or there is not the disease, and $U - A - B$, i.e., the set of all patients for whom it is unknown if there is or there is not the disease. Applying rough set theory, we can compute the lower and the upper approximation of the above mentioned sets of patients from U. For example, the set of patients x from U for whom all the other patients with the same symptoms have the disease constitutes the lower approximation of A denoted by $\underline{R}A$. The set of all patients x from U for whom there is at least one patient with the same symptoms that has the disease constitutes, instead, the upper approximation of A denoted by $\overline{R}A$. Within the Pawlak-Brouwer-Zadeh lattice, the rough approximation can be obtained through the application of the Pawlak operator that assigns pair $(\underline{R}A, \underline{R}B)$ to pair (A, B), that is $(A, B)^L = (\underline{R}A, \underline{R}B)$. Remembering that $\overline{R}A = U - \underline{R}(U - A)$, using the Pawlak operator and the other operators of the Pawlak-Brouwer-Zadeh lattice we can obtain the upper approximation of set A starting from pair (A, B) as follows: $(A, B)^{-\approx L\approx} = (U - \underline{R}(U - A), \underline{R}(U - A)) = (\overline{R}(A), \underline{R}(U - A))$. Analogously, starting from pair (A, B), one can obtain all the following rough approximations: $\underline{R}A, \underline{R}B, \underline{R}(U - A - B), \underline{R}(U - A), \underline{R}(U - B), \underline{R}(A \cup B), \overline{R}A, \overline{R}B, \overline{R}(U - A - B)$. Using these rough approximations we can define all the truth values of the seven-valued logic as follows:

- set of patients from U for whom proposition $DISx$ is true: $\underline{R}A$,
- set of patients from U for whom proposition $DISx$ is sometimes true: $\underline{R}(U - B) \cap \overline{R}A \cap \overline{R}(U - A - B)$,
- set of patients from U for whom proposition $DISx$ is unknown: $\underline{R}(U - A - B)$,
- set of patients from U for whom proposition $DISx$ is contradictory: $\underline{R}(A \cup B) \cap \overline{R}A \cap \overline{R}B$,
- set of patients from U for whom proposition $DISx$ is fully contradictory: $\overline{R}A \cap \overline{R}B \cap \overline{R}(U - A - B)$,
- set of patients from U for whom proposition $DISx$ is sometimes false: $\underline{R}(U - A) \cap \overline{R}B \cap \overline{R}(U - A - B)$,
- set of patients from U for whom proposition $DISx$ is false: $\underline{R}B$.

Using the truth values of the seven-valued logic and the operators of the Pawlak-Brouwer-Zadeh lattice, one can reconstruct all the other logics derived from the aggregation of some of the seven truth values. For example, the truth values of the above mentioned two-valued logic for decisions about the treatment can be formulated as follows:

- set of patients from U for whom proposition $DISx$ is true or sometimes true:
  $\underline{R}A \cup (\underline{R}(U - B) \cap \overline{R}A \cap \overline{R}(U - A - B)) = \underline{R}(U - B) \cap \overline{R}A$, and
- the other cases: $\underline{R}(U - A - B) \cup (\underline{R}(A \cup B) \cap \overline{R}A \cap \overline{R}B) \cup (\overline{R}A \cap \overline{R}B \cap \overline{R}(U - A - B)) \cup (\underline{R}(U - A) \cap \overline{R}B \cap \overline{R}(U - A - B)) \cup \underline{R}B = U - \underline{R}(U - B) = \overline{R}B \cup \underline{R}(U - A - B)$.

The framework we have discussed is also related to three-way decision which is an interesting research line that has been deeply investigated in recent years (see [19] for a survey). In fact, the three basic truth values - true, false, unknown - define a three-valued logic, and the derived seven-valued logic is based on the non-empty subsets of {true, false, unknown}.

This paper presents all the above considerations formally and in detail. It is organized as follows. In the next section, we recall the Pawlak-Brouwer-Zadeh lattice and the basic elements of rough set theory. In the following section, we present the seven-valued logic and the other logics that can be derived from it. The last section collects conclusions.

2   The Pawlak-Brouwer-Zadeh Distributive de Morgan Lattices and Indiscernibility-Based Rough Set Theory

This section recalls the Pawlak-Brouwer-Zadeh distributive De Morgan lattices [10] and shows it as an abstract model of the classical rough set model based on indiscernibility.

A system $\langle \Sigma, \wedge, \vee, ', ^\sim, 0, 1 \rangle$ is a quasi-Brouwer-Zadeh distributive lattice [7] if the following properties (1)–(4) hold:

(1) Σ is a distributive lattice with respect to the join and the meet operations $\vee, \wedge$ whose induced partial order relation is

$$a \leq b \text{ iff } a = a \wedge b \text{ (equivalently } b = a \vee b)$$

Moreover, it is required that Σ is bounded by the least element 0 and the greatest element 1:

$$\forall a \in \Sigma, \quad 0 \leq a \leq 1$$

(2) The unary operation $' : \Sigma \to \Sigma$ is a Kleene (also Zadeh or fuzzy) complementation. In other words, for arbitrary $a, b \in \Sigma$,

(K1) $a'' = a$,
(K2) $(a \vee b)' = a' \wedge b'$,
(K3) $a \wedge a' \leq b \vee b'$.

(3) The unary operation $^\sim : \Sigma \to \Sigma$ is a Brouwer (or intuitionistic) complementation. In other words, for arbitrary $a, b \in \Sigma$,

(B1) $a \wedge a^{\sim\sim} = a$,

(B2) $(a \vee b)^\sim = a^\sim \wedge b^\sim$,

(B3) $a \wedge a^\sim = 0$.

(4) The two complementations are linked by the interconnection rule which must hold for arbitrary $a \in \Sigma$:

(in) $a^\sim \leq a'$.

A structure $\langle \Sigma, \wedge, \vee, ', \sim, 0, 1 \rangle$ is a Brouwer-Zadeh distributive lattice if it is a quasi-Brouwer-Zadeh distributive lattice satisfying the stronger interconnection rule:

(s-in) $a^{\sim\sim} = a^{\sim\prime}$.

A Brouwer-Zadeh distributive lattice satisfying the $\vee$ De Morgan property:

(B2a) $(a \wedge b)^\sim = a^\sim \vee b^\sim$

is called a De Morgan Brouwer-Zadeh distributive lattice.

An approximation operator, called Pawlak operator [10], on a De Morgan Brouwer-Zadeh distributive lattice is an unary operation $^A : \Sigma \to \Sigma$ for which the following properties hold: for $a, b \in \Sigma$

A1) $a^{A\prime} = a^{\prime A}$;

A2) $a \leq b$ implies $b^{A\sim} \leq a^{A\sim}$;

A3) $a^{A\sim} \leq a^\sim$;

A4) $0^A = 0$;

A5) $a^\sim = b^\sim$ implies $a^A \wedge b^A = (a \wedge b)^A$;

A6) $a^A \vee b^A \leq (a \vee b)^A$;

A7) $a^{AA} = a^A$;

A8) $a^{A\sim A} = a^{A\sim}$;

A9) $(a^A \wedge b^A)^A = a^A \wedge b^A$.

2.1   Pawlak-Brouwer-Zadeh Lattices and Rough Set Theory

A *knowledge base* $K = (U, R)$ is a relational system where $U \neq \emptyset$ is a finite set called the *universe* and R is an equivalence relation on U. For any $x \in U$, $[x]_R$ is its equivalence class. The quotient set U/R is composed of all the equivalence classes of R on U. Given the knowledge base $K = (U, R)$, one can associate the two subsets $\underline{R}X$ and $\overline{R}X$ to each subset $X \subseteq U$:

$$\underline{R}X = \{x \in U : [x]_R \subseteq X\},$$

$$\overline{R}X = \{x \in U : [x]_R \cap X \neq \emptyset\}.$$

$\underline{R}X$ and $\overline{R}X$ are called the lower and the upper approximation of X, respectively.

Let us consider the set of all pairs $\langle A, B \rangle$ such that $A, B \subseteq U$ and $A \cap B = \emptyset$. We denote by 3^U the set of these pairs, i.e.,

$$3^U = \{\langle A, B \rangle : A, B \subseteq U \text{ and } A \cap B = \emptyset\}.$$

Given a knowledge base $K = (U, R)$, we can define an unary operator $^L : 3^U \to 3^U$, as follows: for any $\langle A, B \rangle \in 3^U$

$$\langle A, B \rangle^L = \langle \underline{R}A, \underline{R}B \rangle.$$

Let us consider the following operations on 3^U:

$$\begin{aligned}
\langle A, B \rangle \sqcap \langle C, D \rangle &= \langle A \cap C, B \cup D \rangle, \\
\langle A, B \rangle \sqcup \langle C, D \rangle &= \langle A \cup C, B \cap D \rangle, \\
\langle A, B \rangle^- &= \langle B, A \rangle, \\
\langle A, B \rangle^\approx &= \langle B, U - B \rangle.
\end{aligned} \tag{1}$$

The following results hold [10].

Proposition 1. The structure $\langle 3^U, \sqcap, \sqcup, ^-, ^\approx, ^L, \langle \emptyset, U \rangle, \langle U, \emptyset \rangle \rangle$ is a Pawlak-Brouwer-Zadeh lattice.

Proposition 2. For every Pawlak-Brouwer-Zadeh lattice $\mathcal{L}_{PBZ} = \langle \Sigma, \wedge, \vee, ', \sim, ^A, 0, 1 \rangle$, satisfying the condition

(P) there exists $c \in \Sigma$ for which $c = c'$,

there is a knowledge base $K = (U, R)$ such that the structure

$$RS_{PBZ}(U, R) = \langle 3^U, \sqcap, \sqcup, ^-, ^\approx, ^L, \langle \emptyset, U \rangle, \langle U, \emptyset \rangle \rangle$$

is isomorphic to $\mathcal{L}_{PBZ}$.

3   The Seven-Valued Logic of the Pawlak-Brouwer-Zadeh Lattice

In the following, we shall identify a set $S \subseteq U$ with the pair $\langle S, U - S \rangle$. Given a knowledge base $K = (U, R)$, for each pair $a = \langle A, B \rangle, A, B \subseteq U, A \cap B = \emptyset$, the following sets can be considered:

– *the true part* of $\langle A, B \rangle$:
  $\mathbf{T}(A, B)$
  $= \{x \in U : [x]_R \subseteq A\} = \underline{R}A = \langle A, B \rangle^{L-\approx} = a^{A/\sim}$,
– *the sometimes true part* of $\langle A, B \rangle$:
  $\mathbf{s}T(A, B)$
  $= \{x \in U : [x]_R \subseteq (U - B), [x]_R \cap A \neq \emptyset \text{ and } [x]_R \cap (U - A - B) \neq \emptyset\}$
  $= \underline{R}(U - B) \cap \overline{R}A \cap \overline{R}(U - A - B)$
  $= \langle A, B \rangle^{\approx L\approx} \sqcap \left(\langle A, B \rangle^{\approx -} \sqcap \langle A, B \rangle^{-\approx -} \right)^{L\approx -} \sqcap \langle A, B \rangle^{-\approx L\approx -}$
  $= a^{\sim A\sim} \wedge (a^{\sim \prime} \wedge a'^{\sim \prime})^{A\sim \prime} \wedge a'^{\sim A\sim \prime}$,
– *the unknown part* of $\langle A, B \rangle$:
  $\mathbf{U}(A, B)$
  $= \{x \in U : [x]_R \subseteq U - A - B\} = \underline{R}(U - A - B) =$
  $\left(\langle A, B \rangle^{\approx -} \sqcap \langle A, B \rangle^{-\approx -} \right)^{L-\approx}$
  $= (a^{\sim \prime} \wedge a'^{\sim \prime})^{A/\sim}$,

- *the contradictory part* of $\langle A, B \rangle$:
  $$\mathbf{K}(A, B)$$
  $$= \{x \in U : [x]_R \subseteq A \cup B, [x]_R \cap B \neq \emptyset \text{ and } [x]_R \cap A \neq \emptyset\}$$
  $$= \underline{R}(A \cup B) \cap \overline{R}A \cap \overline{R}B =$$
  $$= \left(\langle A, B \rangle \sqcup \langle A, B \rangle^-\right)^{L-\approx} \sqcap \langle A, B \rangle^{-\approx L\approx-} \sqcap \langle A, B \rangle^{\approx L\approx-}$$
  $$= (a \vee a')^{A'\sim} \wedge a'^{\sim A\sim\prime} \wedge a^{\sim A\sim\prime},$$

- *the fully contradictory part* of $\langle A, B \rangle$:
  $$\mathbf{f}K(A, B)$$
  $$= \{x \in U : [x]_R \cap A \neq \emptyset, [x]_R \cap B \neq \emptyset \text{ and } [x]_R \cap (U - A - B) \neq \emptyset\}$$
  $$= \overline{R}A \cap \overline{R}B \cap \overline{R}(U - A - B)$$
  $$= \langle A, B \rangle^{-\approx L\approx-} \sqcap \langle A, B \rangle^{\approx L\approx-} \sqcap \left(\langle A, B \rangle^{\approx-} \wedge \langle A, B \rangle^{-\approx-}\right)^{L\approx-}$$
  $$= a'^{\sim A\sim\prime} \wedge a^{\sim A\sim\prime} \wedge (a^{\sim\prime} \wedge a'^{\sim\prime})^{A\sim\prime},$$

- *the sometimes false part* of $\langle A, B \rangle$:
  $$\mathbf{s}F(A, B)$$
  $$= \{x \in U : [x]_R \subseteq (U - A), [x]_R \cap B \neq \emptyset \text{ and } [x]_R \cap (U - A - B) \neq \emptyset\} =$$
  $$\underline{R}(U - A) \cap \overline{R}B \cap \overline{R}(U - A - B)$$
  $$= \langle A, B \rangle^{-\approx L\approx} \sqcap \left(\langle A, B \rangle^{\approx-} \sqcap \langle A, B \rangle^{-\approx-}\right)^{L\approx-} \sqcap \langle A, B \rangle^{-\approx L\approx-}$$
  $$= a'^{\sim A\sim} \wedge (a^{\sim\prime} \wedge a'^{\sim\prime})^{A\sim\prime} \wedge a'^{\sim A\sim\prime},$$

- *the false part* of $\langle A, B \rangle$:
  $$\mathbf{F}(A, B)$$
  $$= \{x \in U : [x]_R \subseteq B\} = \underline{R}B = \langle A, B \rangle^{L\approx} = a^{A\sim}.$$

The truth values of the seven-valued logic can be represented by the lattice in Fig. 1.

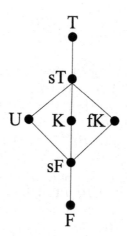

Fig. 1. Seven-valued logic truth value lattice

224   S. Greco and R. Słowiński

Let us remark that truth value operators of the seven-valued logic can be characterized in terms of upper approximations $\overline{R}A, \overline{R}B$ and $\overline{R}(U - A - B)$ as follows: for all $x \in U$

- $x \in \mathbf{T}(A, B) = x \in \overline{R}A \wedge x \notin \overline{R}B \wedge x \notin \overline{R}(U - A - B)$,
- $x \in \mathbf{s}T(A, B) = x \in \overline{R}A \wedge x \notin \overline{R}B \wedge x \in \overline{R}(U - A - B)$,
- $x \in \mathbf{U}(A, B) = x \notin \overline{R}A \wedge x \notin \overline{R}B \wedge x \in \overline{R}(U - A - B)$,
- $x \in \mathbf{K}(A, B) = x \in \overline{R}A \wedge x \in \overline{R}B \wedge x \notin \overline{R}(U - A - B)$,
- $x \in \mathbf{f}K(A, B) = x \in \overline{R}A \wedge x \in \overline{R}B \wedge x \in \overline{R}(U - A - B)$,
- $x \in \mathbf{s}F(A, B) = x \notin \overline{R}A \wedge x \in \overline{R}B \wedge x \in \overline{R}(U - A - B)$,
- $x \in \mathbf{F}(A, B) = x \notin \overline{R}A \wedge x \in \overline{R}B \wedge x \notin \overline{R}(U - A - B)$.

Observe that for all $A, B \subseteq U, A \cap B = \emptyset$,

$$\mathbf{T}(A, B) \cup \mathbf{s}T(A, B) \cup \mathbf{U}(A, B) \cup \mathbf{K}(A, B) \cup \mathbf{f}K(A, B) \cup \mathbf{s}F(A, B) \cup \mathbf{F}(A, B) = U$$

and for all pairs $(\mathbf{O}_1, \mathbf{O}_2)$ with $\mathbf{O}_1, \mathbf{O}_2 \in \mathcal{O} = \{\mathbf{T}, \mathbf{s}T, \mathbf{U}, \mathbf{K}, \mathbf{f}K, \mathbf{s}F, \mathbf{F}\}$ with $\mathbf{O}_1 \neq \mathbf{O}_2$

$$\mathbf{O}_1(A, B) \cap \mathbf{O}_2(A, B) = \emptyset.$$

Several aggregations of the truth value operators in $\mathcal{O}$ are interesting. Among them, the following upward and downward aggregations are particularly interesting:

- *the at least true part* of $\langle A, B \rangle$: $\mathbf{T}^{\uparrow}(A, B) = \mathbf{T}(A, B)$,
- *the at least sometimes true part* of $\langle A, B \rangle$:
  $\mathbf{s}T^{\uparrow}(A, B) = \mathbf{s}T(A, B) \cup \mathbf{T}(A, B)$
  $= \{x \in U : [x]_R \subseteq U - B \text{ and } [x]_R \cap A \neq \emptyset\} = \underline{R}(U - B) \cap \overline{R}A$
  $= \langle A, B \rangle^{\approx L \approx} \sqcap \langle A, B \rangle^{-\approx L \approx -} = a^{\sim A \sim} \wedge a'^{\sim A \sim '}$,
- *the at least unknown part* of $\langle A, B \rangle$:
  $\mathbf{U}^{\uparrow}(A, B) = \mathbf{U}(A, B) \cup \mathbf{s}T(A, B) \cup \mathbf{T}(A, B)$
  $= \{x \in U : [x]_R \subseteq U - B\} = \underline{R}(U - B)$
  $= \langle A, B \rangle^{\approx L \approx} = a^{\sim A \sim}$,
- *the at least contradictory part* of $\langle A, B \rangle$:
  $\mathbf{K}^{\uparrow}(A, B) = \mathbf{K}(A, B) \cup \mathbf{s}T(A, B) \cup \mathbf{T}(A, B)$
  $= \{x \in U : ([x]_R \subseteq (A \cup B) \text{ or } [x]_R \subseteq (U - B)) \text{ and } [x]_R \cap A \neq \emptyset\}$
  $= (\underline{R}(A \cup B) \cup \underline{R}(U - B)) \cap \overline{R}A$
  $= \left((\langle A, B \rangle \sqcup \langle A, B \rangle^-)^{L-\approx} \sqcup \langle A, B \rangle^{\approx L \approx} \right) \sqcap \langle A, B \rangle^{-\approx L \approx -}$
  $= ((a \vee a')^{A \sim} \vee a^{\sim A \sim}) \vee a'^{\sim A \sim '}$,
- *the at least fully contradictory part* of $\langle A, B \rangle$:
  $\mathbf{f}K^{\uparrow}(A, B) = \mathbf{f}K(A, B) \cup \mathbf{s}T(A, B) \cup \mathbf{T}(A, B)$
  $= \{x \in U : [x]_R \subseteq A \text{ or } ([x]_R \cap A \neq \emptyset \text{ and } [x]_R \cap (U - A - B) \neq \emptyset\}$
  $= \underline{R}A \cup (\overline{R}A \cap \overline{R}(U - A - B))$
  $= \langle A, B \rangle^{L-\approx} \sqcup (\langle A, B \rangle^{-\approx L \approx -} \sqcap \left(\langle A, B \rangle^{\approx -} \sqcap \langle A, B \rangle^{-\approx -} \right)^{L \approx -})$
  $= a^{A \sim} \vee (a'^{\sim A \sim '} \wedge (a^{\sim '} \wedge a'^{\sim '})^{A \sim '})$,

– *the at least sometimes false part* of $\langle A, B \rangle$:
$$\mathbf{s}F^{\uparrow}(A, B) = \mathbf{s}F(A, B) \cup \mathbf{U}(A, B) \cup \mathbf{K}(A, B) \cup \mathbf{f}K(A, B) \cup \mathbf{s}T(A, B) \cup \mathbf{T}(A, B)$$
$$= \{x \in U : [x]_R \cap (U - B) \neq \emptyset\} = \overline{R}(U - B) = \langle A, B \rangle^{L \approx -} = a^{A \sim \prime},$$
– *the at least false part* of $\langle A, B \rangle$:
$$\mathbf{F}^{\uparrow}(A, B) = U = \langle U, \emptyset \rangle = 1,$$
– *the at most false part* of $\langle A, B \rangle$: $\mathbf{F}^{\downarrow}(A, B) = \mathbf{F}(A, B)$,
– *the at most sometimes false part* of $\langle A, B \rangle$:
$$\mathbf{s}T^{\downarrow}(A, B) = \mathbf{s}F(A, B) \cup \mathbf{F}(A, B)$$
$$= \{x \in U : [x]_R \subseteq U - A \text{ and } [x]_R \cap B \neq \emptyset\} = \underline{R}(U - A) \cap \overline{R}F$$
$$= \langle A, B \rangle^{-\approx L \approx} \sqcap \langle A, B \rangle^{\approx L \approx -} = a^{\prime \sim A \sim} \wedge a^{\sim A \sim \prime},$$
– *the at most unknown part* of $\langle A, B \rangle$:
$$\mathbf{U}^{\downarrow}(A, B) = \mathbf{U}(A, B) \cup \mathbf{s}F(A, B) \cup \mathbf{F}(A, B)$$
$$= \{x \in U : [x]_R \subseteq U - A\} = \underline{R}(U - A)$$
$$= \langle A, B \rangle^{-\approx L \approx} = a^{\prime \sim A \sim},$$
– *the at most contradictory part* of $\langle A, B \rangle$:
$$\mathbf{K}^{\downarrow}(A, B) = \mathbf{K}(A, B) \cup \mathbf{s}F(A, B) \cup \mathbf{F}(A, B)$$
$$= \{x \in U : ([x]_R \subseteq (A \cup B) \text{ or } [x]_R \subseteq (U - A)) \text{ and } [x]_R \cap B \neq \emptyset\}$$
$$= (\underline{R}(A \cup B) \cup \underline{R}(U - A)) \cap \overline{R}B$$
$$= \left(((\langle A, B \rangle \sqcup \langle A, B \rangle^{-})^{L - \approx} \sqcup \langle A, B \rangle^{-\approx L \approx} \right) \sqcap \langle A, B \rangle^{\approx L \approx -}$$
$$= \left((a \vee a^{\prime})^{A \prime \sim} \vee a^{\prime \sim A \sim} \right) \vee a^{\sim A \sim \prime},$$
– *the at most fully contradictory part* of $\langle A, B \rangle$:
$$\mathbf{f}K^{\downarrow}(A, B) = \mathbf{f}K(A, B) \cup \mathbf{s}F(A, B) \cup \mathbf{F}(A, B)$$
$$= \{x \in U : [x]_R \subseteq B \text{ or } ([x]_R \cap B \neq \emptyset \text{ and } [x]_R \cap (U - A - B) \neq \emptyset)\}$$
$$= \underline{R}B \cup (\overline{R}B \cap \overline{R}(U - A - B))$$
$$= \langle A, B \rangle^{L \approx} \sqcup (\langle A, B \rangle^{\approx L \approx -} \sqcap \left(\langle A, B \rangle^{\approx -} \sqcap \langle A, B \rangle^{-\approx -} \right)^{L \approx -}$$
$$= a^{A \sim} \vee \left(a^{\sim A \sim \prime} \wedge (a^{\sim \prime} \wedge a^{\prime \sim \prime})^{A \sim \prime} \right),$$
– *the at most sometimes true part* of $\langle A, B \rangle$:
$$\mathbf{s}T^{\downarrow}(A, B) = \mathbf{s}T(A, B) \cup \mathbf{U}(A, B) \cup \mathbf{K}(A, B) \cup \mathbf{f}K(A, B) \cup \mathbf{s}F(A, B) \cup \mathbf{F}(A, B)$$
$$= \{x \in U : [x]_R \cap (U - A) \neq \emptyset\} = \overline{R}(U - A) = \langle A, B \rangle^{L - \approx} = a^{A \prime \sim},$$
– *the at most true part* of $\langle A, B \rangle$:
$$\mathbf{T}^{\downarrow}(A, B) = U = \langle U, \emptyset \rangle = 1.$$

Let us consider the set of the upward truth value operators
$$\mathcal{O}^{\uparrow} = \{\mathbf{T}^{\uparrow}, \mathbf{s}T^{\uparrow}, \mathbf{U}^{\uparrow}, \mathbf{K}^{\uparrow}, \mathbf{f}K^{\uparrow}, \mathbf{s}F^{\uparrow}, \mathbf{F}^{\uparrow}\}$$
and the set of the downward truth value operators
$$\mathcal{O}^{\downarrow} = \{\mathbf{T}^{\downarrow}, \mathbf{s}T^{\downarrow}, \mathbf{U}^{\downarrow}, \mathbf{K}^{\downarrow}, \mathbf{f}K^{\downarrow}, \mathbf{s}F^{\downarrow}, \mathbf{F}^{\downarrow}\}.$$
On the basis of $\mathcal{O}^{\uparrow}$ and $\mathcal{O}^{\downarrow}$ one n-valued logic with respect to the knowledge base $K = (U, R)$ is defined by the set of truth value operators $(O_1, \ldots, O_n)$ such that, for all $A, B \subseteq U, A \cap B = \emptyset$, we have

– for all $i = 1, \ldots, n$
  - either $O_i(A, B) = M_1(A, B) \cup \ldots \cup M_k(A, B)$ with $M_1, \ldots, M_k \in \mathcal{O}^{\uparrow}$ or
  - $O_i(A, B) = M_1(A, B) \cup \ldots \cup M_k(A, B)$ with $M_1, \ldots, M_k \in \mathcal{O}^{\downarrow}$ or

- $O_i(A, B) = \left(M_1(A, B) \cup \ldots \cup M_h(A, B) \right) \cap \left(M_{h+1}(A, B) \cup \ldots \cup M_k(A, B) \right)$ with $M_1, \ldots, M_h \in \mathcal{O}^\uparrow$ and $M_{h+1}, \ldots, M_k \in \mathcal{O}^\downarrow$,

and
- for all $O_i, O_j \in (O_1, \ldots O_n)$, $O_i(A, B) \cap O_j(A, B) = \emptyset$,
- $\bigcup_{i=1}^n O_i(A, B) = U$.

Among the many posible logics that can be defined in this way, consider the Belnap four-valued logic [2] defined by the set of truth value operators

$$(\mathbf{T}_{Belnap}, \mathbf{U}_{Belnap}, \mathbf{K}_{Belnap}, \mathbf{F}_{Belnap})$$

with

- $\mathbf{T}_{Belnap}(A, B) = sT^\uparrow(A, B) = \mathbf{T}(A, B) \cup sT(A, B) = \overline{R}A \cap \underline{R}(U - B)$,
- $\mathbf{U}_{Belnap}(A, B) = \mathbf{U}(A, B) = \mathbf{U}^\uparrow(A, B) \cap \mathbf{U}^\downarrow(A, B) = \underline{R}(U - A - B)$,
- $\mathbf{K}_{Belnap}(A, B) = (\mathbf{K}^\uparrow(A, B) \cup \mathbf{f}K^\uparrow(A, B)) \cap (\mathbf{K}^\downarrow(A, B) \cup \mathbf{f}K^\downarrow(A, B)) = \mathbf{K}(A, B) \cup \mathbf{f}K(A, B) = \overline{R}A \cap \overline{R}B$,
- $\mathbf{F}_{Belnap}(A, B) = sF^\uparrow(A, B) = \mathbf{F}(A, B) \cup sF(A, B) = \overline{R}B \cap \underline{R}(U - A)$.

The above Belnap four-valued logic can be interpreted in terms of rough approximations as follows. Consider the knowledge base $K = (U, R)$, $x \in U$, and the concept $\langle A, B \rangle \in 3^U$. For all $x \in U$ we have

- $[x]_R \cap A \neq \emptyset$, i.e., $x \in \overline{R}A$, is an argument for truth,
- $[x]_R \cap B \neq \emptyset$, i.e., $x \in \overline{R}B$, is an argument for falsehood.

Consequently,

- if $x \in \overline{R}A$ and $x \notin \overline{R}B$, there are arguments for truth and there are no arguments for falsehood, so that $x \in \mathbf{T}_{Belnap}(A, B)$,
- if $x \notin \overline{R}A$ and $x \notin \overline{R}B$ (which is equivalent to $x \in \underline{R}(U - A - B)$), there are no arguments for truth and there are no arguments for falsehood, so that $x \in \mathbf{U}_{Belnap}(A, B)$,
- if $x \in \overline{R}A$ and $x \in \overline{R}B$, there are arguments for truth and there are arguments for falsehood, so that $x \in \mathbf{K}_{Belnap}(A, B)$,
- if $x \notin \overline{R}A$ and $x \in \overline{R}B$, there are no arguments for truth and there are arguments for falsehood, so that $x \in \mathbf{F}_{Belnap}(A, B)$.

Comparing the truth value operators of seven-valued logic, expressed in terms of upper approximations $\overline{R}A, \overline{R}B$ and $\overline{R}(U - A - B)$, with the Belnap's truth value operators of the four-valued logic reveals that the latter is different because it does not consider $\overline{R}(U - A - B)$.

4   Conclusions

The seven-valued logic considered in this paper naturally arises within the rough set framework, allowing to distinguish vagueness due to imprecision from ambiguity due to coarseness. We discussed the usefulness of this seven-valued logic for reasoning about data. We showed that the Pawlak-Brouwer-Zadeh lattice is the proper algebraic structure for this seven-valued logic. We proposed also a general framework permitting to obtain other interesting many-valued logics by aggregation of truth value operators in the basic seven-valued logic. We plan to continue our research in this direction, investigating typical rough set topics, such as calculation of reducts and rule induction. We intend also to extend the seven-valued logic to reasoning about ordered data using the dominance-based rough set approach [8] and the related bipolar Pawlak-Brouwer-Zadeh lattice [9,11]. Another interesting line of research we want to pursue is related to investigation of connections with other algebra models for rough sets such as Nelson algebra, Heyting algebra, Łukasiewicz algebra, Stone algebra and so on (see, e.g., Chap. 12 in [17]). We also propose to study the relations between the rough set approach we adopted to obtain the seven-valued logic and the other logics that can be derived from it, and the approach presented in [14] that derives four-valued logic from variable precision rough set model [20]. Finally, with respect to the seven-valued logic we proposed, we plan to investigate the tableau calculi for deduction systems, taking into account also soundness and completeness.

Acknowledgments. Salvatore Greco wishes to acknowledge the support of the Ministero dell'Istruzione, dell'Università e della Ricerca (MIUR) - PRIN 2017, project "Multiple Criteria Decision Analysis and Multiple Criteria Decision Theory", grant 2017CY2NCA. The research of Roman Słowiński was supported by the SBAD funding from the Polish Ministry of Education and Science. This research also contributes to the PNRR GRInS Project.

References

1. Belnap, N.D.: How a computer should think. In: Ryle, G. (ed.) Contemporary Aspects of Philosophy, pp. 30–56. Oriel Press, Boston (1976)
2. Belnap, N.D.: A useful four-valued logic. In: Dunn, J.M., Epstein, G. (eds.) Modern uses of multiple-valued logic, vol. 2, pp. 5–37. Springer, Dordrecht (1977). https://doi.org/10.1007/978-94-010-1161-7_2
3. Burch, G.B.: Seven-valued logic in Jain philosophy. Int. Phil. Q. **4**(1), 68–93 (1964)
4. Cattaneo, G.: Generalized rough sets (preclusivity fuzzy-intuitionistic (BZ) lattices). Stud. Logica. **58**, 47–77 (1997)
5. Cattaneo, G., Algebraic methods for rough approximation spaces by lattice interior–osure operations. In: Mani, A., Cattaneo, G., Düntsch, I. (eds.) Algebraic Methods in General Rough Sets, pp. 13–56. Birkhäuser, Cham (2018)
6. Cattaneo, G., Ciucci, D.: Algebraic structures for rough sets. Trans. Rough Sets II, LNCS **3135**, 208–252 (2004)
7. Cattaneo, G., Nisticò, G.: Brouwer-Zadeh posets and three-valued Łukasiewicz posets. Fuzzy Sets Syst. **33**, 165–190 (1989)

8. Greco, S., Matarazzo, B., Słowiński, R.: Rough set theory for multicriteria decision analysis. Eur. J. Oper. Res. **129**, 1–47 (2001)
9. Greco, S., Matarazzo, B., Słowiński, R.: The bipolar complemented de morgan brouwer-zadeh distributive lattice as an algebraic structure for the dominance-based rough set approach. Fund. Inf. **115**, 25–56 (2012)
10. Greco, S., Matarazzo, B., Słowiński, R.: Distinguishing vagueness from ambiguity by means of Pawlak-Brouwer-Zadeh lattices. In: Greco, S., Bouchon-Meunier, B., Coletti, G., Fedrizzi, M., Matarazzo, B., Yager, R.R. (eds.) IPMU 2012. CCIS, vol. 297, pp. 624–632. Springer, Heidelberg (2012). https://doi.org/10.1007/978-3-642-31709-5_63
11. Greco, S., Matarazzo, B., Słowiński, R.: Distinguishing vagueness from ambiguity in dominance-based rough set approach by means of a bipolar pawlak-brouwer-zadeh lattice. In: Polkowski, L., et al. (eds.) IJCRS 2017. LNCS (LNAI), vol. 10314, pp. 81–93. Springer, Cham (2017). https://doi.org/10.1007/978-3-319-60840-2_6
12. Greco, S., Matarazzo, B., Słowiński, R.: Distinguishing vagueness from ambiguity in rough set approximations. Int. J. Uncertain. Fuzziness Knowl.-Based Syst. **26**(Suppl. 2), 89–125 (2018)
13. Miller, G.A.: The magical number seven, plus or minus two: some limits on our capacity for processing information. Psychol. Rev. **63**(2), 81 (1956)
14. Nakayama, Y., Akama, S., Murai, T.: Four-valued tableau calculi for decision logic of rough set. Procedia Comput. Sci. **126**, 383–392 (2018)
15. Pawlak, Z.: Rough Sets. Int. J. Comput. Inf. Sci. **11**, 341–356 (1982)
16. Pawlak, Z.: Rough Sets. Kluwer, Dordrecht (1991)
17. Polkowski, L.: Rough Sets. Physica-Verlag, Heidelberg (2002)
18. Priest, G.: Jaina logic: a contemporary perspective. Hist. Phil. Logic **29**(3), 263–278 (2008)
19. Yao, Y.: The geometry of three-way decision. Appl. Intell. **51**(9), 6298–6325 (2021)
20. Ziarko, W.: Variable precision rough set model. J. Comput. Syst. Sci. **46**, 39–59 (1993)

Three-way Decisions

Three-Way Conflict Analysis
for Three-Valued Situation Tables
with Rankings and Reference Tuples

Shuting Liu[1], Mengjun Hu[2], Zhifei Zhang[3], and Guangming Lang[1(✉)]

[1] School of Mathematics and Statistics, Changsha University of Science and
Technology, Changsha 410114, Hunan, China
langguangming1984@126.com
[2] Department of Mathematics and Computing Science, Saint Mary's University,
Halifax, NS B3H 3C3, Canada
[3] Department of Computer Science and Technology, Tongji University,
Shanghai 201804, China

Abstract. Nowadays, the researches of conflict analysis are increasing
with the development of three-way decisions, and three-way decisions
with rankings and reference tuples provides new perspectives for conflict
analysis. In this paper, we first divide a whole set of issues into a bundle
of supported issues and a bundle of non-supported issues, and a bundle of
opposed issues and a bundle of non-opposed issues, respectively, from the
perspectives of support and opposition. Accordingly, we give two ranking orders and two reference tuples. Then, we put forward alliance and
conflict measures by considering weights of issues, and develop models of
three-way conflict analysis with ideas of rankings and reference tuples.
Finally, we show how to compute the alliance, neutral and conflict coalitions with the proposed models. It provides an attempt to study conflict
problems with thoughts of rankings and reference tuples.

Keywords: Conflict analysis · Ranking order · Reference tuple ·
Three-way decisions

1 Introduction

Three-way decisions, given by Yiyu Yao [23] in 2010, is a philosophy of thinking
and working with the thought of threes. It divides an object set into three regions
by using evaluation functions. Accordingly, three disjoint regions correspond to
three different actions. After that, researchers [5,6,21,25–27,29] have investigated and enriched three-way decisions in theoretical and application aspects.

Conflict analysis aims to study the essence of conflicts and give feasible strategies for solving conflicts. For example, Pawlak [11,12] depicted conflict problems
by a three-valued situation table, and defined alliance, conflict and neutral relations between two agents. Afterwards, Deja [1] took rough sets to construct models of conflict analysis and studied their applications. Recently, three-way conflict analysis [2–4,7–10,13,15–17,22,24,28,30] has attracted increasing amounts

© The Author(s), under exclusive license to Springer Nature Switzerland AG 2023
A. Campagner et al. (Eds.): IJCRS 2023, LNAI 14481, pp. 231–245, 2023.
https://doi.org/10.1007/978-3-031-50959-9_16

of attention. For example, Du et al. [2] designed effective models of conflict analysis by using Pythagorean fuzzy information. Feng, Yang and Guo [3] clustered agents whose attitudes are described by dual hesitant fuzzy numbers. Hu [4] unified conflict analysis models within the framework of subsethood measures. Lang and Yao [8] studied conflict problems by changing situation tables into formal contexts. Li, Qiao and Ding [9] supplied conflict resolutions by using q-rung orthopair fuzzy information. Suo and Yang [15] clustered agent pairs when attitudes of agents on issues are missing or lost. Yao [24] put forward three-way conflict analysis by combining three-way decisions with conflict analysis. Zhi, Li and Li [30] performed multi-level conflict analysis so as to build the maximal coalitions.

Another research direction is three-way decisions with rankings and reference tuples. In the year of 2021, Xu, Jia and Li [18] provided matching functions by combining a ranking of an attribute set and a reference tuple, and designed a two-universe model of three-way decisions with matching functions. Xu, Jia and Li [19] developed a generalized three-way decision model by using rankings and reference tuples, and studied how to get the optimal division. Xu, Jia and Li [20] introduced two three-way decision models for hybrid information tables by using rankings and references, and showed how to get the local and global optimal divisions. In three-way decisions with rankings and references, it first divides an attribute set into two disjoint parts, and gives a ranking order on the two disjoint parts and a reference tuple. After that, it discusses the relationship between an object and the reference tuple, and trisects an object set into three disjoint parts. That is, the two disjoint subsets of attributes help us to discuss the relationship between an object and the reference tuple from two different perspectives. To solve conflicts, Sun, Ma and Zhao [14] designed models of conflict analysis based on rough sets on two universes. When it divides a set of agents into three disjoint parts, they only consider the relationship between agents and a subset of issues, and does not consider the relationship between agents and its complement. If we neglect the relationship between agents and the rest issues, it will result in some imprecise partitions of the set of agents. It motivates us to divide a whole set of issues into a bundle of supported issues and a bundle of non-supported issues, and a bundle of opposed issues and a bundle of non-opposed issues, respectively, in terms of support and opposition. The two divisions of the set of issues help us to describe the relationship between an agent and the reference tuple from two different perspectives. That is, three-way decisions with rankings and reference tuples gives a new and useful tool for conflict analysis.

For conflict problems depicted by three-valued situation tables, we design three-way conflict analysis models with thoughts of rankings and reference tuples. The contributions are briefly summarized as follows:

(1) From the perspective of support, we divide a whole set of issues into a bundle of supported issues and a bundle of non-supported issues. Similarly, from the perspective of opposition, a set of issues is divided into a bundle of opposed issues and a bundle of non-opposed issues. That is, a whole set of issues is divided into two disjoint parts in terms of support and opposition.

Correspondingly, we define two ranking orders for the two divisions of the set of issues and the support and opposition reference tuples.

(2) We define alliance and conflict measures by considering weights of issues, two ranking orders, and the support and opposition reference tuples. Afterwards, we discuss the relationship between an agent and the support and opposition references. Afterwards, we develop two three-way conflict analysis models by considering rankings and reference tuples, and design two algorithms for dividing an agent set with regard to the support and opposition references. We employ an example to show how to compute the alliance, neutral and conflict coalitions with the two proposed models.

The rest of this paper is listed as follows: Sect. 2 reviews three-way decisions with rankings and reference tuples. Section 3 provides three-way conflict analysis models by considering rankings and reference tuples. Section 4 shows how to use the proposed models with an example. Section 5 gives the conclusion.

2   Preliminaries

In this section, we recall three-way decisions by using rankings and references.

Definition 1 *(Xu [18], 2021). An information table is a quadruple $S = (U, C, V, f)$, where U is an object set, C is an attribute set, and $V = \bigcup \{V_c \mid c \in C\}$, in which V_c stands for a range of attribute values of all objects on c, and $f : U \times C \to V$.*

For any $c \in C$, if the set of attribute values V_c only contains two values 0 and 1, we refer to this type of information tables as two-valued information tables. That is, all objects take attribute values from the set $\{0, 1\}$.

Definition 2 *(Xu [18], 2021). For $S = (U, C, V, f)$, a ranking order $\preceq_r$ divides set C into disjoint sets C_h and C_l such that $C = C_h \cup C_l$ and $C_l \preceq_r C_h$, in which C_h and C_l stand for the high-level and low-level sets of attributes, respectively.*

The set C is divided into two disjoint parts C_h and C_l by the ranking order $\preceq_r$, and $C_l \preceq_r C_h$ means that the ranking of C_h is higher than that of C_l.

Definition 3 *(Xu [18], 2021). For $S = (U, C, V, f)$, an m-tuple $\mathbf{x_r} = (f(x_r, c_1), f(x_r, c_2), \ldots, f(x_r, c_m))$ is referred to as a reference tuple on C if: (1) $\forall c_i \in C$, $f(x_r, c_i) \in \{0, 1\}$; (2) $f(x_r, c_i) = f(x_r, c_j)$ if $\{c_i, c_j\} \subseteq C_h$ or $\{c_i, c_j\} \subseteq C_l$, where $1 \leq i, j \leq m$.*

Example 1. We give an information table with a reference tuple by Table 1, in which $U = \{x_1, x_2, x_3, x_4, x_5, x_6\}$, and $C = \{c_1, c_2, c_3, c_4, c_5\}$. It divides set C into $C_h = \{c_2, c_3, c_4\}$ and $C_l = \{c_1, c_6\}$, and shows a reference tuple $\mathbf{x_r} = (f(x_r, c_1), f(x_r, c_2), f(x_r, c_3), f(x_r, c_4), f(x_r, c_5)) = (0, 1, 1, 1, 0)$.

Table 1. An Information Table.

U	C				
	c_1	c_2	c_3	c_4	c_5
x_1	1	1	1	1	1
x_2	0	0	1	1	1
x_3	1	1	1	1	0
x_4	0	1	1	0	1
x_5	1	1	1	1	1
x_6	0	1	1	0	1
x_r	0	1	1	1	0

In Table 1, Xu et al. [18] divided all attributes into desirable and undesirable attributes, in which c_2, c_3 and c_4 are desirable attributes, and c_1 and c_5 are undesirable attributes. Therefore, the ranking of $\{c_2, c_3, c_4\}$ is higher than that of $\{c_1, c_5\}$ towards the degree of desirable.

Definition 4 *(Xu [18], 2021). For $S = (U, C, V, f)$, and a reference tuple $\mathbf{x_r} = (f(x_r, c_1), f(x_r, c_2), ..., f(x_r, c_m))$, the matching degrees of x and x_r with respect to C_h and C_l are given by:*

$$M[(x, x_r), C_h] = \frac{\sum_{c_i \in C_h} m[(x, x_r), c_i]}{\#(C_h)},$$

$$M[(x, x_r), C_l] = \frac{\sum_{c_i \in C_l} m[(x, x_r), c_i]}{\#(C_l)},$$

where

$$m[(x, x_r), c_i] = \begin{cases} 1, & f(x, c_i) = f(x_r, c_i), \\ 0, & f(x, c_i) \neq f(x_r, c_i), \end{cases}$$

and $\#(C_h)$ and $\#(C_l)$ stand for the cardinalities of C_h and C_l, respectively.

In practice, there are some special ranking orders such as $C_h = \emptyset \wedge C_l = C$ and $C_h = C \wedge C_l = \emptyset$. For the two ranking orders, we have the matching degrees between $x \in U$ and x_r as follows: (1) if $C_h = \emptyset$ and $C_l = C$, then $M[(x, x_r), C_h] = 1$; (2) if $C_h = C$ and $C_l = \emptyset$, then $M[(x, x_r), C_l] = 0$.

Definition 5 *(Xu [18], 2021). For $S = (U, C, V, f)$, a reference tuple $\mathbf{x_r} = (f(x_r, c_1), f(x_r, c_2), ..., f(x_r, c_m))$, a ranking order $\preceq_r$ with $C_l \preceq_r C_h$, and two thresholds α and β such that $0 \leq \alpha, \beta \leq 1$, the positive, negative and boundary regions are given by:*

$$POS_{(\alpha,\beta)}(U, \mathbf{x_r}) = \{x \in U \mid M[(x, x_r), C_h] \geq \alpha \wedge M[(x, x_r), C_l] > \beta\},$$
$$NEG_{(\alpha,\beta)}(U, \mathbf{x_r}) = \{x \in U \mid M[(x, x_r), C_h] < \alpha \wedge M[(x, x_r), C_l] \leq \beta\},$$
$$BND_{(\alpha,\beta)}(U, \mathbf{x_r}) = \{x \in U \mid M[(x, x_r), C_h] \geq \alpha \wedge M[(x, x_r), C_l] \leq \beta\} \cup$$
$$\{x \in U \mid M[(x, x_r), C_h] < \alpha \wedge M[(x, x_r), C_l] > \beta\}.$$

An object set is divided into three disjoint regions with two matching functions. Inspired by three-way decisions with rankings and reference tuples, we will study conflict problems by considering ranking orders and reference tuples.

3 Three-Way Conflict Analysis Models by Considering Rankings and Reference Tuples

3.1 Two Rankings of Issue Sets and Two Reference Tuples

In the year of 1998, Pawlak [12] depicted conflict problems by three-valued situation tables.

Definition 6 *(Pawlak [12], 1998). A triplet $S = (A, I, r)$ is a three-valued situation table, in which A is an agent set, I is an issue set, and $r : A \times I \rightarrow \{+, 0, -\}$. Here, $r(a, i) = +$ implies that a supports i; $r(a, i) = 0$ implies that a holds neutral attitude on i; $r(a, i) = -$ implies that a opposes i.*

Example 2 (Pawlak [12], 1998). Table 2 depicts the Middle East conflict, in which a_1-Israel, a_2-Egypt, a_3-Palsetine, a_4-Jordan, a_5-Syria and a_6-Saudi Arabia; i_1-Autonomous Palestinian state on the West Bank and Gaza, i_2-Israeli military outpost along the Jordan River, i_3-Israel retains East Jerusalem, i_4-Military outposts on the Golan Heights and i_5-Arab countries grant citizenship to Palestinians who choose to remain within their borders.

Table 2. The Middle East Conflict [12].

A	i_1	i_2	i_3	i_4	i_5
a_1	−	+	+	+	+
a_2	+	0	−	−	−
a_3	+	−	−	−	0
a_4	0	−	+	0	−
a_5	+	−	−	−	−
a_6	0	+	−	0	+

Definition 7. *For a three-valued situation table $S = (A, I, r)$,*

(1) *a ranking order $\preceq_s$ divides the issue set I into two disjoint subsets I_s and $I_{\bar{s}}$ such that $I = I_s \cup I_{\bar{s}}$ and $I_{\bar{s}} \preceq_s I_s$, in which I_s and $I_{\bar{s}}$ are called a bundle of supported issues and a bundle of non-supported issues, respectively.*

(2) *a ranking order $\preceq_o$ divides the issue set I into two disjoint subsets I_o and $I_{\bar{o}}$ such that $I = I_o \cup I_{\bar{o}}$ and $I_{\bar{o}} \preceq_o I_o$, in which I_o and $I_{\bar{o}}$ are called a bundle of opposed issues and a bundle of non-opposed issues, respectively.*

We divide a whole set of issues into a bundle of supported issues and a bundle of non-supported issues, and the ranking of the bundle of supported issues is higher than the bundle of non-supported issues from the perspective of support. Moreover, we divide a whole set of issues into a bundle of opposed issues and a bundle of non-opposed issues, and the ranking of the bundle of non-opposed issues is lower than the bundle of opposed issues from the perspective of opposition.

Definition 8. *For a three-valued situation table* $S = (A, I, r)$,

(1) *an m-tuple* $\mathbf{a_r^s} = (r(a_r^s, i_1), r(a_r^s, i_2), ..., r(a_r^s, i_m))$ *is called a support reference tuple on I if it satisfies: (1)* $\forall i \in I_s$, $r(a_r^s, i) = +$*; (2)* $\forall i \in I_{\bar{s}}$, $r(a_r^s, i) \in \{0, -\}$*. We refer to a_r^s as a support reference.*
(2) *an m-tuple* $\mathbf{a_r^o} = (r(a_r^o, i_1), r(a_r^o, i_2), ..., r(a_r^o, i_m))$ *is called an opposition reference tuple on I if it satisfies: (1)* $\forall i \in I_o$, $r(a_r^o, i) = -$*; (2)* $\forall i \in I_{\bar{o}}$, $r(a_r^o, i) \in \{+, 0\}$*. We refer to a_r^o as an opposition reference.*

The support reference divides a set of issues into a bundle of supported issues and a bundle of non-supported issues. That is, the support reference corresponds to a supported plan that takes care of the supported issues and does not care of the non-supported issues by an expert or a government. Furthermore, the opposition reference divides a set of issues into a bundle of opposed issues and a bundle of non-opposed issues. That is, the opposition reference corresponds to an opposed plan that takes care of the opposed issues and does not care of the non-opposed issues by an expert or a government.

Example 3. By Example 2, we give the support and opposition references by Tables 3 as follows:

(1) By Definition 7(1), we have the supported issues $I_s = \{i_1, i_3\}$, and the non-supported issues $I_{\bar{s}} = \{i_2, i_4, i_5\}$. After that, by Definition 8(1), we have the support reference tuple $\mathbf{a_r^s} = (+, 0, +, -, 0)$. That is, the support reference a_r^s supports issues i_1 and i_3; but the support reference a_r^s does not support issues i_2, i_4 and i_5.
(2) By Definition 7(2), we give the opposed issues $I_o = \{i_1, i_3\}$, and the non-opposed issues $I_{\bar{o}} = \{i_2, i_4, i_5\}$. Afterwards, by Definition 8(2), we have the opposition reference tuple $\mathbf{a_r^o} = (-, 0, -, +, 0)$. That is, the opposition reference a_r^o opposes issues i_1 and i_3; but the opposition reference a_r^o does not oppose issues i_2, i_4 and i_5.

3.2   Alliance and Conflict Measures with Weight Factors

For a three-valued situation table, there are nine pairs of attitudes, namely, $(+, +), (+, 0), (+, -), (0, -), (0, 0), (0, +), (-, -), (-, 0)$ and $(-, +)$, in total.

Definition 9 *(Lang [7], 2021). For a three-valued situation table* $S = (A, I, r)$,

Table 3. A Situation Table With the Support and Opposition References.

A	I				
	i_1	i_2	i_3	i_4	i_5
a_1	$-$	$+$	$+$	$+$	$+$
a_2	$+$	0	$-$	$-$	$-$
a_3	$+$	$-$	$-$	$-$	0
a_4	0	$-$	$+$	0	$-$
a_5	$+$	$-$	$-$	$-$	$-$
a_6	0	$+$	$-$	0	$+$
a_r^s	$+$	0	$+$	$-$	0
a_r^o	$-$	0	$-$	$+$	0

(1) *a function $e^{\mathrm{a}} : \{+,0,-\} \times \{+,0,-\} \longrightarrow [0,1]$ is referred to as an alliance measure when:*

(a) $e^{\mathrm{a}}(+,+) \geq e^{\mathrm{a}}(0,0) > e^{\mathrm{a}}(+,0) = e^{\mathrm{a}}(0,+) > e^{\mathrm{a}}(+,-) = e^{\mathrm{a}}(-,+),$

(b) $e^{\mathrm{a}}(+,+) \geq e^{\mathrm{a}}(0,0) > e^{\mathrm{a}}(-,0) = e^{\mathrm{a}}(0,-) > e^{\mathrm{a}}(+,-) = e^{\mathrm{a}}(-,+),$

(c) $e^{\mathrm{a}}(-,-) \geq e^{\mathrm{a}}(0,0) > e^{\mathrm{a}}(+,0) = e^{\mathrm{a}}(0,+) > e^{\mathrm{a}}(+,-) = e^{\mathrm{a}}(-,+),$

(d) $e^{\mathrm{a}}(-,-) \geq e^{\mathrm{a}}(0,0) > e^{\mathrm{a}}(-,0) = e^{\mathrm{a}}(0,-) > e^{\mathrm{a}}(+,-) = e^{\mathrm{a}}(-,+).$

(2) *a function $e^{\mathrm{c}} : \{+,0,-\} \times \{+,0,-\} \longrightarrow [0,1]$ is referred to as a conflict measure when:*

(a) $e^{\mathrm{c}}(+,+) \leq e^{\mathrm{c}}(0,0) < e^{\mathrm{c}}(+,0) = e^{\mathrm{c}}(0,+) < e^{\mathrm{c}}(+,-) = e^{\mathrm{c}}(-,+),$

(b) $e^{\mathrm{c}}(+,+) \leq e^{\mathrm{c}}(0,0) < e^{\mathrm{c}}(-,0) = e^{\mathrm{c}}(0,-) < e^{\mathrm{c}}(+,-) = e^{\mathrm{c}}(-,+),$

(c) $e^{\mathrm{c}}(-,-) \leq e^{\mathrm{c}}(0,0) < e^{\mathrm{c}}(+,0) = e^{\mathrm{c}}(0,+) < e^{\mathrm{c}}(+,-) = e^{\mathrm{c}}(-,+),$

(d) $e^{\mathrm{c}}(-,-) \leq e^{\mathrm{c}}(0,0) < e^{\mathrm{c}}(-,0) = e^{\mathrm{c}}(0,-) < e^{\mathrm{c}}(+,-) = e^{\mathrm{c}}(-,+).$

Example 4. Yao [24] gives a function $d_i : A \times A \longrightarrow \{0, 0.5, 1\}$ with regard to $i \in I$ on A: for $x, y \in A$,

$$d_i(x,y) = \frac{|r(x,i) - r(y,i)|}{2},$$

where $|\cdot|$ denotes the absolute value. After that, we provide a pair of alliance and conflict measures $e^{\mathrm{a}} : \{+,0,-\} \times \{+,0,-\} \longrightarrow \{0, 0.5, 1\}$ and $e^{\mathrm{c}} : \{+,0,-\} \times \{+,0,-\} \longrightarrow \{0, 0.5, 1\}$ by Tables 4 and 5, respectively.

Table 4. The Alliance Measure $e^{\mathrm{a}}(\cdot,\cdot)$.

$e^{\mathrm{a}}(\cdot,\cdot)$	$+$	0	$-$
$+$	1.0	0.5	0.0
0	0.5	1.0	0.5
$-$	0.0	0.5	1.0

Table 5. The Conflict Measure $e^c(\cdot, \cdot)$.

$e^c(\cdot, \cdot)$	$+$	0	$-$
$+$	0.0	0.5	1.0
0	0.5	0.0	0.5
$-$	1.0	0.5	0.0

For conflict problems, there are different weights for different issues, and it must consider weights of issues when designing alliance and conflict measures.

Definition 10 *(Lang [7], 2021). For a three-valued situation table $S = (A, I, r)$,*

(1) *let w_i be the weight of $i \in I$ for alliance, in which $\sum_{i \in I} w_i = 1$, and $0 \leq w_i \leq 1$. The conditional weight $w(i|J)$ for alliance with regard to $J \subseteq I$ is given by:*

$$w(i|J) = \frac{w_i}{\sum_{j \in J} w_j};$$

(2) *let κ_i be the weight of $i \in I$ for conflict, in which $\sum_{i \in I} \kappa_i = 1$, and $0 \leq \kappa_i \leq 1$. The conditional weight $\kappa(i|J)$ of $i \in J$ for conflict with regard to $J \subseteq I$ is given by:*

$$\kappa(i|J) = \frac{\kappa_i}{\sum_{j \in J} \kappa_j}.$$

The conditional weight $w(i|J)$ means the weight of an issue i with regard to alliance, and the conditional weight $\kappa(i|J)$ means the weight of an issue i with regard to conflict.

Definition 11. *For a three-valued situation table $S = (A, I, r)$,*

(1) *the alliance measures $E^a[(a, a_r^s), I_s]$ and $E^a[(a, a_r^s), I_{\bar{s}}]$ with respect to I_s and $I_{\bar{s}}$, respectively, are defined by:*

$$E^a[(a, a_r^s), I_s] = \sum_{i \in I_s} [w(i|I_s) \times e_i^a(a, a_r^s)],$$

$$E^a[(a, a_r^s), I_{\bar{s}}] = \sum_{i \in I_{\bar{s}}} [w(i|I_{\bar{s}}) \times e_i^a(a, a_r^s)];$$

(2) *the conflict measures $E^c[(a, a_r^o), I_o]$ and $E^c[(a, a_r^s), I_{\bar{o}}]$ with respect to I_o and $I_{\bar{o}}$, respectively, are defined by:*

$$E^c[(a, a_r^o), I_o] = \sum_{i \in I_o} [\kappa(i|I_o) \times e_i^c(a, a_r^o)],$$

$$E^c[(a, a_r^o), I_{\bar{o}}] = \sum_{i \in I_{\bar{o}}} [\kappa(i|I_{\bar{o}}) \times e_i^c(a, a_r^o)].$$

The alliance measures $E^a[(a, a_r^s), I_s]$ and $E^a[(a, a_r^s), I_{\bar{s}}]$ depict the matching degrees between $a \in A$ and a_r^s with respect to I_s and $I_{\bar{s}}$, respectively. In addition, if $I_s = \emptyset$ and $I_{\bar{s}} = I$, then $E^a[(a, a_r^s), I_s] = 1$; if $I_s = I$ and $I_{\bar{s}} = \emptyset$, then $E^a[(a, a_r^s), I_{\bar{s}}] = 0$. Therefore, we have $0 \leq E^a[(a, a_r^s), I_s] \leq 1$ and $0 \leq E^a[(a, a_r^s), I_{\bar{s}}] \leq 1$. Furthermore, we use $1 - E^c[(a, a_r^o), I_o]$ and $1 - E^c[(a, a_r^o), I_{\bar{o}}]$ to depict the matching degrees between $a \in A$ and a_r^o with respect to I_o and $I_{\bar{o}}$, respectively. In addition, if $I_o = \emptyset$ and $I_{\bar{o}} = I$, then $E^c[(a, a_r^o), I_o] = 1$; if $I_o = I$ and $I_{\bar{o}} = \emptyset$, then $E^c[(a, a_r^o), I_{\bar{o}}] = 0$. Therefore, we have $0 \leq E^c[(a, a_r^o), I_o] \leq 1$ and $0 \leq E^c[(a, a_r^o), I_{\bar{o}}] \leq 1$.

Definition 12. *For a three-valued situation table $S = (A, I, r)$, two thresholds $0 \leq \alpha, \beta \leq 1$,*

(1) *the alliance coalition, neutral coalition and conflict coalition of the support reference a_r^s with respect to I_s and $I_{\bar{s}}$ are given by:*

$$AL_{(\alpha,\beta)}^s(A, a_r^s) = \{a \in A \mid E^a[(a, a_r^s), I_s] \geq \alpha \wedge E^a[(a, a_r^s), I_{\bar{s}}] > \beta\},$$
$$CO_{(\alpha,\beta)}^s(A, a_r^s) = \{a \in A \mid E^a[(a, a_r^s), I_s] < \alpha \wedge E^a[(a, a_r^s), I_{\bar{s}}] \leq \beta\},$$
$$NE_{(\alpha,\beta)}^s(A, a_r^s) = \{a \in A \mid E^a[(a, a_r^s), I_s] \geq \alpha \wedge E^a[(a, a_r^s), I_{\bar{s}}] \leq \beta\} \cup$$
$$\{a \in A \mid E^a[(a, a_r^s), I_s] < \alpha \wedge E^a[(a, a_r^s), I_{\bar{s}}] > \beta\};$$

(2) *the alliance coalition, neutral coalition and conflict coalition of the opposition reference a_r^o with respect to I_o and $I_{\bar{o}}$ are given by:*

$$AL_{(\alpha,\beta)}^o(A, a_r^o) = \{a \in A \mid 1 - E^c[(a, a_r^o), I_o] \geq \alpha \wedge 1 - E^c[(a, a_r^o), I_{\bar{o}}] > \beta\},$$
$$CO_{(\alpha,\beta)}^o(A, a_r^o) = \{a \in A \mid 1 - E^c[(a, a_r^o), I_o] < \alpha \wedge 1 - E^c[(a, a_r^o), I_{\bar{o}}] \leq \beta\},$$
$$NE_{(\alpha,\beta)}^o(A, a_r^o) = \{a \in A \mid 1 - E^c[(a, a_r^o), I_o] \geq \alpha \wedge 1 - E^c[(a, a_r^o), I_{\bar{o}}] \leq \beta\} \cup$$
$$\{a \in A \mid 1 - E^c[(a, a_r^o), I_o] < \alpha \wedge 1 - E^c[(a, a_r^o), I_{\bar{o}}] > \beta\}.$$

We divide a set of agents into alliance coalition $AL_{(\alpha,\beta)}^s(A, a_r^s)$, neutral coalition $NE_{(\alpha,\beta)}^s(A, a_r^s)$ and conflict coalition $CO_{(\alpha,\beta)}^s(A, a_r^s)$ of the support reference a_r^s. Furthermore, we divide a set of agents into alliance coalition $AL_{(\alpha,\beta)}^o(A, a_r^o)$, neutral coalition $NE_{(\alpha,\beta)}^o(A, a_r^o)$ and conflict coalition $CO_{(\alpha,\beta)}^o(A, a_r^o)$ of the opposition reference a_r^o.

4   An Example

Follows, we introduce the instances of three-way conflict analysis model with support, opposition rankings and reference tuples.

Next, we employ an example to show how to compute the alliance, neutral and conflict coalitions with the proposed models.

Example 5 (Continuation from Example 2). By using Tables 6 and 7, we show how to compute the alliance, neutral and conflict coalitions of the support and opposition references as follows:

Algorithm 1. The algorithm of constructing the alliance coalition, neutral coalition and conflict coalition by using the support ranking and reference tuple.

Input: $S = (A, I, r)$ and $a_{\mathbf{r}}^{\mathbf{s}}$;
Output: $AL_{(\alpha,\beta)}^{\mathbf{s}}(A, a_{\mathbf{r}}^{\mathbf{s}})$, $NE_{(\alpha,\beta)}^{\mathbf{s}}(A, a_{\mathbf{r}}^{\mathbf{s}})$ and $CO_{(\alpha,\beta)}^{\mathbf{s}}(A, a_{\mathbf{r}}^{\mathbf{s}})$.

1: Determine alliance degrees of nine pairs of attitudes;
2: Divide issue set I into a bundle of supported issues $I_{\mathbf{s}}$ and a bundle of non-supported issues $I_{\bar{\mathbf{s}}}$;
3: Compute weights $\omega(i|I_{\mathbf{s}})$ and $\omega(i|I_{\bar{\mathbf{s}}})$ for $i \in I$;
4: Calculate alliance degrees $E^{\mathbf{a}}[(a, a_{\mathbf{r}}^{\mathbf{s}}), I_{\mathbf{s}}]$ and $E^{\mathbf{a}}[(a, a_{\mathbf{r}}^{\mathbf{s}}), I_{\bar{\mathbf{s}}}]$ between $a \in A$ and $a_{\mathbf{r}}^{\mathbf{s}}$;
5: Construct the alliance, neutral and conflict coalitions $AL_{(\alpha,\beta)}^{\mathbf{s}}(A, a_{\mathbf{r}}^{\mathbf{s}})$, $NE_{(\alpha,\beta)}^{\mathbf{s}}(A, a_{\mathbf{r}}^{\mathbf{s}})$ and $CO_{(\alpha,\beta)}^{\mathbf{s}}(A, a_{\mathbf{r}}^{\mathbf{s}})$ with respect to $a_{\mathbf{r}}^{\mathbf{s}}$;
6: Output $AL_{(\alpha,\beta)}^{\mathbf{s}}(A, a_{\mathbf{r}}^{\mathbf{s}})$, $NE_{(\alpha,\beta)}^{\mathbf{s}}(A, a_{\mathbf{r}}^{\mathbf{s}})$ and $CO_{(\alpha,\beta)}^{\mathbf{s}}(A, a_{\mathbf{r}}^{\mathbf{s}})$.

Algorithm 2. The algorithm of computing alliance coalition, neutral coalition and conflict coalition by using the opposition ranking and reference tuple.

Input: $S = (A, I, r)$ and $a_{\mathbf{r}}^{\mathbf{o}}$;
Output: $AL_{(\alpha,\beta)}^{\mathbf{o}}(A, a_{\mathbf{r}}^{\mathbf{o}})$, $NE_{(\alpha,\beta)}^{\mathbf{o}}(A, a_{\mathbf{r}}^{\mathbf{o}})$ and $CO_{(\alpha,\beta)}^{\mathbf{o}}(A, a_{\mathbf{r}}^{\mathbf{o}})$.

1: Determine conflict degrees of nine pairs of attitudes;
2: Divide issue set I into a bundle of opposed issues $I_{\mathbf{o}}$ and a bundle of non-opposed issues $I_{\bar{\sigma}}$;
3: Compute weights $\omega(i|I_{\mathbf{o}})$ and $\omega(i|I_{\bar{\sigma}})$ for $i \in I$;
4: Calculate two conflict degrees $E^{\mathbf{c}}[(a, a_{\mathbf{r}}^{\mathbf{o}}), I_{\mathbf{o}}]$ and $E^{\mathbf{c}}[(a, a_{\mathbf{r}}^{\mathbf{o}}), I_{\bar{\sigma}}]$ between $a \in A$ and $a_{\mathbf{r}}^{\mathbf{o}}$;
5: Construct the alliance, neutral and conflict coalitions $AL_{(\alpha,\beta)}^{\mathbf{o}}(A, a_{\mathbf{r}}^{\mathbf{o}})$, $NE_{(\alpha,\beta)}^{\mathbf{o}}(A, a_{\mathbf{r}}^{\mathbf{o}})$ and $CO_{(\alpha,\beta)}^{\mathbf{o}}(A, a_{\mathbf{r}}^{\mathbf{o}})$ with respect to $a_{\mathbf{r}}^{\mathbf{o}}$;
6: Output $AL_{(\alpha,\beta)}^{\mathbf{o}}(A, a_{\mathbf{r}}^{\mathbf{o}})$, $NE_{(\alpha,\beta)}^{\mathbf{o}}(A, a_{\mathbf{r}}^{\mathbf{o}})$ and $CO_{(\alpha,\beta)}^{\mathbf{o}}(A, a_{\mathbf{r}}^{\mathbf{o}})$.

Table 6. The Alliance Measure $e^{\mathbf{a}}(\cdot, \cdot)$ for Table 3.

$e^{\mathbf{a}}(\cdot, \cdot)$	+	0	−
+	1	1/3	0
0	1/3	2/3	1/3
−	0	1/3	5/6

Table 7. The Conflict Measure $e^{\mathbf{c}}(\cdot, \cdot)$ for Table 3.

$e^{\mathbf{c}}(\cdot, \cdot)$	+	0	−
+	0	1/2	1
0	1/2	1/3	1/2
−	1	1/2	1/4

(1) Table 3 gives a support reference a_r^s. According to Definition 9, by taking $w_{i_1} = w_{i_2} = 0.1, w_{i_3} = w_{i_5} = 0.3, w_{i_4} = 0.2, I_s = \{i_1, i_3\}$ and $I_{\bar{s}} = \{i_2, i_4, i_5\}$, we have alliance degrees between a_i $(1 \leq i \leq 6)$ and a_r^s with respect to I_s and $I_{\bar{s}}$:

$$E^a[(a_1, a_r^s), I_s] = \frac{0.1}{0.4} \times 0 + \frac{0.3}{0.4} \times 1 = \frac{3}{4},$$

$$E^a[(a_2, a_r^s), I_s] = \frac{0.1}{0.4} \times 1 + \frac{0.3}{0.4} \times 0 = \frac{1}{4},$$

$$E^a[(a_3, a_r^s), I_s] = \frac{0.1}{0.4} \times 1 + \frac{0.3}{0.4} \times 0 = \frac{1}{4},$$

$$E^a[(a_4, a_r^s), I_s] = \frac{0.1}{0.4} \times \frac{1}{3} + \frac{0.3}{0.4} \times 1 = \frac{5}{6},$$

$$E^a[(a_5, a_r^s), I_s] = \frac{0.1}{0.4} \times 1 + \frac{0.3}{0.4} \times 0 = \frac{1}{4},$$

$$E^a[(a_6, a_r^s), I_s] = \frac{0.1}{0.4} \times \frac{1}{3} + \frac{0.3}{0.4} \times 0 = \frac{1}{12};$$

$$E^a[(a_1, a_r^s), I_{\bar{s}}] = \frac{0.1}{0.6} \times \frac{1}{3} + \frac{0.2}{0.6} \times 0 + \frac{0.3}{0.6} \times \frac{1}{3} = \frac{2}{9},$$

$$E^a[(a_2, a_r^s), I_{\bar{s}}] = \frac{0.1}{0.6} \times \frac{2}{3} + \frac{0.2}{0.6} \times \frac{5}{6} + \frac{0.3}{0.6} \times \frac{1}{3} = \frac{5}{9},$$

$$E^a[(a_3, a_r^s), I_{\bar{s}}] = \frac{0.1}{0.6} \times \frac{1}{3} + \frac{0.2}{0.6} \times \frac{5}{6} + \frac{0.3}{0.6} \times \frac{1}{3} = \frac{2}{3},$$

$$E^a[(a_4, a_r^s), I_{\bar{s}}] = \frac{0.1}{0.6} \times \frac{1}{3} + \frac{0.2}{0.6} \times \frac{1}{3} + \frac{0.3}{0.6} \times \frac{1}{3} = \frac{1}{3},$$

$$E^a[(a_5, a_r^s), I_{\bar{s}}] = \frac{0.1}{0.6} \times \frac{1}{3} + \frac{0.2}{0.6} \times \frac{5}{6} + \frac{0.3}{0.6} \times \frac{1}{3} = \frac{1}{2},$$

$$E^a[(a_6, a_r^s), I_{\bar{s}}] = \frac{0.1}{0.6} \times \frac{1}{3} + \frac{0.2}{0.6} \times \frac{1}{3} + \frac{0.3}{0.6} \times \frac{1}{3} = \frac{1}{3}.$$

After that, by taking $\alpha = \frac{1}{12}$ and $\beta = \frac{2}{9}$, we have $AL_{(\alpha,\beta)}^s(A, a_r^s) = \{a_2, a_3, a_4, a_5, a_6\}$, $NE_{(\alpha,\beta)}^s(A, a_r^s) = \{a_1\}$ and $CO_{(\alpha,\beta)}^s(A, a_r^s) = \emptyset$. Meanwhile, by taking other 19 pairs of thresholds, we compute three coalitions of the support reference with respect to I_s and $I_{\bar{s}}$ and show all results by Table 8.

(2) Table 3 displays an opposition reference a_r^o. According to Definition 11, by taking $\kappa_{i_1} = \kappa_{i_2} = 0.1, \kappa_{i_3} = \kappa_{i_5} = 0.3, \kappa_{i_4} = 0.2, I_o = \{i_1, i_3\}$ and $I_{\bar{o}} = \{i_2, i_4, i_5\}$, we have conflict degrees between a_i $(1 \leq i \leq 6)$ and a_r^o with respect to I_o and $I_{\bar{o}}$:

Table 8. The Alliance, Conflict and Neutral Coalitions of a_r^s in Middle East Conflicts.

(α,β)	$AL_{(\alpha,\beta)}^s(A,a_r^s)$	$CO_{(\alpha,\beta)}^s(A,a_r^s)$	$NE_{(\alpha,\beta)}^s(A,a_r^s)$
$(\frac{1}{12},\frac{2}{9})$	$\{a_2,a_3,a_4,a_5,a_6\}$	$\emptyset$	$\{a_1\}$
$(\frac{1}{12},\frac{1}{3})$	$\{a_2,a_3,a_5\}$	$\emptyset$	$\{a_1,a_4,a_6\}$
$(\frac{1}{12},\frac{1}{2})$	$\{a_2,a_3\}$	$\emptyset$	$\{a_1,a_4,a_5,a_6\}$
$(\frac{1}{12},\frac{5}{9})$	$\{a_3\}$	$\emptyset$	$\{a_1,a_2,a_4,a_5,a_6\}$
$(\frac{1}{12},\frac{2}{3})$	$\emptyset$	$\emptyset$	$\{a_1,a_2,a_3,a_4,a_5,a_6\}$
$(\frac{1}{4},\frac{2}{9})$	$\{a_2,a_3,a_4,a_5\}$	$\emptyset$	$\{a_1,a_6\}$
$(\frac{1}{4},\frac{1}{3})$	$\{a_2,a_3,a_5\}$	$\{a_6\}$	$\{a_1,a_4\}$
$(\frac{1}{4},\frac{1}{2})$	$\{a_2,a_3\}$	$\{a_6\}$	$\{a_1,a_4,a_5\}$
$(\frac{1}{4},\frac{5}{9})$	$\{a_3\}$	$\{a_6\}$	$\{a_1,a_2,a_4,a_5\}$
$(\frac{1}{4},\frac{2}{3})$	$\emptyset$	$\{a_6\}$	$\{a_1,a_2,a_3,a_4,a_5\}$
$(\frac{3}{4},\frac{2}{9})$	$\{a_4\}$	$\emptyset$	$\{a_1,a_2,a_3,a_5,a_6\}$
$(\frac{3}{4},\frac{1}{3})$	$\emptyset$	$\{a_6\}$	$\{a_1,a_2,a_3,a_4,a_5\}$
$(\frac{3}{4},\frac{1}{2})$	$\emptyset$	$\{a_5,a_6\}$	$\{a_1,a_2,a_3,a_4\}$
$(\frac{3}{4},\frac{5}{9})$	$\emptyset$	$\{a_2,a_5,a_6\}$	$\{a_1,a_3,a_4\}$
$(\frac{3}{4},\frac{2}{3})$	$\emptyset$	$\{a_2,a_3,a_5,a_6\}$	$\{a_1,a_4\}$
$(\frac{5}{6},\frac{2}{9})$	$\{a_4\}$	$\{a_1\}$	$\{a_2,a_3,a_5,a_6\}$
$(\frac{5}{6},\frac{1}{3})$	$\emptyset$	$\{a_1,a_6\}$	$\{a_2,a_3,a_4,a_5\}$
$(\frac{5}{6},\frac{1}{2})$	$\emptyset$	$\{a_1,a_5,a_6\}$	$\{a_2,a_3,a_4\}$
$(\frac{5}{6},\frac{5}{9})$	$\emptyset$	$\{a_1,a_2,a_5,a_6\}$	$\{a_3,a_4\}$
$(\frac{5}{6},\frac{2}{3})$	$\emptyset$	$\{a_1,a_2,a_3,a_5,a_6\}$	$\{a_4\}$

$$E^c[(a_1,a_r^o),I_o]=\frac{0.1}{0.4}\times\frac{1}{4}+\frac{0.3}{0.4}\times1=\frac{13}{16},$$

$$E^c[(a_2,a_r^o),I_o]=\frac{0.1}{0.4}\times1+\frac{0.3}{0.4}\times\frac{1}{4}=\frac{7}{16},$$

$$E^c[(a_3,a_r^o),I_o]=\frac{0.1}{0.4}\times1+\frac{0.3}{0.4}\times\frac{1}{4}=\frac{7}{16},$$

$$E^c[(a_4,a_r^o),I_o]=\frac{0.1}{0.4}\times\frac{1}{2}+\frac{0.3}{0.4}\times1=\frac{7}{8},$$

$$E^c[(a_5,a_r^o),I_o]=\frac{0.1}{0.4}\times1+\frac{0.3}{0.4}\times\frac{1}{4}=\frac{7}{16},$$

$$E^c[(a_6,a_r^o),I_o]=\frac{0.1}{0.4}\times\frac{1}{2}+\frac{0.3}{0.4}\times\frac{1}{4}=\frac{5}{16};$$

$$E^c[(a_1,a_r^o),I_{\overline{o}}]=\frac{0.1}{0.6}\times\frac{1}{2}+\frac{0.2}{0.6}\times0+\frac{0.3}{0.6}\times\frac{1}{2}=\frac{1}{3},$$

$$E^c[(a_2,a_r^o),I_{\overline{o}}]=\frac{0.1}{0.6}\times\frac{1}{3}+\frac{0.2}{0.6}\times1+\frac{0.3}{0.6}\times\frac{1}{2}=\frac{23}{36},$$

$$E^c[(a_3,a_r^o),I_{\overline{o}}]=\frac{0.1}{0.6}\times\frac{1}{2}+\frac{0.2}{0.6}\times1+\frac{0.3}{0.6}\times\frac{1}{3}=\frac{7}{12},$$

$$E^c[(a_4,a_r^o),I_{\overline{o}}]=\frac{0.1}{0.6}\times\frac{1}{2}+\frac{0.2}{0.6}\times\frac{1}{2}+\frac{0.3}{0.6}\times\frac{1}{2}=\frac{1}{2},$$

$$E^c[(a_5,a_r^o),I_{\overline{o}}]=\frac{0.1}{0.6}\times\frac{1}{2}+\frac{0.2}{0.6}\times1+\frac{0.3}{0.6}\times\frac{1}{2}=\frac{2}{3},$$

$$E^c[(a_6,a_r^o),I_{\overline{o}}]=\frac{0.1}{0.6}\times\frac{1}{2}+\frac{0.2}{0.6}\times\frac{1}{2}+\frac{0.3}{0.6}\times\frac{1}{2}=\frac{1}{2}.$$

After that, by taking $\alpha = \frac18$ and $\beta = \frac13$, we have $AL^o_{(\alpha,\beta)}(A,a^o_r) = \{a_1,a_2,a_3,a_4,a_6\}$, $NE^o_{(\alpha,\beta)}(A,a^o_r) = \emptyset$ and $CO^o_{(\alpha,\beta)}(A,a^o_r) = \{a_5\}$. Meanwhile, we take other 19 pairs of thresholds, and compute the three coalitions of the opposition reference with respect to I_o and $I_{\bar{o}}$. We also show all results by Table 9.

Table 9. The Alliance, Conflict and Neutral Coalitions of a^o_r in Middle East Conflicts.

(α,β)	$AL^o_{(\alpha,\beta)}(A,a^o_r)$	$CO^o_{(\alpha,\beta)}(A,a^o_r)$	$NE^o_{(\alpha,\beta)}(A,a^o_r)$
$(\frac18,\frac13)$	$\{a_1,a_2,a_3,a_4,a_6\}$	$\emptyset$	$\{a_5\}$
$(\frac18,\frac{13}{36})$	$\{a_1,a_3,a_4,a_6\}$	$\emptyset$	$\{a_2,a_5\}$
$(\frac18,\frac{5}{12})$	$\{a_1,a_4,a_6\}$	$\emptyset$	$\{a_2,a_3,a_5\}$
$(\frac18,\frac12)$	$\{a_1\}$	$\emptyset$	$\{a_2,a_3,a_4,a_5,a_6\}$
$(\frac18,\frac23)$	$\emptyset$	$\emptyset$	$\{a_1,a_2,a_3,a_4,a_5,a_6\}$
$(\frac{3}{16},\frac13)$	$\{a_1,a_2,a_3,a_6\}$	$\emptyset$	$\{a_4,a_5\}$
$(\frac{3}{16},\frac{13}{36})$	$\{a_1,a_3,a_6\}$	$\emptyset$	$\{a_2,a_4,a_5\}$
$(\frac{3}{16},\frac{5}{12})$	$\{a_1,a_6\}$	$\emptyset$	$\{a_2,a_3,a_4,a_5\}$
$(\frac{3}{16},\frac12)$	$\{a_1\}$	$\{a_4\}$	$\{a_2,a_3,a_5,a_6\}$
$(\frac{3}{16},\frac23)$	$\emptyset$	$\{a_4\}$	$\{a_1,a_2,a_3,a_5,a_6\}$
$(\frac{9}{16},\frac13)$	$\{a_2,a_3,a_6\}$	$\emptyset$	$\{a_1,a_4,a_5\}$
$(\frac{9}{16},\frac{13}{36})$	$\{a_3,a_6\}$	$\emptyset$	$\{a_1,a_2,a_4,a_5\}$
$(\frac{9}{16},\frac{5}{12})$	$\{a_6\}$	$\emptyset$	$\{a_1,a_2,a_3,a_4,a_5\}$
$(\frac{9}{16},\frac12)$	$\emptyset$	$\{a_4\}$	$\{a_1,a_2,a_3,a_5,a_6\}$
$(\frac{9}{16},\frac23)$	$\emptyset$	$\{a_1,a_4\}$	$\{a_2,a_3,a_5,a_6\}$
$(\frac{11}{16},\frac13)$	$\{a_6\}$	$\{a_5\}$	$\{a_1,a_2,a_3,a_4\}$
$(\frac{11}{16},\frac{13}{36})$	$\{a_6\}$	$\{a_2,a_5\}$	$\{a_1,a_3,a_4\}$
$(\frac{11}{16},\frac{5}{12})$	$\{a_6\}$	$\{a_2,a_3,a_5\}$	$\{a_1,a_4\}$
$(\frac{11}{16},\frac12)$	$\emptyset$	$\{a_2,a_3,a_4,a_5\}$	$\{a_1,a_6\}$
$(\frac{11}{16},\frac23)$	$\emptyset$	$\{a_1,a_2,a_3,a_4,a_5\}$	$\{a_6\}$

5   Conclusion and Future Work

In this paper, we provided models of conflict analysis with rankings and reference tuples. First, we defined two ranking orders, in which one divides a whole set of issues into a bundle of supported issues and a bundle of non-supported issues, the other divides a whole set of issues into a bundle of opposed issues and a bundle of non-opposed issues, and the support and opposition reference tuples. Second, we provided alliance and conflict measures by considering weights of issues, and constructed alliance, neutral and conflict coalitions towards the support and opposition references. Finally, we employed an example to show how to compute the alliance, neutral and conflict coalitions with the proposed models.

In the future, we will design effective alliance and conflict measures with thoughts of rankings and reference tuples. Furthermore, we will develop effective models for analyzing conflict problems described by hybrid situation tables.

Acknowledgements. This work is supported by the National Natural Science Foundation of China (No. 62076040), Hunan Provincial Natural Science Foundation of China (No. 2020JJ3034), the Scientific Research Fund of Hunan Provincial Education Department (No. 22A0233), the Scientific Research Fund of Chongqing Key Laboratory of Computational Intelligence (No. 2020FF04), the Graduate Research Innovation Project of Hunan Province (No. CX20220952).

References

1. Deja, R.: Conflict Analysis, Rough Set Methods and Applications, Studies in Fuzziness and Soft Computing, pp. 491–520 (2000)
2. Du, J.L., Liu, S.F., Liu, Y., Yi, J.H.: A novel approach to three-way conflict analysis and resolution with Pythagorean fuzzy information. Inf. Sci. **584**, 65–88 (2022)
3. Feng, X.F., Yang, H.L., Guo, Z.L.: Three-way conflict analysis in dual hesitant fuzzy situation tables. Int. J. Approx. Reason. **154**, 109–132 (2023)
4. Hu, M.J.: Modeling relationships in three-way conflict analysis with subsethood measures. Knowl.-Based Syst. **260**, 110131 (2023)
5. Jiang, C.M., Duan, Y., Guo, D.D.: Effectiveness measure in change-based three-way decision. Soft. Comput. **27**, 2783–2793 (2023)
6. Jiang, C.M., Xu, R.Y., Wang, P.X.: Measuring effectiveness of movement-based three-way decision using fuzzy Markov model. Int. J. Approx. Reason. **152**, 456–469 (2023)
7. Lang, G.M., Yao, Y.Y.: New measures of alliance and conflict for three-way conflict analysis. Int. J. Approx. Reason. **132**, 49–69 (2021)
8. Lang, G.M., Yao, Y.Y.: Formal concept analysis perspectives on three-way conflict analysis. Int. J. Approx. Reason. **152**, 160–182 (2023)
9. Li, T.B., Qiao, J.S., Ding, W.P.: Three-way conflict analysis and resolution based on q-rung orthopair fuzzy information. Inf. Sci. **638**, 118959 (2023)
10. Luo, J.F., Hu, M.J., Lang, G.M., Qin, K.Y.: Three-way conflict analysis based on alliance and conflict functions. Inf. Sci. **594**, 322–359 (2022)
11. Pawlak, Z.: On conflicts. Int. J. Man Mach. Stud. **21**, 127–134 (1984)
12. Pawlak, Z.: An inquiry into anatomy of conflicts. Inf. Sci. **109**, 65–78 (1998)
13. Sun, B.Z., Chen, X.T., Zhang, L.Y., Ma, W.M.: Three-way decision making approach to conflict analysis and resolution using probabilistic rough set over two universes. Inf. Sci. **507**, 809–822 (2020)
14. Sun, B.Z., Ma, W.M., Zhao, H.Y.: Rough set-based conflict analysis model and method over two universes. Inf. Sci. **372**, 111–125 (2016)
15. Suo, L.W.Q., Yang, H.L.: Three-way conflict analysis based on incomplete situation tables: a tentative study. Int. J. Approx. Reason. **145**, 51–74 (2022)
16. Wang, T.X., Huang, B., Li, H.X., Liu, D., Yu, H.: Three-way decision for probabilistic linguistic conflict analysis via compounded risk preference. Inf. Sci. **631**, 65–90 (2023)
17. Wang, L., Pei, Z., Qin, K.: A novel conflict analysis model based on the formal concept analysis. Appl. Intell. (2022). https://doi.org/10.1007/s10489-022-04051-9
18. Xu, W.Y., Jia, B., Li, X.N.: A two-universe model of three-way decision with ranking and reference tuple. Inf. Sci. **581**, 808–839 (2021)
19. Xu, W.Y., Jia, B., Li, X.N.: A generalized model of three-way decision with ranking and reference tuple. Int. J. Approx. Reason. **144**, 51–68 (2022)

20. Xu, W.Y., Yan, Y.C., Li, X.N.: Three-way decision with ranking and reference tuple on information tables. Inf. Sci. **613**, 682–716 (2022)
21. Yao, J.T., Medina, J., Zhang, Y., Slezak, D.: Formal concept analysis, rough sets, and three-way decisions. Int. J. Approx. Reason. **140**, 1–6 (2022)
22. Yang, H.L., Wang, Y., Guo, Z.L.: Three-way conflict analysis based on hybrid situation tables. Inf. Sci. **628**, 522–541 (2023)
23. Yao, Y.Y.: Three-way decisions with probabilistic rough sets. Inf. Sci. **180**, 341–353 (2010)
24. Yao, Y.Y.: Three-way conflict analysis: reformulations and extensions of the Pawlak model. Knowl.-Based Syst. **180**, 26–37 (2019)
25. Yao, Y.Y.: Three-way granular computing, rough sets, and formal concept analysis. Int. J. Approx. Reason. **116**, 106–125 (2020)
26. Yu, H., Wang, X.C., Wang, G.Y., Zeng, X.H.: An active three-way clustering method via low-rank matrices for multi-view data. Inf. Sci. **507**, 823–839 (2020)
27. Zhan, J.M., Wang, J.J., Ding, W.P., Yao, Y.Y.: Three-way behavioral decision making with hesitant fuzzy information systems: survey and challenges. IEEE/CAA J. Autom. Sin. **10**(2), 330–350 (2022)
28. Zhang, X.Y., Chen, J.: Three-hierarchical three-way decision models for conflict analysis: a qualitative improvement and a quantitative extension. Inf. Sci. **587**, 485–514 (2022)
29. Zhang, H.Y., Yang, S.Y.: Three-way group decisions with interval-valued decision-theoretic rough sets based on aggregating inclusion measures. Int. J. Approx. Reason. **110**, 31–45 (2019)
30. Zhi, H.L., Li, J.H., Li, Y.N.: Multi-level conflict analysis based on fuzzy formal contexts. IEEE Trans. Fuzzy Syst. **30**(12), 5128–5142 (2022)

Three-Way Social Network Analysis: Triadic Measures at Three Levels

Yingxiao Chen[1,2(✉)], Yiyu Yao[2], and Ping Zhu[1,3]

[1] School of Science, Beijing University of Posts and Telecommunications,
Beijing 100876, China
`yingx_cbupt@126.com, pzhubupt@bupt.edu.cn`
[2] Department of Computer Science, University of Regina, Regina, SK S4S 0A2,
Canada
`yiyu.yao@uregina.ca`
[3] Key Laboratory of Mathematics and Information Networks (Beijing University of
Posts and Telecommunications), Ministry of Education, Beijing, China

Abstract. Three-way decision, as thinking in threes, realizes the power of triads and has been successfully applied across diverse fields. Due to the important role played by triads, the basic ideas of triadic thinking appear in many studies on social network analysis. While measures based on the use of dyads (i.e., edges), the use of triads (i.e., triangles) has not received its due attention. This paper explores the value of triads in defining and interpreting measures in social network. We present an in-depth examination at the node, community, and network three levels. We propose a set of triadic measures at each level. These new measures contributes to a more comprehensive understanding of the structures and dynamics of social networks.

Keywords: Three-way decision · Social network analysis · Triads

1 Introduction

Three-way decision has emerged as a versatile concept with applications in various fields, encompassing thinking, problem-solving, and computing in threes or triads [23–25]. It is a human-cognitive approach that fosters simplicity while expressing complex patterns. Over time, it has found diverse applications, ranging from three-way classification [9,14,29,30], three-way clustering [1,10,17,28], three-way conflict analysis [4,12,13,15,22], three-way recommendation systems [2,21,26,27], and so on.

It can be observed that triadic thinking is also very common in social network analysis. In the domain of social networks, a "triad" refers to a group of three nodes and the connections between them. Triads, as underscored by

This work was partially supported by the National Natural Science Foundation of China (NO. 62172048), China Scholarship Council (NO. 202006470014), and a Discovery Grant from NSERC, Canada.

Logan et al. [11], serve as a vital structural element in social networks, bringing to the table benefits like stability, innovation, and scalability. Triads play a crucial role in connecting small groups and tribes, enabling the formation of larger and more scalable networks [11]: "The triad is so powerful that it can link tribes together, creating an unlimited capacity for scalability". As a result, the presence and prevalence of triads can serve as a reliable indicator of a group's developmental stage. In well-developed groups, triads serve as fundamental building blocks, fostering large, robust, dynamic, and growing networks of tribes. These groups are characterized by vibrancy, strong values, and a collective effort where leadership is both given and taken. On the contrary, less developed communities may rely more on dyads, which consist of two interconnected nodes. However, an overabundance of dyads can lead to inevitable communication problems [11].

The study of triads in social networks has a rich history, with notable contributions like Georg Simmel's "triadic closure" [18]. Simmel's observation of the tendency for a new connection to form between two nodes if both are already connected to a third has provided profound insights into network growth, enabling the understanding and prediction of network expansion. Moreover, social network analysis categorizes its study into various levels, including the individual actor level, dyad level, triad level, subgroup level, and global level [19]. Triads play a crucial role as a level or view of network analysis, with concepts like transitivity being effectively captured and analyzed through the triad census [8].

Directed networks often rely on triads for structural analysis. The presence of different types of triads in a directed network can provide insights into its overall structure, enabling the classification of networks based on models like balance, clusterability, ranked clusters, transitivity, hierarchical clusters, and no balance [20]. Triads also play a prominent role in Heider's balance theory [7], where they are viewed as 3-cycles in a signed graph representing social networks. This theory has significantly influenced the analysis of signed social networks, offering insights into relationships characterized by agreement, support, disagreement, or hostility. In addition to the aforementioned triad-based social network analysis, studies also employ triads for community detection [3,16,32], link prediction [6], and various other purposes.

Despite the significant importance of triads in social networks, measures based on them have not received as much comprehensive exploration as dyad-based ones in simple networks, which are typically characterized as undirected, unsigned, and unweighted. In node role analysis, the traditional dyadic measures focuse only on the number of nodes directly connected to a given node in the network. While this can reflect the direct influence of a node, in complex social networks, interactions between nodes are not solely based on direct connections. Triad-based measures can describe the triangular relationships formed by nodes in the network, revealing their roles and significance in complex interactions. Interestingly, certain dyad-based measures can be expressed in terms of triads. For example, the local clustering coefficient of a node can be computed as the ratio of the edges between the node's neighbors to the total possible edges among those neighbors. This coefficient can also be interpreted as the ratio of closed triangles formed by the node to the number of open triangles. This observation further underscores the significance of triads in network analysis and their far-reaching implications for understanding the structure and dynamics of social networks.

By considering triadic relationships, researchers can obtain valuable insights into the complexity and behavior of various social systems. Aligned with the principles of triadic thinking, the objective of this paper is to measure social networks from three distinct levels: the node level, the community (subgraph) level, and the network level.

2 A Tri-Level Framework of Triadic Measures

In this section, we present a tri-level framework for analyzing social networks, which comprises node level, community level, and network level analysis. In the following three subsections, we construct triadic measures at each of these levels, respectively.

Figure 1 depicts the tri-level framework. At the node level, we establish three types of triadic measures of centrality, drawing inspiration from dyad-based node degree centrality. The first triadic centrality is derived from a probabilistic perspective, representing the ratio of the number of triangles formed by a node to the total possible triangles in the network. The second centrality is computed as the proportion of the number of triangles formed by a node to the sum of triangles formed by all nodes in the network. The third centrality is based on the total number of triangles in the network, reflecting the ratio of the number of triangles formed by a node to the overall number of triangles in the network. Moving to the community level, triadic measures of importance are calculated by averaging the triadic centralities of nodes within each community. Subsequently, the triadic measures of community-importance of the entire network are measured by calculating the average importance of all communities within the network.

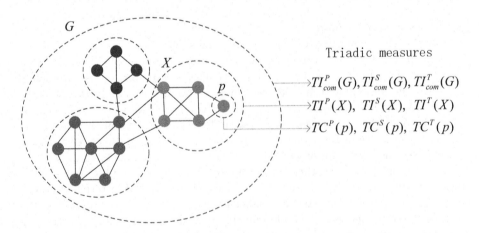

Fig. 1. Triadic measures at three levels

2.1   Triadic Measures of Centrality at the Node-Level

This subsection focuses on measuring nodes through the computation of triads. Before delving into the details, we introduce four heterogeneous types of triads in simple networks. The first type, referred to as the "Empty" type (T_0), represents triads where no edges connect the three nodes. The second type is the "One-edge" type (T_1), where only one edge exists within the triad. The third type, known as "Two-edge" (T_2), consists of triads with two edges. Finally, we have the "Triangle" (T_3) type, which represents triads with three edges forming a complete subgraph, also known as a 3-clique. Figure 2 visually illustrates these triad types.

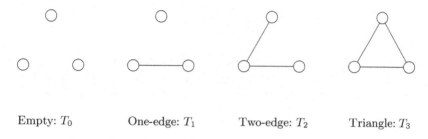

Empty: T_0 One-edge: T_1 Two-edge: T_2 Triangle: T_3

Fig. 2. Types of triads in simple network

When analyzing network nodes, a commonly used measure is degree centrality, proposed by Freeman [5]. It measures the number of connections or the degree of connectivity that a node has with other nodes in the network. A higher degree centrality implies a greater number of connections, indicating that the node holds a more important or central position within the network. In a simple network, a node's degree refers to the number of edges directly connected to that node. The degree centrality formula is as follows:

Definition 1 ([5]). Let $G = (V, E)$ be a simple network with node set V and edge set E. For any node $p \in V$, the degree centrality of p is:

$$DC(p) = \frac{|N(p)|}{|V| - 1},$$

where $N(p) = \{q \in V \mid (p, q) \in E\}$ is the neighbor set of p.

The degree centrality metric evaluates the proportion of a node's neighbors to the maximal potential neighbors within the network, considering a dyadic perspective. Analogously, the significance of nodes can be assessed through triadic frameworks. Sociologically speaking, it is posited that nodes encompassing a larger quantity of triads are predisposed to assume leadership capacities [11]. For triadic network analyses, let us define the set of triads, which comprises any three distinct nodes in a network $G = (V, E)$, as $\mathscr{G} = \{T = \{p, q, h\} \mid p, q, h \in V$

are distinct nodes}. This set is referred to as the triadic granulation of V. Based on $\mathscr{G}$, we propose the following triadic centralities to measure the importance of nodes:

Definition 2. Let $G = (V, E)$ be a simple network and $\mathscr{G}$ be the triadic granulation of V. For any $p \in V$, the three types of triadic centralities of p are defined as:

$$TC^P(p) = \frac{|\{T_3 \in \mathscr{G} \mid p \in T_3\}|}{\{T_0 \in \mathscr{G} \mid T_0 = \{p, q, h\} \wedge q, h \in V\}},$$

$$TC^S(p) = \frac{|\{T_3 \in \mathscr{G} \mid p \in T_3\}|}{\sum_{q \in V} |\{T_3 \in \mathscr{G} \mid q \in T_3\}|},$$

$$TC^T(p) = \frac{|\{T_3 \in \mathscr{G} \mid p \in T_3\}|}{|\{T_3 \in \mathscr{G}\}|}.$$

For simplicity, let's denote the number of Triangle-type triads that a node belongs to as $\triangle_{num}(p)$ and the number of triangles in the network G as $\triangle_{num}(G)$. Consequently, the above formulas can be simplified as follows:

$$TC^P(p) = \frac{2\triangle_{num}(p)}{(|V|)(|V| - 1)},$$

$$TC^S(p) = \frac{\triangle_{num}(p)}{\sum_{q \in V} \triangle_{num}(q)},$$

$$TC^T(p) = \frac{\triangle_{num}(p)}{\triangle_{num}(G)}.$$

The triadic centralities provide insights into different aspects of a node's importance: $TC^P(p)$ represents the ratio of the number of triangles formed by node p to the number of possible triangles it can form with any two nodes in G; $TC^S(p)$ quantifies the ratio of the number of triangles formed by node p to the sum of triangles formed by all nodes in the network; $TC^T(p)$ measures the ratio of the number of triangles formed by node p to the total number of triangles in the network.

2.2   Triadic Measures of Importance at the Community-Level

Communities in networks are commonly observed, even though their definition may not be precise. Typically, internal connections within communities are dense, while connections between communities are sparse. In the preceding subsection, we introduced three distinct triadic centralities for individual nodes, which can reflect the importance of nodes. When examining communities within a network, it becomes natural to assess their importance based on the nodes they comprise. Therefore, we use the average of the triadic centralities of the nodes in the community as the triadic measures of importance of the community.

Definition 3. Let $\mathscr{G} = \{T = \{p, q, h\} \mid p, q, h \in V$ are distinct nodes$\}$ be the triadic granular network of $G = (V, E)$. For any community $X \subseteq V$, the three types of triadic measures of importance of X are given by:

$$TI^P(X) = \frac{\sum\limits_{p \in X} TC^P(p)}{|X|} = \frac{2 \sum\limits_{p \in X} \triangle_{num}(p)}{|X||V|(|V| - 1)},$$

$$TI^S(X) = \frac{\sum\limits_{p \in X} TC^S(p)}{|X|} = \frac{\sum\limits_{p \in X} \triangle_{num}(p)}{|X|(\sum\limits_{q \in V} \triangle_{num}(q))},$$

$$TI^T(X) = \frac{\sum\limits_{p \in X} TC^T(p)}{|X|} = \frac{\sum\limits_{p \in X} \triangle_{num}(p)}{|X|\triangle_{num}(G)}.$$

These triadic measures of importance offer different perspectives on the importance and density of triangles within a community. Each measure is proportional to the total number of triangles formed by the nodes in the community. However, they differ in their normalization factors: $TI^P(X)$ reflects the ratio of the total number of triangles formed by nodes in the community to both the size of the community (the number of nodes in the community) and the size of the entire network (the number of nodes in the network); $TI^S(X)$ reflects the ratio of the total number of triangles formed by nodes in the community to the size of the community and the total number of triangles formed by all nodes in the entire network; $TI^T(X)$ reflects the ratio of the number of triangles formed by nodes in the community to the size of the community and the total number of triangles in the entire network.

The triadic measures of importance of a community provide valuable insights beyond what density alone can reveal. Density is measured by the ratio of the number of edges in the community to the number of possible edges, and it indicates the level of interconnectedness within the community. However, triadic measures we proposed offer a deeper understanding by considering the formation of triangles within the community. A community's triadic measures reflect the quality of the community from the perspective of the relative quantity of triangles formed by its members. The number of triangles formed by nodes reflects the importance of nodes, and the average importance of nodes contained in a community reflects the importance of the community.

2.3   Triadic Measures of Community-Importance at the Network-Level

In this subsection, we adopt a community-centric approach to assess the network, enabling a comprehensive analysis from a hierarchical perspective. The measures presented in this paper adhere to a structured hierarchy, starting with node-level computations, followed by community-level measures derived from the node-level data, and culminating in network-level measures based on the community-level analysis. Specifically, we compute the triadic measures of community-importance for a network by averaging the importance of individual communities as follows.

Definition 4. Suppose that $\mathscr{G} = \{T = \{p, q, h\} \mid p, q, h \in V$ are distinct nodes$\}$ is the triadic granular network of $G = (V, E)$ and $V = \bigcup_{i=1}^{n} X_i$, where $X_i \neq \emptyset$ is the i-th community. The triadic measures of community-importance of G are given by:

$$TI_{com}^P(G) = \frac{\sum_{i=1}^{n} TI^P(X_i)}{n} = \sum_{i=1}^{n} \frac{2 \sum_{p \in X_i} \triangle_{num}(p)}{n |X_i| |V| (|V| - 1)},$$

$$TI_{com}^S(G) = \frac{\sum_{i=1}^{n} TI^S(X_i)}{n} = \sum_{i=1}^{n} \frac{\sum_{p \in X_i} \triangle_{num}(p)}{n |X_i| (\sum_{q \in V} \triangle_{num}(q))},$$

$$TI_{com}^T(G) = \frac{\sum_{i=1}^{n} TI^T(X_i)}{n} = \sum_{i=1}^{n} \frac{\sum_{p \in X_i} \triangle_{num}(p)}{n |X_i| \triangle_{num}(G)}.$$

The proposed triadic measures of the network are obtained through a hierarchical evaluation, starting with the analysis of communities and then nodes within these communities. This approach enables a comprehensive assessment of importance at various levels, yielding valuable insights into the network's structure. According to Definition 4, all triadic measures of the network are proportional to the number of triangles formed by nodes within the communities. However, each centrality measure (TI_{com}^P, TI_{com}^S, and TI_{com}^T) has distinct characteristics. The first one, TI_{com}^P, is inversely proportional to the community size and the network size. The second one, TI_{com}^S, is inversely proportional to the community size and the sum of the number of triangles formed by nodes in the entire network. Lastly, the third one, TI_{com}^T, is inversely proportional to the community size and the total number of triangles in the network. These variations in the triadic measures provide diverse perspectives on the importance of nodes and communities within the network, capturing different aspects of their triadic relationships.

In this section, we have introduced the measurements of triadic centralities $TC^P(p)$, $TC^S(p)$, and $TC^T(p)$ at the node-level, each offering a distinct perspective on a node's importance through its involvement in forming triangles. Building upon these node-level measures, we defined the triadic measures of importance: $TI^P(X)$, $TI^S(X)$, and $TI^T(X)$ at the community-level. These community-level measures represent the average triadic centralities of nodes within a community, offering valuable insights into the triadic characteristics of the communities. Finally, we proposed the triadic measures of community-importance: $TI_{com}^P(G)$, $TI_{com}^S(G)$, and $TI_{com}^T(G)$ at the network-level. These network-level measures provide an overview of the overall triadic characteristics based on the communities within the network, offering a comprehensive understanding of the network's structure and behavior.

3   Two Examples

In this section, we present two examples to illustrate the concepts discussed in the paper. The first example is a demonstration network shown in Fig. 1, and the second one is the well-known karate network [31]. The descriptions of these two networks are provided as follows (Table 1):

Table 1. Description of two networks

	Number of nodes	Number of communities
Network of Fig. 1	16	3
Karate network	34	2

We calculate the degree centrality and triadic centralities for the network shown in Fig. 1 and visualize the results in Fig. 3, with larger nodes represent higher centrality values. Additionally, we perform a similar analysis on the nodes in the karate network. Figure 4 presents the visualization of the nodes sorted based on both degree centrality and triadic centralities, where larger values correspond to larger nodes. A comparison between Fig. 3a and Fig. 3b reveals significant variations in the triadic centralities among nodes with the same degree. Some nodes exhibit high degree centrality but low triadic centralities, suggesting that they have extensive connections across communities but relatively infrequent connections within communities. This observation emphasizes the importance of considering both degree centrality and triadic centralities to gain a comprehensive understanding of a node's importance and its role in the network's structure.

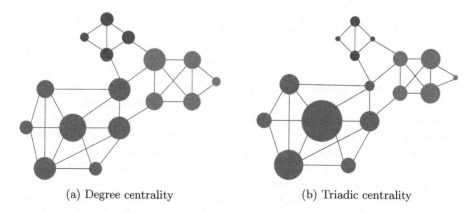

(a) Degree centrality   (b) Triadic centrality

Fig. 3. Visualization of nodes' centralities in Fig. 1

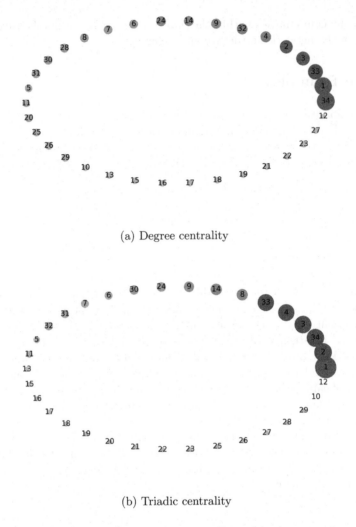

(a) Degree centrality

(b) Triadic centrality

Fig. 4. Visualization of nodes' centralities in karate network

Figure 4 highlights the important nodes identified by both degree and triadic centralities. The key findings reveal that triadic centrality assigns higher importance to node 4 compared to degree centrality. Node 1, serving as the instructor of the karate club, is identified as the most crucial node by triadic centrality, considering his central role in the network. Although node 34, the president, has a higher degree than node 1, triadic centrality recognizes the instructor's significance due to his closer relationships with other members, reflected in his involvement in more triangles. Node 2 emerges as the next significant node, being regarded as the closest partner of node 1. Node 2's participation in two 5-cliques and three triangles, all involving node 1, contributes to its importance according to triadic centrality. Following node 2, node 34, the club's president, holds a prominent position. Nodes 3 and 4 are also identified as significant partners of the instructor, forming two 5-cliques and one triangle each with node 1. Moreover, node 33 is recognized as the closest companion of the president, forming two 4-cliques and six triangles with it, further emphasizing its importance by triadic centrality.

The analysis emphasizes that node 1's significance is further enhanced when considering its partnerships with other important nodes (2, 3, and 4) according to triadic centrality, which is not evident when solely relying on degree centrality. This underlines the importance of considering not just direct connections but also the indirect relationships formed through triangles. Additionally, both nodes 10 and 12 are identified as the least important nodes under triadic centrality, which is reasonable given their lack of involvement in forming triangles. Node 10's limited connections (only two edges) and absence of triangles contribute to its lower importance when compared to nodes that participate in triangles, such as nodes 13, 15–19, 21, 22, 23, and 27. In fact, node 10 is a neutral node, who has no factional affiliation regarding the instructor or the president [31], suggesting that it does not strongly align with either group and does not play a pivotal role in the network's division.

Let's visualize the karate network using Fig. 5. The communities on the left and right are centered on node 1 (the instructor) and node 34 (the president), containing 16 nodes and 18 nodes, respectively. We draw the triangles formed by each node, where the blue lines represent the triangles between the communities, and the black lines represent the triangles within the communities. We denote these two communities as X_1 and X_{34}, respectively. The sum of triangles formed by nodes in X_1 and X_{34} is 91 and 57, respectively. Now, let's calculate the triadic measures of importance of the communities in the network shown in Fig. 1 and the karate network, and display the results in Table 2.

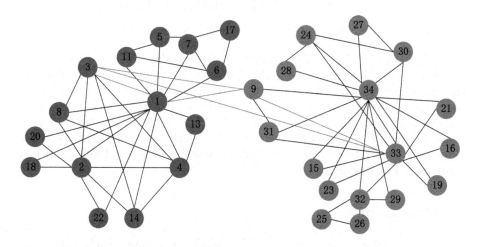

Fig. 5. Nodes' triangles in two communities of karate network

Table 2. Triadic measures of importance of communities in two networks

	Community	TC^P	TC^S	TC^T
Network of Fig. 1	Blue	0.0125	0.0341	0.0882
	Green	0.0249	0.0682	0.1765
	Red	0.0356	0.0974	0.2521
Karate network	X_1	0.0101	0.0384	0.1264
	X_{34}	0.0056	0.0214	0.0704

Figure 1 and Fig. 5 clearly illustrate that triangles formed by nodes within a community tend to be more prevalent. In Fig. 1, triangles are exclusively formed by nodes within communities. In Fig. 5, only two triangles span across different communities, namely {3, 9, 33} and {3, 9, 1}. This observation suggests that members within a community tend to establish closer and more stable relationships with each other compared to those outside the community. Analyzing Fig. 5 and Table 2, it becomes evident that despite the instructor-centered community having relatively fewer nodes, its high triadic measures of importance highlight the significance of its members. Notably, the instructor's supporters perceive him as a fatherly figure who acted as their spiritual and physical mentor [31], further affirming the community's greater stability. Node 9, initially part of the president's camp, chose to join the instructor's community to avoid starting from scratch (White Belt exam) for his promotion [31]. This fact illustrates the relative instability of the president-centered community.

Let us label the networks in Fig. 1 and the karate network as G_1 and G_2, respectively. The triadic measures of the two networks are computed as follows:

$$TI_{com}^P(G_1) = 0.0252, \quad TI_{com}^S(G_1) = 0.0666, \quad TI_{com}^T(G_1) = 0.1723$$
$$TI_{com}^P(G_2) = 0.0079, \quad TI_{com}^S(G_2) = 0.0299, \quad TI_{com}^T(G_2) = 0.0984$$

The larger value of $TI_{com}^P(G_1)$ indicates that the nodes within communities in G_1 form more triangles relative to the size of the communities and the network size. Similarly, the larger value of $TI_{com}^S(G_1)$ indicates that the nodes within communities in G_1 form more triangles relative to the size of the communities and the sum of triangles formed by all nodes in the network. Finally, the larger value of $TI_{com}^T(G_1)$ indicates that the nodes within communities of G_1 form more triangles relative to the size of the communities and the total number of triangles in the network.

4   Conclusion

This paper explores the application of triads-based network analysis to comprehend the structural characteristics of social networks. Triads, representing three-node subgraphs, offer valuable insights into local and global connectivity patterns within a network. Leveraging triads, we hierarchically measured networks at the node, community, and network levels.

The triad-based network analysis opens promising avenues for further research in three-way social network analysis. Indeed, our hierarchical approach has allowed us to construct a set of triad-based measures at different levels, enabling us to gain valuable insights into social networks. However, there are still many unexplored triad-based measures that hold the potential for further investigation within the node-level, community-level, and network-level analyses. Additionally, when constructing higher-level measures based on triads, there are several lower-level choices and approaches that researchers can consider. One such example is measuring the network from the node level using triads, such as calculating the density of triads in the network. Although this paper focused on simple networks, future research could extend these methodologies to more complex and dynamic networks, including directed networks, signed networks, temporal networks, and multiplex networks. Overall, triads-based network analysis holds the potential to become an essential tool in social network research, aiding in the discovery of underlying structures and dynamics within complex social systems.

References

1. Ali, B., Azam, N., Yao, J.: A three-way clustering approach using image enhancement operations. Int. J. Approx. Reason. **149**, 1–38 (2022)
2. Chen, Y., Zhu, P.: Three-way recommendation for a node and a community on social networks. Int. J. Mach. Learn. Cybern. **13**(10), 2909–2927 (2022)
3. Fagnan, J., Zaïane, O., Barbosa, D.: Using triads to identify local community structure in social networks. In: 2014 IEEE/ACM International Conference on Advances in Social Networks Analysis and Mining (ASONAM 2014), pp. 108–112. IEEE (2014)
4. Feng, X., Yang, H., Guo, Z.: Three-way conflict analysis in dual hesitant fuzzy situation tables. Int. J. Approx. Reason. **154**, 109–132 (2023)
5. Freeman, L.C.: A set of measures of centrality based on betweenness. Sociometry **40**(1), 35–41 (1977)
6. Gou, F., Wu, J.: Triad link prediction method based on the evolutionary analysis with IoT in opportunistic social networks. Comput. Commun. **181**, 143–155 (2022)
7. Heider, F.: The Psychology of Interpersonal Relations. Psychology Press, London (1982)
8. Holland, P.W., Leinhardt, S.: Local structure in social networks. Sociol. Methodol. **7**, 1–45 (1976)
9. Hu, M.: Three-way Bayesian confirmation in classifications. Cogn. Comput. **14**, 2020–2039 (2022)
10. Jia, X., Rao, Y., Li, W., Yang, S., Yu, H.: An automatic three-way clustering method based on sample similarity. Int. J. Mach. Learn. Cybern. **12**, 1545–1556 (2021)
11. Logan, D., King, J., Fischer-Wright, H.: Tribal Leadership: Leveraging Natural Groups to Build a Thriving Organization. Harper Business (2011). ISBN: 9780061251320

12. Lang, G., Yao, Y.: New measures of alliance and conflict for three-way conflict analysis. Int. J. Approx. Reason. **132**, 49–69 (2021)
13. Li, X., Yang, Y., Yi, H., Yu, Q.: Conflict analysis based on three-way decision for trapezoidal fuzzy information systems. Int. J. Mach. Learn. Cybern. **13**, 929–945 (2022)
14. Liu, D.: The effectiveness of three-way classification with interpretable perspective. Inf. Sci. **567**, 237–255 (2021)
15. Luo, J., Hu, M., Lang, G., Yang, X., Qin, K.: Three-way conflict analysis based on alliance and conflict functions. Inf. Sci. **594**, 322–359 (2022)
16. Rezvani, M., Liang, W., Liu, C., Yu, J.X.: Efficient detection of overlapping communities using asymmetric triangle cuts. IEEE Trans. Knowl. Data Eng. **30**(11), 2093–2105 (2018)
17. Shah, A., Azam, N., Alanazi, E., Yao, J.: Image blurring and sharpening inspired three-way clustering approach. Appl. Intell. **52**(15), 18131–18155 (2022)
18. Simmel, G.: The Sociology of Georg Simmel, vol. 92892. Simon and Schuster, New York (1950)
19. Wasserman, S., Faust, K.: Social network analysis: methods and applications (1994)
20. Wouter, D.N., Mrvar, A., Batagelj, V.: Ranking, Structural Analysis in the Social Sciences, pp. 244–268, 3 edn. Cambridge University Press, Cambridge (2018)
21. Xu, Y., Gu, S., Li, H., Min, F.: A hybrid approach to three-way conversational recommendation. Soft Comput. **26**(24), 13885–13897 (2022)
22. Yang, H., Wang, Y., Guo, Z.: Three-way conflict analysis based on hybrid situation tables. Inf. Sci. **628**, 522–541 (2023)
23. Yao, Y.: The geometry of three-way decision. Appl. Intell. **51**(9), 6298–6325 (2021)
24. Yao, Y.: Symbols-meaning-value (SMV) space as a basis for a conceptual model of data science. Int. J. Approx. Reason. **144**, 113–128 (2022)
25. Yao, Y.: The Dao of three-way decision and three-world thinking. Int. J. Approx. Reason. **162**, 109032 (2023)
26. Ye, X., Liu, D.: An interpretable sequential three-way recommendation based on collaborative topic regression. Expert Syst. Appl. **168**, 114454 (2021)
27. Ye, X., Liu, D., Li, T.: Multi-granularity sequential three-way recommendation based on collaborative deep learning. Int. J. Approx. Reason. **152**, 434–455 (2023)
28. Yu, H.: A framework of three-way cluster analysis. In: Polkowski, L., et al. (eds.) IJCRS 2017. LNCS (LNAI), vol. 10314, pp. 300–312. Springer, Cham (2017). https://doi.org/10.1007/978-3-319-60840-2_22
29. Yue, X., Chen, Y., Miao, D., Fujita, H.: Fuzzy neighborhood covering for three-way classification. Inf. Sci. **507**, 795–808 (2020)
30. Yue, X., Chen, Y., Yuan, B., Lv, Y.: Three-way image classification with evidential deep convolutional neural networks. Cogn. Comput. 1–13 (2021)
31. Zachary, W.W.: An information flow model for conflict and fission in small groups. J. Anthropol. Res. **33**(4), 452–473 (1977)
32. Zhang, Z., Cui, L., Pan, Z., Fang, A., Zhang, H.: A triad percolation method for detecting communities in social networks. Data Sci. J. **17**(30), 1–12 (2018)

Cognitive and Social Decision Making: Three-Way Decision Perspectives

Yiyu Yao[ID] and JingTao Yao[(⊠)][ID]

Department of Computer Science, University of Regina, Regina, SK S4S 0A2, Canada
{yiyu.yao,jingtao.yao}@uregina.ca

Abstract. This paper introduces a new research direction named cognitive and social decision making (CSDM). From a three-way decision perspective, we discuss the main issues of CSDM, including the research scope, problems, and challenges. We adopt the notion of Symbols-meaning-value spaces as a basis for studying CSDM. We examine three perspectives on CSDM, namely, (1) a research framework consisting of the philosophy, theory, and practice of CSDM, (2) a social hierarchy consisting of the three levels of individual, community, and society, and (3) an intelligence and intelligent systems view consisting of human intelligence, machine intelligence, and human-machine co-intelligence.

Keywords: Cognitive computing · collective intelligence · decision-making · granular computing · social computing · three-way decision

1 Introduction

In 2023, the journal of "Cognitive Computation," published by Springer, introduced a new section called Cognitive and Social Decision Making (CSDM) with the following description[1]:

> "Information seeking, knowledge learning, and decision-making play fundamental roles in cognitive computation. This special section focuses on multidisciplinary scientific studies of cognitive aspects and complex social environments of decision-making including their interactions with other cognitive tasks. The main aim is to introduce and promote an innovative and timely new subfield of Cognitive Computing titled: Cognitive and Social Decision Making (CSDM)."

As indicated by the list of topics, the CSDM Section aims to connect researchers from multiple communities, including decision science, intelligent data analytics, rough sets, cognitive computing, granular computing, social computing, and three-way decision [5,16,29]:

---

[1] https://www.springer.com/journal/12559/updates/18820792.

A. Campagner et al. (Eds.): IJCRS 2023, LNAI 14481, pp. 259–269, 2023.
https://doi.org/10.1007/978-3-031-50959-9_18

- Decision-making with granular computing and rough sets;
- Cognitive group decision-making;
- Neuroscience of decision-making;
- Three-way decision making;
- Social cognition and computing;
- Explainable social artificial-intelligence;
- Human-machine co-intelligence;
- Social and complex network analysis;
- Game-theoretic and information-theoretic models.

The phrase "decision making" may be viewed as an umbrella term that covers all aspects and activities related to, or supporting, decision-making. It focuses on the theory, methods, and processes of decision making both in general and from two distinctive particular angles. One is the cognitive basis of and cognitive computing approaches to decision making, and the other is the social environments for and social computing approaches to decision making. CSDM offers a triadic view for studying decision making, consisting of the cognitive basis, the computational methods, and the social contexts.

While the cognitive and social aspects of decision making have been studies extensively, a combination of the two offers a new viewpoint and a new research direction for decision making. There have been several initiatives along this direction of research. For example, a special issue on "Granular Computing and Three-way Decisions for Cognitive Analytics" in *Cognitive Computation* is devoted to the philosophy, theories, and applications of the three paradigms of computing, namely, cognitive computing, granular computing, and three-way computing [18]. Another special issue on "Uncertainty and Three-way Decision in Data Science" in the *International Journal of Approximate Reasoning* focuses on decision-making under uncertainty in data science [17]. Papers published in the two special issues address many research problems related to the above listed topics. Inspired by human cognitive processes and human intelligence, the introduction of "Cognitive and Social Decision Making" is based on existing research results in data-driven and knowledge-guided individual/social decision-making under uncertainty. In order to realize its potential value, the main objective of this position paper is to promote research on cognitive and social decision making. By recognizing the large scope and a wide spectrum of topics related to cognitive and social decision making, we will focus only on three-way decision perspectives by using several triadic structures.

We organize the discussion into three parts. In Sect. 2, we briefly review the basic ideas of three-way decision and two specific models, namely, tri-level thinking and symbols-meaning-value (SMV) spaces. In Sect. 3, based on the notion of SMV spaces, we discuss three perspectives on studies of CSDM. In Sect. 4, we summarize the main ideas and point out research issues and challenges.

2 Three-Way Decision, Tri-Level Thinking, and SMV Spaces

This section presents an overview of three-way decision and two particular models, namely, tri-level thinking and symbols-meaning-value or SMV spaces.

2.1 An Overview of Three-Way Decision

The theory of three-way decision concerns thinking, problem-solving, and computing in threes [19,20,23]. The theory is motivated by a common human practice of using triadic structures and patterns (i.e., what we do), rests on a solid cognitive basis (i.e., why we do), and follows human ways to problem solving (i.e., how we do). Humans and scientists particularly have an intriguing preference for a ternary patterned theory, model, or explanation of reality [10]. A universal practice across different cultures is using triads for perceiving, understanding, interpreting, and representing the reality [3,13]. For example, triadic structures are abundant in many legends, stories, fictional and non-fictional writings, and scientific documents. Many authors have built theories, frameworks, and models around tripartite conceptions [15].

A possible explanation of our preference for triadic thinking may be offered based on the concept of cognitive load [24]. On the one hand, three seems to the maximum number of things that our brain can handle without much deliberate and conscious effort. An important result from cognitive psychology is that humans can only hold up a few things in the short-term working memory [2,7]. The number is in the range from two to nine and three is a pivoting one. On the other hand, three seems to be the minimum number of things for our brain to form any useful and meaningful patterns. Our cognitive ability to form patterns is crucial for us to make sense of the reality and our experiences. Triadic structures are useful for their simplicity, complexity, and richness. A tripartite theory, model, or method has the cognitive advantages of simple-to-understand, easy-to-remember, and practical-to-use.

In understanding, describing, and solving real-world problems, we often use triadic structures and patterns as a cognitive aid metaphorically, conceptually, and physically. Inspired by human ways to problem solving, three-way decision systematically explores triadic structures, including triangles, tri-segment lines, tri-level hierarchies, three overlapping circles, concentric three circles, and many others [23]. In other words, triads with structures are a basic notion of three-way decision. By attaching specific interpretations and meanings to various triads, we can obtain different models and modes of three-way decision.

2.2 Tri-Level Thinking and SMV Spaces

Tri-level thinking uses a three-level hierarchy depicted in Fig. 1(a), consisting of the top, middle, and bottom levels [21]. There is a control-support relationship between two adjacent levels. In a top-down direction, a higher level is more

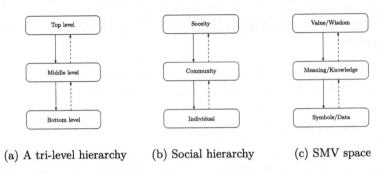

(a) A tri-level hierarchy   (b) Social hierarchy   (c) SMV space

Fig. 1. Tri-level thinking

abstract, of higher granularity, and determines a realization or implementation at a lower level, as indicated by a solid arrowed line. The same abstraction of a higher level may have multiple implementations at a lower level. In a bottom-up direction, a lower level is more concrete, of lower granularity, and supports a higher level, as indicated by a dashed arrowed line. There may exist emergent properties at a higher level that cannot be explained from the properties of entities at a lower level. Emergent properties arise from the interactions of entities at a lower level.

We may interpret the three levels of Fig. 1(a) in many different ways, such as the level of abstraction, the level of governing, the level of operation, the level of scales, and so on. The use of the tri-level hierarchy in fact appears across many disciplines and fields [21]. One example of tri-level hierarchy relevant to cognitive and social decision making is the social hierarchy of Fig. 1(a), which reflects an understanding with respect to the three micro-, meso-, and macro-scales [11]. The bottom level focuses on individuals, the middle level on local communities, and the top level on the human society as a whole. The three levels may correspond to the personal, group, and social computing and decision making. Each level addresses problems and answers questions at a different scale, from the three individual, local, and global points of view.

Another tri-level hierarchy relevant to cognitive and social decision making is the notion of SMV spaces [25, 26], as depicted in Fig. 1(c). The idea of SMV spaces derives from an integration of three powerful triads. The first triad is a division of the communication problems into three categories by Shannon and Weaver [14]. The bottom level concerns the technical problem of transmitting symbols. The middle level deals with the semantic problem of making sense of the transmitted symbols. The top level is about the effectiveness problem regarding whether the received meaning affects conduct of the receiver in the desired way. The second triad is the data-information-knowledge-wisdom hierarchy in information science and management science [1, 4, 12], which is simplified into three levels of data, knowledge, and wisdom by treating information as a kind of weak knowledge [25]. The third triad is the trilogy of perception-cognition-action in psychology [6, 8], which may be metaphorically interpreted as human seeing, knowing, and doing. We can easily observe close connections between the corresponding levels used the three triads. In some sense, the three levels may be interpreted as the input-process-output layers, corresponding to data collection and preparation, knowledge discovery and learning, and wise actions. It is reasonable to claim that the notion of SMV spaces captures the essence of the three triads, namely, the symbols-semantics-effectiveness aspects of a message, the data-knowledge-wisdom hierarchy, and the perception-cognition-action trilogy. Data are raw symbols from observation and perception, knowledge is meaning of data, which is discovered from data and summarizes and theorizes the observations, and wisdom is the wise use of knowledge and appropriate action which actualizes the value and power of knowledge. Thus, SMV spaces serve as a fundamental construct for understanding, describing, and modeling human and intelligent machine problem-solving, behaviours, and actions.

3   Three Perspectives on Cognitive and Social Decision Making

Based on the notion of SMV spaces and other triadic structures introduced in the previous section, we discuss three perspectives on cognitive and social decision, namely, a framework for studying the subject matters of CSMD, a social hierarchy relevant to CSMD, and applications of CSMD for studying human-machine co-intelligence.

3.1   Research Framework Perspective

The notion of SMV spaces provides a conceptual construct for CSDM. According to SMV spaces, we study CSDM at the three levels of symbols, meaning, and value. By following the principles of three-way decision as thinking in threes, we may approach any particular fields from roughly three main angles, namely, the philosophy, theory, and practice perspectives [22,23]. By combining the notion of SMV spaces and the philosophy-theory-practice triad, we immediately arrive at a 3×3 research framework, which is depicted in Table 1. The 3×3 framework was first introduced for building an architecture of explanation for explainable artificial intelligence or XAI, in which the last three columns are labelled by Why, What, and How [24]. The column labelled by "Philosophy" was used for constructing a conceptual model of data science [25].

Table 1. A framework for studying CSDM

	Philosophy	Theory	Practice
Value/Widsom	Wisdom is value	Decision making	Doing
Meaning/Knowledge	Knowledge is power	Knowledge learning	Knowing
Symbols/Data	Data are resources	Information seeking	Seeing

The 3×3 framework focuses on nine different topics at three levels, with three related issues at each level, namely, the philosophical position, the theoretical formulation, and the practical applications[2]. At the bottom level, we treat data as resources; the main theoretical investigations and tasks including data collection, processing, preparation, and information seeking; the level may be metaphorically viewed as the practice of seeing. At the middle level, we focus on the potential power of knowledge; the research questions are related to knowledge discovery, learning, and cognitive processing; the level is about knowing. At the bottom level, we turn our attention to value of knowledge through wise use of knowledge and knowledge supported the right course of actions; we study issues and activities related to decision making; the level concentrates on doing[3].

The framework offers three modes of operations: (1) a bottom-up evidence-based and data-driven knowledge discovery and decision making and action, (2) a top-down decision-guided leaning and information seeking and data collection, and (3) a middle-out knowledge-based decision making and action, as well as knowledge-guided information seeking and data collection. The three modes, in fact, work together iteratively [25].

3.2   Social Hierarchy Perspective

A person exists as a unique individual, as well as a member in communities and the entire human society. The same may be said of an intelligent machine or system. Studies of CSDM may explore the individual, community, and society three levels. In this case, we have a large numbers of individuals at the bottom level, a small number of communities at the middle, and the human society at the top. In the bottom-up direction, individuals form various communities, they in turn form the human society. In the top-down direction, the collective value, belief, and consciousness of the human society determine that of the communities, they in turn determine that of individuals. With respect to SMV spaces, individuals, communities, and human society have their own SMV spaces. Figure 2 describes a tri-level hierarchy with each level characterized by different types of SMV spaces. Studies of CSDM may be formulated according to the interactions of SMV spaces in the three levels.

---

[2] In formulating the Data-Knowledge-Wisdom or DKW hierarchy, we typically take the middle level as information/knowledge [21]. Sometimes, it might also be meaningful to take the bottom level as data/information. It is motivated by an observation that data and information are sometimes closely tied together and there does not exist a clear cut between data and information. Data/information may treated as the input to humans or machines. This alternative understanding is consistent with the earlier quoted description of cognitive and social decision in terms of "information seeking, knowledge learning, and decision-making," which is summarized in the column labeled by "Theory" in Table 1.

[3] We want to clarify that the questions/issues in the cells of Table 1 should be read liberally rather than literately. Depending on a particular context, it is possible to apply the same 3×3 framework, but using a set of different questions/issues.

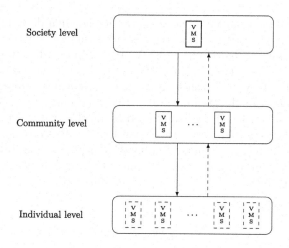

Fig. 2. Individual-community-society tri-level framework

At the bottom level, SMV spaces of individuals, either humans or machines, may be different. The differences stem from the available data, the past experience, the knowledge of an individual. The behaviour and action of an individual reflect the values and beliefs of the individual. At this level, we focus on the cognitive preference, cognitive functions, and cognitive decision-making processes of the individuals. At the middle level, we focus on communities consisting of multiple individuals. The SMV space of a community emerges from the SMV spaces of its members, as well as the relationships and interactions. That is, the SMV space of a community is based on the SMV spaces of its members on the one hand and above these SMV spaces on the other hand. The research questions at the middle levels concern the collective intelligence, collective beliefs, collective values, and collective consciousness, as well as other emergent community based cognitive functions and processes, social cognitive decision making, and cognitive social decision making. The top society level may be similarly explained with respect to the SMV spaces of multiple communities.

A main advantage of using a hierarchy is that it allows investigations on multiple levels. We may use the tri-level hierarchy to study both individual cognitive decision making and social cognitive decision making, as well as their top-down control and bottom-up support.

3.3 Intelligence and Intelligent Systems Perspective

Some of the purposes of studying CSDM are to develop cognitive principles, methods, and tools to empower human beings, and to design and implement intelligent machines by using the same principles. For such purposes, we may cast research of CSDM from a third angle based on the notion of human-machine co-intelligence [26].

Figure 3, adapted from Yiyu Yao [26], gives a triangular description of human-machine co-intelligence. It is assumed that both human intelligence and machine intelligence are interpreted based on SMV spaces. In one direction indicated by the arrowed line, human intelligence guides machine intelligence. In the other direction indicated by the dashed arrowed line, machine intelligence support human intelligence. In this guiding-supporting relationship, human intelligence is augmented, and enriched, and supported by machines. It exploits the human intuition, insights, strategical thinking, moral, and values on one hand, and the machine large volume of memory, computational power, speed, other physical advantages on the other. The combination of human intelligence and machine intelligence gives rise to human-machine co-intelligence.

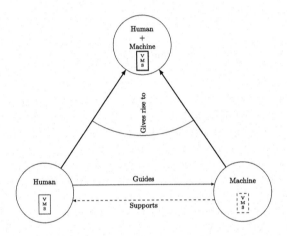

Fig. 3. Human-machine symbiosis triangle (adapted from [23])

Similar to the individual-community-society tri-level hierarchy, we explain the triangular architecture of human-machine intelligence based on the notion of SMV spaces. It is assumed that SMV spaces are a good conceptual construction for understanding and interpreting intelligence. The intelligence of any intelligent beings, either human or machine, must contain explanations at the symbols/data, meaning/knowledge, and wisdom/value three levels. They properly reflect what intelligent beings see, know, and do. Intelligence involves transforming from observations and experience into knowledge, and turning knowledge into the right course of action. That is, an intelligence being should at least be able to receive information from its environment and observe the reality, to learn from experience and adapts itself, and to act according to its knowledge and beliefs. This triadic view of intelligence may deserve further attention and articulation.

Human-machine co-intelligence is the third intelligence emerged from human-machine symbiosis. It is not about the replacement one of by the other, nor the competition of the two; it is more about a seamless integration of the two so that we can take the advantages of both human intelligence and machine intelligence.

Research on CSDM must consider human-machine co-intelligence, in order to have a superior human-guided, machine-supported cognitive decision process in a right social context. In other words, the success of CSDM may lie on the development of human-machine symbiosis that produces the third intelligence through human-machine collaboration, cooperation, coevolution, and co-creation.

4   Concluding Remarks and Future Research Challenges

In this paper, we have made an attempt to draw a research landscape and action plan for studying cognitive and social decision making or CSDM. We must acknowledge that CSDM is a complex multifaceted research field with many challenges. This paper only scratches the surface and many questions remain unanswered. To a large extent, the discussed three perspectives are based on intelligent speculations and high level conceptualization. The ideas of the paper must be further explored and articulated.

For future research, we can briefly mention three directions:

1. A full multidisciplinary study of CSDM. The notion of human-machine co-intelligence may serve as a good starting point. By treating intelligence as a basic concept, we at least discuss the goals or end results for studying CSDM, that is, developing theories, methods, and tools to realize, enhance, and enrich human intelligence, machine intelligence, and human-machine co-intelligence. The next steps may be worked towards achieving such goals.
2. Synthesis of existing results for CSMD. Although the notion of "cognitive and social decision-making" may be a new one, it contents have in fact been investigated in different fields and by many authors. The research challenges may not be the creation of new theories, but more the integration of existing results in a new context. This new context is the cognitive and social aspects of decision making. A combination of a cognitive and a social dimension may lead to new research problems.
3. New triadic structures for CSDM. The introduction of CSMD is based on a triad consisting of cognitive science, social environments, and decision science. In this paper, we have used a number of other triadic structures. Due to the power of triadic structures in theory building, it may be useful to search and examine other new triadic structures for CSDM.

In summary, this paper is perhaps more about raising questions than offering answers. In the beginning of a new field, asking the right questions may greatly increase the probability of its success. Researchers in the rough sets [9], granular computing [28], and three-way decision [23,27] communities may give beautiful answers to our questions, which motivated us writing this position paper.

Acknowledgements. This work was partially supported by Discovery Grants from NSERC, Canada. We thank the anonymous reviewers for their careful reading of our manuscript and their many insightful comments and suggestions.

References

1. Ackoff, R.L.: From data to wisdom. J. Appl. Syst. Anal. **16**, 3–9 (1989)
2. Cowan, N.: The magical number 4 in short-term memory: a reconsideration of mental storage capacity. Behav. Brain Sci. **24**, 87–185 (2000)
3. Dundes, A.: The number three in American culture. In: Dundes, A. (ed.) Every Man His Way: Readings in Cultural Anthropology, pp. 401–424. Prentice-Hall, Englewood Cliffs (1968)
4. Frické, M.: The knowledge pyramid: a critique of the DIKW hierarchy. J. Inf. Sci. **35**, 131–142 (2008)
5. Herbert, J.P., Yao, J.T.: Game-theoretic rough sets. Fund. Inform. **108**, 267–286 (2011)
6. Hilgard, E.R.: The trilogy of mind: cognition, affection, and conation. J. Hist. Behav. Sci. **16**, 107–117 (1980)
7. Miller, G.A.: The magical number seven, plus or minus two: some limits on our capacity for processing information. Psychol. Rev. **63**, 81–97 (1956)
8. Nanay, B.: Between Perception and Action. Oxford University Press, Oxford (2013)
9. Pawlak, Z.: Rough sets. Int. J. Comput. Inf. Sci. **11**, 341–356 (1982)
10. Pogliani, L., Klein, D.J., Balaban, A.T.: Does science also prefer a ternary pattern? Int. J. Math. Educ. Sci. Technol. **37**, 379–399 (2006)
11. Redhead, D., Power, E.A.: Social hierarchies and social networks in humans. Philos. Trans. R. Soc. B **377**(1845), 20200440 (2022)
12. Rowley, J.: The wisdom hierarchy: representations of the DIKW hierarchy. J. Inf. Sci. **33**, 163–180 (2007)
13. Schneider, M.S.: A Beginner's Guide to Constructing the Universe: The Mathematical Archetypes of Nature, Art, and Science. Harper, New York (1994)
14. Shannon, C.E., Weaver, W.: The Mathematical Theory of Communication. The University of Illinois Press, Urbana (1949)
15. Watson, P.: Ideas: A History, from Fire to Freud. Weidenfeld & Nicolson, London (2005)
16. Yao, J.T., Vasilakos, A.V., Pedrycz, W.: Granular computing: perspectives and challenges. IEEE Trans. Cybern. **43**(6), 1977–1989 (2013)
17. Yao, J.T., Cornelis, C., Wang, G.Y., Yao, Y.Y.: Uncertainty and three-way decision in data science. Int. J. Approx. Reason. **162**, 109024 (2023)
18. Yao, J.T., Yao, Y.Y., Ciucci, D., Huang, K.Z.: Granular computing and three-way decisions for cognitive analytics. Cogn. Comput. **14**, 1801–1804 (2022)
19. Yao, Y.Y.: Three-way decisions and cognitive computing. Cogn. Comput. **8**, 543–554 (2016)
20. Yao, Y.Y.: Three-way decision and granular computing. Int. J. Approx. Reason. **103**, 107–123 (2018)
21. Yao, Y.Y.: Tri-level thinking: models of three-way decision. Int. J. Mach. Learn. Cybern. **11**, 947–959 (2020)
22. Yao, Y.Y.: Three-way granular computing, rough sets, and formal concept analysis. Int. J. Approx. Reason. **116**, 106–125 (2020)
23. Yao, Y.Y.: The geometry of three-way decision. Appl. Intell. **51**, 6298–6325 (2021)
24. Yao, Y.Y.: Three-way decision, three-world conception, and explainable AI. In: Yao, J., Fujita, H., Yue, X., Miao, D., Grzymala-Busse, J., Li, F. (eds.) Rough Sets. IJCRS 2022. LNCS, vol. 13633, pp. 39-53. Springer, Cham (2022). https://doi.org/10.1007/978-3-031-21244-4_4

25. Yao, Y.Y.: Symbols-Meaning-Value (SMV) space as a basis for a conceptual model of data science. Int. J. Approx. Reason. **144**, 113–128 (2022)
26. Yao, Y.Y.: Human-machine co-intelligence through symbiosis in the SMV space. Appl. Intell. **53**, 2777–2797 (2023)
27. Yao, Y.Y.: The Dao of three-way decision and three-world thinking. Int. J. Approx. Reason. **162**, 109032 (2023)
28. Zadeh, L.A.: Towards a theory of fuzzy information granulation and its centrality in human reasoning and fuzzy logic. Fuzzy Sets Syst. **90**(2), 111–127 (1997)
29. Zhang, Y., Yao, J.T.: Game theoretic approach to shadowed sets: a three-way tradeoff perspective. Inf. Sci. **507**, 540–552 (2020)

New Models of Three-Way Conflict Analysis for Incomplete Situation Tables

Chengmei Lin, Qimei Xiao, Huiying Yu, and Guangming Lang$^{(\boxtimes)}$

School of Mathematics and Statistics, Changsha University of Science and
Technology, Changsha 410114, Hunan, China
langguangming1984@126.com

Abstract. For conflict problems, attitudes of agents on issues are often
lost due to some mistakes, and trisecting a set of agents is an important
research topic of conflict analysis, and three-way decisions with rankings and references provides an effective method for trisecting a set of
agents. In this paper, we divide a set of issues into two disjoint parts from
different perspectives, and give the support and opposition rankings of
issues and the support and opposition reference tuples for an incomplete situation table. Then, we design an alliance measure with regard
to an issue by a transition probability function, and develop an additive
alliance measure regarding multiple issues with conditional weights of
issues. Afterwards, we take the additive alliance measure to trisect a set
of agents towards multiple issues, and give three types of decision rules
by considering the weights of agents. Finally, we design an algorithm for
deriving three types of decision rules, and use an example to show how
to make decisions with the proposed model.

Keywords: Alliance measure · Conflict analysis · Incomplete
three-valued situation table · Ranking · Reference tuple

1 Introduction

The theory of three-way decisions, proposed by Yiyu Yao [19] in the year of
2010, is a kind of granular computing method of problem solving and information
processing with an idea of threes. Afterwards, it has attracted more and more
attention and has been widely developed in theoretical and application aspects
such as three-way clustering [1,4], three-way concept analysis [2,25], three-way
classification [3,21], three-way recommendation [14,24] and three-way decision
making [22,23]. Recently, Xu, Jia and Li [15] proposed concepts of rankings and
references for two-valued information tables, and trisected a set of objects into
three disjoint parts regarding multiple attributes with matching functions. After
that, Xu, Jia and Li [16] developed a generalized model of three-way decision
with rankings and references and designed an algorithm to compute the optimal
trisection of a set of objects in finite steps. Xu, Yao and Li [17] provided two
models of three-way decisions with ranking and references on hybrid information

tables, and put forward a measure of trisections by the positive, negative and boundary probabilities. The researches of three-way decisions with rankings and references are increasing with the development of three-way decisions.

Another topic of three-way decisions is three-way conflict analysis that studies conflict problems with three-way decisions. In fact, conflict is a kind of antagonistic relationships among agents, and studying the essence of conflicts and finding feasible strategies for solving conflicts are two main topics of conflict analysis. Initially, Pawlak [11] defined the alliance, neutrality and conflict relations between two agents with an auxiliary function, and studied conflict problems with rough set theory and graph theory. After that, many scholars [6–10,12,13,18,20] have taken three-way decisions to study conflict problems. For instance, Lang et al. [6] introduced the probabilistic model of conflict analysis by extending the Pawlak conflict analysis model on the basis of decision-theoretic rough sets and three-way decision theory, and used a pair of thresholds instead of the threshold 0.5 in the Pawlak model to define the alliance, conflict and neutrality relationship. Luo et al. [10] divided an auxiliary function into an alliance function and a conflict function so as to depict the relationship between two agents from alliance and conflict aspects. Sun et al. [12] designed two-universe models of three-way conflict analysis with probabilistic rough set theory. Yang and Suo [13] discussed the relationship between two agents in three types of incomplete situation tables when attitudes or opinions of agents on some issues are lost or missing. Yao [20] reformulated the Pawlak's model to three-way conflict models by adopting distance functions.

In practical situations, there are many incomplete three-valued situation tables to depict conflict problems. Now, there are two methods of treating the loss attitude when constructing alliance and conflict measures in incomplete three-valued situation tables. One is to use the attitude that occurs with the most frequency instead of the missing value, and the other is to use the probability that an attitude occurs. Both methods can not describe the relationship between two agents precisely. Furthermore, when all models of three-way conflict analysis discuss the relationship between two agents towards a subset of issues, they do not consider the relationship between two agents towards the complement of the subset of issues. Actually, it helps us to discuss the relationship between two agents if we consider a subset of issues and its complement at the same time. It motivates us to design models of three-way conflict analysis for incomplete three-valued situation tables with the thoughts of rankings and reference tuples. The contributions of this work are briefly listed as follows:

(1) We divide a set of issues into two disjoint parts with regard to the support and opposition, and define the support and opposition rankings of issues and the support and opposition references for incomplete situation tables. After that, we give the transition probability function to depict the transition probability between the unknown attitude and the support, neutrality and conflict attitudes.

(2) We define an alliance measure for incomplete situation tables regarding an issue with the transition probability function, and develop an additive

alliance measure towards multiple issues with the conditional weights of issues. Afterwards, we trisect a set of agents into three disjoint parts, namely, the alliance, neutrality and conflict coalitions, with the additive alliance measure and derive three types of decision rules with the weights of agents for decision making.

(3) We design an algorithm of trisecting a set of agents into the alliance, neutrality and conflict coalitions with the support and opposition rankings and the support and opposition references, and take an example to illustrate that how to trisect a set of agents into the alliance, neutrality and conflict coalitions and derive three types of decision rules for making decisions.

The rest of this paper is organized as follows. In Sect. 2, we review the basic concepts of three-way decisions with rankings and references. Section 3 develops ranking of issues and reference tuples of agents, and provides two models of three-way conflict analysis with alliance measures. Section 4 shows how to apply the proposed model to derive decision rules. Section 5 concludes the work of this paper.

2   Preliminaries

In this section, we recall three-way decisions with rankings and references.

Definition 1 *(Pawlak [11], 1984). A quadruple $S = (U, C, V, f)$ is an information table, where $U = \{x_1, x_2, \ldots, x_n\}$ is a non-empty finite set of objects, $C = \{c_1, c_2, \ldots, c_m\}$ is a non-empty finite set of attributes, $V = \bigcup\{V_c \mid c \in C\}$, where V_c is the set of values of attribute c on all objects, and f is a function from $U \times C$ into V.*

We refer information tables with two attribute values to as two-valued information tables, and most of three-way decision models with ranking and reference tuple are developed for two-valued information tables.

Definition 2 *(Xu et al. [16], 2022). Let C be a non-empty finite attribute set. A ranking order $\prec_r$ of C divides the set C into two disjoint subsets C_r^1 and C_r^2 such that $C_r = \{C_r^1, C_r^2\}$ and $C_r^2 \prec_r C_r^1$, where C_r^1 and C_r^2 represent the sets of attributes in rank 1 and rank 2, respectively.*

We can divide a set of attributes into two disjoint subsets by the ranking order, and define the ranking order according to the weights of attributes and the preferences of decision makers.

Definition 3 *(Xu et al. [16], 2022). Let $C = \{c_1, c_2, \ldots, c_m\}$ and $\prec_r$ be a ranking order of C. An m-tuple $\mathbf{x_r} = (f(x_r, c_1), f(x_r, c_2), \ldots, f(x_r, c_m))$ is called a reference tuple on $C_{\mathbf{r}}$ if it satisfies: (1) $f(x_r, c_k) \in \{0, 1\}$; (2) $f(x_r, c_i) = f(x_r, c_j)$ when $c_i \in C_r^1 \wedge c_j \in C_r^1$ or $c_i \in C_r^2 \wedge c_j \in C_r^2$, where $k, i, j \in \{1, 2, \ldots, m\}$.*

Definition 4 *(Xu et al. [16], 2022). Let $S = (U, C, V, f)$ be an information table, a ranking order $\prec_r$ divides the set C into two disjoint subsets C_r^1 and C_r^2, a reference tuple $\mathbf{x_r} = (f(x_r, c_1), f(x_r, c_2), \ldots, f(x_r, c_m))$. Then the positive, negative and boundary regions with respect to $\prec_r$ and $\mathbf{x_r}$ are defined by:*

$$POS_{(\alpha,\beta)}(U) = \{x \in U \mid M_{C_r^1}(x, x_r) \geq \alpha \wedge M_{C_r^2}(x, x_r) > \beta\},$$
$$NEG_{(\alpha,\beta)}(U) = \{x \in U \mid M_{C_r^1}(x, x_r) < \alpha \wedge M_{C_r^2}(x, x_r) \leq \beta\},$$
$$BND_{(\alpha,\beta)}(U) = \{x \in U \mid M_{C_r^1}(x, x_r) \geq \alpha \wedge M_{C_r^2}(x, x_r) \leq \beta\}$$
$$\vee \{x \in U \mid M_{C_r^1}(x, x_r) < \alpha \wedge M_{C_r^2}(x, x_r) > \beta\},$$

where

$$M_C(x, x_r) = \frac{\sum_{c_i \in C} m_{c_i}(x, x_r)}{|C|},$$

$$m_{c_i}(x, x_r) = \begin{cases} 1, & f(x, c_i) = f(x_r, c_i), \\ 0, & f(x, c_i) \neq f(x_r, c_i). \end{cases}$$

3   Three-Way Conflict Analysis Models for Incomplete Information Situation Tables

In this section, we design two models of three-way conflict analysis for incomplete situation tables.

Definition 5 *(Yang et al. [5], 1998). A triple $S = (A, I, r)$ is called an incomplete situation table, where $A = \{a_1, a_2, \ldots, a_n\}$ is a non-empty finite set of agents, $I = \{i_1, i_2, \ldots, i_m\}$ is a non-empty finite set of issues, the function $r : A \times I \longrightarrow \{-, 0, +, *\}$, where $r(a, i) = +$ means that the agent a is positive about the issue i, $r(a, i) = -$ means that the agent a is negative about the issue i, $r(a, i) = 0$ means that the agent a is neutral about the issue i, and $r(a, i) = *$ means that the attitude of agent a on issue i is lost or unknown.*

If all attitudes or opinions of agents on issues take the values from the set $\{-, 0, +\}$, then we refer it to as a complete situation table.

Definition 6. *Let $S = (A, I, r)$ be an incomplete situation table,*

(1) *the support ranking order $\prec_s$ of I divides the set I into two disjoint subsets I_s and $I_{\bar{s}}$ such that $\mathbf{I_s} = \{I_s, I_{\bar{s}}\}$ and $I_{\bar{s}} \prec_s I_s$, where I_s and $I_{\bar{s}}$ represent the bundles of supported issues and non-supported issues, respectively.*

(2) *the opposition ranking order $\prec_o$ of I divides the set I into two disjoint subsets I_o and $I_{\bar{o}}$ such that $\mathbf{I_o} = \{I_o, I_{\bar{o}}\}$ and $I_{\bar{o}} \prec_o I_o$, where I_o and $I_{\bar{o}}$ represent the bundles of opposed issues and non-opposed issues, respectively.*

From the perspective of support, we divide the set of issues into a bundle of supported issues and a bundle of non-supported issues. From the perspective of opposition, we divide the set of issues into a bundle of opposed issues and a bundle of non-opposed issues.

Definition 7. *Let* $S = (A, I, r)$ *be an incomplete situation table,*

(1) *An m-tuple* $\mathbf{a_r^s} = (r(a_r^s, i_1), r(a_r^s, i_2), \ldots, r(a_r^s, i_m))$ *is called a support reference tuple on* $\mathbf{I_s}$ *if it satisfies: (1)* $r(a_r^s, i_k) \in \{-, 0, +\}$; *(2)* $r(a_r^s, i_j) = r(a_r^s, i_k)$ *when* $i_j \in I_s \wedge i_k \in I_s$ *or* $i_j \in I_{\overline{s}} \wedge i_k \in I_{\overline{s}}$, *where* $j, k \in \{1, 2, \ldots, m\}$.

(2) *An m-tuple* $\mathbf{a_r^o} = (r(a_r^o, i_1), r(a_r^o, i_2), \ldots, r(a_r^o, i_m))$ *is called an opposition reference tuple on* $\mathbf{I_o}$ *if it satisfies: (1)* $r(a_r^o, i_k) \in \{-, 0, +\}$; *(2)* $r(a_r^o, i_j) = r(a_r^o, i_k)$ *when* $i_j \in I_o \wedge i_k \in I_o$ *or* $i_j \in I_{\overline{o}} \wedge i_k \in I_{\overline{o}}$, *where* $j, k \in \{1, 2, \ldots, m\}$.

Definition 8. *Let* $S = (A, I, r)$ *be an incomplete situation table, if the function* $P : \{*\} \times \{-, 0, +\} \longrightarrow [0, 1]$ *satisfies:*

$$(a)\ P(*, -) \geq 0, P(*, 0) \geq 0, P(*, +) \geq 0;$$
$$(b)\ P(*, -) + P(*, 0) + P(*, +) = 1,$$

then P *is called a transition probability function.*

In Definition 8, $P(*, -)$ means the probability that the unknown opinion $*$ is replaced by the opposition attitude $-$; $P(*, 0)$ means the probability that the unknown opinion $*$ is replaced by the neutral attitude 0; $P(*, +)$ means the probability that the unknown opinion $*$ is replaced by the support attitude $+$. For simplicity, we denote $P(*, -), P(*, 0)$ and $P(*, +)$ as P_{*-}, P_{*0} and P_{*+}, respectively.

Definition 9. *Suppose* $S = (A, I, r)$ *is an incomplete situation table, for an issue* $i \in I$, *the function* $e_i^a : \{-, 0, +, *\} \times \{-, 0, +, *\} \longrightarrow [0, 1]$ *towards* $i \in I$ *is called an alliance measure if it satisfies:*

(1) $e_i^a(+, +) \geq e_i^a(0, 0) > e_i^a(+, 0) = e_i^a(0, +) > e_i^a(+, -) = e_i^a(-, +);$

(2) $e_i^a(+, +) \geq e_i^a(0, 0) > e_i^a(-, 0) = e_i^a(0, -) > e_i^a(+, -) = e_i^a(-, +);$

(3) $e_i^a(-, -) \geq e_i^a(0, 0) > e_i^a(+, 0) = e_i^a(0, +) > e_i^a(+, -) = e_i^a(-, +);$

(4) $e_i^a(-, -) \geq e_i^a(0, 0) > e_i^a(-, 0) = e_i^a(0, -) > e_i^a(+, -) = e_i^a(-, +);$

(5) $e_i^a(+, *) = e_i^a(*, +) = P_{*-} \times e_i^a(+, -) + P_{*0} \times e_i^a(+, 0) + P_{*+} \times e_i^a(+, +);$

(6) $e_i^a(-, *) = e_i^a(*, -) = P_{*-} \times e_i^a(-, -) + P_{*0} \times e_i^a(-, 0) + P_{*+} \times e_i^a(-, +);$

(7) $e_i^a(0, *) = e_i^a(*, 0) = P_{*-} \times e_i^a(0, -) + P_{*0} \times e_i^a(0, 0) + P_{*+} \times e_i^a(0, +);$

(8) $e_i^a(*, *) = P_{*-}P_{*-} \times e_i^a(-, -) + P_{*-}P_{*0} \times e_i^a(-, 0) + P_{*-}P_{*+} \times e_i^a(-, +)$
$\qquad + P_{*0}P_{*-} \times e_i^a(0, -) + P_{*0}P_{*0} \times e_i^a(0, 0) + P_{*0}P_{*+} \times e_i^a(0, +)$
$\qquad + P_{*+}P_{*-} \times e_i^a(+, -) + P_{*+}P_{*0} \times e_i^a(+, 0) + P_{*+}P_{*+} \times e_i^a(+, +).$

Definition 10. *Suppose* $S = (A, I, r)$ *is an incomplete situation table,* w_i *is the weight of* $i \in I$, *where* $\sum_{i \in I} w_i = 1$, *and* $0 \leq w_i \leq 1$. *For* $\emptyset \neq J \subseteq I$, *the conditional weight* $w(i|J)$ *is defined by:*

$$w(i|J) = \frac{w_i}{\sum_{j \in J} w_j}.$$

Definition 11. *Suppose $S = (A, I, r)$ is an incomplete situation table, for $\emptyset \neq J \subseteq I$, an additive alliance measure $E_J^{\mathbf{a}} : A \times A \to [0, 1]$ towards multiple issues J is defined by:*

$$E_J^{\mathbf{a}}(a, b) = \sum_{i \in J} [\omega(i|J) \times e_i^{\mathbf{a}}(r(a, i), r(b, i))].$$

Definition 12. *Suppose $S = (A, I, r)$ is an incomplete situation table,*

(1) *the support reference tuple $\mathbf{a_r^s} = (r(a_r^s, i_1), r(a_r^s, i_2), \ldots, r(a_r^s, i_m))$, I_s and $I_{\bar{s}}$ are the bundles of supported issues and non-supported issues, respectively, the additive alliance measures $E_{I_s}^{\mathbf{a}}(a, a_r^s)$ and $E_{I_{\bar{s}}}^{\mathbf{a}}(a, a_r^s)$ with respect to I_s and $I_{\bar{s}}$, respectively, are defined by:*

$$E_{I_s}^{\mathbf{a}}(a, a_r^s) = \sum_{i \in I_s} [\omega(i|I_s) \times e_i^{\mathbf{a}}(r(a, i), r(a_r^s, i))];$$

$$E_{I_{\bar{s}}}^{\mathbf{a}}(a, a_r^s) = \sum_{i \in I_{\bar{s}}} [\omega(i|I_{\bar{s}}) \times e_i^{\mathbf{a}}(r(a, i), r(a_r^s, i))];$$

(2) *the opposition reference tuple $\mathbf{a_r^o} = (r(a_r^o, i_1), r(a_r^o, i_2), \ldots, r(a_r^o, i_m))$, I_o and $I_{\bar{o}}$ are the bundles of opposed issues and non-opposed issues, respectively, the additive alliance measures $E_{I_o}^{\mathbf{a}}(a, a_r^o)$ and $E_{I_{\bar{o}}}^{\mathbf{a}}(a, a_r^o)$ with respect to I_o and $I_{\bar{o}}$, respectively, are defined by:*

$$E_{I_o}^{\mathbf{a}}(a, a_r^o) = \sum_{i \in I_o} [\omega(i|I_o) \times e_i^{\mathbf{a}}(r(a, i), r(a_r^o, i))];$$

$$E_{I_{\bar{o}}}^{\mathbf{a}}(a, a_r^o) = \sum_{i \in I_{\bar{o}}} [\omega(i|I_{\bar{o}}) \times e_i^{\mathbf{a}}(r(a, i), r(a_r^o, i))];$$

Definition 13. *Suppose $S = (A, I, r)$ is an incomplete situation table, two pairs of thresholds (α_s, β_s) and (α_o, β_o) such that $0 \leq \alpha_s, \beta_s, \alpha_o, \beta_o \leq 1$.*

(1) *we define the alliance, neutrality and conflict coalitions with respect to the support ranking and reference tuple by:*

$$AL_{(\alpha_s, \beta_s)}^s(A) = \{a \in A | E_{I_s}^{\mathbf{a}}(a, a_r^s) \geq \alpha_s \wedge E_{I_{\bar{s}}}^{\mathbf{a}}(a, a_r^s) > \beta_s\};$$

$$CO_{(\alpha_s, \beta_s)}^s(A) = \{a \in A | E_{I_s}^{\mathbf{a}}(a, a_r^s) < \alpha_s \wedge E_{I_{\bar{s}}}^{\mathbf{a}}(a, a_r^s) \leq \beta_s\};$$

$$NE_{(\alpha_s, \beta_s)}^s(A) = (AL_{(\alpha_s, \beta_s)}^s(A) \cup CO_{(\alpha_s, \beta_s)}^s(A))^c;$$

$$= \{a \in A | [E_{I_s}^{\mathbf{a}}(a, a_r^s) \geq \alpha_s \wedge E_{I_{\bar{s}}}^{\mathbf{a}}(a, a_r^s) \leq \beta_s]$$
$$\vee [E_{I_s}^{\mathbf{a}}(a, a_r^s) < \alpha_s \wedge E_{I_{\bar{s}}}^{\mathbf{a}}(a, a_r^s) > \beta_s]\}.$$

(2) *we define the alliance, neutrality and conflict coalitions with respect to the opposition ranking and reference tuple by:*

$$AL_{(\alpha_o, \beta_o)}^o(A) = \{a \in A | E_{I_o}^{\mathbf{a}}(a, a_r^o) \geq \alpha_o \wedge E_{I_{\bar{o}}}^{\mathbf{a}}(a, a_r) > \beta_o\};$$

$$CO_{(\alpha_o, \beta_o)}^o(A) = \{a \in A | E_{I_o}^{\mathbf{a}}(a, a_r^o) < \alpha_o \wedge E_{I_{\bar{o}}}^{\mathbf{a}}(a, a_r^o) \leq \beta_o\};$$

$$NE_{(\alpha_o, \beta_o)}^o(A) = (AL_{(\alpha_o, \beta_o)}^o(A) \cup CO_{(\alpha_o, \beta_o)}^o(A))^c;$$

$$= \{a \in A | [E_{I_o}^{\mathbf{a}}(a, a_r^o) \geq \alpha_o \wedge E_{I_{\bar{o}}}^{\mathbf{a}}(a, a_r^o) \leq \beta_o]$$
$$\vee [E_{I_o}^{\mathbf{a}}(a, a_r^o) < \alpha^o \wedge E_{I_{\bar{o}}}^{\mathbf{a}}(a, a_r^o) > \beta_o]\}.$$

According to Definition 13(1), we trisect the set of agents into three disjoint parts $AL^s_{(\alpha_s,\beta_s)}(A)$, $CO^s_{(\alpha_s,\beta_s)}(A)$ and $NE^s_{(\alpha_s,\beta_s)}(A)$ with regard to the support reference tuple. Moreover, by Definition 13(2), we trisect the set of agents into three disjoint parts $AL^o_{(\alpha_o,\beta_o)}(A)$, $CO^o_{(\alpha_o,\beta_o)}(A)$ and $NE^o_{(\alpha_o,\beta_o)}(A)$ with regard to the opposition reference tuple.

Theorem 1. *Suppose $S = (A, I, r)$ is an incomplete situation table, two pairs of thresholds (α_1, β_1) and (α_2, β_2) such that $0 \le \alpha_1, \beta_1, \alpha_2, \beta_2 \le 1$, and $\star \in \{s, o\}$.*

(1) *if $\alpha_1 = \alpha_2$, and $\beta_1 \le \beta_2$, then*

$$AL^\star_{(\alpha_2,\beta_2)}(A) \subseteq AL^\star_{(\alpha_1,\beta_1)}(A),$$
$$CO^\star_{(\alpha_1,\beta_1)}(A) \subseteq CO^\star_{(\alpha_2,\beta_2)}(A),$$
$$NE^\star_{(\alpha_1,\beta_1)}(A) \subseteq NE^\star_{(\alpha_2,\beta_2)}(A);$$

(2) *if $\beta_1 = \beta_2$, and $\alpha_1 \le \alpha_2$, then*

$$AL^\star_{(\alpha_2,\beta_2)}(A) \subseteq AL^\star_{(\alpha_1,\beta_1)}(A),$$
$$CO^\star_{(\alpha_1,\beta_1)}(A) \subseteq CO^\star_{(\alpha_2,\beta_2)}(A),$$
$$NE^\star_{(\alpha_1,\beta_1)}(A) \subseteq NE^\star_{(\alpha_2,\beta_2)}(A);$$

(3) *if $\alpha_1 \le \alpha_2$, and $\beta_1 \le \beta_2$, then*

$$AL^\star_{(\alpha_1,\beta_1)}(A) \subseteq AL^\star_{(\alpha_2,\beta_2)}(A),$$
$$CO^\star_{(\alpha_2,\beta_2)}(A) \subseteq CO^\star_{(\alpha_1,\beta_1)}(A);$$

(4) *if $\alpha_1 \ge \alpha_2$, and $\beta_1 \ge \beta_2$, then*

$$AL^\star_{(\alpha_2,\beta_2)}(A) \subseteq AL^\star_{(\alpha_1,\beta_1)}(A),$$
$$CO^\star_{(\alpha_1,\beta_1)}(A) \subseteq CO^\star_{(\alpha_2,\beta_2)}(A).$$

Actually, the support and opposition ranking orderings depict the preferences of decision makers for issues, and the support and opposition reference tuples depict proposals with preferences of decision makers. That is, the decision makers hope that the proposal depicted by the support reference tuple can be passed, and the proposal depicted by the opposition reference tuple can be rejected.

Theorem 2. *Suppose $S = (A, I, r)$ is an incomplete situation table, ϕ_a is the weight of $a \in A$ such that $\sum_{a \in A} \phi_a = 1$, and $0 \le \phi_a \le 1$, two pairs of thresholds (μ_s, ν_s) and (μ_o, ν_o) such that $0 \le \nu_s, \mu_s \le 1$ and $0 \le \nu_o, \mu_o \le 1$,*

(1) *if the support reference tuple* $\mathbf{a_r^s}$ *depicts a proposal,*

(**P**) $\displaystyle\sum_{a\in AL^s_{(\alpha_s,\beta_s)}(A)} \phi_a \geq \mu_s \wedge \sum_{a\in CO^s_{(\alpha_s,\beta_s)}(A)} \phi_a \leq \nu_s$

$\implies$ *the proposal is passed;*

(**R**) $\displaystyle\sum_{a\in AL^s_{(\alpha_s,\beta_s)}(A)} \phi_a < \mu_s \wedge \sum_{a\in CO^s_{(\alpha_s,\beta_s)}(A)} \phi_a > \nu_s$

$\implies$ *the proposal is rejected;*

(**N**) $\left[\displaystyle\sum_{a\in AL^s_{(\alpha_s,\beta_s)}(A)} \phi_a \geq \mu_s \wedge \sum_{a\in CO^s_{(\alpha_s,\beta_s)}(A)} \phi_a > \nu_s\right]$

$\vee \left[\displaystyle\sum_{a\in AL^s_{(\alpha_s,\beta_s)}(A)} \phi_a < \mu_s \wedge \sum_{a\in CO^s_{(\alpha_s,\beta_s)}(A)} \phi_a \leq \nu_s\right]$

$\implies$ *the proposal is delayed.*

(2) *if the opposition reference tuple* $\mathbf{a_r^o}$ *depicts a proposal,*

(**P**) $\displaystyle\sum_{a\in AL^o_{(\alpha_o,\beta_o)}(A)} \phi_a \geq \mu_o \wedge \sum_{a\in CO^o_{(\alpha_o,\beta_o)}(A)} \phi_a \leq \nu_o$

$\implies$ *the proposal is passed;*

(**R**) $\displaystyle\sum_{a\in AL^o_{(\alpha_o,\beta_o)}(A)} \phi_a < \mu_o \wedge \sum_{a\in CO^o_{(\alpha_o,\beta_o)}(A)} \phi_a > \nu_o$

$\implies$ *the proposal is rejected;*

(**N**) $\left[\displaystyle\sum_{a\in AL^o_{(\alpha_o,\beta_o)}(A)} \phi_a \geq \mu_o \wedge \sum_{a\in CO^o_{(\alpha_o,\beta_o)}(A)} \phi_a > \nu_o\right]$

$\vee \left[\displaystyle\sum_{a\in AL^o_{(\alpha_o,\beta_o)}(A)} \phi_a < \mu_o \wedge \sum_{a\in CO^o_{(\alpha_o,\beta_o)}(A)} \phi_a \leq \nu_o\right]$

$\implies$ *the proposal is delayed.*

Next, we provide an algorithm for obtaining decision rules with the support and opposition rankings and the support and opposition reference tuples.

Algorithm 1. The algorithm of constructing decision rules with the rankings and references for incomplete situation tables.

Input: $S = (A, I, r)$, $\mathbf{a_r^s}$, $\mathbf{a_r^o}$, P_{*-}, P_{*0}, P_{*+}, ω_i, ϕ_a, (α_s, β_s), (α_o, β_o), (μ_s, ν_s) and (μ_o, ν_o);
Output: Identify that the proposals given by $\mathbf{a_r^s}$ and $\mathbf{a_r^o}$ are passed, delayed or rejected;

1: Determine the alliance degrees of between an agent and the support (respectively, opposition) reference towards an issue;
2: Compute the conditional weights $\omega(i|I_s)$, $\omega(i|I_{\bar{s}})$, $\omega(i|I_o)$ and $\omega(i|I_{\bar{o}})$;
3: Calculate the additive alliance measures $E_{I_s}^{\mathbf{a}}(a, a_r^s)$, $E_{I_{\bar{s}}}^{\mathbf{a}}(a, a_r^s)$, $E_{I_o}^{\mathbf{a}}(a, a_r^o)$ and $E_{I_{\bar{o}}}^{\mathbf{a}}(a, a_r^o)$ for $a \in A$;
4: Construct $AL_{(\alpha_s,\beta_s)}^s(A)$, $CO_{(\alpha_s,\beta_s)}^s(A)$, $NE_{(\alpha_s,\beta_s)}^s(A)$, $AL_{(\alpha_o,\beta_o)}^o(A)$, $CO_{(\alpha_o,\beta_o)}^o(A)$ and $NE_{(\alpha_o,\beta_o)}^o(A)$;
5: Identify that the proposal is passed, delayed or rejected.

4 An Application of the Three-Way Conflict Analysis Model

In this section, a polling question is represented by Table 1, where the agent set $A = \{a_1, a_2, \ldots, a_6\}$ represents six voters, and the issue set $I = \{i_1, i_2, \ldots, i_5\}$ represents five candidates, $r(a, i) = +$ means that the voter a supports the candidate i, $r(a, i) = -$ means that the voter a opposes the candidate i, $r(a, i) = 0$ means that the voter a is neutral about the candidate i, and $r(a, i) = *$ means that the attitude of voter a on candidate i is unknown, the reference tuples $\mathbf{a_r^s}$ and $\mathbf{a_r^o}$ depict proposals with preferences of decision makers. Under the background of incomplete situation table, we study the alliance, neutrality and conflict relationship between voters and decision-makers, and judge whether the reference tuple passes.

Table 1. An incomplete situation table with two reference tuples.

A	i_1	i_2	i_3	i_4	i_5
a_1	$-$	$+$	$+$	$+$	$+$
a_2	$*$	0	$-$	$*$	$-$
a_3	$+$	$-$	$-$	$-$	0
a_4	0	$-$	$+$	$*$	$-$
a_5	$+$	$*$	$-$	$-$	$-$
a_6	0	$+$	$*$	0	$+$
a_r^s	$+$	$-$	$-$	0	$+$
a_r^o	0	$-$	$-$	$+$	$+$

Example 1. For simplicity, by taking the function $e_i^a(r(a, i), r(b, i)) = 1 - \frac{|r(a,i)-r(b,i)|}{2}$, and $P_{*-} = P_{*0} = P_{*+} = \frac{1}{3}$, we have an alliance measure towards an issue:

$$e_i^{\mathbf{a}}(-,-) = e_i^{\mathbf{a}}(0,0) = e_i^{\mathbf{a}}(+,+) = 1;$$

$$e_i^{\mathbf{a}}(-,0) = e_i^{\mathbf{a}}(0,-) = e_i^{\mathbf{a}}(+,0) = e_i^{\mathbf{a}}(0,+) = \frac{1}{2};$$

$$e_i^{\mathbf{a}}(-,+) = e_i^{\mathbf{a}}(+,-) = 0;$$

$$e_i^{\mathbf{a}}(+,*) = e_i^{a}(*,+) = e_i^{\mathbf{a}}(-,*) = e_i^{\mathbf{a}}(*,-) = \frac{1}{2};$$

$$e_i^{\mathbf{a}}(0,*) = e_i^{\mathbf{a}}(*,0) = \frac{2}{3}; e_i^{\mathbf{a}}(*,*) = \frac{5}{9}.$$

(1) From Table 1, we have the support reference tuple $\mathbf{a_r^s} = (+,-,-,0,+)$. Obviously, $I_s = \{i_1, i_5\}$ and $I_{\overline{s}} = \{i_2, i_3, i_4\}$. By taking $\omega_{i_1} = \frac{3}{10}$, $\omega_{i_2} = \frac{1}{10}$, $\omega_{i_3} = \frac{2}{10}$, $\omega_{i_4} = \frac{2}{10}$ and $\omega_{i_5} = \frac{2}{10}$, we have:

$$E_{I_s}^{\mathbf{a}}(a_r^s, a_1) = \frac{2}{5}; E_{I_{\overline{s}}}^{\mathbf{a}}(a_r^s, a_1) = \frac{1}{5};$$

$$E_{I_s}^{\mathbf{a}}(a_r^s, a_2) = \frac{3}{10}; E_{I_{\overline{s}}}^{\mathbf{a}}(a_r^s, a_2) = \frac{23}{30};$$

$$E_{I_s}^{\mathbf{a}}(a_r^s, a_3) = \frac{4}{5}; E_{I_{\overline{s}}}^{\mathbf{a}}(a_r^s, a_3) = \frac{4}{5};$$

$$E_{I_s}^{\mathbf{a}}(a_r^s, a_4) = \frac{3}{10}; E_{I_{\overline{s}}}^{\mathbf{a}}(a_r^s, a_4) = \frac{7}{15};$$

$$E_{I_s}^{\mathbf{a}}(a_r^s, a_5) = \frac{3}{5}; E_{I_{\overline{s}}}^{\mathbf{a}}(a_r^s, a_5) = \frac{7}{10};$$

$$E_{I_s}^{\mathbf{a}}(a_r^s, a_6) = \frac{7}{10}; E_{I_{\overline{s}}}^{\mathbf{a}}(a_r^s, a_6) = \frac{3}{5}.$$

Then, by Definition 13, we have the alliance, neutrality and conflict coalitions:

$$AL_{(\frac{3}{5},\frac{1}{2})}^s(A) = \{a_3, a_5, a_6\}; CO_{(\frac{3}{5},\frac{1}{2})}^s(A) = \{a_1, a_4\}; NE_{(\frac{3}{5},\frac{1}{2})}^s(A) = \{a_2\}.$$

Finally, by taking $\phi_{a_1} = 0.2$, $\phi_{a_2} = 0.1$, $\phi_{a_3} = 0.2$, $\phi_{a_4} = 0.2$, $\phi_{a_5} = 0.2$, $\phi_{a_6} = 0.1$, $\mu_s = 0.5$ and $\nu_s = 0.4$, we get:

$$\sum_{a \in AL_{(\frac{3}{5},\frac{1}{2})}^s(A)} \phi_a = \phi_{a_3} + \phi_{a_5} + \phi_{a_6} = 0.5;$$

$$\sum_{a \in CO_{(\frac{3}{5},\frac{1}{2})}^s(A)} \phi_a = \phi_1 + \phi_4 = 0.3.$$

Therefore, the proposal depicted by the support reference tuple $\mathbf{a_r^s}$ is passed. This means that decision-makers' proposals are accepted by voters.

(2) From Table 1, we have the opposition reference tuple $\mathbf{a_r^o} = (0, -, -, +, +)$. Obviously, $I_0 = \{i_2, i_3\}$ and $I_{\overline{s}} = \{i_1, i_4, i_5\}$. By taking $\omega_{i_1} = \frac{3}{10}, \omega_{i_2} = \frac{1}{10}, \omega_{i_3} = \frac{2}{10}, \omega_{i_4} = \frac{2}{10}$ and $\omega_{i_5} = \frac{2}{10}$, we have:

$$E_{I_o}^{\mathbf{a}}(a_r^o, a_1) = 0; E_{I_{\overline{s}}}^{\mathbf{a}}(a_r^o, a_1) = \frac{11}{14};$$

$$E_{I_o}^{\mathbf{a}}(a_r^o, a_2) = \frac{5}{6}; E_{I_{\overline{s}}}^{\mathbf{a}}(a_r^o, a_2) = \frac{3}{7};$$

$$E_{I_o}^{\mathbf{a}}(a_r^o, a_3) = 1; E_{I_{\overline{s}}}^{\mathbf{a}}(a_r^o, a_3) = \frac{5}{14};$$

$$E_{I_o}^{\mathbf{a}}(a_r^o, a_4) = \frac{1}{3}; E_{I_{\overline{s}}}^{\mathbf{a}}(a_r^o, a_4) = \frac{4}{7};$$

$$E_{I_o}^{\mathbf{a}}(a_r^o, a_5) = \frac{5}{6}; E_{I_{\overline{s}}}^{\mathbf{a}}(a_r^o, a_5) = \frac{3}{14};$$

$$E_{I_o}^{\mathbf{a}}(a_r^o, a_6) = \frac{1}{3}; E_{I_{\overline{s}}}^{\mathbf{a}}(a_r^o, a_6) = \frac{6}{7}.$$

Then, by Definition 13, by taking $\alpha_o = \frac{1}{2}$ and $\beta_o = \frac{1}{3}$, we have:

$$AL_{(\frac{1}{2},\frac{1}{3})}^o(A) = \{a_2, a_3\}; CO_{(\frac{1}{2},\frac{1}{3})}^o(A) = \{\emptyset\}; NE_{(\frac{1}{2},\frac{1}{3})}^o(A) = \{a_1, a_4, a_5, a_6\}.$$

Finally, by taking $\phi_{a_1} = 0.2$, $\phi_{a_2} = 0.1$, $\phi_{a_3} = 0.2$, $\phi_{a_4} = 0.2$, $\phi_{a_5} = 0.2$, $\phi_{a_6} = 0.1$, $\mu_o = 0.5$ and $\nu_o = 0.4$, we get:

$$\sum_{a \in AL_{(\frac{1}{2},\frac{1}{3})}^o(A)} \phi_a = \phi_2 + \phi_3 = 0.3,$$

$$\sum_{a \in CO_{(\frac{1}{2},\frac{1}{3})}^o(A)} \phi_a = 0.$$

Therefore, the proposal depicted by the opposition reference tuple $\mathbf{a_r^o}$ is rejected. This means that decision-makers' proposal is rejected by voters.

5 Conclusion

In this paper, we have provided the support and opposition ranking orders, and the support and opposition reference tuples for an incomplete situation table. Then, we have defined an alliance measure regarding an issue with the transition probability function and designed an additive alliance measure towards multiple issues. After that, we have defined the alliance, neutrality and conflict coalitions with respect to the reference tuples, and derived decision rules with the weights of agents. Finally, we have shown that how to apply the proposed models to derive decision rules with an example.

In the future, we will study how to construct alliance, neutrality and conflict coalitions and derive decision rules for complex incomplete situation tables with rankings and reference tuples. Furthermore, we will investigate that how to find feasible strategy for solving conflicts with the thoughts of rankings and references.

Acknowledgements. This work is supported by the National Natural Science Foundation of China (No. 62076040), Hunan Provincial Natural Science Foundation of China (No. 2020JJ3034), the Scientific Research Fund of Hunan Provincial Education Department (No. 22A0233), the Scientific Research Fund of Chongqing Key Laboratory of Computational Intelligence (No. 2020FF04), the Graduate Research Innovation Project of Hunan Province (No. CX20220952).

References

1. Ali, B., Azam, N., Yao, J.T.: A three-way clustering approach using image enhancement operations. Int. J. Approx. Reason. **149**, 1–38 (2022)
2. Gaeta, A., Loia, V., Orciuoli, F., Parente, M.: Spatial and temporal reasoning with granular computing and three way formal concept analysis. Granul. Comput. **6**, 797–813 (2021)
3. Han, X.Y., Zhu, X.B., Pedrycz, W., Li, Z.W.: A three-way classification with fuzzy decision trees. Appl. Soft Comput. **132**, 109788 (2023)
4. Khan, G.A., Hu, J., Li, T., Diallo, B., Zhao, Y.: Multi-view low rank sparse representation method for three-way clustering. Int. J. Mach. Learn. Cybern. **13**, 233–253 (2021)
5. Kryszkiewicz, M.: Rough set approach to incomplete information systems. Inf. Sci. **112**, 39–49 (1998)
6. Lang, G.M., Mao, D.Q., Cai, M.G.: Three-way decision approaches to conflict analysis using decision-theoretic rough set theory. Inf. Sci. **406–407**, 185–207 (2017)
7. Lang, G.M.: A general conflict analysis model based on three-way decision. Int. J. Mach. Learn. Cybern. **11**, 1083–1094 (2020)
8. Lang, G.M., Mao, D.Q., Fujita, H.: Three-way gruop conflict analysis based on Pythagorean fuzzy set theory. IEEE Trans. Fuzzy Syst. **28**, 447–461 (2020)
9. Lang, G.M., Yao, Y.Y.: New measures of alliance and conflict for three-way conflict analysis. Int. J. Approx. Reason. **132**, 49–69 (2021)
10. Luo, J.F., Hu, M.J., Lang, G.M., Yang, X., Qin, K.Y.: Three-way conflict analysis based on alliance and conflict functions. Inf. Sci. **594**, 322–359 (2022)
11. Pawlak, Z.: On conflicts. Int. J. Man Mach. Stud. **21**, 127–134 (1984)
12. Sun, B.Z., Chen, X.T., Zhang, L.Y., Ma, W.M.: Three-way decision making approach to conflict analysis and resolution using probabilistic rough set over two universes. Inf. Sci. **807**, 809–822 (2020)
13. Suo, L.W.Q., Yang, H.L.: Three-way conflict analysis based on incomplete situation tables: a tentative study. Int. J. Approx. Reason. **145**, 51–74 (2022)
14. Xu, Y.Y., Gu, S.M., Li, H.X., Min, F.: A hybrid approach to three-way conversational recommendation. Soft Comput. **26**, 13885–13897 (2022)
15. Xu, W.Y., Jia, B., Li, X.N.: A two-universe model of three-way decision with ranking and reference tuple. Inf. Sci. **581**, 808–839 (2021)
16. Xu, W.Y., Jia, B., Li, X.N.: A generalized model of three-way decision with ranking and reference tuple. Int. J. Approx. Reason. **144**, 51–68 (2022)
17. Xu, W.Y., Yan, Y.C., Li, X.N.: Three-way decision with ranking and reference tuple on information tables. Inf. Sci. **613**, 682–716 (2022)
18. Yang, H.L., Wang, Y., Guo, Z.L.: Three-way conflict analysis based on hybrid situation tables. Inf. Sci. **628**, 522–541 (2023)
19. Yao, Y.Y.: Three-way decision with probabilistic rough sets. Inf. Sci. **180**, 341–353 (2010)

20. Yao, Y.Y.: Three-way conflict analysis: reformulations and extensions of the Pawlak model. Knowl.-Based Syst. **180**, 26–37 (2019)
21. Yue, X.D., Liu, S.W., Qian, Q., Miao, D.Q., Gao, C.: Semi-supervised shadowed sets for three-way classification on partial labeled data. Inf. Sci. **607**, 1372–1390 (2022)
22. Zhan, J.M., Jiang, H.B., Yao, Y.Y.: Three-way multiattribute decision-making based on outranking relations. IEEE Trans. Fuzzy Syst. **29**, 2844–2858 (2021)
23. Zhan, J.M., Wang, J.J., Ding, W.P., Yao, Y.Y.: Three-way behavioral decision making with hesitant fuzzy information systems: survey and challenges. IEEE/CAA J. Autom. Sin. **10**, 330–350 (2023)
24. Zhang, H.R., Min, F., Shi, B.: Regression-based three-way recommendation. Inf. Sci. **378**, 444–461 (2017)
25. Zhi, H.L., Qi, J.J., Qian, T., Ren, R.: Conflict analysis under one-vote veto based on approximate three-way concept lattice. Inf. Sci. **516**, 316–330 (2020)

Granular-Ball Three-Way Decision

Xin Yang[1][(✉)], Yanhua Li[1], Shuyin Xia[2], Xiaoyu Lian[2], Guoyin Wang[2],
and Tianrui Li[3]

[1] School of Computing and Artificial Intelligence, Southwestern University of
Finance and Economics, Chengdu 611130, China
yangxin@swufe.edu.cn
[2] Chongqing Key Laboratory of Computational Intelligence, Key Laboratory of Big
Data Intelligent Computing, Key Laboratory of Cyberspace Big Data Intelligent
Security, Ministry of Education, Chongqing University of Posts and
Telecommunications, Chongqing 400065, China
{xiasy,wanggy}@cqupt.edu.cn
[3] School of Computing and Artificial Intelligence, Southwest Jiaotong University,
Chengdu 611756, China
trli@swjtu.edu.cn

Abstract. By thinking, information processing and decision-making in threes, the idea, theory and methods of three-way decision have been successfully applied to various domains. However, the current three-way decision has two following limitations. On the one hand, the narrow three-way decision associated with rough sets either has trouble processing continuous data or fails to represent knowledge by equivalence classes. On the other hand, the inputs of generalized three-way decision are individual objects rather than equivalence classes, which reduces the decision efficiency. To this end, we try to integrate efficient granular-ball computing into three-way decision. Firstly, we propose a novel model, i.e., granular-ball three-way decision to improve the efficiency and robustness of three-way decision. Secondly, sequential three-way decision based on granular-ball is presented to investigate the appropriate multi-granularity structures and represent the same object at different granularities. Finally, we analyze the advantages of granular-balls to strengthen the real-world applications of three-way decision.

Keywords: Three-way Decision · Sequential Three-way Decision ·
Granular-ball Computing · Multi-granularity · Robustness

1 Introduction

Three-way decision (3WD) is a useful method to address the complex human cognitive problem with uncertain and insufficient information [1,32], especially in the open dynamic environment [28,29]. Compared to the traditional two-way decision, the third option in 3WD, i.e., the noncommitment decision provides more chance to reduce the decision risk when we consider the cost and uncertainty of problem-solving. Hence, many "three-way+methods" are introduced in

A. Campagner et al. (Eds.): IJCRS 2023, LNAI 14481, pp. 283–295, 2023.
https://doi.org/10.1007/978-3-031-50959-9_20

last ten years, such as three-way approximation [35], three-way concept analysis [7], three-way clustering [17,37], three-way classifications [5], behavioral theories-based 3WD [18,38], etc. Moreover, 3WD is also applied into various domains [27], such as medical diagnosis [18], stock prediction [30], recommendation system [36], credit evaluation [9], attribute reduction [4], etc.

Originally derived from rough sets, the narrow 3WD is gradually developed based on the extended models of Palwak rough sets [31], probabilistic rough sets, and decision-theoretic rough sets [32]. However, the continuous data is difficult to handle in the above three kinds of 3WD. To solve this problem, Hu et al. [6] proposed the neighborhood rough sets by using a neighborhood relation instead of an equivalence relation, which can handle continuous data but fail to represent equivalence classes due to "heterogeneity transmission" [22]. Then, the above narrow 3WD methods are extended into generalized 3WD without the concept of equivalence classes, for example, Anwar Shah et al. [13] studied an ensemble face recognition mechanism based on 3WD, the inputs are single objects rather than equivalence classes, which makes the decision less efficient. More, to improve the decision accuracy and save cost, Yao [33] proposed the sequential three-way decision (S3WD) and pointed out that the challenge of this method lies in how to construct reasonable multi-granularity structures and how to represent the same object at different granularities.

The granular-ball computing is an efficient tool to solve the aforementioned issues in 3WD. Wang and Xia et al. [21] used hypersphere as a "granularity" to represent the dataset and proposed a granular-ball computing method, which only needs two features for any dimension: center and radius. Since granular-ball computing was put forth, it has obtained expansive attention and acquired abundant achievement, such as granular-ball rough set [22], granular-ball classification [20,21,23], granular-ball clustering [25,26], granular-ball feature selection [12], granular-ball attribute reduction [3,15], granular-ball evolutionary computation [24] and granular-ball neural networks [14]. Notably, the granular-ball rough set [22] unified the Pawlak rough set and the neighborhood rough set, enabling it to handle continuous data and use equivalence classes for knowledge representation. Granular-ball computing is based on balls instead of sample points, which greatly reduces the number of input training samples and thus improves efficiency. The radius of a granular-ball naturally reflects its granularity, which helps S3WD to construct an appropriate multi-grained structure adaptively and represent the same objects at multi-level. Additionally, objects with noise are less likely to impact granular-balls characterized by coarse-grained features, underscoring the robustness of granular-ball computing. Given the clear advantages that granular-balls offer in addressing the complexities of 3WD, this paper integrates granular-balls into the 3WD framework. Consequently, we propose both the granular-ball three-way decision method (GB3WD) and the granular-ball sequential three-way decision method (GBS3WD).

The main contributions of this paper are shown as follows:

- In this paper, we integrate granular-ball into 3WD for the first time and propose granular-ball three-way decision methods, which are efficient, robust, and capable of handling continuous data.
- We propose a granular-ball sequential three-way decision method, the characteristics of granular-ball contribute to establishing a more appropriate multi-granularity structure and multiple representations of objects, which in turn leads to more appropriate 3WD boundaries.
- We study how to utilize the advantages of granular-ball to strengthen the applications of 3WD.

2   Preliminaries

This section briefly reviews some basic concepts with respect to 3WD and granular-ball rough set.

2.1   Three-Way Decision

Suppose a universe U, X is a concept. In the narrow 3WD based on rough sets, equivalence classes $[x]$ are obtained by equivalence relation. By the upper and lower approximations of the concept X, we divided the universe U into three regions, i.e., positive region $POS(X)$, boundary region $BND(X)$, and negative region $NEG(X)$, which represent the accept, noncommitment and reject decisions, respectively. The probabilistic rough set-based 3WD method makes decisions according to the conditional probability $Pr(X|[x])$ of the equivalence class $[x]$ belonging to X and the decision thresholds α and β. If $Pr(X|[x]) \geq \alpha$, then we accept $x \in X$; if $\beta < Pr(X|[x]) < \alpha$, then we delay to decision; if $Pr(X|[x]) \leq \beta$, then we reject $x \in X$. To obtain appropriate decision thresholds α and β, Yao [34] introduced the Bayesian decision process into decision-theoretic rough sets based on 3WD.

Moreover, as the classical dynamic models of 3WD, sequential three-way decision (S3WD) is introduced to deal with the multiple stages decision-making by constructing the multilevel granular structures [33]. At the begining, objects are assigned into $POS(X_i)$, $BND(X_i)$ and $NEG(X_i)$ at the coarser granularity. Then, for the objects in $BND(X_i)$, we make further decisions at the finer granularity with more detailed information. In contrast to making decisions at the single granularity, S3WD can be used to balance the decision quality and decision cost. The construction of multilevel granularities and the different representations of objects are two key issues in S3WD.

2.2   Granular-Ball Rough Set

Human cognition possesses a cognitive mechanism known as "global priority" [2], which enables the processing of information input based on coarse-grained details, thus providing adaptive multi-grained descriptive capabilities. Building upon this theory, Wang and Xia [16,21] proposed multi-granularity granular-ball computing by using the granular-ball to cover sample points. This approach replaces individual sample points with the granular-ball as inputs, significantly reducing the number of required training samples. Additionally, the coarse-grained nature of the granular-ball ensures that they are less susceptible to the influence of fine-grained sample points, thereby enhancing the algorithm's robustness. Moreover, each granular-ball $GB = \{x_i, i = 1, 2, \cdots, N\}$ can be expressed using only two features: its center C and radius r, applicable across any dimension. The expression is as follows.

$$C = \frac{1}{N} \sum_{i=1}^{N} x_i,$$

$$r = \frac{1}{N} \sum_{i=1}^{N} \| x_i - C \|, \tag{1}$$

where N denotes the number of samples in the granular-ball, $\| x_i - C \|$ is the distance between x_i and center C. The size of the radius indicates the granularity of different thicknesses. Larger radii result in fewer granular-balls, indicating a coarser level of granularity. More efficient granular-ball computation contributes to improved algorithm robustness. Currently, this method has been successfully employed in the field of rough sets. While the Pawlawk rough set utilizes equivalence classes for knowledge representation, it cannot handle continuous data. Conversely, neighborhood rough sets can address continuous data, but encounter the challenge of "heterogeneous transmission", hindering knowledge representation. To overcome these limitations, Xia et al. [22] introduced granular-balls into rough set theory, proposing granular-ball rough sets. This framework allows for the processing of continuous data while utilizing equivalence classes for knowledge representation. The specific models are defined and described as follows.

Definition 1 [22]. *Let $U = \{x_1, x_2, \cdots, x_n\}$ is a non-empty finite set of real space. $\forall \ x_i \in U$, a granular-ball GB_j is defined as:*

$$GB_j = \{x \mid x \in U, \| x, c_j \| \leq r_j\}, \tag{2}$$

where c_j and r_j denote the center and radius of GB_j, respectively. Obviously, the larger the radius r_j of the granular-ball, the coarser the granularity size, and vice versa, the finer the granularity size.

Definition 2 [22]. *Let $\langle U, A, V, f \rangle$ be an information system, U is the set of objects, A denotes the set of all attributes, V is the values of attributes, and f denotes a mapping function that $f : U \times A \rightarrow V$. $\forall \ x, y \in U$ and $B \subseteq A$, the indiscernible granular-ball relation $INDGB(B)$ of the attribute set B is defined as:*

$$INDGB(B) = \{(x, y) \in U^2 | f(x, a) = f(y, a) = GB, \forall a \in B\}, \tag{3}$$

where a is an attribute of B, If $(x, y) \in INDGB(B)$, then x and y are indiscernible according to attribute set B, denoted as $x \sim y$. In granular-ball rough set, $INDGB(B)$ denotes an equivalence relation on U, which *can create a partition of U, denoted as U/GB(B). An element $[x]_{GB(B)} = \{y \in U | (x,y) \in INDGB(B)\}$ in U/GB(B) represents an equivalence class generated by granular-ball computing.*

Definition 3 [22]. *Let $\langle U, A, V, f \rangle$ be an information system. For $\forall\ B \in A$, GBR_B denotes a corresponding relation on U. $\forall\ X \in U$, the upper and lower approximations of X based on attribute set B can be described as follows:*

$$\overline{GBR_B}X = \cup \left\{ [x]_B \in U/GB(B) | [x]_{GB(B)} \cap X \neq \emptyset \right\},$$
$$\underline{GBR_B}X = \cup \left\{ [x]_B \in U/GB(B) | [x]_{GB(B)} \subseteq X \right\}. \tag{4}$$

Granular-ball rough sets can simultaneously handle continuous data and represent knowledge by equivalence class. Combined with granular-ball computing, granular-ball rough sets adapt to different data distributions flexibly by using granules with different radii, and avoid the propagation of heterogeneity caused by the overlap between the positive regional neighborhoods of different labels in the neighborhood rough set. In addition, it can achieve higher accuracy than feature selection made on a coarse-grained basis.

3   Granular-Ball Three-Way Decision Methods

3.1   Motivation

The idea of 3WD has been widely used in real-world application. Traditional 3WD based on classical rough sets can only process discrete data and have trouble dealing with continuous data. Although neighborhood rough sets can address continuous data, their upper and lower approximations are constructed by sample points rather than equivalence classes, resulting in a loss of knowledge representation capability [22]. Thus, 3WD based on neighborhood rough sets cannot effectively represent equivalence classes. The aforementioned narrow 3WD faces the challenges of processing continuous data or partitioning equivalence classes. However, continuous data prevalence in the real world and the discretization of it will loss of crucial information. In addition, decisions without equivalence classes may increase computational complexity and reduce efficiency. The granular-ball [22] containing a set of points is expressed by the centers and radii. Granular-ball rough sets can perform equivalence class partitioning on continuous data, which is a unify of classical rough sets and neighborhood rough sets. The equivalence classes of granular-ball rough sets have no overlapping, which helps 3WD to obtain appropriate conditional probabilities and decision thresholds, making it easier to find the decision boundary of 3WD.

In addition, the generalized 3WD methods take outside the concept of rough sets, thereby having no equivalence classes. The decision objects of generalized 3WD are individual objects rather than equivalence classes, which may lower the efficiency. By integrating granular-ball computing with generalized 3WD, the inputs are replaced by granular-balls composed of multiple samples instead of individual samples. This will greatly reduce computational costs and improve the decision efficiency.

More importantly, Yao [33] pointed out that S3WD faces the challenges of constructing multiple levels of granularity and multiple representations of the same object. In granular-ball computing, the radius of granular-ball reflects its granularity, where a larger radius represents a coarser granularity ball, while a smaller radius represents a finer granularity ball. Furthermore, by adaptively changing the center and radius, granular-balls can achieve multiple representations of objects. In other words, granular-balls facilitate the construction of multi-granularity structures for S3WD and enable multiple representations of the same objects. Additionally, granular-ball computing is robust, efficient, and interpretive.

Therefore, this work introduces granular-ball into 3WD and S3WD, respectively, and proposes granular-ball three-way decision method and granular-ball sequential three-way decision method.

3.2   Granular-Ball Three-Way Decision (GB3WD)

In this section, we propose granular-ball three-way decision method (GB3WD). The core of GB3WD includes two stages: the generation of granular-balls and the three-way decision of granular-ball equivalence classes. In the following, we introduce granular-ball to several narrow 3WD methods and establish corresponding GB3WD methods, respectively.

Suppose U denotes the whole objects that need to be processed, the granular-balls GB_j $(j = 1, 2, \cdots, n)$ are generated based on k-means. Then, the equivalence classes $[x]_{GB_j}$ with respect to the attribute set are generated by granular-ball computing.

3WD Based on Granular-Ball Rough Set. In 3WD based on Pawlak rough sets [11], the positive region, boundary region, and negative region of 3WD are induced by upper and lower approximation of X. By introducing granular-ball into 3WD, the three regions of three-way decision based on granular-ball rough set are shown as follows:

$$
\begin{aligned}
POS(X) &= \underline{GBR}(X) = \cup \{x \in U \mid [x]_{GB} \subseteq X\}, \\
BND(X) &= \overline{GBR}(X) - \underline{GBR}(X) \\
&= \cup \{x \in U \mid [x]_{GB} \cap X \neq \emptyset \wedge [x]_{GB} \nsubseteq X\}, \\
NEG(X) &= U - \overline{GBR}(X) = \cup \{x \in U \mid [x]_{GB} \cap X = \emptyset\}.
\end{aligned}
\tag{5}
$$

The rough membership function based on granular-ball equivalence class is defined as follows:

$$
\mu_A = Pr(X|[x]_{GB}) = \frac{|X \cap [x]_{GB}|}{|[x]_{GB}|},
\tag{6}
$$

where $|\cdot|$ represents the cardinality of a set, the conditional probability of $[x]_{GB}$ is denoted by $Pr(X|[x]_{GB})$. Then, we construct the new rules for three regions as follows:

$$POS(X) = \{x \in U \mid Pr(X|[x]_{GB}) = 1\},$$
$$BND(X) = \{x \in U \mid 0 < Pr(X|[x]_{GB}) < 1\}, \tag{7}$$
$$NEG(X) = \{x \in U \mid Pr(X|[x]_{GB}) = 0\}.$$

3WD Based on Probabilistic Granular-Ball Rough Set. The 3WD based on granular-ball rough set has some limitations due to its too strict rules on conditional probability. To this end, 3WD based on the probabilistic granular-ball rough set is proposed to solve this issue, the decision rules are shown as follows:

$$POS(X) = \{x \in U \mid Pr(X|[x]_{GB}) \geq \alpha\},$$
$$BND(X) = \{x \in U \mid \beta < Pr(X|[x]_{GB}) < \alpha\}, \tag{8}$$
$$NEG(X) = \{x \in U \mid Pr(X|[x]_{GB}) \leq \beta\}.$$

3WD Based on Decision-Theoretic Granular-Ball Rough Set. Decision-theoretic rough set was proposed by Yao [34], which introduced Bayesian decision process to calculate the decision thresholds α and β. By integrating with granular-ball, we establish 3WD based on the decision-theoretic granular-ball rough set.

Table 1. The loss function matrix.

	X	$\neg X$
a_P	$\lambda_{PP(GB)}$	$\lambda_{PN(GB)}$
a_B	$\lambda_{BP(GB)}$	$\lambda_{BN(GB)}$
a_N	$\lambda_{NP(GB)}$	$\lambda_{NN(GB)}$

Suppose $\Omega = \{X, \neg X\}$ represents the set of states expressing whether an object x belongs to X or not. The set of actions $A = \{a_P, a_B, a_N\}$ denotes three actions of classifying x into $POS(X)$, $BND(X)$ and $NEG(X)$, respectively. Table 1 shows the loss function matrix of different actions under different states. $\lambda_{PP(GB)}, \lambda_{BP(GB)}$ and $\lambda_{NP(GB)}$ represent the loss induced by actions a_P, a_B and a_N, respectively, when $x \in X$. Similarly, $\lambda_{PN(GB)}, \lambda_{BN(GB)}$ and $\lambda_{NN(GB)}$ represent the loss induced by actions a_P, a_B and a_N, respectively, when $x \notin X$. $Pr(X \mid [x]_{GB})$ denotes the probability of the granular-ball equivalence class $[x]_{GB}$ belongs to X. In the Bayesian process, the expected loss is derived by linearly weighting the losses under different states alongside their corresponding conditional probabilities. Therefore, the expected loss $R(a_i \mid [x]_{GB})$

$(i = P, B, N)$ for taking different actions are shown as follows:

$$R(a_P \mid [x]_{GB}) = \lambda_{PP} Pr\left(X \mid [x]_{GB}\right) + \lambda_{PN} Pr\left(\neg X \mid [x]_{GB}\right),$$
$$R(a_B \mid [x]_{GB}) = \lambda_{BP} Pr\left(X \mid [x]_{GB}\right) + \lambda_{BN} Pr\left(\neg X \mid [x]_{GB}\right), \qquad (9)$$
$$R(a_N \mid [x]_{GB}) = \lambda_{NP} Pr\left(X \mid [x]_{GB}\right) + \lambda_{NN} Pr\left(\neg X \mid [x]_{GB}\right).$$

Based on the Bayesian minimum risk principle, the decision rules can be expressed as follows:

$$If\ R(a_P|[x]_{GB}) \leq R(a_B|[x]_{GB}) \& R(a_P|[x]_{GB}) \leq R(a_N|[x]_{GB}), x \in POS(X),$$
$$If\ R(a_B|[x]_{GB}) \leq R(a_P|[x]_{GB}) \& R(a_B|[x]_{GB}) \leq R(a_N|[x]_{GB}), x \in BND(X),$$
$$If\ R(a_N|[x]_{GB}) \leq R(a_P|[x]_{GB}) \& R(a_N|[x]_{GB}) \leq R(a_B|[x]_{GB}), x \in NEG(X).$$
$$(10)$$

In practice, assuming $\lambda_{PP(GB)} \leq \lambda_{BP(GB)} \leq \lambda_{NP(GB)}$ and $\lambda_{NN(GB)} \leq \lambda_{BN(GB)} \leq \lambda_{PN(GB)}$, the decision rules can be simplified as follows:

$$If\ Pr\left(X \mid [x]_{GB}\right) \geq \alpha, then\ x \in POS(X),$$
$$If\ \beta < Pr\left(X \mid [x]_{GB}\right) < \alpha, then\ x \in BND(X), \qquad (11)$$
$$If\ Pr\left(X \mid [x]_{GB}\right) \leq \beta, then\ x \in NEG(X),$$

where α and β can be computed as follows:

$$\alpha = \frac{\lambda_{PN_{GB}} - \lambda_{BN_{GB}}}{(\lambda_{PN_{GB}} - \lambda_{BN_{GB}}) + (\lambda_{BP_{GB}} - \lambda_{PP_{GB}})},$$
$$\qquad (12)$$
$$\beta = \frac{\lambda_{BN_{GB}} - \lambda_{NN_{GB}}}{(\lambda_{BN_{GB}} - \lambda_{NN_{GB}}) + (\lambda_{NP_{GB}} - \lambda_{BP_{GB}})}.$$

Compared with 3WD based on Pawlak rough set, GB3WD is capable of describing continuous data. In contrast to 3WD based on neighborhood rough set, GB3WD can effectively represent equivalence classes. Additionally, granular-balls have good performance in eliminating the influence of noise. Therefore, our GB3WD can get more accurate conditional probabilities and decision thresholds, which in turn leads to more appropriate decision boundaries.

3.3   Granular-Ball Sequential Three-Way Decision (GBS3WD)

S3WD involves multiple steps of 3WD by constructing a hierarchical structure with multiple levels of granularity. S3WD allows for varying descriptions of objects across different granular levels. Granular-balls offer natural representations of different granularities through different centers and radii, thus forming a multi-granularity structure. Within different granular levels, the same object may belong to different granular-ball equivalence classes, resulting in diverse representations of the same object. These inherent properties of granular-ball significantly contribute to the effectiveness of S3WD. In this section, we propose a granular-ball sequential three-way decision method (GBS3WD).

Assuming U is a finite nonempty set of objects x, there are n levels of granularity from coarser to finer in GBS3WD. We generate g_i granular-balls $GB_{i,j}$ ($j = 1, \cdots, g_i$) at level i, where $i = 1, \cdots, n$ denotes the level of granularity. $g_i = g_1, \cdots, g_n$ express the number of granular-balls generated at level i, then $[x]_{GB_{n,.}} \subseteq \cdots \subseteq [x]_{GB_{2,.}} \subseteq [x]_{GB_{1,.}}$. At level i, the granular-ball equivalence classes $[x]_{GB_{ij}}$ are created by the partition U_i/GB_i, and the GB3WD divides $[x]_{GB_{ij}}$ into $POS(U_i)$, $BND(U_i)$ and $NEG(U_i)$.

Let $U_1 = U$, $U_{i+1} = BND(U_i)$, which means that we deal with the objects in $BND(U_i)$ at level $i+1$. The granularity of level $i+1$ is finer than that of level i, which is reflected in the number and radius of granular-balls at each level. Let b_i denote the number of granular-ball equivalence classes that are partitioned into the $BND(U_i)$ at granular level i, g_{i+1} express the number of granular-balls generated at level $i+1$, then, $b_i < g_{i+1}$. Accordingly, the granular-balls generated at the finer levels have smaller radii. Concretely , suppose the object x belongs to $[x]_{GB_{i,.}}$ at level i and $[x]_{GB_{i,.}}$ are divided into $BND(U_i)$, if the object x belongs to $[x]_{GB_{i+1,.}}$ at level $i + 1$, we can get the conclusion of $r(GB_{i+1,.}) \leq r(GB_{i,.})$. The three regions at level i are defined as follows:

$$POS(U_i) = \left\{ x \in U_i \mid Pr(X|[x]_{GB_{ij}}) \geq \alpha_i \right\},$$
$$BND(U_i) = \left\{ x \in U_i \mid \beta_i < Pr(X|[x]_{GB_{ij}}) < \alpha_i \right\}, \tag{13}$$
$$NEG(U_i) = \left\{ x \in U_i \mid Pr(X|[x]_{GB_{ij}}) \leq \beta_i \right\},$$

where α_i and β_i denotes the decision thresholds at level i, $0 \leq \beta_1 \leq \cdots \leq \beta_n < \alpha_n \leq \cdots \leq \alpha_1 \leq 1$.

The core process of GBS3WD is shown in Fig. 1. **Firstly**, at level 1, granular-balls are generated based on all objects in U by *k-means* and all objects are divided into granular-ball equivalence classes $[x]_{GB_{1,1}} \cdots [x]_{GB_{1,13}}$. **Then**, these granular-ball equivalence classes are classified into $POS(U_1)$, $BND(U_1)$, and $NEG(U_1)$ through GB3WD. **Subsequently**, the granular-balls in $BND(U_1)$ are further split, generating a greater number of smaller granular-balls with smaller radii, which leads to the formation of smaller granular-ball equivalence classes $[x]_{GB_{2,1}} \cdots [x]_{GB_{2,12}}$. **Further**, we carry out GB3WD on these smaller granular-ball equivalence classes. By continuously applying multi-step GB3WD, the granular-ball equivalence classes are successively classified into positive and negative regions. When the granular-balls can no longer be split into smaller ones, the two-way decision is employed to classify them.

4 The Application of Granular-Ball Three-Way Decision

3WD has extensive application across diverse domains, such as three-way face recognition [8], three-way credit evaluation [10], three-way investment decision [19], etc. As granular-balls can represent multi-granularity structures naturally, as well as granular-ball computing is efficient, robust, and interpretable, it will definitely improve the performances of 3WD in various areas.

Introducing granular-ball into the three-way face recognition method will improve its robustness and interoperability. Employing multi-granularity

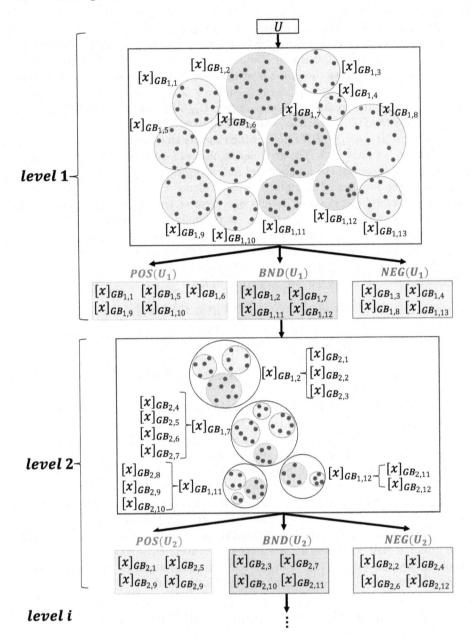

Fig. 1. Granular-ball Sequential Three-way Decision.

granular-balls, we can effectively depict human faces. Notably informative regions such as the eyes are represented using fine-grained smaller granular-balls with smaller radii. Conversely, less informative areas like the forehead are represented using coarse-grained granular-balls with larger radii. This approach enhances the interpretability of the three-way face recognition method. Furthermore, employing granular-balls for facial representation enables the method to disregard the impact of noise. For instance, spots on the cheek are excluded from the input, contributing to the robustness of granular-ball three-way face recognition.

In three-way credit evaluation, granular-ball computing helps construct a multi-granularity structure. In the beginning, all credit objects are represented by granular-balls with larger radii, and GB3WD is made. Then, the granular-ball equivalence classes have been divided into boundary regions in the last stage are split into granular-balls with smaller radius, and further GB3WD are made on them. The radius of granular-ball was used to adjust the granularity, which helps the three-way credit evaluation method to construct more reasonable granularity.

Existing three-way investment decision methods perform 3WD on individual objects one by one. By introducing granular-ball computing into it, similar investment projects are initially grouped within the same granular-ball equivalence class. Subsequently, the 3WD is carried out for each granular-ball equivalence class. The granular-ball three-way investment decision method represents more investment projects by fewer granular-balls, thereby elevating decision efficiency significantly.

5   Conclusion

In this paper, the models of GB3WD and GBS3WD are presented by the integration of granular-ball computing and 3WD. The advantages of granular-ball can improve the weakness of 3WD in terms of the multi-granularity structures, robustness, and interpretability. In future work, the open dynamic environment can be considered in GB3WD and GBS3WD. It is necessary to explore how to continually mine the boundary region and reduce uncertainty in such a framework. The knowledge transfer associated with granular-ball is also a novel topic to improve the robustness and effectiveness of GB3WD and GBS3WD.

Acknowledgements. This work was supported by the Natural Science Foundation of Sichuan Province (No. 2022NSFSC0528), the Sichuan Science and Technology Program (No. 2022ZYD0113), Jiaozi Institute of Fintech Innovation, Southwestern University of Finance and Economics (Nos. kjcgzh20230103, kjcgzh20230201), the Fundamental Research Funds for the Central Universities (No. JBK2307055).

References

1. Campagner, A., Cabitza, F., Ciucci, D.: Three-way decision for handling uncertainty in machine learning: a narrative review. In: Bello, R., Miao, D., Falcon, R., Nakata, M., Rosete, A., Ciucci, D. (eds.) IJCRS 2020. LNCS (LNAI), vol. 12179, pp. 137–152. Springer, Cham (2020). https://doi.org/10.1007/978-3-030-52705-1_10

2. Chen, L.: Topological structure in visual perception. Science **218**(4573), 699–700 (1982)
3. Chen, Y., Wang, P., Yang, X., Mi, J., Liu, D.: Granular ball guided selector for attribute reduction. Knowl.-Based Syst. **229**, 107326 (2021)
4. Fang, Y., Cao, X.M., Wang, X., Min, F.: Three-way sampling for rapid attribute reduction. Inf. Sci. **609**, 26–45 (2022)
5. Han, X.Y., Zhu, X.B., Pedrycz, W., Li, Z.W.: A three-way classification with fuzzy decision trees. Appl. Soft Comput. **132**, 109788 (2023)
6. Hua, H.Q., Ren, Y.D., Xia, X.Z.: Numerical attribute reduction based on neighborhood granulation and rough approximation. J. Softw. **19**(3), 640–649 (2008)
7. Huang, C.C., Li, J.H., Mei, C.L., Wu, W.Z.: Three-way concept learning based on cognitive operators: an information fusion viewpoint. Int. J. Approx. Reason. **83**, 218–242 (2017)
8. Li, H.X., Zhang, L.B., Huang, B., Zhou, X.Z.: Sequential three-way decision and granulation for cost-sensitive face recognition. Knowl.-Based Syst. **91**, 241–251 (2016)
9. Li, W., Yang, B.: Three-way decisions with fuzzy probabilistic covering-based rough sets and their applications in credit evaluation. Appl. Soft Comput. **136**, 110144 (2023)
10. Maldonado, S., Peters, G., Weber, R.: Credit scoring using three-way decisions with probabilistic rough sets. Inf. Sci. **507**, 700–714 (2020)
11. Pawlak, Z.: Rough sets. Int. J. Comput. Inf. Sci. **11**, 341–356 (1982)
12. Qian, W., Xu, F., Huang, J., Qian, J.: A novel granular ball computing-based fuzzy rough set for feature selection in label distribution learning. Knowl.-Based Syst. **278**, 110898 (2023)
13. Shah, A., Ali, B., Habib, M., Frnda, J., Ullah, I., Anwar, M.S.: An ensemble face recognition mechanism based on three-way decisions. J. King Saud Univ.-Comput. Inf. Sci. **35**(4), 196–208 (2023)
14. Shu Yin, X., Da Wei, D., Long, Y., Li, Z., Danf, L., Guoy, W., et al.: Graph-based representation for image based on granular-ball. arXiv preprint arXiv:2303.02388 (2023)
15. Song, M., Chen, J., Song, J., Xu, T., Fan, Y.: Forward greedy searching to κ-reduct based on granular ball. Symmetry **15**(5), 996 (2023)
16. Wang, G.Y.: DGCC: data-driven granular cognitive computing. Granul. Comput. **2**(4), 343–355 (2017)
17. Wang, P.X., Yao, Y.Y.: CE3: a three-way clustering method based on mathematical morphology. Knowl.-Based Syst. **155**, 54–65 (2018)
18. Wang, W.J., Zhan, J.M., Zhang, C., Herrera-Viedma, E., Kou, G.: A regret-theory-based three-way decision method with a priori probability tolerance dominance relation in fuzzy incomplete information systems. Inf. Fusion **89**, 382–396 (2023)
19. Wang, X.H., Wang, B., Liu, S., Li, H.X., Wang, T.X., Watada, J.: Fuzzy portfolio selection based on three-way decision and cumulative prospect theory. Int. J. Mach. Learn. Cybernet. **13**(1), 293–308 (2022)
20. Xia, S.Y., Dai, X.C., Wang, G.Y., Gao, X.B., Giem, E.: An efficient and adaptive granular-ball generation method in classification problem. IEEE Trans. Neural Netw. Learn. Syst. 1–13 (2022)
21. Xia, S.Y., Liu, Y.S., Ding, X., Wang, G.Y., Yu, H., Luo, Y.G.: Granular ball computing classifiers for efficient, scalable and robust learning. Inf. Sci. **483**, 136–152 (2019)
22. Xia, S.Y., et al.: A unified granular-ball learning model of Pawlak rough set and neighborhood rough set. arXiv preprint arXiv:2201.03349 (2022)

23. Xia, S.Y., Wang, G.Y., Gao, X.B., Peng, X.L.: GBSVM: granular-ball support vector machine. arXiv preprint arXiv:2210.03120 (2022)
24. Xia, S.Y., Zheng, S.Y., Wang, G.Y., Gao, X.B., Wang, B.G.: Granular ball sampling for noisy label classification or imbalanced classification. IEEE Trans. Neural Netw. Learn. Syst. **34**, 2144–2155 (2021)
25. Xie, J., Kong, W.Y., Xia, S.Y., Wang, G.Y., Gao, X.B.: An efficient spectral clustering algorithm based on granular-ball. IEEE Trans. Knowl. Data Eng. **35**, 9743–9753 (2023)
26. Xie, J., Xia, S.Y., Wang, G.Y., Gao, X.B.: GBMST: an efficient minimum spanning tree clustering based on granular-ball. arXiv e-prints, pp. arXiv–2303 (2023)
27. Yang, X., Li, Y.H., Li, T.R.: A review of sequential three-way decision and multi-granularity learning. Int. J. Approx. Reason. **152**, 414–433 (2022)
28. Yang, X., Li, Y.J., Liu, D., Li, T.R.: Hierarchical fuzzy rough approximations with three-way multigranularity learning. IEEE Trans. Fuzzy Syst. **30**(9), 3486–3500 (2021)
29. Yang, X., Li, Y.J., Meng, D., Yang, Y.X., Liu, D., Li, T.R.: Three-way multi-granularity learning towards open topic classification. Inf. Sci. **585**, 41–57 (2022)
30. Yang, X., Loua, M.A., Wu, M.J., Huang, L., Gao, Q.: Multi-granularity stock prediction with sequential three-way decisions. Inf. Sci. **621**, 524–544 (2023)
31. Yao, Y.: Three-way decision: an interpretation of rules in rough set theory. In: Wen, P., Li, Y., Polkowski, L., Yao, Y., Tsumoto, S., Wang, G. (eds.) RSKT 2009. LNCS (LNAI), vol. 5589, pp. 642–649. Springer, Heidelberg (2009). https://doi.org/10.1007/978-3-642-02962-2_81
32. Yao, Y.Y.: Three-way decisions with probabilistic rough sets. Inf. Sci. **180**(3), 341–353 (2010)
33. Yao, Y.: Granular computing and sequential three-way decisions. In: Lingras, P., Wolski, M., Cornelis, C., Mitra, S., Wasilewski, P. (eds.) RSKT 2013. LNCS (LNAI), vol. 8171, pp. 16–27. Springer, Heidelberg (2013). https://doi.org/10.1007/978-3-642-41299-8_3
34. Yao, Y.Y., Wong, S.K.M.: A decision theoretic framework for approximating concepts. Int. J. Man Mach. Stud. **37**(6), 793–809 (1992)
35. Yao, Y.Y., Yang, J.L.: Granular fuzzy sets and three-way approximations of fuzzy sets. Int. J. Approx. Reason. **161**, 109003 (2023)
36. Ye, X.Q., Liu, D., Li, T.R.: Multi-granularity sequential three-way recommendation based on collaborative deep learning. Int. J. Approx. Reason. **152**, 434–455 (2023)
37. Yu, H.: Three-way decisions and three-way clustering. In: Nguyen, H.S., Ha, Q.-T., Li, T., Przybyła-Kasperek, M. (eds.) IJCRS 2018. LNCS (LNAI), vol. 11103, pp. 13–28. Springer, Cham (2018). https://doi.org/10.1007/978-3-319-99368-3_2
38. Zhong, Y.H., Li, Y.H., Yang, Y., Li, T., Jia, Y.L.: An improved three-way decision model based on prospect theory. Int. J. Approx. Reason. **142**, 109–129 (2022)

Granular Models

Unsupervised KeyPhrase Extraction Based on Multi-granular Semantics Feature Fusion

Jie Chen[1,2,3], Hainan Hu[1,2,3], Shu Zhao[1,2,3]([⊠]), and Yanping Zhang[1,2,3]

[1] Key Laboratory of Intelligent Computing and Signal Processing, Ministry of Education, Beijing 230601, Anhui, People's Republic of China
zhaoshuzs2002@hotmail.com
[2] School of Computer Science and Technology, Anhui University, Hefei 230601, Anhui, People's Republic of China
[3] Information Materials and Intelligent Sensing Laboratory of Anhui Province, Beijing 230601, Anhui, People's Republic of China

Abstract. In Unsupervised Keyphrase Extraction (UKE) tasks, candidate phrases are ranked based on their similarity to the document embedding. However, This method assumes that every document focuses on only one topic. As a result, it can be difficult to distinguish the significance of potential keyphrases among different topics. Hence, it is necessary to discover a method for acquiring diversified topic information to obtain accurate key phrases. In this paper, we propose a new unsupervised key phrase extraction method (MSFFUKE) that utilizes multi-granularity semantic feature fusion. We first cluster phrases into different clusters through granulation, calculate the semantic similarity between phrases and each cluster, and take the mean to obtain the semantic features of topic granularity. Then, we obtain semantic features of phrase granularity based on the degree centrality of candidate phrases in the graph structure. Finally, we integrate semantic features of different granularity to sort candidate phrases. Three public benchmarks (Inspec, DUC 2001, SemEval 2010) are used to evaluate our model and compared it to the most advanced models currently available. The results demonstrate that our model performs better than most models and can generalize well when processing input documents from various domains and of different lengths. Another ablation study indicates that both topic granularity semantic features and phrase granularity semantic features are crucial for unsupervised keyphrase extraction tasks.

Keywords: Cluster · Unsupervised Keyphrase · Topic information · Multi-granular Semantics Feature Fusion

1 Introduction

Identifying key phrases from a document to succinctly describe its main content is a critical task known as keyphrase extraction [16,17]. Over the years, many

A. Campagner et al. (Eds.): IJCRS 2023, LNAI 14481, pp. 299–310, 2023.
https://doi.org/10.1007/978-3-031-50959-9_21

excellent algorithms have been developed for this task, which is divided into two categories: supervised and unsupervised methods. The unsupervised methods are studied more often because they come with a low computational cost. They generally extract phrases in a more general and adaptive way than the supervised methods.

The UKE model [8,19,24] has gained extensive research in keyphrase extraction with the advancement of pre-trained language models [15]. Popular UKE models use pre-trained language models to embed text and calculate semantic similarity. Most embedding-based methods score candidate phrases by jointly modeling global and local contexts. [23] The global context is calculated by measuring the similarity between candidate phrases and the document vector. The local context is calculated by measuring the similarity between candidate phrases. However, when computing the global score, they often treat the document as having only one topic, which may not fully capture the topic diversity of candidate phrases [13]. In Fig. 1, it is clear that diversifying topics is crucial for effective keyphrase extraction. The nodes represent candidate phrase embeddings and black nodes represent document embeddings, while each black circle indicates a thematic semantic feature. The nodes in the same black circle are related to a topic in the document. Meanwhile, the nodes in the red circle

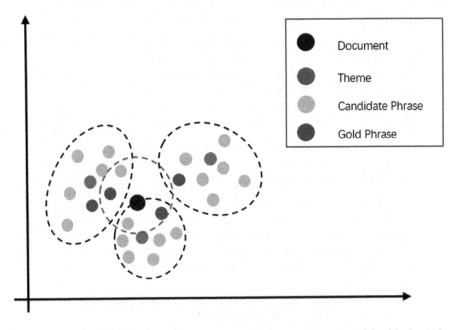

Fig. 1. The nodes in this diagram are potential phrases, represented by black circles for document embeddings. Additionally, black circles indicate semantic features for phrase granularity, while red circles indicate phrases that are similar to single-topic documents. (Color figure online)

are related to a single-topic document. Suppose we only calculate the similarity between candidate phrases and single-topic documents. In that case, the model will tend to select nodes in the red circle, ignoring the topic semantic features of each of the three topics. To accurately obtain keyphrases, to ensure that our candidate phrases are more representative, we must obtain multiple topics to gather the semantic features of their granularity.

To address this, we propose a new unsupervised keyphrase extraction method that fuses multi-granular semantic features. We calculate the semantic features of candidate phrases from both phrase and topic granularity and fuse them. From the topic perspective, we cluster phrases into different clusters to represent different topics. Then, we calculate the similarity between each phrase and each topic to obtain the importance score of each phrase in different topics and take the average score under multiple topics to obtain the thematic semantic features of each phrase. From the phrase perspective, we create a graph structure from the document, with phrases as vertices and edges representing their similarity. Then, we use a graph-based centrality calculation method to calculate the semantic features of phrases at phrase granularity [8]. Finally, we integrate the semantic features of phrase granularity and topic granularity to rank keyphrases. Our method outperforms most existing competitors on three datasets and even outperforms the latest SOTA on individual datasets. Our main contributions are summarized as follows:

(1) We propose a new unsupervised keyphrase extraction model that extracts key phrases by integrating semantic features of phrase granularity and topic granularity.
(2) We obtain different topics by embedding candidate phrases through clustering and calculate the similarity between phrases and topics to obtain semantic features of topic granularity.
(3) Our method outperforms most existing competitors on three datasets and even outperforms the latest SOTA on individual datasets.

2   Related Work

Unsupervised Keyphrase Extraction. There are four types of traditional unsupervised keyphrase phrase extraction. The first type uses statistical models [4] to extract keyphrases by analyzing word frequency, location, or language features. The second type is based on topic models and extracts keyphrases by studying the probability distribution of the document. The third type is based on graph models, which was the most popular method in the early days. This method represents the document as a graph where words or phrases are nodes, and the edges between them are weighted based on their similarity.

In the early stages, TextRank [10] utilized node ranking in a graph to extract keyphrase phrases. Following this, SingleRank [20] used the co-occurrence of words as edge weights. TopicRank [3] grouped potential keyphrases and assigned a significance score to each topic. [2] suggested a multipart ranking approach by incorporating local information within a multipart graph structure. More recently, [21] analyzed nine different centrality measures to determine the most effective combination of word ranking for keyphrase phrase extraction. Additionally, [6] provided a quantitative analysis of statistical and graph-based term weighting schemes for keyphrase phrase extraction. Finally, the fourth type is based on embedding models [11,19,24], which have demonstrated strong performance with the advancement of representation learning.

Specifically, EmbedRank [1] extracts keyphrase by calculating the similarity between the embedded candidate keyphrase and the document.SIFRank [19]extends EmbedRank by replacing static embeddings with deep contextual representations. [8,16,24] solved the problem of length mismatch between candidate keyphrases and the document by calculating the similarity between the masked document and the source document. These existing models only ignore local information by calculating the similarity between the document and the candidate keyphrase.

The process of extracting keyphrases involves various methods. JointGL [8] calculates the similarity between potential keyphrases and the entire document, while also using boundary-aware centrality to determine local significance. Meanwhile, in [16], the global similarity is based on highlighted representations in the title, and local significance is determined by centrality. Other approaches utilize pre-trained language models, as seen in [24], which studied self-attention and cross-attention for unsupervised keyphrase extraction. In this paper, we propose a new method for keyphrase phrase extraction based on multi-granularity. Instead of treating documents as a single topic, we use a different approach. We divide them into multiple topics using clustering, calculate the semantic features of each topic, and then combine the semantic features of each phrase to extract keyphrases.

Text Embedding Models. Currently, the majority of unsupervised methods for extracting key phrases utilize pre-trained language models to acquire an embedded representation of the text. Further work is then conducted on the resulting embedded representation, which may include similarity calculation, clustering, and other related tasks. [12] A pre-trained language model is a model trained on large-scale unlabeled corpora to learn prior knowledge, which is then fine-tuned for downstream tasks. It can provide high-quality natural language embeddings for unsupervised tasks, which is different from static word embeddings such as Word2Vec, GloVe, and FastText [15].

Pre-trained language models can dynamically encode words or sentences using contextual information to solve the Out-of-Vocabulary (OOV) problem. That is, when the model encounters a word or sentence that has never appeared in the training set, it can dynamically generate its vector representation based on contextual information, rather than simply treating it as an unknown word or sentence [14]. Additionally, pre-trained language models can provide document-level or sentence-level embeddings with more semantic information than Sen2Vec or Doc2Vec. ELMoadopts a Bi-LSTM structure and concatenates forward and backward information to capture bidirectional information [15]. BERT is a bidirectional transformer-based pre-trained language model. Compared with concatenating bidirectional information, BERT can capture better contextual information. There are also many other pre-trained language models, such as RoBERTa [9], and XLNet [22]. In this paper, we chose the most commonly used BERT and obtained the vector representations of phrases through joint token embeddings.

3   Methodology

3.1   Data Preprocessing

The model framework is illustrated in Fig. 2. To obtain potential keyphrases from documents, we employed the approach used in previous research [8]. We utilized *Stanford CoreNLP Tools*[1] to label the documents and apply Part of Speech (POS) tagging to label them. Next, we used the *NLTK toolkit*[2] to extract candidate phrases based on part of speech labels using regular expressions. We retained only noun phrases (NP) consisting of zero or more adjectives and combinations of one or more nouns [8,16].

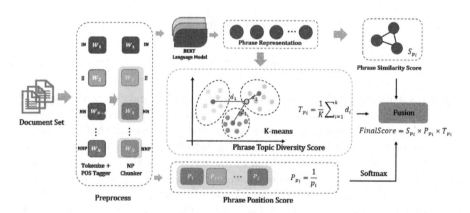

Fig. 2. The core architecture of the proposed MSFFUKE

3.2   Topic and Phrase Vector Representation

We first preprocessed the document D, which was then tokenized into individual words represented by tokens $D = \{w_1, \ldots, w_i\}$ and a set of candidate keyphrases $P = \{p_1, \ldots, p_i\}$. To obtain vector representations of the words in the document, we used the pre-trained language model BERT [5] as the embedding layer, resulting in $H = \{h_{w_1}, \ldots, h_{w_i}\}$ where h_{w_i} refers to the i-th word in the document. Next, we used word embedding to obtain candidate keyphrase representations.

To obtain the central semantics of candidate keyphrases, we applied the maximum pool operation to obtain their representations. This method is both simple and effective, and the calculation is as follows:

$$h_{p_i} = \text{Max-Pooling}(h_{w_j}, \ldots, h_{w_k}) \tag{1}$$

Here, h_{p_i} represents the representation of the i-th candidate keyphrase, and (j-i+1) represents the length of p_i. Specifically, h_k represents the words associated with the candidate keyphrase p_i in the document. Meanwhile, we used the *K-means* clustering operation to obtain the topic representation $\{h_{t_1}, \ldots, h_{t_i}\}$ of the document.

$$\{h_{t_1}, \ldots, h_{t_i}\} = K\text{-}means(h_{p_1}, \ldots, h_{p_i}) \tag{2}$$

3.3   Semantic Features of Topic Granularity

In order to gauge the variety of topics covered by candidate phrases, we assess their similarity to each individual topic and then determine the average. To achieve this, we utilize the Taxicab geometry formula (3) (4) which calculates topic diversity based on experience.

$$h_{d_i} = \frac{1}{|h_{t_i} - h_{p_k}|_1} \tag{3}$$

$$H_{p_i} = \frac{1}{n} \sum_{i=1}^{n} h_{d_i} \tag{4}$$

H_{d_i} represents the similarity between candidate words and one of the topics.

When writing news and scientific articles, authors typically place important information at the beginning or front of the text. This means that the position of words plays a crucial role in identifying keyphrases and can provide useful indicators. To determine the weight of words, authors use the sum of their inverse positions in the document. For instance, words in positions 1, 2, and 5 have a weight of $Pos_i = 1 + 1/2 + 1/5 = 1.7$. Building on previous research [8,16], we incorporate position regularization, where $Pos_i = \text{softmax}(e^{\frac{1}{i}})$ and ρ_i is the regularization factor for the i-th candidate phrase. This allows us to recalculate the weighted topic diversity correlation, $\hat{H}_{p_i}$, as follows:

$$\hat{H}_{p_i} = H_{p_i} * Pos_i \tag{5}$$

3.4   Semantic Features of Phrase Granularity

In general, the diversity of topics in a phrase is calculated independently for each candidate phrase and its topic information. This means it cannot determine which candidate phrases are better than others. To identify the most important candidate phrases, we determine the score of phrase granularity by factoring in the semantic similarity between them and other phrases. Building on previous research [8], we consider phrases that appear at the beginning or end of a document to be more significant than others. Therefore, we use boundary-aware centrality to calculate the importance of candidate phrases when computing phrase similarity scores.

$$H_{p_i}^p = \sum_{d_b(i)<d_b(j)} \max(e_{ij} - \theta, 0) + \lambda \sum_{d_b(i)\geq d_b(j)} \max(e_{ij} - \theta, 0) \qquad (6)$$

We determine how close nodes are to document boundaries using the formula db(i) = the smaller value of either i or α(n-i), with n representing the number of candidate phrases and α being a hyper-parameter that determines the importance of the start and end of the document. If db(i) is smaller than db(j), node i is closer to the boundary than node j. To calculate the centrality of node i, we reduce the contribution of node j. The similarity between two candidate phrases is represented by e_{ij}, and we decrease the impact of phrases using λ. To filter out noise from nodes that are significantly different from node i, we use θ where $\theta = \beta(\max(e_{ij}) - \min(e_{ij}))$. All values of e_{ij} that are less than θ are set to zero to remove their influence on centrality. We control the filter boundary using the hyper-parameter β.

3.5   Multi-granular Semantic Feature Fusion

To determine the final importance score for each candidate, we combine phrase granularity semantic features and topic granularity semantic features using a simple multiplication method.

$$M_i = \hat{H}_{p_i} * H_{p_i}^p \qquad (7)$$

where M_i indicates the importance score of the i-th candidate phrase. Then, we rank all candidates with their importance score M_i and extract top-ranked k phrases as keyphrases of the source document.

4   Experiments

4.1   Datasets and Evaluation Metrics

This paper conducts experiments on three benchmark and popularly used keyphrase datasets, which include DUC 2001, Inspec, and SemEval 2010. The Inspec dataset consists of 2,000 short documents from scientific journal abstracts. We follow previous works [8,16] to use 500 test documents and the version of uncontrolled annotated keyphrases as ground truth. The DUC 2001 dataset is a collection of 308 long-length news articles with an average of 828.4 tokens.

The SemEval 2010 dataset contains ACM full-length papers. In our experiments, we use the 100 test documents and the combined set of author-annotated and reader-annotated keyphrases. [8,16] We follow the common practice and evaluate the performance of our models in terms of f-measure at the top N

Table 1. Comparison of our models with other baselines.

Models	DUC 2001			Inspec			SemEval 2010		
	F1@5	F1@10	F1@15	F1@5	F1@10	F1@15	F1@5	F1@10	F1@15
Statistical-based Models									
TF-IDF [18]	9.21	10.63	11.06	11.28	13.88	13.83	2.81	3.48	3.91
YAKE [4]	12.27	14.37	14.76	18.08	19.62	20.11	11.76	14.4	15.19
Graph-based Models									
TextRank [10]	11.80	18.28	20.22	27.04	25.08	36.65	3.80	5.38	7.65
SingleRank [20]	20.43	25.59	25.70	27.79	34.46	36.05	5.90	9.02	10.58
TopicRank [3]	21.56	23.12	20.87	25.38	28.46	29.49	12.12	12.90	13.54
PositionRank [6]	23.35	28.57	28.60	28.12	32.87	33.32	9.84	13.34	14.33
MultipartiteRank [2]	23.20	25.00	25.24	25.96	29.57	30.85	12.13	13.79	14.92
Embedding-based Models									
SIFRank [19]	24.27	27.43	27.86	29.11	38.80	39.59	–	–	–
JointGL [8]	28.62	35.52	36.29	32.61	40.17	41.09	13.02	19.35	21.72
MDERank [24]	23.31	26.65	26.42	27.85	34.36	36.4	13.05	18.27	20.35
PromptRank [7]	27.39	31.59	31.01	31.73	37.88	38.17	**17.24**	**20.66**	21.35
Our Model									
MSFFUKE	**31.01**	**37.17**	**37.98**	**33.51**	**41.75**	**42.72**	13.31	20.54	**23.31**

keyphrases (F1@N), and apply stemming to both extracted keyphrases and gold truth. Specifically, we report F1@5, F1@10, and F1@15 of each model on three datasets. We adopt the pre-trained language model BERT [5] as the backbone of our model, initialized from their pre-trained weights.

In our experiments, λ is set to 0.9 for three benchmark datasets.

4.2   Overall Performance and Hyperparameter Settings

In Table 1, we present the performance results of both our model and the baseline on three benchmark datasets: DUC 2001, Inspect, and SemEval 2010. To begin with, we compare our approach with traditional statistical methods such as TF-IDF [18] and YAKE [4]. In addition, we also compare our model with five strong graph-based ranking methods. The first of its kind, TextRank [10], uses the co-occurrence of words to convert text to graph and employs PageRank to rank phrases.

In the field of graph-based keyphrase extraction, several models have been proposed to improve the accuracy of the process. SingleRank [20] utilizes a sliding window approach for graph construction, while TopicRank [3] considers the distribution of topics for keyphrase extraction. PositionRank [6] uses position information to weigh the importance of phrases, and MultipartiteRank [2] splits the entire graph into sub-graphs and ranks them using graph theory. Additionally, we have compared five cutting-edge embedding-based models. SIFRank [19] improves upon EmbedRank by utilizing contextualized embeddings obtained from a pre-trained language model. JointGL [8] sorts candidate phrases by jointly modeling global and local information, while MEDRank [24] calculates the similarity of phrases with documents by masking candidate phrases. HGUKE [16] uses highlighted information to guide the calculation of similarity between phrases and documents.

Our model has demonstrated excellent performance on F1@5, F1@10, and F1@15 evaluation metrics, outperforming current state-of-the-art models. The effectiveness of predicting candidate phrases has been proven by calculating the correlation between phrases and documents from different topics perspectives. The multi-granular feature fusion has resulted in a good performance in predicting candidate phrases. Our model's backbone is the pre-trained language model BERT [5], initialized from their pre-trained weights. During our experiments, we set hyperparameters to $\alpha = 1.0$, $\beta = 0.2$, $\lambda = 0.9$.

We conducted an ablation study to evaluate the contribution of topic granularity and phrase granularity in our model. The results, shown in Table 2, indicate that both types of granularity are significant. Additionally, our results demonstrate that the concept of multiple topics is crucial for unsupervised keyphrase extraction, as illustrated in Fig. 3. These findings have significant implications for future work.

Table 2. The results of ablation experiments on three datasets.

Models	DUC 2001			Inspec			SemEval 2010		
	F1@5	F1@10	F1@15	F1@5	F1@10	F1@15	F1@5	F1@10	F1@15
MSFFUKE	31.01	37.17	37.98	33.51	41.75	42.72	13.31	20.54	23.31
-Topic	22.20	34.83	36.50	32.75	40.88	41.59	12.43	18.89	21.78
-Phrase	21.48	27.85	29.26	27.50	36.60	38.45	12.25	15.89	18.41

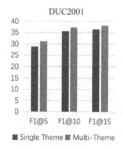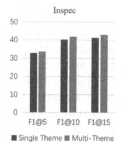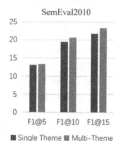

Fig. 3. The results of Keyphrase extraction through single and multiple topics.

4.3   Impact of the Number of Topics

In this section, we will explore how different topics affect the results. Based on the data in Table 3, it can be observed that the model performs the best when there are three topics. This suggests that having multiple topics is beneficial for UKE.

Table 3. The results of Different Topics on three datasets.

Different Topics	DUC 2001			Inspec			SemEval 2010		
	F1@5	F1@10	F1@15	F1@5	F1@10	F1@15	F1@5	F1@10	F1@15
Topic = 1	30.74	37.01	37.71	**34.10**	41.45	42.77	13.32	19.44	21.37
Topic = 2	29.69	36.30	37.02	33.30	41.45	**42.90**	**13.96**	20.43	23.03
Topic = 3	**31.01**	**37.17**	**37.98**	33.51	**41.75**	42.72	13.31	**20.54**	**23.31**

4.4 Impact of Different Similarity Measures

Our approach involves utilizing the Manhattan Distance to gauge the similarity of text between various topics and potential phrases. Additionally, we aim to apply multiple measures to determine the thematic variety of these phrases. The outcomes of these various similarity measures are presented in Table 4. It is evident that the Taxicab geometry distance provides a significant advantage.

Table 4. The results of Different Similarity Measures on three datasets.

Different Similarity Measures	DUC 2001			Inspec			SemEval 2010		
	F1@5	F1@10	F1@15	F1@5	F1@10	F1@15	F1@5	F1@10	F1@15
Cosine Similarity	29.98	35.81	37.84	33.37	40.90	42.13	12.46	19.37	22.40
Euclidean Distance	25.31	32.37	34.57	32.52	39.66	41.25	11.00	17.92	20.17
Manhattan Distance	**31.01**	**37.17**	**37.98**	**33.51**	**41.75**	**42.72**	**13.31**	**20.54**	**23.31**

5 Conclusion

This paper presents a method to enhance unsupervised keyphrase extraction using multi-granularity semantic features based on embedding. Our proposed method, Multi-granularity Feature Semantic Fusion Unsupervised Keyphrase Extraction (MSFFUKE), processes phrase documents as multiple topics instead of a single topic to calculate topic granularity semantic features. The semantic features of phrase granularity are then fused to select relevant candidate phrases. Our experiments show that MSFFUKE performs better than most state-of-the-art unsupervised baselines. Further research could explore the use of different structural information and statistical features of documents to enhance the performance of unsupervised keyphrase extraction.

Acknowledgements. This work was supported by the Major Program of the National Natural Science Foundation of China (Grant No.61876001, 61876157), the National Social Science Foundation of China (GrantNo.18ZDA032), the Natural Science Foundation for the Higher Education Institutions of Anhui Province of China (KJ2021A0039).

References

1. Bennani-Smires, K., Musat, C., Hossmann, A., Baeriswyl, M., Jaggi, M.: Simple unsupervised keyphrase extraction using sentence embeddings. arXiv preprint arXiv:1801.04470 (2018)
2. Boudin, F.: Unsupervised keyphrase extraction with multipartite graphs. arXiv preprint arXiv:1803.08721 (2018)
3. Bougouin, A., Boudin, F., Daille, B.: Topicrank: graph-based topic ranking for keyphrase extraction. In: International Joint Conference on Natural Language Processing (IJCNLP), pp. 543–551 (2013)
4. Campos, R., Mangaravite, V., Pasquali, A., Jorge, A.M., Nunes, C., Jatowt, A.: YAKE! Collection-independent automatic keyword extractor. In: Pasi, G., Piwowarski, B., Azzopardi, L., Hanbury, A. (eds.) ECIR 2018. LNCS, vol. 10772, pp. 806–810. Springer, Cham (2018). https://doi.org/10.1007/978-3-319-76941-7_80
5. Devlin, J., Chang, M.W., Lee, K., Toutanova, K.: Bert: pre-training of deep bidirectional transformers for language understanding. arXiv preprint arXiv:1810.04805 (2018)
6. Florescu, C., Caragea, C.: Positionrank: an unsupervised approach to keyphrase extraction from scholarly documents. In: Proceedings of the 55th Annual Meeting of the Association for Computational Linguistics (Volume 1: Long Papers), pp. 1105–1115 (2017)
7. Kong, A., et al.: Promptrank: unsupervised keyphrase extraction using prompt. ACL (2023)
8. Liang, X., Wu, S., Li, M., Li, Z.: Unsupervised keyphrase extraction by jointly modeling local and global context. arXiv preprint arXiv:2109.07293 (2021)
9. Liu, Y., et al.: Roberta: a robustly optimized bert pretraining approach. arXiv preprint arXiv:1907.11692 (2019)
10. Mihalcea, R., Tarau, P.: Textrank: bringing order into text. In: Proceedings of the 2004 Conference on Empirical Methods in Natural Language Processing, pp. 404–411 (2004)
11. Papagiannopoulou, E., Tsoumakas, G.: Local word vectors guiding keyphrase extraction. Inf. Process. Manag. 54(6), 888–902 (2018)
12. Sarwar, T.B., Noor, N.M., Miah, M.S.U.: Evaluating keyphrase extraction algorithms for finding similar news articles using lexical similarity calculation and semantic relatedness measurement by word embedding. PeerJ Comput. Sci. 8, e1024 (2022)
13. Schopf, T., Klimek, S., Matthes, F.: Patternrank: leveraging pretrained language models and part of speech for unsupervised keyphrase extraction. arXiv preprint arXiv:2210.05245 (2022)
14. Song, M., Feng, Y., Jing, L.: Hyperbolic relevance matching for neural keyphrase extraction. arXiv preprint arXiv:2205.02047 (2022)
15. Song, M., Feng, Y., Jing, L.: A survey on recent advances in keyphrase extraction from pre-trained language models. Find. Assoc. Comput. Linguist. EACL 2023, 2108–2119 (2023)
16. Song, M., Liu, H., Feng, Y., Jing, L.: Improving embedding-based unsupervised keyphrase extraction by incorporating structural information. ACL Finds (2023)
17. Song, M., Xiao, L., Jing, L.: Learning to extract from multiple perspectives for neural keyphrase extraction. Comput. Speech Lang. 81, 101502 (2023)

18. Sparck Jones, K.: A statistical interpretation of term specificity and its application in retrieval. J. Doc. **28**(1), 11–21 (1972)
19. Sun, Y., Qiu, H., Zheng, Y., Wang, Z., Zhang, C.: SIFRank: a new baseline for unsupervised keyphrase extraction based on pre-trained language model. IEEE Access **8**, 10896–10906 (2020)
20. Wan, X., Xiao, J.: Single document keyphrase extraction using neighborhood knowledge. In: AAAI, vol. 8, pp. 855–860 (2008)
21. Wang, R., Liu, W., McDonald, C.: Corpus-independent generic keyphrase extraction using word embedding vectors. In: Software Engineering Research Conference, vol. 39, pp. 1–8 (2014)
22. Yang, Z., Dai, Z., Yang, Y., Carbonell, J., Salakhutdinov, R.R., Le, Q.V.: Xlnet: generalized autoregressive pretraining for language understanding. Adv. Neural Inf. Process. Syst. **32** (2019)
23. Zhang, C., Zhao, L., Zhao, M., Zhang, Y.: Enhancing keyphrase extraction from academic articles with their reference information. Scientometrics **127**(2), 703–731 (2022)
24. Zhang, L., et al.: Mderank: a masked document embedding rank approach for unsupervised keyphrase extraction. arXiv preprint arXiv:2110.06651 (2021)

Multi-granularity Feature Fusion for Transformer-Based Single Object Tracking

Ziye Wang[✉] and Duoqian Miao

Department of Computer Science and Technology, Tongji University, No. 4800,
Cao'an Highway, Jiading District, Shanghai 201804, People's Republic of China
{yeziwang,dqmiao}@tongji.edu.cn

Abstract. The recently developed transformer has been largely explored in the research field of computer vision and especially improve the performance of single object tracking. However, the majority of current efforts concentrate on combining and enhancing convolutional neural network (CNN)-generated features and cannot fully excavating the potential of transformer. Motivated by this, we introduce multi-granularity theory into the pure transformer-based single object tracker and design a multi-granularity feature fusion module. With a view to fuse the feature of different granularity and enhance the feature representation, we design the double-branch transformer feature extractor and utilize cross-attention mechanism to fuse the feature. In our extensive experiments on multiple tracking benchmarks, including OTB2015, VOT2020, TrackingNet, GOT-10k, LaSOT, our proposed method named MGTT, the results could demonstrate that the proposed tracker achieves better performance than multiple state-of-the-art trackers.

Keywords: Computer vision · Single object tracking · Multi granularity · Rough set · Transformer

1 Introduction

Visual object tracking has a wide range of applications in fields including military guidance, video surveillance, unmanned driving, robot vision, and medical diagnosis, which is a fundamental task in computer vision research area. The aim of single object tracking is to predict the location and shape of the target which is given in the first frame of the video in each video frame. Since the advent of deep learning, single object tracking has advanced significantly [1–7]. Because of its ability to simulate long-term dependencies, the recent Transformer [8] has greatly pushed the state-of-the-art in tracking. However, occlusion, deformation, fast motion, illumination variation et.al are crucial tracking issues. Although numerous efforts have been made recently [9,10], creating a high-accuracy and real-time tracker is still a difficult issue. A significant improvement has been made in the capability for sequence-to-sequence modeling in NLP

© The Author(s), under exclusive license to Springer Nature Switzerland AG 2023
A. Campagner et al. (Eds.): IJCRS 2023, LNAI 14481, pp. 311–323, 2023.
https://doi.org/10.1007/978-3-031-50959-9_22

——— Ours   - - - - Siam R-CNN   — — TransT   Gt bbox

Fig. 1. The comparison of our proposed tracker MGTT with Siam R-CNN and TransT.

applications [11] thanks to the unique transformer design [8]. The vision community is particularly interested in learning if transformers can effectively compete with the leading convolutional neural network-based architectures (CNNs) in vision tasks [12–14], as a result of transformers' tremendous success in natural language processing (NLP). However, current approaches [15–20] typically focus on employing Transformer to incorporate and enhance features produced by convolutional neural networks (CNNs), such as ResNet [13]. Although these hybrid techniques show promise, they are less scalable computationally than strictly attention-based transformers. To solve this problem, a pure transformer framework named SwinTrack is proposed, which, in comparison to pure CNN-based [1,2] and hybrid CNN-Transformer [15–17] frameworks, enables better interactions within the feature learning of template and search region and their fusion, resulting to more reliable performance. However, transformer's potential to be used for feature representation learning is still mostly undeveloped. Granular computing [21–32] is an effective method that utilizes granulation tools to decompose complex problems into multiple relatively simple subproblems for solution. The multi granularity theory can be seen as a partitioning method for combined features which is widely utilized in the field of computer vision [33–37]. The typical single object tracking pipeline [15–17,38] extracts and aggregates features to obtain a single feature vector as the target representation and obviously has some limitations. Additionally, it is capable to capturing discriminative variables and semantics at various granularities (regions of various sizes). Effective techniques to investigate these hierarchical traits are lacking, nevertheless. Inspired by the success of multi-granularity or multi-scale based deep convolutional networks vision methods, we introduce the theory of multi granularity into the transformer-based single object tracking to make the feature representation more precise, discriminative and comprehensive. In this paper, We address the afore-mentioned challenges by developing a dual-branch transformer that combines image patches of varying widths to provide stronger visual features for single object tracking. We name our approach MGTT (Multi-Granularity Transformer based object Tracking). The proposed object tracking architecture consists of the multi-granularity transformer-based feature extractor, the multi-granularity feature fusion module and the head network. The framework of the proposed MGTT method is demonstrated in Fig. 2. We evaluate our method in various large-scale benchmarks including OTB100 [39] VOT2020 [40], TrackingNet [41], GOT-10k [42] and LaSOT [43,44] to verify the effectiveness of our MGTT. In Fig. 1, it's noticeable that our method can lead more precise and robust tracking performance.

In summary, this work has three main contributions as shown as follows.

- We propose a novel MGTT, a transformer based architecture dedicated to single object tracking. It consists of the multi-granularity transformer-based feature extractor, the multi-granularity feature fusion module and the head network. It is capable to combining multi-granularity feature of the target to enhance the tracking performance.
- We develop the feature-fusion module, enabling to extract better feature representations by introducing multi-granularity theory. Compared with the single-granularity based feature, our method adaptively establishes associa-tions between distant features.
- The proposed tracker achieves state-of-the-art performance on numerous tracking benchmarks, especially on some large scale datasets such as Track-ingNet, GOT-10k and LaSOT, while meets the real-time-requirements.

2   Related Work

2.1   Transformer in Language and Vision

Transformer is a language modeling architecture that was initially pitched by Vaswani et al. [8] for the machine translation task. Transformer reads a sequence as its input and determines the dependencies between each element. Transformer is intrinsically good at capturing global information in sequential data thanks to its property. Transformer has recently demonstrated their significant potential in tasks involving vision, such as picture classification [45], object identification [46], semantic segmentation [47], multiple object tracking [48,49], etc. Our approach draws inspiration from the recently published DETR, however there are a few key changes. (1) The tasks under study are dissimilar. This effort focuses on object tracking while DETR is intended for object detection. (2) There are several network inputs. While DETR requires the entire image as input, ours is a triplet made up of one search region and two templates. Their backbone features are first flattened, then they are concatenated, and then they are transmitted to the encoder. (3) The training strategies and query design are different. When DETR is training, it does 100 object queries and use the Hungarian method to compare predictions to actual data. In contrast, our approach doesn't use the Hungarian algorithm and only ever matches one query with the actual data. (4) There are different bounding box heads. DETR predicts boxes using a three-layer perceptron.

2.2   Transformer in Tracking

Motivated by the success of transformer in other research fields, researchers have applied transformer for visual tracking. Two of the fundamental recent exemplary works on transformer tracking are TransTrack [49] and TrackFormer [48]. TransTrack takes the image features of both the current and the previous frame as the input and contains two decoder which, respectively, accept the learnt object questions and the inquiries from the previous frame as input. Differently, TrackFormer only takes the feature of the current frame as the input and contains only one decoder where there is interaction between the track queries from the previous frame and the learnt object queries. Some other methods utilize transformer to improve and fuse the features [15] and utilize temporal information to achieve spatial-temporal feature augmentation [16,50] to increase tracking reliability. Unlike all aforementioned methods, SwinTrack [38] is a pure transformer-based tracking approach where Transformer is employed to execute feature fusion and representation learning, allowing the development of better features for robust tracking.

2.3   Multi Granularity

At present, the methods of granular computing [51–55] mainly include three categories, namely word Theory of computation, rough set theory [56–58] and quotient space theory. Multi-granularity computing is an effective new model to solve the problem of data-knowledge fusion in intelligent decision-making based on big data. This theory also has very wide applications in the field of computer vision. For instance, in the convolutional neural network, a variety of new approaches use different layers in a ConvNet to improve detection and segmentation such as FCN [59], HyperNet [60], ParseNet [61]. U-Net [62] and SharpMask [63] for segmentation, Recombinator networks [64] for face detection, and Stacked Hourglass networks [65] for keypoint estimation are recent methods that exploit lateral/skip connections to associate low-level feature maps across resolutions and semantic levels.

3   Proposed Method

In this section, we now present the proposed multi-granularity transformer-based tracking method, named MGTT as shown in Fig. 2. For clarity, we first introduce the overview the of tracking framework architecture in Sect. 3.1. The proposed framework contains three main components: the multi-granularity transformer-based feature extractor, the multi-granularity feature fusion module and the head network. Each part mentioned above will be detailed described in the following sections.

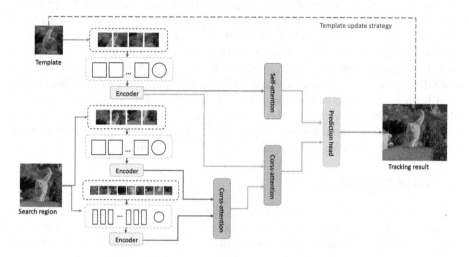

Fig. 2. Architecture of our tracking framework MGTT.

3.1   Overall Architecture

ViT's accuracy and complexity are affected by the granularity of the patch size; fine-grained patch size allows ViT to perform better but results in increased FLOPs and memory usage. This is the driving force behind our suggested method, which aims to balance complexity and utilize the benefits of more fine-grained patch sizes. In order to combine transformer feature of different granularity, we introduce a transformer feature extractor with a double branch to extract multi-granularity features and operate them at a different strategy. As illustrated in the Fig. 2, the proposed method could be divided into three components. The first one is a multi-granularity transformer-based feature extractor, which utilize transformer as a backbone. This extractor could be separated into two branches: a L-branch and a M-branch. In the L-branch, the search region is first cropped into 4 parts and then serve as the input of encoder. In the M-branch, the search region is cropped into 9 parts. After that, we utilize cross-attention mechanism to fuse the multi-granularity features to obtain better feature representation. Finally, we input the fusion feature and the feature extract from the template into the prediction head network and attain the tracking result. We also apply the template update strategy to avoid error accumulation.

3.2   Multi-granularity Feature Fusion Module

In order to enhance the feature representation and improve the tracking accuracy, we introduce the cross-attention mechanism and design a feature fusion module to fuse the multi-granularity feature extracts from the abovementioned transformer-based feature extractor. In particular, we first employ the CLS token at each branch as an agent to exchange information among the patch tokens from the other branch and then back project it to its own branch in order to fuse multi-scale features more effectively and efficiently. Interacting with the patch tokens at the other branch enables the inclusion of information at a different scale because the CLS token already learns abstract information among all patch tokens in its own branch. The CLS token interacts with its own patch tokens once more at the next transformer encoder after the fusion with other branch tokens. Here, it is able to transmit the knowledge it has learnt from the other branch to its own patch tokens, enhancing the representation of each patch token.

4   Experiments

This section initially provides technical details as well as the results of our MGTT tracker on numerous benchmarks, with comparisons to state-of-the-art approaches. Then, abolition studies are presented to examine the effects of the suggested networks' essential components. In order to highlight our method's superiority, we also provide the outcomes of other prospective frameworks and compare them to ours.

4.1   Implementation Details

The backbone we utilize in our proposed method is Swin Transformer-Tiny which is initialized with the parameters pretrained Imagenet-22k. We train our model by the train-splits of COCO [66], TrackingNet [41], GOT-10k [42] (For fair comparison, 1,000 videos are eliminated following [16]) and LaSOT [43,44]. To build picture pairings for COCO detection datasets, we perform several alterations to the source image. To increase the size of the training set, popular data augmentation techniques (such as translation and brightness jitter) are used. The search region patch and template patch have sizes of 224×224 and 112×112, respectively. AdamW [67] is used to optimize the model, which has a learning rate of 5e-4 and a weight decay of 1e-4. The backbone's learning rate is set to 5e-5. Our trackers are implemented using Python 3.6 and pytorch 1.5.1.

4.2   Comparisons on Multiple Benchmarks

In this subsection, we compare our MGTT tracking method with numerus state-of-the-art trackers (SiamR-CNN, KYS, Ocean, ATOM, SiamFC, SiamFC++, SiamRPN, SiamAttn, SiamCar, TransT, STARK, KeepTrack, Swintrack). In Table 1, 2, 3, 4 and 5, we present extensive comparison results for the large-scale OTB2015 [39], VOT2020 [40], TrackingNet [41], GOT-10k [42] and LaSOT [43, 44] datasets.

Otb2015. The Otb2015 [38] is a well-known tracking benchmark that contains 100 difficult video sequences. With an AUC score of 70.8, the MGTT tracker easily beats all other examined trackers on the OTB2015 benchmark, as shown in Table 1.

Table 1. Comparison with state-of-the-art trackers on the Otb2015 test set [39] in terms of success (AUC score) and. The best two results are shown in red and blue fonts.

	SiamFC [1]	MDNet [68]	SiamRPN++ [2]	Ocean [6]	ECO [69]	LightTrack [70]	ATOM [5]	TransT [15]	STARK [16]	SwinTrack [38]	MGTT
AUC	58.3	65.0	68.7	68.4	66.6	65.4	66.3	69.5	68.3	70.2	70.8

VOT2020. VOT2020 [39] comprises of 60 lengthy films in which target items frequently disappear and reappear. Furthermore, trackers are expected to disclose the target's confidence score. Precision (Pr) and recall (Re) are calculated using a set of confidence levels. Respectively, MGTT beat most of previous techniques, as shown in Table 2.

Table 2. Comparison with state-of-the-art trackers on the VOT2020 test set [40] in terms of accuracy (A), robustness (R), and expected average overlapn (EAO). The best two results are shown in red and blue fonts.

	SiamFC [1]	DiMP [71]	ATOM [5]	Ocean [6]	STARK [43]	UPDT [72]	ToMP [38]	SwinTrack	MGTT
A	0.600	0.457	0.462	0.693	0.481	0.465	0.453	0.471	0.482
R	0.234	0.734	0.734	0.754	0.775	0.755	0.814	0.775	0.784
EAO	0.414	0.274	0.271	0.430	0.308	0.278	0.309	0.302	0.270

TrackingNet. TrackingNet [41] is a large-scale short-term tracking benchmark with a test set of 511 video sequences and avliable ground-truth bounding box which is released recently. Table 3 shows that MGTT outperform SiamFC++ in AUC by 5.8% respectively. MGTT gets the better AUC of 81.5% and outperforming Siam R-CNN by 0.4

GOT-10k. The GOT-10k [42] dataset has 10,000 training sequences and 180 testing sequences. We adhere to the refined protocol given in [42] and send the tracking outputs to the official evaluation site. The obtained results (AO and SRT) are then reported in Table 4. The MGTT method achieve the best results. In the key AO statistic, the MGTT approach outperforms SwinTRACK by 0.2%. The result in Table 4 shows that MGTT has a better performance than most of other state-of-the-art trackers on this benchmark.

318   Z. Wang and D. Miao

Table 3. Comparison with state-of-the-art trackers on the TrackingNet test set [41] in terms of success (AUC score), precision (P), and normalized precision (P_{NORM}). The best two results are shown in red and blue fonts.

	MDNet [68]	SiamRPN++ [2]	SiamAttn [66]	SiamFC++ [3]	KYS [73]	Siam R-CNN [74]	TransT [4]	STARK [5]	SwinTrack [38]	MGTT
AUC	60.6	73.3	75.2	75.4	74.0	81.2	81.4	82.0	81.1	81.5
P_{NORM}	70.5	80.0	81.7	80.0	80.0	85.4	86.7	86.9	-	-
P	56.5	69.4	-	70.5	68.8	80.0	80.3	-	78.4	78.3

Table 4. Comparison with state-of-the-art trackers on the GOT-10k test set [42] in terms of average overlap (AO), $SR_{0.5}$, and $SR_{0.75}$. The best two results are shown in red and blue fonts.

	SiamFC [1]	MDNet [68]	SiamRPN++ [2]	SiamCAR [75]	SiamFC++ [3]	Ocean [6]	Siam R-CNN [74]	TransT [15]	STARK [16]	SwinTrack [38]	MGTT
AO	34.8	29.9	51.7	56.9	59.5	61.1	64.9	67.1	68.8	71.3	71.5
$SR_{0.5}$	35.3	30.3	61.6	67.0	69.5	72.1	72.8	76.8	78.1	81.9	82.1
$SR_{0.75}$	9.8	9.9	32.5	41.5	47.9	47.3	59.7	60.9	64.1	64.5	64.3

LaSOT. LaSOT [43] is a current large-scale long-term tracking benchmark with high-quality annotations that includes 1120 training videos and 280 testing videos. It is actually more difficult than the prior short-term tracking datasets. The ability to deal with major target appearance changes using temporal context and geographical information is crucial in this dataset. Table 1 summarizes the precision and success of cutting-edge approaches. As shown in the table, our suggested approach MGTT (67.4 AUC) has competitive performance and outperforms other competing trackers.

Table 5. Comparison with state-of-the-art trackers on the LaSOT test set [43] in terms of success (AUC score), precision (P), and normalized precision (P_{NORM}). The best two results are shown in red and blue fonts.

	MDNet [68]	SiamRPN++ [2]	SiamAttn [66]	SiamFC++ [3]	Ocean [6]	Siam R-CNN [74]	TransT [15]	STARK [16]	SwinTrack [38]	MGTT
AUC	39.7	49.6	56.0	54.4	56.0	64.8	64.9	67.1	67.2	67.4
P_{NORM}	46.0	56.9	64.0	62.3	65.1	72.2	73.8	77.0	-	76.8
P	37.3	49.1	-	54.7	56.6	-	69.0	-	70.8	71.0

5 Conclusions

In this work, we present MGTT, a multi-granularity transformer-based tracker, to improve the prediction accuracy for single object tracking. In MGTT, we utilize cross-attention mechanism to combine the multi-granularity features which are extracted from two branches to effectively enhance the feature representation. Extensive experiments can demonstrate the effectiveness of our proposed architecture and our method perform better that multiple of state-of-the-art algorithms in numerus benchmarks. While our proposed work only scratch the surface of the search region of single object tracking, we anticipate that multi-granularity theory will be applied more widely in many search regions. We also hope that our work will encourage and assist new research.

Acknowledgements. This work is supported in part by the National Key Research and Development Plan under Grant No. 2022YFB3104700, the National Science Foundation of China under Grant No. 61976158 and No. 62376198, the National Science Foundation of China under Grant No. 62076182. This paper is partially supported by the Jiangxi "Double Thousand Plan", and the National Natural Science Foundation of China (Serial No. 62163016), and the Jiangxi Provincial natural science fund (No. 20212ACB202001) and the National Natural Science Foundation of China No. 62006172.

References

1. Bertinetto, L., Valmadre, J., Henriques, J.F., Vedaldi, A., Torr, P.H.S.: Fully-convolutional Siamese networks for object tracking. In: Hua, G., Jégou, H. (eds.) ECCV 2016. LNCS, vol. 9914, pp. 850–865. Springer, Cham (2016). https://doi.org/10.1007/978-3-319-48881-3_56

2. Li, B., Yan, J., Wu, W., Zhu, Z., Hu, X.: High performance visual tracking with Siamese region proposal network. In: Proceedings of the IEEE Conference on Computer Vision and Pattern Recognition, pp. 8971–8980 (2018a)

3. Yinda, X., Wang, Z., Li, Z., Yuan, Y., Gang, Y.: SiamFC++: towards robust and accurate visual tracking with target estimation guidelines. In: Proceedings of the AAAI Conference on Artificial Intelligence, vol. 34, pp. 12549–12556 (2020)

4. Noor, S., Waqas, M., Saleem, M.I., Minhas, H.N.: Automatic object tracking and segmentation using unsupervised SiamMask. IEEE Access **9**, 106550–106559 (2021)

5. Danelljan, M., Bhat, G., Khan, F.S., Felsberg, M.: Atom: accurate tracking by overlap maximization. In: Proceedings of the IEEE/CVF Conference on Computer Vision and Pattern Recognition, pp. 4660–4669 (2019)

6. Zhang, Z., Peng, H., Fu, J., Li, B., Hu, W.: Ocean: object-aware anchor-free tracking. In: Vedaldi, A., Bischof, H., Brox, T., Frahm, J.-M. (eds.) ECCV 2020. LNCS, vol. 12366, pp. 771–787. Springer, Cham (2020). https://doi.org/10.1007/978-3-030-58589-1_46

7. Cucci, D.A., Matteucci, M., Bascetta, L.: Pose tracking and sensor self-calibration for an all-terrain autonomous vehicle. IFAC-PapersOnLine **49**(15), 25–31 (2016)

8. Vaswani, A., et al.: Attention is all you need. Adv. Neural Inf. Process. Syst. **30** (2017)

9. Li, P., Wang, D., Wang, L., Lu, H.: Deep visual tracking: review and experimental comparison. Pattern Recognit. **76**, 323–338 (2018b)

10. Marvasti-Zadeh, S.M., Cheng, L., Ghanei-Yakhdan, H., Kasaei, S.: Deep learning for visual tracking: a comprehensive survey. IEEE Trans. Intell. Transp. Syst. **23**(5), 3943–3968 (2021)

11. Devlin, J., Chang, M.W., Lee, K., Toutanova, K.: Bert: pre-training of deep bidirectional transformers for language understanding. arXiv preprint arXiv:1810.04805 (2018)
12. Girshick, R.: Fast r-cnn. In: Proceedings of the IEEE International Conference on Computer Vision, pp. 1440–1448 (2015)
13. He, K., Zhang, X., Ren, S., Sun, J.: Deep residual learning for image recognition. In: Proceedings of the IEEE Conference on Computer Vision and Pattern Recognition, pp. 770–778 (2016)
14. Tan, M., Le, Q.: Efficientnet: rethinking model scaling for convolutional neural networks. In: International Conference on Machine Learning, pp. 6105–6114. PMLR (2019)
15. Chen, X., Yan, B., Zhu, J., Wang, D., Yang, X., Lu, H.: Transformer tracking. In: Proceedings of the IEEE/CVF Conference on Computer Vision and Pattern Recognition, pp. 8126–8135 (2021a)
16. Yan, B., Peng, H., Fu, J., Wang, D., Lu, H.: Learning spatio-temporal transformer for visual tracking. In: Proceedings of the IEEE/CVF International Conference on Computer Vision, pp. 10448–10457 (2021a)
17. Wang, N., Zhou, W., Wang, J., Li, H.: Transformer meets tracker: exploiting temporal context for robust visual tracking. In: Proceedings of the IEEE/CVF Conference on Computer Vision and Pattern Recognition, pp. 1571–1580 (2021a)
18. Bello, I., Zoph, B., Vaswani, A., Shlens, J., Le, Q.V.: Attention augmented convolutional networks. In: Proceedings of the IEEE/CVF International Conference on Computer Vision, pp. 3286–3295 (2019)
19. Ramachandran, P., Parmar, N., Vaswani, A., Bello, I., Levskaya, A., Shlens, J.: Stand-alone self-attention in vision models. Adv. Neural Inf. Process. Syst. **32** (2019)
20. Srinivas, A., Lin, T.Y., Parmar, N., Shlens, J., Abbeel, P., Vaswani, A.: Bottleneck transformers for visual recognition. In: Proceedings of the IEEE/CVF Conference on Computer Vision and Pattern Recognition, pp. 16519–16529 (2021)
21. Li, J., Huang, C., Qi, J., Qian, Y., Liu, W.: Three-way cognitive concept learning via multi-granularity. Inf. Sci. **378**, 244–263 (2017a)
22. Herrera, F., Herrera-Viedma, E., Martınez, L.: A fusion approach for managing multi-granularity linguistic term sets in decision making. Fuzzy Sets Syst. **114**(1), 43–58 (2000)
23. Yao, Y.: Perspectives of granular computing. In: 2005 IEEE International Conference on Granular Computing, vol. 1 (2005)
24. Qian, Y., Liang, J., Yao, Y., Dang, C.: MGRS: a multi-granulation rough set. Inf. Sci. **180**(6), 949–970 (2010)
25. Yao, J.T., Vasilakos, A.V., Pedrycz, W.: Granular computing: perspectives and challenges. IEEE Trans. Cybern. **43**(6), 1977–1989 (2013)
26. Yao, J.T., Yao, Y.Y.: Induction of classification rules by granular computing. In: Alpigini, J.J., Peters, J.F., Skowron, A., Zhong, N. (eds.) RSCTC 2002. LNCS (LNAI), vol. 2475, pp. 331–338. Springer, Heidelberg (2002). https://doi.org/10.1007/3-540-45813-1_43
27. Yao, J.T.: A ten-year review of granular computing. In: 2007 IEEE International Conference on Granular Computing (GRC 2007), p. 734. IEEE (2007)
28. Li, F., Miao, D., Pedrycz, W.: Granular multi-label feature selection based on mutual information. Pattern Recognit. **67**, 410–423 (2017b)
29. Zhang, X., Miao, D., Liu, C., Le, M.: Constructive methods of rough approximation operators and multigranulation rough sets. Knowl.-Based Syst. **91**, 114–125 (2016)

30. Miao, D.Q., Wang, G.Y., Liu, Q., Lin, T.Y., Yao, Y.Y.: Granular computing: past, present and future prospects (2007)
31. Wang, Z., Miao, D., Zhao, C., Luo, S., Wei, Z.: A robust long-term pedestrian tracking-by-detection algorithm based on three-way decision. In: Mihálydeák, T., et al. (eds.) IJCRS 2019. LNCS (LNAI), vol. 11499, pp. 522–533. Springer, Cham (2019). https://doi.org/10.1007/978-3-030-22815-6_40
32. Wang, Z.Y., Miao, D.Q., Zhao, C.R., Luo, S., Wei, Z.H.: Pedestrian tracking and detection combined algorithm based on multi-granularity features. Comput. Res. Dev. **57**, 996–1002 (2020)
33. Ruoyi, D., Xie, J., Ma, Z., Chang, D., Song, Y.-Z., Guo, J.: Progressive learning of category-consistent multi-granularity features for fine-grained visual classification. IEEE Trans. Pattern Anal. Mach. Intell. **44**(12), 9521–9535 (2021)
34. Li, J., Zhang, S., Huang, T.: Multi-scale 3D convolution network for video based person re-identification. In: Proceedings of the AAAI Conference on Artificial Intelligence, vol. 33, pp. 8618–8625 (2019)
35. Chen, C.F.R., Fan, Q., Panda, R.: Crossvit: cross-attention multi-scale vision transformer for image classification. In: Proceedings of the IEEE/CVF International Conference on Computer Vision, pp. 357–366 (2021b)
36. Zhang, Z., Lan, C., Zeng, W., Chen, Z.: Multi-granularity reference-aided attentive feature aggregation for video-based person re-identification. In: Proceedings of the IEEE/CVF Conference on Computer Vision and Pattern Recognition, pp. 10407–10416 (2020b)
37. Lin, T.-Y., Dollár, P., Girshick, R., He, K.: Feature pyramid networks for object detection. In: Proceedings of the IEEE Conference on Computer Vision and Pattern Recognition, pp. 2117–2125 (2017)
38. Lin, L., Fan, H., Zhang, Z., Yong, X., Ling, H.: Swintrack: a simple and strong baseline for transformer tracking. Adv. Neural Inf. Process. Syst. **35**, 16743–16754 (2022)
39. Wu, Y., Lim, J., Yang, M.H.: Online object tracking: a benchmark. In: Proceedings of the IEEE Conference on Computer Vision and Pattern Recognition, pp. 2411–2418 (2013)
40. Kristan, M., et al.: The eighth visual object tracking VOT2020 challenge results. In: Bartoli, A., Fusiello, A. (eds.) ECCV 2020. LNCS, vol. 12539, pp. 547–601. Springer, Cham (2020). https://doi.org/10.1007/978-3-030-68238-5_39
41. Müller, M., Bibi, A., Giancola, S., Alsubaihi, S., Ghanem, B.: TrackingNet: a large-scale dataset and benchmark for object tracking in the wild. In: Ferrari, V., Hebert, M., Sminchisescu, C., Weiss, Y. (eds.) ECCV 2018. LNCS, vol. 11205, pp. 310–327. Springer, Cham (2018). https://doi.org/10.1007/978-3-030-01246-5_19
42. Huang, L., Zhao, X., Huang, K.: Got-10k: a large high-diversity benchmark for generic object tracking in the wild. IEEE Trans. Pattern Anal. Mach. Intell. **43**(5), 1562–1577 (2019)
43. Fan, H., et al.: Lasot: a high-quality benchmark for large-scale single object tracking. In: Proceedings of the IEEE/CVF Conference on Computer Vision and Pattern Recognition, pp. 5374–5383 (2019)
44. Fan, H., et al.: Lasot: a high-quality large-scale single object tracking benchmark. Int. J. Comput. Vis. **129**, 439–461 (2021)
45. Dosovitskiy, A., et al.: An image is worth 16 × 16 words: transformers for image recognition at scale. arXiv preprint arXiv:2010.11929 (2020)
46. Zheng, M., et al.: End-to-end object detection with adaptive clustering transformer. arXiv preprint arXiv:2011.09315 (2020)

47. Wang, H., Zhu, Y., Adam, H., Yuille, A., Chen, L.C. Max-deeplab: end-to-end panoptic segmentation with mask transformers. In: Proceedings of the IEEE/CVF Conference on Computer Vision and Pattern Recognition, pp. 5463–5474 (2021b)
48. Meinhardt, T., Kirillov, A., Leal-Taixe, L., Feichtenhofer, C.: Trackformer: multi-object tracking with transformers. In: Proceedings of the IEEE/CVF Conference on Computer Vision and Pattern Recognition, pp. 8844–8854 (2022)
49. Sun, P., et al.: Transtrack: multiple object tracking with transformer. arXiv preprint arXiv:2012.15460 (2020)
50. Wang, Z., Miao, D.: Spatial-temporal single object tracking with three-way decision theory. Int. J. Approx. Reason. **154**, 38–47 (2023)
51. Yao, Y., Zhong, N.: Granular computing (2008)
52. Wang, Z., Shi, C., Wei, L., Yao, Y.: Tri-granularity attribute reduction of three-way concept lattices. Knowl.-Based Syst. 110762 (2023)
53. Chen, Y., Zhu, P., Li, Q., Yao, Y.: Granularity-driven trisecting-and-learning models for interval-valued rule induction. Appl. Intell. 1–23 (2023)
54. Deng, W., Wang, G., Zhang, X., Ji, X., Li, G.: A multi-granularity combined prediction model based on fuzzy trend forecasting and particle swarm techniques. Neurocomputing **173**, 1671–1682 (2016)
55. Liu, K., Li, T., Yang, X., Ju, H., Yang, X., Liu, D.: Feature selection in threes: neighborhood relevancy, redundancy, and granularity interactivity. Appl. Soft Comput. 110679 (2023)
56. Pawlak, Z.: Rough sets. Int. J. Comput. Inf. Sci. **11**, 341–356 (1982)
57. Stepaniuk, J., Skowron, A.: Three-way approximation of decision granules based on the rough set approach. Int. J. Approx. Reason. **155**, 1–16 (2023)
58. Janusz, A., Zalewska, A., Wawrowski, Ł, Biczyk, P., Ludziejewski, J., Sikora, M., et al.: Brightbox-a rough set based technology for diagnosing mistakes of machine learning models. Appl. Soft Comput. **141**, 110285 (2023)
59. Long, J., Shelhamer, E., Darrell, T.: Fully convolutional networks for semantic segmentation. In: Proceedings of the IEEE Conference on Computer Vision and Pattern Recognition, pp. 3431–3440 (2015)
60. Kong, T., Yao, A., Chen, Y., Sun, F.: Hypernet: towards accurate region proposal generation and joint object detection. In: Proceedings of the IEEE Conference on Computer Vision and Pattern Recognition, pp. 845–853 (2016)
61. Liu, W., Rabinovich, A., Berg, A.C.: Parsenet: looking wider to see better. In: ICLR Workshop. Cited on, p. 111 (2016)
62. Ronneberger, O., Fischer, P., Brox, T.: U-Net: convolutional networks for biomedical image segmentation. In: Navab, N., Hornegger, J., Wells, W.M., Frangi, A.F. (eds.) MICCAI 2015. LNCS, vol. 9351, pp. 234–241. Springer, Cham (2015). https://doi.org/10.1007/978-3-319-24574-4_28
63. Pinheiro, P.O., Lin, T.-Y., Collobert, R., Dollár, P.: Learning to refine object segments. In: Leibe, B., Matas, J., Sebe, N., Welling, M. (eds.) ECCV 2016. LNCS, vol. 9905, pp. 75–91. Springer, Cham (2016). https://doi.org/10.1007/978-3-319-46448-0_5
64. Honari, S., Yosinski, J., Vincent, P., Pal, C.: Recombinator networks: learning coarse-to-fine feature aggregation. In: Proceedings of the IEEE Conference on Computer Vision and Pattern Recognition, pp. 5743–5752 (2016)
65. Newell, A., Yang, K., Deng, J.: Stacked hourglass networks for human pose estimation. In: Leibe, B., Matas, J., Sebe, N., Welling, M. (eds.) ECCV 2016. LNCS, vol. 9912, pp. 483–499. Springer, Cham (2016). https://doi.org/10.1007/978-3-319-46484-8_29

66. Lin, T.-Y., et al.: Microsoft COCO: common objects in context. In: Fleet, D., Pajdla, T., Schiele, B., Tuytelaars, T. (eds.) ECCV 2014. LNCS, vol. 8693, pp. 740–755. Springer, Cham (2014). https://doi.org/10.1007/978-3-319-10602-1_48

67. ILoshchilov, I., Hutter, F.: Decoupled weight decay regularization. arXiv preprint arXiv:1711.05101 (2017)

68. Zhang, Z., Xie, Y., Xing, F., McGough, M., Yang, L.: MDNet: a semantically and visually interpretable medical image diagnosis network. In: Proceedings of the IEEE Conference on Computer Vision and Pattern Recognition, pp. 6428–6436 (2017)

69. Danelljan, M., Bhat, G., Shahbaz Khan, F., Felsberg, M.: Eco: efficient convolution operators for tracking. In: Proceedings of the IEEE Conference on Computer Vision and Pattern Recognition, pp. 6638–6646 (2017)

70. Yan, B., Peng, H., Wu, K., Wang, D., Fu, J., Lu, H.: Lighttrack: finding lightweight neural networks for object tracking via one-shot architecture search. In: Proceedings of the IEEE/CVF Conference on Computer Vision and Pattern Recognition, pp. 15180–15189 (2021b)

71. Bhat, G., Danelljan, M., Gool, L.V., Timofte, R.: Learning discriminative model prediction for tracking. In: Proceedings of the IEEE/CVF International Conference on Computer Vision, pp. 6182–6191 (2019)

72. Bhat, G., Johnander, J., Danelljan, M., Khan, F.S., Felsberg, M.: Unveiling the power of deep tracking. In: Ferrari, V., Hebert, M., Sminchisescu, C., Weiss, Y. (eds.) ECCV 2018. LNCS, vol. 11206, pp. 493–509. Springer, Cham (2018). https://doi.org/10.1007/978-3-030-01216-8_30

73. Yu, Y., Xiong, Y., Huang, W., Scott, M.R.: Deformable Siamese attention networks for visual object tracking. In: Proceedings of the IEEE/CVF Conference on Computer Vision and Pattern Recognition, pp. 6728–6737 (2020)

74. Voigtlaender, P., Luiten, J., Torr, P.H., Leibe, B.: Siam R-CNN: visual tracking by re-detection. In: Proceedings of the IEEE/CVF Conference on Computer Vision and Pattern Recognition, pp. 6578–6588 (2020)

75. Guo, D., Wang, J., Cui, Y., Wang, Z., Chen, S.: Siamcar: siamese fully convolutional classification and regression for visual tracking. In: Proceedings of the IEEE/CVF Conference on Computer Vision and Pattern Recognition, pp. 6269–6277 (2020)

A Multi-granularity Network for Time Series Forecasting on Multivariate Time Series Data

Zongqiang Wang[1], Yan Xian[1], Guoyin Wang[2], and Hong Yu[1,2]([✉])

[1] Chongqing Key Laboratory of Computational Intelligence, Chongqing University of Posts and Telecommunications, Chongqing, China
[2] Key Laboratory of Cyberspace Big Data Intelligent Security, (Chongqing University of Posts and Telecommunications), Ministry of Education, Chongqing, China
yuhong@cqupt.edu.cn

Abstract. Multivariate time series forecasting is a significant research problem in many fields, such as economics, finance, and transportation. The main challenge faced by current time series forecasting models is effectively capturing the information embedded in different temporal patterns. However, most existing methods analyze forecasting at a single time granularity. To alleviate this issue, we propose a novel approach called the multi-granularity hierarchical temporal forecasting network (MGTNet), which integrates three temporal convolution kernels based on the principles of granular computing. Specifically, the data from fine-grained, meso-grained, and coarse-grained are processed, respectively. Then, temporal dependence and variable dependence analyses are performed at different levels according to distinct granularities. Finally, they are stacked to form a complete multi-granularity network. The experimental results demonstrate that the proposed MGTNet outperforms a bunch of compared methods in terms of RSE, CORR on Traffic, Electricity and Solar-Energy datasets.

Keywords: Multi-granularity · Granular computing · Hierarchy network · Multivariate time series

1 Introduction

In everyday life, multivariate time series (MTS) data are in many fields including household electricity consumption, solar power generation, highway traffic flow. The complex relationships within multivariate time series data are difficult to accurately capture, such as the traffic department tracking traffic flow at multiple intersections. This makes multivariable time series forecasting very challenging [26].

Time series forecasting has a rich history, with statistical-based methods playing a crucial role. For instance, the autoregressive sliding average (ARIMA) and smoothing exponential are mainly used for univariate time series forecasting

A. Campagner et al. (Eds.): IJCRS 2023, LNAI 14481, pp. 324–338, 2023.
https://doi.org/10.1007/978-3-031-50959-9_23

[2]. However, these methods treat each time series as an individual entity and fail to fully utilize the information of variables in multivariate time series. To solve the MTS problem, the VAR model was proposed [14], which is an autoregressive model that extends from univariate to vector scale. The VAR models struggle to capture nonlinear relationships within high-dimensional data, where dependencies across time and correlations among variables are nonlinear. Since the multivariate time series data is nonlinear, which both the dependencies in the time dimension and the correlation among multivariates. Therefore, this paper exploits the powerful nonlinear modeling capability of deep learning to enhance multivariate time series forecasting.

Recently, deep learning-based methods have been widely used for time series forecasting owing to their ability to capture nonlinear relationships and the increasing availability of nonlinear and non-stationary time series data [20]. There are many deep learning methods have been applied to time series forecasting, such as recurrent neural networks (RNNs) [16], long short-term memory networks (LSTMs) [6], convolutional neural networks (CNNs) [12], and attention mechanisms [19].

For the specific problem of multivariate time series forecasting, Lai et al. [9] proposed LSTNet , which combines LSTM and CNN to analyze the intricate periodic patterns present in time series data. These patterns are complex and subtle, consisting mainly of short-term patterns, long-term patterns, and their mixtures. For example, the traffic dataset contains daily and weekly patterns, which representing short-term and long-term patterns, respectively. The former describing morning and evening peaks, while the latter reflecting weekday and weekend patterns. The capacity to detect these two trends is critical in time series forecasting models. In their study, a novel deep learning model is proposed specifically for the MTS forecasting problem, where convolutional and recurrent neural networks are used to capture short-term and long-term dependence patterns between multiple variables, respectively. However, the main drawback of LSTNet is that the length of the "recurrent jumps" should be predetermined based on the dataset before training. To address this issue, Shih et al. [17] introduced a dual self-attentive network for multivariate time series. This network utilizes both CNNs and local CNNs to extract periodicity and nonperiodicity effectively. Since LSTNet performs poorly on non-periodic datasets, Huang et al. [7] proposed a dual self-attentive network for multivariate time series, mainly employing all CNNs and local CNNs for extracting periodicity and non-periodicity. For the complex dependencies between multiple variables in time series and mixed dependencies in time dimension, Song and Fujimura [18] proposed a combinatorial model for capturing long-term and short-term correlations in multivariate time series forecasting, the model leverages stacked inflated convolution and recursive units to capture long-term and short-term correlations in the data, enabling the capturing of complex patterns of mixed long-term and short-term dependencies.

Although the above methods have been successfully applied to multivariate time series forecasting, they do not fully consider the temporal patterns in the time series. However, the relationship between different temporal patterns is very significant. Therefore, it is important to consider multiple temporal perspectives when predicting a point in time. As Yao introduced the granular computing [23], which involves the use of granules and multiple levels of granularity to analyze problems from different perspectives. Yao proposed [25] that artificial intelligence perspectives on granular computing of hierarchical problem solving reveals some important aspects of granular computing. For example, communicating up and down the different levels of granularity and switching between differing granularity. The methodology of hierarchical problem solving in artificial intelligence and other fields can be adapted for granular computing.

Therefore, we propose a multi-granular hierarchical time series network, which draws inspiration from granular computing while consider the unique properties of time series data. Our goal is to address certain limitations overlooked in previous work and provide some new problem solving ideas for the field. The main contributions of this work can be summarized as follows:

- Based on the idea of granular computing, a hierarchical multi-granularity temporal forecasting model is proposed to extract different temporal correlations from various temporal granularities, thereby providing multi-level support for prediction results.
- The concept of temporal convolution kernel is proposed to solve the problem from fine, meso and coarse, which is a kind of micro-detail to global research. This approach provide some new directions for the meso concept.
- It is experimentally demonstrated that we achieve advanced performance on three real-world multivariate datasets, and the successful application also advances the exploration of multiple granularity levels in time series forecasting.

The remainder of this paper is organized as follows. Section 2 analyzes the related work on multivariate time series forecasting. Section 3 introduce our proposed method. Section 4 describes the comparative experiment and ablation experiment. Section 5 conclude this paper.

2   Related Work

This section reviews the development of multivariate time series forecasting, gives a brief review of the granular computing, and describes the temporal granularity of its own design to further illustrate the details of the work.

2.1   Multivariate Time Series Forecasting

Time series forecasting has been an area of interest for researchers due to its wide range of applications, and various forecasting methods have been proposed one after another. Among all the methods, statistical methods have a large share, and one of the most popular models is the autoregressive integrated moving average model (also known as ARIMA model), which contains autoregressive, moving average and autoregressive moving average [2]. The success of ARIMA model is due to its robustness to non-stationary data and interpretability of statistical features. However, ARIMA is more suitable for univariate time series forecasting due to the high computational cost required by the model. To solving the MTS problem, the VAR model was proposed [14], which is an autoregressive model extended from univariate to vector scale. It is widely used to solve the MTS problem, and many VAR-based models have been constructed [35], including Varmax [13], elliptical VAR [15], etc. VAR-based models perform poorly when encountering high-dimensional data, and VAR models are proned to overfitting when dealing with high-dimensional data. However, the temporal and variable dependencies are mostly nonlinear relationships, which are difficult to capture by statistical models. Some machine learning based models such as support vector machines [3] and neural networks were subsequently proposed to capture the nonlinear relationships, where the forecasting problem is treated as a regression problem and kernel methods are applied to the models to increase their ability to handle nonlinearities. However, each MTS dataset possesses different nonlinear relationships, and existing methods like support vector machines are limited in addressing these diverse nonlinearities.

Recently, deep neural networks have been widely used in time series forecasting, which performance has been significantly improved [21]. Recurrent neural network RNN [16], long short-term memory LSTM [6] and gated recurrent unit GRU [12] are exactly specialized sequence modeling networks. CNNs plays a great role in feature extraction. The following work is built and used for multivariate time series forecasting based on the above networks. For example, Lai et al. analyzed that periodic patterns in time series data are complex and subtle, consisting of short-term patterns, long-term patterns, and their combinations, the work called LSTNet-skip [9]. Shih et al. [17] proposed a new attention concept in their work called TPA-LSTM. This concept computes attention values of different variables at each time step, enabling the model to extract temporal patterns without relying on predefined time steps. Huang et al. constructed a DSANet network using local CNN and global CNN to capture local and global dependencies [7]. Song et al. extracted complex patterns of mixed long-term and short-term dependencies between multiple variables using stacked expanded convolutional kernel recursion units [18]. Although the above methods have achieved good results in multivariate time-series forecasting, none of them fully consider the different temporal patterns in the time series. For LSTnet and MDTNet networks although long-term and short-term time patterns are analyzed to some extent, this method does not fully consider the complete time patterns in the time series and may not adequately mine temporal correlations. Therefore, we want to analyze the same forecasting task at multiple levels from different time levels, and the idea of granular computing is exactly suitable for this scenario.

2.2   Theory and Application of Granular Computing

The basic components of the granular computing include three main parts: granules, granules layers and granules structures. granules is the most basic element of the granular computing model and is the original language of the granular computing model [27]. The totality of all granules obtained according to some practically required granulation criterion constitutes a granular layer, an abstracted description of the problem space. According to some relation or operator, the corresponding granules produced. The granules in the same layer tend to have the same certain property or function within them. Different degrees of granulation result in different granules layers being produced for the same problem space. The internal structure of a granular layer is the structure of the thesis domain consisting of the individual granules on that granular layer and the interrelationships between the granules. One granularity criterion corresponds to one granules layer, different granularity criteria correspond to multiple granules layers. The different layer respond people looking at the problem, understanding the problem, and solving the problem from different perspectives and sides. The interconnections between all granules layers form a relational structure called granules structure [24].

As presented, in Yao's article on the processing and interpretation of time series, understanding the use of granules. From human perspective, we can describe time series in a semiqualitative manner by pointing at specific segments of such signals. We always granulate all phenomena that are understandable to human beings regardless of the original signals being discrete or analog [23].

The following is description the granulation of time: time is another important and omnipresent variable that is subject to granulation. We use seconds, minutes, days, months, and years. Depending upon a specific problem we have in mind and who the user is, the size of information granules (time intervals) could vary quite significantly. To the high-level management, time intervals of quarters of year or a few years could be meaningful temporal information granules on the basis of which one develops any predictive model [23]. From the above it can be seen that for time series we can use the principle of granular computing to analyze the forecasting problem from multi perspectives, and granular computing is a feasible direction to help us consider different temporal patterns from time series.

In recent years, the granular computing has some research in the area of time series forecasting [1]. Yang et al. [22] achieved long-term forecasting by structuring numerical time series into granular time series. Ma et al. [11] proposed a method for smoothing the original time series into granular time series using a sliding window strategy using the idea of granular computing. The method has achieved better results in long-term forecasting, not only with the ability to avoid cumulative errors, but also with better explanatory power. Li et al. [10] proposed a new granulation algorithm to demonstrate the feasibility of the granulation idea in short-term forecasting, which is centered on the basic rules and is free from external disturbances and successfully applied on short-term forecasting. Kouloumpris et al. [8] based on an aggregation method with hourly granularity captured better temporal patterns on short-term forecasts in energy, making short-term forecasts more accurate. Hao et al. [5] used the fuzzy grain method to granulate the interval values and proposed the concept of dynamic fuzzy information grain to establish a long-term forecasting model, which provides a new way of thinking for long-term forecasting.

2.3   Construction of Time Granules

For multivariate time series data, we need to analyze prediction tasks from different temporal patterns. Time series data has special temporal relationships, so we must strictly follow the temporal nature of the construction of time granules to avoid the risk of data leakage. Also to perform analysis from different temporal patterns we need to use different temporal convolution kernels for granules construction. In this paper, three temporal convolutional kernels are constructed and analysis of time series prediction tasks at three levels.

Theorem 1. *Three sets of temporal convolution kernels are used $l_1 \times 1$, $l_2 \times 1$ and $l_3 \times 1$, then the three temporal convolution kernels are granularized over the time series X to extract the temporal granularity, producing temporal granularity as $G_1(X)$, $G_2(X)$, and $G_3(X)$, respectively.*

Because the three temporal convolution kernels are of different lengths and produce temporal granularity of different sizes, analyzing temporal correlations from different temporal granularity is different to analyzing prediction results from different temporal patterns. This allows the final result to contain information from multiple temporal patterns. For the sake of convenience, we give an example of the extraction process of a temporal convolution kernel.

$$X = \begin{bmatrix} x_{11} & x_{12} & \cdots & x_{1\,T} \\ x_{21} & x_{22} & \cdots & x_{2\,T} \\ \vdots & \vdots & \cdots & \vdots \\ x_{D1} & x_{D2} & \cdots & x_{DT} \end{bmatrix} \overset{\text{Granulation}}{=} \begin{bmatrix} g_{11} & g_{12} & \cdots & x_{1\,T-l+1} \\ g_{21} & g_{22} & \cdots & g_{2\,T-l+1} \\ \vdots & \vdots & \cdots & \vdots \\ g_{D1} & g_{D2} & \cdots & g_{DT-l+1} \end{bmatrix} = G(X) \quad (1)$$

For the above X is the multivariate temporal data, where x is the specific value of the time scale represented, for example x_{DT} represents the specific value of the variable D at time T. Then we will get the $G(X)$ matrix by temporal granularity of the temporal convolution kernel. g granularity in the $G(X)$ matrix is calculated by the temporal convolution kernel. For example, a specific example: a convolution kernel $l_1 = [a_1, a_2, a_3, a_4, a_5, a_6]$ of length 6, where a_1 to a_6 are time-specific convolution kernel values, which will be updated with each iteration of deep learning. Extraction is performed on the X data by a temporal convolution kernel, which will first extract the T timescales of the first row of variables sequentially, sliding backward one bit at a time, and sequentially sliding the extracted time granules until it reaches the end of the slide. The computation is done for $g_{11} = x_{11} \times a_1 + x_{12} \times a_2 + x_{13} \times a_3 + x_{14} \times a_4 + x_{15} \times a_5 + x_{16} \times a_6$, then ditto for $g_{22} = x_{22} \times a_1 + x_{23} \times a_2 + x_{24} \times a_3 + x_{25} \times a_4 + x_{26} \times a_5 + x_{27} \times a_6$, each temporal convolutional kernel extracts all D variables and then combines the final result into $G(X)$. Finally we extract all the variables of the time series data by three temporal convolution kernels of different length. $l_1 \times 1$, $l_2 \times 1$ and $l_3 \times 1$ to get three different temporal granularity $G_1(X)$, $G_2(X)$, and $G_3(X)$, respectively. Then for the three temporal granularity generated we used different methods for learning, trying to mine useful information from the different temporal patterns.

3   Proposed Method

In this section, a complete description of the proposed model is given. The model is mainly divided into three modules:the temporal granulation module,the temporal forecasting module and the granules layer fusion module.

Figure 1 shows the general model structure. The model extracts three different granularity levels of temporal patterns from the original data using three temporal convolution kernels. These granularity levels are referred to as fine-grained, meso-grained, and coarse-grained, respectively. For the fine and meso granularity levels, we use LSTMs to find the temporal relationships between each temporal granularity, and then utilize the self-attentive network to identify the relationships between different variables. As for the coarse-grained level, we use a $T \times 1$ temporal convolution kernel for temporal granularity extraction, where the length of the kernel matches the length of the entire data. Consequently, each variable is extracted only once, resulting in a temporal granularity that can also be referred to as global granularity. To learn the relationships between different variables, N groups of simultaneous extraction are employed for each temporal convolution kernel. Thus, the extracted temporal granularity will form N groups and subsequently learn.

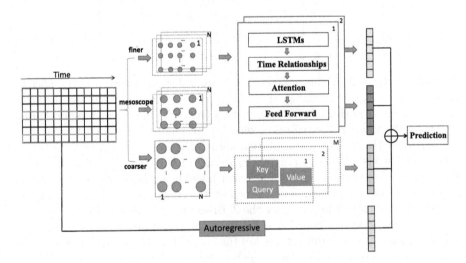

Fig. 1. General Framework Diagram

3.1   Temporal Granulation Module

After three different lengths of temporal convolution kernels, three different temporal patterns of temporal granularity are produced. Following the extraction of N sets of temporal convolution kernels, there are N layers of each kind of temporal granule as shown in Fig. 2. For the coarse-grained level, where the extraction is performed globally, only one layer of $D \times N$ granularity is generated. For fine-grained and meso-grained, we use a long-and short-term memory network to predict the temporal granularity in each layer. The relationship between multiple temporal granularity is mapped to a temporal granules of size $D \times 1$, allowing the results of the N layers are integrated into a single $D \times N$ granules layer for computational convenience. Finally, the output size of the temporal granulation module is guaranteed to be consistent. The specific process is shown in the following Fig. 2.

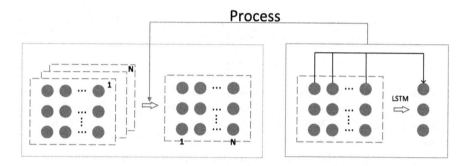

Fig. 2. Mining the relationship between time granules

The long short-term memory network is a kind of artificial recurrent neural network used in deep learning, it can deal with a single number of points or a sequence segment. In our context, the main purpose of introducing the LSTM network is to obtain the predicted output h_t of a time series and the corresponding hidden state. Then, we calculate the weights between the hidden layer and the previous time window according to the corresponding hidden state. These weights are utilized to adjust the time weight matrix, enabling the capture of more useful time information. The specific formula of long short-term memory network is as follows:

$$h_t, c_t = F(h_{t-1}, c_{t-1}, x_t) \tag{2}$$

$$i_t = \sigma(U_i x_t + W_i h_{t-1}) \tag{3}$$

$$f_t = \sigma(U_f x_t + W_f h_{t-1}) \tag{4}$$

$$o_t = \sigma(U_o x_t + W_o h_{t-1}) \tag{5}$$

$$c_t = f_t \odot c_{t-1} + i_t \odot tanh(U_g x_t + W_g h_{t-1}) \tag{6}$$

$$h_t = o_t \odot tanh(c_t) \tag{7}$$

3.2 Temporal Forecasting Module

Each of the three different temporal patterns is processed differently. To account for the nonlinear nature, we utilize nonlinear neural networks to handle all temporal modes. Specifically, we designed a linear AR module for linear balancing. For the granularity matrix of $D \times N$ obtained from the temporal granularity module, a multi-headed attention network is used for learning. In the attention module, query, key and value vectors are used to dig deeper into the hidden layers of the inputs. Subsequently, the outputs are weighted using a certain weighting relation, which enables the emphasis on valuable information. To compute the weight for each position, we calculate the inner product between the query of other positions in the time series and the key of the current position. The resulting weight is then multiplied by the value to obtain the final output. This output is then forwarded to generate the output of the current module. The relationship between query, key, and value is defined as follows:

$$Z = softmax(\frac{Q(K)^T}{\sqrt{d}})V \tag{8}$$

$$\hat{G}(X) = ReLU(ZW_1 + b_1)W_2 + b_2 \tag{9}$$

where W represents the weight matrix, b is the offset, and the final resulting $\hat{G}(X)$ is the corresponding forecasting vector value learned from the multi-headed attention network. The three predictions obtained from each of the three temporal models are denoted as $\hat{G}_1(X) = X^{g_1}_{T+h} \in R^D$, $\hat{G}_2(X) = X^{g_2}_{T+h} \in R^D$ and $\hat{G}_3(X) = X^{g_3}_{T+h} \in R^D$. Due to the nonlinearity of the convolutional and self-attentive components, the scale of the neural network output is insensitive to the scale of the input. To address this drawback, the final forecasting result as a mixture of linear and nonlinear components. In addition to the nonlinear components introduced above, the classical AR model is treated as a linear component. The forecasting of the AR component can be represented as $h_L = X^L_{T+h} \in R^D$.

3.3 Granules Layer Fusion Module

For the four vectors obtained from the forecasting module and analyze the problem at different levels, we perform vector fusion. The advantage of this method is that it allows the granular layers to learn from each other during the network learning process, resulting in more diverse final results. The fusion process is achieved using the following formula:

$$F(\hat{G}_1(X), \hat{G}_2(X), \hat{G}_3(X), h_L) = X_{T+h} \tag{10}$$

The equation presented above represents the function F, which can involve various calculations such as summation. In this equation, h_L represents the result of the linear forecasting layer, and $\hat{G}_1(X), \hat{G}_2(X), \hat{G}_3(X)$ is the prediction result of time granules $G_1(X), G_2(X)$ and $G_3(X)$, respectively. The X_{T+h} represents the final output result after the fusion process.

4   Experiments

In all experiments, we used a consistent learning rate of 0.001 and employed the Adam optimizer for training the model. To assess the performance of model, Empirical Correlation Coefficient (CORR) and Root Relative Squared Error(RSE) are utilized as evaluation metrics. We compare with four distinct methods on three datasets. Additionally, the effectiveness of multigranularity is further illustrated by ablation experiments.

4.1   Datasets and Baseline

For a more comprehensive validation, we conducted experiments on three distinct datasets, which represent different domain applications in electricity, transportation, and solar energy. The descriptions of the datasets are shown in Table 1.

- **Traffic:** for the Traffic dataset is 48 months (2015–2016) of hourly data collected by the California Department of transportation. The data depicts the road occupancy (between 0 and 1) measured by different sensors on freeways in the San Francisco Bay area.
- **Solar-Energy:** which records solar production in 2006, sampled every 10 min from 137 PV plants in Alabama.
- **Electricity:** which records electricity consumption (kWh) for n=321 customers every 15 min from 2012 to 2014. We transformed the data to reflect hourly consumption.

Table 1. Details of the datasets.

Datasets	Objects	Dimensions	Time Interval
Electricity	26304	321	1 h
Traffic	17544	862	1 h
Solar-Energy	52560	137	10 min

For the above dataset we divide it and the division ratio is 60%, 20% and 20% for training set, validation set and testing set. Also we normalized the data to between 0 and 1.

Five benchmark models are selected for performance comparison:

- **AR:** [4] denotes the autoregression model, which is the most commonly used machine learning algorithm for linear multi time series forecasting.
- **LSTNet-skip:** [9] uses a combination of CNNs and RNNs to extract local and long-term dependencies and introduces the recurrent-skip component to alleviate the problem that RNNs cannot capture long-term dependencies.

- **TPA-LSTM:** [17] denotes an attention-based RNN, which uses using a set of filters to extract time-invariant temporal patterns, similar to transforming time series data into its "frequency domain".
- **DSANet:** [7] utilizes a combination of CNN and self-attention to capture local and global dependencies
- **MDTNet-direct:** [18] is a model for extracting mixed long-term and short-term dependencies between multiple variables using stacked dilated convolution and using recurrent neural networks.

4.2   Analysis of Experimental Results

In this experiment, CORR and RSE were used as evaluation metrics to assess the experiment. A higher value of CORR indicates better performance, while a lower value of RSE also indicates better performance. After conducting the tests on the three datasets, we obtained the following results:

In Table 2, 3 and 4, we present a summary of the experimental results. The prediction ranges of 3,6,12 and 24. which means that the time-step prediction ranges from 3 to 24 h for electricity and traffic, and from 30 min to 4 h for the solar energy. Here, the best results are highlighted in bold and the prediction task becomes more difficult as the prediction range increases. It is observed that our method outperforms the rest of the methods on the Traffic dataset and the best performance is achieved for both the metrics RSE and CORR. The proposed MGTNet framework achieves the best performance on the Solar-Energy dataset at 3, 6, and 12 steps and the third performance at 24 steps. The state-of-the-art performance is achieved on the Electricity dataset for both 3 and 6 step lengths, and we maintain the second effect on 12 step length. The above experiments demonstrate the effectiveness of our proposed multi-granular hierarchical time series network. It further demonstrates the complexity of the idea of granular computing, which has a multi-layered role for mining different patterns in time series.

Table 2. Comparison results on Traffic.

Databaset	Traffic							
	3		6		12		24	
Method/Metrics	RSE	CORR	RSE	CORR	RSE	CORR	RSE	CORR
AR	0.4777	0.7752	0.6218	0.7568	0.6252	0.7544	0.6293	0.7519
LSTNet-skip	0.4487	0.8721	0.4893	0.8690	0.4950	0.8614	0.4963	0.8588
TPA-LSTM	0.4487	0.8812	0.4658	0.8717	0.4641	0.8717	0.4765	0.8629
DSANet	0.4456	0.8829	0.4779	0.8642	0.4789	0.8659	0.4983	0.8541
MDTNet-direct	0.4513	0.8809	0.4754	0.8698	0.4710	0.8774	0.4851	0.8631
ours	**0.4396**	**0.8905**	**0.4539**	**0.8835**	**0.4632**	**0.8782**	**0.4757**	**0.8684**

Table 3. Comparison results on Solar-Energy.

Databaset	Solar-Energy							
	3		6		12		24	
Method/Metrics	RSE	CORR	RSE	CORR	RSE	CORR	RSE	CORR
AR	0.2435	0.9710	0.3790	0.9263	0.5911	0.8107	0.8699	0.5314
LSTNet-skip	0.1843	0.9843	0.2559	0.9690	0.3254	0.9467	0.4643	0.8870
TPA-LSTM	0.1843	0.9850	0.2347	0.9742	0.3234	0.9487	0.4389	0.9081
DSANet	0.1816	0.9862	0.2310	0.9733	0.3255	0.9553	**0.4313**	**0.9156**
MDTNet-direct	0.1805	0.9841	0.2336	0.9725	0.3236	0.9467	0.4357	0.9097
ours	**0.1794**	**0.9869**	**0.2252**	**0.9809**	**0.3202**	**0.9548**	0.4363	0.9127

Our proposed MGTNet model is a multi-granularity hierarchical network, where the G1 module represents the module with temporal convolution kernel l_1, and similarly the G2 module represents the module with temporal convolution kernel l_2 and the G3 module represents the module with temporal convolution kernel l_3. In order to prove the effectiveness of these modules, we designed ablation experiments. Based on MGTNet, some components were removed to obtain three new algorithms in MGTNet/G1, MGTNet/G2 and MGTNet/G3 respectively. These newly algorithms are then compared with MGTNet. MGTNet/G1 denotes the MGTNet framework with the G1 component removed, that is, G1 is not involved in the operation, and the rest of the components remain similar.

Table 5 shows the ablation results on Traffic. It can be seen that the overall performance of the proposed MGTNet method decreases regardless of which component is discarded. The results demonstrate the effectiveness of our multi-perspective temporal model, and the success of the temporal kernel extraction method in capturing temporal relationships. This further confirms the efficacy of granular computing in temporal forecasting tasks and the effectiveness of our multi-granularity analysis method.

Table 4. Comparison results on Electricity.

Databaset	Electricity							
	3		6		12		24	
Method/Metrics	RSE	CORR	RSE	CORR	RSE	CORR	RSE	CORR
AR	0.0995	0.8845	0.1035	0.8632	0.1050	0.8591	0.1054	0.8595
LSTNet-skip	0.0864	0.9283	0.0931	0.9135	0.1007	0.9077	0.1007	0.9119
TPA-LSTM	0.0823	0.9429	0.0916	0.9337	0.0964	0.9250	0.1006	0.9133
DSANet	0.0783	0.9517	0.0874	0.9463	0.0990	0.9271	0.1040	**0.9209**
MDTNet-direct	0.0821	0.9472	0.0889	0.9331	**0.0945**	0.9267	**0.0981**	0.9140
ours	**0.0768**	**0.9598**	**0.0862**	**0.9529**	0.0962	**0.9294**	0.1027	0.9158

Table 5. The ablation experiment results on Traffic.

Traffic		Methods			
horizon	Metrics	MGTNet/G1	MGTNet/G2	MGTNet/G3	Ours
3	RSE	0.4558	0.4519	0.4583	**0.4396**
	CORR	0.8782	0.8780	0.8790	**0.8905**
6	RSE	0.4605	0.4568	0.4729	**0.4539**
	CORR	0.8761	0.8759	0.8716	**0.8835**
12	RSE	0.4678	0.4691	0.4774	**0.4632**
	CORR	0.8727	0.8675	0.8674	**0.8782**
24	RSE	0.4904	0.4896	0.4987	**0.4757**
	CORR	0.8589	0.8631	0.8560	**0.8684**

5 Conclusions

In this paper, we propose a multi-granular hierarchical time series forecasting network based on the idea of granular computation. This approach addresses the limitations of existing multivariate time series forecasting models by using multiple temporal models to incorporate complex relationships within the time series data. The experimental results demonstrate that our proposed method outperforms the comparison methods, leading to overall improved performance. The ablation experiments further highlight the significant role of multi-granularity in addressing the forecasting problem, providing evidence for the importance of considering the problem from multiple levels. Moreover, the effectiveness of our coarse-to-fine multi-granularity analysis method is also confirmed. It also further shows that this coarse-to-fine multi-granularity analysis method is effective. In our future work, we will further investigate the method of solving time series forecasting with granular computing and rough set.

Acknowledgements. This work was supported in part by the National Natural Science Foundation of China (62136002, 62233018, 62221005), and the Natural Science Foundation of Chongqing (cstc2022ycjh-bgzxm0004).

References

1. Bargiela, A., Pedrycz, W.: Granular computing. In: HANDBOOK ON COMPUTER LEARNING AND INTELLIGENCE: Volume 2: Deep Learning, Intelligent Control and Evolutionary Computation, pp. 97–132. World Scientific (2022)
2. Box, G.E., Jenkins, G.M., Reinsel, G.C., Ljung, G.M.: Time Series Analysis: Forecasting and Control. Wiley, Hoboken (2015)
3. Cao, L.J., Tay, F.E.H.: Support vector machine with adaptive parameters in financial time series forecasting. IEEE Trans. Neural Networks **14**(6), 1506–1518 (2003)
4. Hamilton, J.: Time Series Analysis. Princeton University Press, Princeton (1994)

5. Hao, Y., Jiang, S., Yu, F., Zeng, W., Wang, X., Yang, X.: Linear dynamic fuzzy granule based long-term forecasting model of interval-valued time series. Inf. Sci. **586**, 563–595 (2022)
6. Hochreiter, S., Schmidhuber, J.: Long short-term memory. Neural Comput. **9**(8), 1735–1780 (1997)
7. Huang, S., Wang, D., Wu, X., Tang, A.: Dsanet: dual self-attention network for multivariate time series forecasting. In: Proceedings of the 28th ACM International Conference on Information and Knowledge Management, pp. 2129–2132 (2019)
8. Kouloumpris, E., Konstantinou, A., Karlos, S., Tsoumakas, G., Vlahavas, I.: Short-term load forecasting with clustered hybrid models based on hour granularity. In: Proceedings of the 12th Hellenic Conference on Artificial Intelligence, pp. 1–10 (2022)
9. Lai, G., Chang, W.C., Yang, Y., Liu, H.: Modeling long-and short-term temporal patterns with deep neural networks. In: The 41st International ACM SIGIR Conference on Research & Development in Information Retrieval, pp. 95–104 (2018)
10. Li, F., Tang, Y., Yu, F., Pedrycz, W., Liu, Y., Zeng, W.: Multilinear-trend fuzzy information granule-based short-term forecasting for time series. IEEE Trans. Fuzzy Syst. **30**(8), 3360–3372 (2021)
11. Ma, C., Zhang, L., Pedrycz, W., Lu, W.: The long-term prediction of time series: a granular computing-based design approach. IEEE Trans. Syst. Man, Cybern. Syst. **52**(10), 6326–6338 (2022)
12. Ma, Q., Chen, E., Lin, Z., Yan, J., Yu, Z., Ng, W.W.: Convolutional multitimescale echo state network. IEEE Trans. Cybern. **51**(3), 1613–1625 (2019)
13. Östermark, R., Saxén, H.: Varmax-modelling of blast furnace process variables. Eur. J. Oper. Res. **90**(1), 85–101 (1996)
14. Phillips, P.C.: Fully modified least squares and vector autoregression. Econometrica J. Econometric Soc. **63**, 1023–1078 (1995)
15. Qiu, H., Xu, S., Han, F., Liu, H., Caffo, B.: Robust estimation of transition matrices in high dimensional heavy-tailed vector autoregressive processes. In: International Conference on Machine Learning, pp. 1843–1851. PMLR (2015)
16. Rumelhart, D.E., Hinton, G.E., Williams, R.J.: Learning representations by back-propagating errors. Nature **323**(6088), 533–536 (1986)
17. Shih, S.Y., Sun, F.K., Lee, H.Y.: Temporal pattern attention for multivariate time series forecasting. Mach. Learn. **108**, 1421–1441 (2019)
18. Song, W., Fujimura, S.: Capturing combination patterns of long-and short-term dependencies in multivariate time series forecasting. Neurocomputing **464**, 72–82 (2021)
19. Vaswani, A., et al.: Attention is all you need. In: Advances in Neural Information Processing Systems, vol. 30 (2017)
20. Wang, X., Cai, Z., Luo, Y., Wen, Z., Ying, S.: Long time series deep forecasting with multiscale feature extraction and seq2seq attention mechanism. Neural Process. Lett. **54**(4), 3443–3466 (2022)
21. Wang, X., Liu, H., Yang, Z., Du, J., Dong, X.: Cnformer: a convolutional transformer with decomposition for long-term multivariate time series forecasting. Appl. Intell. 1–15 (2023)
22. Yang, X., Yu, F., Pedrycz, W.: Long-term forecasting of time series based on linear fuzzy information granules and fuzzy inference system. Int. J. Approximate Reasoning **81**, 1–27 (2017)
23. Yao, J.T., Vasilakos, A.V., Pedrycz, W.: Granular computing: perspectives and challenges. IEEE Trans. Cybern. **43**(6), 1977–1989 (2013)

24. Yao, Y.: Granular computing for data mining. In: Data Mining, Intrusion Detection, Information Assurance, and Data Networks Security 2006, vol. 6241, pp. 44–55. SPIE (2006)
25. Yao, Y.: Artificial intelligence perspectives on granular computing. In: Pedrycz, W., Chen, S.M. (eds.) Granular Computing and Intelligent Systems. Intelligent Systems Reference Library, vol. 13, pp. 17–34. Springer, Heidelberg (2011). https://doi.org/10.1007/978-3-642-19820-5_2
26. Yin, C., Dai, Q.: A deep multivariate time series multistep forecasting network. Appl. Intell. **52**(8), 8956–8974 (2022)
27. Zheng, Z., Hu, H., Shi, Z.: Tolerance granular space and its applications. In: 2005 IEEE International Conference on Granular Computing, vol. 1, pp. 367–372. IEEE (2005)

Adaptive Multi-granularity Aggregation Transformer for Image Captioning

Daitianxia Li, Ye Wang, and Qun Liu$^{(\boxtimes)}$

Chongqing Key Laboratory of Computational Intelligence, Chongqing University
of Posts and Telecommunications, Chongqing 400065, China
liuqun@cqupt.edu.cn

Abstract. In image captioning, images often contain complex scenes where features at a single granularity level fail to capture all the visual information. For instance, grid features of an image provide spatial details but lack an understanding of semantic objects. Therefore, it is necessary to fuse the multi-granularity features of an image for a comprehensive representation. In this paper, we propose an adaptive multi-granularity aggregation transformer that integrates grid, region and global features of image. In contrast to previous approaches that rely on single-feature or two-feature representation, our approach integrates features of different granularity levels, which overcomes the incompleteness of traditional visual information characterization. Specifically, we construct an encoder with a multi-granularity feature enhancement module that explores intrinsic relationships between different features to reduce the redundancy of feature representation. We also design a multi-granularity feature adaptive fusion module to adjust the attention of features at different scales, enhancing cross-modal inference ability. Experiments on the MSCOCO dataset demonstrate that our model achieves superior performance, with a CIDEr score of 138.6 on the "Karpathy" split, surpassing the state-of-the-art fusion model by 2.5 points.

Keywords: Image captioning · Multi-granularity features fusion

1 Introduction

Images are a widely used medium for conveying information, containing rich visual content. Image captioning is the task of automatically generating natural language descriptions by extracting features from the visual information in an image. The main challenge of image captioning lies not only in comprehending the visual content of the image but also in generating language that conforms to its visual semantics.

At first, researchers in image captioning utilize pre-trained Convolutional Neural Networks (CNNs) as encoders to extract image grid features, which represent the visual information. The advantage of this method is that the grid features capture all contents of the given image in a fragmented manner. This

ⓒ The Author(s), under exclusive license to Springer Nature Switzerland AG 2023
A. Campagner et al. (Eds.): IJCRS 2023, LNAI 14481, pp. 339–353, 2023.
https://doi.org/10.1007/978-3-031-50959-9_24

method is also easy to implement and expand. Nonetheless, grid features only concentrate on local image regions and fail to comprehend semantic objects in the image, resulting in generated captions that may lack coherence and completeness. Consequently, researchers seek more effective methods to extract image information to enhance image captioning performance.

Anderson et al. [1] first use image visual object region-level features extracted by pre-trained Faster R-CNN as image information. Region features offer object-level information, as most salient regions in the image can be recognized and represented by feature vectors. These region features can be regarded as the semantic representation of objects. The region features greatly reduce the difficulty of visual semantic embedding, and improve the quality and reliability of image captioning. Grid-level features extracted by CNNs are gradually being discarded as a result. However, relying solely on region features has limitations, as they may not capture fine-grained details or provide spatial relationships between different rgion-level features, which negatively impacts decoding ability. These are precisely the advantages that grid features can furnish.

Features of different granularity capture various ranges between local and global information, imparting different levels of visual details. Therefore, grid and region features are essentially complementary, where grid features provide low-level spatial context and detail information, while region features offer high-level object semantics.

In addition, Global features in images that capture overall information with higher-level abstraction and integration are often neglected. They are widely applied in tasks such as image classification, retrieval, and recognition. Global features are also indispensable in image captioning, as they offer semantic information about the whole image, such as content, scene, and emotion. Incorporating global features can enhance the model's understanding of image content and semantics, leading to more accurate captions. To leverage the advantages of multi-granularity visual features, we propose a method that combines these features to obtain comprehensive and accurate visual information. In contrast to previous methodologies that rely on a single or dual feature representation, we integrate grid, region, and global features to create a comprehensive and precise visual representation. This multi-granularity fusion allows us to extract rich and diverse visual cues, leading to improved caption generation performance.

Combining features of different granularity to represent images is both crucial and challenging. Due to redundancy and irrelevant information within features at different scales, feature representations can become overly complex after fusion. Therefore, enhancing the features before fusing them at multiple scales is a natural idea. Note that this enhancement does not supplement the feature information. Instead, it reduces redundancy within the features by extracting more critical features, to enhance feature representation ability. Additionally, features of different granularity may have varying importance in generating each word. We propose a multi-granularity feature fusion mechanism that jointly models grid and region features. This mechanism adaptively determines the weight of different scale information at each time step, achieving effective feature fusion.

In this paper, we propose an adaptive multi-granularity aggregation transformer for image captioning, which utilizes the complementary advantages of multi-scale features. Our model, illustrated in Fig. 1, processes image grid and region features with the multi-granularity feature enhancement module to capture local and global information and explore intrinsic feature relationships. Simultaneously, image global features representing the semantic information of the entire image are fused with the text for effective semantic guidance during decoding. Secondly, the multi-granularity feature adaptive fusion module investigates the relationships between grid and region features to enhance complex cross-modal reasoning ability through their respective strengths.

Our contributions are summarized as follows:

- For a more comprehensive understanding of images in terms of content and semantics, we introduce a multi-granularity feature representation method. By integrating of image global features into the image captioning, the model's ability to perceive and understand the objects, scenes, and other elements within the image is enhanced.
- To reduce the feature redundancy, we propose a multi-granularity feature enhancement module that effectively extracts local and global information from different granularities of features such as grid features and region features, resulting in more powerful image feature representations.
- To achieve the complementary advantages between multi-scale features, we design a multi-granularity feature adaptive fusion module to dynamically adjust the weights of different granularity visual features at each time step to generate more accurate captions.
- We conduct extensive experiments on the publicly available MSCOCO dataset to demonstrate that our proposed AMGAT outperforms state-of-the-art methods, and is capable of generating more natural linguistic expressions.

2 Related Work

The encoder-decoder architecture is a generalized learning framework for image captioning, which encodes an image into a sequence of vectors for generating descriptive text. It is evident that the development of image captioning relies on the advancement of computer vision technology. Different pre-trained image models can be utilized to extract visual features of various granularity, which can improve the quality and accuracy of the generated captions.

In previous studies, pre-trained convolutional neural networks (CNN), such as VGG and ResNet, are used as encoders to extract fixed-length grid-level features from images. The pioneering work by Vinyals et al. [15] employs CNNs to encode images into grid features and generated captions using the LSTM network. Subsequently, Xu et al. [20] design soft and hard attention mechanisms to assign different weights to different image grid features, allowing the model to focus on the important parts of the image. Cornia et al. [2] introduce the use of cross-layer grid connections in the decoder to extract information between grid features at different levels, improving the quality and accuracy of generated image captions. Wang et al. [16] utilize a window-based multi-head attention to model the interaction between grid features, effectively promoting the interaction and fusion of information between different grid features. Clearly, utilizing grid features is a flexible and effective approach for image captioning.

With the continuous development of object detection technology, researchers find that object-level region features extracted by Faster R-CNN are more consistent with human visual perception than grid-based features. This lead to region-based features becoming a typical method for feature extraction in subsequent image captioning. Anderson et al. [1] are the first to use region features to provide object-level information to the decoder, which significantly improved the model's performance. Huang et al. [9] design an adaptive attention time (AAT) mechanism that can dynamically adjust the many-to-many relationships between image regions and different caption words to better capture the semantic relationship between images and text. Pan et al. [12] propose an X-Linear attention that fully uses bilinear pooling to obtain attention between words and visual objects, improving the model's ability to model complex relationships between images and text. Herdate et al. [6] propose an image transformer that uses object IoU to calculate the relative spatial relationships between image regions, allowing the model to better understand the relationships between objects in the image. All of these methods demonstrate that region features can more accurately describe objects in the image and improve the quality and accuracy of image captioning, compared to encoding methods based on the entire image.

In recent years, more and more researchers begin to explore the fusion of grid and region features [7]. This is because the extraction of single-granularity features has some limitations. Xian et al. [19] integrate grid features into region features using a visual global adaptive attention module to learn complete semantic information. Hu et al. [7] explore the inherent properties between different features and map them to joint representations, promoting the model's decoding ability. Zhang et al. [23] construct a novel cross-attention mechanism to align different types of visual features and establish connections between different features, thereby improving the model's comprehensive performance. Wu et al. [17] represent both grid features and region features as graph nodes and use a joint graph for modeling, achieving the transfer of information between features. These methods not only improve the accuracy of image captioning but also help the model better understand objects and scenes in the image, thereby improving the model's visual reasoning ability.

3   Method

In this paper, we introduce an adaptive multi-granularity aggregation transformer for image captioning. The overall structure is illustrated in Fig. 1. In the following sections, we will introduce our proposed AMGAT model in detail.

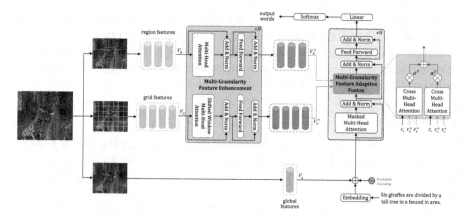

Fig. 1. The overview of our proposed model AMGAT.

3.1 Attention Mechanism

The transformer has become a mainstream method for image captioning due to the powerful modeling ability of the Multi-Head Attention mechanism (MA). This mechanism can better capture the semantic information. In addition, the independent nature of the MA mechanism allows for parallel attention computations, greatly improving computational efficiency. Our model utilizes the MA mechanism and its variant Shifted Window MSA (SW-MSA) from Swin-Transformer. Specifically, the attention module calculates weights via dot product operation, obtains attention values through weighted summation and uses them as the similarity scoring function, as formulated below:

$$\text{MA}\left(Q,K,V\right) = \text{Concat}\left(head_1, head_2, ..., head_h\right)W_o \quad (1)$$

$$head_i = \text{Att}\left(Q_i, K_i, V_i\right) \quad (2)$$

$$\text{Att}\left(Q_i, K_i, V_i\right) = \text{softmax}\left(\frac{Q_i K_i^T}{\sqrt{s}}\right)V_i \quad (3)$$

where $head_i$ represents the attention result of the i-th head, $i = 1, 2, ..., h$. W_o is the linear transformation matrix. s is a scaling factor. Q_i, K_i, V_i represents the i-th sub-sequence of Q, K and V. Note that when Q, K and V are the same sequence, this attention mechanism is called Multi-Head Self-Attention (MSA), otherwise it is called Multi-Head Cross-Attention (MCA).

SW-MSA divides the input sequence into fixed-size windows and applies MSA in each window to reduce computational complexity while preserving local information. It incorporates a shifted window operation to improve the modeling of relationships between positions in the input sequence, as expressed below:

$$\text{SW} - \text{MSA}\left(Q,K,V\right) = \text{Merge}\left(window_1, window_2, ..., window_n\right)W_w \quad (4)$$

$$window_i = MSA\left(Q_i^W, K_i^W, V_i^W\right), i = 1, 2, ..., w \quad (5)$$

where $window_i$ represents the attention result of the i-th window, and w is the number of windows. The Merge($\cdot$) operation merges the output representations of different partitions into a global output sequence in the shifted window partitioning scheme. W_w is the linear transformation matrix, Q_i^W, K_i^W and V_i^W represent the i-th window of Q, K and V, respectively.

3.2 Encoder

We use Faster R-CNN based on object detection to extract the region features $V_R = \{v_{r_1}, v_{r_2}, ..., v_{r_m}\}$ of a given image, Swin-Transformer to obtain the grid features $V_G = \{v_{g_1}, v_{g_2}, ..., v_{g_n}\}$, where m and n are the number of region features and grid features. They are mapped to a dimension D. We adopt average pooling of grid features V_G as the global feature of the image, which serves as the input to the subsequent decoder. The calculation formula is as follows:

$$V_g = \frac{1}{n} \sum_{i=1}^{n} v_{g_i} \qquad (6)$$

The image regions and grid features are fed into a multi-granularity feature enhancement module for encoding. As shown in Fig. 1, the module comprises N stacked blocks, each with an attention layer, a feedforward layer and a fully connected layer. It is worth noting that SW-MSA requires inputs of the same length, but the number of region features in each image is not fixed. Therefore, we input image grid features into SW-MSA and region features into MSA. In the $(l+1)$-th block, it can be represented as follows:

$$\tilde{V}_G^{l+1} = \text{SW} - \text{MSA}\left(V_G^l, V_G^l, V_G^l\right) \qquad (7)$$

$$\tilde{V}_R^{l+1} = \text{MSA}\left(V_R^l, V_R^l, V_R^l\right) \qquad (8)$$

where V^l represents the output of the l-th block. The input of the first block are $V_G^0 = V_G$, $V_R^0 = V_R$.

Then, residual connections and layer normalization are performed. After N blocks, the features of different scales are encoded into V_G^N and V_R^N by self-attention and used for subsequent multi-scale feature fusion is performed.

3.3 Decoder

We construct an adaptive semantic-guided decoder to fuse multi-granularity features from the encoder in the transformer-based decoder, achieving a balance of information across different scales of the image. We first fuse image global feature with textual context information for richer context and better semantic guidance. At time step t, the $(l+1)$-th decoder block is expressed as follows:

$$\hat{c}_t^{l+1} = \text{LayerNorm}\left(\text{ReLU}\left(W_c \left[H_{<t}^l; V_g\right]\right) + V_g\right) \qquad (9)$$

where W_c is a learnable parameter matrix, and $H_{<t}^l$ is the output of the l-th layer of the decoder. It is worth noting that the input of the first block is $H_{<t}^0 = [e_0, e_1, ..., e_{t-1}]$, $e_i \in R^D$ represents the word embedding.

Next, context information is input to masked MSA with residual connections and layer normalization, as shown in the following formula:

$$c_t^{l+1} = \text{LayerNorm}\left(\text{MSA}\left(\hat{c}_t^{l+1}, \hat{c}_t^{l+1}, \hat{c}_t^{l+1}\right) + \hat{c}_t^{l+1}\right) \tag{10}$$

To generate the corresponding output word, the model selects appropriate feature information based on the word type generated at the current time step. We introduce a gating mechanism that dynamically distributes weights between region and grid features for a suitable visual representation. Contextual information is used as the query to perform multi-head cross-attention on region and grid features, as shown in Fig. 1, and defined as follows:

$$h_G^{l+1} = \text{MCA}\left(c_t^{l+1}, V_G^N, V_G^N\right) \tag{11}$$

$$h_R^{l+1} = \text{MCA}\left(c_t^{l+1}, V_R^N, V_R^N\right) \tag{12}$$

where h_G^{l+1}, h_R^{l+1} representing the image grid and region features attended to after being guided by contextual information.

We learn adaptive weights for balancing different granularity of feature information from context. The weight formula for grid features is defined as follows:

$$\alpha^G = \sigma\left(c_t^{l+1} W_\alpha\right) \tag{13}$$

where σ is a sigmoid function and W_α is the learnable weight matrix.

We compute the weights of the region features in a simple way as follows:

$$\alpha^R = 1 - \alpha^G \tag{14}$$

The information h_V^{l+1}, which adaptively incorporates multiple granularities, is obtained:

$$h_V^{l+1} = \alpha^G \otimes h_G^{l+1} + \alpha^R \otimes h_R^{l+1} \tag{15}$$

where $\otimes$ denotes element-by-element multiplication. h_V^{l+1} is fed into subsequent layers for further decoding to obtain the output H^{l+1} of the $l+1$ layer.

Finally, the output of the last block of the decoder $H_{<t}^N$ is used to predict words w_t by softmax.

3.4   Objective

Firstly, the cross-entropy, which is commonly used in image captioning, is used as the loss function in the form shown below:

$$\mathcal{L}_{XE}\left(\theta\right) = -\sum_{t=1}^{T} \log\left(p_\theta\left(y_t^* | y_{1:t-1}^*\right)\right) \tag{16}$$

where $y_{1:T}^*$ is the real sequence of the target language, θ indicating the parameters that the model needs to be trained.

Next, the CIDEr is further optimized according to the self-critical sequence training (SCST) strategy with the following formula:

$$\mathcal{L}_{RL}\left(\theta\right) = -\mathrm{E}_{y_{1:T} \sim p_\theta}\left[r\left(y_{1:T}\right)\right] \tag{17}$$

where the reward $r(\cdot)$ denotes the CIDEr score of the sentence. We can approximate the gradient of as:

$$\nabla_\theta \mathcal{L}_{RL}\left(\theta\right) \approx -\left(r\left(y_{1:T}^{S}\right) - r\left(y_{1:T}^{*}\right)\right) \nabla_\theta \log p_\theta\left(y_{1:T}^{S}\right) \tag{18}$$

where $y_{1:T}^{S}$ represents the sampled sentence, $y_{1:T}^{*}$ is the baseline reward of the sentence generated by greedy decoding.

4   Experiments

4.1   Experiment Setup and Evaluation Metrics

We evaluate model on the MSCOCO 2014 dataset, which has been widely used as a standard dataset for image captioning. The MSCOCO dataset comprises 123,287 images, each with at least five manually annotated descriptions. Sentences are preprocessed to lowercase, and words that appear less than six times are filtered out to construct a vocabulary. We follow the "Karpathy" split, using 113,287 images for training, 5,000 for validation, and another 5,000 for testing.

We use pre-trained Faster R-CNN on the VG dataset to extract region features, taking the 2048-dimensional features after the first FC layer. Grid features, sized at $12 \times 12 \times 1536$, are extracted using Swin-Transformer. All features are projected onto the same 512-dimensional space. Additionally, The encoder and decoder consist of 3 blocks each, with 8 transformer heads. For training, we follow the standard process for image captioning. XE training uses a warm-up (10,000 iterations) learning rate scheduling strategy [14] and train for 20 epochs. RL training [13] uses a fixed learning rate of 5×10^{-6} and trained for 30 epochs. Both stages use the Adam optimizer and beam search with a beam size of 5.

We select BLEU, METEOR, ROUGE, CIDEr and SPICE as metrics to evaluate caption quality. These metrics are widely recognized as common objective evaluation standards in image captioning [2,13,16]. Note that B-1, B-4, M, R, C, and S denote BLEU-1, BLEU-4, METEOR, ROUGE, CIDEr, and SPICE in Tables 1, 2, 3 and 4 in the paper.

4.2   Ablation Experiments

In this paper, we conduct several ablation experiments to measure the impact of each design in the model on performance. Note that we do not utilize a reinforcement learning-based training strategy in this section, as all of our models can be further improved using RL-based methods to enhance overall performance.

To better understand the role of multi-granularity features, Table 1 reports the results using different image features. "Region" and "Grid" indicate that

Table 1. Performance comparison using image features of different granularity.

Features	B-1	B-4	M	R	C	S
Grid(G)	77.8	36.9	28.9	57.7	122.4	21.8
Region(R)	77.7	37.3	28.8	57.6	121.1	21.9
G+R(MGAT)	78.4	37.9	**29.2**	**58.3**	124.5	22.3
AMGAT	**78.5**	**38.5**	**29.2**	**58.3**	**126.7**	**22.4**

only image region features or grid features are utilized in the model, without the multi-granularity feature adaptive fusion module. "G+R" represents the feature addition of region and grid features instead of the multi-granularity feature adaptive fusion module. This approach simply performs a straightforward fusion of multi-granularity features, namely "MGAT". AMGAT is our final proposed model, which not only uses region and grid features, but also constructs a gate mechanism to adaptively fuse them. Note that global image features are used to supplement context information in all experiments, and the other model settings are identical. From the table, it can be seen that both MGAT and AMGAT using features at two different granularities perform better than the models using only one feature (the first and second rows), which demonstrates the benefits of multi-scale features in improving caption generation quality. Furthermore, comparing the last two rows in Table 1, we can see that the CIDEr score increased from 124.5 to 126.7. The 2.2 point improvement of AMGAT over MGAT in CIDEr validates the effectiveness of our proposed multi-granularity feature adaptive fusion module. This adaptation enhances the model's ability to model the importance and correlations between different modalities, effectively integrating both features and thus improving the model's performance.

Table 2. Performance comparison of different window size ws and shift size ss.

ws	ss	B-1	B-4	M	R	C	S
12	0	78.1	38.0	29.0	**58.3**	123.5	22.1
6	0	78.2	38.0	29.1	58.1	124.6	22.2
6	3	**78.5**	**38.5**	**29.2**	**58.3**	**126.7**	**22.4**
4	0	78.3	38.0	29.1	58.2	124.2	21.9
4	2	78.1	37.7	**29.2**	58.1	124.2	22.0

We investigate the impact of different window sizes (ws) and shift sizes (ss) in SW-MSA on model performance, as shown in Table 2. When ss is set to 0, the window remains unshifted. Notably, the size of the region features are 12×12. When the window size is set to 12, SW-MSA degenerates into ordinary Multi-Head Self-Attention (MSA). We can see that the model performs best when $ws = 6$ and $ss = 3$, which is natural. Compared to using MSA ($ws = 12$,

$ss = 0$), the SW-MSA mechanism can utilize sliding windows of different sizes to process features of different scales, thus better capturing feature information of different scales. The model with $ss = 0$ lacks cross-window connections and only consider local information within each window, disregarding global information across different windows. For the model with $ws = 4$ and $ss = 2$, its small window size only captures local semantic information, unable to get long-range dependency relationships, affecting model performance. Considering all factors, setting $ws = 6$ and $ss = 3$ can increase the model's receptive field, capture more detailed semantic information, as well as improve the model's understanding of global information to obtain the best performance.

4.3   Comparative Experiments

We compare our model AMGAT with state-of-the-art models. The experimental results are shown in Table 3, 4.

Table 3. Performance comparison of the proposed model with the model using only a single feature.

Model	Cross-Entropy Loss						CIDEr Score Optimization					
	B-1	B-4	M	R	C	S	B-1	B-4	M	R	C	S
models using grid features												
SCST [13]	-	30.0	25.9	53.4	99.4	-	-	34.2	26.7	55.7	114.0	-
M^2-Transformer [2]	-	-	-	-	-	-	80.8	39.1	29.2	58.6	131.2	22.6
LSTNet [11]	-	-	-	-	-	-	81.5	40.3	29.6	59.4	134.8	23.1
RSTNet [24]							81.8	40.1	29.8	59.5	135.6	23.0
ViTCAP [4]	-	36.3	29.3	58.1	125.2	22.6	-	41.2	30.1	60.1	138.1	24.1
PureT [16]	77.8	36.9	28.9	57.7	122.4	21.8	82.1	40.9	**30.2**	60.1	138.2	24.2
models using region features												
Up-Down [1]	77.2	36.2	27.0	56.4	113.5	20.3	79.8	36.3	27.7	56.9	120.1	21.4
SGAE [21]	77.6	36.9	27.7	57.2	117.3	21.3	80.8	38.4	28.4	58.6	127.8	22.1
AOANet [8]	77.4	37.2	28.4	57.5	119.8	21.3	80.2	38.9	29.2	58.8	129.8	22.4
X-Transformer [12]	77.3	37.0	28.7	57.5	120.0	21.8	80.9	39.7	29.5	59.1	132.8	23.4
A^2-Transformer [5]	78.6	38.2	29.2	58.3	125.0	22.1	81.5	39.8	29.6	59.1	133.9	23.0
TCIC [3]	**78.8**	**39.1**	29.1	**58.5**	123.9	22.2	81.8	40.8	29.5	59.2	135.3	22.5
AMGAT	**78.8**	38.5	**29.3**	**58.5**	**126.7**	**22.4**	**82.3**	**41.2**	**30.2**	**60.3**	**138.6**	**24.3**

In our study, we compare the results of the AMGAT model with some models that use only a single feature, demonstrating the advantages of the AMGAT model. Specifically, we report the comparison results of the following models in Table 3: SCST [13], M^2-Transformer [2], LSTNet [11], RSTNet [24], ViTCAP [4], and PureT [16]. These models use grid features as the image representation and generate description sentences through feature interaction. In addition, we also compare Up-Down [1], SGAE [21], AOANet [8], X-Transformer [12],

A^2-Transformer [5], and TCIC [3]. These models focus on salient object region features and explore the relationship between region features to generate caption. It can be observed that our AMGAT model achieves the best results on all metrics after being trained with Cross-Entropy Loss and optimized with CIDEr. Compared to using a single image feature representation, our model not only considers grid and region features but also introduces global image features, achieving better performance on all metrics. For models using grid features, AMGAT achieves an average improvement of 6.6 on the CIDEr metric, as the region features can capture salient objects in the images. AMGAT achieves an average improvement of 8.7 on CIDEr when compared to models using region features, due to the fact that the grid features can provide the local structure and details of the images. The experimental results demonstrate the huge potential of fusing multiple features in image captioning tasks. Different features may have interactions and dependencies between them, and integrating multiple features can facilitate the interaction and relationship modeling among them.

Table 4. Performance comparison of the proposed model with the model integrating multiple features.

Model	Cross-Entropy Loss						CIDEr Score Optimization					
	B-1	B-4	M	R	C	S	B-1	B-4	M	R	C	S
Double-stream GCN [18]	76.8	36.2	27.8	56.8	115.6	-	80.4	38.2	28.5	58.2	126.4	-
DLCT [10]	-	-	-	-	-	-	81.4	39.8	29.5	59.1	133.8	23.0
DGET [19]	-	-	-	-	-	-	81.3	40.3	29.2	59.4	132.4	23.5
TSFNet [7]	78.9	39.3	28.8	58.6	121.4	22.3	81.7	40.3	29.8	59.6	133.5	23.4
AS-Transformer [22]	-	38.3	29.0	58.1	123.6	22.3	-	41.0	29.8	60.0	136.1	23.8
AMGAT	**78.5**	**38.5**	**29.2**	**58.3**	**126.7**	**22.4**	**82.3**	**41.2**	**30.2**	**60.3**	**138.6**	**24.1**

To further demonstrate the effectiveness of our multi-feature fusion approach, We compare the AMGAT model with some models that integrate multiple features, and the results are presented in Table 4. Double-stream GCN [18] processes grid and region features using GCN, combines the feature information of each node with that of its surrounding nodes to produce more discriminative node feature representations. DLCT [10] embeds geometric information to align and enhance grid and region features and proposes a local constraint cross-attention mechanism to address the semantic noise problem caused by direct fusion of these two features. DGET [19] views grid features as visual global information and adaptively fuses them into region features at each layer to enhance visual information. TSFNet [7] uses the cascading representation of grid, region, and scene graph features in image captioning and guides caption generation through joint attention. Clearly, our model consistently outperforms other models in both the cross-entropy training phase and the reinforcement learning phase. After two-stage training, our model achieves high scores of 82.3, 41.2, 30.2, 60.3, 138.6, and 24.1 on BLEU-1, BLEU-4, METEOR, ROUGE, CIDEr, and

GT1: a lady in a hat standing in front of a big cake.
GT2: a woman wearing a hat while looking at a cake.
GT3: a woman in a silly hat cuts the cake.
AoANet: a woman standing in front of a cake with a knife.
PureT: a woman cutting a birthday cake on a table.
AMGAT: a women wearing a hat standing next to a table with a cake.

GT1: the man is sleeping in a bed near two cats.
GT2: a man sleeping in a bed with two cats.
GT3: a man is sleeping next to a couple of cats.
AoANet: a man and a cat are laying on a bed.
PureT: a man laying on a bed with two cats.
AMGAT: a man sleeping in a bed with two cats.

GT1: a surfer in the ocean on the crest of a wave.
GT2: a man on a surf board rides a wave.
GT3: a man that is on a surfboard in the water.
AoANet: a man riding a surfboard on top of a wave.
PureT: a man riding a wave on top of a surfboard.
AMGAT: a man riding a wave on a surfboard in the ocean.

GT1: a dog holding a yellow frisbee in it's mouth.
GT2: a dog is holding a yellow disc in its mouth.
GT3: a brown and white dog is holding a yellow frisbee.
AoANet: a dog with a frisbee in its mouth.
PureT: a dog with a yellow frisbee in its mouth.
AMGAT: a brown and white dog holding a yellow frisbee in its mouth.

Fig. 2. Examples of image captioning results on the MSCOCO dataset.

SPICE, respectively. While the state-of-the-art multi-feature fusion model AS-Transformer [22] also improves multi-head attention by adaptively adjusting the weights of grid and region features, our model incorporates global image features as context information, which can provide global semantic information and guide the decoding process more effectively. AMGAT achieves a significant improvement of 2.5 in CIDEr compared to AS-Transformer, demonstrating the advantage of our fusion method. By fusing global image features and textual features to guide the entire decoding process, including enhancing self-attention within grid and region features and interacting between the two features, the fusion of multi-granularity features is beneficial for improving the model's ability to handle complex scenes and enhance cross-modal reasoning ability, resulting in more accurate and natural captions.

4.4   Qualitative Experiments

In Fig. 2, We show some examples of AMGAT and comparison models generated on the MSCOCO 2014 dataset. The ground truth is represented as GT. As shown in the first example, AMGAT accurately generates the noun "hat" compared to other models, benefiting from the semantic information provided by region features. This indicates the effective utilization of precise semantic words in our proposed model. In the second example, AMGAT not only correctly generates "two cats" but also predicts the verb "sleep" based on contextual information. This is because our model incorporates global image features and can grasp the semantic information of the entire image, which helps the model understand the content and semantics of the image. Similarly, the "ocean" in the third example is not fully captured by the object detector but visible in the image. In addition, in the last example, our model clearly captures additional fine-grained information and generates a more descriptive caption. These examples demonstrate the superiority of AMGAT in image caption generation.

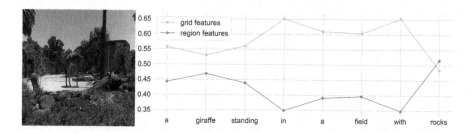

Fig. 3. The contribution of the learned attention weights to word prediction.

To visually illustrate the effectiveness of the multi-granularity feature adaptive fusion module, we show in Fig. 3 the contribution of the learned attention weights to word prediction. Different feature types exhibit varying contributions to word prediction based on the word type. Specifically, when generating prepositions like "in" and "with", region features are not important, resulting in lower weights compared to grid features. However, when predicting nouns, the weight of region features will increase due to their richer semantic information, facilitating improved predictions. Especially for the word "rocks", the contribution of region features is particularly significant. This indicates that our model has successfully learned to dynamically allocate feature weights when generating different words, thereby achieving better prediction performance.

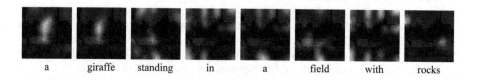

Fig. 4. Visualization of grid features attention maps for each word generation.

In addition, we visualize how our model utilizes grid features during the decoding process. We average the attention weights of the eight heads of the cross-attention layer over the grid features in the last decoding block at each time step, as shown in Fig. 4. The results show that our AMGAT model can correctly focus on the corresponding grid regions and achieve excellent performance in generating words. For example, when predicting "rocks", our model does not blindly follow the region features just because their weights are high, but correctly focuses attention on the rocks in the image.

5   Conclusion

In this paper, we propose the Adaptive Multi-Granularity Aggregation Transformer (AMGAT) to improve transformer-based image captioning models. AMGAT incorporates global, grid, and region features of images to enhance the model's cross-modal reasoning ability. To obtain targeted feature representations by leveraging the interaction of the multi-granularity features, we introduce a multi-granularity feature enhancement module that applies self-attention to features of different granularities. In addition, we propose a novel decoder that utilizes a fusion representation of global image features and textual information. This fusion representation serves as context-guided information during decoding, facilitating more accurate caption generation. Furthermore, we design a multi-granularity feature adaptive fusion module that dynamically allocates attention weights between grid and region features based on guidance information, improving the model's ability to handle complex scenes and tasks. AMGAT achieves state-of-the-art performance across all evaluation metrics, confirming the proposed approach's superiority in improving the image captioning task.

References

1. Anderson, P., et al.: Bottom-up and top-down attention for image captioning and visual question answering. In: Proceedings of the IEEE Conference on Computer Vision and Pattern Recognition, pp. 6077–6086 (2018)
2. Cornia, M., Stefanini, M., Baraldi, L., Cucchiara, R.: Meshed-memory transformer for image captioning. In: Proceedings of the IEEE/CVF Conference on Computer Vision and Pattern Recognition, pp. 10578–10587 (2020)
3. Fan, Z., et al.: TCIC: theme concepts learning cross language and vision for image captioning. arXiv preprint arXiv:2106.10936 (2021)
4. Fang, Z., et al.: Injecting semantic concepts into end-to-end image captioning. In: Proceedings of the IEEE/CVF Conference on Computer Vision and Pattern Recognition, pp. 18009–18019 (2022)
5. Fei, Z.: Attention-aligned transformer for image captioning. In: proceedings of the AAAI Conference on Artificial Intelligence, vol. 36, pp. 607–615 (2022)
6. Herdade, S., Kappeler, A., Boakye, K., Soares, J.: Image captioning: transforming objects into words. In: Advances in Neural Information Processing Systems, vol. 32 (2019)

7. Hu, N., Ming, Y., Fan, C., Feng, F., Lyu, B.: TSFNet: triple-steam image captioning. IEEE Trans. Multimedia (2022)
8. Huang, L., Wang, W., Chen, J., Wei, X.Y.: Attention on attention for image captioning. In: Proceedings of the IEEE/CVF International Conference on Computer Vision, pp. 4634–4643 (2019)
9. Huang, L., Wang, W., Xia, Y., Chen, J.: Adaptively aligned image captioning via adaptive attention time. In: Advances in Neural Information Processing Systems, vol. 32 (2019)
10. Luo, Y., et al.: Dual-level collaborative transformer for image captioning. In: Proceedings of the AAAI Conference on Artificial Intelligence, vol. 35, pp. 2286–2293 (2021)
11. Ma, Y., Ji, J., Sun, X., Zhou, Y., Ji, R.: Towards local visual modeling for image captioning. Pattern Recogn. **138**, 109420 (2023)
12. Pan, Y., Yao, T., Li, Y., Mei, T.: X-linear attention networks for image captioning. In: Proceedings of the IEEE/CVF Conference on Computer Vision and Pattern Recognition, pp. 10971–10980 (2020)
13. Rennie, S.J., Marcheret, E., Mroueh, Y., Ross, J., Goel, V.: Self-critical sequence training for image captioning. In: Proceedings of the IEEE Conference on Computer Vision and Pattern Recognition, pp. 7008–7024 (2017)
14. Vaswani, A., et al.: Attention is all you need. In: Advances in Neural Information Processing Systems, vol. 30 (2017)
15. Vinyals, O., Toshev, A., Bengio, S., Erhan, D.: Show and tell: a neural image caption generator. In: Proceedings of the IEEE Conference on Computer Vision and Pattern Recognition, pp. 3156–3164 (2015)
16. Wang, Y., Xu, J., Sun, Y.: End-to-end transformer based model for image captioning. In: Proceedings of the AAAI Conference on Artificial Intelligence, vol. 36, pp. 2585–2594 (2022)
17. Wu, D., Li, H., Gu, C., Guo, L., Liu, H.: Improving fusion of region features and grid features via two-step interaction for image-text retrieval. In: Proceedings of the 30th ACM International Conference on Multimedia, pp. 5055–5064 (2022)
18. Wu, L., Xu, M., Sang, L., Yao, T., Mei, T.: Noise augmented double-stream graph convolutional networks for image captioning. IEEE Trans. Circ. Syst. Video Technol. **31**(8), 3118–3127 (2020)
19. Xian, T., Li, Z., Zhang, C., Ma, H.: Dual global enhanced transformer for image captioning. Neural Netw. **148**, 129–141 (2022)
20. Xu, K., et al.: Show, attend and tell: Neural image caption generation with visual attention. In: International Conference on Machine Learning, pp. 2048–2057. PMLR (2015)
21. Yang, X., Tang, K., Zhang, H., Cai, J.: Auto-encoding scene graphs for image captioning. In: Proceedings of the IEEE/CVF Conference on Computer Vision and Pattern Recognition, pp. 10685–10694 (2019)
22. Zhang, J., Fang, Z., Sun, H., Wang, Z.: Adaptive semantic-enhanced transformer for image captioning. IEEE Trans. Neural Netw. Learn. Syst. (2022)
23. Zhang, J., Xie, Y., Ding, W., Wang, Z.: Cross on cross attention: Deep fusion transformer for image captioning. IEEE Trans. Circ. Syst. Video Technol. (2023)
24. Zhang, X., et al.: RSTNet: captioning with adaptive attention on visual and non-visual words. In: Proceedings of the IEEE/CVF Conference on Computer Vision and Pattern Recognition, pp. 15465–15474 (2021)

A Causal Disentangled Multi-granularity Graph Classification Method

Yuan Li[1,2], Li Liu[1,2], Penggang Chen[1,2], Youmin Zhang[1,2], and Guoyin Wang[1,2(✉)]

[1] Chongqing Key Laboratory of Computational Intelligence, Chongqing, China
D190201011@stu.cqupt.edu.cn, {liliu,wanggy}@cqupt.edu.cn
[2] Key Laboratory of Cyberspace Big Data Intelligent Security,
Ministry of Education, Chongqing University of Posts and Telecommunications,
Chongqing 400065, The People's Republic of China

Abstract. Graph data widely exists in real life, with large amounts of data and complex structures. It is necessary to map graph data to low-dimensional embedding. Graph classification, a critical graph task, mainly relies on identifying the important substructures within the graph. At present, some graph classification methods do not combine the multi-granularity characteristics of graph data. This lack of granularity distinction in modeling leads to a conflation of key information and false correlations within the model. So, achieving the desired goal of a credible and interpretable model becomes challenging. This paper proposes a causal disentangled multi-granularity graph representation learning method (CDM-GNN) to solve this challenge. The CDM-GNN model disentangles the important substructures and bias parts within the graph from a multi-granularity perspective. The disentanglement of the CDM-GNN model reveals important and bias parts, forming the foundation for its classification task, specifically, model interpretations. The CDM-GNN model exhibits strong classification performance and generates explanatory outcomes aligning with human cognitive patterns. In order to verify the effectiveness of the model, this paper compares the three real-world datasets MUTAG, PTC, and IMDM-M. Six state-of-the-art models, namely GCN, GAT, Top-k, ASAPool, SUGAR, and SAT are employed for comparison purposes. Additionally, a qualitative analysis of the interpretation results is conducted.

Keywords: Multi-granularity · Interpretability · Explainable AI · Causal disentanglement · Graph classification

1 Introduction

Graph data is characterized by complex structures and vast amounts of data, widely prevalent in our daily lives. Therefore, it is crucial to map graph data into low-dimensional embedding. Among various graph downstream tasks, graph classification is an essential task. Examples of such tasks include superpixel graph classification [8], molecular graph property prediction [10], and more.

© The Author(s), under exclusive license to Springer Nature Switzerland AG 2023
A. Campagner et al. (Eds.): IJCRS 2023, LNAI 14481, pp. 354–368, 2023.
https://doi.org/10.1007/978-3-031-50959-9_25

In the research on graph classification, representation learning is an important approach for data analysis. Graph data, inherently possessing multiple levels of granularity, comprises nodes at a fine-grained level, coarser-grained substructures. For example, in the Tox21 dataset, atoms are fine-grained, functional groups are coarse-grained. This dataset compounds containing the azo functional group that are associated with carcinogenic and mutagenic properties [5]. In graph classification, however, traditional representation learning methods overlook the multi-granularity of graph data, and these methods' interpretability is limited. In graph classification, the outcome is primarily determined by certain important substructures [13]. Indeed, current graph classification methods do not consider the inherent multi-granularity nature of graph data. This lack of granularity differentiation during modeling leads to the mixing of critical information and false correlations within the models. As a result, it becomes challenging to accurately distinguish and achieve the goal of building interpretable models.

In summary, there is a need to construct a substructure recognition model that takes into account multi-granularity graph data for modeling the graph representations. Therefore, this paper attempts to build a causal disentangled GNN model based on the idea of multi-granularity [21]. This model can disentangle the important substructures and bias parts in the graph from the multi-granularity perspective, and then conduct representation learning and classification for the entire graph.

Specifically, this paper designs a causal disentangled multi-granularity graph representation learning method (CDM-GNN). First, from a fine-grained perspective, this paper uses the feature and topological information of nodes to build a mask describing the closeness between nodes. Next, from a coarser-grained perspective, CDM-GNN uses this mask to disentangle the important substructures and bias parts. It obtains the reason for learning the current representation, which is the interpretable result. Subsequently, the masked graph is input into each slice layer to disentangle the key substructures from the false association relationship. Increase the depth of the model, gradually transitioning from fine-grained to coarse-grained, and expanding global information. Finally, this paper learns an adaptive weight for each layer of slice results and adaptively fuses the results of each layer. Then it obtains the final representation for graph classification.

The main contributions of this work are summarized as follows:

1. The CDM-GNN considers the multi-granularity characteristics of graph data. It models through granularity transformation, fully taking into account the information at different granularities and their fusion.
2. The proposed model is capable of disentangling the key substructures and bais parts associative relationships in the graph while providing corresponding explanations.
3. Compared with six state-of-the-art models, namely GCN, GAT, Top-k, ASAPool, SUGAR, and SAT in MUTAG, PTC, and IMDB-M, the CDM-GNN model achieves better graph classification results.

2   Related Work

2.1   Graph Classification Representation Learning

To obtain continuous low-dimensional embedding, graph representation learning aims to map non-Euclidean data into a low-dimensional representation space. For graph classification, there are two categories of research methods. One category is similarity-based graph classification methods, including graph kernel methods and graph matching methods. However, these methods are often inflexible and computationally expensive. In these methods, the process of graph feature extraction and graph classification is independent, which limits optimization for specific tasks. The other category is based on GNNs. When applied to graph classification problems, GCN [11] and GAT [20] perform graph classification through convolution and pooling operations. Pooling is the process of graph coarsening, where the operation progressively aggregates fine-grained nodes. Subsequent research has also introduced changes to the pooling operation. For example, SAT [2] proposes a graph transformer method used in pooling.

2.2   Disentangled Learning

The idea of disentangling initially originated from Bengio et al. [1] and is primarily focused on computer vision [9]. However, some researchers have extended this concept to graphs. DisenGCN model [15] introduces a neighborhood routing mechanism to disentangle the various latent factors behind interactions in the graph. The IPGDN model [12], based on DisenGCN, add the Hilbert Schmidt Independence Criterion (HSIC) to further enhance the independence between different modules. Based on the routing mechanism, the authors demonstrate the user-item relationship at the granularity of user intent and disentangle these intents in the representations of users and items [22]. However, this method mainly focuses on bipartite graphs and may not be suitable for more complex graph structures. It lacks scalability. In the context of knowledge graphs with richer types of relationships, some works consider leveraging relationship information in the process of disentangled representation learning. For example, they guide the disentangled representation of entity nodes based on the semantics of relationships [24]. However, these methods overlook the information from different types of relationship edges. These studies have successfully achieved disentangled. But, their primary emphasis is on manipulating the intermediate hidden layer states, which poses challenges in comprehending the structure of the graph. Consequently, their interpretability from a human perspective is limited.

2.3   Interpretability Method

Despite the excellent representation capability of GNN models, their learning process is often opaque and difficult for humans to understand. To address this issue, some researchers have proposed post-hoc methods for explaining. GNNExplainer model [23] learns masks for the adjacency matrix and node features to identify important substructures. The PGExplainer model [14] attempts to learn an MLP function to mask edges in the graph, incorporating sparsity and continuity constraints in the model to obtain the final explanations. The post-hoc methods for explaining can only understand the model, not adjust the model. Another category of GNN explanation methods involves constructing self-explainable GNNs. Compared with post-hoc methods, this kind of model not only provides predictions but also offers explanations for the reasons behind those predictions. It can guide the model to some extent. Some self-explainable models require prior knowledge. For example, KerGNNs [6] is a subgraph-based node aggregation algorithm that manually constructs graph kernel functions to compare the similarity between graph filters and input subgraphs. The trained graph filters are also visualized and used as the model's explanation, which is then integrated into the GNN. In addition to explaining isomorphic GNN models, researchers have also explored self-explainable models for heterogeneous GNN models, such as Knowledge Router [3]. However, these self-explainable GNNs have not extensively considered the issue of the multi-granularity structure of graphs.

3 Preliminaries

3.1 Notations

Let $G = (V, E)$ be a graph, where V is the node set, and E is the edge set. The $A \in \{0,1\}^{|N| \times |N|}$ is defined as the adjacency matrix of graph G. If there is an edge between node i and node j, the $A_{ij} = 1$; otherwise, $A_{ij} = 0$. The X is defined as the features matrix. $X \in R^{|V| \times F}$ represents the features of each node, where F denotes the feature dimension for each node. The neighbour of node i is N_i. The true label set is Y.

3.2 A Causal View on GNNs

We analyze this problem using the Structural Causal Model (SCM). Figure 1 illustrates the five components. Z: input graph data, Y: labels, B: bias part in the graph, C: important substructures, and E: learned embedding by GNN model.

$C \rightarrow Z \leftarrow B$: Z is composed of B and C.

$C \rightarrow Z \rightarrow E \rightarrow Y$: The structure of C is learned through GNNs and represented as E. Then, establish a causal correlation between C and Y.

$B \rightarrow Z \rightarrow E \rightarrow Y$: Due to the confusion between B and C, it affects the representation E obtained from GNNs, which also impacts the prediction of Y. Consequently, a spurious correlation is formed, leading to misleading predictions.

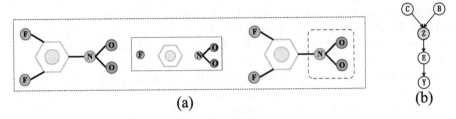

Fig. 1. (a) A real example of MUTAG dataset. A molecular diagram consists of a carbon ring, F atomic, and NO_2. The NO_2 atomic group determines to have mutagenicity, while others (bias part) do not determine this property. (b) Causal view of graph classification. The NO_2 atomic group is the C, and others are the bias parts B.

4 Proposed Method

This paper employs a multi-granularity approach for modeling. The CDM-GNN is introduced with an overall framework illustrated in Fig. 2.

4.1 Fine-Grained Closeness Mask

Based on the multi-granularity characteristics of graph data, this paper first considers modeling the nodes at a fine-grained level to capture the closeness between nodes, forming a mask matrix.

Given a graph G, the attention values are calculated based on the feature similarity between node i and node j from a fine-grained perspective.

$$e_{ij} = \alpha^T \left(W_{feat} \cdot x_i \parallel W_{feat} \cdot x_j \right) \tag{1}$$

where W_{feat} is the learnable parameters. Then the calculated attention values are normalized.

$$\overline{e_{ij}} = \frac{exp\left(LeakyRelu\left(e_{ij}\right)\right)}{\sum_{m \in N_i} exp\left(LeakyRelu\left(e_{im}\right)\right)} \tag{2}$$

At the fine-grained level, we calculate the interaction between the structures of node i and j and use $Stru$ as the attention value of the topological structure to describe the relationship between nodes.

$$Stru_{ij} = \frac{\sum_{p \in N_i \cup N_j} min\left(\omega_{ip}, \omega_{jp}\right)}{\sum_{p \in N_i \cup N_j} max\left(\omega_{ip}, \omega_{jp}\right)} \tag{3}$$

Here, We use restart random walks to describe the degree of structural similarity between the center node i and other nodes p. Particles start from the center node i and randomly walk to their neighbors p and $p \in N_i$. At each step, there is a certain probability of returning to the center node i. After t iterations, the probability vector of visiting the neighbors around node i is obtained.

$$\omega_{ip}^{t+1} = q \cdot \tilde{A}\omega_{ip}^{(t)} + (1 - q) \cdot vec_i \tag{4}$$

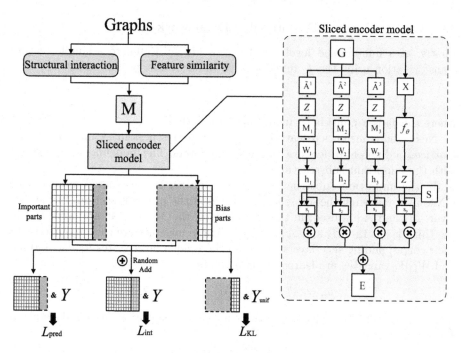

Fig. 2. The model framework of CDM-GNN.

where $\tilde{A} = D^{-0.5} A D^{-0.5}$ and D is the degree matrix. q is the probability of restarting the random walk. vec_i is a one-hot vector where the center node i is assigned 1. Others are assigned 0.

The probability vector is proportional to the edge weights. The higher the probability, the larger the edge weight. This probability vector is used as the weight vector. When $t \to \infty$, the vector converges to the following equation:

$$\omega_{ip} = (1 - q) \cdot \left(I - \tilde{A}\right)^{-1} \cdot e_i \tag{5}$$

In Eq. (3), ω_{ip} represents the weight of node p ($p \in Ni$). After normalizing $Stru_{ij}$, we obtain the following expression:

$$\overline{Stru_{ij}} = \frac{exp\left(Stru_{ij}\right)}{\sum_{m \in N_i} exp\left(Stru_{im}\right)} \tag{6}$$

We integrate e_{ij} and $Stru_{ij}$ to obtain M_{ij}, resulting in the formation of matrix M:

$$M_{ij} = \frac{\overline{e_{ij}} + \overline{Stru_{ij}}}{2} \tag{7}$$

At a fine-grained level, this paper describes the closeness between each node from its features and structure. And it fuses them to form a mask. The mask part is an important substructure, while others are the bias parts.

4.2   Coarse-Grained Disentangled Framework

Firstly, at a fine-grained level, the features are subjected to a simple feature transformation using the f_θ function with the learnable parameters θ.

$$Z = f_\theta(X) \tag{8}$$

where the f_θ is a multi-layer neural network with θ.

Next, the transformed features, fine-grained closeness mask, and adjacency matrices of different orders are sent into slice layers. The modeling process starts with the important substructures.

In the important substructures, the M matrix is constructed in each slice GNN layer following the approach described in the previous section. In the first layer, M_1 and $\tilde{A}$ are used, in the second layer, the M_2 and $\tilde{A}^2$ are used, and in the third layer, the M_3 and $\tilde{A}^3$ are used. These matrices are then inputted into their corresponding slice layers to obtain the hidden layer states h_1, h_2, and h_3, and W_1, W_2, and W_3 are learnable parameters.

$$h_n = \tilde{A}^n \cdot Z \cdot M_n \cdot W_n, \qquad n = 1, 2, 3 \tag{9}$$

The CDM-GNN model stacks the transformed feature results and obtained hidden layer states:

$$H = stack(Z, h_1, h_2, h_3) \tag{10}$$

This model adaptive learns the weight S for each slice layer:

$$S = reshape(\sigma(Hs)) \tag{11}$$

$$E = squeeze(SH) \tag{12}$$

For the graph G_k, this model performs pooling on the obtained embedding to obtain the representation of Emb_k:

$$E_k = pooling(E) \tag{13}$$

For bias parts, the CDM-GNN obtains the status of hidden layer $\overline{h_1}$, $\overline{h_2}$, and $\overline{h_3}$. In this model, we share these learnable parameters W_1, W_2, and W_3.

$$\overline{h_n} = \tilde{A}^n \cdot Z \cdot (1 - M_n) \cdot W_n, \qquad n = 1, 2, 3 \tag{14}$$

Similarly, this model stacks the transformed feature results and obtained hidden layer states:

$$\overline{H} = stack(Z, \overline{h_1}, \overline{h_2}, \overline{h_3}) \tag{15}$$

Then, the CDM-GNN model adaptive learns the weight $\overline{S}$ of each slice layer and gets the final embedding $\overline{Emb_k}$:

$$\overline{S} = reshape(\sigma(H\overline{s})) \tag{16}$$

$$\overline{E} = squeeze(\overline{S}H) \tag{17}$$

$$\overline{E_k} = pooling(\overline{E}) \tag{18}$$

4.3   Causal Distangled Learning

This paper aims to train K graphs by the causal components of their representations to enable the CDM-GNN model to classify correctly. To achieve this,
CDM-GNN employs the supervised classification cross-entropy loss as follows:

$$Y'_k = softmax\left(E_k\right) \tag{19}$$

$$L_{pred} = -\sum_k Y_k^T log\left(Y'_k\right) \tag{20}$$

For the representation of bias parts, it should not affect the classification
results. Therefore, the prediction results of the bias parts should be evenly distributed across all categories. Using uniform distribution to help learn the representation of bias parts as follows:

$$L_{KL} = \sum_k KL\left(Y_{unif}, \overline{E}_k\right) \tag{21}$$

where KL means the KL-Divergence, Y_{unif} denotes the uniform distribution.

In order to reduce the interference of bias parts, the representation of bias
parts is recombined with the causal parts to construct intervention terms. The
CDM-GNN redefines the $\oplus$ function. It adds the bias representation of random
disturbance back to the corresponding positions of the causal part representation
to obtain the constructed intervention term E_k^{int}:

$$E_k^{int} = E_k \oplus \overline{E}_k \tag{22}$$

In this case, the CDM-GNN can get the correct classification results by the
representation of the causal part. The loss function is as follows:

$$Y'^{int}_k = softmax\left(E_k^{int}\right) \tag{23}$$

$$L_{int} = -\sum_k Y_k^T log\left(Y'^{int}_k\right) \tag{24}$$

The overall loss function is as follows:

$$Loss = \frac{1}{|K|}\left(\alpha L_{pred} + \beta L_{KL} + \gamma L_{int}\right) \tag{25}$$

where α, β, and γ are hyper-parameters that determine the strength of disentanglement and causal influences.

5   Experiment

5.1   Datasets

The commonly-used datasets for graph classification are summarized in Table 1.

- MUTAG [4]: This dataset contains 188 compounds marked according to whether it has a mutagenic effect on a bacterium.
- PTC [19]: This dataset contains 344 organic molecules marked according to their carcinogenicity on male mice.
- IMDB-M [17]: This dataset is a movie collaboration dataset marked according to the genre an ego-network belongs to (romance, action, and science).

Table 1. Statistics of datasets

Dataset	Graphs	Classes	Avg.$|V|$	Avg.$|E|$
MUTAG	188	2	17.93	19.79
PTC	344	2	14.29	14.69
IMDB-M	1500	3	13.00	65.94

5.2 Baselines

This paper compares the proposed CDM-GNN model with several state-of-the-art methods, which are summarized as follows:

- GCN [11]: It is a semi-supervised graph convolution network model for graph embedding.
- GAT [20]: It is a graph neural network model which employs an attention mechanism to obtain graph representations.
- Top-K [7]: It is a graph representation method that adaptively selects some critical nodes to form smaller subgraphs based on their importance vectors.
- ASAPool [16]: It is a graph neural network model that utilizes attention mechanisms to capture the importance of nodes and pools subgraphs into a coarse graph through learnable sparse soft clustering allocation for graph representation.
- SUGAR [18]: It is a graph representation method that first samples some subgraphs. And then it uses the DQN algorithm to select top-k key subgraphs as representative abstractions of the entire graph.
- SAT [2]: It is a graph transformer method that incorporates the structure explicitly. Before calculating attention, it fuses structural information into the original self-attention by extracting k-hop subgraphs or k-subtrees on each node.

5.3 Performance on Real-World Graphs

In the graph classification task, this paper adopts accuracy as the evaluation metric to measure the performance of different models.

From Table 2, it can be observed that the proposed CDM-GNN model achieves the best graph classification results on the MUTAG, PTC, and IMDB-M datasets. Specifically, compared with the GCN and GAT, the CDM-GNN has an improvement of about 5% on the MUTAG dataset in terms of graph classification accuracy. Compared to Top-K and ASAPool, which are specifically designed for pooling operations, CDM-GNN demonstrates improvements ranging from 12% to 20%. In comparison to methods focusing on substructure extraction, SUGAR and SAT, CDM-GNN shows improvements of 3% to 7%. On the PTC dataset, the CDM-GNN also shows a notable improvement and achieves accurate graph classification. On the multi-class IMDB-M dataset, the proposed CDM-GNN achieves about 5.4% improvement over GCN and GAT. When compared to some pooling methods, CDM-GNN shows accuracy improvements ranging from 5.5% to 15%. In the substructure research, CDM-GNN demonstrates accuracy improvements of 4% to 9%. These results highlight the superior graph classification performance of the CDM-GNN model across different datasets. It surpasses other popular models and specialized approaches for pooling or subgraph analysis.

Table 2. The Accuracy of graph classification

Models	MUTAG	PTC	IMDB-M
GCN	0.8924	0.5726	0.5700
GAT	0.8994	0.5944	0.5810
Top-K Pool	0.7291	0.5721	0.4836
ASAPool	0.8211	0.5677	0.5794
SUGAR	0.8660	0.5821	0.5988
SAT	0.9030	0.6070	0.5329
Ours	**0.9474**	**0.6154**	**0.6348**

5.4 Visualization Results

To demonstrate the interpretability of the CDM-GNN method, this paper conducts a qualitative analysis. According to the existing chemical knowledge, people know that the MUTAG dataset determines whether it has a mutagenic effect by judging whether the molecule contains the substructure of NO_2 or NH_2. Therefore, this paper visualizes the MUTAG dataset to observe the interpretability of the CDM-GNN model and perform qualitative analysis. If CDM-GNN model can recognize the important substructure NO_2 and disentangle this part with bias parts, then it indicates that the model has better interpretability. This model can obtain classification and interpretation results that are consistent with human cognition and prior knowledge.

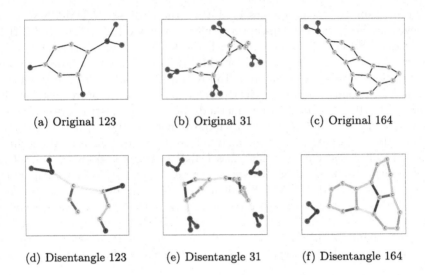

(a) Original 123   (b) Original 31   (c) Original 164

(d) Disentangle 123   (e) Disentangle 31   (f) Disentangle 164

Fig. 3. The visualization results on the MUTAG dataset.

In this study, the edges of the graph are colored based on the values of the mask, where darker colors indicate greater importance. In the visualized results, we can observe that the edges of important substructures NO_2 are colored darker, indicating their higher weights and greater significance in the classification of the MUTAG dataset. The edges connecting the important substructures NO_2 and the bias parts are colored lighter, indicating that the CDM-GNN model has successfully disentangled the important substructures from the bias parts. From Fig. 3 (a)(b)(d)(e), it can be observed that regardless of whether the graph contains one or multiple NO_2 substructures, the CDM-GNN can recognize them and successfully disentangle them from the carbon rings. Even in cases where there are multiple carbon rings in the graph, like Fig. 3 (c)(f), the CDM-GNN model can still identify the important NO_2 parts. This indicates the model has the ability to capture and distinguish the relevant features, enabling it to recognize and disentangle the important NO_2 parts from other structural components.

5.5 Ablation Experiment

Here, this paper discusses the role of three losses in learning for the CDM-GNN model. From Fig. 4, It can be observed that when this model only uses the L_{KL}, L_{int}, and L_{pred}, the L_{pred} plays a more important role in this model training. The CDM-GNN model is designed by incorporating downstream graph classification tasks. The L_{pred} plays a leading role in guiding this model to achieve better differentiation among different categories of graph data. This way, the CDM-GNN model focuses on the goal of graph classification and progressively improves its performance. The L_{KL} is mainly aimed at reducing the interference caused by the bias component. Without this loss, the classification results are the most significant decrease in accuracy. The L_{int} is to consider the decoupling of causality. If the CDM-GNN model is without L_{int}, it also leads to a decrease in accuracy. When three losses are used together, the model can achieve the best effect. It has better graph classification results and can separate the important substructure and the bias parts.

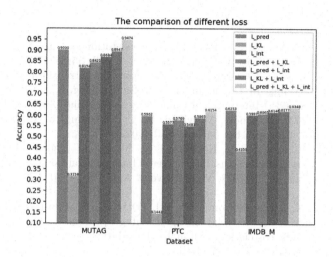

Fig. 4. The ablation results about the Loss on MUTAG, PTC, and IMDB-M dataset

6 Conclusion

This paper introduces a causal disentangled multi-granularity graph representation learning method, namely CDM-GNN. It is primarily based on the idea of multi-granularity, aiming to identify important substructures and bias parts and disentangle them in the graph classification tasks. The proposed method achieves favorable classification performance and provides qualitative interpretability of the results. This technology can be extended for researching drug molecule properties and facilitating new drug development. Identifying important substructures that influence drug properties within molecules and their disentanglement can aid in exploring drug molecule properties and innovating drug development based on these substructures. This paper has not yet explored additional downstream graph tasks, such as node classification and link prediction. Future research could explore the extension of the concept of causal disentangled multi-granularity to these graph-related tasks. Subsequent research also can prioritize in-depth discussions of interpretable quantitative assessments for self-explainable models.

Acknowledgments. This work is supported by the National Natural Science Foundation of China (Nos. 62221005, 61936001, 61806031), Natural Science Foundation of Chongqing, China (Nos. cstc2019jcyj-cxttX0002, cstc2021ycjh-bgzxm0013), Project of Chongqing Municipal Education Commission, China (No. HZ2021008), and Doctoral Innovation Talent Program of Chongqing University of Posts and Telecommunications, China (Nos. BYJS202108, BYJS202209, BYJS202118).

References

1. Bengio, Y., Courville, A.C., Vincent, P.: Representation learning: a review and new perspectives. IEEE Trans. Pattern Anal. Mach. Intell. **35**(8), 1798–1828 (2013)
2. Chen, D., O'Bray, L., Borgwardt, K.M.: Structure-aware transformer for graph representation learning. In: Chaudhuri, K., Jegelka, S., Song, L., Szepesvári, C., Niu, G., Sabato, S. (eds.) International Conference on Machine Learning, ICML 2022, 17–23 July 2022, Baltimore, Maryland, USA. Proceedings of Machine Learning Research, vol. 162, pp. 3469–3489. PMLR (2022)
3. Cucala, D.J.T., Grau, B.C., Kostylev, E.V., Motik, B.: Explainable GNN-based models over knowledge graphs. In: The Tenth International Conference on Learning Representations, ICLR 2022, Virtual Event, April 25–29, 2022. OpenReview.net (2022)
4. Debnath, A.K., Lopez de Compadre, R.L., Debnath, G., Shusterman, A.J., Hansch, C.: Structure-activity relationship of mutagenic aromatic and heteroaromatic nitro compounds. correlation with molecular orbital energies and hydrophobicity. J. Med. Chem. **34**(2), 786–797 (1991)
5. Fang, Y., et al.: Knowledge graph-enhanced molecular contrastive learning with functional prompt. Nat. Mach. Intell. 1–12 (2023)
6. Feng, A., You, C., Wang, S., Tassiulas, L.: Kergnns: Interpretable graph neural networks with graph kernels. In: Thirty-Sixth AAAI Conference on Artificial Intelligence, AAAI 2022, Thirty-Fourth Conference on Innovative Applications of Artificial Intelligence, IAAI 2022, The Twelveth Symposium on Educational Advances in Artificial Intelligence, EAAI 2022 Virtual Event, February 22 - March 1, 2022, pp. 6614–6622. AAAI Press (2022)
7. Gao, H., Ji, S.: Graph u-nets. In: Chaudhuri, K., Salakhutdinov, R. (eds.) Proceedings of the 36th International Conference on Machine Learning, ICML 2019, 9–15 June 2019, Long Beach, California, USA. Proceedings of Machine Learning Research, vol. 97, pp. 2083–2092. PMLR (2019)
8. Hendrycks, D., Dietterich, T.G.: Benchmarking neural network robustness to common corruptions and perturbations. In: 7th International Conference on Learning Representations, ICLR 2019, New Orleans, LA, USA, May 6–9, 2019. OpenReview.net (2019)
9. Hsieh, J., Liu, B., Huang, D., Fei-Fei, L., Niebles, J.C.: Learning to decompose and disentangle representations for video prediction. In: Bengio, S., Wallach, H.M., Larochelle, H., Grauman, K., Cesa-Bianchi, N., Garnett, R. (eds.) Advances in Neural Information Processing Systems 31: Annual Conference on Neural Information Processing Systems 2018, NeurIPS 2018, December 3–8, 2018, Montréal, Canada, pp. 515–524 (2018)
10. Hu, W., et al.: Open graph benchmark: Datasets for machine learning on graphs. In: Larochelle, H., Ranzato, M., Hadsell, R., Balcan, M., Lin, H. (eds.) Advances in Neural Information Processing Systems, vol. 33: Annual Conference on Neural

Information Processing Systems 2020, NeurIPS 2020, December 6–12, 2020, virtual (2020)

11. Kipf, T.N., Welling, M.: Semi-supervised classification with graph convolutional networks. In: 5th International Conference on Learning Representations, ICLR 2017, Toulon, France, April 24–26, 2017, Conference Track Proceedings. OpenReview.net (2017)

12. Liu, Y., Wang, X., Wu, S., Xiao, Z.: Independence promoted graph disentangled networks. In: The Thirty-Fourth AAAI Conference on Artificial Intelligence, AAAI 2020, The Thirty-Second Innovative Applications of Artificial Intelligence Conference, IAAI 2020, The Tenth AAAI Symposium on Educational Advances in Artificial Intelligence, EAAI 2020, New York, NY, USA, February 7–12, 2020, pp. 4916–4923. AAAI Press (2020)

13. Lucic, A., Ter Hoeve, M.A., Tolomei, G., De Rijke, M., Silvestri, F.: Cfgnnexplainer: counterfactual explanations for graph neural networks. In: Camps-Valls, G., Ruiz, F.J.R., Valera, I. (eds.) International Conference on Artificial Intelligence and Statistics, AISTATS 2022, 28–30 March 2022, Virtual Event. Proceedings of Machine Learning Research, vol. 151, pp. 4499–4511. PMLR (2022)

14. Luo, D., et al.: Parameterized explainer for graph neural network. In: Larochelle, H., Ranzato, M., Hadsell, R., Balcan, M., Lin, H. (eds.) Advances in Neural Information Processing Systems 33: Annual Conference on Neural Information Processing Systems 2020, NeurIPS 2020, December 6–12, 2020, virtual (2020)

15. Ma, J., Cui, P., Kuang, K., Wang, X., Zhu, W.: Disentangled graph convolutional networks. In: Chaudhuri, K., Salakhutdinov, R. (eds.) Proceedings of the 36th International Conference on Machine Learning, ICML 2019, 9–15 June 2019, Long Beach, California, USA. Proceedings of Machine Learning Research, vol. 97, pp. 4212–4221. PMLR (2019)

16. Ranjan, E., Sanyal, S., Talukdar, P.P.: ASAP: adaptive structure aware pooling for learning hierarchical graph representations. In: The Thirty-Fourth AAAI Conference on Artificial Intelligence, AAAI 2020, The Thirty-Second Innovative Applications of Artificial Intelligence Conference, IAAI 2020, The Tenth AAAI Symposium on Educational Advances in Artificial Intelligence, EAAI 2020, New York, NY, USA, February 7–12, 2020, pp. 5470–5477. AAAI Press (2020)

17. Rossi, R.A., Ahmed, N.K.: The network data repository with interactive graph analytics and visualization. In: Bonet, B., Koenig, S. (eds.) Proceedings of the Twenty-Ninth AAAI Conference on Artificial Intelligence, January 25–30, 2015, Austin, Texas, USA, pp. 4292–4293. AAAI Press (2015)

18. Sun, Q., et al.: SUGAR: subgraph neural network with reinforcement pooling and self-supervised mutual information mechanism. In: Leskovec, J., Grobelnik, M., Najork, M., Tang, J., Zia, L. (eds.) WWW '21: The Web Conference 2021, Virtual Event / Ljubljana, Slovenia, April 19–23, 2021, pp. 2081–2091. ACM / IW3C2 (2021)

19. Toivonen, H., Srinivasan, A., King, R.D., Kramer, S., Helma, C.: Statistical evaluation of the predictive toxicology challenge 2000–2001. Bioinformatics 19(10), 1183–1193 (2003)

20. Velickovic, P., Cucurull, G., Casanova, A., Romero, A., Liò, P., Bengio, Y.: Graph attention networks. In: 6th International Conference on Learning Representations, ICLR 2018, Vancouver, BC, Canada, April 30 - May 3, 2018, Conference Track Proceedings. OpenReview.net (2018)

21. Wang, G.: DGCC: data-driven granular cognitive computing. Granular Comput. 2(4), 343–355 (2017)

22. Wang, X., Jin, H., Zhang, A., He, X., Xu, T., Chua, T.: Disentangled graph collaborative filtering. In: Huang, J.X., Chang, Y., Cheng, X., Kamps, J., Murdock, V., Wen, J., Liu, Y. (eds.) Proceedings of the 43rd International ACM SIGIR Conference on Research and Development in Information Retrieval, SIGIR 2020, Virtual Event, China, July 25–30, 2020, pp. 1001–1010. ACM (2020)

23. Ying, Z., Bourgeois, D., You, J., Zitnik, M., Leskovec, J.: Gnnexplainer: generating explanations for graph neural networks. In: Wallach, H.M., Larochelle, H., Beygelzimer, A., d'Alché-Buc, F., Fox, E.B., Garnett, R. (eds.) Advances in Neural Information Processing Systems 32: Annual Conference on Neural Information Processing Systems 2019, NeurIPS 2019, December 8–14, 2019, Vancouver, BC, Canada, pp. 9240–9251 (2019)

24. Zhang, S., Rao, X., Tay, Y., Zhang, C.: Knowledge router: Learning disentangled representations for knowledge graphs. In: Proceedings of the 2021 Conference of the North American Chapter of the Association for Computational Linguistics: Human Language Technologies, pp. 1–10 (2021)

Distances and Similarities

Towards ML Explainability with Rough Sets, Clustering, and Dimensionality Reduction

Marek Grzegorowski[1]([✉]) [iD], Andrzej Janusz[1,2] [iD], Grzegorz Śliwa[3],
Łukasz Marcinowski[3], and Andrzej Skowron[4] [iD]

[1] Institute of Informatics, University of Warsaw, Banacha 2, 02-097 Warsaw, Poland
{M.Grzegorowski,janusza}@mimuw.edu.pl
[2] QED Software, Mazowiecka 11/49, 00-052 Warsaw, Poland
[3] FitFood, Solskiego 11/28, 31-216 Kraków, Poland
{g.sliwa,l.marcinowski}@fitfoodpoland.pl
[4] Systems Research Institute, Polish Academy of Sciences, Nawelska 6, 01-447
Warszawa, Poland
https://qed.pl, https://fitboxy.com

Abstract. This study discusses some essential problems of explainable machine learning applications in the FMCG market. The solution combines several machine learning techniques, including clustering, dimensionality reduction, rough set reducts, and rule-based explanations. We propose a novel approach to improve human-computer interaction with the XAI prototype method by generating human-readable cluster descriptions, emphasizing each cluster's most discernible characteristics. To evaluate our method, we refer to the challenging task of demand prediction. The results confirmed that we could achieve five times better work performance without losing quality.

Keywords: RST · decision reducts · DAAR · XAI · FMCG

1 Introduction

Despite the growing popularity of machine learning, such solutions are often incomprehensible to employees and difficult to control [24]. They usually try to mimic the employees' activities, striving to replace them, less often supporting their work. This causes reluctance and concern among staff at various levels. The anxiety that "artificial intelligence" may take over jobs [6], compounded by concerns of the complexity of machine learning tools resulting in general misunderstanding [15], are the key issues holding back the popularization and implementation of machine learning [1].

The presented study concerns the application of ML to optimize the efficiency of operational processes. Our goal is to provide an auxiliary tool that facilitates

---

Research co-funded by Polish National Centre for Research and Development (NCBiR) grant no. POIR.01.01.01-00-0963/19-00 and by Polish National Science Centre (NCN) grant no. 2018/31/N/ST6/00610.

the work of experts, allowing them to achieve better results. In the article, we propose a new approach to the semi-automatic prescription of future demand based on unsupervised clustering techniques and cluster prototypes, considered as a prediction. The solution is focused not only on prediction accuracy but also on stability in time and, foremost interpretability of results. We achieve this by providing two-dimensional visualization and human-readable cluster descriptions emphasizing the most significant similarity characteristics of clustered objects. The conducted experimental evaluation confirmed that the proposed approach achieved a fair trade-off between ML performance and interpretability [3].

The conducted research confirmed that, without a significant loss of the quality of predictions, we could operate on points of sale (PoS) in an aggregate manner, reducing the amount of work needed to prepare delivery plans in the fast-moving consumer goods (FMCG) industry [9]. The challenge here is to properly aggregate the points of sale in such a way as to minimize variety in purchasing patterns. We want to achieve such a granulation that groups together PoSs with similar sales patterns. We ensure that the results of the clustering are understandable for the team of experts by referring to XAI prototypes [12], which explicitly define tasks for employees. Furthermore, apart from the unambiguousness of the tasks, we proposed an innovative method of generating human-readable cluster descriptions inspired by feature ranking [1], based on reduction algorithms from the rough set theory (RST) [20,21] - this way, obtaining a good understanding of each cluster most discernible characteristics. In the frame of this study, we evaluate real data collected from several hundreds of FitBoxY.com vending machines [9]. The main contributions of the paper are:

1. Visual clustering stability assessment by the 2D projections.
2. Novel approach to improve interpretability based on reducts from RST.
3. Experimental study on the real data and two different data representations.

The rest of the paper is organized as follows. In Sect. 2, we review the related literature. Section 3 provides the essential preliminary knowledge. In Sect. 4, we present the solution. Section 5 presents the case study and experimental evaluation. Finally, in Sect. 6, we conclude the paper.

2   Related Works

Food production is a complex process under high uncertainty resulting in differences between planned and actual demand. Considering the short shelf-life of many products that may result in unnecessary food waste, the accurate prediction of the future demand at each point of sale is highly important [27]. It is particularly interesting to prepare such a delivery plan for each vending machine, the realization of which will bring maximum profit and minimize food waste at the same time. One of the ways is to predict demand with ML models.

Intrinsically interpretable, simple models are often less accurate than more sophisticated methods [1]. On the other end, more complex multivariate methods, like the random forest, boosting models, or deep neural networks, suffer from the lack of interpretability [24]. Typically, time series collected from vending machines are very short and scattered between many PoS and products, and ML models need to handle the cold-start problem [14].

Maintaining trustworthy human-computer collaboration is a vital research topic, and we may refer to the number of well-established decision support methods [16]. Among the plethora of ML explainability-related methods [2], in our case, a particularly interesting are post-hoc model agnostic approaches [1]. The example-based explanations, for instance, are explicitly inspired by the cognitive science of human reasoning, which is often prototype-based. For explaining text clusters, keyword extraction seems to be a feasible approach [22], but this method is not applicable in the general case. Other methods capable of explaining the clusters' similarities are based on variable rankings [5,29].

In the context of PoS clustering, we require finding a set of the most relevant differences between objects from different clusters. Therefore, the application of RST-based reduction methods [10,13] to facilitate this process, we find a promising approach. Furthermore, considering the variability of sales patterns in time impacting the cluster structure, it is also worth paying attention to the stability of clustering and explanations and various approaches to visual explanation techniques [2], particularly 2D projections.

3 Preliminary Knowledge

3.1 Rudiments of Rough Sets

Rough set theory as a whole provides a formalism for reasoning about imperfect data, handling such problems as data veracity, uncertainty, or incompleteness [20,21]. In RST, we assume that the whole available information about an object $u \in U$ is represented in a structure called an information system – a tuple (U, A), where U is a finite, non-empty set of objects, and A is a finite, non-empty set of attributes. Let us distinguish a decision attribute, which defines a partitioning of U into disjoint sets representing decision classes [8]. An information system with a specified decision attribute is called a decision table and is denoted by $\mathbb{S} = (U, A \cup \{d\})$, $A \cap \{d\} = \emptyset$. For a given $\mathbb{S}$, one considers functions $a : U \rightarrow V_a$, $a \in A$, where V_a is the set of values of a. Such functions allow us to represent $\mathbb{S}$ as a table with rows labeled by objects, columns labeled by attributes, and cells corresponding to pairs (u, a) assigned with values $a(u) \in V_a$.

Typically, some attributes in A may be dispensable or could be irrelevant from the point of view of a given problem corresponding to the decision attribute d. In such situations, A-based information about objects in U may be simplified. Selecting informative sets of attributes is conducted by referring to the notion of a reduct [21], i.e., an irreducible subset of attributes $R \subseteq A$ that, for a given decision table $\mathbb{S} = (U, A \cup \{d\})$, determines d, denoted as $R \Rightarrow d$.

There are plenty of interpretations of the reduct definition and their approximate interpretations [8,20]. Criteria for calculating approximate decision reducts are usually based on functions evaluating degrees of decision information induced by attribute subsets and thresholds for values of those functions' specifying which

---

Algorithm 1: DAAR reduct calculation

Input: $\mathbb{S} = (U, A \cup \{d\})$; $\phi_d : 2^A \to \mathbb{R}$; $p_{probe} \in [0,1)$; $mTry$;
Output: DAAR reduct R of $\mathbb{S}$

1 **begin**

2     $R := \emptyset$; $\phi_{max} := -\infty$; $stopFlag := FALSE$;

3   **while** $stopFlag = FALSE$ **do**

4   Sample $A' \subseteq A \setminus R$ that satisfy $|A'| == mTry$;

5   **foreach** $a \in A' \setminus R$ **do**

6           $R' := R \cup \{a\}$;

7   **if** $\phi_d(R') > \phi_{max}$ **then**

8              $\phi_{max} := \phi_d(R')$; $a_{best} := a$;

9   **end**

10   **end**

11   **if** $P\left(\phi_d\big(R \cup \{a_{best}\}\big) \leq \phi_d\big(R \cup \{\hat{a}_{best}\}\big)\right) \leq p_{probe}$ **then**

12           $R := R \cup \{a_{best}\}$;

13   **end**

14   **else**

15           $stopFlag := TRUE$;

16   **end**

17   **end**

18   **foreach** $a \in R$ **do**

19        $AR' := R \setminus \{a\}$;

20   **if** $\phi_d(R') \geq \phi_d(R)$ **then**

21           $AR := R'$;

22   **end**

23   **end**

24   **return** R;

25 **end**

---

of those subsets are good enough. Such an approach may lead us to obtain subsets of attributes that are less accurate than exact reducts but could be preferred in some real-life applications to deal with large or noisy data, ultimately leading to smaller data representations [10].

RST reducts are often hybridized with other methods [25]. For instance, Algorithm 1, is a combination of iterative filter-based feature selection with statistical significance tests based on random probes, here, RST-based feature elimination is applied to calculate dynamically adjusted approximate reducts (DAAR) [13]. Algorithm 1 operated on a given decision table $\mathbb{S}$, attribute subset quality measure $\phi_d : 2^A \to \mathbb{R}$, probability threshold of adding irrelevant attribute p_{probe}, and assumed attribute sample size $mTry$. This concept is applied in this study to determine the most distinguishing attributes of PoS as a special implementation of the variable importance XAI method [5]. However, the proposed approach does not rank features but provides a complete subset of descriptive attributes.

3.2   Distance Based Clustering Methods

In the conducted research, we verified several distance-based clustering methods. The two flat clustering algorithms that we used in our experiments are *kmeans* and *pam* [11]. In the first one, the initial cluster centers are selected randomly, and then iteratively refined by assigning all instances to the closest cluster center and computing new centers by averaging the corresponding instances. In this method, the final cluster center does not have to coincide with an actual data instance, thus the representative of a cluster is chosen as the instance closest to the final cluster center. The second method, *partitioning around medoid*, works in a similar way as *kmeans*, but in each iteration, the new cluster center (medoid) is chosen as the instance with the smallest sum of distances to other cluster members. Hence, after the final clusters are found, their medoids can be used as the most representative instances.

Agglomerative clustering algorithms create a hierarchy of data clusters by starting from singleton groupings, and iteratively merging the two closest groups into a bigger cluster [14]. This process stops when all data instances are merged into a single cluster (a bottom-up approach). To measure the proximity (dissimilarity) between groups, agglomerative clustering algorithms use so-called, linkage functions. In our experiments, we used *single_linkage*, *complete_linkage*, and *ward_linkage*. The first one (*single_linkage*) defines the dissimilarity between two groups as the smallest distance between any two instances from those groups. Analogically, the second function (*complete_linkage*) asserts the proximity of groups as the largest distance between any two instances. The last function (*ward_linkage*), also called Ward's minimum variance method, associates the dissimilarity between two groups with the sum of squared distances between all pairs of instances from those groups [7].

The divisive approach to hierarchic clustering can be regarded as the opposite of the agglomerative method. It starts by placing all data instances into a single group, and then recursively dividing a group with the largest diameter into two groups whose diameters are possibly small. In our experiments, we used *diana_linkage* method [7], which defines the cluster's diameter as the largest distance between any two members of a group. The clustering algorithm stops when each of the resulting groups consists of only one data instance.

3.3   Dimensionality Reduction

Dimensionality reduction techniques are useful for 2D visualizations of high-dimensional data [30]. Given a matrix $D^{m \times m}$ containing distance between pairs of m objects and a number of target dimensions, multidimensional scaling places each object into low-dimensional space in a way that preserves pairwise distances. In genetics and microbiology, dimensionality reduction is used for visualizing the data with t-distributed stochastic neighbor embedding [17]. There are a plethora of dimensionality reduction methods, yet the most prominent and broadly used in this context are uniform manifold approximation and projection (UMAP) and principal component analysis (PCA).

The central idea of PCA is to reduce the dimensionality of a dataset $\mathbf{A} \in \mathbb{R}^{m \times n}$, which contains a potentially large number of interrelated variables, retaining as much as possible of the variation in the data. This is achieved by transforming the original representation to a new set of variables, so-called principal components, which are uncorrelated and ordered in a way the first few retain most of the variation from the original variables. In the first step, the data in matrix $\mathbf{A} \in \mathbb{R}^{m \times n}$ is centered by subtracting it with matrix $\mathbf{A}_{mean}$ of mean vectors for each column. The next step is to calculate the co-variance matrix $\mathbf{C} \in \mathbb{R}^{n \times n}$ for the columns (features) in table $\mathbf{B} = \left(\vec{b_1} \cdots \vec{b_n} \right)$ as $\mathbf{C} = \frac{1}{m} \mathbf{B}^T \mathbf{B}$. We can calculate eigenvectors, and the corresponding eigenvalues, for matrix $\mathbf{C}$, such as $\mathbf{CW} = \mathbf{W\Lambda}$, where matrix $\mathbf{W}$ contains eigenvectors, a diagonal matrix $\mathbf{\Lambda}$ contains eigenvalues. For the purpose of dimensionality reduction, we can project the data points onto the first k principal components, i.e., the truncating matrix $\mathbf{W}$ to only k most significant features ($\mathbf{W}_k$) and projecting the original data $\mathbf{A_k} = \mathbf{AW}_k$ retaining enough variance. The first principal component is the direction in feature space along which projections of observations have the largest variance. The second principal component is the direction which maximizes variance among all directions orthogonal to the first one, etc.

UMAP returns a low-dimensional graph that preserves relationships from the high-dimensional dataset, hence, is useful for identifying similarities and outliers in data since similar samples tend to be grouped together. There are two main steps. The first corresponds to learning the structure of the manifold (high dimensional data). The second focuses on finding a low-dimensional representation. This can be achieved by constructing a neighbor graph based on calculating the similarity score for each point and its nearest neighbors [18]. After learning the approximate manifold and constructing the data graph in the high dimensional space, UMAP starts constructing a graph in a lower dimensional embedding space similar to the graph in the input space. UMAP uses fuzzy cross-entropy to minimize the difference between two distributions and to retain similarities of points in the low-dimensional embedding space.

4   Solution Overview

The introduced mechanism is based on the clustering of similar objects and selecting the central element as the representative. The method complies with the assumptions of Industry 5.0 with the central role of humans in the ML-driven automation of industrial processes [9, 28]. To ensure the process is understandable to the expert, we introduce two complementary mechanisms. The first is based on instance-based explainability (so-called prototypes). The second is the generation of cluster descriptions based on their most characteristic features. The developed technique has mechanisms for tracking changes in data by projecting them onto two-dimensional space. The flowchart of our approach is depicted in Fig. 1.

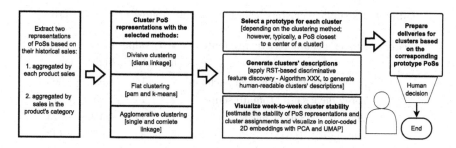

Fig. 1. Flowchart of the proposed method.

4.1 Data Representation and Clustering

In the first step, the developed solution requires data ingestion and integration, leading to an appropriate vectorized representation. Next, we need to decide on the distance metric. In the presented study, we decided to build two alternative representations. One is based on the historical sales aggregated by each product offered in a PoS. The second one aggregated sales by category (cf. Sect. 5.1). To make those two representations comparable, we used the Manhattan distance.

We verified several clustering methods to group similar points of sale. It was also vital to select the most representative PoS (prototype) for each cluster. The two flat clustering algorithms we used are *kmeans* and *pam* [11]. For *kmeans* the prototypes are chosen as the instances closest to the clusters' centers. Whereas for *pam*, we use medoids. Agglomerative clustering, we used *single_linkage*, *complete_linkage*, and *ward_linkage*. We used also the *diana_linkage* method [7] which defines the cluster's diameter as the largest distance between any two members of a group. The details of each method are described in Sect. 3.2.

Additionally, we used two random methods, i.e., *random_random* and *random_custom* as a reference for the more sophisticated approaches. For both methods, the division into clusters is performed at random. The difference refers to the way we select the representative vending machine - *random_custom* method selects the machine closest to the cluster's center.

4.2 RST-Based Discriminative Features Discovery

The important stage in our method is to indicate which attributes are sufficient to distinguish objects from different clusters. The proposed method is based on the concept of decision reduct derived from RST (cf. Sect. 3.1), i.e., an irreducible subset of attributes that brings enough information to distinguish objects of different classes. In our study, we discuss two approaches to reduct computation - *global* and *local*.

In the *global* approach, we compute a single decision reduct that discerns PoSs belonging to different clusters. To achieve that, we use the greedy local discretization method described in [23] coupled with the DAAR computation [13]. This way, we may identify a single set of attributes to quantify differences

Table 1. Exemplary clusters' characteristic attributes - global discern.

cluster		cluster's characteristic attributes	attr. value		ratio	
no.	size		high	low	high	low
4	3	7prev days sales 'Chicken Sechuan'	2	1	1.99	1.865
		7prev days sales 'Tomato soup'	2	1	1.95	1.515
		sales in cat.'Other meals'	19	11	1.459	1.323
		sales in cat.'Pasta'	6	3	2.977	2.143
		sales in cat.'Snacks'	3	1	1.206	3.32
5	51	7prev days sales 'Chicken Sechuan'	2	1	10.115	2.604
		7prev days sales 'Tomato soup'	2	1	19.926	2.837
		sales in cat.'Other meals'	19	11	14.907	2.205
		sales in cat.'Pasta'	6	3	10.039	4.613
		sales in cat.'Snacks'	3	1	6.084	1.472

in sales patterns from all clusters. In the subsequent step, for each cluster C, attribute in the reduct a, and its discretized value v, we estimate the *lift* of a rule $u \in C \implies a(u) = v$. With this information, we may construct natural language descriptions of clusters such as those listed below. Of course, in practice, we may want to create the descriptions using only the values exceeding the required lift threshold to indicate only the most relevant cluster characteristics.

In the *local* method, we independently compute decision reducts for each cluster. Reducts are computed with the same algorithm as in the global approach. However, instead of discriminating all PoSs from all clusters, they focus on a single (corresponding) cluster only. In this way, we obtain a different DAAR reduct for each cluster. Since such a reduct is specialized in capturing the most discriminating factors of the corresponding group of PoSs, it allows identifying the attribute values with greater lift coefficients and creating even more meaningful descriptions in natural language.

Tables 1 and 2 present the exemplary outcomes of the two considered approaches to constructing cluster descriptions for the investigated case study, based on the global and local reduction methods, respectively. The global method results in high consistency in the description of all clusters because they are based on the same attributes. The only difference between groups of PoS is in the ratios that reflect the lift coefficients of the corresponding attribute value.

For those users who prefer to investigate more pronounced differences between clusters, we suggest using the second method,i.e., based on the local approach. In which case, the reducts may differ in their attributes, their discretization, and resulting lift values, an example of which can be seen in Table 2.

4.3 Human-Readable Clusters' Descriptions

Indicating a particular cluster's most central object as a task to be performed - in the discussed case study, a point of sale that needs to be completed - is a

Table 2. Exemplary clusters' characteristic attributes - local discern.

cluster		cluster's characteristic attributes	attr. value		ratio	
no.	size		high	low	high	low
1	2	sales in cat.'Other meals'	18	17	1.349	3.478
		total sales	54	45	3.181	1.497
6	30	sales in cat.'Other meals'	8.5	5	5.244	2.399
		total sales	56	38	3.54	1.01
9	23	7prev days sales 'Chicken Sechuan'	3	2	1.185	1.85
		total sales	68	50	1.706	2.359
13	7	7prev days sales 'Tomato soup'	6	3	13.3	9.198

well-defined task that meets the assumptions of Industry 5.0. However, it may be unclear to users why the particular object (points of sale) were considered similar. This may be even more confusing given that the allocation to the clusters may vary in time (cf. discussion on stability in Sect. 4.4). To mitigate this problem, we generate a human-readable description of a given cluster, emphasizing its most individual characteristics.

The idea is to prepare such clusters' descriptions that emphasize the properties of objects belonging to them that make them stand out from other clusters. To highlight such characteristics, we apply RST reduction algorithms. Afterward, we use simple rule-based text formatting to make the prepared descriptions more self-descriptive. Below, we present exemplary descriptions for selected clusters for the global and local approaches, which are generated on data from Tables 1 and 2. For the global method, cluster no. 5, and the lift threshold of 10.0, the generated cluster descriptions are as follows:

1. PoSs from cluster 5 are 10.1 times more likely to have greater sales in *seven previous days* than 2 for *'Chicken Sechuan'* than other PoSs
2. PoSs from cluster 5 are 19.9 times more likely to have greater sales in *seven previous days* than 2 for *'Tomato soup'* than other PoSs
3. PoSs from cluster 5 are 14.9 times more likely to have greater sales in *seven previous days* than 19 in the category *'Other meals'* than other PoSs, etc.

For local approach, cluster no. 6, and lift greater than 3.0, descriptions are:

1. PoSs from cluster 6 are 5.244 times more likely to have greater sales in *seven previous days* than 8 in the category *'Other meals'* than other PoSs
2. PoSs from cluster 6 are 3.540 times more likely to have greater *total sales* than 56 in *seven previous days* than other PoSs

It is worth noting that both methods are performed after the clustering is completed. They are independent of each other, i.e., we can generate both local and global descriptions simultaneously, and it only depends on users, which would be more understandable to them. Observably, there is still a potential for further improvement in acquiring human-readable cluster descriptions. By introducing a dialogue with experts we could acquire ontologies of concepts [4] to support more granular descriptions. It would be also valuable to operate on more natural terms like 'few', or 'many' instead of raw numerics.

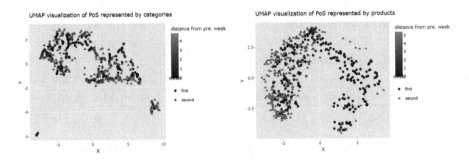

Fig. 2. 2D projection with UMAP that emphasizes the distance between the particular PoS in the first and second week.

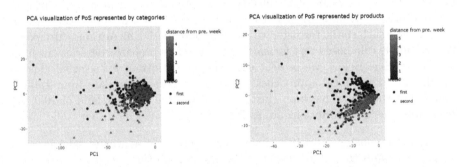

Fig. 3. 2D projection with PCA that emphasizes the distance between the particular PoS in the first and second week.

4.4   Clusters' Stability Visualisations

Users expect both adaptability and transparency of decision support systems, so observability of changes in the system's behavior is essential. The developed solution adapts to changes in customer preferences. Such changes are reflected in the data representation, thereby causing the PoS vector representation (based on purchasing patterns) to change significantly over time. This, in turn, influences the clustering, which may vary from week to week.

We can observe these regularities by projecting vectors representing PoS into two-dimensional space. Figure 2 shows 2D-embedding with UMAP for *cat_data* and *prod_data* representations. Observably, the category-based representation is more stable, i.e., the points representing each PoS for two consecutive weeks in data form a more compact structure. We can clearly see this regularity in Figs. 2 and 3 for two dimensionality reduction methods, namely UMAP and PCA. We represent each PoS from the first week with circles and those from the second week as triangles. The distances between the same points from the first and second week are emphasized with the color depth, i.e., the further the city-block distance between vectors representing PoS for the second week from the first one, the brighter the triangle.

4.5   Clusters' Prototypes

In the proposed framework, we assume that to optimally reflect customers' needs, the menu should be prepared by human experts. However, we would like to avoid preparing individual menus for hundreds of locations. The gist is to group PoS based on the historically observed customers' behavior (reflected in product purchases) and prepare the menu for the whole clusters of similar locations. For that reason, we first cluster similar PoS together, and then we select the most representative one. By describing a cluster by its most representative PoS, we implement the instance-based XAI method [2]. Furthermore, assuming the average cluster size of X, we ensure that experts have X-times less work. In our study, we aim at $X \geq 5$. We also validate if such an approach could bring satisfactory results by a data mining investigation of real data collected from FitBoxY.com - presented in the subsequent Sect. 5.

5   Experimental Evaluation

5.1   Data

The dataset used in the experiments was obtained from FitBoxY.com vending machines and contains the sales history of FitFoodPoland.pl products collected between June 21, 2017, and May 21, 2021, thus it covers the first three waves of the COVID-19 pandemic. The test data covers the period between December 2, 2019, and May 21, 2021, so they cover the first year of the pandemic and a period of a few months before the COVID-19 outbreak.

For the purpose of the experiments, there were two versions of the dataset prepared, namely: *prod_data* and *cat_data*. The first (i.e., *prod_data*), encodes weekly sales at each PoS as a vector of all available products, for each, indicating the quantity of one-week sales. The second version (i.e., *cat_data*), was created by aggregating products into seven categories, i.e., breakfast, small lunch dishes, pasta, etc. Additionally, we considered each PoS in time as a separate instance. In other words, having two selling points: A and B in weeks: t and t', the input to clustering was: $A_t, A_{t'}, B_t, B_{t'}$. Such a representation allowed us to take into account that customers' behavior may vary in time in a given location.

5.2   Experiment Flow

We perform the following steps for both category and product data representations for each week in the test data, i.e., 2019-12-02, 2019-12-09, 2019-12-16, ..., 2021-05-31. We select all the data before the test week according to the logic of the following query:

```
SELECT t.product, t.PoS, t.week_no,
       t.year, count(*) as sales_qty
FROM transactions t
WHERE t.date < TEST_WEEK_START_DATE_PARAM
GROUP BY t.product, t.PoS, t.week_no, t.year;
```

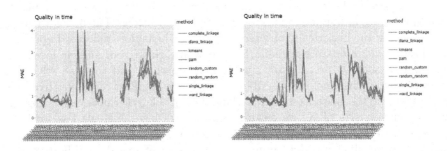

Fig. 4. MAE for *cat_data prod_data*

For the *cat_data* the query is similar. The only difference is to replace *t.product* with *t.product.category*. In the next step, we build vectors of products and categories with the values of 'sales_qty', which are thereafter clustered with the selected clustering methods (cf. Sect. 4.1). For each cluster C, we choose its most representative *PoS*, e.g., the one with the smallest distance to the cluster center (cf. Sect. 3.2). Similarly, we process the test data.

In the prediction phase, each PoS is assigned to the closest cluster. As the delivery prediction for the test week, we consider products delivered to the most representative PoS of the corresponding cluster. We compare such recommendations with the ground truth, i.e., the real sales of the products at the PoS during the next seven days. We assess the quality with mean average error (MAE).

5.3   Results

The conducted experiments revealed that the representation of sales points by their weekly sales for each product (*prod_data*) performs better than a more concise representation based on 7 selected product categories (*cat_data*). The highest quality was achieved by the PAM algorithm, with MAE of 1.151801 and an average cluster size of approximately 5 vending machines. In the case of category-based representation, the error was slightly higher. The best performing was hierarchical clustering, i.e., complete_linkage method, which achieved MEA equals 1.257736. So, the proposed method results in approx. one product mismatch in a week-long prediction horizon providing a significant acceleration of work. The detailed experiment outcomes are presented in Table 3.

Table 3. MAE of demand prediction - *prod_data* vs. *cat_data*.

Method	prod_data		cat_data	
	mean	std_dev	mean	std_dev
kmeans	1.1734	0.6641	1.2732	0.707
pam	1.1518	0.6555	1.3041	0.7191
single_linkage	1.1595	0.6685	1.4079	0.74
complete_linkage	1.1786	0.6705	1.2577	0.6662
ward_linkage	1.1667	0.6477	1.2808	0.678
diana_linkage	1.1698	0.6732	1.2865	0.6901

In Fig. 4, we see the fluctuation of the error in time. The empty parts correspond to lockdowns related to COVID-19. That sales patterns corresponding to several weeks just after (or just before) the lockdowns were hard to predict. Not matching historically observed behavior, leading to significantly higher error. As the pandemic continued, the greater amount of collected data and various patterns allowed us to represent such a situation better. The error level observed at the end of the chart is only slightly higher than the pre-pandemic one.

When assessing the clustering stability in time, the category-based representations are more stable overall. Since the sum of all attribute values in both representations is the same, we may compare their distributions of week-to-week distances. The city-block distance between points of sale represented by categories is relatively small, with a mean of 18.85, a median of 15, and 123 at max. Observably lower than for product-based representation, 41.49, 37, and 225, respectively. The overall PoS spread for the two consecutive weeks is similar for the first representation and completely different for the second.

6 Summary

In the article, we introduce an end-to-end framework enabling the application of soft computing methods to optimize the efficiency of operational processes in an interpretable way. For that purpose, we considered clustering techniques on data representations that encode customer purchasing patterns. In the presented case study of the FMCG market, we showed that it is possible to operate on whole groups of PoSs, significantly reducing the work required to prepare delivery plans resulting in a fivefold increase in work efficiency.

One of the possible extensions of the proposed method is related to interaction with experts [8]. Introducing a dialogue with experts would allow us to give more natural, i.e., human-readable names for the clusters relying on the acquired ontologies of concepts [4]. To further improve clusters' descriptions and support dialogue with experts, instead of referring to raw numerics like 9 or 10, it would be vital to operate with more natural terms like 'often', 'rarely', 'intensive', 'few', or 'many'. In the search for customer behavior patterns the promising area for future research is related to granulation techniques provided by, e.g., RST in the context of process mining [26] as well as for the discovery of process models from sample data and domain knowledge [19].

References

1. Adadi, A., Berrada, M.: Peeking inside the black-box: a survey on Explainable Artificial Intelligence (XAI). IEEE Access **6**, 52138–52160 (2018). https://doi.org/10.1109/ACCESS.2018.2870052
2. Barredo Arrieta, A., et al.: Explainable Artificial Intelligence (XAI): concepts, taxonomies, opportunities and challenges toward responsible AI. Inf. Fusion **58**, 82–115 (2020). https://doi.org/10.1016/j.inffus.2019.12.012
3. Baryannis, G., Dani, S., Antoniou, G.: Predicting supply chain risks using machine learning: the trade-off between performance and interpretability. Futur. Gener. Comput. Syst. **101**, 993–1004 (2019). https://doi.org/10.1016/j.future.2019.07.059
4. Dutta, S., Skowron, A.: Concepts approximation through dialogue with user. In: Mihálydeák, T., et al. (eds.) IJCRS 2019. LNCS (LNAI), vol. 11499, pp. 295–311. Springer, Cham (2019). https://doi.org/10.1007/978-3-030-22815-6_23
5. Fisher, A., Rudin, C., Dominici, F.: All models are wrong, but many are useful: learning a variable's importance by studying an entire class of prediction models simultaneously. J. Mach. Learn. Res. **20**, 177:1–177:81 (2019)
6. Frey, C.B., Osborne, M.A.: The future of employment: how susceptible are jobs to computerisation? Technol. Forecast. Soc. Chang. **114**, 254–280 (2017). https://doi.org/10.1016/j.techfore.2016.08.019
7. Goy, S., Coors, V., Finn, D.: Grouping techniques for building stock analysis: a comparative case study. Energy Build. **236**, 110754 (2021). https://doi.org/10.1016/j.enbuild.2021.110754
8. Grzegorowski, M.: Selected aspects of interactive feature extraction. In: Peters, J.F., Skowron, A., Bhaumik, R.N., Ramanna, S. (eds.) Transactions on Rough Sets XXIII. LNCS, vol. 13610, pp. 121–287. Springer, Heidelberg (2022). https://doi.org/10.1007/978-3-662-66544-2_8
9. Grzegorowski, M., Litwin, J., Wnuk, M., Pabis, M., Marcinowski, L.: Survival-based feature extraction - application in supply management for dispersed vending machines. IEEE Trans. Ind. Inform. **19**(3), 3331–3340 (2023). https://doi.org/10.1109/TII.2022.3178547
10. Grzegorowski, M., Ślęzak, D.: On resilient feature selection: computational foundations of r-C-reducts. Inf. Sci. **499**, 25–44 (2019). https://doi.org/10.1016/j.ins.2019.05.041
11. Guo, X., Lin, H., Wu, Y., Peng, M.: A new data clustering strategy for enhancing mutual privacy in healthcare IoT systems. Futur. Gener. Comput. Syst. **113**, 407–417 (2020). https://doi.org/10.1016/j.future.2020.07.023

12. Heide, N.F., Muller, E., Petereit, J., Heizmann, M.: X^3SEG: model-agnostic explanations for the semantic segmentation of 3D point clouds with prototypes and criticism. In: 2021 IEEE International Conference on Image Processing (ICIP), pp. 3687–3691 (2021). https://doi.org/10.1109/ICIP42928.2021.9506624

13. Janusz, A., Ślęzak, D.: Computation of approximate reducts with dynamically adjusted approximation threshold. In: Esposito, F., Pivert, O., Hacid, M.-S., Raś, Z.W., Ferilli, S. (eds.) ISMIS 2015. LNCS (LNAI), vol. 9384, pp. 19–28. Springer, Cham (2015). https://doi.org/10.1007/978-3-319-25252-0_3

14. Kannout, E., Grodzki, M., Grzegorowski, M.: Towards addressing item cold-start problem in collaborative filtering by embedding agglomerative clustering and FP-growth into the recommendation system, vol. 2023 OnLine-First (2023). https://doi.org/10.2298/CSIS221116052K

15. Khan, I.A., et al.: XSRU-IoMT: explainable simple recurrent units for threat detection in internet of medical things networks. Futur. Gener. Comput. Syst. **127**, 181–193 (2022). https://doi.org/10.1016/j.future.2021.09.010

16. Khan, S.A., Naim, I., Kusi-Sarpong, S., Gupta, H., Idrisi, A.R.: A knowledge-based experts' system for evaluation of digital supply chain readiness. Knowl.-Based Syst. **228**, 107262 (2021). https://doi.org/10.1016/j.knosys.2021.107262

17. Kobak, D., Berens, P.: The art of using t-SNE for single-cell transcriptomics. Nat. Commun. **10**, 5416 (2019). https://doi.org/10.1038/s41467-019-13056-x

18. McInnes, L., Healy, J., Melville, J.: UMAP: uniform manifold approximation and projection for dimension reduction (2018). https://doi.org/10.48550/arXiv.1802.03426

19. Nguyen, H.S., Jankowski, A., Peters, J.F., Skowron, A., Stepaniuk, J., Szczuka, M.: Discovery of process models from data and domain knowledge: a rough-granular approach. IGI Glob. (2010). https://doi.org/10.4018/978-1-60566-324-1.ch002

20. Pawlak, Z., Skowron, A.: Rudiments of rough sets. Inf. Sci. **177**(1), 3–27 (2007). https://doi.org/10.1016/j.ins.2006.06.003

21. Pawlak, Z.: Rough sets. Int. J. Comput. Inform. Sci. **11**, 341–356 (1982). https://doi.org/10.1007/BF01001956

22. Penta, A., Pal, A.: What is this cluster about? Explaining textual clusters by extracting relevant keywords. Knowl.-Based Syst. **229**, 107342 (2021). https://doi.org/10.1016/j.knosys.2021.107342

23. Riza, L.S., et al.: Implementing algorithms of rough set theory and fuzzy rough set theory in the R package 'RoughSets'. Inf. Sci. **287**, 68–89 (2014). https://doi.org/10.1016/j.ins.2014.07.029

24. Rudin, C.: Please stop explaining black box models for high stakes decisions. CoRR abs/1811.10154 (2018). arxiv.org/abs/1811.10154

25. Stawicki, S., Ślęzak, D., Janusz, A., Widz, S.: Decision bireducts and decision reducts - a comparison. Int. J. Approx. Reason. **84**, 75–109 (2017). https://doi.org/10.1016/j.ijar.2017.02.007

26. Suraj, Z.: Discovering concurrent process models in data: a rough set approach. In: Sakai, H., Chakraborty, M.K., Hassanien, A.E., Ślęzak, D., Zhu, W. (eds.) RSFDGrC 2009. LNCS (LNAI), vol. 5908, pp. 12–19. Springer, Heidelberg (2009). https://doi.org/10.1007/978-3-642-10646-0_2

27. Tarallo, E., Akabane, G.K., Shimabukuro, C.I., Mello, J., Amancio, D.: Machine learning in predicting demand for fast-moving consumer goods: an exploratory research. IFAC-PapersOnLine **52**(13), 737–742 (2019). https://doi.org/10.1016/j.ifacol.2019.11.203. 9th IFAC Conference on Manufacturing Modelling, Management and Control MIM 2019

28. Xu, X., Lu, Y., Vogel-Heuser, B., Wang, L.: Industry 4.0 and Industry 5.0 - inception, conception and perception. J. Manuf. Syst. **61**, 530–535 (2021). https://doi.org/10.1016/j.jmsy.2021.10.006
29. Zhang, C.X., Zhang, J.S., Yin, Q.Y.: A ranking-based strategy to prune variable selection ensembles. Knowl.-Based Syst. **125**, 13–25 (2017). https://doi.org/10.1016/j.knosys.2017.03.031
30. Zong, W., Chow, Y., Susilo, W.: Interactive three-dimensional visualization of network intrusion detection data for machine learning. Future Gener. Comput. Syst. **102**, 292–306 (2020). https://doi.org/10.1016/j.future.2019.07.045

Decision Rule Clustering—Comparison of the Algorithms

Agnieszka Nowak-Brzezińska[iD] and Igor Gaibei[(✉)][iD]

Institute of Computer Science, Faculty of Science and Technology,
University of Silesia, Bankowa 12, 40-007 Katowice, Poland
{agnieszka.nowak-brzezinska,igor.gaibei}@us.edu.pl

Abstract. In this paper, we present the complexity of decision rule clustering. When the rules are first clustered, then in the inference process we do review only the representatives of the rule clusters. This shortens the inference time significantly, because we search only k rule cluster representatives instead of n rules, where $k << n$. The main goal of the research was to examine the two well-known clustering algorithms: the *K-means* and the *AHC*, in the context of rule-based knowledge representation. We tested different clustering approaches, distance measures, clustering methods, and values for the parameter representing the number of created rule clusters. We studied the clustering time and cluster quality indices. This paper is the first step of a more extensive study. After we have checked which algorithm clustering the rules faster in the knowledge base, we will propose our own version of the inference algorithm for rule clusters, a modification of the classic forward chaining process (on rules). Next, we will carry out experiments that are a continuation of those carried out for this work. These experiments will focus on analyzing the times of the classical inference process and its modification and the efficiency of inference, which will be measured, among others, by the frequency of successful conclusions of inference for both versions of inference algorithms. In this way, we will check whether, by clustering the rules and generating the conclusions on clusters of rules while significantly reducing the reasoning time, we can maintain high efficiency of reasoning.

Keywords: rule clustering · rule-based knowledge base · inference algorithm

1 Introduction

Decision support systems are starting to become our everyday reality and are indeed our future. We are using intelligent applications more and more often to get expert knowledge in a given field. Such intelligent applications support our decisions or imitate our consultation with a domain expert (access to human expert knowledge is difficult and often impossible). Such a solution can guarantee us access to specialist knowledge without restrictions. It is enough to build a

A. Campagner et al. (Eds.): IJCRS 2023, LNAI 14481, pp. 387–401, 2023.
https://doi.org/10.1007/978-3-031-50959-9_27

system that we will equip with inference algorithms (reasoning conducted by an expert who draws conclusions, i.e., derives new knowledge based on the information he has and his knowledge and experience). Such a system must also be equipped with expert knowledge. This knowledge can be provided to us by domain experts or automatically generated by dedicated algorithms that analyze the collected data and look for patterns, the so-called *rules*. Of course, domain knowledge can take different forms, however, rule-based representation of knowledge is the most natural (the form of cause-and-effect chains $IF - THEN$ also known as production rules). The reasoning process conducted by the decision support system represents logical thinking of a human equipped with knowledge and experience. This process is called *inference*. There are two inference algorithms: *forward chaining* (from premises to conclusions) and *backward chaining* (from hypothesis/conclusion to premises). In this work, we have only dealt with the first method. In the literature, it is often called data-driven inference. The forward chaining process is based on activating only those rules whose premises are true (they are facts stored in memory). If more than one rule can be activated at a given moment, we select one of them using appropriate rule selection strategies. Activation leads to adding the conclusion of such a rule to the fact base and blocking this rule before the next activation. When more than one rule be activated, these rule selection strategies can significantly affect inference efficiency. Hence, research in this area is necessary. It is easy to see that if a knowledge base contains many rules (hundreds or thousands), then the time needed to analyze such a knowledge base is long. The more rules the longer inference time. Our idea is to cluster the rules that are similar to each other, hoping that when we divide such a large rule set into clusters of similar rules, we reduce the inference time significantly. Therefore, the aim of this work is a comparative analysis of the classical non-hierarchical algorithm (K-means) with the classical hierarchical algorithm (AHC) in terms of clustering time (i.e., the time of creating a knowledge base with the structure of rule clusters), the quality (cohesion and separation) of the created rule clusters and the inference time. Mostly we are interested in checking the differences in the inference time of the classical inference algorithm (based on single rules) and the proposed rule cluster-based inference algorithm. The proposed approach reduces inference time significantly as only cluster representatives are reviewed instead of every rule in the entire knowledge base. This research only checks the inference time and recency rule selection strategy as the simplest (and the most intuitive). Other strategies will be verified in the next stage of the research.

1.1   The Structure of the Article

In Sect. 2 we present the research results in the given topic from our perspective and knowledge. Section 3 contains both the definition of rule representation and brief introduction to algorithms that allow us to generate such rules automatically from data. It also contains a short description of the *RSES* system, which can be used for such a rule induction process based on the *Rough Set Theory* approach [6]. Clustering algorithms that we used in this research are presented in Sect. 4. The methodology, the data source description and information about the programming environment used in the research are included in Sect. 5. Section 6 contains the selected results in which we may compare the clustering time or other clustering parameters and their impact on the inference process. The research presented in this paper constitutes the first significant part of the research, which in effect, is to allow the implementation of classical inference algorithms and its modification, based on the structure of rule clusters proposed in this work. Then it will be possible to compare the different created rule cluster structures (using different clustering algorithms and different methods of rule cluster representation) and their impact on the efficiency of reasoning. Therefore, in summary (Sect. 7), we evaluate only the first stage of the research and refer to the next stage.

2   State of Art

In the literature you can find many papers on either the comparison of the *K-means* and the *AHC* algorithm, the use of different distance measures or methods of combining clusters, or methods of analyzing the quality of clustering. In [1], the authors discuss and compare clustering algorithms and methods of cluster quality assessment (F-measure, Entropy) for different values of the number of clusters. In [2], the authors compare the clustering times for *AHC* and *K-means*. However, in their research, they do not consider the rule-based knowledge representation in the data as well as the study of the quality of clusters. Comparison of using the *K-means* and the *AHC* algorithm in terms of the number of groups, objects in groups, number of iterations, clustering time for small and large data sets was presented in [3]. However, the aspect of cluster quality research was not included there. In the paper [4] it is presented comparison of dozens of different approaches based on clustering but without research details. Although, it is impossible to find papers that would combine these issues in one study. In the authors' previous research [5] clustering algorithms, outlier detection algorithms and methods for assessing the quality of created clusters were included but never before did the authors merge all the issues in one research.

3   Rules as Knowledge Representation: The Definition and Algorithm Generation

The rule with pure $IF - THEN$ representation, containing only premises and a conclusion, is called the production rule. However, there are various types of rule representations, for example, association rules or certainty factor rules. Rules can be fed directly by a domain expert or induced from the data gathered in a given domain for a certain period of time. The process of knowledge acquisition from a domain expert is time consuming and requires the experts to explain their knowledge and reasoning. Sometimes the knowledge acquired from many experts is inconsistent which, in turn, requires methods of dealing with inconsistent knowledge. In case of rules generated out of the data many different methods can be used to build rule-based knowledge bases. There are well-known algorithms for rule generation from so-called decision tables with the use of the *Rough Set Theory* as well as the algorithms for generating *association rules* and *decision trees*.

3.1   RSES System and LEM2 Algorithm for Rule Induction

In this paper, the authors focus on knowledge representation in the form of *rules* generated automatically from data with the use of *LEM2* algorithm described in [6]. *Rough Set Theory* allows to deal with inconsistency in knowledge gathered in decision tables, to reduce unnecessary attributes and to generate the set of decision rules from original data. For a given data table (also known as a decision table), a set $KB = \{r_1, r_2, \ldots, r_N\}$ of N rules as the $premise_1$ & ... & $premise_m$ → $conclusion$ is generated and for each rule $r_i \in KB$, a numerical value is added depending on its cover within a set of objects. As an example let us take a dataset used for contact lenses fitting, which contains 24 instances, described by 4 nominal attributes[1] and a decision attribute with 3 classes[2].

The piece of the original dataset is as follows:

```
1 1 1 1 1 3
2 1 1 1 2 2
...
24 3 2 2 2 3
```

Using the RSES system with the *LEM2* algorithm implementation the knowledge base with 5 rules has been achieved. The source file of the knowledge base is as follows:

```
RULE_SET lenses
ATTRIBUTES 5
age symbolic
...
contact-lenses symbolic
DECISION_VALUES 3
none
...
RULES 5
(tear-prod-rate=reduced)=>(contact-lenses=none[12]) 12
...
(spectacle-prescrip=myope)&(astigmatism=no)&
(tear-prod-rate=normal) &(age=young)=> (contact-lenses=soft[1]) 1
```

---

[1] Age of the patient: (1) young, (2) pre-presbyopic, (3) presbyopic, spectacle prescription: (1) myope, (2) hypermetrope, astigmatic: (1) no, (2) yes and tear production rate: (1) reduced, (2) normal.

[2] 1 : hard contact lenses, 2: soft contact lenses and 3:no contact lenses. Class distribution is following: 1: 4, 2: 5 and 3: 15.

The rule: (tear-prod-rate=reduced)=> (contact-lenses=none[12]) should be read as: *if* tear-prod-rate=reduced *then* contact-lenses=none which is covered by 12 of instances in the original dataset (50% of instances cover this rule). When the size of input data (the ones that rules are to be generated from) increases, the number of generated rules does too. Let us look at the diabetes data set [7]. It contains the data for 768 objects described with 8 continuous attributes. The objects are divided into two decision classes where 1 means "tested positive for diabetes" and covers 268 objects and 0 means the opposite and covers 500 instances. Processing the data with *LEM2* and *RSES* which contains an implementation of the *LEM2* algorithm, 490 rules have been created. For the nursery dataset which originally contains 12960 instances described with 9 conditional attributes and a decisional attribute for which there are 5 possible values, 867 rules have been generated.

3.2   Rule Clusters

When looking globally at a knowledge base with rules, it turns out that a knowledge base might in theory contain a large number of short rules (with one premise or few), but also some rules described with a large number of premises where only a few premises differ for some clusters of rules. Figure 1 presents an example of rule cluster structure.

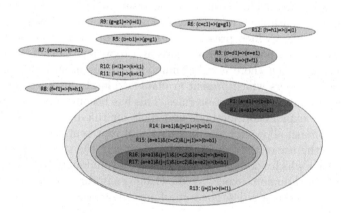

Fig. 1. The example of rule cluster structure.

The example shows a case where seventeen rules in the knowledge base have been divided into many small clusters, including one largest cluster composed of 6 small clusters containing the rules R1, R2, R13...R17, and the remaining small clusters containing one or two rules in them. It should be remembered that the structure of many small clusters requires reviewing each such cluster in the inference process (in fact, the representative of such a cluster is reviewed). If we create fewer clusters but contain more rules inside, then we only review the representatives of these few clusters, which means a shorter analysis time. Domain knowledge determines the distribution of rules in clusters. If we have a lot of rules with very similar premises, then clustering algorithms will tend to create a small number of clusters which in turn contain a large number of similar rules in the middle. Many small clusters should be expected when the knowledge base consists of many unrelated rules. When there are many rules in a cluster, it is more difficult to create coherent group representatives. This will then translate into difficulties finding a cluster where the rules we are potentially looking to activate are located. The process from creating a knowledge base with singular rules, through these rules clustering, to practical reasoning algorithm is, therefore, very complex. For the effectiveness of the inference processes, decision support systems founded on rule-based knowledge representation should be equipped with rule management mechanisms. In other words, they are methods and tools which help effectively review the rules and quickly find those to be activated. One of the possibilities is the clustering of similar rules. In the literature on the subject, this issue has been extensively described and most of the time it focuses on cluster analysis [8].

4 Clustering Algorithms

Clustering is one of the methods which allow to effectively manage huge datasets. Among the available clustering techniques, *non-hierarchical* and *hierarchical* methods can be used. The subject of the research in this paper is representation of specific data such as rule-based knowledge bases. Even though there are numerous papers which present rule representation as decision tables and association rules and provide the methods and tools for their effective management (especially when there is a big number of rules in such sets), so far we have not found papers which present available exploration tools which can deal with big sets of data for production rules. This has become our main motivation for research on exploration methods and tools for rule-based knowledge bases. This new approach allows for the designation of representatives for the created clusters with the use of the generalization approach, the specification approach or an approach which combines both. This would undoubtedly provide a massive support for a domain expert or knowledge engineer who can improve their knowledge on the domain described in a knowledge base.

4.1 Distance Measures

Good-quality clustering requires the created groups to be as internally homogenous and externally distinct as possible. It is essential to use the proper distance or similarity measure in accordance with the given data type (quantitative, qualitative, or binary). Having two rules r_i and r_j in a multidimensional space with the dimension p ($p = 1, 2, \ldots, m$), the distance between these objects can be determined as *Euclidean, Chebyshev* or *Manhattan*. The *Euclidean* distance is defined as $d(r_i, r_j) = \sqrt{\sum_{p=1}^{m}(r_{i_p} - r_{j_p})^2}$, the *Chebyshev* as $d(r_i, r_j) = \max_p(|r_{i_p} - r_{j_p}|)$ and the *Manhattan* as $d(r_i, r_j) = \sum_p |r_{i_p} - r_{j_p}|$.

4.2   Clustering Algorithms: Hierarchical *AHC* vs Partitional *K-means*

Hierarchical clustering is an algorithm that creates a tree of clusters by identifying and merging similar objects (rules in our case). It can be performed in an agglomerate or divisive mode. Agglomerative clustering starts with each observation being its own cluster. They merge into subgroups as we move up the tree. Divisive clustering starts with one cluster of all observations. The cluster is split into subgroups as we move down the tree. The clustering algorithms used are the two most popular cluster analysis algorithms. It is known from the literature that the K-means algorithm should lead to the formation of a certain number of clusters in a short time. However, with no certainty as to the optimal quality of the obtained division. In turn, the hierarchical algorithm should allow the achievement of consistent clusters of objects, but at the cost of long clustering time. Rules are peculiar data, and the effectiveness of clustering them needs to be studied. These classic literature approaches should be translated into rules. In the Agglomerative Hierarchical Clustering (*AHC*), we compute the distance between pairs of rule clusters using the following popular methods: *Single Linkage* (*SL*), *Complete Linkage* (*CL*), *Average Linkage* (*AL*) [8]. The *AHC* algorithm works as follows:

1. In the first step, each rule constitutes a separate cluster. So there are $k = N$ rule clusters, and we must calculate the distance between each pair of rule clusters.
2. Find and join the two most similar rule clusters reducing the number of rule clusters by one.
3. Repeat the second step until obtaining the declared final number of rule clusters (k) or combining all rules into one big cluster.

In each iteration of the *K-means* algorithm, we try to divide N original rules into k rule clusters so well that each rule belongs to the cluster to which it is most similar. The main idea of the algorithm is as follows:

1. Select the number of rule clusters (k) and assign k hypothetical centers.
2. For each rule the nearest cluster center is determined.
3. The rule cluster representative is created - which is an average value for each attribute of a rule cluster representative (conditional and decisional). If there are qualitative attributes (instead od quantitative) then the rule cluster representative vector is a mode value.
4. The center of the rule cluster is shifted to its centroid. Then, the centroid becomes the center of the new rule cluster.
5. The 3rd and 4th steps repeat iteratively.
6. The algorithm ends when no rule cluster changes occur at some iteration.

When analyzing both algorithms, we notice that the *AHC* algorithm seems to be more resistant to outliers in rules. Unfortunately, this algorithm, in turn, requires more memory occupation.

4.3 Clustering Quality Indices

The study of the quality of rule clusters consists in identifying such a partition among many possible partitions of rules into clusters that provides the greatest possible internal consistency of rule clusters (cohesion) and the greatest possible separability of rule clusters between themselves (separation). In this work, the *Dunn* and *Davies-Bouldin* indices have been used to validate the quality of clustering. The *Dunn* Index defines compact rule clusters, the elements (rules) of which are well-grouped together, and the clusters themselves are located as far away from each other as possible. The *Dunn* index attempts to identify those rule clusters partitions that are compact and well separated. The higher its values the better quality. The *Dunn* index for k clusters is defined as $D(u) = \min_{1 \le i \le k}\{\min_{1 \le i \le k, j \ne i}\{\frac{\delta(X_i, X_j)}{\max_{1 \le c \le k}\{\Delta(X_c)\}}\}\}$ where $\delta(X_i, X_j)$ is the inter-cluster distance between rule cluster centroids (representatives) C_i and C_j, and $\Delta(X_c)$ is the intra-cluster distance of cluster X_c. The index itself is sensitive to noise and outliers in rules. The quality of clustering performed using the quantitative and qualitative characteristics of the dataset is shown by the *Davies-Bouldin* index. Since clusters must be compact and well-separated, the lowest possible index value means a high-quality clustering. The *Davies - Bouldin* index for k clusters is defined as $DB(u) = \frac{1}{k}\sum_{i=1}^{k}\max_{i \ne j}\{\frac{\Delta(X_i) + \Delta(X_j)}{(\delta(X_i, X_j))}\}$ for $\Delta(X_i)$ being the average distance between rules within rule cluster X_i ($\Delta(X_j)$ respectively for rule cluster X_j). It determines an average similarity between each individual rule cluster and the rule cluster closest to it. This index attempts to minimize the average distance between each rule cluster and the one most similar to it.

5 Methodology

The experiments proceeded as follows. The source data set is loaded into the *RSES* tool, where decision rules are generated using the *LEM2* algorithm. *LEM2* (Learning from Examples Module, version 2) is the algorithm for rule induction. It uses local coverings, yet is based on global approximations. The *LEM2* uses an idea of blocks of attribute-value pairs and explores the search space of attribute-value pairs. In general, *LEM2* computes a local covering and then converts it into a rule set. These rules are then loaded in the *Python* environment in a customized software that runs the selected clustering algorithm (*K-means* or *AHC*) with the selected clustering parameters (distance measure, number of clusters, clustering method).

5.1 The Course of the Experiments

Seven hundred fifty-six experiments have been carried out for each of the four bases (see Table 1 for the details). Where did the number 756 come from? We perform clustering sequentially for each algorithm (*K-means*, *AHC*) for $k = 2, 3, \ldots, 22$. For each algorithm, we perform clustering using one of the three distance measures, successively *Euclidean*, *Chebyshev*, and *Manhattan*.

This already makes each algorithm run $3 * 21$ times, so 63 times. We repeat each algorithm for three different inputs: the conditions alone, the conclusions alone, and the conditions and conclusions of the rules together. This triples the number of our experiments for each algorithm - $63 * 3 = 189$. We repeat the *AHC* algorithm for three different clustering methods, i.e., *SL*, *CL*, and *AL* $(189 * 3 = 567)$. This means that for each knowledge base (and we used four different knowledge bases), we performed 189 experiments (for *K-means*) and 567 (for *AHC*), so a total of 756. For four knowledge bases, this gives a total of 3024 experiments. The course of the experiments is shown in the Fig. 2.

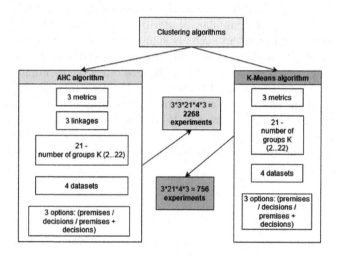

Fig. 2. The course of the experiments.

5.2 Environment

The runtime for the experiments had the following configuration: Spyder compiler with *Python* version 3.9 from the Anaconda platform. The computer parameters on which all experiments were carried out are as follows: Intel Core i5-7500K, 16 Gb RAM. The following libraries were used: *Pandas* for data processing and analysis and *NumPy* for basic operations on n-arrays and matrices. Finally, we used the *RSES* system and the *LEM2* algorithm to generate rules, although we also checked the *exhaustive* algorithm. Unfortunately, for a small number of attributes, this algorithm generated rules at a similar time as the *LEM2*. Still, for larger data sets (number of attributes > 40), after an hour of calculations, not even 1% of the rules were created, and the algorithm crashed.

5.3 Data Description

In the experiments, we included factual knowledge bases with different structures. The structure of these sources is presented in Table 1.

Table 1. Data description

KB	Items	Attributes	Rules	Data source description
kb_1	4435	37	937	Statlog landsat satellite [10]
kb_2	7027	65	4125	Polish companies bankruptcy [11]
kb_3	527	38	123	Water treatment plant [12]
kb_4	17898	9	6432	Pulsar candidates [13]

The number of attributes in the analyzed knowledge bases ranged from 9 to 65. It can be seen that the number of attributes of about 40 (and most of the analyzed knowledge bases had such) is significant and will undoubtedly affect the time of clustering and the process of creating rules. It will also affect the process of creating cluster representatives. The size range of the number of objects subject to rule induction ranged from several hundred to several thousand. In turn, the size of the set of rules created from these source data sets ranged from several hundred to several thousand. If we divide several thousand rules into several groups and, in the inference process, we analyze the similarity of only these groups to the set of facts; then we will certainly be able to shorten the inference time significantly.

6   Experiments

In the experiments, we focused on comparing the clustering algorithms known in the literature, i.e., the *K-means* and the *AHC* algorithms, for knowledge-based rules. It should be emphasized again that rules are complex data structures. It is not apparent that known algorithms that are well adapted to the analyzing typical data structures (tables with numerical data) will adapt equally well to such complex structures as rules.

Table 2 shows a comparison of the *K-means* and the *AHC* algorithm in terms of the number of experiments performed and the average rule clustering time (including the standard deviation together with the minimum and maximum values). The algorithms were compared with the Student's T-test. It can be seen there are statistically significant differences between the algorithms both in terms

Table 2. Comparison of clustering algorithms

Algorithm	# exp	Clustering time [s]
		$Avg \pm std \, [min - max]$
K-means	756	$0.113 \pm 0.138 \, [0.002 - 0.667]$
AHC	2268	$0.282 \pm 0.294 \, [0.001 - 0.943]$
$p = 0.000000$		

of clustering time (the *AHC* clustering time is more than twice as long as compared to the *K-means* algorithm). Table 3 compares three of the most popular methods for clustering according to the *AHC* algorithm only. The comparison is made in terms of the number of experiments performed and the average rule clustering time (including the standard deviation, as well as the minimum and maximum values) It was expected (research conducted earlier in the literature on the subject was confirmed) that the *SL* method would be performed in the shortest time and the *CL* method is the longest. These are statistically significant differences. We also compared the distance measures used in the rule clustering process. It can be seen, in Table 4, that using three different distance measures does not provide statistically significant differences in the clustering time. In our research, we were also interested in examining the correlation between the size of the data and the time of clustering or the values of cluster quality indices. The results are included in Table 5. We expected a positive correlation between the input dataset size and the clustering time. The more objects in the data set or the more attributes in the data set, the longer the clustering time, but it can

Table 3. Comparison of clustering methods (for *AHC* algorithm)

Clustering method	Clustering time [s]
	$Avg \pm std\,[min - max]$
SL	$0.207 \pm 0.218\,[0.001 - 0.608]$
CL	$0.325 \pm 0.325\,[0.001 - 0.913]$
AL	$0.315 \pm 0.311\,[0.001 - 0.943]$
$p = 0.000000$	

Table 4. Comparison of distance measures (for *AHC* algorithm)

Distance measure	Clustering time [s]
	$Avg \pm std\,[min - max]$
Euclidean	$0.251 \pm 0.286\,[0.0013 - 0.943]$
Chebyshev	$0.235 \pm 0.265\,[0.0013 - 0.778]$
Manhattan	$0.234 \pm 0.269\,[0.0013 - 0.833]$
$p > 0.05$	

Table 5. Analysis of the correlation between the size of the data and the parameters of the clustering efficiency

	Clustering time [s]	Dunn index	Davies-Bouldin index
# items	$r = 0.5598^*$	$r = -0.0021; p > 0.05$	$r = -0.3473^*$
# Attr	$r = 0.1018^*$	$r = 0.0017; p > 0.05$	$r = 0.2432^*$
# Rules	$r = 0.7205^*$	$r = -0.0018; p > 0.05$	$r = -0.3670^*$

be noticed that this correlation is significant for the number of objects subject to rule creation or the number of rules (∗ signifies a statistically significant correlation, i.e. p < 0.05.). The correlation between the number of attributes and the clustering time is not that significant (of course, it is positive, i.e. the more attributes, the longer the rule clustering time). The most important part of our recent research was analyzing the inference processes and assessing the results in the context of inference time and success. We were interested in answering the following questions: which clustering algorithm allows us to finish the inference process in the shortest time, or which clustering methods or distance measure results in finishing the inference process in the shortest time? The other aspect is assessing which clustering parameters (algorithm, clustering method, or distance measure) result in successfully finishing the inference process. Inference success means that during the inference process, at least one rule was fired (activated) and its conclusion was added to the database of new inferred facts. If no rule could be fired during the inference process, it means that the inference failed. Firing a rule means finding such a rule, in which the premises (conditions) cover the input facts. Fact coverage is checked by calculating the similarity of facts to rules in the rule cluster. The results are included in Tables 6, 7, 8, and 9. There is a statistically significant difference (based on the Chi-square Test) between the *K-means* and the *AHC* algorithm. It can be seen that in the case of the *AHC* algorithm, the inference process is more often successful than in the case of the *K-means* algorithm. Our explanation for this situation is as follows: clustering the rules using the *AHC* algorithm creates a more consistent structure of rule clusters, making the inference process more likely to succeed. We also compared clustering methods like *SL*, *CL*, and *AL* methods in the case of the *AHC* algorithm. We may see in Table 7 that there is a statistically significant difference between these three methods in the context of finishing the inference process successfully. The *SL* method allows us to finish the inference process successfully more often than the two other methods. The explanation is rela-

Table 6. The comparison of clustering algorithms in the context of a successfully finished inference process

Algorithm	Success frequency	Failure frequency	p value
K-means	21.83%	78.17%	p = 0.00002
AHC	29.89%	70.11%	

Table 7. The comparison of clustering methods in the context of a successfully finished inference process (for *AHC* algorithm)

Clustering method	Success frequency	Failure frequency	p value
SL	35.85%	64.15%	p = 0.00001
CL	24.47%	75.53%	
AL	29.37%	70.63%	

Table 8. The comparison of distance measures in the context of a successfully finished inference process (for AHC algorithm)

Clustering method	Success frequency	Failure frequency	p value
Euclidean	26.59%	73.41%	p = 0.02768
Chebyshev	30.95%	69.05%	
Manhattan	26.09%	73.91%	

Table 9. The comparison of clustering algorithms in the context of a successfully finished inference process - separately for each knowledge base

KB	Algorithm	Success	False	p value
kb_1	*K-means*	10.58%	89.42%	p > 0.05
	AHC	10.23%	89.77%	
kb_2	*K-means*	70.90%	29.10%	p = 0.00000
	AHC	47.44%	52.56%	
kb_3	*K-means*	0%	100%	—
	AHC	0%	100%	
kb_4	*K-means*	47.62%	52.38%	p = 0.02833
	AHC	56.79%	43.21%	

tively simple. The *SL* method builds a chaining structure, meaning that most nodes are connected together. Therefore, we will probably not omit the relevant rule cluster in the inference process. The distance measure used in the clustering process also influences the result of the inference process. We examined three distance measures *Euclidean, Chebyshev* and *Manhattan* for *AHC* algorithm in the context of finishing the inference process successfully. Table 8 presents the result of a comparison of those three measures. The *Chebyshev* measure allows us to finish the inference process successfully, the most often (it is a statistically significant difference). The most interesting results are presented in Table 9. We see the comparison of clustering algorithms in the context of a successfully finished inference process - separately for each knowledge base. We examined an inference process for each of the analyzed knowledge bases. We may see that a lot depends on the given knowledge base. There is a knowledge base (kb_3) for which every inference process finished unsuccessfully, no matter which clustering algorithm we chose and which other parameters. Probably the structure of this specific knowledge base creates a complicated rule cluster structure that causes problems during the inference process to find a relevant rule cluster and fire a proper decision rule. However, there is also a knowledge base (kb_2) in which the inference process finishes successfully in more than 70% of all experiments. It proves that, undoubtedly, we need additional research in this area. We need to find a way to create the optimal rule cluster structure for every knowledge base and form their optimal representatives. Only then may we expect a successfully finished inference process.

7 Summary

In this paper, we have attempted to present the complexity of clustering decision rules. We group rules to shorten the reasoning process carried out by the decision support system. Such a system, theoretically designed for rules in the knowledge base, would have to go through each rule, to select which rules to activate. When the rules are first grouped and the clusters created for them together with the representatives, then in the inference process, we will review not the rules but the representatives of the rule clusters. This will significantly shorten the inference time because usually $k << n$, where k is the number of rule clusters and n is the number of rules. The course of the experiments was as follows. We generated rule-based knowledge bases with the $IF - THEN$ structure for four entire data sets, from different areas of life, with different structures and sizes of input data. Then, the rules created this way were grouped using two clustering algorithms known in the literature, i.e. *K-means* and *AHC*. We tested different distance measures, different clustering methods (for the *AHC* algorithm), and different values for the parameter representing the number of rule groups created (from 2 to 22). We studied the clustering time and cluster quality indices, the *Dunn* and the *Davies-Bouldin* indexes (known in the literature). The results showed a statistically significantly twice as long grouping time for the *AHC* compared to the *K-means*. We also showed a statistically significant correlation between the size of the input data and the clustering time. The research also confirmed that combining clusters based on considering the nearest neighbor allows for a statistically significantly shorter clustering time than the other two methods, i.e. average or complete. There was no statistically significant difference in the clustering time between the distance measures used for clustering (*Euclidean*, *Chebyshev*, or *Manhattan*). This paper is the first step of a more extensive study. We have checked which algorithm clustering the rules faster in the knowledge base, we proposed our own version of the inference algorithm for rule clusters, a modification of the classic forward chaining process (on rules). We have carried out experiments which focus on analyzing the time of the inference process based on rule cluster structure. We also analyze the efficiency of inference, measured by the frequency of successful conclusions of inference algorithm. In this way, we can check whether, by clustering the rules and generating the conclusions on clusters of rules while significantly reducing the reasoning time, we can maintain high efficiency of reasoning.

References

1. Steinbach, M.S., Karypis, G., Kumar, V.: A Comparison of Document Clustering Techniques. Department of Computer Science and Engineering, Computer Science (2000)
2. Karthikeyan, B., George, D.J., Manikandan, G., Thomas, T.: A comparative study on k-means clustering and agglomerative hierarchical clustering. Int. J. Emerg. Trends Eng. Res. 8(5) (2020). https://doi.org/10.30534/ijeter/2020/20852020

3. Saleena, T.S., Sathish, A.J., Joseph, A.: Comparison of k-means algorithm and hierarchical algorithm using Weka tool. Int. J. Adv. Res. Comput. Commun. Eng. IJARCCE **7**(7), 74–79 (2018)
4. Jabbar, A.M.: Local and global outlier detection algorithms in unsupervised approach: a review. Iraqi J. Electr. Electron. Eng. Coll. Eng. **17**(1), 1–12 (2021)
5. Nowak-Brzezińska, A., Horyń, C.: Outliers in rules - the comparision of LOF, COF and K-means algorithms. Procedia Comput. Sci. **176**, 1420–1429 (2020)
6. Grzymała-Busse, J.W.: A comparison of rule induction using feature selection and the LEM2 algorithm. In: Stańczyk, U., Jain, L.C. (eds.) Feature Selection for Data and Pattern Recognition. SCI, vol. 584, pp. 163–176. Springer, Heidelberg (2015). https://doi.org/10.1007/978-3-662-45620-0_8
7. Machine Learning Repository. www.archive.ics.uci.edu/ml/datasets/diabetes.com. Accessed Dec 2022
8. Kaufman, L., Rousseeuw, P.J.: Finding Groups in Data: An Introduction to Cluster Analysis. Wiley, Hoboken (2005)
9. Legany, C., Juhasz, S., Babos, A.: Cluster validity measurement techniques. In: Knowledge Engineering and Data Bases, WSEAS, USA, pp. 388–393 (2006)
10. Machine Learning Repository. www.archive-beta.ics.uci.edu/dataset/146/statlog+landsat+satellite. Accessed Dec 2022
11. Machine Learning Repository. www.archive-beta.ics.uci.edu/dataset/365/polish+companies+bankruptcy+data. Accessed Dec 2022
12. Machine Learning Repository. www.archive-beta.ics.uci.edu/dataset/106/water+treatment+plant. Accessed Dec 2022
13. Machine Learning Repository. www.archive-beta.ics.uci.edu/dataset/372/htru2. Accessed Dec 2022

Classifying Token Frequencies Using Angular Minkowski p-Distance

Oliver Urs Lenz[✉] and Chris Cornelis

Research Group for Computational Web Intelligence, Department of Applied
Mathematics, Computer Science and Statistics, Ghent University, Ghent, Belgium
{oliver.lenz,chris.cornelis}@ugent.be

Abstract. Angular Minkowski p-distance is a dissimilarity measure that
is obtained by replacing Euclidean distance in the definition of cosine
dissimilarity with other Minkowski p-distances. Cosine dissimilarity is
frequently used with datasets containing token frequencies, and angular
Minkowski p-distance may potentially be an even better choice for certain tasks. In a case study based on the *20-newsgroups* dataset, we evaluate classification performance for classical weighted nearest neighbours,
as well as fuzzy rough nearest neighbours. In addition, we analyse the
relationship between the hyperparameter p, the dimensionality m of the
dataset, the number of neighbours k, the choice of weights and the choice
of classifier. We conclude that it is possible to obtain substantially higher
classification performance with angular Minkowski p-distance with suitable values for p than with classical cosine dissimilarity.

Keywords: Cosine dissimilarity · Fuzzy rough sets · Minkowski
distance · Nearest Neighbours

1 Introduction

Cosine (dis)similarity [12,13] is a popular measure for data that can be charac-
terised by a collection of token frequencies, such as texts, because it only takes
into account the relative frequency of each token. Cosine dissimilarity is particularly relevant for distance-based algorithms like classical (weighted) nearest
neighbours (NN) and fuzzy rough nearest neighbours (FRNN). In the latter case,
cosine dissimilarity has been used to detect emotions, hate speech and irony in
tweets [9].

A common way to calculate cosine dissimilarity is to normalise each record
(consisting of a number of frequencies) by dividing it by its Euclidean norm, and
then considering the squared Euclidean distance between normalised records.
Euclidean distance can be seen as a special case of a larger family of Minkowski
p-distances (namely the case $p = 2$). It has previously been argued that in
high-dimensional spaces, classification performance can be improved by using
Minkowski p-distance with fractional values for p between 0 and 1 [1].

A. Campagner et al. (Eds.): IJCRS 2023, LNAI 14481, pp. 402–413, 2023.
https://doi.org/10.1007/978-3-031-50959-9_28

In light of this, we propose *angular Minkowski p-distance*: a natural generalisation of cosine dissimilarity obtained by substituting other Minkowski p-distances into its definition. The present paper is a case study of angular Minkowski p-distance using the well-known *20-newsgroups* classification dataset. In particular, we investigate the relationship between the hyperparameter p, the dimensionality m, the number of neighbours k, and the choice of classification algorithm and weights.

To the best of our knowledge, this topic has only been touched upon once before in the literature. Unlike the present paper, the authors of [5] do not evaluate classification performance directly, but rather the more abstract notion of 'neighbourhood homogeneity', and they only consider a limited number of values for p and m.

The remainder of this paper is organised as follows. In Sect. 2, we motivate and define angular Minkowski p-distance. In Sect. 3, we recall the definitions of NN and FRNN classification. Then, in Sect. 4, we describe our experiment, and in Sect. 5 we present and analyse our results, before concluding in Sect. 6.

2 Angular Minkowski p-Distance

In this section, we will work in a general m-dimensional real vector space $\mathbb{R}^m$, for some $m \in \mathbb{N}$.

The cosine similarity between any two points $x, y \in \mathbb{R}^m$ is defined as the cosine of the angle θ between x and y. We obtain the cosine dissimilarity by subtracting the cosine similarity from 1. Defined thus, cosine similarity and dissimilarity take values in, respectively, $[-1, 1]$ and $[0, 2]$. However, when all records are located in $\mathbb{R}^m_{\geq 0}$, such as token frequencies, both measures take values in $[0, 1]$.

It is a well-known fact that cosine dissimilarity is proportional to the squared Euclidean distance between x and y once these points have been normalised by their Euclidean norm (note that $\cdot$ denotes the vector in-product):

$$
\begin{aligned}
1 - \cos\theta &= 1 - \frac{x \cdot y}{|x|\,|y|} \\
&= 1 - \frac{x}{|x|} \cdot \frac{y}{|y|} \\
&= \frac{1}{2}\left|\frac{x}{|x|}\right| + \frac{1}{2}\left|\frac{y}{|y|}\right| - \frac{x}{|x|} \cdot \frac{y}{|y|} \\
&= \frac{1}{2}\left(\frac{x}{|x|} \cdot \frac{x}{|x|} + \frac{y}{|y|} \cdot \frac{y}{|y|} - 2\frac{x}{|x|} \cdot \frac{y}{|y|}\right) \\
&= \frac{1}{2}\left(\frac{x}{|x|} - \frac{y}{|y|}\right)^2
\end{aligned}
\tag{1}
$$

The Euclidean norm is the special case $p = 2$ of the more general Minkowski p-size, defined for any $x \in \mathbb{R}^m$ as:

$$
|x|_p = \left(\sum |x_i^p|\right)^{\frac{1}{p}},
\tag{2}
$$

where p is allowed to be any positive real number. Note that this is only a norm for $p \geq 1$. The Minkowski p-distance between any two $x, y \in \mathbb{R}^m$ is defined as the p-size of their difference $|y - x|_p$. This is a metric if $p \geq 1$.

Similarly, we can also view the squared Euclidean norm (distance) as the special case $p = 2$ of the *rootless* Minkowski p-size (distance), defined for any $x \in \mathbb{R}^m$ as:

$$|x|_p^p = \sum |x_i^p|, \tag{3}$$

The rootless p-size is not a norm for any p (other than $p = 1$, for which it coincides with the ordinary 1-norm); rootless p-distance is a metric for $p \leq 1$.

With these definitions in place, we can define the angular Minkowski p-distance between any two vectors $x, y \in \mathbb{R}^m$ as:

$$\left| \frac{y}{|y|_p} - \frac{x}{|x|_p} \right|_p . \tag{4}$$

as well as their rootless angular Minkowski p-distance:

$$\left| \frac{y}{|y|_p} - \frac{x}{|x|_p} \right|_p^p . \tag{5}$$

Thus, cosine dissimilarity corresponds to rootless angular Minkowski 2-distance, and we can consider angular Minkowski p-distance with different values for p as alternatives to cosine dissimilarity.

3   Classical and Fuzzy Rough Nearest Neighbour Classification

We will now briefly review the definition of classical weighted nearest neighbour (NN) classification [2–4] and fuzzy rough nearest neighbour classification (FRNN) [7,10]. Both approaches require a choice of a dissimilarity measure, weights, and a positive integer k determining the number of nearest neighbours to be considered. In what follows, we will specify the class prediction that each method makes for a test instance y, given a training set X and a decision class $C \subseteq X$.

3.1   Nearest Neighbour Classification

For NN, let x_i be the ith nearest neighbour of y in X. Then the class score for C is given by:

$$\sum_{i \leq k | x_i \in C} w_i \bigg/ \sum_{i \leq k} w_i \tag{6}$$

where w_i is the weight attributed to the ith nearest neighbour of y. Two popular choices [2,3] for the weights are linear distance weights:

$$w_i = \begin{cases} \dfrac{d_k - d_i}{d_k - d_1} & k > 1; \\ 1 & k = 1, \end{cases}$$
(7)

and reciprocally linear distance weights:

$$w_i = \frac{1}{d_i},$$
(8)

where d_i is the distance between y and x_i.

3.2 Fuzzy Rough Nearest Neighbour Classification

Properly speaking, FRNN consists of two different classifiers, the upper and the lower approximation, which can be combined to form the mean approximation. For the upper approximation, let d_i be the distance between y and its ith nearest neighbour in C. Then the class score for C is given by:

$$\overline{C}(y) = \sum_{i \leq k} w_i \cdot \min(0, 1 - d_i/2).$$
(9)

For the lower approximation, let d_i be the distance between y and its ith nearest neighbour in $X \backslash C$. Then the class score for C is given by:

$$\underline{C}(y) = \sum_{i \leq k} w_i \cdot \max(d_i/2, 1).$$
(10)

For the mean approximation, the class score for C is given by:

$$\left(\overline{C}(y) + \underline{C}(y) \right) / 2.$$
(11)

In the definition of both the upper and the lower approximation, $\langle w_i \rangle_{i \leq k}$ is a weight vector of values in $[0,1]$ that sum to 1. As with NN, two popular weight choices are linear weights:

$$w_i = \frac{2(k + 1 - i)}{k(k + 1)},$$
(12)

and reciprocally linear weights:

$$w_i = \frac{1}{i \cdot \sum_{i \leq k} \frac{1}{i}}.$$
(13)

4   Experimental Setup

To evaluate angular Minkowski p-distance, we conduct a case study on the well known text dataset *20-newsgroups* [8]. Originally, this contained 20 000 usenet posts from 20 different newsgroups (1000 each) from the period February-May 1993, and was collected by Ken Lang. We use the version of this dataset provided by the Python machine learning library *scikit-learn* [11], which comprises a training set (11 314 records) and a test set (7532 records, consisting of later posts than those in the training set), preprocessed to remove headers, footers and quotes.

We first convert each text into a set of words, defined as any sequence of at least two alphanumeric characters separated by non-alphanumeric characters, regardless of case. Next, we count the word frequencies per text and transform this into an m-dimensional dataset by selecting the top-m overall most frequent words, and discarding the rest.

In order to evaluate the behaviour of NN and FRNN with angular Minkowski p-distance, we systematically vary different values for p, m as well as the number of nearest neighbours k. In the case of FRNN, we consider the upper, lower and mean approximations separately. For both NN and FRNN, we will consider linear and reciprocally linear weights, as described in Sect. 3.

For p, we consider all multiples of 0.1 in the range of $[0.1, 4]$, centred on the canonical values of 1 and 2. Since k and m encode magnitudes, we investigate them on a logarithmic scale, with values corresponding to powers of 2 in the range of, respectively, $[1, 256]$ and $[2, 4096]$.

We measure classification performance using the area under the receiver operator characteristic (AUROC) [6].

5   Results

Figures 1 and 2 display AUROC as a function of dimensionality (the number of most frequent tokens taken into consideration) and as a function of p, for $k = 256$. There are a few things to be noted from these response curves:

- The choice of weights doesn't appear to play a role in the overall behaviour of these response curves.
- The response curves are substantially smoother for the upper approximation than for the lower approximation and for NN. The mean approximation appears to inherit some of this smoothness from the upper approximation. This qualitative difference is somewhat surprising, but it can perhaps be explained by the fact that for the upper approximation, neighbours are drawn from a uniform concept (each decision class), whereas for the lower approximation and NN, neighbours are drawn from across decision classes.
- The upper approximation is a better classifier (in terms of AUROC) than the lower approximation and NN for the *20-newsgroups* dataset. Given the relatively poor performance of the lower approximation, it is surprising that the mean approximation produces even better results than the upper approximation.

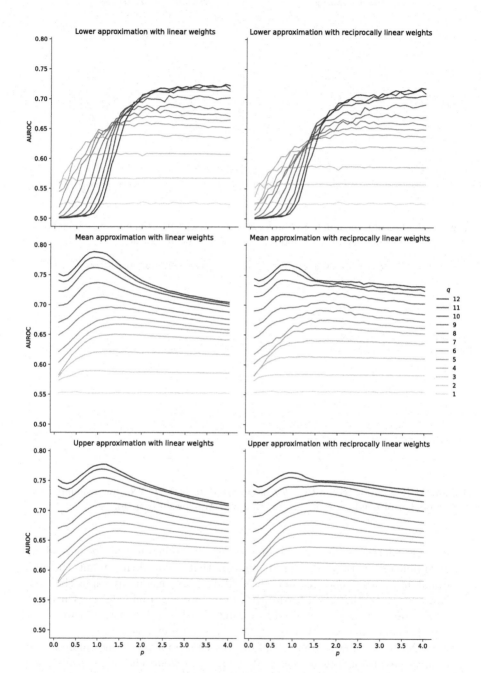

Fig. 1. AUROC obtained on the *20-newsgroups* dataset with FRNN, number of neighbours $k = 256$, dimensionality $m = 2^q$ and angular Minkowski p-distance.

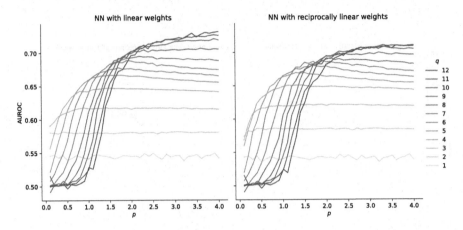

Fig. 2. AUROC obtained on the *20-newsgroups* dataset with NN, number of neighbours $k = 256$, dimensionality $m = 2^q$ and angular Minkowski p-distance.

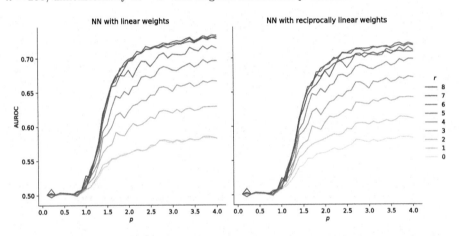

Fig. 3. AUROC obtained on the *20-newsgroups* dataset with NN, dimensionality $m = 4096$, number of neighbours $k = 2^r$ and angular Minkowski p-distance.

- AUROC increases with dimensionality, but the difference between 2048 and 4096 dimensions is quite small. It appears that up until that point, the additional information encoded in each additional dimension outweighs the noise. Note, however, that even before that point, we get diminishing returns. For each subsequent curve we need to double the dimensionality, and we obtain a performance increase that is smaller than the previous one.
- For NN and the lower approximation, the choice for p becomes more important as dimensionality increases. Not only is a good choice for p necessary to make use of the potential performance increase from adding more dimensions, choosing p poorly can actually cause performance to decrease with dimensionality.

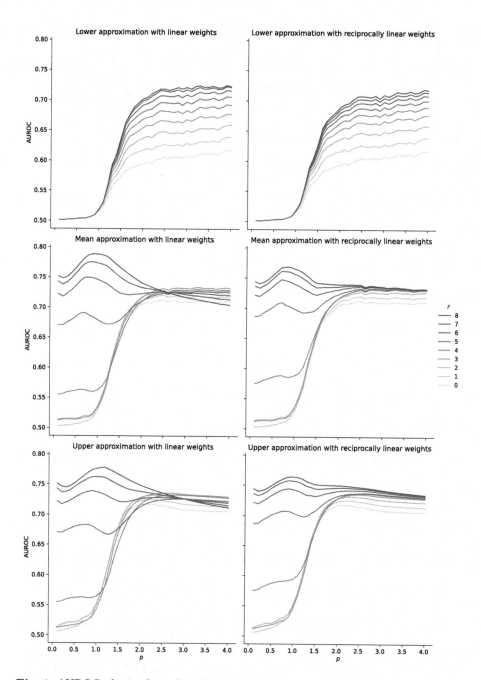

Fig. 4. AUROC obtained on the *20-newsgroups* dataset with FRNN, dimensionality $m = 4096$, number of neighbours $k = 2^r$ and angular Minkowski p-distance.

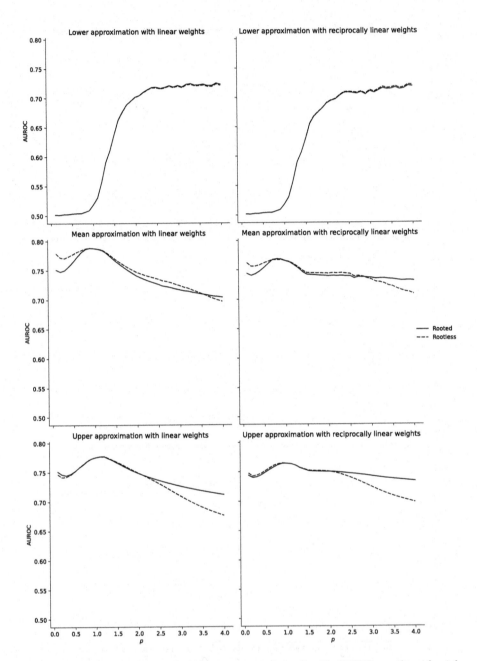

Fig. 5. AUROC obtained on the *20-newsgroups* dataset with FRNN, number of neighbours $k = 256$, dimensionality $m = 4096$ and rooted and rootless angular Minkowski p-distance.

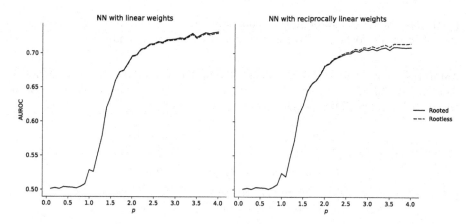

Fig. 6. AUROC obtained on the *20-newsgroups* dataset with NN, number of neighbours $k = 256$, dimensionality $m = 4096$ and rooted and rootless angular Minkowski p-distance.

- There is a marked difference with respect to the optimal values for p between the different classifiers. For NN and the lower approximation, higher values appear to be better within the range $[0.1, 4]$ that we have investigated, albeit with diminishing returns. For the upper and mean approximations, the optimum is located near $p = 1$ for high dimensionalities.

As mentioned above, Figs. 1 and 2 reflect a choice of the number of neighbours $k = 256$. The effect of k on performance is illustrated in Figs. 3 and 4, for $m = 4096$.

- For NN and the lower approximation, the overall behaviour of the response curve does not change with k. Higher values for k lead to higher AUROC, and within the range of investigated values, the relationship appears to be similar to the relationship between AUROC and m: each doubling of k leads to an increase in AUROC that is slightly smaller than the previous increase. From $k = 128$ to $k = 256$, the increase is already quite small.
- In contrast, for the upper and mean approximations, AUROC starts out quite high for high values of p, and increases only little thereafter. However, from $k = 8$ upwards, AUROC starts to strongly increase for lower values of p, eventually surpassing the AUROC obtained with higher values of p from $k = 64$ upwards. This means that the good performance of the mean and upper approximations around $p = 1$ is only realised for high values of k.

Finally, we may also ask whether it makes a difference whether we use rooted ('ordinary') or rootless angular Minkowski p-distance. The results discussed above were obtained using rooted angular Minkowski p-distance. It turns out that using rootless angular Minkowski p-distance, which generalises cosine dissimilarity more closely, does not make much difference (Figs. 5 and 6). In particular, there is (by definition) no difference for $p = 1$, which maximises classification performance for the upper and mean approximations.

Table 1. Highest AUROC and corresponding value for p obtained on the *20-newsgroups* dataset, with linear weights, number of neighbours $k = 256$, dimensionality $m = 4096$ and rooted angular Minkowski p-distance.

Classifier	p	AUROC
NN	4.0	0.731
FRNN (lower approximation)	3.9	0.725
FRNN (mean approximation)	0.9	0.788
FRNN (upper approximation)	1.1	0.777

In summary (Table 1), we obtain the best classification performance on the *20-newsgroups* dataset with the upper and mean approximation and angular Minkowski p-distance with values of p around 1, but only when k is high enough (≥ 64).

6 Conclusion

We have presented angular Minkowski p-distance, a generalisation of the popular cosine (dis)similarity measure. In an exploratory case study of the large *20-newsgroups* text dataset, we showed that the choice of p can have a large effect on classification performance, and in particular that the right choice of p can increase classification performance over cosine dissimilarity (which corresponds to $p = 2$).

We have also examined the interaction between p and the dimensionality m of a dataset, the choice of classification algorithm (NN or FRNN), the choice of weights (linear or reciprocally linear), and the choice of the number of neighbours k. We found that while the choice of weights was not important, the best value for p can depend on m, k and the classification algorithm. Under optimal circumstances (high k and high m), the best-performing values for p are in the neighbourhood of 1 (FRNN with upper or mean approximation) and around 4 (NN and FRNN with lower approximation).

A major advantage of angular Minkowski p-distance is that it is defined in terms of ordinary Minkowski p-distance, which is widely available. Thus, angular Minkowski p-distance does not require any dedicated implementation and can easily be used in experiments by other researchers.

The most important open question to be investigated in future experiments is to which extent these results generalise to other text datasets, as well as to other datasets containing token frequencies. Depending on the outcome of these experiments, it may be possible to formulate more general conclusions about the best choice for p, or we may be forced to conclude that this is a hyperparameter that must be optimised for each individual dataset.

Acknowledgements. The research reported in this paper was conducted with the financial support of the Odysseus programme of the Research Foundation – Flanders (FWO).

References

1. Aggarwal, C.C., Hinneburg, A., Keim, D.A.: On the surprising behavior of distance metrics in high dimensional space. In: Van den Bussche, J., Vianu, V. (eds.) ICDT 2001. LNCS, vol. 1973, pp. 420–434. Springer, Heidelberg (2001). https://doi.org/10.1007/3-540-44503-x_27
2. Dudani, S.A.: An experimental study of moment methods for automatic identification of three-dimensional objects from television images. Ph.D. thesis, The Ohio State University (1973)
3. Dudani, S.A.: The distance-weighted k-nearest-neighbor rule. IEEE Trans. Syst. Man Cybern. **6**(4), 325–327 (1976)
4. Fix, E., Hodges, Jr, J.: Discriminatory analysis — nonparametric discrimination: Consistency properties. Technical report 21-49-004, USAF School of Aviation Medicine, Randolph Field, Texas (1951). https://apps.dtic.mil/sti/citations/ADA800276
5. France, S.L., Carroll, J.D., Xiong, H.: Distance metrics for high dimensional nearest neighborhood recovery: compression and normalization. Inf. Sci. **184**(1), 92–110 (2012)
6. Hand, D.J., Till, R.J.: A simple generalisation of the area under the ROC curve for multiple class classification problems. Mach. Learn. **45**(2), 171–186 (2001)
7. Jensen, R., Cornelis, C.: A new approach to fuzzy-rough nearest neighbour classification. In: Chan, C.-C., Grzymala-Busse, J.W., Ziarko, W.P. (eds.) RSCTC 2008. LNCS (LNAI), vol. 5306, pp. 310–319. Springer, Heidelberg (2008). https://doi.org/10.1007/978-3-540-88425-5_32
8. Joachims, T.: A probabilistic analysis of the Rocchio algorithm with TFIDF for text categorization. Technical report CMS-CS-96-118, Carnegie Mellon University, School of Computer Science, Pittsburgh (1996)
9. Kaminska, O., Cornelis, C., Hoste, V.: Fuzzy rough nearest neighbour methods for detecting emotions, hate speech and irony. Inf. Sci. **625**, 521–535 (2023)
10. Lenz, O.U.: Fuzzy rough nearest neighbour classification on real-life datasets. Doctoral thesis, Universiteit Gent (2023)
11. Pedregosa, F., et al.: Scikit-learn: machine learning in python. J. Mach. Learn. Res. **12**(85), 2825–2830 (2011)
12. Rosner, B.S.: A new scaling technique for absolute judgments. Psychometrika **21**(4), 377–381 (1956)
13. Salton, G.: Some experiments in the generation of word and document associations. In: Proceedings of the 1962 Fall Joint Computer Conference. AFIPS Conference Proceedings, vol. 22, pp. 234–250. Spartan Books (1962)

On kNN Class Weights for Optimising
G-Mean and F1-Score

Grzegorz Góra[1]($\boxtimes$) and Andrzej Skowron[2]

[1] University of Warsaw, Stefana Banacha 2, 02-097 Warszawa, Poland
`ggora@mimuw.edu.pl`
[2] Systems Research Institute PAS, Newelska 6, 01-447 Warszawa, Poland
`skowron@mimuw.edu.pl`

Abstract. We present two novel theorems that allow for estimating the weight parameter in (weighted) kNN while dealing with imbalanced data. More precisely, the theorems for G-mean and F1-score are presented. The theorems assume 'totally random' distribution i.e. lack of dependency between features and class value.

These results can be used for setting the default weights of classes for kNN-type classifiers, e.g. for imbalanced learning problems. Moreover, these theorems taken together illustrate the fact that without a precise specification of the particular performance measure we are interested in, the 'best classifier' term can be ambiguous or even misleading.

Keywords: k Nearest Neighbours · Classification · Supervised Learning · Imbalanced Data · Performance Measures · G-mean · F1-score · Borderline Examples

1 Introduction

Machine learning (ML) algorithms construct from training sets *classifiers* that provide decisions for test objects. Well known ML methods are kNN algorithms. Recently, much scientific effort has been put into supervised learning that concerns learning from so-called *imbalanced data* [7]. For the classification task with a binary decision, which we focus on in the paper, there is just one class of special importance. It is referred to as the *minority class* and the other one as the *majority class*.

For such data sets, kNN may require using different weights for the objects from minority class and majority class (see e.g. [3]). There exist also other approaches of using weights for kNN approach (see e.g. [14]). In paper, we consider only the case of classes weights for kNN (for binary classification task) with weights being searched globally. Optimal weight might be found (according to the chosen performance measure) during learning phase. It may be interesting, even for artificially defined data sets, to find optimal weights for chosen performance measures. Below, we explain why considering 'totally random' distribution can be interesting from the practical point of view, i.e. can help to understand the problems that can arise in the construction of classifiers for real data sets.

Generally instances can be categorised into 3 main categories: safe examples, borderline examples, outlier examples [11]. In this paper, we exclude outlier examples from our considerations as another problem. In this case, one can divide the space of instances into two main regions: safe regions and borderline regions. It is known [12] that the problem is most difficult in borderline regions. Essentially, this is where the quality of the classifiers makes some win out over others. These borderline regions are worth focusing on the most (for both balanced and imbalanced data). Let us focus on just one such region in the data set.

In particular, in the paper we consider 'totally random' data set, i.e. where there is no dependence of objects descriptions and its decision. Such data sets may occur in practice, at least as a part of real-life data sets, e.g. as borderline regions [11]. Of course, borderline regions are not usually 'totally random'. However, for some data sets may occur some borderline regions 'totally random' or at least to some extent random.

Even for the last case the problem presented in the paper holds (to some extent). We want to show that it may happen (on real-life data sets) that one classifier wins with another not because it is 'better' but only due to the performance measure we assess them with. On the other hand the provided results can be a base in searching for optimal weight parameter in case when borderline regions are close to 'totally random'.

Thus, it is interesting to find in this case optimal weights for *class based weighted kNN* for the chosen performance measures. We focus on two widely used performance measures for imbalanced data, namely G-mean and F1-score[1].

It should be noted that the essence of the problem we consider is not based on the assumption that the distribution of data is known. Certainly, if a priori it is known that this distribution is random one can use the approach based on MLE (Maximum Likelihood Estimation) or MAP (Maximum A Posteriori). We consider the situation when one is trying to solve a real-life problem with unknown distribution. What if by chance this distribution (of part of data set) is close to random distribution? We try to highlight the problem which can occur in real-life classification problems which a researcher should take into account.

The considered issue relates to the problem of optimisation and evaluation using different performance measures, in particular for imbalanced data (see e.g. [13]). One can also treat the paper as a step toward answering the following question: What do learning systems learn when they are trained with random labels (see e.g. [9])? The issue also somehow relates to the well-known so-called 'no free lunch theorem' (see [15]).

Presented in the paper weighted kNN is not a new method. Rather it can be starting point for new methods (and in fact it was; see e.g. [4,5]). The presented theorems (and experiments) may accelerate the process of searching for the optimal values of weights by starting the process from the default values

---

[1] It corresponds to the case F_β-*measure* when $\beta = 1$.

Algorithm 1: class based weighted kNN

Input: test example tst, training set $trnSet$, integer $k > 0$, real $p \in [0, 1]$, metric ϱ

$neighbourSet$ = the set of k training examples that are most similar to tst
according to metric ϱ

$supportSet(d_{min}) = \emptyset$; $supportSet(d_{maj}) = \emptyset$

for all $trn \in neighbourSet$ **do**

   $v = d(trn)$

   $supportSet(v) = supportSet(v) \cup \{trn\}$

end for

$p_{current} = \frac{|supportSet(d_{min})|}{|neighbourSet|}$

if $p_{current} > p$ **then**

   return d_{min}

else

   return d_{maj}

end if

---

derived from the theorems. Such approximation of the optimal values of weights can be also used for algorithms combining kNN with other approaches. The paper is related to rough sets indirectly as kNN may be combined with rule-based classifiers based on the rough set approach for imbalanced data problems and then the presented theorems can somehow relate to such cases. The paper reports several results from the PhD thesis [5] that have not yet been published and extends them.

2 Basic Notions

The acceptable performance measures for imbalanced data are among others *F1-score* and *G-mean*. They are composed out of the sub-measures Sensitivity, Specificity and Precision (see e.g. [7]).

In Algorithm 1 we present *class based weighted kNN* (in short *weighted kNN*) which is of our interest. In the algorithm the tie-breaking procedure is settled in favour of the majority class. This algorithm is equivalent to assigning weights to minority and majority examples with values $1 - p$ and p, respectively in standard kNN algorithm. In particular, for parameter $p = \frac{1}{2}$ it becomes standard kNN algorithm.

The default candidate for the parameter p of Algorithm 1 in the case of G-mean is the percentage of the size of the minority class from the size of the whole data set. Theorem 1 together with the discussion that follows it can be treated as an intuitive explanation of why this default choice can be really good.

It is well known that not all examples in data sets are equally difficult to classify. In [11], four types of example are identified regarding their difficulty of classification. We focus on two of them, namely *safe* and *borderline* examples. The *safe* examples are those which lay in the interior of the homogeneous regions of the minority class (or the majority class). The *borderline* examples are those which are located close to the boundary between two classes and are quite 'mixed'. *Borderline regions* contain only borderline examples; *safe regions* contain only safe examples. In *safe regions* it is easy to classify examples. However, in regions of borderline examples the classification becomes harder. We restrict our considerations to the case consisting of borderline examples only which additionally are 'totally mixed' (with fixed imbalance ratio).

We consider $\mathbf{X} = X \times V_d$, where X is the space of vectors of values of conditional attributes, and V_d the binary value set of the decision attribute d, i.e. $V_d = \{d_{maj}, d_{min}\}$, where $d_{maj} = 0$ and $d_{min} = 1$. Any *example (object)* is a pair $(x, d) \in X \times V_d$. Any training set *trnSet* of the length n is a sequence of examples, i.e. $trnSet = (z_1, \ldots, z_n)$, where $z_i \in X \times V_d$ for $i = 1, \ldots, n$.

By $z \sim \mathcal{D}$ we denote random sampling of z from a set Z according to $\mathcal{D}$, where $\mathcal{D}$ is a probability distribution over Z. Usually, we denote by $\mathcal{D}$ a probability distribution over the set $\mathbf{X} = X \times V_d$. By $trnSet \sim \mathcal{D}^n$ we denote random sampling of the training set *trnSet* of size n, where each example from *trnSet* is sampled independently using the same distribution $\mathcal{D}$.

By $\mathbf{E}R$ we denote the expected value of the given random variable R. The subscript of $\mathbf{E}$ in $\mathbf{E}_{z \sim \mathcal{D}} R(z)$ is used to indicate that sampling of z is according to the probability distribution $\mathcal{D}$. Analogously, we denote by $\Pr_{z \sim \mathcal{D}}(Event(z))$ the probability of the event *Event*, where sampling of z is according to the probability distribution $\mathcal{D}$.

The Accuracy of a given classifier C is equal to the probability that this classifier correctly classifies any test example (see e.g. [10]), i.e. $Acc(C) = \Pr_{(x,d) \sim \mathcal{D}}(C(x) = d) = \mathbf{E}_{(x,d) \sim \mathcal{D}}(I(C(x) = d))$, where $C(x)$ is the decision assigned to x by the classifier C, d is the correct decision on x and $I(\cdot)$ is the indicator function (equal to 1, if the condition in the argument is satisfied and 0, otherwise)[2]. For calculating G-mean we need Sensitivity (called also Accuracy for Positive Class or Recall) and Specificity (called also Accuracy for Negative class):

$$Acc_{min}(C) = \Pr_{(x,d) \sim \mathcal{D}}(C(x) = d \mid d = d_{min}),$$

$$Acc_{maj}(C) = \Pr_{(x,d) \sim \mathcal{D}}(C(x) = d \mid d = d_{maj}).$$

For calculating F1-score, we need Sensitivity and Precision. Precision is the conditional probability that the classification is correct provided that the classifier predicts the positive (minority) class:

$$Prec(C) = \Pr_{(x,d) \sim \mathcal{D}}(C(x) = d \mid C(x) = d_{min}).$$

However, we are interested in computing Accuracy of a learning algorithm $Alg(trnSet)$ constructing a classifier from a given training set *trnSet* Formally, Accuracy should be averaged over all possible training data sets of fixed size n (see e.g. [10]), i.e. we need to calculate $AvgAcc(Alg) =$

---

[2] $I(C(x) = d) = 1 - L(C(x), d)$, where L is the 0–1 loss function (equal to 0, if C correctly classifies the given example (x, d) and 1, otherwise).

$\mathbf{E}_{trnSet \sim \mathcal{D}^n} Acc(Alg(trnSet))$
$= \mathbf{E}_{trnSet \sim \mathcal{D}^n} \Pr_{(x,d) \sim \mathcal{D}}(Alg(trnSet)(x) = d)$
$= \mathbf{E}_{trnSet \sim \mathcal{D}^n} \mathbf{E}_{(x,d) \sim \mathcal{D}} I(Alg(trnSet)(x) = d)^3$. Analogously, for calculating measures related to G-mean and F1-score we need the measures presented above averaged over all possible training data sets of fixed size n. Hence, we introduce:

$$AvgAcc_{min}(Alg) = \mathbf{E}_{trnSet \sim \mathcal{D}^n} Acc_{min}(Alg(trnSet)),$$
$$AvgAcc_{maj}(Alg) = \mathbf{E}_{trnSet \sim \mathcal{D}^n} Acc_{maj}(Alg(trnSet)),$$
$$AvgPrec(Alg) = \mathbf{E}_{trnSet \sim \mathcal{D}^n} Prec(Alg(trnSet)).$$

For each test example $tst = (x, d)$, any sequence of training examples $trnSet = ((x_1, d_1), \ldots, (x_n, d_n))$, and any pseudometric ϱ, let $\pi_1(x), \ldots, \pi_n(x)$ be the permutation of $\{1, \ldots, n\}$ reordering $(x_1, \ldots, x_n)$ according to $\varrho(x, x_i)$, as follows

$$\varrho(x, x_{\pi_i(x)}) \leq \varrho(x, x_{\pi_{i+1}(x)}), \text{ for each } i \in \{1, \ldots, n-1\}.$$

Without loss of generality for our considerations, one can assume that the permutation is determined uniquely. It should be noted that ϱ may depend on $trnSet$.

As it was mentioned, we consider the 'totally random' distribution over set $\mathbf{X}$. Intuitively, it means that for this distribution the decisions of examples for the majority and minority classes are 'totally mixed' (with fixed imbalance ratio) without any dependence on values of conditional attributes. Formally, this means that the distribution $\mathcal{D}$ over $\mathbf{X}$ can be expressed as the product of independent distributions $\mathcal{D}_{\mathcal{X}}$ and $\mathcal{D}_{\mathcal{V}}$ over X and V_d, respectively, i.e. $\mathcal{D} = \mathcal{D}_{\mathcal{X}} \times \mathcal{D}_{\mathcal{V}}$.

In our following considerations, Alg will be interpreted as the class based weighted kNN learning algorithm (Algorithm 1) for the fixed k. However it is treated as random variable. It is parametrised by $p \in [0, 1]$. Hence, $AvgAcc_{min}$, $AvgAcc_{maj}$ and $AvgPrec$ are functions of p.

3   Theoretical Results

3.1   Optimal Classes Weights for G-mean

Now, we present a theorem which roughly says that the optimal value for the parameter p in the case of G-mean for the class based weighted kNN (Algorithm 1) under the assumption of the 'totally random' distribution is very close to the percentage of the size of the minority class from the size of the whole data set.

Theorem 1. *(version for G-mean) Let $k, n \in \mathbb{N}$, $k \leq n$, $q \in (0, 1)$ be given constants. Let $p \in [0, 1]$ be a parameter. Let $\mathcal{D}$ be a distribution over $\mathbf{X} = X \times V_d$*

---

[3] Formally, $\Pr_{(x,d) \sim \mathcal{D}}(Alg(trnSet)(x) = d)$ is a random variable, where $trnSet$ is fixed. It can also be seen as the conditional probability on $\mathcal{D}^n \times \mathcal{D}$ given a training set $trnSet$, i.e. $\Pr_{trnSet \sim \mathcal{D}^n, (x,d) \sim \mathcal{D}}(Alg(trnSet)(x) = d \mid trnSet)$.

such that $\mathcal{D} = \mathcal{D}_{\mathcal{X}} \times \mathcal{D}_{\mathcal{Y}}$, where $\mathcal{D}_{\mathcal{X}}$ in any distribution over X and $\mathcal{D}_{\mathcal{Y}}$ is the Bernoulli(q) distribution taking values $d_{min} = 1$ with probability q and $d_{maj} = 0$ with probability $1 - q$. Let $tst = (x, d) \sim \mathcal{D}$, $trnSet = ((x_1, d_1), \ldots, (x_n, d_n)) \sim \mathcal{D}^n$. Let D_i be a random variable equal to $d_{\pi_i(x)}$, i.e. the decision of the i-th nearest neighbour (from trnSet) to x. Let us consider the random variable Alg with arguments trnSet and x taking the decision on the basis of values $D_1(trnSet, x), D_2(trnSet, x), \ldots, D_k(trnSet, x)$ defined as follows

$$Alg(trnSet)(x) = \begin{cases} d_{min} \text{ if } \frac{1}{k} \sum_{i=1}^{k} D_i(trnSet, x) > p, \\ d_{maj} \text{ otherwise.} \end{cases}$$

Let us consider the function[4]
$$AvgGmean(p) = \sqrt{AvgAcc_{min}(Alg(p)) \cdot AvgAcc_{maj}(Alg(p))}.$$
If we consider all the values p_{opt} such that the function $AvgGmean(p)$ takes the maximal value at p_{opt}, then

$$\inf_{p_{opt}} |p_{opt} - q| \leq \frac{\ln 2}{k}.$$

Proof. For any fixed $trnSet$ we have $Acc_{min}(Alg(trnSet)) = \Pr_{(x,d) \sim \mathcal{D}}(Alg(trnSet)(x) = d \mid d = d_{min}) = \Pr_{(x,d) \sim \mathcal{D}}(Alg(trnSet)(x) = d_{min} \mid d = d_{min})$

$$= \Pr_{(x,d) \sim \mathcal{D}}(Alg(trnSet)(x) = d_{min}) \tag{1}$$

$$= \Pr_{x \sim \mathcal{D}_{\mathcal{X}}}(Alg(trnSet)(x) = d_{min}) \tag{2}$$

Eq. 1 follows from the fact that events $Alg(trnSet)(x) = d_{min}$ and $d = d_{min}$ are independent. Equation 2 follows from the fact that $\mathcal{D} = \mathcal{D}_{\mathcal{X}} \times \mathcal{D}_{\mathcal{Y}}$.
Analogously, we have

$$Acc_{maj}(Alg(trnSet)) = \Pr_{x \sim \mathcal{D}_{\mathcal{X}}}(Alg(trnSet)(x) = d_{maj}).$$

For any $trnSet$ we have $Acc_{min}(Alg(trnSet)) + Acc_{maj}(Alg(trnSet)) = P_{x \sim \mathcal{D}_{\mathcal{X}}}(Alg(trnSet)(x) = d_{min}) + P_{x \sim \mathcal{D}_{\mathcal{X}}}(Alg(trnSet)(x) = d_{maj}) = 1$. Hence, we also have $AvgAcc_{min}(Alg) + AvgAcc_{maj}(Alg) = \mathbf{E}_{trnSet \sim \mathcal{D}^n} Acc_{min}(Alg(trnSet)) + \mathbf{E}_{trnSet \sim \mathcal{D}^n} Acc_{maj}(Alg(trnSet)) = 1$.
Thus, we have

$$AvgGmean(p) = \sqrt{AvgAcc_{maj}(Alg(p)) \cdot (1 - AvgAcc_{maj}(Alg(p)))}.$$

---

[4] Here and later for F1-score, we determine the average performance using averaged values of the submeasures. One could also consider computing the average of these performance measures directly, what seems more appropriate. In such case the analogous computations of the proofs seem to be much more hard mathematical task.

The root square function is monotonic and under the root square we have a quadratic function of $AvgAcc_{maj}$, which achieves the maximal value for $AvgAcc_{maj} = \frac{1}{2}$. This quadratic function is symmetrical (around $\frac{1}{2}$) and monotonically increasing up to $AvgAcc_{maj} = \frac{1}{2}$ and from that point monotonically decreasing. Thus, $AvgGmean(p)$ achieves the maximal value for any p_{opt} such that: $p_{opt} \in \arg\min_{p \in [0,1]} |AvgAcc_{maj}(Alg(p)) - \frac{1}{2}|$.

It should be noted that the above formula not defines the optimal value of p uniquely; we obtain the set of optimal values of p. We consider all such optimal values p_{opt}. Later on we prove that the set of all optimal values p_{opt} is 'close' to the value q.

We have $AvgAcc_{maj}(Alg)$

$$= \mathbf{E}_{trnSet \sim \mathcal{D}^n} Acc_{maj}(Alg(trnSet)) = \mathbf{E}_{trnSet \sim \mathcal{D}^n} \Pr_{x \sim \mathcal{D}_{\mathcal{X}}} (Alg(trnSet)(x) = d_{maj})$$

$$= \mathbf{E}_{trnSet \sim \mathcal{D}^n} \mathbf{E}_{x \sim \mathcal{D}_{\mathcal{X}}} I(Alg(trnSet)(x) = d_{maj}) \tag{3}$$

$$= \mathbf{E}_{trnSet \sim \mathcal{D}^n} \mathbf{E}_{(x,d) \sim \mathcal{D}} I(Alg(trnSet)(x) = d_{maj})$$

$$= \mathbf{E}_{trnSet \sim \mathcal{D}^n, (x,d) \sim \mathcal{D}} I(Alg(trnSet)(x) = d_{maj})$$

$$= \Pr_{trnSet \sim \mathcal{D}^n, (x,d) \sim \mathcal{D}} (Alg(trnSet)(x) = d_{maj}) \tag{4}$$

$$= \Pr_{trnSet \sim \mathcal{D}^n, (x,d) \sim \mathcal{D}} (\sum_{i=1}^{k} D_i \le pk) = F_{B(k,q)}(pk), \tag{5}$$

where $B(k,q)$ denotes the binomial distribution and $F_{B(k,q)}(v)$ its cumulative distribution function at point v, i.e. $F_{B(k,q)}(v) = \sum_{i=0}^{\lfloor v \rfloor} \binom{k}{i} q^i (1-q)^{(k-i)}$, $\lfloor v \rfloor$ is the 'floor' under v, i.e. the greatest integer less than or equal to v.

Equations 3 and 4 follow from the definition of indicator function. Equation 5 follows from the definition of Alg. Any permutation of training examples (formally, random variables) does not change their distribution and independence, thus for any $1 \le i \le k$, $D_i \sim Bernoulli(q)$ (D_i are identically distributed) and D_i are (mutually) independent. Thus, the probability in Eq. 5 is the cumulative distribution function of binomial distribution $B(k,q)$ at point pk. This implies the last equation.

From the previous considerations we obtain the set of all optimal values p_{opt} which satisfy:

$$p_{opt} \in \arg\min_{p \in [0,1]} |F_{B(k,q)}(pk) - \frac{1}{2}|.$$

Let us denote by $\tilde{p}_{opt}$ the smallest optimal value p_{opt}. Then $\tilde{p}_{opt}k$ is an integer value since the cumulative binomial distribution function is a step function with jumps in integer values and constant between them. First, let us consider a specific case when the cumulative distribution function achieves the optimal value (i.e. the closest to $\frac{1}{2}$) at both integer values $\tilde{p}_{opt}k$ and $\tilde{p}_{opt}k + 1$. Then all the optimal values p_{opt} form the interval $[\tilde{p}_{opt}, \tilde{p}_{opt} + \frac{2}{k})$ since the optimal values p_{opt} are contained in the sum of two intervals for which $F_{B(k,q)}$ is closest to $\frac{1}{2}$, i.e. $[\tilde{p}_{opt}, \tilde{p}_{opt} + \frac{1}{k})$ and $[\tilde{p}_{opt} + \frac{1}{k}, \tilde{p}_{opt} + \frac{2}{k})$. We will return to this case at the end of the proof. From now on, we consider the opposite case. Then all the optimal values p_{opt} form the interval $[\tilde{p}_{opt}, \tilde{p}_{opt} + \frac{1}{k})$.

For the need of the considerations that follow, it is worthwhile to recall the definition of the median. The median of the distribution induced by a random variable R is any real number m that satisfies the inequalities: $\Pr(R \leq m) \geq \frac{1}{2}$ and $\Pr(R \geq m) \geq \frac{1}{2}$.

In the three cases considered below it will be convenient to denote by e the value $\tilde{p}_{opt} k = \lfloor \tilde{p}_{opt} k \rfloor$ and by D the random variable with distribution $B(k,q)$.

The first case: the optimal (i.e. the closest to $\frac{1}{2}$) value $F_{B(k,q)}(\tilde{p}_{opt}k)$ is equal to $\frac{1}{2}$. Then $P(D \leq e) = F_{B(k,q)}(e) = \frac{1}{2}$; $P(D \geq e) > P(D > e) = 1 - P(D \leq e) = \frac{1}{2}$. Hence, e is the median of $B(k,q)$.

The second case: the optimal value $F_{B(k,q)}(\tilde{p}_{opt}k)$ is less than $\frac{1}{2}$. Then $P(D \leq e) = F_{B(k,q)}(e) < \frac{1}{2}$. Then $P(D \leq e + 1) = F_{B(k,q)}(e + 1) > \frac{1}{2}$ (otherwise $F_{B(k,q)}(e)$ would not be the optimal value). We also have $Pr(D > e) = 1 - P(D \leq e) > \frac{1}{2}$. Thus, $Pr(D \geq e + 1) = Pr(D > e) > \frac{1}{2}$. It implies that $e + 1$ is the median of $B(k,q)$.

The third case: the (optimal) value $F_{B(k,q)}(\tilde{p}_{opt}k)$ is greater than $\frac{1}{2}$. Then $P(D \leq e) = F_{B(k,q)}(e) > \frac{1}{2}$. We also have $P(D < e) = P(D \leq e - 1) < \frac{1}{2}$ (otherwise $F_{B(k,q)}(e)$ would not be the optimal value). Then $P(D \geq e) = 1 - P(D < e) > \frac{1}{2}$. Hence, that e is the median of $B(k,q)$.

To sum up, we have shown that for all $\tilde{p}_{opt}$, we have that $e = \tilde{p}_{opt}k$ or $e + 1$ is the median of $B(k,q)$.

On the other hand in [6] it is shown that any median M of $B(k,q)$ cannot be 'far' from its mean value $\mu = kq$. More precisely, the distance between M and μ can be at most $\ln 2$, i.e.:

$$|M - \mu| \leq \ln 2. \tag{6}$$

This implies that for any $\tilde{p}_{opt}$ we have (respectively for the cases when e is the median or $e+1$ is the median) $|e - kq| \leq \ln 2$ or $|e+1 - kq| \leq \ln 2$. Thus, $|\tilde{p}_{opt}k - kq| \leq \ln 2$ or $|\tilde{p}_{opt}k+1-kq| \leq \ln 2$. Hence, $|\tilde{p}_{opt} - q| \leq \frac{\ln 2}{k}$ or $|\tilde{p}_{opt} + \frac{1}{k} - q| \leq \frac{\ln 2}{k}$.

Let us recall that all the optimal values p_{opt} form the interval $[\tilde{p}_{opt}, \tilde{p}_{opt} + \frac{1}{k})$. Therefore, either (in the case when e is the median) the beginning of this interval is distanced from q not more than $\frac{\ln 2}{k}$ or (in the case when $e + 1$ is the median) the end of it is distanced from q not more than $\frac{\ln 2}{k}$. Thus,

$$\inf_{p_{opt}} |p_{opt} - q| \leq \frac{\ln 2}{k}.$$

We still have to prove the theorem for the specific case when all the optimal values p_{opt} form the interval $[\tilde{p}_{opt}, \tilde{p}_{opt} + \frac{2}{k})$. Then from the above considerations, it is easy to see that in such case the value q belongs to this interval. Hence, it belongs to the set of optimal values p_{opt}.

$\square$

We used in the proof of the above theorem the best possible approximation between the median and the mean (independent of q and k) of the binomial distribution [6]. However, it should be noted that we do not want to look for the maximal distance between median and mean, but the distance between q and the interval of optimal values p_{opt}. In particular, in many cases this interval contains the value q.

Intuitively, the algorithm Alg in Theorem 1 represents the weighted kNN algorithm (Algorithm 1) with a fixed parameter k (given in the assumption of the theorem), and with a data set represented only by a single region with the high degree of overlapping between classes, i.e. only borderline examples occur in the data set under consideration The function $AvgGmean(p)$ represents the G-mean for weighted kNN (with fixed parameters as described above) for different values of the parameter p. The theorem roughly says that the maximal G-mean value for weighted kNN is achieved for p equal roughly to q (the percentage of the size of the minority class). For example, without going to the technical details, the theorem says that if there are 5% of examples from the minority class mixed totally randomly with examples from the majority class, then the optimal value for the parameter p for weighted kNN is achieved for $p = 5\%$.

Obviously, in practice, data sets contain not only borderline examples but also safe examples. Thus, it would be valuable for applications to formulate and prove more general theorem for borderline and safe regions. We leave it for future work. However, below we give an intuitive explanation that, roughly speaking, the theorem's conclusions will remain true in such more general situation.

First, let us assume that there are some other regions with borderline examples with the same overlapping level of the minority and majority classes (formally, distributed randomly with the same parameters of Bernoulli distribution). If one adds such regions, the conclusion of the theorem will also hold since it can be treated as one region of borderline examples.

Second, let us consider the case when both borderline and safe examples occur in data. In this case, one can divide the whole space of examples into the safe region (consisting of the safe regions of the majority class and the safe regions of the minority class) and the borderline region (consisting of borderline regions in different areas of data). Let us assume that all examples from the safe region are correctly classified by the algorithm (independently of the parameter p)[5]. Let us also assume that the global percentage of the minority class is the same as the percentage of the minority class in the borderline region[6]. Under these assumptions, it is easy to check that the optimal parameter p will be the same as in the theorem's conclusion (i.e. the optimal parameter p for the borderline region).

This shows that, in a sense, only regions with borderline examples are important to focus on in order to achieve the high G-mean value.

---

[5] In practice, such an assumption is satisfied for a wide range of values of the parameter p.

[6] This assumption is important only from the technical point of view. If it is not satisfied we should consider the percentage of the minority class coming from borderline region (value q from theorem).

In the case of dealing with real-life data sets, even if borderline examples are 'totally mixed', the given above assumptions may be not satisfied. For instance, some examples treated as safe examples can be misclassified for the optimal value of the parameter p, the borderline regions can have different percentage of the minority class, or the global percentage can be different from the percentage of borderline regions. However, if the given above assumptions are 'roughly' satisfied, then the optimal parameter p can be only 'slightly' different from that given in the theorem.

3.2 Optimal Classes Weights for F1-score

Now, we present the theorem which roughly says that the optimal value of the parameter p in the case of F1-score for the weighted kNN algorithm and 'totally random' distribution is 0.

Theorem 2. *(version for F1-score) Under the assumptions of Theorem 1 let us consider the function $AvgF1score(p) = H(AvgAcc_{min}(Alg(p)), AvgPrec(Alg(p)))$, where $H(\cdot, \cdot)$ is the function of harmonic mean of its arguments, i.e. $H(a, b) = \frac{2}{a^{-1}+b^{-1}}$.*
Then the function $AvgF1score(p)$ takes the maximal value at

$$p_{opt} \in \left[0, \frac{1}{k}\right).$$

Proof. Let us recall that by $F_{B(k,q)}$ we denote the cumulative binomial distribution function. $AvgAcc_{min}(Alg)$

$$= 1 - AvgAcc_{maj}(Alg) = 1 - \Pr_{(x,d)\sim\mathcal{D}, S\sim\mathcal{D}^m}\left(\sum_{i=1}^{k} D_i \le pk\right) = 1 - F_{B(k,q)}(pk).$$

Both the first and second equation come from the proof of Theorem 1. The third equation follows from the fact that the probability in the previous equation is equal to the cumulative distribution function of the binomial distribution $B(k, q)$ at point pk (for details see the proof of Theorem 1).

We also have $AvgPrec(Alg(p))$

$$= \mathbf{E}_{trnSet\sim\mathcal{D}^n}\mathbf{E}_{(x,d)\sim\mathcal{D}}(I(Alg(trnSet)(x) = d) \mid Alg(trnSet)(x) = d_{min})$$

$$= \mathbf{E}_{trnSet\sim\mathcal{D}^n}\mathbf{E}_{(x,d)\sim\mathcal{D}}(I(d_{min} = d) \mid Alg(trnSet)(x) = d_{min})$$

$$= \mathbf{E}_{trnSet\sim\mathcal{D}^n}\mathbf{E}_{(x,d)\sim\mathcal{D}}I(d_{min} = d) \tag{7}$$

$$= \mathbf{E}_{trnSet\sim\mathcal{D}^n}\mathbf{E}_{x\sim\mathcal{D}_\mathcal{X}}\mathbf{E}_{d\sim\mathcal{D}_\mathcal{Y}}I(d_{min} = d) \tag{8}$$

$$= \mathbf{E}_{trnSet\sim\mathcal{D}^n}\mathbf{E}_{x\sim\mathcal{D}_\mathcal{X}}\Pr_{d\sim\mathcal{D}_\mathcal{Y}}(d = d_{min}) \tag{9}$$

$$= \mathbf{E}_{trnSet\sim\mathcal{D}^n}\mathbf{E}_{x\sim\mathcal{D}_\mathcal{X}}q = q. \tag{10}$$

Eq. 7 follows from the fact that events $d_{min} = d$ and $Alg(trnSet)(x) = d_{min}$ are independent (for any fixed $trnSet$). Equation 8 follows from the fact that

$\mathcal{D} = \mathcal{D_X} \times \mathcal{D_Y}$. Equation 9 follows from definition of the indicator function. Equation 10 follows from the fact that $\mathcal{D_Y}$ is the $Bernoulli(q)$ distribution.

Thus, we have $AvgF1score(p) = H(AvgPrec(Alg(p)), AvgAcc_{min}(Alg(p)))$ $= H(q, 1 - F_{B(k,q)}(pk))$. The first argument of H with respect to p is constant and H is monotonically increasing function of the second argument. Thus, the function $AvgF1score(p)$ takes the maximal value at those values of p for which the function $1 - F_{B(k,q)}(pk)$ takes the maximal value. Hence, the function $AvgF1score(p)$ takes the maximal value at

$$p_{opt} \in \underset{p \in [0,1]}{\arg \min} \, F_{B(k,q)}(pk).$$

Every cumulative distribution function is non-decreasing. Thus, the function $AvgF1score(p)$ takes the maximal value at p_{opt} such that $F_{B(k,q)}(p_{opt}k) = 0$. From definition of $F_{B(k,q)}(pk) = 0$ we have $\lfloor p_{opt}k \rfloor = 0$. Thus $p_{opt} \in [0, \frac{1}{k})$.

□

Intuitively, function $AvgF1score(p)$ represents the F1-score for weighted kNN (with fixed parameters as described above) for different values of the parameter p. Intuitively, the theorem says that the maximal value of F1-score for weighted kNN is achieved for p equal roughly to 0. This relates to the algorithm classifying examples to the minority class if at least one minority example occurs in the neighbourhood. This is intuitively clear because for F1-score we need to balance between Precision and Sensitivity. Precision is constant for totally random examples, i.e. in a sense, it does not depend on algorithm. Thus, to maximise the F1-score one needs to maximise Sensitivity. It is done by setting the minimal possible value of the parameter p. This is related to classifying all objects to the minority class (excluding only the situations such that all neighbours of a given test object are from the majority class).

It can seem strange that the set of optimal values p_{opt} does not depend on the value of q. For example, both for q close to 0 and close to 1 the optimal value does not change. However, it should be observed that F1-score is the harmonic mean of Sensitivity and Precision. Thus, in a sense, this performance measure 'favours' one class, that is the minority class. This measure does not balance between classifying to the minority class and the majority class, but rather between classifying to the minority class and quality of this classification, i.e. Precision. Hence, if Precision is constant (which is the case when classes are 'totally mixed' with the fixed imbalance ratio), then to maximise F1-score one should choose such p that the classifier chooses the minority class as often as possible. In fact, p close to 0 relates to this case. Irrespective of the value of q, it is better to classify examples to the minority class (if it is only possible). This is an intuitive explanation of the above theorem. However, it is worth pointing out that for practical data sets Precision is not constant.

Analogously as for the previous theorem (for G-mean), it would be more relevant for practical applications to formulate and prove more general theorem with borderline and safe regions. Again, we leave it for future work. The given previously intuitive explanation that the theorem for G-mean can be easily generalised for such a case would not work for F1-score. This is due to the fact that safe examples from the majority class could be misclassified for p close to 0 (which is the value close to the optimal values of p from the theorem for F1-score). In consequence, Precision would be not constant (would depend on p). Then, the optimal values of p in such case could be greater than 0 and should be recalculated for the generalised theorem for F1-score.

3.3   Discussion

We show use of these theorem in two ways. First, one can possibly accelerate the process of searching for the optimal values by searching them close to the values derived from the theorems. In fact we show that it holds for G-mean even for real-life data sets (see Sect. 4). Such approximation of the optimal values of p can be also used for algorithms combining kNN with other approaches (see [4,5]).

Second, these theorems with the following considerations may be a warning against drawing too hasty conclusions from experiments comparing algorithms on G-mean or F1-score. Comparing the optimal values for G-mean and F1-score for the case when classes are 'totally mixed' one can see that the optimal values for different performance measures can be very different. In fact, in the described situation we do not optimise the parameter p according to the given data (since as high randomness occurs one can deduce nothing) but to the selected performance measure. In this sense, these theorems illustrate that in some specific situations learning algorithms may rather 'learn' optimisation measure more than useful relations between conditional attributes and decision. One should be aware of that.

Also, these theorems lead to another interesting observation. To be specific, consider 'random' data set with the percentage of the minority class (i.e. the value of q from the assumptions of the theorems) equal to 0.3 and $k = 50$ (size of the considered neighbourhood). Then, these theorems show that for a given data set (in our case, 'random' data set), the optimal classifiers from a given class of classifiers may be significantly different (in respect to classification) depending on the performance measure relative to which the optimal classifier is selected. Moreover, it can be easily calculated (using formulas from the proofs of the theorems) in the considered case that the assessments of these two optimal classifiers are significantly different depending on the performance measure used for the assessment. For one performance measure, the first optimal classifier is much better than the second one; and for another performance measure, vice versa (the second one is much better than the first one). It should be noted that weighted kNN with two different setting of p are examples of two classifiers. But the discussion may be true for other pairs of classifiers. These observations may help to understand that the 'best' classifier selection may strongly depend on the chosen performance measure. Also, it shows that without a precise specification of what particular performance measure we optimise, the 'best classifier' term can be ambiguous or even misleading.

How can it impact real-life problem? Let us assume that one is assuming that there exist dependencies between conditional features and class value but it is just the case that the distribution is random (what is not known for the analyst). By applying different methods the analyst can obtain results leading to conclusion that some methods lead to better results for F1-score and other for G-mean. Our results are explaining this. In the case of random distribution learning of dependencies (which do not exist!) is not realised but it is optimised the adjusting of induced classifiers to the quality measure which we would like to use for evaluation of classifiers. Hence, in the real-life problem the winner will be not 'better' algorithm but the one better adjusted to a given measure.

4   Experiments for G-mean

In Subsect. 3.1, we presented intuitive arguments suggesting that the theorem for G-mean may possibly roughly hold for data sets with safe regions (assuming that borderline regions are roughly 'totally random'). In this section, we report results of experiments aiming to check whether the conclusions of the theorem roughly hold for real life data sets. In other words, we need to check whether the optimal parameter p for weighted kNN would be close to the percentage of the minority class in data sets (being an approximation of the real q from the theorem).

In weighted kNN we used the metric ϱ being sum of metrics for all attributes. For any symbolic attributes we compute SVDM metric [2]; and for numerical attributes we use Euclidean metric on $\mathbb{R}$ normalised by the distance of maximal and minimal values in data set (on that attribute).

From now on, we will present estimations of the classification quality (relative to G-mean) depending on parameter p (sometimes on two parameters k and p) of the weighted kNN algorithm. The classification quality (in the function of parameters) was computed using the leave-one-out method applied to the whole data set. This allowed us to obtain the best possible approximation of the optimal p (or the pair of optimal values of parameters k and p) for data sets.

Data sets used in experiments are based mainly on UCI Machine Learning Repository [8].[7] Data sets containing originally more than two classes were transformed into the binary classification task by choosing one small class or joining several small classes into one (minority) class; other classes were joined into another (majority) class. We have selected 20 fairly diverse imbalanced data sets considering many aspects related to difficulty of imbalanced data classification (see [12]).

Experiments were performed for three cases: fixed $k = 30$, fixed $k = 20$ and optimal k. In the last case we search for the pair of values of parameters maximising G-mean. The used set of possible values for parameter p was $\{0.00, 0.01, 0.02, \ldots 0.5\}$.[8] In case with optimal k, the used set of possible values

---

[7] Only *mammography* data set is not publicly available and was supported by Nitesh Chawla [1].

[8] It is reasonable that for imbalanced data p should be not greater than 0.5 since minority class should have greater weight than majority class.

for parameter k was $\{1, 2, \ldots, 100\}$. Δ is the absolute difference between p_{min} and p_{opt} (from the respective case).

Table 1. Experiments comparing percent of minority class rounded to two decimals (p_{min}) and optimal p (p_{opt}) for three cases: fixed $k = 30$, fixed $k = 20$ and optimal k (optimal k, k_{opt} is also presented). Δ is the absolute difference between p_{min} and p_{opt} (from the respective case).

dataset	$k = 30$			$k = 20$		$k = k_{opt}$			dataset	$k = 30$			$k = 20$		$k = k_{opt}$		
	p_{min}	p_{opt}	Δ	p_{opt}	Δ	k_{opt}	p_{opt}	Δ		p_{min}	p_{opt}	Δ	p_{opt}	Δ	k_{opt}	p_{opt}	Δ
abalone	.08	.08	0	.08	0	31	.08	0	hepatitis	.21	.1	.11	.15	.06	42	.16	.05
balance-scale	.08	.04	.04	.11	.03	11	.11	.03	ionosphere	.36	.16	.2	.15	.21	5	.2	.16
breast-cancer	.3	.3	0	.3	0	82	.28	.02	mammography	.02	.02	0	.02	0	100	.03	.01
breast-w	.34	.06	.28	.17	.17	10	.3	.04	new-thyroid	.16	.13	.03	.16	0	60	.13	.03
car	.04	.21	.17	.21	.17	63	.33	.29	nursery	.03	.35	.32	.29	.26	1	.03	0
cleveland	.12	.12	0	.12	0	34	.11	.01	pima	.35	.35	0	.35	0	40	.36	.01
credit-g	.3	.3	0	.31	.01	52	.3	0	postoperative	.27	.26	.01	.25	.02	11	.27	0
ecoli	.1	.21	.11	.26	.16	20	.26	.16	transfusion	.24	.24	0	.24	0	26	.24	0
glass	.08	.08	0	.08	0	24	.09	.01	vehicle	.24	.34	.1	.31	.07	10	.31	.07
haberman	.26	.22	.04	.26	0	48	.25	.01	yeast-ME2	.03	.03	0	.05	.02	48	.03	0

In Table 1, we present the absolute difference between the optimal parameter p (maximising G-mean) for the considered data set and the percentage of the minority class, p_{min}. If there were several optimal values, we chose the one that is closest to p_{min} (it is enough for p_{min} to be close to any true optimal value to satisfy the theorem). If there were many optimal values of k we took for presenting the smallest one.

First, let us consider the case with fixed $k = 30$. For 9 data sets (out of 20), values p_{min} and p_{opt} coincide, i.e. $p_{min} = p_{opt}$ ($\Delta = 0$). For 13 data sets, $\Delta < 0.05$. For the remaining 7 data sets, the difference is greater than 0.1 (differences are relatively large).

Second, let us consider the case with fixed $k = 20$. For 9 data sets, values p_{min} and p_{opt} coincide ($\Delta = 0$). For 13 data sets, $\Delta < 0.05$. For 15 data sets, $\Delta < 0.1$. For the remaining 5 data sets, the difference is 0.16 or more (differences are relatively large).

Third, let us consider the case with using optimal k for any data set ($k = k_{opt}$). For 6 data sets, values p_{min} and p_{opt} coincide. For 15 data sets, $\Delta < 0.05$. For 17 data sets, $\Delta < 0.1$. For the remaining 3 data sets, the differences are relatively large (0.16 or more).

To sum up, for first and second case, for roughly half of data sets results are similar and show surprising equality of p_{min} and p_{opt}. This is strong argument that the proved theorem may be used for quick approximation of optimal value of p. For the case with optimal k, although for less data sets we have equality, for almost all data sets the approximation is relatively not large. In practice we would like to find not only optimal p, but simultaneously pair of optimal values of k and p. In this case, approximation of optimal p is remarkably good by using the proved theorem.

For $k = 30$ (and $k = k_{opt}$), we also analysed the graphs of G-mean as function of parameter p. Generally, there are two types of graphs among analysed data sets. The representatives of these types are presented in Fig. 1(a) and Fig. 1(b). On the first graph the optimum is 'evident'. On the second graph around the optimum there exist 'smooth area'. Generally, we found that for the first type of graphs the p_{min} and p_{opt} are equal or close to each other. For the second type of graphs the p_{min} and p_{opt} it may not hold. Of course, these are only intuitions, which could bring in future further conclusions or ideas.

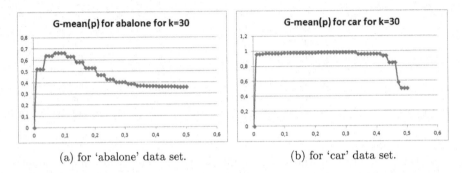

(a) for 'abalone' data set.   (b) for 'car' data set.

Fig. 1. Graphs representing G-mean for weighted kNN for two exemplary data sets as a function of parameter p with fixed $k = 30$.

4.1 Hypothesis About 'Totally Random' borderline regions

Apart of approximation of value of p_{opt}, the results of theorem for G-mean and provided experiments allow us to formulate the hypothesis that maybe for those data sets for which Δ is 0 or close to 0, the borderline regions are really (or 'close' to) 'totally random'. Of course, this needs further investigation. But if this is true then strong implications come from that. Then the considerations in Subsect. 3.3 would be true for some considered real-life data sets.

5 Conclusions

For G-mean and F1-score, we proved two theorems providing estimates of the optimal degree of importance of the minority class (weight for the minority class) for weighted kNN under the assumption of a 'totally random' distribution. These estimates are faster alternatives than solutions obtained by parameter learning and can be used for setting the default value for the appropriate parameter in weighted kNN or other analogous approaches. We experimentally justified for G-mean that conclusions of these theorems may be applied to the construction of weighted kNN classifiers for real-life data sets. Moreover, an interesting conclusion follows from these theorems. Namely, for a certain class of classifiers, the optimal one might be significantly different (relative to classification) for two different performance measures. Additionally, the assessments of such two optimal classifiers may be significantly different depending on the performance measure used for the assessment. The practical implication for real-life classifications is that without a precise specification of the particular performance measure we are interested in, the 'best classifier' term can be ambiguous or even misleading. Our experiments lead to the hypothesis that it may be the case for real-life data sets.

References

1. Chawla, N.V., Bowyer, K.W., Hall, L.O., Kegelmeyer, W.P.: SMOTE: synthetic minority over-sampling technique. J. Artif. Intell. Res. **16**, 321–357 (2002). https://doi.org/10.1613/jair.953
2. Domingos, P.: Unifying instance-based and rule-based induction. Mach. Learn. **24**(2), 141–168 (1996). https://doi.org/10.1007/BF00058656
3. Dubey, H., Pudi, V.: Class based weighted k-nearest neighbor over imbalance dataset. In: Pei, J., Tseng, V.S., Cao, L., Motoda, H., Xu, G. (eds.) PAKDD 2013. LNCS (LNAI), vol. 7819, pp. 305–316. Springer, Heidelberg (2013). https://doi.org/10.1007/978-3-642-37456-2_26
4. G. Góra, Skowron, A., Wojna, A.: Explainability in RIONA algorithm combining rule induction and instance-based learning. In: Ganzha, M., Maciaszek, L.A., Paprzycki, M., Ślęzak, D. (eds.) Proceedings of the 18th Conference on Computer Science and Intelligence Systems, FedCSIS 2023, Warsaw, Poland, September 17–20, 2023. Annals of Computer Science and Information Systems, vol. 31, pp. 485–496. IEEE (2023). https://annals-csis.org/proceedings/2023/
5. Góra, G.: Combining instance-based learning and rule-based methods for imbalanced data. Ph.D. thesis, University of Warsaw, Warsaw (2022). https://www.mimuw.edu.pl/sites/default/files/gora_grzegorz_rozprawa_doktorska.pdf
6. Hamza, K.: The smallest uniform upper bound on the distance between the mean and the median of the binomial and Poisson distributions. Stat. Probab. Lett. **23**(1), 21–25 (1995). https://doi.org/10.1016/0167-7152(94)00090-U
7. He, H., Ma, Y.: Imbalanced Learning: Foundations, Algorithms, and Applications. Wiley-IEEE Press, Piscataway, NJ, 1st edn. (2013). https://doi.org/10.1002/9781118646106
8. Lichman, M.: UCI Machine Learning Repository (2013). http://archive.ics.uci.edu/ml
9. Maennel, H., et al.: What do neural networks learn when trained with random labels? (2020). https://doi.org/10.48550/ARXIV.2006.10455
10. Mitchell, T.M.: Machine Learning. McGraw-Hill, New York (1997)
11. Napierała, K.: Improving Rule Classifiers For Imbalanced Data. Ph.D. thesis, Poznań University of Technology, Poznań (2012)
12. Napierała, K., Stefanowski, J.: Types of minority class examples and their influence on learning classifiers from imbalanced data. J. Intell. Inf. Syst. **46**(3), 563–597 (2016). https://doi.org/10.1007/s10844-015-0368-1

13. Raeder, T., Forman, G., Chawla, N.V.: Learning from imbalanced data: evaluation matters. In: Holmes, D.E., Jain, L.C. (eds.) Data Mining: Foundations and Intelligent Paradigms. igms. Intelligent Systems Reference Library, vol. 23, pp. 315–331. Springer, Heidelberg (2012). https://doi.org/10.1007/978-3-642-23166-7_12
14. Wang, L., Khan, L., Thuraisingham, B.: An effective evidence theory based k-nearest neighbor (KNN) classification. In: 2008 IEEE/WIC/ACM International Conference on Web Intelligence and Intelligent Agent Technology, vol. 1, pp. 797–801. IEEE (2008)
15. Wolpert, D.H.: The supervised learning no-free-lunch theorems. In: Roy, R., Klöppen, M., Ovaska, S., Furuhashi, T., Hoffmann, F. (eds.) Soft Computing and Industry, pp. 25–42. Springer, London (2002). https://doi.org/10.1007/978-1-4471-0123-9_3

Searching of Potentially Anomalous Signals in Cosmic-Ray Particle Tracks Images Using Rough k-Means Clustering Combined with Eigendecomposition-Derived Embedding

Tomasz Hachaj$^{(\boxtimes)}$ ⓘ, Marcin Piekarczyk ⓘ, and Jarosław Wąs ⓘ

Faculty of Electrical Engineering, Automatics, Computer Science and Biomedical Engineering, AGH University of Krakow, Al. Mickiewicza 30, 30-059 Krakow, Poland
{tomasz.hachaj,marcin.piekarczyk,jaroslaw.was}@agh.edu.pl

Abstract. Our work presents the application of the rough sets method in the field of astrophysics for the analysis of observational data recorded by the Cosmic Ray Extremely Distributed Observatory (CREDO) project infrastructure. CREDO research has produced huge datasets that are not well yet studied in terms of the information they contain, including specific anomalous observations, which are of particular interest to physicists and other scientists. From the pool of data available for analysis registered under CREDO infrastructure, containing approximately 10^7 of events, a set of 10^4 of samples was selected. We have applied eigendecomposition-derived embedding limiting data to 62 dimensions (95% of variance). We have adapted rough k-means algorithm for the purpose of anomalies detection task. We have validated our approach on various configurations of adaptable parameters of the proposed algorithm. The potential anomalies retrieved with the proposed algorithm have morphological features consistent with what a human expert would expect from anomalous signals in this case. The source codes and data of our experiments are available for download to make research reproducible.

Keywords: Cosmic-ray particle · Anomalies detection · Rough sets · Rough k-means · CMOS detectors · Eigendecomposition

1 Introduction

Rough sets [27] is one of the methodologies that can be used for unsupervised learning [34] and especially clusterization [28,39]. Using rough sets in the clustering process can enhance expressive and algorithmic capabilities allowing to work with lower and upper approximations of clusters sets [2,30,33,35]. If we treat the problem of anomaly detection as a problem of finding outliers in unlabelled datasets, we can apply rough set clustering algorithms to solve it. Such

A. Campagner et al. (Eds.): IJCRS 2023, LNAI 14481, pp. 431–445, 2023.
https://doi.org/10.1007/978-3-031-50959-9_30

rough set applications have been used, among others, for: anomaly detection in electric smart grids [32], intrusion traces of "sendmail" daemon process [42], Border Gateway Protocol (BGP) analysis [19], computer network systems analysis [9], IoT anomaly usage [25] as well as demonstrated their effectiveness on a variety of test data [20,41]. Methods using rough sets have been used effectively to detect anomalies in specific applications. The solutions proposed in the literature are particularly concentrated in applications where categorical data dominate [37,43], but not only, as models designed to operate on numerical or mixed data can also be pointed out [7,21,23,25,32]. Models have also been proposed in g-base, which clustering approaches to anomaly detection are extended and supported by the use of rough set methodology, e.g. [11,12], where the authors use fuzzy C-means (FCM) or [25], where the density-based clustering (DBSCAN) is supported by information reduction based on rough sets.

Our work presents the application of the rough sets method in the field of astrophysics for the analysis of observational data recorded by the Cosmic Ray Extremely Distributed Observatory (CREDO) project infrastructure [8,17]. The idea of CREDO project is to detect very high-energy cosmic rays by analyzing secondary showers of particles reaching the earth's surface. The analysis of supermassive cosmic ray particles offers a chance to understand fundamental physical theories about Dark Energy and Dark Matter [17]. CREDO implements the concept of a global distributed system of detectors consisting of various types of devices, including mobile detectors based on smartphones and other detectors based on CMOS image matrices [8,18,31]. In principle, this allows for effective scaling, using the citizen science paradigm, of the range covered by observations in contrary to stationary observatories located in a specific spatially limited location, such as the Pierre Auger Observatory [3,10,13], IceCube [1,5] and Baikal-GVD at Lake Baikal [4,36].

CREDO research has produced huge datasets that are not well yet studied in terms of the information they contain, including specific anomalous observations, which are of particular interest to physicists. In order to analyze the dataset for detection of potential anomalies, efficient and effective algorithms from the field of signal processing and data science are required. To our knowledge, the results presented in this work are the first practical application of rough set-based methodology to analyze observational data obtained by the Cosmic Ray Extremely Distributed Observatory.

2   Material and Methods

2.1   Dataset

We used a representative subset of observations archived under CREDO as the basis for the experimental verification of the hypotheses considered in this paper and for the presentation of the computational results. The dataset used by us was selected in such a way as to reflect well the internal diversity of the observed signals. From the pool of data available for analysis registered under CREDO infrastructure, containing approximately 10^7 of events, a set of 10^4 of samples was selected. For the purposes of the work, we used only image data and omitted all other metadata. The obtained dataset contains all known types of observed signals recorded by the CREDO detectors [8], i.e. dots, tracks, worms, as well as other atypical observations that meet the criteria for recognizing them as potential cosmic-ray particle tracks. The shape morphology of cosmic-ray particle tracks is a factor that is taken into account when determining the type of signal. The subset does not contain clearly incorrect signals resulting from various types of measurement errors, which are referred to in the nomenclature as artefacts [8,29]. Examples of signal types appearing in the CREDO database are shown in Fig. 1.

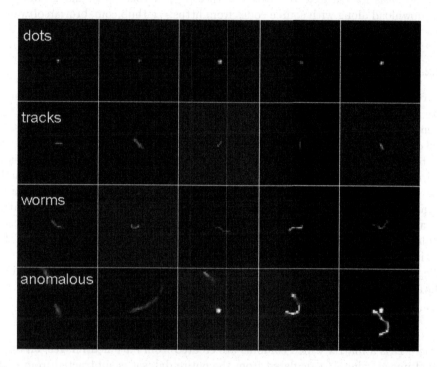

Fig. 1. Examples of types of signals in the CREDO observation database. The illustration shows the basic classes of useful signals acquired using CMOS sensors. Artifacts resulting from measurement errors and incorrect calibration were omitted as irrelevant to the research issues of the article. Signal types are described in the first column.

The selection of samples was carried out in several stages. First, automatic methods were used for initial identification of signals [6] and effective filtering out of artifacts [29]. Then, the obtained data was manually verified for consistency by a human annotator and thus additionally cleaned. As a result, a set of image observations with labels of signal classes was obtained.

Due to the specificity of recording traces of high-energy cosmic ray particles [16] and the limitations of both automatic and human classification, unusual observations (anomalies) may occur in basically every class of recorded signals that are initially annotated. In other words, there is no certainty which objects in a given class may constitute unusual observations (anomalies) from the point of view of morphological and physical interpretation [17]. Therefore, the labels assigned to the samples are not useful in this problem and have not been used.

From the technical side, all selected events in the prepared dataset are represented by RGB images with a resolution of 60×60 pixels. They do not contain any additional metadata in the volume as well as labels. The subset of the CREDO data we use in this work consists of 13804 instances. This dataset will be the basis for calculations in the further part of this study and is available for download along with source codes from https://github.com/browarsoftware/particle_pars.

2.2   Features Generation

In order to be able to effectively compare image data, it is necessary to generate an appropriate embedding that preserves the interrelationships between the elements of the dataset. In this case the embedding should allow us to search for objects that differ in some way from other typical cosmic-ray particle tracks in terms of morphology. Knowing this, we should base embedding on statistical relationships in the dataset, such as analysis of variance. Effective methods for generating embedding using analysis of variance are algorithms similar to Eigenfaces [15,38]. In this approach, the feature vector of images is generated as a linear combination of image coordinates after projecting them into a space which coordinate system is calculated on the basis of variance analysis of the entire data set. The axes of the new system follow the principal components of the covariance matrix COV created from the individual vectors of the original data.

$$COV = \frac{1}{s}D^T D \tag{1}$$

where D is a matrix which columns are created from flattened images, an averaged image value M calculated from the entire dataset is subtracted from each image.

$$D = [I_1 - M, ..., I_s - M] \tag{2}$$

where I_1 is the first image from the dataset, there are s images in total.

The entire analysis is performed using the well-known Principal Components Analysis (PCA) approach. A very important fact is that image analysis based on eigendecomposition with PCA is very sensitive to even small variance distortions, so before applying it to an image dataset it is necessary to normalize dataset (to perform so called image aligning). In the case of our dataset, aligning consisted of translating the center of mass of the image so that it is at the center of the image, and rotating the image so that the axis relative to which the variance of non-zero pixels has the largest value becomes the axis parallel to the horizontal axis. This is also done using PCA, calculated on each image from the dataset separately.

As a result of applying eigendecompostion of the COV matrix, we get a new coordinate system in which our embedding will be expressed. Each of the axes of this system can be interpreted in a similar way to the Eigenfaces approach. The axes responsible for the higher variance of the data have the characteristics of components responsible for the low-frequency deviation from the average image and further axes are responsible for high-frequency deviations. The value of the mean image from our dataset and selected components visualized as 2D images are shown in Fig. 2. The images in our dataset have a resolution of 60×60 so the corresponding vectors have 3600 dimensions. After PCA analysis, we limited the number of dimensions to express 95% of the variance. In the case of our dataset, these were the first 62 dimensions. Thus, in the rest of the paper we will work on the 62-dimensional embedding of our image dataset.

It is also possible to perform feature extraction and dimensionality reduction using deep learning approach using a deep encoder-decoder (E-D) architecture [26,40]. To do this, an E-D network is trained as an autoencoder. Latent space of such trained network is used to generate low dimensional embedding similar to one calculated by PCA.

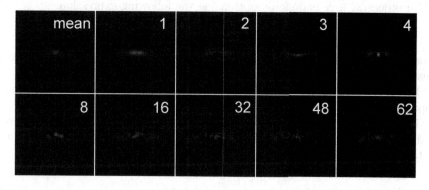

Fig. 2. This figure shows the results of applying PCA analysis to the dataset described in Sect. 2.1. In the upper left corner is the mean image calculated from all images in the dataset. We also show the two-dimensional interpretation of the 1, 2, 3, 4, 8, 16, 42, 48, and 62 axes of the coordinate system defined using PCA. We limited embedding to 62 dimensions (95% of the variance).

2.3   Potential Anomalies Detection

Using the embedding described in Sect. 2.2, anomalous images can be defined as those that are relatively far given the Euclidean metric from the other images. In other words, we want to find such images whose embedding will have the maximum distance from the other objects in the set. We know about the image dataset that it contains several classes of objects, morphologically different from each other (see Fig. 1). Similar objects will form clusters. An additional issue that we discussed before in Sect. 2.1 is that it is difficult to unambiguously determine exactly where in space the boundary between objects that belong to the classes dots, tracks, worms should be defined and which objects might be counted as anomalous images. This has already been pointed out by preparing manual annotations for the CREDO dataset, in which a group of annotators, through a blind voting process, determined to which class each object belongs [29]. As a result of this process, the experiment described in [29] eliminated those objects to which the annotators were uncertain about the class to which they belonged. In our case, of course, we do not make such a selection but we use the entire dataset. Thus, it is natural that if we want to perform unsupervised analysis of dataset, which contains sets of objects against which even human annotators cannot make an unambiguous decision it is reasonable to use an approach that allows modelling uncertainty in the decision-making process. An approach that allows modelling of uncertainty in the clustering process is rough k-means clustering [22]. The algorithm described in [22] adds a number of improvements to the classic k-means algorithm introduced in [14,24]. The object's cluster membership is defined using rough set methodology.

Let us assume that objects are represented by n-dimensional vectors and are contained in the set X. In the classical approach, finding the nearest centroid for an object $x_j \in X$ is done by optimizing the following expression:

$$d(x_j)_{min} = min_{i \in k} E(x_j, c_i) \tag{3}$$

where k is the number of clusters represented by centroids, c_i is the centroid of cluster C_i with index i and E is the Euclidean metric.

Let us assume that $\overline{C_i}$ is upper approximation of cluster C_i and $\underline{C_i}$ is lower approximation of cluster C_i.

In rough k-means the object belongs not only to the closest cluster in terms of distance to the centroid, but also to all other clusters to whose centroids the distance satisfies the condition:

$$\frac{d(x_j)_{min}}{E(x_j, c_l)} \leqslant t, l \neq i \tag{4}$$

where t is the threshold of the method and l is the index of the centroid that does not minimize (3). If $t \leqslant 1$ then rough k-means is performed like a classical k-means. If $t > 1$ then:

- if (4) is satisfied $x_j \in \overline{C_i}, x_j \in \overline{C_l}$ - that means x_j belongs to at least two upper approximations of clusters: ($\overline{C_i}$ and all $\overline{C_i}$ that satisfies (4)),
- if (4) is not satisfied $x_j \in \overline{C_i}, x_j \in \underline{C_i}$

In rough k-means, it is also necessary to modify the centroid updating algorithm, which takes the form of a weighted sum:

$$c_m = w_{upper} \frac{\sum v \in \overline{C_m}}{|\overline{C_m}|} + w_{lower} \frac{\sum v \in \underline{C_m}}{|\underline{C_m}|} \tag{5}$$

where $|\overline{C_m}|$ is cardinal number of set $\overline{C_m}$ and $w_{upper} + w_{lower} = 1$. If $|\overline{C_m}| = 0$ or $|\underline{C_m}| = 0$ then only the component in which the denominator is non-zero is taken into the sum.

To use rough k-means to find anomalies, we need to calculate the distances of each element in the set to the centroid of cluster to which this element has been assigned. Since we want to take into account the uncertainty of assignment to a cluster we use the upper approximation of each cluster. For this reason, each object can be assigned to more than one cluster. Suppose we want to identify p anomalous elements in our dataset. To do this, we order the objects belonging to the upper approximation of each cluster by their distance from the centroid of the cluster they belong to, and take p unique outermost elements.

Data: X—dataset described in Section 2.2,;
$k, t, w_{upper}, w_{lower}$—parameters of rough k-means algorithm described in Section 2.3,;
p—number of potentially anomalous objects to be returned
Result: R—set of p potentially anomalous objects
```
// perform rough k-means
```
$C \leftarrow \mathrm{km}(X, k, t, w_{upper}, w_{lower})$;
```
// calculate distances from centroid, assign a pair: element
    and its distance to centroid to set P
```
$P \leftarrow \varnothing$;
for $x_i \in X$ **do**
    for $c_j \in C$ **do**
        if $x_i \in \overline{C_j}$ **then**
            $P \leftarrow P \cup \{(x_i, E(x_i, c_j)\}$;
        end
    end
end
```
// sort P by distances in descending order
```
$P_{sorted} \leftarrow sort(P)$;
```
// get first p unique objects xi from Psorted
```
$R \leftarrow \varnothing$;
$s \leftarrow 0$;
while $|R| < p$ **do**
    if $P_{sorted}[s] \notin R$ **then**
        $R \leftarrow \{P_{sorted}[s]\}$;
    end
end

Algorithm 1: Algorithm for finding anomalous objects using rough k-means clustering

---

There are three operations in the proposed solution that have significant computational complexity. These are: calculation of the covariance matrix, PCA solved with Singular Value Decomposition (SVD) and rough k-means. These algorithms are performed sequentially one by one. Assuming no parallel computation, the computational complexity is $\mathcal{O}(n^3)$ where n is proportional to the number of elements in the dataset and the resolution of the images.

3   Results

In order to test the proposed method, we prepared its implementation in Python 3.X. We used the packages numpy 1.22, opencv-python 4.5 and the modified package https://github.com/geofizx/rough-clustering so that it can work with Python 3.X. The source codes and data are available for download from the address https://github.com/browarsoftware/particle_pars.

We performed a series of computational experiments involving clustering of the set described in Sect. 2.1. We performed embedding of the set using the method described in Sect. 2.2. Potentially anomalous images were detected using the method described in Sect. 2.3. To ensure the reproducibility of the experiment, the pseudorandom number generator had a seed set. We tested the application of the rough k-means algorithm on the following parameter ranges: $k \in \{3, 4, 5, 6, 7, 8, 9, 10\}$ and $t \in \{1.0, 1.25, 1.5, 1.75, 2.0\}$. The parameters w_{upper} and w_{lower} were set to values of 0.9 and 0.1, respectively. The range of parameters was chosen experimentally so that the method would achieve convergence at $\varepsilon = 10^{-4}$. In order to test the effect of different values of t and k on the set of potential anomalies returned by the method for each rough k-means configuration, we returned a set of R (see Algorithm 1) with a count of 70 objects (about 0.5% of the dataset size). We performed a comparison of the R sets so obtained using Jaccard index. The results are shown in Figs. 3 and 4.

In Figs. 5 and 6 we have presented a visualization of the results of the proposed algorithm for the $(k = 4, t = 1.25, w_{lower} = 0.9, w_{upper} = 0.1)$ and $(k = 4, t = 1)$. The second case is the classic k-means algorithm. The count of R has been set to 25.

4   Discussion

Based on the results shown in Figs. 3 and 4, it can be seen that increasing the parameter k (number of clusters) changes the level of diversity of the returned set of potential anomalies. This manifests itself by decreasing the value of the Jaccard index. In the case of $(k = 10, t > 1.25)$ the Jaccard index is at 0.5. This means that with a relatively large number of clusters, the parameter t plays an increasingly important role in the designation of objects as potential anomalies.

The potential anomalies retrieved with the proposed algorithm, examples of which we have shown in Fig. 5, have morphological features consistent with what a human expert would expect from anomalous signals in this case. They deviate significantly from the visual features typical of the known classes of dots, tracks and worms observations. In contrast, Fig. 6, which shows the results obtained with the classical k-means algorithm, contains single copies of objects that appear to represent well-known classes: dots (fourth row, third column) and tracks (fourth row, fifth column). Analyses carried out in this area have shown

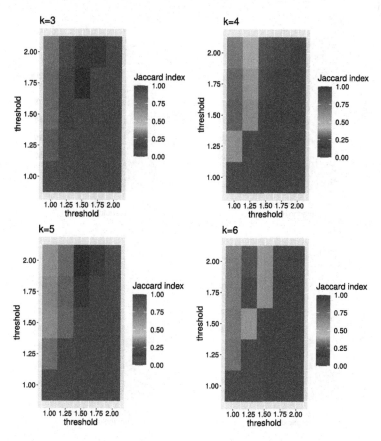

Fig. 3. Comparison of intersecting sets of detected anomalies in terms of the Jaccard coefficient for a fixed $w_{lower} = 0.9$ and a various number of clusters in the range $k \in \{3, 4, 5, 6\}$.

that the spatial distribution of image feature vectors is built in such a way that in some cases we can obtain coherently, clusters composed of non-anomalous objects of relatively large radius. Classical clustering (not using rough sets) does not take into account uncertainty in the assignment of objects to individual clusters and objects are assigned uniquely to a single cluster. In such a case, when searching for objects on the boundary regions of clusters, far from the centroids, objects from clusters that represent unambiguously typical signals may also be indicated as potential anomalies. For this reason, modeling the search for potential anomalies using only crisp sets may yield unsatisfactory results. If, on the other hand, we apply uncertainty modeling using rough sets, then the phenomenon of cluster boundary blurring appears, the properties of which can be controlled using the parameters of the rough k-means algorithm. In practice, this results in the fact that a given object can belong to several clusters in different degrees while being in their upper approximations. This increase the capabilities

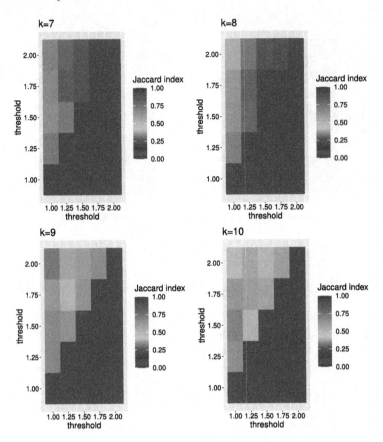

Fig. 4. Comparison of intersecting sets of detected anomalies in terms of the Jaccard coefficient for a fixed $w_{lower} = 0.9$ and a various number of clusters in the range $k \in \{7, 8, 9, 10\}$.

of potential anomalies detection and, as our experiments show, compensates the disadvantages present in classic k-means. We performed a similar visual analysis for the other configurations of rough k-means obtaining analogous results. In conclusion, basing potential anomalies detection on the proposed approach using PCA for embedding generation and rough sets for uncertainty modeling proves to be an effective approach yielding the expected results.

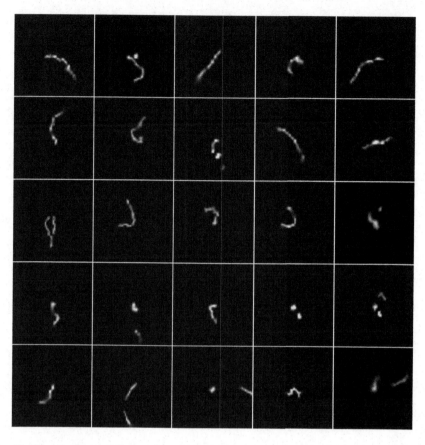

Fig. 5. Visualization of the results of the proposed algorithm for rough k-means with parameters ($k = 4, t = 1.25, w_{lower} = 0.9, w_{upper} = 0.1$). The count of R has been set to 25.

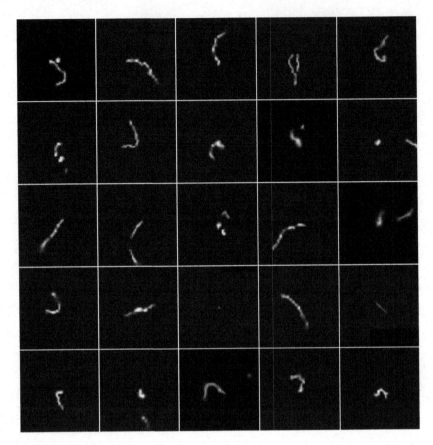

Fig. 6. Visualization of the results of the proposed algorithm for classical k-means ($k = 4$). The count of R has been set to 25.

5 Conclusion

Based on the experiments carried out and the discussion of their results in Sect. 4, we can conclude that the algorithm proposed in this work is an effective approach for finding potential anomalies in cosmic-ray particle tracks images. This is very promising for research in the areas of astrophysics and astronomy, where detectors based on CMOS or similar sensors are used. The approach presented in this work will be further tested for its deployment into the image processing pipeline in research conducted for the CREDO project.

It seems that our proposed solution might not be limited to such a specific problems as the analysis of cosmic rays tracks. The universal features of the proposed rough set-based method for finding anomalies and its efficiency give basis to test its usability also for other types of datasets, not only images. Note that it is very important to select an appropriate set of features describing the objects. In the case of images, the PCA-based method used in this work may be effective.

An open topic for further research is the analysis of the effect of rough k-means parameters on the speed of computation, and whether anomalous signals will not be grouped into clusters with low number of elements as the number of clusters increases. It would also be useful to automatically determine the suboptimal number of clusters that should be used for a given dataset.

References

1. Aartsen, M.G., et al.: The IceCube neutrino observatory: instrumentation and online systems. J. Instrum. **12**(03), P03012 (2017)
2. Afridi, M.K., Azam, N., Yao, J., Alanazi, E.: A three-way clustering approach for handling missing data using GTRS. Int. J. Approximate Reasoning **98**, 11–24 (2018)
3. Allekotte, I., et al.: The surface detector system of the pierre auger observatory. Nucl. Instrum. Methods Phys. Res. Sect. A **586**(3), 409–420 (2008)
4. Avrorin, A., et al.: Baikal-GVD: status and prospects. In: EPJ Web of Conferences, vol. 191, p. 01006. EDP Sciences (2018)
5. Avrorin, A., et al.: Deep-underwater Cherenkov detector in lake Baikal. J. Exp. Theor. Phys. **134**(4), 399–416 (2022)
6. Bar, O., et al.: Zernike moment based classification of cosmic ray candidate hits from CMOs sensors. Sensors **21**(22), 7718 (2021). https://doi.org/10.3390/s21227718. https://www.mdpi.com/1424-8220/21/22/7718
7. Bhuyan, M.H., Bhattacharyya, D.K., Kalita, J.K.: Network anomaly detection: methods, systems and tools. IEEE Commun. Surv. Tutorials **16**(1), 303–336 (2013)
8. Bibrzycki, Ł., et al.: Towards a global cosmic ray sensor network: Credo detector as the first open-source mobile application enabling detection of penetrating radiation. Symmetry **12**(11), 1802 (2020). https://doi.org/10.3390/sym12111802, https://www.mdpi.com/2073-8994/12/11/1802
9. Cai, Z., Guan, X., Shao, P., Peng, Q., Sun, G.: A rough set theory based method for anomaly intrusion detection in computer network systems. Expert. Syst. **20**(5), 251–259 (2003)
10. Cataldi, G., et al.: The upgrade of the Pierre auger observatory with the scintillator surface detector. Proc. Sci. **395**, 251 (2022). https://doi.org/10.22323/1.395.0251
11. Chimphlee, W., Abdullah, A.H., Sap, M.N.M., Chimphlee, S., Srinoy, S.: Unsupervised clustering methods for identifying rare events in anomaly detection. Eng. Technol. **2**, 1 (2005)
12. Chimphlee, W., Abdullah, A.H., Sap, M.N.M., Srinoy, S., Chimphlee, S.: Anomaly-based intrusion detection using fuzzy rough clustering. In: 2006 International Conference on Hybrid Information Technology, vol. 1, pp. 329–334. IEEE (2006)
13. Collaboration, P.A., et al.: The pierre auger cosmic ray observatory. Nucl. Instrum. Methods Phys. Res., Sect. A **798**, 172–213 (2015)
14. Forgey, E.: Cluster analysis of multivariate data: efficiency vs. interpretability of classification. Biometrics **21**(3), 768–769 (1965)
15. Hachaj, T., Koptyra, K., Ogiela, M.R.: Eigenfaces-based steganography. Entropy **23**(3) (2021). https://doi.org/10.3390/e23030273, https://www.mdpi.com/1099-4300/23/3/273

16. Hachaj, T., Piekarczyk, M.: The practice of detecting potential cosmic rays using CMOs cameras: hardware and algorithms. Sensors **23**(10), 4858 (2023). https://doi.org/10.3390/s23104858, https://www.mdpi.com/1424-8220/23/10/4858

17. Homola, P., Beznosko, D., Bhatta, G., Bibrzycki, Ł., et al.: Cosmic-ray extremely distributed observatory. Symmetry **12**(11), 1835 (2020). https://doi.org/10.3390/sym12111835, https://www.mdpi.com/2073-8994/12/11/1835

18. Karbowiak, M., et al.: Small shower array for education purposes-the credo-maze project. Proc. Sci. **395**, 199 (2021)

19. Li, Y., et al.: Classification of bgp anomalies using decision trees and fuzzy rough sets. In: 2014 IEEE International Conference on Systems, Man, and Cybernetics (SMC), pp. 1312–1317 (2014). https://doi.org/10.1109/SMC.2014.6974096

20. Lin, T.: Anomaly detection. In: Proceedings New Security Paradigms Workshop, pp. 44–53 (1994). https://doi.org/10.1109/NSPW.1994.656226

21. Lin, T.: Anomaly detection. In: Proceedings New Security Paradigms Workshop, pp. 44–53. IEEE (1994)

22. Lingras, P., Peters, G.: Applying rough set concepts to clustering. In: Peters, G., Lingras, P., Slezak, D., Yao, Y. (eds.) Rough Sets: Selected Methods and Applications in Management and Engineering. Advanced Information and Knowledge Processing, pp. 23–37. Springer, London (2012). https://doi.org/10.1007/978-1-4471-2760-4_2

23. Liu, H., Zhou, J., Li, H.: Using rough sets to improve the high-dimensional data anomaly detection method based on extended isolation forest. In: 2023 26th International Conference on Computer Supported Cooperative Work in Design (CSCWD), pp. 231–236. IEEE (2023)

24. Lloyd, S.: Least squares quantization in PCM. IEEE Trans. Inf. Theory **28**(2), 129–137 (1982)

25. Mazarbhuiya, F.A.: Detecting anomaly using neighborhood rough set based classification approach (2022). Available at SSRN 4124453

26. Pang, G., Shen, C., Cao, L., Hengel, A.V.D.: Deep learning for anomaly detection: a review. ACM Comput. Surv. (CSUR) **54**(2), 1–38 (2021)

27. Pawlak, Z.: Rough sets. Int. J. Comput. Inf. Sci. **11**, 341–356 (1982)

28. Peters, J.F., Skowron, A., Suraj, Z., Rzasa, W., Borkowski, M.: Clustering: a rough set approach to constructing information granules. In: Soft Computing and Distributed Processing, Proceedings of 6th International Conference, SCDP, vol. 5761 (2002)

29. Piekarczyk, M., Bar, O., Bibrzycki, Ł., Niedźwiecki, M., et al.: CNN-based classifier as an offline trigger for the CREDO experiment. Sensors **21**(14), 4804 (2021). https://doi.org/10.3390/s21144804, https://www.mdpi.com/1424-8220/21/14/4804

30. Pięta, P., Szmuc, T.: Applications of rough sets in big data analysis: an overview. Int. J. Appl. Math. Comput. Sci. **31**(4), 659–683 (2021)

31. Pryga, J., et al.: Analysis of the capability of detection of extensive air showers by simple scintillator detectors. Universe **8**(8), 425 (2022)

32. Rawat, S.S., Polavarapu, V.A., Kumar, V., Aruna, E., Sumathi, V.: Anomaly detection in smart grid using rough set theory and k cross validation. In: 2014 International Conference on Circuits, Power and Computing Technologies [ICCPCT-2014], pp. 479–483. IEEE (2014)

33. Riza, L.S., et al.: Implementing algorithms of rough set theory and fuzzy rough set theory in the r package "roughsets". Inf. Sci. **287**, 68–89 (2014)

34. Skowron, A., Dutta, S.: Rough sets: past, present, and future. Nat. Comput. **17**, 855–876 (2018)

35. Skowron, A., Ślęzak, D.: Rough sets turn 40: From information systems to intelligent systems. In: 2022 17th Conference on Computer Science and Intelligence Systems (FedCSIS), pp. 23–34. IEEE (2022)
36. Stasielak, J., et al.: High-energy neutrino astronomy-baikal-gvd neutrino telescope in lake baikal. Symmetry **13**(3), 377 (2021)
37. Taha, A., Hadi, A.S.: Anomaly detection methods for categorical data: a review. ACM Comput. Surv. (CSUR) **52**(2), 1–35 (2019)
38. Turk, M., Pentland, A.: Face recognition using eigenfaces. In: Proceedings. 1991 IEEE Computer Society Conference on Computer Vision and Pattern Recognition, pp. 586–591 (1991). https://doi.org/10.1109/CVPR.1991.139758
39. Wang, P., Yao, Y.: CE3: a three-way clustering method based on mathematical morphology. Knowl.-Based Syst. **155**, 54–65 (2018)
40. Wei, R., Mahmood, A.: Recent advances in variational autoencoders with representation learning for biomedical informatics: a survey. IEEE Access **9**, 4939–4956 (2021). https://doi.org/10.1109/ACCESS.2020.3048309
41. Yuan, Z., Chen, B., Liu, J., Chen, H., Peng, D., Li, P.: Anomaly detection based on weighted fuzzy-rough density. Appl. Soft Comput. **134**, 109995 (2023). https://doi.org/10.1016/j.asoc.2023.109995, https://www.sciencedirect.com/science/article/pii/S1568494623000133
42. Zeng, F., Yin, K., Chen, M., Wang, X.: A new anomaly detection method based on rough set reduction and hmm. In: 2009 Eighth IEEE/ACIS International Conference on Computer and Information Science, pp. 285–289 (2009). https://doi.org/10.1109/ICIS.2009.140
43. Zeng, F., Yin, K., Chen, M., Wang, X.: A new anomaly detection method based on rough set reduction and hmm. In: 2009 Eighth IEEE/ACIS International Conference on Computer and Information Science, pp. 285–289. IEEE (2009)

Hybrid Approaches

Crisp-Fuzzy Concept Lattice Based on Interval-Valued Fuzzy Sets

Tong-Jun Li[1,2]($\boxtimes$) and Yi-Qian Wang[1]

[1] School of Information Engineering, Zhejiang Ocean University, Zhoushan 316022, Zhejiang, China
ltj72@126.com
[2] Key Laboratory of Oceanographic Big Data Mining and Application of Zhejiang Province, Zhejiang Ocean University, Zhoushan 316022, Zhejiang, China

Abstract. Fuzzy concept lattices can be viewed as the generalizations of the classical concept lattices in fuzzy formal contexts, which is a key issue and a major research direction in knowledge discovery. Crisp-fuzzy concept lattices are special fuzzy concept lattices, the existing crisp-fuzzy concept lattices can be divided into two categories, that is, one is the extension of the classical concept lattice, and the other is based on rough fuzzy approximation operations. In this paper, by combing these two types of crisp-fuzzy concept lattices and using interval-valued fuzzy sets, a novel crisp-fuzzy concept lattice is firstly presented, then the properties of the new model are discussed in detail. From two aspects of granular and algebraic structures, the new concept lattice is compared with two types of existing crisp-fuzzy concept lattices, which shows that the former has obvious advantages over the latter. Therefore, the work has not only enriched the theory of fuzzy concept lattice, but helpful for its application.

Keywords: Crisp-fuzzy concepts · Fuzzy formal contexts · Interval-valued fuzzy sets · Formal concept analysis

1 Introduction

Formal Concept Analysis (FCA), firstly proposed by German mathematician Wille [1] in 1982, serves as an effective tool for data analysis and knowledge discovery. As a significant research direction of artificial intelligence, FCA has been widely used in information retrieval, cognitive concept learning, rule acquisition, and other fields [2–4].

FCA deals with a set of data called a formal context, formal concepts are the primary knowledge units extracted from the formal context, and the relationship among the formal concepts indicates that they form a complete lattice in a mathematical sense, the complete lattice is also called a concept lattice. Each formal concept consists of extension and intension, where the extension is a set of objects covered by this concept, and the intension is a set of attributes shared by all the objects of extension.

A. Campagner et al. (Eds.): IJCRS 2023, LNAI 14481, pp. 449–462, 2023.
https://doi.org/10.1007/978-3-031-50959-9_31

A formal context in the classical FCA is mainly depicted by an ordinary binary relation between the object set and the attribute set, however, in a large number of practical applications, binary fuzzy relations need to be dealt with. Therefore, the extension of the classical concept lattices in fuzzy environments become an important research issue. Thus fuzzy sets were initially brought into FCA by Burusco and Fuentes-Gonzalez [5], so L-fuzzy concept lattices were proposed based on residuated lattice theory. From then on, various fuzzy concept lattice models have been presented successively [6,7]. In order to enhance the practicality of fuzzy concepts, one-sided concept lattices were suggested [8,9], the "one-sided" means that for the extension and intension of each concept, one is a crisp set, and the other a fuzzy set. Many achievements have been made for the one-sided concept lattices [10,11].

As for the generalization of the classical concept lattices, the use of the approximation operations of rough set theory is a better choice. Duntsch and Gediga [12] first put forward a property-oriented concept lattice by a pair of modal operators, Yao [13] further constructed an object-oriented concept lattice, they are known as rough concept lattices collectively. By introducing three-way decision theory [14] into the rough concept lattices, Wei and Qian [15] defined the three-way object-oriented and property-oriented concept lattices. Based on approximation operations, He et al. [16] proposed a property-oriented interval-set concept lattice in an incomplete formal context. Li et al. [17] applied fuzzy rough approximation operations to fuzzy formal contexts, and came up with a crisp-fuzzy concept lattice, which pioneered the application of rough approximation on one-sided concept lattices.

Interval-valued fuzzy sets are extensions of classical fuzzy sets. In many practical matters, it is more advantageous to utilize interval-valued fuzzy sets to describe uncertainty. However, there is little research on fuzzy concept analysis by interval-valued fuzzy sets. This paper introduces interval-valued fuzzy sets into one-sided concept lattices, so a novel crisp-fuzzy concept lattice is proposed in a fuzzy formal context, a main characteristic of which is that the intensions of all concepts are interval-valued fuzzy sets. After analyzing the basic properties of the new concept lattice, the relationships between it and two existing crisp-fuzzy concept lattices are investigated.

2   Preliminary

This section primarily recalls the relevant knowledge of two one-sided concept lattices in fuzzy formal contexts and interval-valued fuzzy sets.

2.1   Existing Crisp-Fuzzy Concepts

Let U be a nonempty and finite set, and the class of all subsets of U and the class of all fuzzy sets on U will be denoted $P(U)$ and $F(U)$, respectively. Here $\widetilde{X} \in F(U)$ mains a mapping from U to [0,1].

Definition 1. *A fuzzy formal context is a triplet $(U, A, \tilde{I})$, in which U is a nonempty and finite set, called the object set, whose elements are known as objects, A is a nonempty and finite set, called the attribute set, whose elements are called attributes, and $\tilde{I}$ is a fuzzy set on $U \times A$, that is, a binary fuzzy relation from U to A.*

For $x \in U$ and $a \in A$, $x\tilde{I}$ is a fuzzy set on A, and $\tilde{I}a$ is a fuzzy set on U, which are defined respectively as:

$$(x\tilde{I})(b) = \tilde{I}(x, b), \quad b \in A; \quad (\tilde{I}a)(y) = \tilde{I}(y, a), \quad y \in U.$$

We assume that the following fuzzy formal context $(U, A, \tilde{I})$ satisfies:

$$x\tilde{I} \neq y\tilde{I}, \quad x, y \in U; \quad \tilde{I}a \neq \tilde{I}b, \quad a, b \in A,$$

and such a fuzzy formal context is said to be clarified.

Example 1. Table 1 shows a fuzzy formal context $(U, A, \tilde{I})$, where

$$U = \{x_1, x_2, x_3, x_4, x_5\}, \quad A = \{a, b, c, d, e\},$$

and the fuzzy relation $\tilde{I}$ can be read from Table 1.

Table 1. A fuzzy formal context $(U, A, \tilde{I})$ in Example 1

	a	b	c	d	e
x_1	0.56	0.20	0.72	0.35	0.52
x_2	0.73	0.83	0.30	0.41	0.29
x_3	0.19	0.48	0.57	0.61	0.00
x_4	0.44	0.67	0.15	0.49	0.36
x_5	0.28	0.59	0.44	0.50	0.17

Krajci and Yahia et al. [8,9] first proposed a kind of crisp-fuzzy concepts in fuzzy formal contexts independently.

For a fuzzy formal context $(U, A, \tilde{I})$, two operators $* : P(U) \to F(A)$ and $* : F(A) \to P(U)$ are defined as follows:

$$X^* = \bigcap_{x \in X} x\tilde{I}, \quad X \in P(U); \quad \tilde{B}^* = \{x \in U \mid \tilde{B} \subseteq x\tilde{I}\}, \quad \tilde{B} \in F(A).$$

For $X \in P(U)$ and $\tilde{B} \in F(A)$, if $X^* = \tilde{B}$, $\tilde{B}^* = X$, then $(X, \tilde{B})$ is called a crisp-fuzzy concept of $(U, A, \tilde{I})$ or a type-I concept, and X and $\tilde{B}$ are known as the extension and intension of $(X, \tilde{B})$, respectively. The set of all type-I concepts is denoted $L^I(\tilde{I})$, the set of the extensions of all type-I concepts is denoted $Ext^I(\tilde{I})$, and the set of the intensions of all type-I concepts is denoted $Int^I(\tilde{I})$.

Table 2. All the type-I concepts of $(U, A, \tilde{I})$ in Example 1

Number	(Extension, Intension)	Number	(Extension, Intension)
FC_1^{I}	$(\{x_1, x_2, x_3, x_4, x_5\}, \{a^{0.19}, b^{0.20}, c^{0.15}, d^{0.35}, e^{0.00}\})$	FC_{13}^{I}	$(\{x_1, x_4\}, \{a^{0.44}, b^{0.20}, c^{0.15}, d^{0.35}, e^{0.36}\})$
FC_2^{I}	$(\{x_1, x_2, x_3, x_5\}, \{a^{0.19}, b^{0.20}, c^{0.30}, d^{0.35}, e^{0.00}\})$	FC_{14}^{I}	$(\{x_1, x_5\}, \{a^{0.28}, b^{0.20}, c^{0.44}, d^{0.35}, e^{0.17}\})$
FC_3^{I}	$(\{x_1, x_2, x_4, x_5\}, \{a^{0.28}, b^{0.20}, c^{0.15}, d^{0.35}, e^{0.17}\})$	FC_{15}^{I}	$(\{x_2, x_4\}, \{a^{0.44}, b^{0.67}, c^{0.15}, d^{0.41}, e^{0.29}\})$
FC_4^{I}	$(\{x_2, x_3, x_4, x_5\}, \{a^{0.19}, b^{0.48}, c^{0.15}, d^{0.41}, e^{0.00}\})$	FC_{16}^{I}	$(\{x_2, x_5\}, \{a^{0.28}, b^{0.59}, c^{0.30}, d^{0.41}, e^{0.17}\})$
FC_5^{I}	$(\{x_1, x_2, x_4\}, \{a^{0.44}, b^{0.20}, c^{0.15}, d^{0.35}, e^{0.29}\})$	FC_{17}^{I}	$(\{x_3, x_5\}, \{a^{0.19}, b^{0.48}, c^{0.44}, d^{0.50}, e^{0.00}\})$
FC_6^{I}	$(\{x_1, x_2, x_5\}, \{a^{0.28}, b^{0.20}, c^{0.30}, d^{0.35}, e^{0.17}\})$	FC_{18}^{I}	$(\{x_4, x_5\}, \{a^{0.28}, b^{0.59}, c^{0.15}, d^{0.49}, e^{0.17}\})$
FC_7^{I}	$(\{x_1, x_3, x_5\}, \{a^{0.19}, b^{0.20}, c^{0.44}, d^{0.35}, e^{0.00}\})$	FC_{19}^{I}	$(\{x_1\}, \{a^{0.56}, b^{0.20}, c^{0.72}, d^{0.35}, e^{0.52}\})$
FC_8^{I}	$(\{x_2, x_3, x_5\}, \{a^{0.19}, b^{0.48}, c^{0.30}, d^{0.41}, e^{0.00}\})$	FC_{20}^{I}	$(\{x_2\}, \{a^{0.73}, b^{0.83}, c^{0.30}, d^{0.41}, e^{0.29}\})$
FC_9^{I}	$(\{x_2, x_4, x_5\}, \{a^{0.28}, b^{0.59}, c^{0.15}, d^{0.41}, e^{0.17}\})$	FC_{21}^{I}	$(\{x_3\}, \{a^{0.19}, b^{0.48}, c^{0.57}, d^{0.61}, e^{0.00}\})$
FC_{10}^{I}	$(\{x_3, x_4, x_5\}, \{a^{0.19}, b^{0.48}, c^{0.15}, d^{0.49}, e^{0.00}\})$	FC_{22}^{I}	$(\{x_4\}, \{a^{0.44}, b^{0.67}, c^{0.15}, d^{0.49}, e^{0.36}\})$
FC_{11}^{I}	$(\{x_1, x_2\}, \{a^{0.56}, b^{0.20}, c^{0.30}, d^{0.35}, e^{0.29}\})$	FC_{23}^{I}	$(\{x_5\}, \{a^{0.28}, b^{0.59}, c^{0.44}, d^{0.50}, e^{0.17}\})$
FC_{12}^{I}	$(\{x_1, x_3\}, \{a^{0.19}, b^{0.20}, c^{0.57}, d^{0.35}, e^{0.00}\})$	FC_{24}^{I}	$(\{\varnothing\}, \{a^{1.00}, b^{1.00}, c^{1.00}, d^{1.00}, e^{1.00}\})$

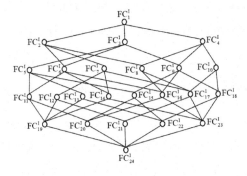

Fig. 1. The concept lattice $L^{\mathrm{I}}(\tilde{I})$ of $(U, A, \tilde{I})$ in Example 1

Example 2. For the fuzzy formal context $(U, A, \tilde{I})$ in Example 1, all the type-I concepts of $(U, A, \tilde{I})$ are listed in Table 2, and Fig. 1 is the concept lattice $L^{\mathrm{I}}(\tilde{I})$.

Based on rough approximation operations, Li and Wu [17] defined another kind of crisp-fuzzy concepts in fuzzy formal contexts.

In a fuzzy formal context $(U, A, \tilde{I})$, two operators $\Diamond : P(U) \to F(A)$ and $\Box : F(A) \to P(U)$ are defined as follows:

$$X^{\Diamond} = \bigcup_{x \in X} x\tilde{I}, \ X \in P(U); \quad \tilde{B}^{\Box} = \{x \in U \mid x\tilde{I} \subseteq \tilde{B}\}, \ \tilde{B} \in F(A).$$

For $X \in P(U)$ and $\tilde{B} \in F(A)$, if $X^{\Diamond} = \tilde{B}$, $\tilde{B}^{\Box} = X$, then $(X, \tilde{B})$ is called a crisp-fuzzy concept of $(U, A, \tilde{I})$ based on rough fuzzy approximations of the first kind or a type-II concept, in which X and $\tilde{B}$ are known as the extension and intension of $(X, \tilde{B})$, respectively. The set of all type-II concepts is denoted $L^{\mathrm{II}}(\tilde{I})$, the set of the extensions of all type-II concepts is denoted $Ext^{\mathrm{II}}(\tilde{I})$, and the set of the intensions of all type-II concepts is denoted $Int^{\mathrm{II}}(\tilde{I})$.

Example 3. For the fuzzy formal context in Example 1, all the type-II concepts of $(U, A, \tilde{I})$ are shown in Table 3, and Fig. 2 depicts the concept lattice $L^{\mathrm{II}}(\tilde{I})$.

Table 3. All the type-II concepts of $(U, A, \tilde{I})$ in Example 1

Number	(Extension, Intension)	Number	(Extension, Intension)
FC_1^{II}	$(\{x_1,x_2,x_3,x_4,x_5\}, \{a^{0.73}, b^{0.83}, c^{0.72}, d^{0.61}, e^{0.52}\})$	FC_{13}^{II}	$(\{x_1,x_4\}, \{a^{0.56}, b^{0.67}, c^{0.72}, d^{0.49}, e^{0.52}\})$
FC_2^{II}	$(\{x_1,x_2,x_4,x_5\}, \{a^{0.73}, b^{0.83}, c^{0.72}, d^{0.50}, e^{0.52}\})$	FC_{14}^{II}	$(\{x_1,x_5\}, \{a^{0.56}, b^{0.59}, c^{0.72}, d^{0.50}, e^{0.52}\})$
FC_3^{II}	$(\{x_1,x_3,x_4,x_5\}, \{a^{0.56}, b^{0.67}, c^{0.72}, d^{0.61}, e^{0.52}\})$	FC_{15}^{II}	$(\{x_2,x_4\}, \{a^{0.73}, b^{0.83}, c^{0.30}, d^{0.49}, e^{0.36}\})$
FC_4^{II}	$(\{x_2,x_3,x_4,x_5\}, \{a^{0.73}, b^{0.83}, c^{0.57}, d^{0.61}, e^{0.36}\})$	FC_{16}^{II}	$(\{x_2,x_5\}, \{a^{0.73}, b^{0.83}, c^{0.44}, d^{0.50}, e^{0.29}\})$
FC_5^{II}	$(\{x_1,x_2,x_4\}, \{a^{0.73}, b^{0.83}, c^{0.72}, d^{0.49}, e^{0.52}\})$	FC_{17}^{II}	$(\{x_3,x_5\}, \{a^{0.28}, b^{0.59}, c^{0.57}, d^{0.61}, e^{0.17}\})$
FC_6^{II}	$(\{x_1,x_3,x_5\}, \{a^{0.56}, b^{0.59}, c^{0.72}, d^{0.61}, e^{0.52}\})$	FC_{18}^{II}	$(\{x_4,x_5\}, \{a^{0.44}, b^{0.67}, c^{0.44}, d^{0.50}, e^{0.36}\})$
FC_7^{II}	$(\{x_1,x_4,x_5\}, \{a^{0.56}, b^{0.67}, c^{0.72}, d^{0.50}, e^{0.52}\})$	FC_{19}^{II}	$(\{x_1\}, \{a^{0.56}, b^{0.20}, c^{0.72}, d^{0.35}, e^{0.52}\})$
FC_8^{II}	$(\{x_2,x_3,x_5\}, \{a^{0.73}, b^{0.83}, c^{0.57}, d^{0.61}, e^{0.29}\})$	FC_{20}^{II}	$(\{x_2\}, \{a^{0.73}, b^{0.83}, c^{0.30}, d^{0.41}, e^{0.29}\})$
FC_9^{II}	$(\{x_2,x_4,x_5\}, \{a^{0.73}, b^{0.83}, c^{0.44}, d^{0.50}, e^{0.36}\})$	FC_{21}^{II}	$(\{x_3\}, \{a^{0.19}, b^{0.48}, c^{0.57}, d^{0.61}, e^{0.00}\})$
FC_{10}^{II}	$(\{x_3,x_4,x_5\}, \{a^{0.44}, b^{0.67}, c^{0.57}, d^{0.61}, e^{0.36}\})$	FC_{22}^{II}	$(\{x_4\}, \{a^{0.44}, b^{0.67}, c^{0.15}, d^{0.49}, e^{0.36}\})$
FC_{11}^{II}	$(\{x_1,x_2\}, \{a^{0.73}, b^{0.83}, c^{0.72}, d^{0.41}, e^{0.52}\})$	FC_{23}^{II}	$(\{x_5\}, \{a^{0.28}, b^{0.59}, c^{0.44}, d^{0.50}, e^{0.17}\})$
FC_{12}^{II}	$(\{x_1,x_3\}, \{a^{0.56}, b^{0.48}, c^{0.72}, d^{0.61}, e^{0.52}\})$	FC_{24}^{II}	$(\{\varnothing\}, \{a^{0.00}, b^{0.00}, c^{0.00}, d^{0.00}, e^{0.00}\})$

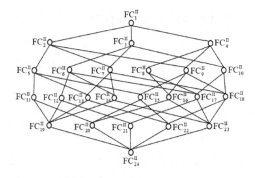

Fig. 2. The concept lattice $L^{II}(\tilde{I})$ of $(U, A, \tilde{I})$ in Example 1

The operators $\Diamond$ and $\Box$ satisfy the properties of upper and lower approximation operators of rough set theory, so they can be considered as two kinds of rough approximation operators, and the corresponding crisp-fuzzy concepts are viewed as a kind of crisp-fuzzy concepts based on rough approximations.

2.2 Interval-Valued Fuzzy Set

The closed interval (also called interval) in the set $\mathbf{R}$ of real numbers is denoted $[a, b]$, where $a, b \in \mathbf{R}$. If $a < b$, then $[a, b]$ represents an ordinary real interval; If $a = b$, then $[a, a]$ degenerates into the real number a; If $a > b$, then $[a, b]$ expresses the empty set $\varnothing$.

In the following, we only consider the intervals contained in the unit interval $I = [0, 1]$, and the class of all such intervals will be denoted $IP([0, 1])$ or $IP(I)$.

Two operations on $IP(I)$ are defined as follows: Let $[a_1, b_1] \in IP(I)$, $[a_2, b_2] \in IP(I)$, then

– the intersection of $[a_1, b_1]$ and $[a_2, b_2]$, denoted $[a_1, b_1] \cap [a_2, b_2]$, is given as:

$$[a_1, b_1] \cap [a_2, b_2] = [a_1 \vee a_2, b_1 \wedge b_2];$$

– the union of $[a_1, b_1]$ and $[a_2, b_2]$, denoted $[a_1, b_1] \cup [a_2, b_2]$, is given as:

$$[a_1, b_1] \cup [a_2, b_2] = [a_1 \wedge a_2, b_1 \vee b_2].$$

where $\wedge$ and $\vee$ represent the operations minimum and maximum on [0,1].

It is easy to verify that the corresponding order relation on $IP(I)$ coincides with the inclusion relation of crisp sets, and still be denoted $\subseteq$.

For the universe of discourse U, an interval-valued fuzzy set $\hat{X}$ on U is defined as a mapping from U to $IP(I)$, i.e. $\hat{X} : U \rightarrow IP(I)$. For $x \in U$, $\hat{X}(x)$ is represented as $[\hat{X}^-(x), \hat{X}^+(x)]$, then $\hat{X}$ can be expressed as $[\hat{X}^-, \hat{X}^+]$. It is clear that $\hat{X}^-$ and $\hat{X}^+$ are two classical fuzzy sets on U with $\hat{X}^- \subseteq \hat{X}^+$. Denote the set of all interval-valued fuzzy sets on U as $IF(U)$.

If $U = \{x_1, x_2, \cdots, x_n\}$ then $\hat{X} \in IF(U)$ can be expressed as

$$\{x_1^{[\hat{X}^-(x_1), \hat{X}^+(x_1)]}, x_2^{[\hat{X}^-(x_2), \hat{X}^+(x_2)]}, \cdots, x_n^{[\hat{X}^-(x_n), \hat{X}^+(x_n)]}\}.$$

Of course, the fuzzy set $\tilde{X} \in F(U)$ should be denoted

$$\{x_1^{\tilde{X}(x_1)}, x_2^{\tilde{X}(x_2)}, \cdots, x_n^{\tilde{X}(x_n)}\}.$$

For any $\tilde{X} \in F(U)$ and $\hat{Y} \in IF(U)$, if $\forall x \in U$, $\tilde{X}(x) \in \hat{Y}(x)$, we say that $\tilde{X}$ belongs to $\hat{Y}$, which is denoted $\tilde{X} \in \hat{Y}$.

Pointwise application of the inclusion relation $\subseteq$, and the intersection $\cap$ and union $\cup$ operations on $IP(I)$ leads to a relation and two operations on $IF(U)$.

Let $\hat{X} \in IF(U)$, $\hat{Y} \in IF(U)$. Then:

– $\hat{X}$ being included in $\hat{Y}$, denoted $\hat{X} \subseteq \hat{Y}$, is defined as:

$$\hat{X}(x) \subseteq \hat{Y}(x), \quad \forall x \in U;$$

– the intersection of $\hat{X}$ and $\hat{Y}$, denoted $\hat{X} \cap \hat{Y}$, is defined as:

$$(\hat{X} \cap \hat{Y})(x) = \hat{X}(x) \cap \hat{Y}(x), \quad \forall x \in U;$$

– the union of $\hat{X}$ and $\hat{Y}$, denoted $\hat{X} \cup \hat{Y}$, is defined as:

$$(\hat{X} \cup \hat{Y})(x) = \hat{X}(x) \cup \hat{Y}(x), \quad \forall x \in U.$$

We can see that for $\tilde{X} \in F(U)$, $\hat{Y}_1, \hat{Y}_2 \in IF(U)$, if $\hat{Y}_1 \subseteq \hat{Y}_2$, then $\tilde{X} \in \hat{Y}_1$ implies $\tilde{X} \in \hat{Y}_2$.

Proposition 1. *Let U be a universe of discourse and $\hat{X}, \hat{Y} \in IF(U)$. Then*

(1) $\hat{X} \subseteq \hat{Y} \Leftrightarrow \hat{Y}^- \subseteq \hat{X}^-$, $\hat{X}^+ \subseteq \hat{Y}^+$;
(2) $(\hat{X} \cap \hat{Y})^- = \hat{X}^- \cup \hat{Y}^-$, $(\hat{X} \cap \hat{Y})^+ = \hat{X}^+ \cap \hat{Y}^+$;
(3) $(\hat{X} \cup \hat{Y})^- = \hat{X}^- \cap \hat{Y}^-$, $(\hat{X} \cup \hat{Y})^+ = \hat{X}^+ \cup \hat{Y}^+$.

Proposition 2. *Let U be a universe of discourse and $\hat{X}, \hat{Y} \in IF(U)$. Then*

(1) $\hat{X} \cap \hat{Y} \subseteq \hat{X}$, $\hat{X} \cap \hat{Y} \subseteq \hat{Y}$;
(2) $\hat{X} \subseteq \hat{X} \cup \hat{Y}$, $\hat{Y} \subseteq \hat{X} \cup \hat{Y}$;
(3) $\hat{X} \subseteq \hat{Y} \Leftrightarrow \hat{X} \cap \hat{Y} = \hat{X} \Leftrightarrow \hat{X} \cup \hat{Y} = \hat{Y}$.

3 Novel Crisp-Fuzzy Concepts

Definition 2. *Let $(U, A, \tilde{I})$ be a fuzzy formal context. For $X \in P(U)$ and $\hat{B} \in IF(A)$, a interval-valued fuzzy set $X^\#$ on A and a crisp subset $\hat{B}^\&$ of U are defined as follows:*

$$X^\#(a) = \left[\bigwedge_{x \in X}(x\tilde{I})(a), \bigvee_{x \in X}(x\tilde{I})(a) \right], \quad \forall a \in A,$$
$$\hat{B}^\& = \{x \in U \mid \forall a \in A, \hat{B}^-(a) \le \tilde{I}(x,a) \le \hat{B}^+(a)\}.$$

The following conclusion follows from Definition 2.

Proposition 3. *Let $(U, A, \tilde{I})$ be a fuzzy formal context, $X \in P(U)$, $\hat{B} \in IF(A)$. Then*

(1) $X^\# = [X^*, X^\diamond]$, i.e. $(X^\#)^- = X^*$, $(X^\#)^+ = X^\diamond$;

(2) $\hat{B}^\& = (\hat{B}^-)^* \cap (\hat{B}^+)^\square$.

Proposition 4. *Let $(U, A, \tilde{I})$ be a fuzzy formal context, $X, X_1, X_2 \in P(U)$, $\hat{B}, \hat{B}_1, \hat{B}_2 \in IF(A)$. Then*

(1) $X_1 \subseteq X_2 \Rightarrow X_1^\# \subseteq X_2^\#$, $\hat{B}_1 \subseteq \hat{B}_2 \Rightarrow \hat{B}_1^\& \subseteq \hat{B}_2^\&$;

(2) $(X_1 \cup X_2)^\# = X_1^\# \cup X_2^\#$, $(X_1 \cap X_2)^\# \subseteq X_1^\# \cap X_2^\#$;

(3) $(\hat{B}_1 \cap \hat{B}_2)^\& = \hat{B}_1^\& \cap \hat{B}_2^\&$, $(\hat{B}_1 \cup \hat{B}_2)^\& \supseteq \hat{B}_1^\& \cup \hat{B}_2^\&$;

(4) $X \subseteq X^{\#\&}$, $\hat{B}^{\&\#} \subseteq \hat{B}$;

(5) $X^{\#\&\#} = X^\#$, $\hat{B}^{\&\#\&} = \hat{B}^\&$;

(6) $X \subseteq \hat{B}^\& \Leftrightarrow X^\# \subseteq \hat{B}$.

Proof. (1) If $X_1 \subseteq X_2$, then $\bigcap_{x \in X_2} x\tilde{I} \subseteq \bigcap_{x \in X_1} x\tilde{I}$ and $\bigcup_{x \in X_1} x\tilde{I} \subseteq \bigcup_{x \in X_2} x\tilde{I}$, i.e.
$(X_2^\#)^- \subseteq (X_1^\#)^-$ and $(X_1^\#)^+ \subseteq (X_2^\#)^+$. Hence $X_1^\# \subseteq X_2^\#$.

If $\hat{B}_1 \subseteq \hat{B}_2$, then for any $x \in U$, $x\tilde{I} \in \hat{B}_1$ implies $x\tilde{I} \in \hat{B}_2$, by Definition 2 we have $\hat{B}_1^\& \subseteq \hat{B}_2^\&$.

(2) According to Definition 2 we obtain

$$(X_1 \cup X_2)^{\#-} = \bigcap_{x \in X_1 \cup X_2} x\tilde{I} = \left(\bigcap_{x \in X_1} x\tilde{I} \right) \cap \left(\bigcap_{x \in X_2} x\tilde{I} \right) = (X_1^\#)^- \cap (X_2^\#)^-,$$
$$(X_1 \cup X_2)^{\#+} = \bigcup_{x \in X_1 \cup X_2} x\tilde{I} = \left(\bigcup_{x \in X_1} x\tilde{I} \right) \cup \left(\bigcup_{x \in X_2} x\tilde{I} \right) = (X_1^\#)^+ \cup (X_2^\#)^+.$$

Hence $(X_1 \cup X_2)^\# = X_1^\# \cup X_2^\#$.

It immediately follows from (1) that $(X_1 \cap X_2)^\# \subseteq X_1^\# \cap X_2^\#$.

(3) For any $x \in U$, we get

$$(\hat{B}_1 \cap \hat{B}_2)^- \subseteq x\tilde{I} \subseteq (\hat{B}_1 \cap \hat{B}_2)^+ \Leftrightarrow \hat{B}_1^- \cup \hat{B}_2^- \subseteq x\tilde{I} \subseteq \hat{B}_1^+ \cap \hat{B}_2^+$$
$$\Leftrightarrow \hat{B}_1^- \subseteq x\tilde{I} \subseteq \hat{B}_1^-, \hat{B}_2^+ \subseteq x\tilde{I} \subseteq \hat{B}_2^+.$$

According to Definition 2, it can be concluded that $(\hat{B}_1 \cap \hat{B}_2)^{\&} = \hat{B}_1^{\&} \cap \hat{B}_2^{\&}$.

It can be directly derived from (1) that $(\hat{B}_1 \cup \hat{B}_2)^{\&} \supseteq \hat{B}_1^{\&} \cup \hat{B}_2^{\&}$.

(4) From $X^{\#} = [(X^{\#})^-, (X^{\#})^+] = \left[\bigwedge_{x \in X}(x\tilde{I}), \bigvee_{x \in X}(x\tilde{I}) \right]$ it follows that for

any $x \in X$, $(X^{\#})^- \subseteq x\tilde{I} \subseteq (X^{\#})^+$, hence $X \subseteq X^{\#\&}$.

According to $\hat{B}^{\&} = \{x \in U \mid \hat{B}^- \subseteq x\tilde{I} \subseteq \hat{B}^+\}$, we assert $\hat{B}^- \subseteq (\hat{B}^{\&\#})^-$, $(\hat{B}^{\&\#})^+ \subseteq \hat{B}^+$, hence $\hat{B}^{\&\#} \subseteq \hat{B}$.

(5) In terms of (1) and (4), we get $X^{\#} \subseteq X^{\#\&\#}$. Substituting $X^{\#}$ for $\hat{B}$ in $\hat{B}^{\&\#} \subseteq \hat{B}$ gives $X^{\#\&\#} \subseteq X^{\#}$. Thus $X^{\#} = X^{\#\&\#}$.

Equally, we can show that $\hat{B}^{\&\#\&} = \hat{B}^{\&}$.

(6) For $X \in P(U)$, $\hat{B} \in IF(A)$, we have

$$X \subseteq \hat{B}^{\&} \Leftrightarrow \forall x \in X(x \in \hat{B}^{\&}) \Leftrightarrow \forall x \in X(\hat{B}^- \subseteq x\tilde{I} \subseteq \hat{B}^+)$$
$$\Leftrightarrow \hat{B}^- \subseteq \bigcap_{x \in X} x\tilde{I} \subseteq \bigcup_{x \in X} x\tilde{I} \subseteq \hat{B}^+ \Leftrightarrow X^{\#} \subseteq \hat{B}.$$

Definition 3. *Let $(U, A, \tilde{I})$ be a fuzzy formal context, $X \in P(U)$, $\hat{B} \in IF(A)$. If $X^{\#} = \hat{B}$, $\hat{B}^{\&} = X$, then $(X, \hat{B})$ is called a crisp-fuzzy concept of $(U, A, \tilde{I})$ based on rough fuzzy approximations of the second kind or a type-III concept, where X and $\hat{B}$ are known as the extension and intension of $(X, \hat{B})$, respectively.*

The set of all type-III concepts of $(U, A, \tilde{I})$ is denoted $L^{III}(\tilde{I})$, the set of the extensions of all type-III concepts is denoted $Ext^{III}(\tilde{I})$, and the set of the intensions of all type-III concepts is denoted $Int^{III}(\tilde{I})$.

A partially ordered relation on $L^{III}(\tilde{I})$ is defined as follows: For $(X_1, \hat{B}_1) \in L^{III}(\tilde{I})$, $(X_2, \hat{B}_2) \in L^{III}(\tilde{I})$, then

$$(X_1, \hat{B}_1) \leq (X_2, \hat{B}_2) \Leftrightarrow X_1 \subseteq X_2$$

The partially ordered relation on $L^{I}(\tilde{I})$ and $L^{II}(\tilde{I})$ is the same as on $L^{III}(\tilde{I})$. With respect to these partially ordered relations, $L^{I}(\tilde{I})$, $L^{II}(\tilde{I})$, and $L^{III}(\tilde{I})$ form three complete lattices, and called type-I, type-II, and type-III concept lattices, respectively. The meet and join operations are list below:

- Type-I concept lattice $L^{I}(\tilde{I})$:

$$(X_1, \tilde{B}_1) \wedge (X_2, \tilde{B}_2) = (X_1 \cap X_2, (X_1 \cap X_2)^*),$$
$$(X_1, \tilde{B}_1) \vee (X_2, \tilde{B}_2) = ((\tilde{B}_1 \cap \tilde{B}_2)^*, \tilde{B}_1 \cap \tilde{B}_2);$$

- Type-II concept lattice $L^{II}(\tilde{I})$:

$$(X_1, \tilde{B}_1) \wedge (X_2, \tilde{B}_2) = (X_1 \cap X_2, (X_1 \cap X_2)^{\diamond}),$$
$$(X_1, \tilde{B}_1) \vee (X_2, \tilde{B}_2) = ((\tilde{B}_1 \cup \tilde{B}_2)^{\square}, \tilde{B}_1 \cup \tilde{B}_2);$$

- Type-III concept lattice $L^{III}(\tilde{I})$:

$$(X_1, \hat{B}_1) \wedge (X_2, \hat{B}_2) = (X_1 \cap X_2, (X_1 \cap X_2)^{\#}),$$
$$(X_1, \hat{B}_1) \vee (X_2, \hat{B}_2) = ((\hat{B}_1 \cup \hat{B}_2)^{\&}, \hat{B}_1 \cup \hat{B}_2).$$

Example 4. For the fuzzy formal context in Example 1, all the type-III concepts of $(U, A, \tilde{I})$ are listed in Table 4, and Fig. 3 illustrates the concept lattice $L^{III}(\tilde{I})$.

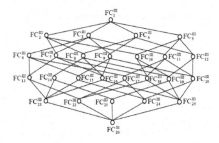

Fig. 3. The concept lattice $L^{III}(\tilde{I})$ of $(U, A, \tilde{I})$ in Example 1

4 Relationships Between Three Concept Lattices

4.1 Granular Structures

Theorem 1. *Let $(U, A, \tilde{I})$ be a fuzzy formal context. Then*

$$Ext^{I}(\tilde{I}) \subseteq Ext^{III}(\tilde{I}), \quad Ext^{II}(\tilde{I}) \subseteq Ext^{III}(\tilde{I}).$$

Proof. For any $(X, \tilde{B}) \in L^{I}(\tilde{I})$, we have $\tilde{B}^* = X$ and $X^* = \tilde{B}$. By Proposition 3 we have $X^{\#} = (X^*, X^{\diamond}) = (\tilde{B}, X^{\diamond}) \in IF(A)$ and $(\tilde{B}, X^{\diamond})^{\&} = \tilde{B}^* \cap X^{\diamond\square}$. According to $X \subseteq X^{\diamond\square}$ and $\tilde{B}^* = X$, we obtain $(\tilde{B}, X^{\diamond})^{\&} = X$. By Definition 3 we get $(X, X^{\#}) \in L^{III}(\tilde{I})$, and conclude $Ext^{I}(\tilde{I}) \subseteq Ext^{III}(\tilde{I})$.

By a similar way we can prove $Ext^{II}(\tilde{I}) \subseteq Ext^{III}(\tilde{I})$.

It should be noted that the inverses of the inequalities in Theorem 1 do not hold, which can be illustrated by Tables 2, 3, and 4.

Theorem 2. *Let $(U, A, \tilde{I})$ be a fuzzy formal context. Then*

$$Int^{I}(\tilde{I}) = \{\hat{B}^- \in F(A) \mid (\hat{B}^-, \hat{B}^+) \in Int^{III}(\tilde{I})\},$$
$$Int^{II}(\tilde{I}) = \{\hat{B}^+ \in F(A) \mid (\hat{B}^-, \hat{B}^+) \in Int^{III}(\tilde{I})\}.$$

Proof. It is clear that for any $X \in P(U)$, $X^* \in Int^{I}(\tilde{I})$. For any $(\hat{B}^-, \hat{B}^+) \in Int^{III}(\tilde{I})$, let X be the corresponding extension, we know $X^* = \hat{B}^-$ and hence $\hat{B}^- \in Int^{I}(\tilde{I})$. Thus

$$\{\hat{B}^- \in F(A) \mid (\hat{B}^-, \hat{B}^+) \in Int^{III}(\tilde{I})\} \subseteq Int^{I}(\tilde{I}).$$

From the proof of Theorem 1 we can see

$$Int^{I}(\tilde{I}) \subseteq \{\hat{B}^- \in F(A) \mid (\hat{B}^-, \hat{B}^+) \in Int^{III}(\tilde{I})\}.$$

We consequently obtain

$$Int^{I}(\tilde{I}) = \{\hat{B}^- \in F(A) \mid (\hat{B}^-, \hat{B}^+) \in Int^{III}(\tilde{I})\}.$$

Similarly, we can prove

$$Int^{II}(\tilde{I}) = \{\hat{B}^+ \in F(A) \mid (\hat{B}^-, \hat{B}^+) \in Int^{III}(\tilde{I})\}.$$

Table 4. All the type-III concepts of $(U, A, \tilde{I})$ in Example 1

$Number$	$(Extension, Intension)$
FC_1^{III}	$(\{x_1, x_2, x_3, x_4, x_5\}, \{a^{[0.19,0.73]}, b^{[0.20,0.83]}, c^{[0.15,0.72]}, d^{[0.35,0.61]}, e^{[0.00,0.52]}\})$
FC_2^{III}	$(\{x_1, x_2, x_3, x_5\}, \{a^{[0.19,0.73]}, b^{[0.20,0.83]}, c^{[0.30,0.72]}, d^{[0.35,0.61]}, e^{[0.00,0.52]}\})$
FC_3^{III}	$(\{x_1, x_2, x_4, x_5\}, \{a^{[0.28,0.73]}, b^{[0.20,0.83]}, c^{[0.15,0.72]}, d^{[0.35,0.50]}, e^{[0.17,0.52]}\})$
FC_4^{III}	$(\{x_1, x_3, x_4, x_5\}, \{a^{[0.19,0.56]}, b^{[0.20,0.67]}, c^{[0.15,0.72]}, d^{[0.35,0.61]}, e^{[0.00,0.52]}\})$
FC_5^{III}	$(\{x_2, x_3, x_4, x_5\}, \{a^{[0.19,0.73]}, b^{[0.48,0.83]}, c^{[0.15,0.57]}, d^{[0.41,0.61]}, e^{[0.00,0.36]}\})$
FC_6^{III}	$(\{x_1, x_2, x_4\}, \{a^{[0.44,0.73]}, b^{[0.20,0.83]}, c^{[0.15,0.72]}, d^{[0.35,0.49]}, e^{[0.29,0.52]}\})$
FC_7^{III}	$(\{x_1, x_2, x_5\}, \{a^{[0.28,0.73]}, b^{[0.20,0.83]}, c^{[0.30,0.72]}, d^{[0.35,0.50]}, e^{[0.17,0.52]}\})$
FC_8^{III}	$(\{x_1, x_3, x_5\}, \{a^{[0.19,0.56]}, b^{[0.20,0.59]}, c^{[0.44,0.72]}, d^{[0.35,0.61]}, e^{[0.00,0.52]}\})$
FC_9^{III}	$(\{x_1, x_4, x_5\}, \{a^{[0.28,0.56]}, b^{[0.20,0.67]}, c^{[0.15,0.72]}, d^{[0.35,0.50]}, e^{[0.17,0.52]}\})$
FC_{10}^{III}	$(\{x_2, x_3, x_5\}, \{a^{[0.19,0.73]}, b^{[0.48,0.83]}, c^{[0.30,0.57]}, d^{[0.41,0.61]}, e^{[0.00,0.29]}\})$
FC_{11}^{III}	$(\{x_2, x_4, x_5\}, \{a^{[0.28,0.73]}, b^{[0.59,0.83]}, c^{[0.15,0.44]}, d^{[0.41,0.50]}, e^{[0.17,0.36]}\})$
FC_{12}^{III}	$(\{x_3, x_4, x_5\}, \{a^{[0.19,0.44]}, b^{[0.48,0.67]}, c^{[0.15,0.57]}, d^{[0.49,0.61]}, e^{[0.00,0.36]}\})$
FC_{13}^{III}	$(\{x_1, x_2\}, \{a^{[0.56,0.73]}, b^{[0.20,0.83]}, c^{[0.30,0.72]}, d^{[0.35,0.41]}, e^{[0.29,0.52]}\})$
FC_{14}^{III}	$(\{x_1, x_3\}, \{a^{[0.19,0.56]}, b^{[0.20,0.48]}, c^{[0.57,0.72]}, d^{[0.35,0.61]}, e^{[0.00,0.52]}\})$
FC_{15}^{III}	$(\{x_1, x_4\}, \{a^{[0.44,0.56]}, b^{[0.20,0.67]}, c^{[0.15,0.72]}, d^{[0.35,0.49]}, e^{[0.36,0.52]}\})$
FC_{16}^{III}	$(\{x_1, x_5\}, \{a^{[0.28,0.56]}, b^{[0.20,0.59]}, c^{[0.44,0.72]}, d^{[0.35,0.50]}, e^{[0.17,0.52]}\})$
FC_{17}^{III}	$(\{x_2, x_4\}, \{a^{[0.44,0.73]}, b^{[0.67,0.83]}, c^{[0.15,0.30]}, d^{[0.41,0.49]}, e^{[0.29,0.36]}\})$
FC_{18}^{III}	$(\{x_2, x_5\}, \{a^{[0.28,0.73]}, b^{[0.59,0.83]}, c^{[0.30,0.44]}, d^{[0.41,0.50]}, e^{[0.17,0.29]}\})$
FC_{19}^{III}	$(\{x_3, x_5\}, \{a^{[0.19,0.28]}, b^{[0.48,0.59]}, c^{[0.44,0.57]}, d^{[0.50,0.61]}, e^{[0.00,0.17]}\})$
FC_{20}^{III}	$(\{x_4, x_5\}, \{a^{[0.28,0.44]}, b^{[0.59,0.67]}, c^{[0.15,0.44]}, d^{[0.49,0.50]}, e^{[0.17,0.36]}\})$
FC_{21}^{III}	$(\{x_1\}, \{a^{[0.56,0.56]}, b^{[0.20,0.20]}, c^{[0.72,0.72]}, d^{[0.35,0.35]}, e^{[0.52,0.52]}\})$
FC_{22}^{III}	$(\{x_2\}, \{a^{[0.73,0.73]}, b^{[0.83,0.83]}, c^{[0.30,0.30]}, d^{[0.41,0.41]}, e^{[0.29,0.29]}\})$
FC_{23}^{III}	$(\{x_3\}, \{a^{[0.19,0.19]}, b^{[0.48,0.48]}, c^{[0.57,0.57]}, d^{[0.61,0.61]}, e^{[0.00,0.00]}\})$
FC_{24}^{III}	$(\{x_4\}, \{a^{[0.44,0.44]}, b^{[0.67,0.67]}, c^{[0.15,0.15]}, d^{[0.49,0.49]}, e^{[0.36,0.36]}\})$
FC_{25}^{III}	$(\{x_5\}, \{a^{[0.28,0.28]}, b^{[0.59,0.59]}, c^{[0.44,0.44]}, d^{[0.50,0.50]}, e^{[0.17,0.17]}\})$
FC_{26}^{III}	$(\{\varnothing\}, \{a^{[0.00,0.00]}, b^{[0.00,0.00]}, c^{[0.00,0.00]}, d^{[0.00,0.00]}, e^{[0.00,0.00]}\})$

The results of Theorems 1 and 2 can be verified by Tables 2, 3 and 4.

Theorems 1 and 2 indicate that for the same fuzzy formal context, the extensions and intensions of all type-I and type-II concepts are hidden in the type-III concept lattice. Further, according to the following two theorems, we know that the type-I and type-II concept lattices can be written out from the type-III concept lattice.

Theorem 3. *Let $(U, A, \tilde{I})$ be a fuzzy formal context, $(X, \hat{B}) \in L^{\mathrm{III}}(\tilde{I})$. Then $(X, \hat{B}^-) \in L^{\mathrm{I}}(\tilde{I})$, if and only if for any $(Y, \hat{C}) \in L^{\mathrm{III}}(\tilde{I})$, it implies $(Y, \hat{C}) = (X, \hat{B})$ that $(X, \hat{B}) \leq (Y, \hat{C})$ and $\hat{C}^- = \hat{B}^-$.*

Proof. For $(X, \hat{B}) \in L^{\mathrm{III}}(\tilde{I})$, if $(X, \hat{B}^-) \in L^{\mathrm{I}}(\tilde{I})$, then $X^* = \hat{B}^-$ and $X^{**} = X$. For any $(Y, \hat{C}) \in L^{\mathrm{III}}(\tilde{I})$, if $(X, \hat{B}) \leq (Y, \hat{C})$ and $\hat{C}^- = \hat{B}^-$, then $X^* = Y^*$ and $X \subseteq Y$. Then we have $X^{**} = X \subseteq Y \subseteq Y^{**}$, and it follows from $X^* = Y^*$ that $X^{**} = Y^{**}$. Therefore $X = Y$, which implies that $(X, \hat{B}) = (Y, \hat{C})$.

Conversely, for $(X, \hat{B}) \in L^{\mathrm{III}}(\tilde{I})$, we assume that for any $(Y, \hat{C}) \in L^{\mathrm{III}}(\tilde{I})$, if $(X, \hat{B}) \leq (Y, \hat{C})$ and $\hat{C}^- = \hat{B}^-$, then $(Y, \hat{C}) = (X, \hat{B})$. In order to prove

$(X, \hat{B}^-) \in L^I(\tilde{I})$, it is only needed to prove $X = (\hat{B}^-)^* = X^{**}$. It is evident that $(X^{**}, X^*) \in L^I(\tilde{I})$, and it is easy to prove $(X^{**}, (X^*, X^{**\diamond})) \in L^{III}(\tilde{I})$. Then we can see that $(X, \hat{B}) \leq (X^{**}, (X^*, X^{**\diamond}))$ and $X^* = \hat{B}^-$, by the assumption we get $(X^{**}, (X^*, X^{**\diamond})) = (X, \hat{B})$, which means $X^{**} = X$.

Analogously, we can prove the next theorem.

Theorem 4. *Let $(U, A, \tilde{I})$ be a fuzzy formal context, $(X, \hat{B}) \in L^{III}(\tilde{I})$. Then $(X, X^\diamond) \in L^{II}(\tilde{I})$, if and only if for any $(Y, \hat{C}) \in L^{III}(\tilde{I})$, it implies $(Y, \hat{C}) = (X, \hat{B})$ that $(X, \hat{B}) \leq (Y, \hat{C})$ and $\hat{C}^+ = \hat{B}^+$.*

On the other hand, for the same fuzzy formal context, by virtue of the following two theorems, we can see that the type-III concept lattice can be generated from the type-I and type-II concept lattices.

Theorem 5. *Let $(U, A, \tilde{I})$ be a fuzzy formal context. Then*

$$Ext^{III}(\tilde{I}) = \{X \cap Y \mid X \in Ext^I(\tilde{I}), Y \in Ext^{II}(\tilde{I})\}.$$

Proof. It's not difficult to prove that $Ext^{III}(\tilde{I}) = \{\tilde{B}^* \cap \tilde{C}^\square \mid \tilde{B}, \tilde{C} \in F(A)\}$. For $X \in Ext^I(\tilde{I})$ and $Y \in Ext^{II}(\tilde{I})$, we have $X = X^{**}$ and $Y = Y^{\diamond\square}$, so $X \cap Y = (X^*)^* \cap (Y^\diamond)^\square = (X^*, Y^\diamond)^\&$, hence $X \cap Y \in Ext^{III}(\tilde{I})$. We get

$$\{X \cap Y \mid X \in Ext^I(\tilde{I}), Y \in Ext^{II}(\tilde{I})\} \subseteq Ext^{III}(\tilde{I}).$$

On the other hand, for $Z \in Ext^{III}(\tilde{I})$, we have $(Z, (Z^*, Z^\diamond)) \in L^{III}(\tilde{I})$ and hence $Z = Z^{**} \cap Z^{\diamond\square}$. It is clear that $Z^{**} \in Ext^I(\tilde{I})$, $Z^{\diamond\square} \in Ext^{II}(\tilde{I})$. Then $Z \in \{X \cap Y \mid X \in Ext^I(\tilde{I}), Y \in Ext^{II}(\tilde{I})\}$. We obtain

$$\{X \cap Y \mid X \in Ext^I(\tilde{I}), Y \in Ext^{II}(\tilde{I})\} \supseteq Ext^{III}(\tilde{I}).$$

Summarizing the results above we can conclude that

$$Ext^{III}(\tilde{I}) = \{X \cap Y \mid X \in Ext^I(\tilde{I}), Y \in Ext^{II}(\tilde{I})\}.$$

Theorem 6. *Let $(U, A, \tilde{I})$ be a fuzzy formal context, $(X, \hat{B}) \in L^{III}(\tilde{I})$. Then*

$$Int^{III}(\tilde{I}) = \left\{ (\tilde{B}, \tilde{C}) \in IF(A) \,\middle|\, \begin{matrix} (X, \tilde{B}) \in L^I(\tilde{I}), (Y, \tilde{C}) \in L^{II}(\tilde{I}), \\ (X \cap Y)^* = \tilde{B}, (X \cap Y)^\diamond = \tilde{C} \end{matrix} \right\}.$$

Proof. For $(\tilde{B}, \tilde{C}) \in Int^{III}(\tilde{I})$, it is clear that $(\tilde{B}^*, \tilde{B}) \in L^I(\tilde{I})$, $(\tilde{C}^\square, \tilde{C}) \in L^{II}(\tilde{I})$, and $(\tilde{B}^* \cap \tilde{C}^\square, (\tilde{B}, \tilde{C})) \in L^{III}(\tilde{I})$, so $(\tilde{B}^* \cap \tilde{C}^\square)^* = \tilde{B}$, $(\tilde{B}^* \cap \tilde{C}^\square)^\diamond = \tilde{C}$. Thus,

$$Int^{III}(\tilde{I}) \subseteq \left\{ (\tilde{B}, \tilde{C}) \in IF(A) \,\middle|\, \begin{matrix} (X, \tilde{B}) \in L^I(\tilde{I}), (Y, \tilde{C}) \in L^{II}(\tilde{I}), \\ (X \cap Y)^* = \tilde{B}, (X \cap Y)^\diamond = \tilde{C} \end{matrix} \right\}.$$

Moreover, for $(X, \tilde{B}) \in L^I(\tilde{I})$ and $(Y, \tilde{C}) \in L^{II}(\tilde{I})$ with $(X \cap Y)^* = \tilde{B}$ and $(X \cap Y)^\diamond = \tilde{C}$, making use of Definitions 2 and 3 we can verify that $(X \cap Y, (\tilde{B}, \tilde{C})) \in L^{III}(\tilde{I})$. Therefore

$$\left\{ (\tilde{B}, \tilde{C}) \in IF(A) \,\middle|\, \begin{matrix} (X, \tilde{B}) \in L^I(\tilde{I}), (Y, \tilde{C}) \in L^{II}(\tilde{I}), \\ (X \cap Y)^* = \tilde{B}, (X \cap Y)^\diamond = \tilde{C} \end{matrix} \right\} \subseteq Int^{III}(\tilde{I}).$$

Based on the inclusions above, we can conclude

$$Int^{III}(\tilde{I}) = \left\{ (\tilde{B}, \tilde{C}) \in IF(A) \,\middle|\, \begin{array}{l} (X, \tilde{B}) \in L^{I}(\tilde{I}), (Y, \tilde{C}) \in L^{II}(\tilde{I}), \\ (X \cap Y)^* = \tilde{B}, (X \cap Y)^{\diamond} = \tilde{C} \end{array} \right\}.$$

By Tables 2, 3 and 4, Theorems 5 and 6 can be checked.

4.2   Algebraic Structures

Theorem 7. *Let $(U, A, \tilde{I})$ be a fuzzy formal context. Then there is a homomorphic mapping from $(L^{I}(\tilde{I}), \wedge)$ to $(L^{III}(\tilde{I}), \wedge)$.*

Proof. For $(X, \tilde{B}) \in L^{I}(\tilde{I})$, let $\varphi((X, \tilde{B})) = (X, X^{\#})$, then it can be checked that $(X, X^{\#}) \in L^{III}(\tilde{I})$, so φ is a mapping from $L^{I}(\tilde{I})$ to $L^{III}(\tilde{I})$. For any $(X_1, \tilde{B}_1), (X_2, \tilde{B}_2) \in L^{I}(\tilde{I})$, we have $(X_1, \tilde{B}_1) \wedge (X_2, \tilde{B}_2) = (X_1 \cap X_2, (X_1 \cap X_2)^*)$ and hence $\varphi((X_1, \tilde{B}_1) \wedge (X_2, \tilde{B}_2)) = (X_1 \cap X_2, (X_1 \cap X_2)^{\#})$. From $\varphi((X_1, \tilde{B}_1)) = (X_1, X_1^{\#})$ and $\varphi((X_2, \tilde{B}_2)) = (X_2, X_2^{\#})$, it follows that $\varphi((X_1, \tilde{B}_1)) \wedge \varphi((X_2, \tilde{B}_2)) = (X_1, X_1^{\#}) \wedge (X_2, X_2^{\#}) = (X_1 \cap X_2, (X_1 \cap X_2)^{\#})$. Therefore $\varphi((X_1, \tilde{B}_1) \wedge (X_2, \tilde{B}_2)) = \varphi((X_1, \tilde{B}_1)) \wedge \varphi((X_2, \tilde{B}_2))$, which means that φ is a homomorphic mapping from $(L^{I}(\tilde{I}), \wedge)$ to $(L^{III}(\tilde{I}), \wedge)$.

Similar to Theorem 7, the following theorem can be proved.

Theorem 8. *Let $(U, A, \tilde{I})$ be a fuzzy formal context. Then there is a homomorphic mapping from $(L^{II}(\tilde{I}), \wedge)$ to $(L^{III}(\tilde{I}), \wedge)$.*

Definition 4. *Let $(U, A, \tilde{I})$ be a fuzzy formal context. For any $(x, a) \in U \times A$, let*

$$\tilde{I}^c(x, a) = 1 - \tilde{I}(x, a)$$

then $\tilde{I}^c$ is a binary fuzzy relation from U to A. The fuzzy formal context $(U, A, \tilde{I}^c)$ is called the complement context of $(U, A, \tilde{I})$.

In the following, for the fuzzy formal context $(U, A, \tilde{I})$, the operators $*$, $\square$, and $\diamond$ are also denoted as $*^{\tilde{I}}$, $\square^{\tilde{I}}$, and $\diamond^{\tilde{I}}$, respectively.

Proposition 5. *Let $(U, A, \tilde{I}^c)$ be the complement context of fuzzy formal context $(U, A, \tilde{I})$. Then for any $X \in P(U)$, $\tilde{B} \in F(A)$, we have $\tilde{B}^{\square_{\tilde{I}^c}} = (\tilde{B}^c)^{*\tilde{I}}$, $\tilde{B}^{*\tilde{I}^c} = (\tilde{B}^c)^{\square_{\tilde{I}}}$, $X^{\diamond_{\tilde{I}^c}} = (X^{*\tilde{I}})^c$, $X^{*\tilde{I}^c} = (X^{\diamond_{\tilde{I}}})^c$. Where $\tilde{B}^c$ denotes the complement of fuzzy set $\tilde{B}$, i.e. $\tilde{B}^c(x) = 1 - \tilde{B}(x)$, $x \in A$, and X^c denotes the complement of crisp subset $X \subseteq U$.*

Theorem 9. *Let $(U, A, \tilde{I})$ be a fuzzy formal context. Then $(X, \tilde{B}) \in L^{I}(\tilde{I})$ if and only if $(X, \tilde{B}^c) \in L^{II}(\tilde{I}^c)$.*

Proof. For any $(X, \tilde{B}) \in L^{II}(\tilde{I}^c)$, we have $X^{\diamond_{\tilde{I}^c}} = \tilde{B}$ and $\tilde{B}^{\square_{\tilde{I}^c}} = X$. From Proposition 3 it follows that $X^{*\tilde{I}} = (X^{\diamond_{\tilde{I}^c}})^c = \tilde{B}^c$ and $(\tilde{B}^c)^{*\tilde{I}} = \tilde{B}^{\square_{\tilde{I}^c}} = X$. Therefore $(X, \tilde{B}^c) \in L^{I}(\tilde{I})$.

Moreover, for any $(X, \tilde{B}) \in L^{I}(\tilde{I})$, we know $X^{*\tilde{I}} = \tilde{B}$ and $\tilde{B}^{*\tilde{I}} = X$. Again by Proposition 3 we have $X^{\diamond_{\tilde{I}^c}} = (X^{*\tilde{I}})^c = \tilde{B}^c$ and $(\tilde{B}^c)^{\square_{\tilde{I}^c}} = \tilde{B}^{*\tilde{I}} = X$. Thus $(X, \tilde{B}^c) \in L^{II}(\tilde{I}^c)$.

The following conclusion can be drawn from Theorem 9.

Corollary 1. *Let $(U, A, \tilde{I})$ be a fuzzy formal context. Then $(X, \tilde{B}) \in L^{\mathrm{II}}(\tilde{I})$ if and only if $(X, \tilde{B}^c) \in L^{\mathrm{I}}(\tilde{I}^c)$.*

Theorem 10. *Let $(U, A, \tilde{I})$ be a fuzzy formal context. Then $L^{\mathrm{I}}(\tilde{I})$ is isomorphic to $L^{\mathrm{II}}(\tilde{I}^c)$, i.e. $L^{\mathrm{I}}(\tilde{I}) \cong L^{\mathrm{II}}(\tilde{I}^c)$.*

Proof. For any $(X, \tilde{B}) \in L^{\mathrm{I}}(\tilde{I})$, let $\varphi((X, \tilde{B})) = (X, \tilde{B}^c)$, then according to Theorem 9 we know that φ is a one-to-one mapping from $L^{\mathrm{I}}(\tilde{I})$ to $L^{\mathrm{II}}(\tilde{I}^c)$, and order-preserving apparently. Thus φ is a isomorphic mapping between $L^{\mathrm{I}}(\tilde{I})$ and $L^{\mathrm{II}}(\tilde{I}^c)$, i.e. $L^{\mathrm{I}}(\tilde{I})$ is isomorphic to $L^{\mathrm{II}}(\tilde{I}^c)$.

Of course, we know that the following corollary holds.

Corollary 2. *Let $(U, A, \tilde{I})$ be a fuzzy formal context. Then $L^{\mathrm{II}}(\tilde{I})$ is isomorphic to $L^{\mathrm{I}}(\tilde{I}^c)$.*

5 Conclusions

Classical concept lattice and rough concept lattices can be generalized into one-sided fuzzy concept lattices in fuzzy formal contexts. Making use of interval-valued fuzzy sets, this paper proposed a new kind of crisp-fuzzy concept lattice. By investigating its properties, it can be found that the operations defining the crisp-fuzzy concepts of the new concept lattice can be viewed as two types of rough fuzzy approximation operations. From two aspects of granular and algebraic structures, a comparison has been made for it and two existing crisp-fuzzy concept lattices. Consequently, it can be understood that two the existing crisp-fuzzy concept lattices are hidden in the new concept lattice, and by synthesizing two existing crisp-fuzzy concept lattices, the new concept lattice can be generated. Therefore, the new concept lattice has more information than each of two existing crisp-fuzzy concept lattices.

Acknowledgements. This work was supported by grants from the National Natural Science Foundation of China (Nos. 61773349, 61976194).

References

1. Ganter, B., Wille, R.: Formal Concept Analysis: Mathematical Foundations. Springer, Heidelberg (1999)
2. Sampath, S., Sprenkle, S., Gibson, E., et al.: Applying concept analysis to user-session-based testing of web applications. IEEE Trans. Software Eng. **33**(10), 643–658 (2007)
3. Yuan, K.H., Xu, W.H., Li, W.T., Ding, W.P.: An incremental learning mechanism for object classification based on progressive fuzzy three-way concept. Inf. Sci. **584**, 127–147 (2022)

4. Niu, J.J., Chen, D.G., Li, J.H., Wang, H.: A dynamic rule-based classification model via granular computing. Inf. Sci. **584**, 325–341 (2022)
5. Burusco, A., Fuentes-Gonzalez, R.: Concept lattices defined from implication operators. Fuzzy Sets Syst. **114**(3), 431–436 (2000)
6. Long, B.H., Xu, W.H.: Fuzzy three-way concept analysis and fuzzy three-way concept lattice. J. Nanjing Univ. (Nat. Sci.) **55**(4), 537–545 (2019)
7. Gao, Y.Q., Ma, J.M.: Variable threshold interval-set concept lattices. J. Nanjing Univ. (Nat. Sci.) **56**(4), 437–444 (2020)
8. Yahia, S.B., Arour, K., Slimani, A., et al.: Discovery of compact rules in relational databases. Inf. Sci. J. **4**(3), 497–511 (2000)
9. Krajci, S.: Cluster based efficient generation of fuzzy concepts. Neural Netw. World **13**(5), 521–530 (2003)
10. Lin, Y.D., Li, J.J., Zhang, C.L.: Attribute reductions of fuzzy-crisp concept lattices based on matrix. Pattern Recogn. Art. Intell. (Chinese J.) **33**(1), 21–31 (2020)
11. Li, T.J., Zhang, X.Y., Wu, W.Z., et al.: Attribute reduction of single-sided fuzzy concept lattice in formal fuzzy contexts. J. Nanjing Univ. (Nat. Sci.) **58**(1), 38–48 (2022)
12. Duntsch, N., Gediga, G.: Modal-style operators in qualitative data analysis. In: Proceedings of 2002 IEEE International Conference on Data Mining, pp. 155–162. IEEE, Maebashi City (2002)
13. YAO, Y.Y.: Concept lattices in rough set theory. In: Proceedings of 2004 Annual Meeting of the North American Fuzzy Information Processing Society, pp. 796–801. IEEE, Banff (2004)
14. Yao, Y.Y.: Three-way decisions with probabilistic rough sets. Inf. Sci. **180**(3), 341–353 (2010)
15. Wei, L., Qian, T.: The three-way object oriented concept lattice and the three-way property oriented concept lattice. In: Proceedings of the International Conference on Machine Learning and Cybernetics, pp. 854–859. IEEE, Guangzhou (2015)
16. He, X.L., Wei, L., Qian, T.: Property oriented interval-set concept lattice. J. Front. Comput. Sci. Technol. **12**(9), 1506–1512 (2018)
17. Li, T.J., Wu, M.R., Wu, W.Z.: Attribute reduction of crisp-fuzzy concept lattices based on rough approximation operations. (accepted by Chinese Journal Of Engineering Mathematics)

Normal Fuzzy Three-Way Decision Based on Prospect Theory

Yanhua Li[1], Jiafen Liu[1], Yihua Zhong[2], and Xin Yang[1(✉)]

[1] School of Computing and Artificial Intelligence, Southwestern University of Finance and Economics, Chengdu 611130, China
{jfliu,yangxin}@swufe.edu.cn
[2] School of Sciences, Southwest Petroleum University, Chengdu 610500, China

Abstract. Prospect theory-based three-way decision has been successfully applied in various fuzzy information systems owing to its excellent performance in expressing the risk attitude of decision makers. However, the current prospect theory-based three-way decisions have two following limitations. On the one hand, they are constrained in processing uncertain continuous data or neglecting the distribution of uncertain fuzzy numbers. On the other hand, the risk attitudes of decision-makers are not considered when calculating the conditional probability. To address the two issues, we propose a normal fuzzy prospect theory-based three-way decision model and a normal fuzzy ideal solution method. First, since normal fuzzy numbers can describe the continuous uncertain data subjected to the normal distribution, we use it to represent the uncertain decision information, i.e., normal fuzzy outcome matrix, normal fuzzy reference points. Then, by integrating prospect theory and TOPSIS, we propose a normal fuzzy ideal solution method to calculate conditional probability, which considers the risk attitudes of decision-makers. Finally, the comparative experiments demonstrate the effectiveness and superiority of our proposal.

Keywords: Three-way decision · Prospect theory · Normal fuzzy number

1 Introduction

Three-way decision (3WD) [17,18] mainly deals with uncertain and incomplete information. 3WD gives the noncommitment decision when the information is inadequate [14,21,23]. In traditional 3WD model [17], the corresponding losses for taking different actions are calculated by Bayesian minimum loss, but the decision-makers' psychological risk attitudes are ignored. Prospect theory points out the "bounded rational" behavior of decision makers, which expresses that people will be risk-averse toward gains and risk chase toward losses [9]. In recent years, prospect theory has been introduced into 3WD to represent the psychological risk attitudes, the achievements include 3WD based on various prospect theories [11,12,22] and prospect theory-based 3WD in diverse fuzzy environment [15,20].

Gu et al. [3] presented a prospect theory-based decision framework under intuitionistic fuzzy environment, Liang et al. [6] studied 3WD in Pythagorean fuzzy environment, they all focused on dealing with discrete uncertain information rather than continuous uncertain information. In practice, there is more continuous uncertain data in real life and the discretization of it will lead to the loss of information. Then, Wang et al. [12] presented the 3WD based on third-generation prospect theory, which transformed Z-numbers into triangular fuzzy numbers to describe decision information. Triangular fuzzy number can describe continuous uncertain data but it neglect the distribution of uncertainty, which results in an insufficiently detailed depiction of uncertainty.

To against the above issues, we observed that normal fuzzy numbers can describe the uncertain continuous information subjected to normal distribution. There are many things that obey normal distribution in human activities and natural environment [16]. For instance, the score of students and the service life of products both obey normal distribution. Therefore, we proposed normal fuzzy prospect theory-based three-way decision (NFP3WD) by describing decision information with normal fuzzy numbers. In addition, previous TOPSIS methods of calculating conditional probability neglected the risk attitudes of decision-makers. To this end, we propose a normal fuzzy ideal solution method to compute conditional probability under normal fuzzy environment without class label. The contributions of this work are expressed as follows:

- We proposed normal fuzzy prospect theory-based three-way decision method (NFP3WD) to handle continuous uncertain information with normal distribution.
- A normal fuzzy ideal solution based on TOPSIS and prospect theory is proposed to compute conditional probability, which includes the risk attitudes of decision-makers.

The rest of this paper is set out as follows. In Sect. 2, we review some fundamental concepts and notations of normal fuzzy numbers, 3WD, and prospect theory. Section 3 proposes a normal fuzzy prospect theory-based three-way decision method. In Sect. 4, we propose a normal fuzzy ideal solution based on TOPSIS and prospect theory. In Sect. 5, we give an illustrative example, then some comparative analyzes are carried out, which verify the effectiveness and superiority of our proposed method. Section 6 summarizes our study.

2 Preliminaries

2.1 Normal Fuzzy Numbers

Definition 1. *[7, 13] Suppose $\tilde{A}$ is a fuzzy number, if $\tilde{A}$ has the following membership function:*

$$\tilde{A}(x) = exp\left\{-\frac{(x-a)^2}{\sigma^2}\right\}, x, a \in R, \sigma > 0, \tag{1}$$

then, $\tilde{A}$ is a normal fuzzy number, represented as $\tilde{A} = (a, \sigma^2)$, and R is a set of real numbers, a is the mean of $\tilde{A}$ and σ^2 denotes variance of $\tilde{A}$. Obviously, when $\sigma = 0$, the normal fuzzy number $\tilde{A} = (a, \sigma^2)$ degenerates to real number a.

Definition 2. [5, 8] $E(\tilde{A})$ is the expectation of normal fuzzy number $\tilde{A}$, which is defined as:

$$E(\tilde{A}) = \frac{\int_{-\infty}^{+\infty} x\tilde{A}(x)dx}{\int_{-\infty}^{+\infty} \tilde{A}(x)dx}. \tag{2}$$

when $\tilde{A} = (a, \sigma^2)$, $E(\tilde{A}) = a$.

Definition 3. [5, 8] Suppose normal fuzzy numbers $\tilde{A} = (a, \sigma_a^2)$, $\tilde{B} = (b, \sigma_b^2)$, we can derive:

(1) if $a > b$, then $\tilde{A} > \tilde{B}$;
(2) if $a = b$, then, when $\sigma_a = \sigma_b$, $\tilde{A} = \tilde{B}$; when $\sigma_a < \sigma_b$, $\tilde{A} > \tilde{B}$;
(3) if $a < b$, then $\tilde{A} < \tilde{B}$.

Definition 4. [2] Suppose normal fuzzy numbers $\tilde{A} = (a, \sigma_a^2)$, $\tilde{B} = (b, \sigma_b^2)$, the distance between $\tilde{A}$ and $\tilde{B}$ is defined as:

$$d(\tilde{A}, \tilde{B}) = \sqrt{(a - b)^2 + \frac{1}{2}(\sigma_a^2 - \sigma_b^2)^2}. \tag{3}$$

2.2 Three-Way Decision

3WD theory divides a universe into three parts reasonably and takes effective strategies to deal with each part [19]. The two states $\Omega = \{C, \neg C\}$ in 3WD indicate that an object x is in a decision class C or not, respectively. There are three actions $\mathcal{A} = \{a_P, a_B, a_N\}$ in 3WD. Taking action a_P denotes that we accept x belongs to C and classify x to positive region POS(C); taking action a_B denotes that we classify x into boundary region BND(C); and taking action a_N denotes that we reject x belongs to C and classify x into negative region NEG(C).

Table 1. Loss function matrix.

	C	$\neg C$
a_P	λ_{PP}	λ_{PN}
a_B	λ_{BP}	λ_{BN}
a_N	λ_{NP}	λ_{NN}

Table 1 shows the different loss functions. When an object $x \in C$, the losses for taking actions a_i ($i = P, B, N$) are λ_{iP}, respectively. When $x \notin C$, the losses for taking actions a_i are λ_{iN}, respectively. Assume that $Pr(C|x)$ and $Pr(\neg C|x)$

represent conditional probability of $x \in C$ and $x \notin C$, respectively. Then, based on Bayesian process, the expected losses for taking three different actions can be calculated [17], and the decision rules are based on the minimum loss. The decision rules are simplified as comparing decision thresholds and conditional probability as follows:

$$If \ Pr(C|x) \geq \alpha, then \ x \in POS(C);$$
$$If \ \beta < Pr(C|x) < \alpha, then \ x \in BND(C); \tag{4}$$
$$If \ Pr(C|x) \leq \beta, then \ x \in NEG(C).$$

2.3   Prospect Theory

Prospect theory, proposed by Kahneman and Tversky [4], describes decision makers' behaviors under uncertainty and risk. Prospect theory integrates decision makers' value perception factor into the decision process, and the risk attitudes of decision makers are evaluated by value function and weight function.

The value function describes decision makers' risk attitudes toward gains and losses, which is an asymmetric S-shaped function. Different decision-makers may have different reference points, and that may lead to different judgments of gains and losses. Decision makers show risk aversion toward gains and risk-chasing toward losses. The value function is shown as follows [10]:

$$v(\Delta z_k) = \begin{cases} (\Delta z_k)^\mu, & \Delta z_k \geq 0 \\ -\theta(-\Delta z_k)^v, & \Delta z_k < 0 \end{cases}, \tag{5}$$

where $\Delta z_k = z_k - z_r$, Δz_k measures the k-th difference between reference point z_r and the k-th outcome z_k. When $\Delta z_k \geq 0$, the observed outcome is considered as a gain relative to the reference point. Conversely, if $\Delta z_k < 0$, it is perceived as a loss.

Prospect theory holds that decision-makers always over-weight small probabilities and under-weight large probabilities [10]. Weight function w_k is a nonlinear transformation of the probability, the weight function given by Tversky and Kahneman [10] is shown as follows:

$$w_k = \begin{cases} w^+(p(\Delta z_k)) = \dfrac{p(\Delta z_k)^\sigma}{(p(\Delta z_k)^\sigma + (1-p(\Delta z_k))^\sigma)^{1/\sigma}} \\ w^-(p(\Delta z_k)) = \dfrac{p(\Delta z_k)^\delta}{(p(\Delta z_k)^\delta + (1-p(\Delta z_k))^\delta)^{1/\delta}} \end{cases}, \tag{6}$$

where w_k represents the decision weight, and $p(\Delta z_k)$ denotes the actual probability of Δz_k. The influence degree of overweighting and underweighting to gains and losses are represented through parameters σ and δ, respectively, and they satisfy $0 < \sigma, \delta < 1$.

Prospect theory holds that people prefer the maximum prospect value [11]. Suppose n denotes the number of outcomes, then, the prospect value function is shown as follows:

$$V = \sum_{k=1}^{n} w_k v(\Delta z_k). \tag{7}$$

3   A Normal Fuzzy Prospect Theory-Based Three-Way Decision Method (NFP3WD)

In this section, we represent the decision information with normal fuzzy numbers and propose the NFP3WD method.

Since prospect theory seeks the maximum prospect value instead of the minimum loss, the losses in 3WD are replaced by outcomes in NFP3WD. The outcome denotes the final state of wealth for taking different actions in different states. In NFP3WD, the outcome matrix is described by normal fuzzy numbers, as shown in Table 2, where $\tilde{Z}_{ij} = (a_{ij}, \sigma_{ij}^2)$ $(i = P, B, N; j = P, N)$ indicates the normal fuzzy outcomes incurred for taking action i in state j.

Table 2. Normal fuzzy outcome matrix.

	C	$\neg C$
a_P	$\tilde{Z}_{PP} = (a_{PP}, \sigma_{PP}^2)$	$\tilde{Z}_{PN} = (a_{PN}, \sigma_{PN}^2)$
a_B	$\tilde{Z}_{BP} = (a_{BP}, \sigma_{BP}^2)$	$\tilde{Z}_{BN} = (a_{BN}, \sigma_{BN}^2)$
a_N	$\tilde{Z}_{NP} = (a_{NP}, \sigma_{NP}^2)$	$\tilde{Z}_{NN} = (a_{NN}, \sigma_{NN}^2)$

Suppose $\tilde{Z}_r = (a_r, \sigma_r^2)$ represents the normal fuzzy reference point of the r-th decision maker, if $\tilde{Z}_{ij} \geq \tilde{Z}_r$, the outcome $\tilde{Z}_{ij}$ is perceived as a gain. Conversely, if $\tilde{Z}_{ij} < \tilde{Z}_r$, it is perceived as a loss. According to Definition 4, the distance between $\tilde{Z}_{ij}$ and $\tilde{Z}_r$ is represented by d_{ij} $(i = P, B, N; j = P, N)$, which can be computed as follows:

$$d_{ij} = d(\tilde{Z}_{ij}, \tilde{Z}_r) = \sqrt{(a_{ij} - a_r)^2 + \frac{1}{2}(\sigma_{ij}^2 - \sigma_r^2)^2}, \ (i = P, B, N; j = P, N). \quad (8)$$

Based on prospect theory, individuals tend to risk aversion to gains and risk chasing toward losses, they tend to be more sensitive to losses compared to gains, which are described by the value function. With the distance d_{ij} between normal fuzzy outcome and reference point obtained as well as the gains and losses judged, the value functions $\tilde{v}_{ij}$ $(i = P, B, N; j = P, N)$ for taking different actions in different states are computed based on Eq. (5), shown as follows:

$$\tilde{v}_{ij} = \begin{cases} (d_{ij})^\mu, & \tilde{Z}_{ij} \geq \tilde{Z}_r, \\ -\theta(d_{ij})^\upsilon, & \tilde{Z}_{ij} < \tilde{Z}_r \end{cases}, \quad (9)$$

where μ, υ and θ are suggested to set $\mu = \upsilon = 0.88$, $\theta = 2.25$ after many psychological experiments [4]. The value function matrix is shown in Table 3.

In this table, the values $\tilde{v}_{PP}$, $\tilde{v}_{BP}$, and $\tilde{v}_{NP}$ represent the value function associated with actions a_P, a_B, and a_N respectively, when $x \in C$. Similarly, the values $\tilde{v}_{PN}$, $\tilde{v}_{BN}$, and $\tilde{v}_{NN}$ represent the value function associated with actions a_P, a_B, and a_N respectively, when $x \notin C$.

Table 3. Value function matrix.

	C	$\neg C$
a_P	$\tilde{v}_{PP}$	$\tilde{v}_{PN}$
a_B	$\tilde{v}_{BP}$	$\tilde{v}_{BN}$
a_N	$\tilde{v}_{NP}$	$\tilde{v}_{NN}$

Next, in the 3WD theory, the conditional probability $Pr(C|x)$ denotes the probability of $x \in C$. $Pr(C|x) + Pr(\neg C|x) = 1$. Prospect theory suggests that the probability should be extended to weight functions for gains and losses. The two different weight functions: $w_i(Pr(C|x))$ ($i = P, B, N$) corresponding to gains and $w_i(Pr(\neg C|x))$ ($i = P, B, N$) corresponding to losses are presented as follows:

$$
\begin{aligned}
w_i(Pr(C|x)) &= \begin{cases} w_i^+(Pr(C|x)), & \tilde{Z}_{iP} \geq \tilde{Z}_r \\ w_i^-(Pr(C|x)), & \tilde{Z}_{iP} < \tilde{Z}_r \end{cases}, \\
w_i(Pr(\neg C|x)) &= \begin{cases} w_i^+(1 - Pr(C|x)), & \tilde{Z}_{iN} \geq \tilde{Z}_r \\ w_i^-(1 - Pr(C|x)), & \tilde{Z}_{iN} < \tilde{Z}_r \end{cases}.
\end{aligned}
\tag{10}
$$

In fact, the weight functions are nonlinear transformations of conditional probabilities. Based on Eq. (6), the detailed calculation of weight function $w_i(Pr(C|x))$ for $Pr(C|x)$ and $w_i(Pr(\neg C|x))$ for $Pr(\neg C|x)$ are presented as follows:

$$
\begin{aligned}
w_i(Pr(C|x)) &= \begin{cases} \frac{Pr(C|x)^\sigma}{((Pr(C|x))^\sigma + (1-Pr(C|x))^\sigma)^{1/\sigma}}, & \tilde{Z}_{iP} \geq \tilde{Z}_r \\ \frac{Pr(C|x)^\delta}{((Pr(C|x))^\delta + (1-Pr(C|x))^\delta)^{1/\delta}}, & \tilde{Z}_{iP} < \tilde{Z}_r \end{cases}, i = P, B, N, \\
w_i(Pr(\neg C|x)) &= \begin{cases} \frac{(1-Pr(C|x))^\sigma}{((1-Pr(C|x))^\sigma + (Pr(C|x))^\sigma)^{1/\sigma}}, & \tilde{Z}_{iN} \geq \tilde{Z}_r \\ \frac{(1-Pr(C|x))^\delta}{((1-Pr(C|x))^\delta + (Pr(C|x))^\delta)^{1/\delta}}, & \tilde{Z}_{iN} < \tilde{Z}_r \end{cases}, i = P, B, N,
\end{aligned}
\tag{11}
$$

where the parameters are suggested to set $\sigma = 0.61$ and $\delta = 0.69$ by Tversky and Kahneman [4] and the settings are extensively used in the studies corresponding to prospect theory.

Subsequently, with value functions and weight functions obtained, based on Eq. (7), the prospect value $\tilde{V}(a_i|x)$ ($i = P, B, N$) of taking actions a_P, a_B, and a_N are calculated as follows:

$$
\begin{aligned}
\tilde{V}(a_P|x) &= \tilde{v}_{PP}w_P(Pr(C|x)) + \tilde{v}_{PN}w_P(Pr(\neg C|x)); \\
\tilde{V}(a_B|x) &= \tilde{v}_{BP}w_B(Pr(C|x)) + \tilde{v}_{BN}w_B(Pr(\neg C|x)); \\
\tilde{V}(a_N|x) &= \tilde{v}_{NP}w_N(Pr(C|x)) + \tilde{v}_{NN}w_N(Pr(\neg C|x)).
\end{aligned}
\tag{12}
$$

Then, the decision rules based on the maximum prospect value are shown as follows:

$$If \ \tilde{V}(a_P|x) \geq \tilde{V}(a_B|x) \ \& \ \tilde{V}(a_P|x) \geq \tilde{V}(a_N|x), \ then \ x \in POS(C);$$

$$If \ \tilde{V}(a_B|x) \geq \tilde{V}(a_P|x) \ \& \ \tilde{V}(a_B|x) \geq \tilde{V}(a_N|x), \ then \ x \in BND(C); \quad (13)$$

$$If \ \tilde{V}(a_N|x) \geq \tilde{V}(a_P|x) \ \& \ \tilde{V}(a_N|x) \geq \tilde{V}(a_B|x), \ then \ x \in NEG(C).$$

In general, the decision rule of three-way decision is simplified to the comparison of decision thresholds and conditional probability $Pr(C|x)$. Wang et al. [11] have proved that the decision thresholds α, β and γ exist and are unique in prospect theory-based three-way decisions. Similarly, the decision thresholds α, β and γ also exist and are unique in our NFP3WD.

Suppose α, β and γ are the intersections between $\tilde{V}(a_P|x)$ and $\tilde{V}(a_B|x)$, $\tilde{V}(a_B|x)$ and $\tilde{V}(a_N|x)$, $\tilde{V}(a_P|x)$ and $\tilde{V}(a_N|x)$, respectively. Let $\tilde{V}_1 = \tilde{V}(a_P|x) - \tilde{V}(a_B|x)$, $\tilde{V}_2 = \tilde{V}(a_B|x) - \tilde{V}(a_N|x)$ and $\tilde{V}_3 = \tilde{V}(a_P|x) - \tilde{V}(a_N|x)$. Then, α, β and γ are the zero points of $\tilde{V}_1$, $\tilde{V}_2$ and $\tilde{V}_3$, respectively. If $\alpha > \beta$, the decision rules are:

$$If \ Pr(C|x) \geq \alpha, then \ x \in POS(C);$$

$$If \ \beta < Pr(C|x) < \alpha \ , then \ x \in BND(C); \quad (14)$$

$$If \ Pr(C|x) \leq \beta, then \ x \in NEG(C).$$

Otherwise, the decision rules are:

$$If \ Pr(C|x) \geq \gamma, then \ x \in POS(C);$$

$$If \ Pr(C|x) < \gamma, then \ x \in NEG(C). \quad (15)$$

4   The Normal Fuzzy Ideal Solutions for NFP3WD

For the information system without class label, the TOPSIS method can calculate the conditional probability [6]. However, traditional TOPSIS method does not consider the psychological risk attitudes. Therefore, we integrated prospect theory with TOPSIS to design a method to compute conditional probability of normal fuzzy system without class label.

Suppose $IS = (U, AT, V, f)$ is a normal fuzzy information system without class label, where $U = \{o_1, o_2, \cdots, o_m\}$ represents the universe with m objects, $AT = \{g_1, g_2, \cdots, g_n\}$ denotes the attribute set of normal fuzzy information system. Then, the weights of attributes are expressed as $\omega = \{\omega_1, \omega_2, \cdots, \omega_n\}^T$, where $\omega_q \ (q = 1, \cdots, n)$ represents the weight of attribute g_q and satisfies $0 \leq \omega_q \leq 1$, $\sum_{q=1}^{n} \omega_q = 1$. In our normal fuzzy information system, let $\tilde{A}_{pq} = (a_{pq}, \sigma_{pq}^2) \ (p = 1, \cdots, m; q = 1, \cdots, n.)$ represents the value of the q-th attribute of the p-th object. The detailed normal fuzzy information system is shown in Table 4.

In Table 4, assume that all the attributes in normal fuzzy information system are positive attributes. To eliminate the dimensional effect of attributes, we

Table 4. Normal fuzzy information system.

	g_1	g_2	$\cdots$	g_n
o_1	$\tilde{A}_{11} = (a_{11}, \sigma_{11}^2)$	$\tilde{A}_{12} = (a_{12}, \sigma_{12}^2)$	$\cdots$	$\tilde{A}_{1n} = (a_{1n}, \sigma_{1n}^2)$
o_2	$\tilde{A}_{21} = (a_{21}, \sigma_{21}^2)$	$\tilde{A}_{22} = (a_{22}, \sigma_{22}^2)$	$\cdots$	$\tilde{A}_{2n} = (a_{2n}, \sigma_{2n}^2)$
$\cdots$	$\cdots$	$\cdots$	$\cdots$	$\cdots$
o_m	$\tilde{A}_{m1} = (a_{m1}, \sigma_{m1}^2)$	$\tilde{A}_{m2} = (a_{m2}, \sigma_{m2}^2)$	$\cdots$	$\tilde{A}_{mn} = (a_{mn}, \sigma_{mn}^2)$

perform the following transformation to standardize the values of the attributes:

$$\tilde{a}_{pq} = \frac{a_{pq}}{\max_{1 \le p \le m} \{a_{pq}\}}, \tilde{\sigma}_{pq}^2 = \frac{\sigma_{pq}^2}{\max_{1 \le p \le m} \{\sigma_{pq}^2\}} \cdot \frac{\sigma_{pq}^2}{a_{pq}}. \tag{16}$$

The standardized attribute value is expressed as $\tilde{B}_{pq} = (\tilde{a}_{pq}, \tilde{\sigma}_{pq}^2)$. In general, TOPSIS chooses the maximum attribute value as the positive ideal solution and the minimum attribute value as the negative ideal solution. In NFP3WD, we determine the two ideal solutions of each attribute by the mean and variance of the normal fuzzy number. More specifically, the maximum mean and the minimum variance of each attribute are selected to constitute the positive ideal solution $\tilde{B}_q^+$. And the minimum mean and the maximum variance of each attribute are selected to construct the negative ideal solution $\tilde{B}_q^-$. Then, the positive ideal solution is expressed as $o^+ = \left\{ \tilde{B}_1^+, \tilde{B}_2^+, \cdots, \tilde{B}_n^+ \right\}$ and the negative ideal solution is expressed as $o^- = \left\{ \tilde{B}_1^-, \tilde{B}_2^-, \cdots, \tilde{B}_n^- \right\}$, where

$$\begin{aligned} \tilde{B}_q^+ &= (\tilde{a}_q^+, \tilde{\sigma}_q^{2+}), \ \tilde{a}_q^+ = max_{1 \le p \le m} \tilde{a}_{pq}, \ \tilde{\sigma}_q^{2+} = min_{1 \le p \le m} \tilde{\sigma}_{pq}^2, \\ \tilde{B}_q^- &= (\tilde{a}_q^-, \tilde{\sigma}_q^{2-}), \ \tilde{a}_q^- = min_{1 \le p \le m} \tilde{a}_{pq}, \ \tilde{\sigma}_q^{2-} = max_{1 \le p \le m} \tilde{\sigma}_{pq}^2. \end{aligned} \tag{17}$$

In the TOPSIS method, with the ideal solutions obtained, the distance between object o_p and o^+ as well as the distance between o_p and o^- can be computed. Then, the conditional probability of the object o_p belonging to C can be calculated based on the above distances.

Based on Eq. (3), the distance $d_{pq}^+ = d(\tilde{B}_{pq}, \tilde{B}_q^+)$ between $\tilde{B}_{pq}$ and $\tilde{B}_q^+$, and the distance $d_{pq}^- = d(\tilde{B}_{pq}, \tilde{B}_q^-)$ between $\tilde{B}_{pq}$ and $\tilde{B}_q^-$ are calculated as follows:

$$\begin{aligned} d_{pq}^+ &= d(\tilde{B}_{pq}, \tilde{B}_q^+) = \sqrt{(\tilde{a}_{pq} - \tilde{a}_q^+)^2 + \frac{1}{2}(\tilde{\sigma}_{pq}^2 - \tilde{\sigma}_q^{2+})^2}, \\ d_{pq}^- &= d(\tilde{B}_{pq}, \tilde{B}_q^-) = \sqrt{(\tilde{a}_{pq} - \tilde{a}_q^-)^2 + \frac{1}{2}(\tilde{\sigma}_{pq}^2 - \tilde{\sigma}_q^{2-})^2}. \end{aligned} \tag{18}$$

However, the distances calculated in TOPSIS do not include the risk attitudes. In prospect theory, value functions are utilized to represent the risk attitudes toward gains and losses. Thus, we utilize the value functions of the original distances as the new distances between objects and ideal solutions.

Algorithm 1: Decision process of NFP3WD with normal fuzzy ideal solution.

Require:

Normal fuzzy outcome matrix in Table 2; The normal fuzzy reference points $\tilde{Z}_r = (a_r, \sigma_r^2)$ of decision makers;

Parameters value in prospect theory: $\mu = \upsilon = 0.88$, $\theta = 2.25$, $\sigma = 0.61$ and $\delta = 0.69$;

Normal fuzzy information system $U = (U, AT, V, f)$ in Table 4;

Ensure:

Decision thresholds α, β and γ of each decision-makers;

Conditional probability $Pr(C|o_p)$ of each object o_p being in the state C;

Decision results on each object of every decision maker.

1: Calculate the distance d_{ij} between $\tilde{Z}_{ij}$ and $\tilde{Z}_r$ by Eq. (8);
2: Calculate the value functions $\tilde{v}_{ij}$ ($i = P, B, N$; $j = P, N$) for taking different actions in different states by Eq. (9);
3: Calculate the weight functions $w_i(Pr(C|x))$ and $w_i(Pr(\neg C|x))$ by Eq. (11);
4: Calculate the prospect value $\tilde{V}(a_i|x)$ ($i = P, B, N$) of taking actions a_P, a_B and a_N by Eq. (12);
5: Calculate α, β and γ by the zero points of $\tilde{V}_1$, $\tilde{V}_2$ and $\tilde{V}_3$.
6: Standardize the normal fuzzy values $\tilde{A}_{pq} = (a_{pq}, \sigma_{pq}^2)$ in Table 4 as $\tilde{B}_{pq} = (\tilde{a}_{pq}, \tilde{\sigma}_{pq}^2)$ by Eq. (16);
7: Determine the two ideal solutions o^+, o^- and see them as positive reference points and negative reference points, respectively.
8: Calculate the distance $d_{pq}^+ = d(\tilde{B}_{pq}, \tilde{B}_q^+)$ between $\tilde{B}_{pq}$ and $\tilde{B}_q^+$, and the distance $d_{pq}^- = d(\tilde{B}_{pq}, \tilde{B}_q^-)$ between $\tilde{B}_{pq}$ and $\tilde{B}_q^-$ by Eq. (18);
9: Calculate the value functions of d_{pq}^+ and d_{pq}^- by Eq. (19);
10: Calculate the new distance between object o_p and o^+, and the new distance between object o_p and o^- by Eq. (20);
11: Obtain the conditional probability of objects in normal fuzzy information system according to the relative closeness by Eq. (21);
12: Obtain the decision results based on by Eq. (14) and Eq. (15).

According to prospect theory, let $o^+ = \left\{ \tilde{B}_1^+, \tilde{B}_2^+, \cdots, \tilde{B}_n^+ \right\}$ be the positive reference point of decision maker and $o^- = \left\{ \tilde{B}_1^-, \tilde{B}_2^-, \cdots, \tilde{B}_n^- \right\}$ be the negative reference point of decision maker. For attribute g_q, let $\tilde{B}_q^+$ and $\tilde{B}_q^-$ be the positive reference point and negative reference point, respectively. Compared to $\tilde{B}_q^+$, g_q represents a loss, and people will show risk chasing toward it conversely, g_q represents a gain in contrast to $\tilde{B}_q^-$, and people will show risk averse toward it.

Then, the value functions of original distance d_{pq}^+ and d_{pq}^- are calculated to represent the new distances that take into account decision-maker's risk attitude. Because all attributes in our normal fuzzy information system are positive attributes, we know that all $\tilde{B}_{pq} \leq \tilde{B}_q^+$ and all $\tilde{B}_{pq} \geq \tilde{B}_q^-$. Thus, the value functions are computed as follows:

$$\tilde{v}_{pq}^+ = -\theta(d_{pq}^+)^\upsilon, \quad \tilde{v}_{pq}^- = (d_{pq}^-)^\mu, \tag{19}$$

where $\tilde{v}_{pq}^+$ denotes the value function of d_{pq}^+, and $\tilde{v}_{pq}^+ \leq 0$; $\tilde{v}_{pq}^-$ denotes the value function of d_{pq}^-, and $\tilde{v}_{pq}^- \geq 0$.

Since each attribute in our information system has different weights, the new distance between object o_p and o^+, as well as the new distance between object o_p and o^- are calculated as follows:

$$\tilde{v}_p^+ = \sum_{q=1}^{n} \omega_q \tilde{v}_{pq}^+, \quad \tilde{v}_p^- = \sum_{q=1}^{n} \omega_q \tilde{v}_{pq}^-, \tag{20}$$

where $\tilde{v}_p^+$ denotes the new distance between object o_p and the positive ideal solution o^+; $\tilde{v}_p^-$ denotes the new distance between o_p and the negative ideal solution o^-.

Since the relative closeness of o_p to o^+ is a good reflection of the conditional probability [1,6]. Thus, we compute the conditional probability by relative closeness as follows:

$$Pr(C|o_p) = RC(o_p) = \frac{\tilde{v}_p^-}{\left|\tilde{v}_p^+\right| + \tilde{v}_p^-}. \tag{21}$$

The whole decision process of NFP3WD with normal fuzzy ideal solution is shown in Fig. 1. Algorithm 1 describes the pseudocode of our proposed methods. In Algorithm 1, the decision thresholds are calculated through steps 1 to 5, conditional probability is calculated through steps 6 to 11, and step 12 obtains the decision results.

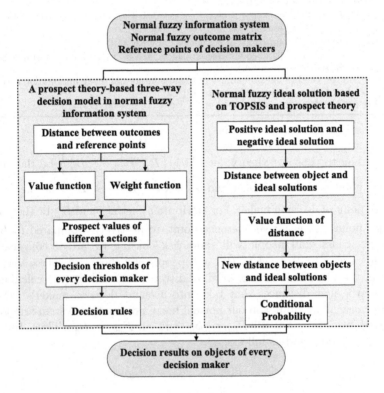

Fig. 1. Decision procedure of NFP3WD with normal fuzzy ideal solution.

5   Illustrative Example

In this section, we apply our proposed methods to make decisions about investment projects.

5.1   Background Description

There are six investment projects represented as $U = \{o_1,\ o_2,\ o_3,\ o_4,\ o_5,\ o_6\}$. These investment projects have four attributes represented as $AT = \{g_1,\ g_2,\ g_3,\ g_4\}$, denote the "Safety", "Efficiency", "Marketing environment" and "Team capability" of investment projects. All four attributes are positive and their weights are $\omega = \{0.1,\ 0.4,\ 0.2,\ 0.3\ \}^T$. The attribute values are described by normal fuzzy numbers, the normal fuzzy information system of the six investment projects is shown in Table 5.

Table 5. Normal fuzzy information system.

	Safety	Efficiency	Marketing environment	Team capability
o_1	$(15, 9)$	$(17, 13)$	$(36, 30)$	$(45, 31)$
o_2	$(16, 7)$	$(22, 11)$	$(43, 24)$	$(52, 31)$
o_3	$(16, 8)$	$(23, 12)$	$(46, 25)$	$(54, 30)$
o_4	$(14, 9)$	$(26, 12)$	$(38, 24)$	$(45, 30)$
o_5	$(13, 8)$	$(22, 11)$	$(38, 22)$	$(44, 35)$
o_6	$(10, 17)$	$(18, 22)$	$(27, 30)$	$(32, 31)$

The normal fuzzy outcome matrix of investment projects is shown in Table 6. There are 10 investors denoted as $E = \{e_1,\ e_2,\ e_3,\ e_4,\ e_5,\ e_6,\ e_7,\ e_8,\ e_9,\ e_{10}\}$. Each investor has different expectations for the outcome of an investment project, which can be denoted by reference points. The normal fuzzy reference points for 10 investors are shown in Table 7.

Table 6. Normal fuzzy outcome matrix.

	C	$\neg C$
a_P	$\tilde{Z}_{PP} = (12, 7)$	$\tilde{Z}_{PN} = (5, 12)$
a_B	$\tilde{Z}_{BP} = (9,\ 8)$	$\tilde{Z}_{BN} = (8,\ 9)$
a_N	$\tilde{Z}_{NP} = (6, 10)$	$\tilde{Z}_{NN} = (11, 6)$

Table 7. Normal fuzzy reference points.

	e_1	e_2	e_3	e_4	e_5	e_6	e_7	e_8	e_9	e_{10}
$\tilde{Z}_r$	$(4, 13)$	$(4, 12)$	$(6, 11)$	$(6, 10)$	$(8, 9)$	$(8, 8)$	$(10, 7)$	$(10, 6)$	$(12, 5)$	$(12, 4)$

In this example, we need to give the actions that each investor should take for each project. Investors are usually bounded-rational when making decisions. Therefore, the above decision-making problem can be solved using our NFP3WD method and normal fuzzy ideal solution method.

5.2   Decision Processes and Decision Results

Based on steps 1 to 5 in Algorithm 1, the decision thresholds are obtained as shown in Table 8.

Table 8. Decision thresholds of NFP3WD.

	e_1	e_2	e_3	e_4	e_5	e_6	e_7	e_8	e_9	e_{10}
α	0.6370	0.6154	0.7929	0.8153	0.8479	0.8156	0.6206	0.5360	0.6296	0.6746
β	0.5354	0.5204	0.5512	0.4776	0.2769	0.3861	0.5153	0.5740	0.5662	0.5628
γ	0.5865	0.5674	0.7029	0.6948	0.5984	0.5945	0.5651	0.5557	0.5953	0.6136

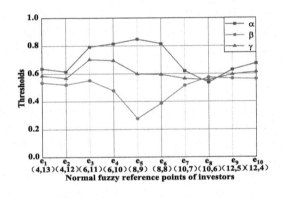

Fig. 2. Decision thresholds of NFP3WD

The figure depicted in Fig. 2 illustrates the changes in decision thresholds as the reference points undergo variation, the reference point of investor increase from e_1 to e_{10}. By Fig. 2, it can be observed that variations in reference points significantly affect α and β, but have little impact on γ. As the reference point increases, α initially increases and then decreases, while β initially decreases and then increases. On the other hand, γ remains relatively stable throughout the variations in reference points.

Then, conditional probabilities of six investment projects are calculated according to steps 6 to 11 in Algorithm 1, as shown in Table 9.

The decision results by step 12 of Algorithm 1 are presented in Table 10. Through Table 10, the decision results will change with the variation of normal

Table 9. Conditional probabilities of projects.

	o_1	o_2	o_3	o_4	o_5	o_6
$Pr(C|o_p)$	0.4008	0.7192	0.8082	0.6785	0.5186	0.0139

fuzzy reference point $\tilde{Z}_r = (a_r, \sigma_r^2)$. In addition, with the increase of normal fuzzy reference points, $BND(C)$ becomes larger first and then gets smaller, while $POS(C)$ and $NEG(C)$ are on the contrary. This variation trend is the same as in the P3WD method [11].

Table 10. Decision results of investment projects.

	$POS(C)$	$BND(C)$	$NEG(C)$
e_1 $(4,13)$	$\{o_2, o_3, o_4\}$	$\emptyset$	$\{o_1, o_5, o_6\}$
e_2 $(4,12)$	$\{o_2, o_3, o_4\}$	$\emptyset$	$\{o_1, o_5, o_6\}$
e_3 $(6,11)$	$\{o_3\}$	$\{o_2, o_4\}$	$\{o_1, o_5, o_6\}$
e_4 $(6,10)$	$\emptyset$	$\{o_2, o_3, o_4, o_5\}$	$\{o_1, o_6\}$
e_5 $(8,9)$	$\emptyset$	$\{o_1, o_2, o_3, o_4, o_5\}$	$\{o_6\}$
e_6 $(8,8)$	$\emptyset$	$\{o_1, o_2, o_3, o_4, o_5\}$	$\{o_6\}$
e_7 $(10,7)$	$\{o_2, o_3, o_4\}$	$\{o_5\}$	$\{o_1, o_6\}$
e_8 $(10,6)$	$\{o_2, o_3, o_4\}$	$\emptyset$	$\{o_1, o_5, o_6\}$
e_9 $(12,5)$	$\{o_2, o_3, o_4\}$	$\emptyset$	$\{o_1, o_5, o_6\}$
e_{10} $(12,4)$	$\{o_2, o_3, o_4\}$	$\emptyset$	$\{o_1, o_5, o_6\}$

5.3 Comparative Analysis

We compare our NFP3WD method with traditional 3WD model [17] and P3WD model [11] by calculating the decision thresholds of investment projects in Sect. 5 using these three methods. The traditional 3WD model [17] makes decisions based on minimum loss, and the loss matrix is represented by crisp numbers. P3WD model [11] incorporates decision-makers risk attitudes using prospect theory, but its outcome matrix is still represented by crisp numbers rather than fuzzy numbers. The calculated decision thresholds of the three methods are shown in Fig. 3.

From Fig. 3, we find that the decision threshold does not change within the ten investors in traditional 3WD model. In the P3WD model, when decision-makers have the same expectation a_r in reference points, the corresponding values of decision thresholds do not change with the change of σ_r^2 in the reference points. While in our NFP3WD method, the decision threshold changes with both the variations of a_r and σ_r^2 in normal fuzzy reference points. These results

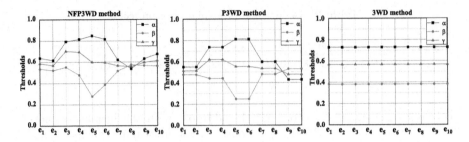

Fig. 3. Decision thresholds of three models.

indicate that our NFP3WD method performs better on considering decision-makers' uncertain preferences.

In addition, our NFP3WD method can handle the continuous uncertain decision information with normal distribution. More, our normal fuzzy ideal solution method combined TOPSIS and prospect theory, which takes into account decision-makers' risk attitudes. The comparison of our proposed method with other methods is shown in Table 11.

Table 11. The comparative analysis with other methods.

	P3WD [11]	Method in [12]	Method in [3]	Ours
Maximum-prospect value	✓	✓	✓	✓
Continuous uncertain data		✓		✓
Distribution of uncertainty				✓
Ideal solution		✓		✓
Risk attitudes in $Pr(X|[x])$				✓

6   Conclusion

In this paper, we present a normal fuzzy prospect theory-based three-way decision method, in which we utilize normal fuzzy numbers subjected to normal distribution to represent the continuous uncertain decision information. The other is that we design a normal fuzzy ideal solution method to estimate conditional probability in normal fuzzy information system without class label, which considers the risk attitudes of decision-makers. In the end, an illustrative example and comparative analysis verify the effectiveness and superiority of our proposed methods.

Acknowledgements. This work was supported by the Natural Science Foundation of Sichuan Province (No. 2022NSFSC0528), the Sichuan Science and Technology Program (No. 2022ZYD0113), Jiaozi Institute of Fintech Innovation, Southwestern University of Finance and Economics (Nos. kjcgzh20230103, kjcgzh20230201), the Fundamental Research Funds for the Central Universities (No. JBK2307055).

References

1. Gao, Y., Li, D.S., Zhong, H.: A novel target threat assessment method based on three-way decisions under intuitionistic fuzzy multi-attribute decision making environment. Eng. Appl. Artif. Intell. **87**, 103276 (2020)
2. Gu, C.-L., Wang, W., Wei, H.-Y.: Regression analysis model based on normal fuzzy numbers. In: Fan, T.-H., Chen, S.-L., Wang, S.-M., Li, Y.-M. (eds.) Quantitative Logic and Soft Computing 2016. AISC, vol. 510, pp. 487–504. Springer, Cham (2017). https://doi.org/10.1007/978-3-319-46206-6_46
3. Gu, J., Wang, Z., Xu, Z., Chen, X.: A decision-making framework based on the prospect theory under an intuitionistic fuzzy environment. Technol. Econ. Dev. Econ. **24**(6), 2374–2396 (2018)
4. Kahneman, D., Tversky, A.: Prospect theory: an analysis of decision under risk. Econometrica **47**(2), 363–391 (1979)
5. Li, A.G., Zhang, Z.H., Meng, Y.: Fuzzy Mathematics and Its Application. Metallurgical Industry Press, Dongcheng (2005)
6. Liang, D.C., Xu, Z.S., Liu, D., Wu, Y.: Method for three-way decisions using ideal TOPSIS solutions at Pythagorean fuzzy information. Inf. Sci. **435**, 282–295 (2018)
7. Peng, Z.Z., Sun, Y.Y.: The Fuzzy Mathematics and Its Application. Wuhan University Press, Wuhan (2007)
8. Sang, G., Wu, T.: Normal fuzzy number multi-attribute decision making model and its application. J. Shanxi Univ. (Nat. Sci. Ed.) **36**(1), 34–39 (2013)
9. Tian, X.L., Xu, Z.S., Gu, J., Herrera-Viedma, E.: How to select a promising enterprise for venture capitalists with prospect theory under intuitionistic fuzzy circumstance? Appl. Soft Comput. **67**, 756–763 (2018)
10. Tversky, A., Kahneman, D.: Advances in prospect theory: cumulative representation of uncertainty. J. Risk Uncertain. **5**(4), 297–323 (1992)
11. Wang, T.X., Li, H.X., Zhou, X.Z., Huang, B., Zhu, H.B.: A prospect theory-based three-way decision model. Knowl.-Based Syst. **203**, 106129 (2020)
12. Wang, T.X., Li, H.X., Zhou, X.Z., Liu, D., Huang, B.: Three-way decision based on third-generation prospect theory with z-numbers. Inf. Sci. **569**, 13–38 (2021)
13. Yang, M.S., Ko, C.H.: On a class of fuzzy c-numbers clustering procedures for fuzzy data. Fuzzy Sets Syst. **84**(1), 49–60 (1996)
14. Yang, X.P., Yao, J.T.: Modelling multi-agent three-way decisions with decision-theoretic rough sets. Fund. Inform. **115**(2–3), 157–171 (2012)
15. Yang, X., Li, Y.H., Li, T.R.: A review of sequential three-way decision and multi-granularity learning. Int. J. Approx. Reason. **152**, 414–433 (2022)
16. Yang, Z.L., Chang, J.P.: Interval-valued Pythagorean normal fuzzy information aggregation operators for multi-attribute decision making. IEEE Access **8**, 51295–51314 (2020)
17. Yao, Y.Y.: Three-way decisions with probabilistic rough sets. Inf. Sci. **180**(3), 341–353 (2010)
18. Yao, Y.Y.: The superiority of three-way decisions in probabilistic rough set models. Inf. Sci. **181**(6), 1080–1096 (2011)

19. Yao, Y.Y.: Interval sets and three-way concept analysis in incomplete contexts. Int. J. Mach. Learn. Cybernet. **8**(1), 3–20 (2017)
20. Zhan, J.M., Wang, J.J., Ding, W.P., Yao, Y.Y.: Three-way behavioral decision making with hesitant fuzzy information systems: survey and challenges. IEEE/CAA J. Automatica Sinica (2022)
21. Zhang, S.C.: Cost-sensitive classification with respect to waiting cost. Knowl.-Based Syst. **23**(5), 369–378 (2010)
22. Zhong, Y., Li, Y., Yang, Y., Li, T., Jia, Y.: An improved three-way decision model based on prospect theory. Int. J. Approx. Reason. **142**, 109–129 (2022)
23. Zhou, B., Yao, Y., Luo, J.: A three-way decision approach to email spam filtering. In: Farzindar, A., Kešelj, V. (eds.) AI 2010. LNCS (LNAI), vol. 6085, pp. 28–39. Springer, Heidelberg (2010). https://doi.org/10.1007/978-3-642-13059-5_6

On Several New Dempster-Shafer-Inspired Uncertainty Measures Applicable for Active Learning

Daniel Kałuża[1,2(✉)] 🆔, Andrzej Janusz[1,2] 🆔, and Dominik Ślęzak[1,2,3] 🆔

[1] Institute of Informatics, University of Warsaw, ul. Banacha 2, 02-097 Warsaw,
Poland
d.kaluza@mimuw.edu.pl
[2] QED Software Sp. z o.o., ul. Miedziana 3A/18, 00-814 Warsaw, Poland
[3] DeepSeas, 12121 Scripps Summit Drive Suite #320, San Diego, CA 92131, USA
https://qed.pl, https://www.deepseas.com

Abstract. Uncertainty-based sampling is one of the most successful and commonly used techniques in active learning. The key element of this approach is an uncertainty function that measures the informativeness of cases from the data pool and is one of the main query selection criteria. In this paper, we investigate the mathematical properties of popular uncertainty functions. We also propose a new family of functions that are inspired by Dempster-Shafer's theory of evidence. Finally, we conduct a series of experiments in which we test the proposed functions against commonly used benchmarks. We argue that our approach is a safe choice for real-life active learning applications.

Keywords: active learning · uncertainty-based sampling · theory of evidence

1 Introduction

Modeling uncertainty based on the probability distribution outputs of machine learning models is a complex task. Typically, the uncertainty is measured with respect to a single decision class [1], which is supposed to be the final model's output for a given input object (usually the class with the maximum posterior probability, sometimes put against the prior). However, in some applications, such as e.g. active learning, it is needed to measure the uncertainty of the whole distributions (to assess the uncertainty of particular objects) rather than focus on the most probable classes.

In this paper, we employ the mathematical apparatus taken from Dempster-Shafer's theory of evidence [16] to tackle this problem. That theory has been used with success in multiple practical applications for reasoning in state of

This research was co-funded by Smart Growth Operational Programme 2014-2020, financed by European Regional Development Fund, in frame of project POIR.01.01.01-00-0213/19, operated by National Centre for Research and Development in Poland.

uncertainty [3] including active learning applications. In that theory, the so-called belief and plausibility functions are considered for arbitrary subsets of values – in our case, the subsets of decision classes. The belief and plausibility functions are calculated from the so-called basic probability assignments ($b.p.a.$'s in short; also called the mass functions). Herein, our goal is to define the b.p.a.'s in such a way that the corresponding belief and plausibility functions describe the (un)certainty of decision making. Existing work in the field of active learning using evidence theory focuses on approaches specialized to particular model families or model ensembles. J. Vandoni et al. [15] used the evidence theory b.p.a. combination rule to aggregate beliefs of multiple SVM classifiers and reason on the uncertainty of such combinations. The authors propose how to incorporate the belief-based entropy, ignorance, and conflict measures into the query-by-committee active learning algorithm to obtain better results for pedestrian detection based on binary classification. P. Hemmer et al. [10] used the evidence theory-based neural network with Dirichlet distribution as the last layer instead of softmax to measure the uncertainty of predictions in the active learning setup. A. Hoarau et al. [11] proposed a novel version of the evidence-based KNN that can work on b.p.a. instead of precise labels. All of the known to us and referenced approaches, if used on hard labels for training, assign mass only to singleton classes or the set of all classes for a single classifier, therefore they do not lead to better expression power than the standard probabilities for the uncertainty-based active learning sample selection. In contrast, in this work, we propose model agnostic mappings that assign mass to multi-element sets, suitable even for non-ensemble models. Proposed approaches might be extended to ensembles using one of the existing combination methods to obtain finer-grained b.p.a and possibly even better uncertainty-measuring capabilities.

Our intuition is that the b.p.a.'s designed for this task should reflect the differences between the most probable and the second/third/etc. most probable decision classes. For example, in the literature there is often used, so-called, margin sampling [13] measure, equal to $p_1 - p_2$, whereby p_1 and p_2 denote the posterior probabilities of the most probable decision class (let us call it class #1) and the second most probable decision class (class #2). This measure is designed to reflect the certainty of reasoning about class #1 given the fact that class #2 is "right behind it". Now, let us consider the third most probable class #3 with probability p_3. We claim that the difference $p_2 - p_3$ should be treated analogously to $p_1 - p_2$, but now in the context of classes #1 and #2 against all the others, and so on.

The above discussion leads us toward defining positive b.p.a.'s only for the subsets of the form {class #1}, {class #1, class #2}, {class #1,class #2,class #3} and so on. In the remainder of this section, i.e., Subsects. 1.1 and 1.2, we formulate definitions and further discuss our intuition on the uncertainty description in the evidence theory. Next, in Sect. 3, we show examples of uncertainty measures that follow this intuition. First, however, in Sect. 2 we provide more basic facts about functions proposed mass functions, later referenced as $m_{\blacktriangle}$ and m_h. Finally, in Sect. 4 we evaluate the proposed uncertainty measures in an active learning experimental setup.

1.1 Definitions

There may be multiple classes with the same posterior probability, therefore we formalize the desired family of such subsets as follows:

Definition 1. *For a given n-dimensional probability distribution $\bar{p}$ on a set of events V and corresponding descending sequence p of all unique positive probability values in that distribution $p_1 > p_2 > ... > p_k > 0$, where $k \leq n$. We define a descending (in sense of inclusion) sequence of sets $X_1^p \subset X_2^p \subset, ... \subset X_k^p$, such that*

$$X_i^p = \{\#j \,|\, \bar{p}_j \geq p_i\} \tag{1}$$

for every i in $1, ..., k$,. We will call those $X_i^p \subseteq V$ sets the layered sets and denote the whole sequence as X^p.

There are different ways of using the quantities of the form $p_1 - p_2$, $p_2 - p_3$, generally $p_i - p_{i+1}$, to provide the subsets $X_1^p, ..., X_k^p$ with their b.p.a.'s. In this paper, we consider two such examples as defined below. The reasons for calling them *pyramidal* and *height ratio* mass functions will become clear in further sections. To simplify further notation, we will consider $p_{k+1} = 0$, which allows for the natural definition of corner cases of mass assigned to X_k^p.

Definition 2. *For a given probability distribution $\bar{p}$ and corresponding layered sets X^p, we define a pyramidal mass function $m_\blacktriangle : 2^V \to [0, 1]$ as:*

$$
\begin{aligned}
m_\blacktriangle(X_i^p) &= (p_i - p_{i+1}) \cdot |X_i^p| &&for\,i = 1, ..., k, \\
m_\blacktriangle(X) &= 0 &&for\ any\ other\,X \subseteq V.
\end{aligned}
\tag{2}
$$

The pyramidal mass function assigns each layered set a value corresponding to the size of this set multiplied by the size of the "step" in the ordered probability distribution, i.e. for the layered set X_i^p, the difference between the i-th largest probability and the next one.

Definition 3. *For a given probability distribution $\bar{p}$ and corresponding layered sets X^p, we define a height ratio mass function $m_h : 2^V \to [0, 1]$ as:*

$$
\begin{aligned}
m_h(X_i^p) &= \frac{(p_i - p_{i+1})}{p_1} &&for\,i = 1, ..., k, \\
m_h(X) &= 0 &&for\ any\ other\,X \subseteq V.
\end{aligned}
\tag{3}
$$

Height-ratio mass function assigns each layered set a value corresponding only to the size of the "step" in the ordered probability distribution normalized by the largest probability in the distribution. It differs from the pyramidal mass as it does not take into account the size of the set in the assigned values, i.e., the value assigned to the set X_i^p is proportional to the difference between i-th largest probability and the next one.

As one can easily check, both $m_\blacktriangle$ and m_h are valid b.p.a.'s as we have $\sum_{X \subseteq V} m_\blacktriangle(X) = \sum_{X \subseteq V} m_h(X) = 1$. Moreover, both of them can be called the certainty assignments, because they label the corresponding subsets with the increasing levels of certainty that it is safe to infer about the values in those subsets.

Let us now recall the formulas for the belief and plausibility functions, defined for an arbitrary b.p.a. m:

$$Bel(X) = \sum_{A|A\subseteq X} m(A) \qquad Pl(X) = 1 - Bel(\overline{X}) = \sum_{A|A\cap X\neq\emptyset} m(A) \qquad (4)$$

where $\overline{X}$ is a complement of set X. These functions, when computed using our $m_{\blacktriangle}$ and m_h, provide two alternative interpretations of the extent in which we are certain about particular subsets of decision values, and on the other hand, we are certain that we do not infer with classes outside those subsets.

1.2 Uncertainty Description in Evidence Theory

Functions Bel and Pl, computed based on b.p.a.'s $m_{\blacktriangle}$ and m_h, can be a useful tool that supports decision making. In particular, the differences of the form $Pl(X) - Bel(X)$ can be regarded as the measures of uncertainty of decision making about $X \subseteq V$. In [14], it was proven that the average value of the differences between Pl and Bel for arbitrary subsets can be expressed in terms of the cardinalities of the sets assigned with b.p.a.'s, namely:

$$\frac{1}{2^{|V|}} \sum_{X\subseteq V} (Pl(X) - Bel(X)) = 1 - \sum_{X\subseteq V} m(X) \cdot 2^{-|X|+1} \qquad (5)$$

This leads us to the first idea of the uncertainty measure which can be used when assessing posterior probability distributions, namely the exponent evidence function introduced later in Eq. (10). However, one may operate with a broader family of uncertainty measures based on such b.p.a.'s, whereby the fundamental intuition is that the uncertainty grows, if high values of b.p.a.'s begin to correspond to the higher-cardinality subsets of decision classes.

2   Interpretation

To better explain the intuition behind the mass functions we visualize them in Fig. 1 and Fig. 2. As visualized in Fig. 1, the $m_{\blacktriangle}$ assigns mass corresponding to horizontal slices of the probability histogram, which might be thought of as layers of a pyramid if we consider ordering of the probabilities with larger values in the middle instead of on the beginning of the chart. The mass from one layer is assigned the set of all classes from the appropriate layered set, i.e., which have bar on the same height or higher than the considered probability value. This is why we call $m_{\blacktriangle}$ the pyramidal mass function.

For a machine learning model prediction this can be viewed as simplification that all of the classes with same posterior probability value are undecidable

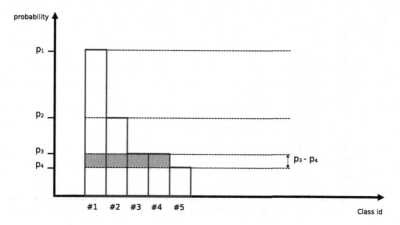

Fig. 1. Pyramidal mass function visualization. In this example a mass marked with green is assigned to the set of classes $X_3^p = \{\#1, \#2, \#3, \#4\}$. It is worth noting that class $\#4$ is included in this set as it has exactly the same probability as class $\#3$.

for the model, i.e. model cannot distinguish in any way to which class from those the object should be assigned. In such abstraction the posterior probabilities returned by the model are viewed only from the perspective of differences. Specifically, the mass assigned to a class $\#1$, may be interpreted as the probability with which this class can be distinguished from the second most probably class $\#2$. If we take for example the following probability distribution $p(\{\#1\}) = 0.4$, $p(\{\#2\}) = 0.2$, $p(\{\#3\}) = p(\{\#4\}) = 0.15$, $p(\{\#5\}) = 0.1$ the mass assigned to the sets are as follows:

$$m_\blacktriangle(\{\#1\}) = 0.2,$$
$$m_\blacktriangle(\{\#1, \#2\}) = 0.1,$$
$$m_\blacktriangle(\{\#1, \#2, \#3, \#4\}) = 0.2,$$
$$m_\blacktriangle(\{\#1, \#2, \#3, \#4, \#5\}) = 0.5.$$

The above example shows, that such construction may lead to counter-intuitively large mass being assigned to sets with many classes in comparison to the mass assigned to the dominant class in original probability distribution. This leads us to the second mass function m_h, which may lead to more intuitive values.

Visualization of the height ratio mass function can be viewed in Fig. 2. m_h assigns mass corresponding to the ratio of the height of the layer and the largest probability. This can also be thought of as the height of one "step" in a sorted histogram to the height of the largest bar. The division by largest probability is in fact a normalization factor to a correct b.p.a. - without it, assigned masses, i.e., $p_i - p_{i+1}$ sum just to the largest probability value p_1. Once again, this mass is assigned to the whole layered set, i.e., all classes that have the same or higher probability. For the previously-mentioned example, $p(\{\#1\}) = 0.4$, $p(\{\#2\}) =$

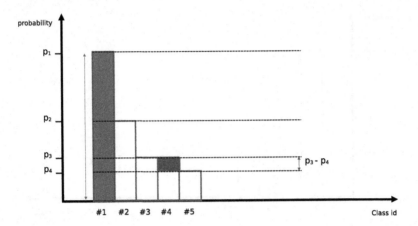

Fig. 2. Height ratio mass function visualization. In this example, a mass marked with blue divided by the area of red is assigned to the set of classes $X_3^p = \{\#1, \#2, \#3, \#4\}$. This might be also viewed as a ratio of blue and red lines with arrows next to the appropriate bars. It is worth noting that class #4 is included in this set as it has exactly the same probability as class #3. (Color figure online)

0.2, $p(\{\#3\}) = p(\{\#4\}) = 0.15$, $p(\{\#5\}) = 0.1$, this leads to the following mass assignments:

$$m_h(\{\#1\}) = 0.5,$$
$$m_h(\{\#1, \#2\}) = 0.125,$$
$$m_h(\{\#1, \#2, \#3, \#4\}) = 0.125,$$
$$m_h(\{\#1, \#2, \#3, \#4, \#5\}) = 0.25.$$

This function does not include the size of the set in the definition, only the difference between the probability values and their relation to the largest probability value. This leads to b.p.a.'s strongly connected to the original distribution, which might be easier to reason on and might lead to more intuitive uncertainty measures.

Moreover, both of the proposed mass functions assign mass to groups of classes, allowing us to reason based on not only the values but also the sizes of the groups, therefore to better model the spread of the distribution.

In most basic 2 class setup, i.e. $V = \{\#1, \#2\}$ proposed mass function formulas lead to the following simplified equations:

$$m_\blacktriangle(\{\#1\}) = p_1 - p_2$$

$$m_h(\{\#1\}) = 1 - \frac{p_2}{p_1}$$

which correspond to the formulas of well-known active learning measures: margin sampling and ratio of confidence [9].

3   Uncertainty Measures

In this section, we display examples of uncertainty measures built upon proposed mass functions. As mass concentrated in sets of smaller cardinalities indicate the confidence, all of the proposed functions base on the mass values and sizes of sets. We have prepared those functions without basing on the internals of our b.p.a. methods therefore we strongly believe that the same functions might be relevant for other b.p.a. At the same time as those are only the examples other interesting uncertainty measures might be constructed with proposed b.p.a. methods. To better show the similarities and differences of the proposed approach we visualize them on 2 dimensional simplex and compare the functions with classical model agnostic measures, i.e. least confidence, margin sampling, entropy and confidence ratio.

3.1   Classical Measures

For simplicity of notation we will denote $\max_i(\overline{p})$ as a i-th maximal value in probability distribution $\overline{p}$.

1. Least confidence
$$f_{least}(\overline{p}) = 1 - \max_1(\overline{p}) \tag{6}$$

2. Margin sampling
$$f_{margin}(\overline{p}) = 1 - (\max_1(\overline{p}) - \max_2(\overline{p})) \tag{7}$$

3. Confidence ratio
$$f_{ratio}(\overline{p}) = \frac{\max_2(\overline{p})}{\max_1(\overline{p})} \tag{8}$$

4. Entropy
$$f_{entropy}(\overline{p}) = -\sum_{\overline{p}_i \in \overline{p}} \overline{p} \log(\overline{p}) \tag{9}$$

The visualization of the values assigned by each of these uncertainty measures for 3 class classification problem in the form of a 2d simplex is available in Fig. 3. All measures obtain largest value in the center of the simplex corresponding to the uniform distribution, which is expected behavior for the balanced problems. Characteristic "crosses" are visible for the ratio and margin measures, which corresponds to the similar uncertainty values for samples in those areas, even if the probability distribution wanders off from the uniform distribution to the edge at right angle.

3.2   Evidence-Based Measures

Evidence based measures make use of both value of mass assigned to a particular set of classes and the cardinality of the set, therefore they have a more natural ability to express the spread of the distribution. Using this fact we propose the following exemplary functions:

1. Exponent evidence

$$f_{exp}(\overline{p}) = 1 - \sum_{\substack{X_i^p \in X^p \\ m(X_i^p) \neq 0}} m(X_i^p) \cdot 2^{-|X_i^p|+1} \tag{10}$$

2. Large exponent evidence

$$f_{large\_exp}(\overline{p}) = 1 - \sum_{\substack{X_i^p \in X^p \\ m(X_i^p) \neq 0}} m(X_i^p) \cdot 4^{-|X_i^p|+1} \tag{11}$$

3. Log divide evidence

$$f_{log\_div}(\overline{p}) = 1 - \sum_{\substack{X_i^p \in X^p \\ m(X_i^p) \neq 0}} \frac{m(X_i^p)}{\log(|X_i^p| + \epsilon)} \tag{12}$$

where $\epsilon = 1e-12$ is small constant to make this function numerically defined for $|X_i^p| = 1$.

4. Log plus evidence

$$f_{log+}(\overline{p}) = \sum_{\substack{X_i^p \in X^p \\ m(X_i^p) \neq 0}} m(X_i^p) \log(|X_i^p| + 1). \tag{13}$$

Those uncertainty measures might be used together with different mass functions and therefore lead to various properties of the active learning selection. The measures are visualized together with pyramidal mass function in Fig. 4 and with height ratio mass function in Fig. 5. It is worth noting that some of the visualization correspond to known classical measures, e.g. Log divide evidence with pyramidal uncertainty looks similar to margin sampling and log divide for height ratio visualization looks similar to confidence ratio. Of course, as the evidence measures take into account all of the classes, this correspondence might not be the case for problems with higher number of classes. Moreover, exponent evidence with height ratio b.p.a. might be viewed as a generalization of margin sampling to include a larger number of classes as it takes into account distances between descending pairs of probabilities with weighting corresponding to the size of the set, which is equivalent to indication which pair is used in the computation. Log plus evidence is a slight modification of one of known entropy counterparts in the Dempster-Shafer theory [7], addition of 1 in the logarithm argument softens the difference between weights of values corresponding to sets of larger sizes.

From our perspective, the exponent height ratio evidence measures should be given particular attention. Based on empirical inspection for a three-class scenario, they have intuitive properties that, in our opinion, a good active learning measure should have, i.e.:

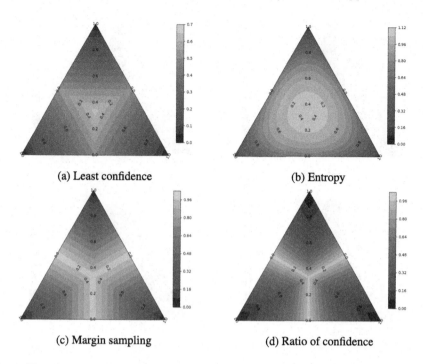

Fig. 3. Visualization of classical uncertainty measures for 3 class scenario presented on 2d simplex. High uncertainty points are indicated with yellow and when the uncertainty decreases color changes to dark blue. Each plot is done with 15 color levels, this allows to examine points with the same value, i.e. on the "lines" better. Numbers rotated to one of the vertices represent the distance from this vertex, which correspond to probability of particular class in this point. (Color figure online)

- The uncertainty values are increasing towards the middle of of the simplex, therefore when we follow the gradient from any point in the simplex we should steadily approach the middle.
- The uncertainty values decrease most toward the nearest simplex vertex.

A final important aspect that should be kept in mind is that proposed mass functions are similar at their core, therefore a mapping from one to another can be easily done in the uncertainty measure formula using the cardinality of the class set and the largest probability.

4 Experiments

To validate the proposed mass functions and based on them uncertainty measures a series of active learning experiments has been performed. A classical active learning setup has been employed, i.e.:

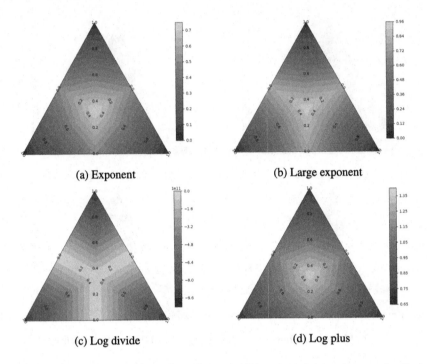

Fig. 4. Visualization of evidence based pyramidal uncertainty measures for 3 class scenario presented on 2d simplex. High uncertainty points are indicated with yellow and when the uncertainty decreases color changes to dark blue. Each plot is done with 15 color levels, this allows to examine points with the same value, i.e. on the "lines" better. Numbers rotated to one of the vertices represent the distance from this vertex, which correspond to probability of particular class in this point. (Color figure online)

1. Each dataset has been split to a training pool and a holdout test set for the purpose of evaluation. The split was performed with stratification based on the label of samples.
2. An initial training set has been chosen at random, but the same one for every uncertainty measure to make the evaluation more deterministic.
3. Machine learning model has been trained on the training set and evaluated on the test set.
4. An iterative active learning process has been simulated for each uncertainty measure:
   (a) One sample, with highest uncertainty, has been chosen in each iteration using uncertainty measure and added to the training set.
   (b) Machine learning model has been trained on the training set.
   (c) Obtained model has been evaluated on the test set.
   (d) The steps 4a–4c are repeated until desired number of samples is chosen.

The parameters of the evaluation for each dataset are available in Table 1. Each dataset has been split in half for the testing purposes. The active learning procedure, from step 2 has been also repeated 10 times to make the evaluation

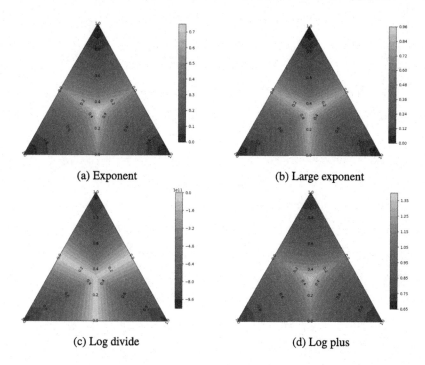

(a) Exponent   (b) Large exponent

(c) Log divide   (d) Log plus

Fig. 5. Visualization of evidence based height ratio uncertainty measures for 3 class scenario presented on 2d simplex. High uncertainty points are indicated with yellow and when the uncertainty decreases color changes to dark blue. Each plot is done with 15 color levels, this allows to examine points with the same value, i.e. on the "lines" better. Numbers rotated to one of the vertices represent the distance from this vertex, which correspond to probability of particular class in this point. (Color figure online)

more independent from random sampling of initial training set. Balanced accuracy has been used as evaluation metric, as it is easily interpretable and suitable for both balanced and imbalanced datasets. For each experiment exactly 100 samples have been drawn in an active manner. RandomForestClassifier from scikit-learn [12], with fixed initial random seed and other parameters set to default values, has been used as the machine learning model.

The following datasets has been used for the evaluation:

- vowel [6] - a dataset with task of recognition of steady state vowels of British English based on features extracted from speech
- pendigits [2] - a dataset of tabular features retrieved from handwritten digits using a pressure sensitive tablet, with the task of digits classification
- letter [8] - a dataset of tabular features retrieved from black-and-white rectangular pixel displays, e.g. statistical moments and edge counts, with the task of letters classification
- car [5] - dataset with hierarchical structure describing car properties with decision problem of car purchase evaluation

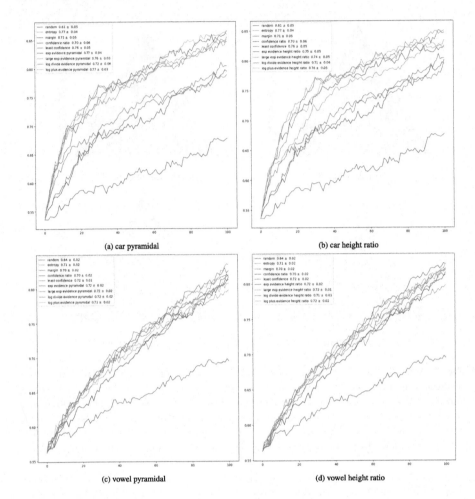

Fig. 6. Visualization of active learning experiments results on vowel and car datasets. The y-axis of each plot features balanced accuracy metric, the x-axis features number of iterations of active learning loop. The plot display average across 10 experiments, values in the legend denote an average across all loop iterations for a given measure and standard deviation across the experiments. To maintain readability of the plots, evidence based uncertainty measures have been divided according to mass function used to obtain the b.p.a. Plots on the left display the pyramidal mass based measures and plots on the right height ratio based. Classical measures have been added to each plot for common reference.

Experiment results are visible in Figs. 6 and 7. The summary of the mean ranks obtained by each uncertainty measure across the datasets is available in Table 2. Most of the proposed evidential uncertainty measures behave similarly. They were worse than margin and ratio of confidence on the *letter* dataset but better than entropy and least confidence. The same case was for the *pendigits*

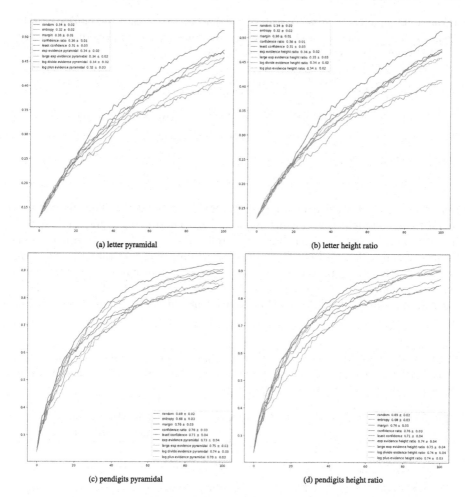

(a) letter pyramidal

(b) letter height ratio

(c) pendigits pyramidal

(d) pendigits height ratio

Fig. 7. Visualization of active learning experiments results on letter and pendigits datasets. The y-axis of each plot features balanced accuracy metric, the x-axis features number of iterations of active learning loop. The plot display average across 10 experiments, values in the legend denote an average across all loop iterations for a given measure and standard deviation across the experiments. To maintain readability of the plots, evidence based uncertainty measures have been divided according to mass function used to obtain the b.p.a. Plots on the left display the pyramidal mass based measures and plots on the right height ratio based. Classical measures have been added to each plot for common reference.

dataset. What is interesting, margin and ratio of confidence behaved exactly the same for this 2 datasets. For the *vowel* dataset, least confidence obtained the best results followed evidence based measures and entropy, with margin and confidence ratio on the other side of the spectrum. In the experiments on the *car* dataset entropy with pyramidal evidence came out with best values, they

Table 1. Parameters of active learning experiments setup. Each row indicates a different dataset. For each dataset exactly 100 iterations have been performed. The size of initial training set depends on the dataset size.

Dataset	# classes	# features	initial size	training pool	test size
vowel	11	12	100	495	495
pendigits	10	16	5	5496	5496
letter	26	16	10	10 000	10 000
car	4	6	100	864	864

were followed with least confidence and height ratio evidence. For this dataset, margin and confidence ratio measures were last. Those results suggest that similar situation as with no free lunch theorem occurs for active learning uncertainty measures, i.e. there might be no uncertainty measure that is always the best. Nevertheless, experiments have indicated that proposed evidential measures, especially pyramidal exp and pyramidal large exp are the safest choice with average rank equal to 5.0, followed by height ratio exp with mean rank 5.25.

Table 2. Average ranks across all datasets obtained by uncertainty measures. The lower rank the better, in case of a tie an average of consequent ranks was given to the measures. The lowest average rank is marked with bold.

Classical Measure	Mean rank	Measure	Pyramidal mean rank	Height ratio Mean rank
Entropy	9.25	Exp	**5.0**	5.25
Least confidence	8.0	Large Exp	**5.0**	6.25
Margin	6.25	Log divide	5.25	7.5
Confidence ratio	6.75	Log plus	8.25	6.75

5   Conclusions

In this paper, we explored the concept of uncertainty-based query selection, a widely used approach in active learning. Our primary focus was on the mathematical properties of uncertainty functions used in this approach. Inspired by Dempster-Shafer's theory of evidence, we introduced a novel family of uncertainty functions. These functions are based on the concept of basic probability assignments (b.p.a.) and belief and plausibility functions derived from them. Our goal was to establish a b.p.a. that accurately reflects the uncertainty of decision-making processes. We also introduce additional examples of uncertainty measures that align with the intuition that uncertainty grows when high b.p.a.

values correspond to larger subsets of decision classes. The paper provides a comprehensive exploration of these measures and their properties.

The proposed uncertainty measures are then put to the test through a series of experiments using active learning benchmarks. The findings suggest that while there is no universally best uncertainty measure, the proposed evidential measures, particularly those labeled as pyramidal exp and pyramidal large exp, consistently perform well across datasets. These measures are identified as reliable choices for active learning applications. It shows that uncertainty functions inspired by the evidence theory are a plausible choice for real-life applications and should be further studied in this context.

Additionally, we have verbalized our intuition describing desired properties of a good uncertainty measure. The formalization of those properties with mathematical apparatus and relevant experiments supportive the intuition remains a future work. To achieve that, one might consider an extension of the lattice and entropy theory described by D. Bianucci and G. Cattaneo [4].

References

1. Agrawal, A., Tripathi, S., Vardhan, M.: Active learning approach using a modified least confidence sampling strategy for named entity recognition. Prog. Artif. Intell. **10**(2), 113–128 (2021). https://doi.org/10.1007/s13748-021-00230-w
2. Alpaydin, E., Alimoglu, F.: Pen-Based Recognition of Handwritten Digits. UCI Machine Learning Repository (1998). https://doi.org/10.24432/C5MG6K
3. Bezerra, E.D.C., Teles, A.S., Coutinho, L.R., da Silva e Silva, F.J.: Dempster-Shafer theory for modeling and treating uncertainty in IoT applications based on complex event processing. Sensors **21**(5), 1863 (2021). https://www.mdpi.com/1424-8220/21/5/1863
4. Bianucci, D., Cattaneo, G.: Information entropy and granulation co–entropy of partitions and coverings: a summary. In: Peters, J.F., Skowron, A., Wolski, M., Chakraborty, M.K., Wu, W.-Z. (eds.) Transactions on Rough Sets X. LNCS, vol. 5656, pp. 15–66. Springer, Heidelberg (2009). https://doi.org/10.1007/978-3-642-03281-3_2
5. Bohanec, M.: Car Evaluation. UCI Machine Learning Repository (1997). https://doi.org/10.24432/C5JP48
6. Deterding, D.H.: Speaker normalisation for automatic speech recognition. Ph.D. thesis, University of Cambridge (1990)
7. Dubois, D., Prade, H.: Properties of measures of information in evidence and possibility theories. Fuzzy Sets Syst. **24**(2), 161–182 (1987). https://www.sciencedirect.com/science/article/pii/0165011487900881
8. Frey, P.W., Slate, D.J.: Letter recognition using Holland-style adaptive classifiers. Mach. Learn. **6**(2), 161–182 (1991). https://doi.org/10.1007/BF00114162
9. Guochen, Z.: Four uncertain sampling methods are superior to random sampling method in classification. In: 2021 2nd International Conference on Artificial Intelligence and Education (ICAIE), pp. 209–212 (2021)
10. Hemmer, P., Kühl, N., SchÖffer, J.: Deal: deep evidential active learning for image classification. In: 2020 19th IEEE International Conference on Machine Learning and Applications (ICMLA), pp. 865–870 (2020)

11. Hoarau, A., Martin, A., Dubois, J.C., Le Gall, Y.: Imperfect labels with belief functions for active learning. In: Le Hégarat-Mascle, S., Bloch, I., Aldea, E. (eds.) BELIEF 2022. LNCS, vol. 13506, pp. 44–53. Springer, Cham (2022). https://doi.org/10.1007/978-3-031-17801-6_5

12. Pedregosa, F., et al.: Scikit-learn: machine learning in Python. J. Mach. Learn. Res. **12**, 2825–2830 (2011)

13. Scheffer, T., Decomain, C., Wrobel, S.: Active hidden Markov models for information extraction. In: Hoffmann, F., Hand, D.J., Adams, N., Fisher, D., Guimaraes, G. (eds.) IDA 2001. LNCS, vol. 2189, pp. 309–318. Springer, Heidelberg (2001). https://doi.org/10.1007/3-540-44816-0_31

14. Ślęzak, D.: Approximate Decision Reducts. Ph.D. Thesis under Supervision of A. Skowron, University of Warsaw, Poland (2002). (in Polish)

15. Vandoni, J., Aldea, E., Le Hégarat-Mascle, S.: Evidential query-by-committee active learning for pedestrian detection in high-density crowds. Int. J. Approx. Reason. **104**, 166–184 (2019). https://www.sciencedirect.com/science/article/pii/S0888613X18303517

16. Yager, R.R., Liu, L.: Classic Works of the Dempster-Shafer Theory of Belief Functions, 1st edn. Springer, Heidelberg (2010). https://doi.org/10.1007/978-3-540-44792-4

Rough Fuzzy Concept Analysis
via Multilattice

Gaël Nguepy Dongmo[1]([✉]) [ID], Blaise Blériot Koguep Njionou[1] [ID],
Léonard Kwuida[2] [ID], and Mathias Onabid[1] [ID]

[1] Department of Mathematics and Computer Science, University of Dschang,
Dschang, Cameroon
dongmogaelc@gmail.com, blaise.koguep@univ-dschang.org
[2] Department of Business, Bern University of Applied Sciences, Bern, Switzerland
leonard.kwuida@bfh.ch

Abstract. J. Medina and J. Ruiz-Calvino presented Fuzzy Formal Concept Analysis via multilattice (M-FCA for short) as a fuzzy generalization of Formal Concept Analysis with multilattice as underlying set of
truth degree. The truth degrees are not always linear or the existence of
a least upper bound of two elements are no longer required, but there
is a possibility of having minimal upper bounds, and dually. Hence, M-
FCA is a more flexible than Fuzzy Formal Concept Analysis (L-FCA).
We investigate formal M-concepts with rough intents and, get a generalization of E. Bartl and J. Konecny work. First, we define appropriate
concept forming operators and give their properties. Next, we show that
the concepts in the new framework form a complete multilattice.

Keywords: Multilattice · Rough set Theory · Formal concept
analysis · Fuzzy sets

1 Introduction

Rough set theory (RST) is a mathematical tool for studying information systems
with inexact, uncertain or fuzzy information. It was developed by Pawlak [27]
in the 80ies. In RST, data are analyzed by approximating sets, usually based on
an equivalence relation.

Fuzzy set theory (FST) is another tool for studying imprecise information. It
evaluates the degree to which an element belongs to a set. The set of membership
degrees could be the real unit interval $[0, 1]$ or even a residuated lattice.

The need to combine RST and FST to handle uncertain or imprecise data
arose almost naturally. The first work in this direction was proposed by Dubois
and Prade [11], where they used the real unit interval as the set of membership
degree. Radzikowska [23] investigated L-rough sets where the set of truth degrees
is a residuated lattice L.

Formal Concept Analysis (FCA) is a tool for data analysis based on lattice
theory. FCA was introduced in 1982 by Rudolph Wille [25]. The mathematical

foundations are documented in [14]. In FCA-stteing, the information is described by a set of objects, a set of attributes and an incidence relation indicating if an object has an attribute. We call this a **formal context**. Formal concepts can then be derived from the formal context and form a complete lattice, when ordered by their subsumption relation.

When analyzing data in real life, we might face situations where the incidence relation is imprecise. Burusco and Fuentes-Gonzáles [5] proposed a fuzzy generalization of FCA, by replacing the incidence relation by a fuzzy relation encoding the vagueness. But some useful properties such as extensity of concept-forming operators were missing in their approach. Bělohlávek [3] developed fuzzy FCA further, using residuated lattices as set of truth values, and a different approach in constructing the concept forming operators. Lattices structures required the existence of least upper bound and greatest lower bound for each pair of elements. This restriction could be weakened to the existence of minimal upper bounds and maximal lower bounds for each pair of elements. That is in fact replacing lattices by multilattices, as introduced in [4,17,18].

In this direction, Medina et al. presented fuzzy logic programming via multilattices [21], after, they have used multilattices as the underlying sets of truth degrees for fuzzy formal concept analysis [20] and made concept-forming operators [19].

The combination of FCA and Rough Set arrive since, not every pair of a set of objects and a set of attribute defines a concept. Moreover we can be faced with a situation where we have a set of attributes and need to find the best concept that approximates theses attributes. For example, when a search engine finds an object, that has certain attributes, he finds objects whose attributes are close to required attributes. That is why, many authors have introduced the notion of approximation in FCA [24]. These notions were later extended in L-FCA [1], and will be extended to M-FCA in this contribution. The aim of this paper, is to extend the work of E. Bartl and J. Konecny [1] by replacing the complete residuated lattice with a multilattice, as truth value structures.

The paper is organized as follows. In Sect. 2 we recall basic notions to make this paper self-contained. Section 3 is devoted to the M-rough formal concept. Our main result shows that the set of $M-$rough concepts is a multilattice. We provide an example to illustrate our construction Sect. 4. Section 5 concludes the paper.

2   Preliminaries

2.1   Multilattice

In this section we introduce several notions from lattice and multilattice theory in order to make our paper self contained. Let $(P, \leq)$ be a poset and $X \subseteq P$. We denote by $U(X)$ (resp. $L(X)$) the set of upper (resp. lower) bounds of X. The **supremum** (resp. **infimum**) of X is the least (resp. greatest) element of $U(X)$ (resp. $L(X)$), whenever it exists. The supremum (resp. infimum) of X is denoted by $\vee X$ or $\sup X$ (resp. $\wedge X$ or $\inf X$). A **lattice** is a poset $(P, \leq)$ in which any pair of elements has a supremum and an infimum. If every subset of P has a supremum and an infimum then $(P, \leq)$ is called a **complete lattice** [10]. A subset $X \subseteq P$ is called a **chain** (resp. **antichain**) if for every $x, y \in X$ we have $x \leq y$ or $y \leq x$ (resp. $x \not\leq y$ and $y \not\leq x$). A poset $(P, \leq)$ is said to be **coherent** if every chain has a supremum and an infinimum [19]. To extend the notion of lattice, Benado [4] introduced multilattices, where he replaces the supremum (resp. infimum) with minimal upper bounds (resp. maximal lower bounds).

A **multisupremum** (resp. **multiinfimum**) of X is a minimal (resp. maximal) element of $U(X)$ (resp. $L(X)$). The set of multisuprema (resp. multiinfima) of X is denoted by $\sqcup X$ (resp. $\sqcap X$). For $a, b \in P$ we simply write $U(a)$, $L(a)$, $a \sqcup b$, $a \sqcap b$ for $U(\{a\})$, $L(\{b\})$, $\sqcup\{a, b\}$, $\sqcap\{a, b\}$, respectively.

Definition 1. *[4, 7] A poset $(M, \leq)$ is called **multilattice** if, for all $a, b \in M$*

- *$c \in M$ is an upper bound of $\{a, b\} \implies \exists d \in a \sqcup b$ such that $d \leq c$, and*
- *$c \in M$ is a lower bound of $\{a, b\} \implies \exists d \in a \sqcap b$ such that $d \geq c$.*

A **complete multilattice** *[21]* is a multilattice $(M, \leq)$ in which $\sqcup X$ and $\sqcap X$ are non-empty for any $X \subseteq M$.

Any lattice $(A, \wedge, \vee)$ is a multilattice since for all $a, b \in A$, $a \sqcap b = \{a \wedge b\}$ and $a \sqcup b = \{a \vee b\}$. Whenever $\sqcap X$ or $\sqcup X$ is a singleton, it is denoted by $\bigwedge X$ or $\bigvee X$. Any complete lattice is also a complete multilattice. A multilattice will be called **pure** if it is not a lattice.

In Fig. 1, a and b do not have supremum because c and d are not comparable. This shows that, M_7 is not a lattice. However, a and b have minimal upper bounds c and d. Therefore M_7 is a multilattice. In fact, any finite (bounded) poset is a (complete) multilattice.

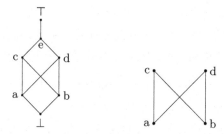

Fig. 1. Left: $(M_7, \leq)$, is an example of a complete and pure multilattice. It will be used as set of truth degrees in our example. Right: An example of multilattice which is not complete.

We now define the adjoint pair, which is a generalization of a t-norm and its implication.

Definition 2. *[12] Let $(P, \leq)$ be a poset. A couple $(\&, \rightarrow)$ of binary operations on P is called **adjoint pair** if*

1. $\&$ is order-preserving in both arguments;
2. $\rightarrow$ is order-preserving in the first argument and reversing in the second argument;
3. $a \leq (c \rightarrow b) \iff (a\&c) \leq b$, for all $a, b, c \in P$.

Some examples of adjoint pairs are the product, Gödel and Lukasiewicz t-norms together with their residuated implications. They are defined on $[0, 1]$ by:

$$x\&_P y = x \cdot y \qquad\qquad z \rightarrow_P x = \min\{1, z/x\}$$

$$x\&_G y = \min\{x, y\} \qquad\qquad z \rightarrow_G x = \begin{cases} 1 & \text{if } x \leq z \\ z & \text{otherwise} \end{cases}$$

$$x\&_L y = \max\{0, x + y - 1\} \qquad z \rightarrow_L x = \min\{1, 1 - x + z\}.$$

Definition 3. *If $(\&, \rightarrow)$ is an adjoint pair on a poset $(P, \leq)$ with $\&$ commutative and associative, with $\top$ a neutral (wrt. $\&$) and top element of P, then we call $(P, \leq, \&, \rightarrow, \top)$ a **pocrim** (partial order commutative residuated integral monoid). If in addition $(P, \leq)$ is a bounded lattice, then we call it a **commutative integral residuated lattice**.*

Proposition 1. *[7] Let $(P, \leq)$ be a poset and $(\&, \rightarrow)$ be an adjoint pair on $(P, \leq)$. If $\&$ is associative and commutative, then, for all $a, b, c \in P$, $\quad a \rightarrow (b \rightarrow c) = (a\&b) \rightarrow c$.*

Definition 4. *[20] Let $(P_1, \leq)$ and $(P_2, \leq)$ be two posets and $^\triangle : P_1 \rightarrow P_2$ and $^\triangledown : P_2 \rightarrow P_1$ be maps. We call $(^\triangle, ^\triangledown)$ a **Galois connection** between P_1 and P_2 if for all $f, f_1, f_2 \in P_1$ and $g, g_1, g_2 \in P_2$ we have:*

- $f_1 \leq f_2 \implies f_2^\triangle \leq f_1^\triangle$
- $g_1 \leq g_2 \implies g_2^\triangledown \leq g_1^\triangledown$
- $f \leq f^{\triangle\triangledown}$ and $g \leq g^{\triangledown\triangle}$.

Let $(P_1, \leq)$ and $(P_2, \leq)$ be two posets and let $(^\triangle, ^\triangledown)$ be a Galois connection between P_1 and P_2. A couple $(f, g) \in P_1 \times P_2$ is called a **concept** if $f^\triangle = g$ and $g^\triangledown = f$.

From now on, we fix the following notations with P_1, P_2, P_3 posets and $\lozenge : P_1 \times P_2 \rightarrow P_3$:

$$X\lozenge m := \{x\lozenge m \mid x \in X\}, \quad \text{where } X \subseteq P_1 \text{ and } m \in P_2$$

$$Y \leq m : \iff y \leq m \text{ for all } y \in Y, \quad \text{where } Y \cup \{m\} \subseteq P_j, 1 \leq j \leq 3.$$

Definition 5. *[19] Let $(M_1, \leq_1)$, $(M_2, \leq_2)$ be two multilattices. Let $(P, \leq)$ be a poset, and $\lozenge : M_1 \times M_2 \rightarrow P$ be a map. We say that $\lozenge$ is:*

(i) *left-continuous in the first argument if $\quad K_1\lozenge m_2 \leq p \implies \sqcup K_1\lozenge m_2 \leq p$ for all $K_1 \subseteq M_1$, $m_2 \in M_2$, and $p \in P$,*

(ii) *left-continuous in the second argument if $\quad m_1\lozenge K_2 \leq p \implies m_1\lozenge \sqcup K_2 \leq p$ for all $K_2 \subseteq M_2$, $m_1 \in M_1$ and $p \in P$,*

(iii) *left-continuous if it is left-continuous in both arguments.*

Dually, we introduce the notion of right-continuity.

Definition 6. *Let* $(M_1, \leq_1)$, $(M_2, \leq_2)$ *be two multilattices. Let* $(P, \leq)$ *be a poset, and* $\Diamond : M_1 \times M_2 \to P$ *be a map. We say that* $\Diamond$ *is:*

(i) *right-continuous in the first argument if* $K_1 \Diamond m_2 \geq p \implies \sqcap K_1 \Diamond m_2 \geq p$
    for all $K_1 \subseteq M_1$, $m_2 \in M_2$, *and* $p \in P$

(ii) *right-continuous in the second argument if* $m_1 \Diamond K_2 \geq p \implies m_1 \Diamond \sqcap K_2 \geq$
    p *for all* $K_2 \subseteq M_2$, $m_1 \in M_1$, *and* $p \in P$

(iii) *right-continuous if it is right-continuous in both arguments.*

Example 1. We consider the multilattice in Fig. 1 and define the map $\Diamond_1 : M_7 \times$

$$M_7 \to M_7, \text{ by } y\Diamond_1 x := \begin{cases} \top & \text{if } y \leq x \\ b & \text{if } y \notin \{\bot, b\} \text{ and } x \in \{\bot, b\} \\ e & \text{if } y = \top \text{ and } x \in \{a, c, d, e\} \\ \top & \text{otherwise} \end{cases} \text{ for all } x, y \in$$

M_7.

$\Diamond_1$ is not right-continuous, because if we consider $K_2 = \{c, d, e, \top\}$, we have, $a\Diamond_1 K_2 \geq a$, but, $\sqcap K_2 = \{a, b\}$ and $a\Diamond_1 \sqcap K_2 \ngeq a$.

The operator $\Diamond_2 : M_7 \times M_7 \to M_7$ defined by,

$$y\Diamond_2 x := \begin{cases} \top & \text{if } y \leq x \\ e & \text{if } y = \top \text{ and } x \in \{a, b, c, d, e\} \\ \top & \text{otherwise} \end{cases} \text{ for all } x, y \in M_7, \text{ is right-}$$

continuous in the second argument.

Remark 1. Let M be a complete multilattice.

1. If a map $\Diamond : M \times M \to M$ is left-continuous and M has a neutral element with respect to $\Diamond$ then, $\sqcup X$ is a singleton for all $X \subseteq M$. In fact, if n is a neutral element of M with respect to $\Diamond$ and $X \subseteq M$ such that $\sqcup X$ is not a singleton, we would have for $a \in \sqcup X$, $X \Diamond n \leq a$ and there would be $b \in \sqcup X$ such that $b \nleq a$. Therefore, $\sqcup X \Diamond n \nleq a$, which is a contradiction to left-continuity.
2. Dually, if a map $\Diamond : M \times M \to M$ is right-continuous, and M has a neutral element with respect to $\Diamond$, then $\sqcap X$ is a singleton for all $X \subseteq M$.

From now, we denote by **M** a complete multilattice equipped with an adjoint pair $(\&, \to)$, where $\&$ is commutative and left-continuous, $\to$ is right-continuous in second argument. We will see how to use a multilattice as underlying set of truth values in fuzzy concept analysis, as proposed in [19, 20] by M. Medina et al.

2.2   M-fuzzy Formal Concept Analysis

Our universe is a non-empty set U. A mapping f assigning to each $y \in U$ a truth degree $f(y) \in M$ is called an **M-fuzzy set** of U. The set of all **M**-fuzzy sets of U is denoted by $\mathbf{M}^U$, and is ordered pointwise:

$$f \leq g \iff f(y) \leq g(y), \text{for all } y \in U.$$

$\left(\mathbf{M}^U, \leq\right)$ is a multilattice [16].

Let X, Y be two non-empty sets, and $I : X \times Y \to \mathbf{M}$ a fuzzy incidence relation. The triple (X, Y, I) is called an **M-fuzzy formal context**, where X and Y are the sets objects and attributes respectively. Let (X, Y, I) be an M-fuzzy formal context. The derivation operators $\uparrow : \mathbf{M}^X \to M^Y$ and $\downarrow : \mathbf{M}^Y \to \mathbf{M}^X$ are defined for $f \in \mathbf{M}^X$ and $g \in \mathbf{M}^Y$ by:

$$f^\uparrow(y) : \bigwedge_{x \in X} \Big(f(x) \to I(x,y)\Big) \quad \text{and} \quad g^\downarrow(x) : \bigwedge_{y \in X} \Big(g(y) \to I(x,y)\Big).$$

An **M-fuzzy formal concept** of the M-fuzzy formal context (X, Y, I) is a pair $(f, g) \in \mathbf{M}^X \times \mathbf{M}^Y$ such that $f^\uparrow = g$ and $g^\downarrow = f$. This concept forming operators were first proposed by Bělohlávek [3], where a residuated lattice were used instead of $\mathbf{M}$.

Proposition 2. *[19,20] Let (X, Y, I) be a $M-$fuzzy formal context, $f, f_1, f_2,$ $f_i \in M^X$ and $g, g_1, g_2, g_i \in M^Y$, $i \in J$, where J is an index set. Then*

  (i) $f_1 \leq f_2 \implies f_2^\uparrow \leq f_1^\uparrow$ *and* $g_1 \leq g_2 \implies g_2^\downarrow \leq g_1^\downarrow$;
  (ii) $f \leq f^{\uparrow\downarrow}$ *and* $g \leq g^{\downarrow\uparrow}$;
  (iii) $f^\uparrow = f^{\uparrow\downarrow\uparrow}$ *and* $g^\downarrow = g^{\downarrow\uparrow\downarrow}$;
  (iv) $\underset{i \in I}{\sqcap}\left(f_i^\uparrow\right) \subseteq \{f^\uparrow, \quad f \in \underset{i \in I}{\sqcup} f_i\}$ *and* $\underset{i \in I}{\sqcap}\left(g_i^\downarrow\right) \subseteq \{g^\downarrow, \quad g \in \underset{i \in I}{\sqcup} g_i\}$.

We denoted by $\mathcal{C}^{\uparrow\downarrow}$ the set of all concepts and by $\leq$ the order relation defined on concepts by: $(f_1, g_1) \leq (f_2, g_2) : \iff f_1 \leq f_2$ (equivalently $g_1 \geq g_2$).

3   Rough Fuzzy Concept Multilattice

We can now define **M**-rough concepts and will prove that the set of all **M**-rough concepts forms a complete multilattice.

3.1   M-Fuzzy Rough Set

Let U be our universe. A **M**-fuzzy relation $R : U \times U \to \mathbf{M}$ is called an **equivalence** relation if R is

- reflexive (i.e. $\forall x \in U : R(x, x) = \top$),
- symmetric (i.e. $\forall x, y \in U : R(x, y) = R(y, x)$)
- transitive (i.e. $\forall x, y, z \in U : R(x, y)\&R(y, z) \leq R(x, z)$)

A pair (U, R) is called **M-fuzzy approximation space** [23].

Theorem 1. *Let $g \in M^U$ and $R \in M^{U^2}$. Then, for any $x \in U$ the following hold:*

  (i) *The set $\{R(x, y) \to g(y); \ y \in U\}$ has an infimum.*
  (ii) *The set $\{R(x, y)\&g(y); \ y \in U\}$ has a supremum.*

Proof. (i) [19].

(ii) Let $a_1, a_2 \in \sqcup\{R(x,y)\&g(y); \; y \in U\}$, then for all $y \in U$ we have $R(x,y)\&g(y) \leq a_1, a_2$, by adjointness condition, $g(y) \leq R(x,y) \rightarrow a_1$, $R(x,y) \rightarrow a_2$. As $\rightarrow$ is second argument right-continuous, for all $r \in a_1 \sqcap a_2$, we have, $g(y) \leq R(x,y) \rightarrow r$ for all $y \in U$. Therefore $g(y)\&R(x,y) \leq r$ for all $y \in U$, hence, $a_1 = r = a_2$, by the minimality of a_1 and a_2. $\square$

For any $g \in \mathbf{M}^U$ the lower and upper approximations of g, respectively denoted by $\underline{g}$ and $\overline{g}$, coincide with lower and upper approximations of g defined in [9] by:

$$\underline{g}(a) := \bigwedge_{y \in U} \Big(R(a,y) \rightarrow g(y)\Big) \text{ and } \overline{g}(a) := \bigvee_{y \in U} \Big(R(a,y)\&g(y)\Big), \text{ for all } a \in U.$$

The pair $< \underline{g}, \overline{g} >$ is called an **M**-fuzzy rough set in (U, R). The set of all **M**-fuzzy rough sets in the **M**-fuzzy approximation space (U, R) is denoted by $\mathcal{RF}(U, \mathbf{M})$. On $\mathcal{RF}(U, \mathbf{M})$ we define an order relation by

$$< \underline{f}, \overline{f} > \; \leq \; < \underline{g}, \overline{g} > \text{ if and only if } \underline{f} \leq \underline{g} \text{ and } \overline{g} \leq \overline{f}$$

Moreover, for any family $\{g_i\}_{i \in J}$ we have:

$$\underset{i \in J}{\sqcap} < \underline{g_i}, \overline{g_i} >= \{< f, h >: \; f \in \sqcap \underline{g_i} \text{ and } h \in \sqcup \overline{g_i}\}$$

and

$$\underset{i \in J}{\sqcup} < \underline{g_i}, \overline{g_i} >= \{< f, h >: \; f \in \sqcup \underline{g_i} \text{ and } h \in \sqcap \overline{g_i}\}.$$

3.2 M-Rough Concept Multilattice

In this subsection, after having introduced the notion of M-rough concept, we will prove that the set of M-rough concepts forms a complete multilattice.

Corollary 1. *Let (X, Y, I) be an $\mathbf{M}$-context, $f \in \mathbf{M}^X$ and $g \in \mathbf{M}^Y$. Then*

1. for $x \in X$, $\underset{y \in Y}{\sqcap}\{g(y) \rightarrow I(x,y)\}$ and $\underset{y \in Y}{\sqcap}\{I(x,y) \rightarrow g(y)\}$ are singletons;

2. for $y \in Y$, $\underset{x \in X}{\sqcap}\{f(x) \rightarrow I(x,y)\}$ and $\underset{x \in X}{\sqcup}\{f(x)\&I(x,y)\}$ are singletons.

Proof. The proof is similar to that Theorem 1. $\square$

We define two pairs of operators $(\uparrow, \downarrow)$ and $(\cap, \cup)$ with $\uparrow, \cap: \mathbf{M}^X \rightarrow \mathbf{M}^Y$ and $\downarrow, \cup: \mathbf{M}^Y \rightarrow \mathbf{M}^X$ by:

$$f^\uparrow(y) = \underset{x \in X}{\wedge}\{f(x) \rightarrow I(x,y)\}, \qquad g^\downarrow(x) = \underset{y \in Y}{\wedge}\{g(y) \rightarrow I(x,y)\},$$

$$f^\cap(y) = \underset{x \in X}{\vee}\{f(x)\&I(x,y)\}, \qquad g^\cup(x) = \underset{y \in Y}{\wedge}\{I(x,y) \rightarrow g(y)\}.$$

Theorem 2. *Let (X, Y, I) be an **M**-context. For all $g, h \in \mathbf{M}^Y$, the set $g^{\downarrow}(x) \sqcap h^{\cup}(x)$ is a singleton for all $x \in X$.*

Proof. Let $x \in X$. Let $a_1, a_2 \in g^{\downarrow}(x) \sqcap h^{\cup}(x)$. Then $a_1, a_2 \leq g^{\downarrow}(x)$ and $a_1, a_2 \leq h^{\cup}(x)$. For all $y \in Y$, we have, $a_1, a_2 \leq g(y) \to I(x, y)$ and $a_1, a_2 \leq I(x, y) \to h(y)$. Hence, $a_1 \& g(y), a_2 \& g(y) \leq I(x, y)$ and $a_1 \& I(x, y), a_2 \& I(x, y) \leq h(y)$, for all $y \in Y$. Since $\&$ is left-continuous, for all $t \in \sqcup \{a_1, a_2\}$, $t \& g(y) \leq I(x, y)$ and $t \& I(x, y) \leq h(y)$, for all $y \in Y$. Therefore, $t \leq g(y) \to I(x, y)$ and $t \leq I(x, y) \to h(y)$, for all $y \in Y$. Hence, $t \leq g^{\downarrow}(x)$ and $t \leq h^{\cup}(x)$. By the maximality of a_1 and a_2, we obtain $a_1 = t = a_2$.

Definition 7. *Let (X, Y, I) be an **M**-context. We define the **M-rough concept-forming operator** as a couple $(^{\triangle}, ^{\triangledown})$ with $^{\triangle} : \mathbf{M}^X \to (\mathbf{M} \times \mathbf{M})^Y$ and $^{\triangledown} : (\mathbf{M} \times \mathbf{M})^Y \to \mathbf{M}^X$ given by*

$$f^{\triangle} := < f^{\uparrow}, f^{\cap} > \quad and \quad < \underline{g}, \overline{g} >^{\triangledown} := (\underline{g})^{\downarrow} \sqcap (\overline{g})^{\cup}$$

for $f \in \mathbf{M}^X, < \underline{g}, \overline{g} > \in (\mathbf{M} \times \mathbf{M})^Y$.
i.e. $\forall y \in Y$, $f^{\triangle}(y) = < f^{\uparrow}(y), f^{\cap}(y) >$ and $\forall x \in X, < \underline{g}, \overline{g} >^{\triangledown}(x) = (\underline{g})^{\downarrow}(x) \sqcap (\overline{g})^{\cup}(x)$.

Proposition 3. *Let (X, Y, I) be an **M**-context. Then $(^{\triangle}, ^{\triangledown})$ forms a Galois connection between $\mathbf{M}^X$ and $(\mathbf{M} \times \mathbf{M})^Y$.*

Proof. Let $f_1, f_2 \in \mathbf{M}^X$ and $< \underline{g_1}, \overline{g_1} >, < \underline{g_2}, \overline{g_2} > \in (\mathbf{M} \times \mathbf{M})^Y$ such that, $f_1 \leq f_2$ and $< \underline{g_1}, \overline{g_1} > \leq < \underline{g_2}, \overline{g_2} >$. Then $\begin{cases} f_2^{\uparrow} \leq f_1^{\uparrow} \text{ and } f_1^{\cap} \leq f_2^{\cap} \\ (\underline{g_2})^{\downarrow} \leq (\underline{g_1})^{\downarrow} \text{ and } (\overline{g_2})^{\cup} \leq (\overline{g_1})^{\cup} \end{cases}$.

It follows that $\begin{cases} f_2^{\triangle} \leq f_1^{\triangle} \\ (\underline{g_2})^{\downarrow} \sqcap (\overline{g_2})^{\cup} \leq (\underline{g_1})^{\downarrow} \sqcap (\overline{g_1})^{\cup} \end{cases}$. Therefore, $^{\triangle}$ and $^{\triangledown}$ are order-reversing.

Let

$f \in \mathbf{M}^X$ and $< \underline{g}, \overline{g} > \in (\mathbf{M} \times \mathbf{M})^Y$. We have, $\begin{cases} f \leq f^{\uparrow\downarrow} \text{ and } f \geq f^{\cap\cup} \\ \underline{g} \leq (\underline{g})^{\downarrow\uparrow} \leq ((\underline{g})^{\downarrow} \sqcap (\overline{g})^{\cup})^{\uparrow} \end{cases}$

and $((\underline{g})^{\downarrow} \sqcap (\overline{g})^{\cup})^{\cap} \leq (\overline{g})^{\cup\cap} \leq \overline{g}$. Hence, $f \leq f^{\triangle\triangledown}$ and $(\underline{g}, \overline{g}) \leq (\underline{g}, \overline{g})^{\triangledown\triangle}$. □

Definition 8. *An **M-rough concept** is a fixed-point of $(^{\triangle}, ^{\triangledown})$. That is a pair $(f, < \underline{g}, \overline{g} >)$ in $\mathbf{M}^X \times (\mathbf{M} \times \mathbf{M})^Y$ such that $f^{\triangle} = < \underline{g}, \overline{g} >$ and $< \underline{g}, \overline{g} >^{\triangledown} = f$. For an **M-rough concept** $(f, < \underline{g}, \overline{g} >)$*

i) $\underline{g}$ is the **lower intent approximation**, i.e. contains the attributes shared by all objects of f

ii) $\overline{g}$ is a **upper intent approximation**, i.e. contains the attributes possessed by at least one object of f.

*Thus, one can consider the two intents to be a **lower and upper approximation** of attributes possessed by f.*

Note 1. The set of all **M**-rough concepts is denoted by $\mathfrak{B}^{\vartriangle\triangledown}(X,Y,I)$. That is

$$\mathfrak{B}^{\vartriangle\triangledown}(X,Y,I) = \{(f,<\underline{g},\overline{g}>) \in \mathbf{M}^X \times (\mathbf{M}\times\mathbf{M})^Y \mid f^\vartriangle =<\underline{g},\overline{g}> \text{ and } <\underline{g},\overline{g}>^\triangledown= f\}.$$

Let (X,Y,I) be an **M**$-$context and $(^\vartriangle,^\triangledown)$ a Galois connection between $\mathbf{M}^X$ and $(\mathbf{M}\times\mathbf{M})^Y$. $\mathfrak{B}^{\vartriangle\triangledown}(X,Y,I)$ can be equipped with a partial order $\leq$ defined by:

$$(f_1,<\underline{g_1},\overline{g_1}>) \leq (f_2,<\underline{g_2},\overline{g_2}>) \text{ iff } f_1 \leq f_2 \; (\text{ iff } <\underline{g_2},\overline{g_2}>\leq<\underline{g_1},\overline{g_2}>).$$

(Indeed, $f_1 \leq f_2$ is equivalent to $f_1^\vartriangle \geq f_2^\vartriangle$ equivalent to $<\underline{g_2},\overline{g_2}>\leq<\underline{g_1},\overline{g_2}>$).

We are going to prove that the set of all **M**-rough concepts is a complete multilattice.

Theorem 3. *Let (X,Y,I) be an **M**-context and let $(^\vartriangle,^\triangledown)$ be a Galois connection between $\mathbf{M}^X$ and $(\mathbf{M}\times\mathbf{M})^Y$.*

(i) If $\left\{\left(f_i,<\underline{g_i},\overline{g_i}>\right)\right\}_{i\in J}$ is a set of concepts, we have:

$$\underset{i\in J}{\sqcap}\{<\underline{g_i},\overline{g_i}>^\triangledown\} \subseteq \left\{\underset{i\in J}{\sqcup}<\underline{g_i},\overline{g_i}>\right\}^\triangledown \text{ and } \underset{i\in J}{\sqcap}\{f_i^\vartriangle\} \subseteq \left\{\underset{i\in J}{\sqcup}f_i\right\}^\vartriangle,$$

where $\left\{\underset{i\in J}{\sqcup}<\underline{g_i},\overline{g_i}>\right\}^\triangledown = \left\{<h,g>^\triangledown| \; <h,g>\in \underset{i\in J}{\sqcup}<\underline{g_i},\overline{g_i}>\right\}$ and $\left\{\underset{i\in J}{\sqcup}f_i\right\}^\vartriangle$ is given similarly.

*(ii) $(\mathfrak{B}^{\vartriangle\triangledown}(X,Y,I),\leq)$ is a complete multilattice called **M-Rough Concept Multilattice**. The multisupremum and multiinfimun of a set of concepts $\left\{\left(f_i,<\underline{g_i},\overline{g_i}>\right)\right\}_{i\in J}$ are given by*

$$\underset{i\in J}{\sqcap}\left(f_i,<\underline{g_i},\overline{g_i}>\right) = \left\{\langle h,h^\vartriangle\rangle \mid h \in \underset{i\in J}{\sqcap}f_i\right\}$$

$$\underset{i\in J}{\sqcup}\left(f_i,<\underline{g_i},\overline{g_i}>\right) = \left\{\left(<h,g>^\triangledown,<h,g>\right) |<h,g>\in \left(\underset{i\in J}{\sqcap}<\underline{g_i},\overline{g_i}>\right)\right\}.$$

Proof. (i) Let $f \in \underset{i\in J}{\sqcap}\{<\underline{g_i},\overline{g_i}>^\triangledown\}$. We have $f \leq<\underline{g_i},\overline{g_i}>^\triangledown$ for every $i \in J$. Then, $<\underline{g_i},\overline{g_i}>^{\triangledown\vartriangle}\leq f^\vartriangle$ that is $<\underline{g_i},\overline{g_i}>=<\underline{g_i},\overline{g_i}>^{\triangledown\vartriangle}\leq f^\vartriangle$. Then there is $<h,g>\in \underset{i\in J}{\sqcup}\{<\underline{g_i},\overline{g_i}>\}$ such that $<h,g>\leq f^\vartriangle$. Hence $f = f^{\vartriangle\triangledown} \leq<h,g>^\triangledown$. But we have $<h,g>\in \underset{i\in J}{\sqcup}\{<\underline{g_i},\overline{g_i}>\}$; i.e. $<\underline{g_i},\overline{g_i}>\leq<h,g>$ for every $i \in J$. Hence, $<h,g>^\triangledown\leq<\underline{g_i},\overline{g_i}>^\triangledown$ for every $i \in J$ so, $<h,g>^\triangledown$ is a lower bound of $\{<\underline{g_i},\overline{g_i}>^\triangledown| i \in J\}$ but $f \in \underset{i\in J}{\sqcap}\{<\underline{g_i},\overline{g_i}>^\triangledown\}$ and $f \leq<h,g>^\triangledown$. Hence by the maximality of f, $f =<h,g>^\triangledown$.
By the same manner we can prove the second one.

(ii) It is obvious to show that for $h \in \underset{i \in J}{\sqcap} f_i$, $< h, h^{\triangle} >$ are maximal element of lower bounds of the set $\left\{ \left(f_i, < \underline{g_i}, \overline{g_i} > \right) \mid i \in J \right\}$ it remains to show that $< h, h^{\triangle} >$ is a concept.

$\underset{i \in J}{\sqcap} \{f_i\} = \underset{i \in J}{\sqcap} \left\{ < \underline{g_i}, \overline{g_i} >^{\triangledown} \right\} \subseteq \left\{ \underset{i \in J}{\sqcup} < \underline{g_i}, \overline{g_i} > \right\}^{\triangledown}$. For $h \in \underset{i \in J}{\sqcap} f_i$ there exists $< f, g > \in \underset{i \in J}{\sqcup} < \underline{g_i}, \overline{g_i} >$ such that $h = < f, g >^{\triangledown}$. Hence $h^{\triangle \triangledown} = < f, g >^{\triangledown \triangle \triangledown} = < f, g >^{\triangledown} = h$ therefore $< h, h^{\triangle} >$ is an **M**-rough concept. the same manner, we can prove the next one.

4 Working Example

When we look a university staff, chancellor is the upper hierarchy of the vice-chancellors who are the upper hierarchies of the deans of faculties, but if we consider only the deans of faculties we cannot say who is the upper hierarchy. Therefore since multilattices better deal better with objects which are incomparable, we will use it to evaluate university staff.

Let us assume that, a vice-chancellor of an university wants to reward his best collaborator among the following persons: the dean of the faculty of science (FS), the dean of the faculty of industrial engineering (FG), the dean of the faculty of agronomy (FA), the head of the department of mathematics (HM), the head of the department of technologies (HT) and the head of the department of computer science (HC). They will be assessed on punctuality (Pu), innovation (In), leadership (Le), grade (Gr) and sociability (So).

The attributes that a staff possesses are assessed by the elements of the ordered structure represented by the Fig. 1.

The incidence relation between each member of staff and attributes is given by the Table 1.

Table 1. Incidence relation

I	Gr	Pu	So	In	Le
FS	d	$\bot$	c	$\top$	a
FI	c	a	d	c	b
FA	$\top$	b	e	d	c
HM	a	e	$\top$	e	d
HC	b	c	d	c	$\bot$
HT	d	d	c	d	a

For example, if we consider Grades, we can say that, some Grades are better than others but we cannot compare Grades of peoples from different Academy (Academy of Science, Academy of Arts, ...), if we see the first column, the Grade of FS and FI are appreciated by the value d and c respectively but are not comparable. Its means that, FS can be an associate professor at the Academy of Science and FI an associate professor at the Academy of Engineering. The Grade of FA is appreciated by the value top to signify that FA has a better appreciation in terms of Grade. Something similar occurs for the others attributes. That is why, we use multilattice as underlying set for evaluating academy staff. Therefore we can look this as a Concept Multilattice and that the Web queries are expressed by the sets of attributes. A Concept is a couple where intent are staff members and extend his performances.

The criteria of the company that will give the award are defined by a mapping g as follows:

$$g(Gr) = e, \ g(Po) = \top, \ g(So) = c, \ g(In) = \top, \ g(Le) = d$$

As we can see, there is nobody who has this criteria! Then, our goal is to find collaborator whose performance better approximate the fuzzy set of attributes specified by the company.

Let us consider the multilattice M_7 and the adjoint pairs & and ← defined in the following tables (Table 2).

Table 2. Of & (left) and → (right) on M_7

&	⊥	a	b	c	d	e	⊤
⊥	⊥	⊥	⊥	⊥	⊥	⊥	⊥
a	⊥	a	⊥	⊥	a	⊥	a
b	⊥	⊥	b	b	⊥	⊥	b
c	⊥	⊥	b	c	⊥	⊥	c
d	⊥	a	⊥	⊥	d	⊥	d
e	⊥	⊥	⊥	⊥	⊥	a	e
⊤	⊥	a	b	c	d	e	⊤

→	⊥	a	b	c	d	e	⊤
⊥	⊤	⊤	⊤	⊤	⊤	⊤	⊤
a	b	⊤	b	⊤	⊤	⊤	⊤
b	a	a	⊤	⊤	⊤	⊤	⊤
c	⊥	a	b	⊤	d	⊤	⊤
d	⊥	a	b	c	⊤	⊤	⊤
e	d	e	d	e	d	⊤	⊤
⊤	⊥	a	b	c	d	e	⊤

We are going to use the notion of lower/upper approximations to propose the staff member that performances are closed to the criteria.

Let $X = \{FS, FI, FA, HM, HC, HT\}$ and $Y = \{Gr, Pu, So, In, Le\}$ be the set of objects and attributes respectively then (X, Y, I) is a M-fuzzy context.

To evaluate the lower and upper approximation, we consider the following M-fuzzy relation

R	Gr	Pu	So	In	Le
Gr	⊤	e	c	d	d
Pu	e	⊤	d	a	⊤
So	c	d	⊤	a	b
In	d	a	a	⊤	a
Le	d	⊤	b	a	⊤

Let us now evaluate the lower approximation of the criteria.

$$\underline{g}(Gr) = \underset{y \in Y}{\wedge} \{g(y) \leftarrow R(Gr, y)\} = \underset{y \in Y}{\wedge} \{e \leftarrow \top, \top \leftarrow e, c \leftarrow c, \top \leftarrow d, d \leftarrow d\}$$
$$= e \quad \underline{g}(Pu) = c, \quad \underline{g}(So) = c, \quad \underline{g}(In) = \top, \quad \underline{g}(Le) = d$$

$$\underline{g} = \{^e/Gr,^c/Pu,^c/So,^\top/In,^d/Le\}.$$

We can also evaluate the upper approximation of g $\overline{g}(Gr) = \underset{y \in Y}{\vee}$
$\{R(Gr, y)\&g(y)\} = \underset{y \in Y}{\vee} \{\top\&e, e\&\top, c\&c, d\&\top, d\&d\} = e$ $\overline{g}(Pu) = \top, \quad \overline{g}(So) = e, \quad \overline{g}(In) = \top, \quad \overline{g}(Le) = \top.$

$$\overline{g} = \{^e/Gr,^\top/Pu,^e/So,^\top/In,^\top/Le.\}$$

We therefore have,

$$\underline{g}^\downarrow(FS) = \underset{y \in Y}{\wedge} \{I(FS, y) \leftarrow \underline{g}(y)\} = \underset{y \in Y}{\wedge} \{d \leftarrow e, \bot \leftarrow c, c \leftarrow c, \top \leftarrow \top, a \leftarrow d\} = \bot$$
$$\underline{g}^\downarrow(FI) = \bot, \quad \underline{g}^\downarrow(FA) = b, \quad \underline{g}^\downarrow(HM) = e, \quad \underline{g}^\downarrow(HC) = \bot, \quad \underline{g}^\downarrow(HT) = a$$
$$(\underline{g})^\downarrow = \{^\bot/FS,^\bot/FI,^b/FA,^e/HM,^\bot/HC,^a/HT\}.$$

$$\overline{g}^\cup(FS) = \underset{y \in Y}{\wedge} \{\overline{g}(y) \leftarrow I(FS, y)\} = inf \{e \leftarrow d, \top \leftarrow \bot, e \leftarrow c, \top \leftarrow \top, \top \leftarrow a\} = \top$$
$$\overline{g}^\cup(FI) = \top, \quad \overline{g}^\cup(FA) = e, \quad \overline{g}^\cup(HM) = e, \quad \overline{g}^\cup(HC) = \top, \quad \overline{g}^\cup(HT) = \top.$$
$$(\overline{g})^\cup = \{^\top/FS,^\top/FI,^e/FA,^e/HM,^\top/HC,^\top/HT\}.$$

We have:

$$\langle \underline{g}, \overline{g} \rangle^\nabla(FS) = (\underline{g}^\downarrow \wedge \overline{g}^\cup)(FS) = \bot, \quad \langle \underline{g}, \overline{g} \rangle^\nabla(FI) = \langle \underline{g}^\downarrow \wedge \overline{g}^\cup \rangle(FI) = \bot,$$
$$\langle \underline{g}, \overline{g} \rangle^\nabla(FA) = (\underline{g}^\downarrow \wedge \overline{g}^\cup)(FA) = b, \quad \langle \underline{g}, \overline{g} \rangle^\nabla(HM) = (\underline{g}^\downarrow \wedge \overline{g}^\cup)(HM) = e,$$
$$\langle \underline{g}, \overline{g} \rangle^\nabla(HC) = (\underline{g}^\downarrow \wedge \overline{g}^\cup)(HC) = \bot, \quad \langle \underline{g}, \overline{g} \rangle^\nabla(HT) = (\underline{g}^\downarrow \wedge \overline{g}^\cup)(HT) = a.$$

Since the truth degree of HM is greater than others, thus the best staff member is the Head of the department of mathematics.

5 Conclusion

In this paper we use multilattice as underlying set of truth degree in rough fuzzy concept analysis. We prove that, the set of all **M**-rough concepts of an **M**-context forms a complete multilattice.

The present results on M-rough concept analysis open ways to continue the research on the practical applications on semantic web, also we will investigate the Algorithm that can generate concepts. Since, we have two intents in each $M-$rough concept, the size of concept multilattice can be very large, then the study of reduction of M-rough concept multilattice via linguistic hedges is also very interesting.

Acknowledgement. The authors would like to thank the reviewers for useful remarks that led to improvement of the presentation of this research.

References

1. Bartl, E., Konecny, J.: Formal L-concepts with rough intents. In: CLA 2014: Proceeding of the 11th International Conference on Concept lattice and their Applications, pp. 207–218 (2014)
2. Bartl, E., Konecny, J.: Using linguistic hedges in L-rough concept analysis. In: CLA 2015, pp. 229–240 (2015). ISBN 978-2-9544948-0-7
3. Bělohlávek, R., Vychodil, V.: What is a fuzzy concept lattice? In: Belohlavek, R., et al. (eds.) CLA, pp. 34–45 (2005)
4. Benado, M.: Les ensembles partiellement ordonnes et théorème de raffinement de Schreier II (Théorie des multistructures). Czechoslovak Math. J. **5**, 308–344 (1955)
5. Burusco, A., Fuentes-González, R.: The study of L-fuzzy concept lattice. Mathware Soft Comput. **3**, 209–218 (1994)
6. Cabrera, I.P., Cordero, P., Gutiérrez, G., Martínez, J., Ojeda-Aciego, M.: Residuated operations in hyperstructures: residuated multilattices. In: 11th International Conference on Computational and Mathematical Methods in Science and Engineering CMMSE, pp. 26–30 (2011)
7. Cabrera, I.P., Cordero, P., Gutiérrez, G., Martínez, J., Ojeda-Aciego, M.: On residuation in multilattices: filters, congruences, and homomorphisms. Fuzzy Sets Syst. **234**, 1–21 (2014)
8. Cabrera, I.P., Cordero, P., Gutiérrez, G., Martínez, J., Ojeda-Aciego, M.: Finitary coalgebraic multisemilattices and multilattices. Appl. Math. Comput. **219**, 31–44 (2012)
9. Cornelis, C., Medina, J., Verbiest, N.: Multi-adjoint fuzzy rough sets: definition, properties and attribute selection. Int. J. Approx. Reason. **55**, 412–426 (2013)
10. Davey, B., Priestley, H.: Introduction to Lattices and Order. Cambridge University Press, Cambridge (2002)
11. Dubois, D., Prade, H.: Rough fuzzy sets and fuzzy rough sets. Int. J. Gen Syst **17**, 191–209 (1990)
12. Eugenia, M., Medina, J., Ramírez, E.: A Comparative study of adjoint triples. Fuzzy Sets Syst. **211**, 1–14 (2013)
13. Formica, A.: Semantic web search based on rough sets and Fuzzy Formal Concept Analysis. Knowl.-Based Syst. **26**, 40–47 (2012)
14. Ganter, B., Wille, R.: Formal Concept Analysis: Mathematical Foundations. Springer, Heindelberg (1999). https://doi.org/10.1007/978-3-642-59830-2
15. Goguen, J.A.: L-fuzzy sets. J. Math. Anal. Appl. **18**, 145–174 (1967)
16. Njionou, B.B.K., Kwuida, L., Lele, C.: Formal concepts and residuation on multilattices. Fundamental Informaticae **188**(4), 1–21 (2023)
17. Martínez, J., Gutiéerrez, G., de Guzmän, I.P., Cordero, P.: Multilattices via multisemilattices. Top. Appl. Theor. Math. Comput. Sci. 238–248 (2001)
18. Martínez, J., Gutiérrez, G., de Guzmän, I.P., Cordero, P.: Generalizations of lattices via non-deterministic operators. Discrete Math. **295**(13), 107–141 (2005)
19. Medina-Moreno, J., Ojeda-Aciego, M., Ruiz-Calviño, J.: Concept-forming operators on multilattices. In: Cellier, P., Distel, F., Ganter, B. (eds.) ICFCA 2013. LNCS (LNAI), vol. 7880, pp. 203–215. Springer, Heidelberg (2013). https://doi.org/10.1007/978-3-642-38317-5_13

20. Medina, J., Ruiz-Calviño, J.: Fuzzy formal concept analysis via multilattices: first prospects and results. In: CLA, pp. 69–79 (2012)
21. Medina, J., Ojeda-Aciego, M., Ruiz-Calviño, J.: Fuzzy logic programming via multilattices. Fuzzy Sets Syst. **158**(6), 674–688 (2007)
22. Poelmans, J., Ignatov, D.I., Kuznetsov, S.O., Dedene, G.: Fuzzy and rough formal concept analysis: a survey. Int. J. Gen Syst **43**(2), 105–134 (2014)
23. Radzikowska, A.M., Kerre, E.E.: Fuzzy rough sets based on residuated lattices. In: Peters, J.F., Skowron, A., Dubois, D., Grzymała-Busse, J.W., Inuiguchi, M., Polkowski, L. (eds.) Transactions on Rough Sets II. LNCS, vol. 3135, pp. 278–296. Springer, Heidelberg (2004). https://doi.org/10.1007/978-3-540-27778-1_14
24. Saquer, J., Deogun, J.S.: Formal rough concept analysis. In: Zhong, N., Skowron, A., Ohsuga, S. (eds.) RSFDGrC 1999. LNCS (LNAI), vol. 1711, pp. 91–99. Springer, Heidelberg (1999). https://doi.org/10.1007/978-3-540-48061-7_13
25. Wille, R.: Restructuring lattice theory: an approach based on hierarchies of concepts. In: Rival (ed.) Ordered Sets, pp. 445–470 (1982)
26. Zadeh, L.A.: Fuzzy sets. Inf. Control **8**, 338–358 (1965)
27. Pawlak, Z.: Rough sets. Int. J. Comput. Inform. Sci. **11**(5), 341–356 (1982)

Applications

Clustering Methods for Adaptive e-Commerce User Interfaces

Adam Wasilewski[1,2](✉) and Mateusz Przyborowski[3,4]

[1] Fast White Cat S.A., Wrocław, Poland
[2] Faculty of Management, Wrocław University of Science and Technology, Wrocław, Poland
`ml.przyborowsk@uw.edu.pl`
[3] QED Software, Warsaw, Poland
[4] Faculty of Mathematics, Informatics, and Mechanics, University of Warsaw, Warsaw, Poland

Abstract. Typical online shops have one interface provided to all users, regardless of their use of the shop. Meanwhile, user behavior varies and therefore different interfaces could be provided to different user groups. Various methods can be used to cluster users, including those using artificial intelligence (AI) methods. AI-based personalization allows e-commerce businesses to provide tailored recommendations to each individual customer based on preferences, purchase history, and behavior on the website. This article presents a study of the impact of an AI-based clustering method on the effectiveness of a dedicated user interface implemented and delivered to the customers of an e-shop. The first study included five methods, and two of them - agglomerative clustering and K-means clustering - were selected for detailed analysis. For both of these methods, an in-depth research was carried out and the impact of the clustering method on the quality of user clusters, as measured by the effectiveness of the dedicated interface in relation to the effectiveness of the default interface, was verified.

Keywords: Clustering · personalisation · user interface · e-commerce

1 Introduction

The rapid growth of e-commerce has led to an increase in the amount of available products, making it more challenging for customers to find products that match their preferences.

Personalization has become a crucial aspect of the e-commerce experience, as it allows customers to receive recommendations that match their interests and needs. One of the most effective ways to achieve personalization is through adaptive user interfaces, which are capable of dynamically adjusting the presentation of content based on the user's preferences and behavior. An adaptive user interface (AUI) can adjust and modify its presentation, functionality, and content to suit the preferences, needs, and abilities of individual users. AUIs can be designed to adapt to various factors such as the user's cognitive abilities, motor skills, language proficiency, cultural background, device type, screen size, and input methods. This type of interface may use various techniques such as machine learning and artificial intelligence (AI). By leveraging AI-based algorithms, businesses can analyze large amounts of data and provide recommendations that are relevant and timely. This, in turn, can lead to increased customer engagement, higher conversion rates, and improved customer satisfaction.

Today's e-businesses make extensive use of product recommendation mechanisms that rely on advanced data analysis, including AI/ML [11,29], but the possibilities are definitely greater. Considerable potential can be seen in personalizing the user interface as well.

In this work, the impact of the clustering method (agglomerative and K-means) on the effectiveness of dedicated UIs in e-commerce was compared. The contributions are:

1. Verification of the four clustering methods used to group e-commerce customers for the purpose of providing them with a dedicated interface.
2. An experimental evaluation of dedicated e-commerce interfaces provided to customers divided by means of two different clustering methods.

In Sect. 2 previous works on personalised e-commerce interfaces and clusterization methods are discussed. Section 3 briefly describes the research method. Experiments conducted are detailed and discussed in Sect. 4. Section 5 concludes the work.

2   Related Work

The problem of adapting the user interface with recommendation systems was addressed before artificial intelligence-based solutions were widely used [15]. Nowadays several AI-based techniques can be used to achieve personalization in the adaptive web shop interface [10], including collaborative filtering. Collaborative filtering is one of the most commonly used techniques and involves analyzing the behavior of similar users to make recommendations [32]. The use of clustering in recommendations systems has been extensively researched [36], and clustering algorithms, including: K-Means [30], Evolutionary Clustering [21], Fuzzy C-means [37], Bi-Clustering [2], and Locality Sensitive Hashing [27], have been widely discussed in the literature.

The adaptive web shop interface, delivered as a result of customer segregation, has been the focus of significant research in recent years, with many studies evaluating the effectiveness of different techniques and approaches.

According to [12,22,24] web mining techniques can be a valuable source of information, which is used to produce personalised e-commerce services. Segmentation of e-commerce customers using data mining techniques was presented in [20,31]. The possibilities of clustering users to divide them into groups with similar preferences are widely discussed [5,35]. Analysis of selected clustering algorithms (hybrid partitioning-based heuristic sequence clustering algorithm inspired from K-medoid, DBSCAN algorithms, a hybrid tree-based sequence clustering algorithm inspired from B-Trees and BIRCH algorithm) for an e-commerce recommendation system was presented in [25]. K-means algorithm can be used for segmenting e-commerce customers to obtain groups of customers with different characteristics [3, 16, 23]. The use of the K-mode clustering algorithm for e-commerce customer segmentation, was described in [13]. The Partition around Medoids (PAM) algorithm was also used to group the services provided to e-commerce customers [9]. Another possibility was presented in [28] - using the K-medoids method for e-commerce customer segmentation can provide valuable insights into customer behavior and preferences, which can be used to develop targeted marketing strategies and personalized product recommendations. A less popular method of clustering e-commerce customers, Fuzzy Temporal Clustering Approach (FTCA), is discussed in [26].

Data mining techniques and analysis of user's behavior, frequency and content may lead to the descriptive characteristics of each user and can be transformed to knowledge about the behavior of user groups. Further analysis and understanding of their characteristics allows for preparation of dedicated interfaces for each group. In [17] research on related e-commerce recommendation technologies and algorithms is described. An e-commerce recommendation system architecture should including key features: income level, online shopping experience, commodity price and quality, quality of services etc. [6].

Using data mining techniques (e.g. customer segmentation) to deliver personalized user interfaces involves identifying groups of customers with similar preferences and behaviors, and then developing user interfaces that are customized to meet the specific needs of each group. By delivering a user interface that is tailored to the preferences of each group, businesses can increase customer engagement and satisfaction.

3   Research Method

The aim of the study was to verify the impact of the clustering method on the effectiveness of the interface provided to the customers of a sample e-commerce.

To verify the hypothesis that the choice of clustering method affects the quality of the adaptable user interface, four clustering methods were analysed and two were selected as the most promising. The methods considered were: the K-means algorithm [14], DBSCAN (density-based spatial clustering of applications with noise) [8], BIRCH (balanced iterative reduction and clustering using hierarchies) [33] and the agglomerative clustering algorithm [1].

The effectiveness of the verified clustering algorithms was measured using several internal quality metrics. Moreover, proposed method for analyzing the resulting user groups and identifying key characteristics of users assigned to each cluster was used to define changes in each variant of dedicated interface. Users who were assigned to selected clusters were provided with interface variants, different from the default interface provided to the other users. Tests were carried out independently for both analysed clustering methods.

Both clustering methods were evaluated for 45 days, divided into 15 three-day research periods (RP). In each RP, an average PCR was calculated for all sessions of customers that had a dedicated interface provided (PCR_v) and for customers that had a default interface provided (PCR_d).

$$PCR_k = \frac{1}{s} \sum_{i=1}^{s} PCR_s \tag{1}$$

$$PCR_v = \frac{1}{k} \sum_{i=1}^{k} PCR_k \tag{2}$$

k is the number of RP, s is the session number of the user to whom the dedicated interface has been provided in the RP k, PCR_s is the calculated PCR metric value for the session s, PCR_k is the average value of the PCR metric for session k

The PCR_d value is calculated analogously, except that PCR metrics are calculated for sessions where the default user interface is provided.

The effectiveness of the dedicated interfaces was compared by measuring the partial conversion rate of customers who were provided with a dedicated interface and customers who were provided with the default interface.

For the purpose of assessing the quality of the dedicated interfaces, a measure of partial conversion rate (PCR) has been defined. This indicator measures user behavior along the purchase path, both for the dedicated interface and for the default interface.

Assuming that the purchase path includes a homepage, a product listing and a product page, the behavior desired from the e-commerce owner's point of view can be defined for each step. For example, it can be assumed that the user from the homepage should go to the product listing or to a specific product card. The PCR can be calculated for different steps of the shopping process, in particular for:

- homepage (expected result: change to listing or to product page),
- listing (expected result: switching to a product page)
- the product page (expected result: addition to basket or transition to another product page).

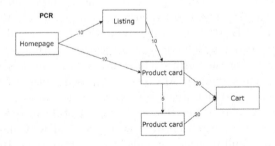

Fig. 1. Calculation of PCR metrics

Each of the expected behaviors is scored and the sum of the scores is the total PCR for the session (Fig. 1). If a user interface variant was changed during the session, the PCR value is counted independently for both interface variants.

Comparison of the PCR values for the both analysed clustering methods allowed to assess the effectiveness of the dedicated interfaces relative to the standard interface and to identify the clustering method that yields better results.

The research was conducted using the platform for self-adaptation of e-commerce interfaces (named AIM^2). It is an intelligent system that allows e-commerce platforms to optimize their user interfaces automatically. The architecture of the platform consists of several components that work together to provide a seamless user experience:

- User model - a representation of the user's behavior and preferences,
- UI Designer - used to generate different versions of the UI, which are tested against the user model to determine which version performs best,
- Adaptation engine - provides UI variants and uses artificial intelligence (AI) algorithms to group customers on the base of actions, events, purchases, and other factors,
- Monitoring - provide feedback on the performance of different UI versions, which is used to improve the effectiveness of the adaptation engine and UI variants.

A pilot version of the AIM^2 platform was implemented in the e-shop of one of the market leaders in sportswear in Central Europe and was the technical basis for the research presented in the article.

4   Result Analysis and Discussion

4.1   Analysis of Selected Clustering Methods

The dataset used for producing the clustering consists of 156 days of user activity, making on average, 5162 visits daily. About 5.67% of the data is directly connected to known customers, i.e. accounts recorded in the database. Detailed logs include information about the device used (type, resolution, browser, etc.), country and language, or previous visits. The most important part of the data is the type of pages that were loaded during users' activity, taking into account such features as the category of the viewed object, time spent, or finalizing activity with a purchase. Furthermore, the dataset has been enriched with additional information about existing products and their categories. Products that are noted during users' activity have been matched to existing historical products, which allows for better segmentation of users at a given product category level.

In order to produce a clustering, the dataset was filtered and aggregated according to the activity of a given user. Each user was represented by a list of activities summarized by a vector of features. This representation transforms logs about website activity into a feature space where distinct users yield distinct points. Each dimension of this space comes from a precalculated feature that concerns information that is useful for further applications, e.g. categories of the most commonly viewed products. Even though some features, such as detailed timestamp data, did allow for more stable user segmentation, they were dismissed as less useful in the case of the considered application.

Four distinct clustering methods were compared: DBSCAN [8], BIRCH [33], Gaussian mixture models (GMM) [34], and agglomerative clustering [1]. Each clustering was evaluated using Variance Ratio Criterion (Calinski-Harabasz, CH score), Eq. 3, and Davies-Bouldin (DB) score, Eq. 4.

$$CH = \frac{(N-k)tr(\sum_{q=1}^{k} n_q(c_q - c_E)(c_q - c_E)^T)}{(k-1)tr(\sum_{q=1}^{k} \sum_{x \in A_q}(x - c_q)(x - c_q)^T)} \tag{3}$$

N is the size of the dataset, k is the number of clusters, A_q is the set of points in cluster q, c_q is the centre of cluster q, n_q is the number of points in cluster q, c_E is the centre of the entire dataset. The CH index is the ratio of the variance between clusters and the sum of the variances within each cluster; it yields higher values for clusters that are better separated and denser.

$$DB = \frac{1}{k} \sum_{i=1}^{k} \max_{i \neq j} \frac{s_i + s_j}{d_{ij}} \tag{4}$$

k is the number of clusters, s_i is the average distance of points from cluster i from its centre, d_{ij} is the distance between the centres of clusters i and j. The DB index yields higher values for clustering where the clusters are wide and overlapping, with their centres remaining rather close to each other. Conversely, lower values may indicate that the clusters are well separated.

1. **DBSCAN**: For epsilon values that created clusters of size between 2 and 10, CH score ranges between 2.7 to 6.8 and DB score ranges from 1.17 to 1.37. Furthermore, cluster entropy remains below 0.15, which indicates that a single cluster dominates the others in terms of size. The mean silhouette score consistently remains negative (<-0.54).
2. **BIRCH**: Clusterings for cluster number between 2 and 10, CH score ranges from 4876 to 6886 and DB score ranges from 0.6 to 0.85. The mean silhouette score indicates 4–5 clusters.
3. **GMM**: Clusterings for cluster number between 2 and 10, CH score ranges from 5250 to 8213 and DB score ranges from 0.62 to 0.76. The mean silhouette score indicates 4–5 clusters.
4. **Agglomerative clustering**: Clusterings for cluster number between 3 and 9, CH score ranges from 4893 to 6886 and DB score ranges from 0.6 to 0.85. The mean silhouette score indicates 4–6 clusters, but CH score and DB score yield better results (6727 and 0.71) for 6 clusters.

The obtained results opt towards agglomerative clustering and Gaussian mixture models. These methods additionally have properties that allow a better interpretation of results in terms of connecting different clusters. Agglomerative clustering produces a dendrogram that asserts relations between clusters at a given cut-off level, therefore giving a method for merging different clusters. The Gaussian mixture model by design provides a method for evaluating the probability that a given point comes from a given cluster. In the experiments, the Gaussian mixture model returned results consistent with the K-means algorithm. Finally, the agglomerative clustering method was selected for further detailed research (Figs. 2 and 3).

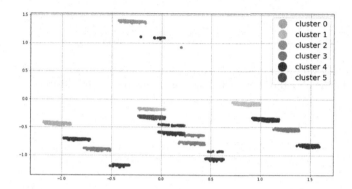

Fig. 2. Scatter plot of the selected clustering along the two most significant dimensions produced by the PCA decomposition of the initial dataset.

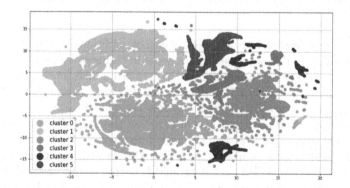

Fig. 3. Scatter plot of the selected clustering along the dimensions of low-dimensional representation produced by the UMAP algorithm [18].

Additional analysis regarding clustering using Shapley values was performed. Using the final data representation and clustering indices as labels, a gradient boosted random forest was fitted using the XGBoost framework [4]. The Shapley values measure the contribution (importance) of different features in influencing models' output. As for the obtained clustering, the Shapley values indicate a strong relationship between users belonging to some clusters and manifesting a

518   A. Wasilewski and M. Przyborowski

particular interest in certain categories of products (Figs. 4 and 5); therefore, these results could be useful in terms of applications.

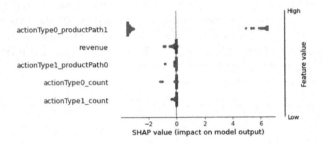

Fig. 4. Shapley values for cluster 2. Browsing products from the *productPath1* category increases the likelihood of being classified as a member of this cluster.

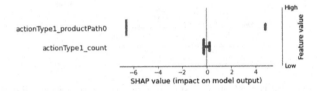

Fig. 5. Shapley values for cluster 4. Browsing products from the *productPath0* category increases the likelihood of being classified as a member of this cluster.

The main application of clustering is based on additional assumptions. A clustering not only should meet the requirements in terms of its regularity but also needs to be periodically updated/recalculated using the latest, rapidly growing transaction dataset. For this reason, the decision was made to also verify in practice an approach that could be less computationally demanding. Initially, Gaussian mixture model was selected as the second clustering method for detailed study, as the clustering results it received were promising for use in providing a dedicated e-commerce interface. Ultimately, for business and technical reasons, it was decided that K-means clustering would be studied in detail alongside agglomerative clustering. The K-means method has lower computational requirements [19], and could potentially yield clustering results similar to those of the GMM method. In practical applications within the AIM^2 platform, the K-means method could be particularly useful for large e-commerces, where the dataset of user behavior would be huge.

4.2 Effectiveness of Clustering-Based Interfaces

Research into the effectiveness of dedicated interfaces was conducted under real-life conditions of a fashion e-shop.

The analysis included 328.721 sessions during tests of the interface delivered after client clustering using the agglomerative clustering method and 208.858 sessions during tests of the interface delivered after using the K-means method. The difference in the number of sessions, with the same research time (15 RPs), was due to differences in e-commerce traffic. The research was not conducted during major promotional and marketing events (e.g. major sales), but minor promotions (e.g. weekend sales) did take place, which was reflected in the number of sessions and in the results. The possible impact of in-store promotional campaigns was offset by calculating relative ratios and relating the effect of the dedicated interface to the effect of the default interface.

The dedicated interface variants were optimised by UX experts based on an analysis of the behavior of two clusters of customers: a) the group with the highest activity in the e-shop and b) the group with the lowest activity in the e-shop. A dedicated user interface was designed for each of these groups and was delivered in subsequent sessions in the e-store to those users whom the clustering method categorised as either group a) or group b). The user behavior investigated by UX experts included the number of actions per session, number of events, sequences of events, orders placed, use of the search engine, selection of product attributes such as size and colour.

Chronologically, the first research was carried out for clustering using the agglomerative clustering method. Based on cluster analysis and customer behavior, it was decided to provide a dedicated interface to 4728 users during the first iteration of the research.

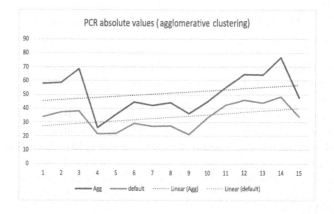

Fig. 6. PCR absolute values for agglomerative clustering

In each RP, the average PCR of users who received a dedicated interface were higher than the average PCR of users who had the default interface provided (Fig. 6).

It can be seen that there is a large variation in the average values of the PCR indicator from period to period, which indicates that the value is influenced by factors such as promotions organised in the e-shop. Nevertheless, a high correlation can also be observed between the metric values for the dedicated and the default interface, indicating that factors not directly related to the interface similarly influence the PCR values.

Analysis of the absolute values of the PCR may be subject to error, due to the seasonality of e-commerce purchases and active promotions. For this reason, the relative PCR values for the dedicated interfaces were calculated by relating them to the PCR values for the default interface (Fig. 7).

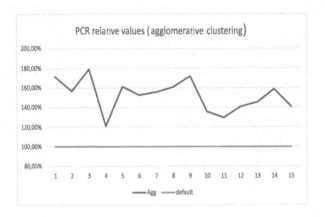

Fig. 7. PCR relative values for dedicated interfaces

Relative PCR values with agglomerative clustering show high inter-period variability, with none of the periods falling below 120% of the default interface value. It took values ranging from 120.64% to 179.09%.

In the second iteration (K-means clustering), the decision was made to provide a dedicated interface to 7425 users. The higher number of users to which a dedicated interface was assigned was due to the splitting of customers into clusters grouping the most active and least active users.

Comparing the absolute values of the PCR for K-means clustering, it can again be seen that they are higher for users provided with a dedicated interface in each RP (Fig. 8).

When users are divided into clusters using the K-means method, similar to agglomerative clustering, a correlation can be observed between the PCR values for the dedicated and default interfaces.

Relative PCR values for K-means clustering (Fig. 9) were below the 120% threshold in 4 cases out of 15, with 2 cases where the values were even below

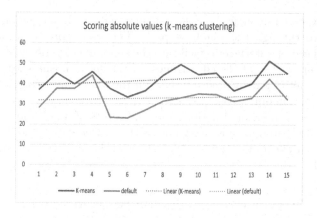

Fig. 8. PCR absolute values for K-means clustering

110% (105.41% and 103.82% in RP 3 and 4). In contrast, the maximum value of this indicator was 160.57%.

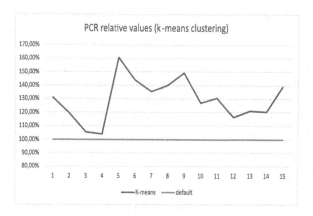

Fig. 9. PCR relative values for dedicated interfaces

If the entire study of each user clustering approach (45 days = 15 RPs) were taken into account, it turns out that for agglomerative clustering, PCR values were on average 51.94% higher than the average values of this indicator for the default interface, while for K-means clustering, users who were provided with a dedicated interface achieved an average PCR value only 28.89% higher than for the default interface.

The results obtained allow to conclude that regardless of the clustering method chosen, the values of PCRs for dedicated interfaces are higher than those for the default interface, but the choice of clustering approach seems to matter for the efficiency of the dedicated interface.

The PCR indicator shows how the behavior of the e-shop's customers changes as they move through the purchase path. This is important information for optimising the shop interface to facilitate the shopping experience. Nevertheless, from the point of view of the effectiveness of the entire business, it is crucial to check the impact of a dedicated interface on the final conversion rate (CR) and the value of the shopping basket (AvB) [7].

In the case of traditional performance indicators of an e-shop, it turned out that a dedicated interface positively affects the conversion rate in each case, but there is no clear effect of a dedicated interface on the average value of a shopping cart (Table 1).

Table 1. Conversion Rate and Average basket Value

Period	Standard		Agglomerative		K-means	
	CR	AvB	CR	AvB	CR	AvB
A1	1.96	63.37	6.36	61.70		
A2	3.76	61.79	8.98	82.76		
K1	3.42	58.27			7.46	68.92
K2	1.90	68.84			6.29	65.29

It is worth noting that the research periods for verification of CR and AvB values were different than for the PCR analysis, but covered the same 45 days for each clustering method. The A1 and K1 periods corresponded to the first 8 periods from PCR studies (for each clustering method separately), while the A2 and K2 periods corresponded to the next 7 periods of the PCR studies.

5   Conclusion

Clustering of users of an e-shop can be done by various methods. Within the study described in the paper, 4 methods were selected for preliminary analysis on the basis of a set of information about the e-shop's customer behavior from 156 days, and based on the results, the two most promising ones - agglomerative clustering and K-means clustering - were selected for further research.

Using both methods, user clustering was performed, and then customers from selected clusters were provided with a dedicated interface for 45 days (divided into 15 research periods).

Using the PCR indicator defined for the study, the impact of the clustering method on the behavior of users provided with a dedicated interface was analysed. The results of the conducted research show a slight advantage for the agglomerative clustering method, but the results obtained with K-means clustering are also satisfactory, as they outperform those obtained for the default interface.

It should also be remembered that agglomerative clustering is more computationally appealing, so the expected clustering time should also be taken into account when deciding on a clustering method for delivering a dedicated e-commerce user interface.

Acknowledgements. Project co-funded by the National Centre for Research and Development under the Sub-Action 1.1.1 of the Operational Programme Intelligent Development 2014-2020.

Data Availability. The datasets used during the current study are available from authors on reasonable request.

References

1. Ah-Pine, J.: An efficient and effective generic agglomerative hierarchical clustering approach. J. Mach. Learn. Res. **19**(1), 1615–1658 (2018)
2. Bansal, S., Baliyan, N.: Bi-MARS: a bi-clustering based memetic algorithm for recommender systems. Appl. Soft Comput. **97**, 106785 (2020). https://doi.org/10.1016/j.asoc.2020.106785
3. Chatterjee, R.P., Deb, K., Banerjee, S., Das, A., Bag, R.: Web mining using k-means clustering and latest substring association rule for e-commerce. J. Mech. Continua Math. Sci. **14**(6), 28–44 (2019). https://doi.org/10.26782/jmcms.2019.12.00003
4. Chen, T., Guestrin, C.: XGBoost: a scalable tree boosting system. In: Proceedings of the 22nd ACM SIGKDD International Conference on Knowledge Discovery and Data Mining, KDD 2016, pp. 785–794. ACM, New York (2016). https://doi.org/10.1145/2939672.2939785
5. Chu, X., Lv, D., Zhao, D.: Personalized e-commerce website construction based on data mining. J. Phys. Conf. Ser. **1345**, 362–376 (2019). https://doi.org/10.1088/1742-6596/1345/5/052038
6. Das, P., Kar, B., Misra, S.N.: Information processing style and e-commerce website design: a clustering-based conjoint approach. J. Humanit. Soc. Sci. Res. **4**, 63–74 (2022). https://doi.org/10.37534/bp.jhssr.2022.v4.nS.id1191.p63
7. Di Fatta, D., Patton, D., Viglia, G.: The determinants of conversion rates in SME e-commerce websites. J. Retail. Consum. Serv. **41**, 161–168 (2018). https://doi.org/10.1016/j.jretconser.2017.12.008
8. Ester, M., Kriegel, H.P., Sander, J., Xu, X.: A density-based algorithm for discovering clusters in large spatial databases with noise. In: Proceedings of the Second International Conference on Knowledge Discovery and Data Mining, pp. 226–231, KDD 1996. AAAI (1996). https://doi.org/10.37394/232020.2022.2.17
9. Gaikwad, D., Lamkuche, H.: Segmentation of services provided by e-commerce platforms using pam clustering. J. Phys. Conf. Ser. **1964**, 1–15 (2021). https://doi.org/10.1088/1742-6596/1964/4/042036
10. Gulzar, Y., Alwan, A.A., Abdullah, R.M., Abualkishik, A.Z., Oumrani, M.: OCA: ordered clustering-based algorithm for e-commerce recommendation system. Sustainability **15**(4), 1–22 (2023). https://doi.org/10.3390/su15042947
11. Guo, S., Zhai, R.: E-commerce precision marketing and consumer behavior models based on IoT clustering algorithm. J. Cases Inf. Technol. **24**(5), 1–21 (2022). https://doi.org/10.4018/JCIT.302244

12. Ji, C., Wang, K., Chen, X.: Design of e-commerce personalized service model based on web mining classification technique. In: 2010 International Conference on Internet Technology and Applications (2010). https://doi.org/10.1109/ITAPP. 2010.5566332
13. Kamthania, D., Pahwa, A., Madhavan, S.S.: Market segmentation analysis and visualization using k-mode clustering algorithm for e-commerce business. J. Comput. Inf. Technol. **26**(1), 57–68 (2018). https://doi.org/10.20532/cit.2018.1003863
14. Kaufman, L., Rousseeuw, P.J.: Finding Groups in Data: An Introduction to Cluster Analysis. Wiley, Hoboken (1990)
15. Kopel, M., Sobecki, J., Wasilewski, A.: Automatic web-based user interface delivery for SOA-based systems. In: Bădică, C., Nguyen, N.T., Brezovan, M. (eds.) ICCCI 2013. LNCS (LNAI), vol. 8083, pp. 110–119. Springer, Heidelberg (2013). https://doi.org/10.1007/978-3-642-40495-5_12
16. Li, Y., Qi, J., Chu, X., Mu, W.: Customer segmentation using k-means clustering and the hybrid particle swarm optimization algorithm. Comput. J. **66**(4), 941–962 (2022). https://doi.org/10.1093/comjnl/bxab206
17. Liu, L.: e-commerce personalized recommendation based on machine learning technology. Mob. Inf. Syst. **2022**, 1–11 (2022). https://doi.org/10.1155/2022/1761579
18. McInnes, L., Healy, J.: UMAP: Uniform Manifold Approximation and Projection for Dimension Reduction. arXiv e-prints (2018)
19. Patel, E., Kushwaha, D.S.: Clustering cloud workloads: K-means vs gaussian mixture model. Procedia Comput. Sci. **171**, 158–167 (2020). https://doi.org/10.1016/j.procs.2020.04.017
20. Punhani, R., Arora, V., Sabitha, A.S., Shukla, V.K.: Segmenting e-commerce customer through data mining techniques. J. Phys. Conf. Ser. **1714**, 1–11 (2020). https://doi.org/10.1088/1742-6596/1714/1/012026
21. Rana, C., Jain, S.K.: An extended evolutionary clustering algorithm for an adaptive recommender system. Soc. Netw. Anal. Min. **4**, 1–13 (2014). https://doi.org/10.1007/s13278-014-0164-x
22. Rao, T.R.K., Knah, S.A., Begum, Z., Divakar, C.: Design of e-commerce personalized service model based on web mining classification technique. In: 2013 IEEE International Conference on Computational Intelligence and Computing Research (2013). https://doi.org/10.1109/ICCIC.2013.6724233
23. Shaik, I., Nittela, S.S., Hiwarkar, T., Nalla, S.: K-means clustering algorithm based on e-commerce big data. Int. J. Innov. Technol. Explor. **8**(11), 1–5 (2019). https://doi.org/10.35940/ijitee.K2121.0981119
24. Shi, Y., Li, X., ZhouGong, S., Li, X., Wang, H.: Precise marketing classification of agricultural products fore-commerce live broadcast platform using clustering. Mob. Inf. Syst. **2022**(03), 1–8 (2022). https://doi.org/10.1155/2022/1062938
25. Singh, H., Kaur, P.: An effective clustering-based web page recommendation framework for e-commerce websites. SN Comput. Sci. **2**, 1–20 (2021). https://doi.org/10.1007/s42979-021-00736-z
26. Sudhamathy, G., Venkateswaran, J.: Fuzzy temporal clustering approach for e-commerce websites. Int. J. Eng. Technol. **4**(3), 119–132 (2012)
27. Wu, Y., Li, Y., Qian, R.: NE-UserCF: collaborative filtering recommender system model based on NMF and E2LSH. Int. J. Perform. Eng. **13**(5), 610–619 (2017). https://doi.org/10.1088/1742-6596/1714/1/012026
28. Wu, Z., Jin, L., Zhao, J., Jing, L., Chen, L.: Research on segmenting e-commerce customer through an improved k-medoids clustering algorithm. Comput. Intell. Neurosci. **2022**, 1–10 (2022). https://doi.org/10.1155/2022/9930613

29. Xu, Q., Wang, J.: A social-aware and mobile computing-based e-commerce product recommendation system. Comput. Intell. Neurosci. **2022**, 1–8 (2022). https://doi.org/10.1155/2022/9501246

30. Yang, F.: A hybrid recommendation algorithm - based intelligent business recommendation system. J. Discrete Math. Sci. Cryptogr. **21**(6), 1317–1322 (2018). https://doi.org/10.1080/09720529.2018.1526408

31. Yang, S., Hou, Y.: A clustering algorithm for cross-border e-commerce customer segmentation. PROOF **2**, 138–145 (2022). https://doi.org/10.37394/232020.2022.2.17

32. Zhang, Q.: Construction of personalized learning platform based on collaborative filtering algorithm. Wirel. Commun. Mob. Comput. **2022**, 1–9 (2022). https://doi.org/10.1155/2022/5878344

33. Zhang, T., Ramakrishnan, R., Livny, M.: Birch: an efficient data clustering method for very large databases. ACM SIGMOD Rec. **25**(2), 103–114 (1996). https://doi.org/10.1145/235968.233324

34. Zhang, Y., et al.: Gaussian mixture model clustering with incomplete data. ACM Trans. Multimedia Comput. Commun. Appl. **17**(1s), 1–14 (2021). https://doi.org/10.1145/3408318

35. Zhao, Y., He, Y., Wutao, Z.: Personalized clustering method of cross-border e-commercetopics based on art algorithm. Math. Probl. Eng. **2**, 1–7 (2022). https://doi.org/10.1155/2022/8190544

36. Zhu, X., Li, Y., Wang, J., Fu, J.: Automatic recommendation of a distance measure for clustering algorithms. ACM Trans. Knowl. Discov. Data **15**, 1–22 (2020). https://doi.org/10.1145/3418228

37. Öner, S.C., Öztayşi, B.: An interval type 2 hesitant fuzzy MCDM approach and a fuzzy C means clustering for retailer clustering. Soft. Comput. **22**, 4971–4987 (2018). https://doi.org/10.1007/s00500-018-3191-0

Application of Federated Learning to Prediction of Patient Mortality in Vasculitis Disease

Jan G. Bazan[1]($\boxtimes$), Pawel Milan[2], Stanislawa Bazan-Socha[3], and Krzysztof Wójcik[3]

[1] Institute of Computer Science, University of Rzeszow, Pigonia 1, 35-310 Rzeszow, Poland
bazan@ur.edu.pl
[2] State University of Applied Sciences in Krosno, Rynek 1, 38-400 Krosno, Poland
pawel.milan@pans.krosno.pl
[3] II Department of Internal Medicine, Jagiellonian University Medical College, M. Jakubowskiego 2, 30-688 Kraków, Poland
{stanislawa.bazan-socha,krzysztof.wojcik}@uj.edu.pl

Abstract. In this paper, we propose two methods of application of federated learning to the construction of classifiers for the analysis of data related to predicting the death of patients suffering from the vasculitis. The paper contains results of experiments on medical data obtained from Second Department of Internal Medicine, Collegium Medicum, Jagiellonian University, Krakow, Poland. In order to evaluate the proposed methods, which are trained on data samples, we compared their functionality with the work results of classical classifiers trained on the entire data. It turned out that the quality of classification of federated learning methods is comparable to the quality of classical methods. This means that access to the whole data is not necessary to construct effective classifiers for the considered decision problem.

Keywords: rough sets · classifiers · federated learning

1 Introduction

Federated learning (see, e.g., [4,12]) is a machine learning technique that trains a model through multiple independent sessions, each using its own dataset. This approach contrasts with classic centralized machine learning techniques where local datasets are combined into one large dataset used to train a global model. One of the kind of models that can be trained using federated learning is classifiers. For the purposes of this paper, we establish that in federated learning for classifiers construction, the data needed to train them is available in multiple places. The place where a global classifier model is created is called the *server*, while individual locations with small data fragments are called *nodes*. Federated classifier training is based on the server sending a request to nodes asking

© The Author(s), under exclusive license to Springer Nature Switzerland AG 2023
A. Campagner et al. (Eds.): IJCRS 2023, LNAI 14481, pp. 526–536, 2023.
https://doi.org/10.1007/978-3-031-50959-9_36

them to participate in the construction of the global classifier using the data available in individual nodes. Next the nodes send back their partial models to server. Finally, the server aggregates partial models, getting a global classifier model. In this paper, we propose two methods of classifiers construction based on federate learning for the analysis of data related to predicting the death of patients suffering from vasculitis, which is a serious disease. The first of these methods uses the aggregation of classifiers from nodes based on the arithmetic mean of weights generated by individual classifiers for the test object. The second method uses a specific node vote on the decision for the test object. The paper contains the results of experiments on medical data obtained from Second Department of Internal Medicine, Collegium Medicum, Jagiellonian University, Krakow, Poland. One of the main goals of the experimental part of the paper was to answer the question whether the classifiers based on federated learning proposed in this paper can match the efficiency of the classifier based on classical learning for the medical data analysed in the paper. It turned out that federated learning methods are comparable in terms of classification quality to classical methods.

2   Classical and Federated Classifier Learning

Classifiers are referred in the literature as decision-making algorithms, classification algorithms or learning algorithms, those can be treated as constructive, approximate descriptions of concepts (decision classes) [9]. The basis for creating (in other words - learning) classifiers are the so-called training data, which is usually in the form of a rectangular table, called a decision table in rough set theory (see, e.g., [11]). In a decision table, all columns except the last one are usually so-called conditional attributes, and the last column is the so-called decision attribute. Classic classifier training is based on the fact that for the entire available set of training data, based on the selected type of model, a classifier is trained that approximates the decision classes represented by the decision attribute which enables determine the value of the decision attribute, also for unknown objects during classifier training.

However, in practical applications related to the construction of classifiers, training data may not come from a single source but from many sources and may be stored in different locations. For example, the data necessary to construct a classifier supporting the treatment of a certain disease may be stored in different hospitals. In addition, data from different locations sometimes cannot be collected in one place, which forces them to be analysed in a dispersed manner. The reasons may be, for example, the following:

- legal provisions - protecting data against data portability, which prevent individual organizations from combining their own users' data to train artificial intelligence models; this is because users live in different parts of the world, so their data is subject to different data protection laws,

- data privacy - e.g., in finance, due to the need to maintain customer privacy, it is not possible for financial institutions to share sensitive data with each other,
- amount of data - some sensors, such as cameras, produce such a large amount of data that it is neither feasible in practice to collect it all in one place.

When data is available in multiple places, we may use the federated learning. In the federated learning, the place where the global classifier model is created in this paper we call as the server, while the individual locations with small pieces of data we call as nodes. As we already wrote in the introduction, classifier training is when the server sends a request to nodes asking them to participate in the construction of a global classifier using the data available at individual node. Next, the nodes send back their partial models to the server. Finally, the server aggregates partial models, obtaining a global classifier model. Depending on the methods of training partial models, methods of aggregation of partial models into a global model and methods for organizing successive learning iterations, you can define many different federated learning methods.

3   Classic Classifiers for the Medical Problem Related to Vasculitis

Vasculitis is defined as inflammation within the walls of blood vessels, often resulting in vessel damage and necrosis. Consequently, there may be bleeding or ischemia of organs or tissues supplied with blood by the damaged vessels (e.g., stroke). This process can involve various tissues and organs; however, it usually affects the kidneys, lungs, upper respiratory tract, skin, nervous system, and eye vessels. Therefore, vasculitis is a challenging disease, which often leads to death despite the applied treatment. That is why scientific research is currently conducted to understand this group of diseases and improve their treatment efficiency. One of the leading centers in Poland where such research is conducted is the Jagiellonian University Medical College (UJCM) in Krakow. As part of the research, UJCM developed a register of primary vasculitides in Poland as part of the Scientific Consortium of the Polish Vasculitis Register (POLVAS), in the construction of which 14 medical centers from Poland are involved. Recently, POLVAS also participated in the European project FAIRVASC together with other medical centers from 8 European countries, providing a new opportunity to understand and treat vasculitides (European Union's Horizon 2020 research project, grant No. 811171). Thanks to the POLVAS register, it is possible to conduct scientific research to improve the effectiveness of vasculitides therapy, such as predicting patient mortality under a particular treatment modality. If it would be possible to develop IT tools to predict mortality risk effectively, then in the case of patients who are expected to die, other or more aggressive treatment modes might be considered to save the patient's life.

The data made available for the experiments for this paper is a decision table with 819 rows (patients), 142 columns constituting conditional attributes and one decision attribute that informs about the possible death of a given patient. In fact, the value of the decision attribute represents the patient's condition after treatment of the last observed exacerbation of the disease. If the patient survived this exacerbation, it is recorded in the database that he is alive. However, it is not known whether the patient will die, e.g., during the next exacerbation of the disease, if any. Long-term patient survival has not yet been analysed because data on this topic is ongoing. Conditional attributes are mostly numerical attributes and describe the current situation of a given patient, including, e.g., age, gender, information about the type of vasculitis detected, the presence of diabetes, treatment applied, results of laboratory tests and many others.

Let us add that the values in the first column (that is in the first conditional attribute) of the decision table uniquely identify the patient and the medical center (hospital) where the patient is treated.

If we ignore the first conditional attribute that is referring to the hospital from which the patient comes, then the medical data described above is in the form of a classic decision table (see [11]). Therefore, we can use here any model of creating a classifier that deals with numerical attributes. One of them is a model based on a random forest [2], which is available in the Scikit-Learn library [13] in the `RandomForestClassifier` class. This model generally has a very good opinion as a useful model for creating efficient classifiers for various data sets. Therefore, for a practical illustration of the approach described in this paper, we decided to use this classifier creation model. However, the medical data analysed in this paper are unbalanced (the decision class about patient death is about 6 times less numerous than the decision class about patient survival). Therefore, in the experiments, we used a random forest model dedicated to creating classifiers for unbalanced data, which is available in the Imbalanced-learn library in the `BalancedRandomForestClassifier` class [5,6]. This library is compatible with the Scikit-Learn interface [13].

4   Classifiers Based on Federated Learning

The data set described in Sect. 3 includes information on patients from 14 medical centers in Poland. Due to their origin, these data could be made available for scientific research conducted in Poland. The FAIRVASC project also collected data from medical centers in other European countries, but due to the legal protection mentioned above (see Sect. 2), they could not be made available for research conducted in Poland. However, the data provided from centers in Poland give the opportunity to test the methods of federated learning for classifiers based on the aggregation of partial models obtained on the basis of data from individual Polish hospitals. In addition, for comparison purposes, a classic classifier can be constructed for the entirety of the data. Thanks to this, it is possible to answer the question whether classifiers based on federated learning proposed in this paper can match the efficiency of the classifier based on classical learning.

It is worth mentioning that many federated learning methods have been described in the literature that could be used here (see, e.g., [10,12]). These approaches are usually based on iterative training of classifiers in nodes and iterative creation of the aggregation mechanism. This is because in practical applications, the data available at the nodes changes. However, in this paper we analyse data that is immutable and therefore do not use iterative learning mechanisms.

Currently, most research on federated learning focuses on training deep learning models with privacy protection (see, e.g., [7,14]). While the deep learning model is very efficient for a range of real federated learning tasks, it can be beaten for tabular data by complex tree models (see, e.g., [15]). Therefore, using the BalancedRandomForestClassifier model from the Imbalanced-learn library to construct classifiers for federated learning nodes makes sense.

Algorithm 1: Classifier based on average weights from nodes

Input:
1. $H_1, ..., H_n$ are hospitals and $\mathbf{A}_1, ..., \mathbf{A}_n$ are decision tables obtained from hospitals $H_1, ..., H_n$, respectively (e.g., $k = 14$),
2. D_0 and D_1 are decision classes represented by values of decision attributes from tables $\mathbf{A}_1, ..., \mathbf{A}_n$ (D_0 and D_1 correspond to the situation when the patient is alive or died, respectively),
3. u is a tested object,
4. t is the weight threshold of belonging to class D_1, used to classify objects in such a way that an object is classified into class D_1, when the weight of belonging u to class D_1 generated by a given classifier is greater than t.

Output: The value of decision attribute for the object u.

1 **begin**
2 | Build classifiers $\mathbf{C}_1, ..., \mathbf{C}_n$ for decision tables $\mathbf{A}_1, ..., \mathbf{A}_n$.
3 | Classify the object u by classifiers $\mathbf{C}_1, ..., \mathbf{C}_n$ to obtain a collection of weights $W_1^{D_1}(u), ..., W_n^{D_1}(u)$ belonging to the class D_1 of the object u.
4 | Assign $mW := \frac{(W_1^{D_1}(u)+...+W_n^{D_1}(u)}{n}$
5 | **if** $mW > t$ **then**
6 | | **return** final decision value corresponding to the class D_1.
7 | **else**
8 | | **return** final decision value corresponding to the class D_0.
9 | **end**
10 **end**

In this paper, we propose two algorithms for federated learning and testing of classifiers. The first one is Algorithm 1. This algorithm first creates classifiers for individual hospitals, and then these classifiers are used to generate weights that indicate the predicted degree of belonging of the u object to the D_1 class. Both of these operations can be performed in nodes, i.e., in individual hospitals. Then, the determined weights are sent to the server where they are aggregated by calculating the arithmetic mean (compare with [10]). Finally, a final decision is generated, with the object classified to D_1, when the aggregated weight is greater than the t threshold. Otherwise, the object is classified to D_0 class.

Algorithm 2: Classifier based on nodes voting

Input:
1. $H_1, ..., H_n$ are hospitals and $\mathbf{A}_1, ..., \mathbf{A}_n$ are decision tables obtained from hospitals $H_1, ..., H_n$, respectively (e.g., $k = 14$),
2. D_0 and D_1 are decision classes represented by values of decision atrributes from tables $\mathbf{A}_1, ..., \mathbf{A}_n$ (D_0 and D_1 correspond to the situation when the patient is alive or died, respectively),
3. u is a tested object,
4. t is the weight threshold of belonging to class D_1, used to classify objects in such a way that an object is classified into class D_1, when the weight of belonging u to class D_1 generated by a given classifier is greater than t.

Output: The value of decision attribute for the object u.

```
 1  begin
 2      Build classifiers C_1, ..., C_n for decision tables A_1, ..., A_n.
 3      Classify the object u by classifiers C_1, ..., C_n to obtain a collection of
        weights W_1^{D_1}(u), ..., W_n^{D_1}(u) belonging to the class D_1 of the object u.
 4      Assign votesD_0 := 0 and votesD_1 := 0
 5      for i := 1 to n do
 6          if W_i^{D_1}(u) > t then
 7              votesD_1 := votesD_1 + 1
 8          else
 9              votesD_0 := votesD_0 + 1
10          end
11      end
12      if votesD_1 > votesD_0 then
13          return final decision value corresponding to the class D_1.
14      else
15          return final decision value corresponding to the class D_0.
16      end
17  end
```

The second of the federated learning and testing algorithms proposed in this paper is Algorithm 2. Similarly to the previous one, this algorithm first creates classifiers for individual hospitals, and then these classifiers are used to generate the weights of object u belonging to class D_1. As before, both of these operations can be performed in nodes, i.e., in hospitals. Then, the determined weights are sent to the server where they are aggregated in order to make a decision for the test object u. However, this is done differently like in Algorithm 1. First, based on the weight threshold t, a decision is made for object u by each of the nodes using the weights provided by the individual nodes. Then the nodes vote on the decision value to be proposed for the object u. The object u is classified into the decision class that received the most votes.

It is easy to see that the above two algorithms are representatives of two different types of algorithms resolving conflicts between classifiers generated in individual nodes. Algorithm 1 is an example of an algorithm that uses a certain established function to aggregate the numerical weight values returned by individual classifiers when classifying a test object. Algorithm 1 uses the arithmetic mean function of the weights, but other functions of this type are possible (see Sect. 6). Algorithm 2, on the other hand, is an example of an algorithm that uses heuristics to resolve conflicts, operating not on the weights returned by the classifiers, but on the decisions generated by individual classifiers. In Algorithm 2, the method of simple voting of individual classifiers was used to select one of two decision classes. However, other heuristics that select the resulting decision class are possible. It is worth adding that the algorithms of both types require a different way of tuning the sensitivity level of the classifier aggregating classifiers from individual nodes. In Algorithm 1, we adjust the sensitivity after aggregating the weights. However, in Algorithm 2, the sensitivity is adjusted when classifying an object by individual classifiers. This is a more flexible approach because you can adjust the sensitivity for individual classifiers separately. However, for the purposes of this paper, we use one sensitivity adjustment parameter for all classifiers generated in individual nodes.

In the above description of the algorithms, we assumed that the test object is classified by classifiers generated in individual nodes. This is because data from nodes often cannot be made available to the server. However, with this assumption, the test object representing the classified patient would have to be sent to each of the nodes (in order to classify it), which is also usually contrary to the protection of personal data. Therefore, in practice, the above approach may look like this: classifiers are created on data in individual nodes, but after training, ready-made classifier models (e.g., a decision tree or a family of decision trees) are sent to the server and can be freely used there to classify test objects. Therefore, neither data from individual nodes is sent to the server nor the test object is sent to the nodes. It is also worth adding that the classifier models generated in the nodes do not contain sensitive data about patients, but only knowledge about classification expressed at a higher level of abstraction than data about patients.

5   Experiments and Results

To verify the effectiveness of our classifiers based on federated learning in relation to classical classifiers, we have implemented our algorithms in Python language using the Scikit-Learn library [13] and the Imbalanced-Learn library [5,6]. The experiments have been performed on the medical data set obtained from Second Department of Internal Medicine, Collegium Medicum, Jagiellonian University, Krakow, Poland (POLVAS register - see Sect. 3).

The aim of our experiments was to answer the question whether the classifiers based on federated learning proposed in this paper can match the efficiency of the classifier based on classical learning.

Table 1. Results of experiments with classic classifier

Weight threshold	Accuracy	Accuracy for D_0	Accuracy for D_1	Difference
0.05	0.136	0.0	1.0	1.0
0.1	0.137	0.001	1.0	0.999
0.15	0.16	0.028	1.0	0.972
0.2	0.204	0.079	1.0	0.921
0.25	0.279	0.165	1.0	0.835
0.3	0.391	0.298	0.982	0.684
0.35	0.484	0.41	0.955	0.545
0.4	0.592	0.539	0.928	0.389
0.45	0.677	0.652	0.838	0.186
0.5	**0.757**	**0.755**	**0.766**	**0.011**
0.55	0.822	0.844	0.676	0.169
0.6	0.858	0.912	0.514	0.399
0.65	0.877	0.953	0.387	0.566
0.7	0.884	0.986	0.234	0.752
0.75	0.877	0.999	0.099	0.899
0.8	0.869	1.0	0.036	0.964
0.85	0.867	1.0	0.018	0.982
0.9	0.866	1.0	0.009	0.991
0.95	0.864	1.0	0.0	1.0

As we have already written, the dataset used in this paper for experiments has 819 rows. In such a situation, the cross-validation method (see, e.g., [9]), is usually used to determine the quality of the created classifiers, because the data set has more than 100 rows. However, the samples of the entire dataset that are available in hospitals are often less than 100 rows, whilst in case of Algorithm 1 and Algorithm 2 classifiers are created for these samples. Therefore, in this paper, the leave-one-out method was proposed for the experimental verification of the proposed classifiers, which is often used when the data is small, i.e., it has less than 100 rows (see, e.g., [9]). The leave-one-out method is a variation of the cross-validation method, where the data partitioning elements are single-element, i.e., an N-element array is divided into N subsets (each subset contains one row).

The following popular measures are used to measure the quality of the proposed methods: "Accuracy" (accuracy for the entire table, that is, the classification accuracy of objects from both class D_0 and class D_1), "Accuracy for D_0" (accuracy for the decision class D_0) i "Accuracy for D_1" (accuracy for the decision class D_1). The parameter "Difference" is also given, which gives the absolute value of the difference between "Accuracy for D_0" and "Accuracy for D_1".

Table 1 shows the results obtained for the classic classifier construction method (without federated learning). In each row with the results, the "Weight

Table 2. Results of experiments with classifier from Algorithm 1

Weight threshold	Global accuracy	Accuracy for D_0	Accuracy for D_1	Difference
0.05	0.136	0.0	1.0	1.0
0.1	0.136	0.0	1.0	1.0
0.15	0.136	0.0	1.0	1.0
0.2	0.136	0.0	1.0	1.0
0.25	0.144	0.01	1.0	0.99
0.3	0.181	0.052	1.0	0.948
0.35	0.291	0.182	0.982	0.8
0.4	0.467	0.393	0.937	0.544
0.45	0.632	0.6	0.838	0.238
0.5	**0.748**	**0.754**	**0.712**	**0.042**
0.55	0.83	0.877	0.532	0.345
0.6	0.87	0.955	0.333	0.621
0.65	0.878	0.987	0.18	0.807
0.7	0.87	0.994	0.081	0.913
0.75	0.867	1.0	0.018	0.982
0.8	0.864	1.0	0.0	1.0
0.85	0.864	1.0	0.0	1.0
0.9	0.864	1.0	0.0	1.0
0.95	0.864	1.0	0.0	1.0

threshold" value, that was used in the experiment from the given row is given in the first column (parameter t also appeared in Algorithms 1 and 2).

Incidentally, each row of the results table corresponds to one point on the so called ROC curve. In practical applications, we select a point on the ROC curve, according to which we calculate the classifiers to be used. In medical applications, we often want accuracy for both D_0 and D_1 decision classes to be similar. From this point of view, the most interesting row in Table 1 is the row with a weight of 0.5. It is easy to see that it concerns the experiment with the smallest value of the Difference" parameter (equals 0.01) and general accuracy 0.757.

However, from a general point of view, the Table 1 is a tool for adjusting the sensitivity and specificity of classifiers. For example, if we wanted the sensitivity of the classifier for the class D_1 to be at least 0.9, we would choose a value of "Weight threshold" equal to 0.4. Then the specificity (accuracy per class D_1) reaches 0.539.

Table 2 shows the results obtained for Algorithm 1. In this case, the most interesting row in Table 2 is the row with a weight of 0.5. It concerns the experiment with the smallest value of the "Difference" parameter and general accuracy of 0.748.

Table 3. Results of experiments with classifier from Algorithm 2

Weight threshold	Global accuracy	Accuracy for D_0	Accuracy for D_1	Difference
0.05	0.136	0.0	1.0	1.0
0.1	0.136	0.0	1.0	1.0
0.15	0.136	0.0	1.0	1.0
0.2	0.139	0.004	1.0	0.996
0.25	0.166	0.035	1.0	0.965
0.3	0.208	0.083	1.0	0.917
0.35	0.315	0.209	0.991	0.782
0.4	0.473	0.405	0.91	0.505
0.45	0.627	0.597	0.82	0.223
0.5	**0.74**	**0.743**	**0.721**	**0.022**
0.55	0.813	0.857	0.532	0.326
0.6	0.859	0.946	0.306	0.64
0.65	0.875	0.986	0.171	0.815
0.7	0.874	0.996	0.099	0.897
0.75	0.867	0.999	0.027	0.972
0.8	0.864	1.0	0.0	1.0
0.85	0.864	1.0	0.0	1.0
0.9	0.864	1.0	0.0	1.0
0.95	0.864	1.0	0.0	1.0

Finally, Table 3 shows the results obtained for Algorithm 2. In this case, the most interesting row in Table 3 is the row with a weight of 0.5. It concerns the experiment with the smallest value of the "Difference" parameter and general accuracy of 0.74.

6   Conclusion

We presented two methods of constructing classifiers based on federated learning. We also experimentally compared their effectiveness with the classical method of creating classifiers based on data related to the diagnosis of patient mortality in vasculitis disease.

Experiments have shown that our two methods are comparable to the results of the classical method but do not match the quality of the classical method of creating a classifier. This is probably due to the fact that classical methods use the entire data, and methods based on federated learning create classifiers only in nodes, and use aggregation methods to combine the knowledge acquired in individual nodes. The slightly lower quality of the proposed methods based on federated learning compared to the classical methods suggests that research on aggregation methods should be continued. Therefore, we plan research in the following three directions:

1. the use of other aggregation methods in Algorithm 1 known from the literature (see [3,8,10]),
2. the use of other node voting methods in Algorithm 2 using more information about the data available in the nodes,
3. the use of additional classifiers to resolve conflicts between nodes.

Acknowledgement. This paper was partially supported by the Centre for Innovation and Transfer of Natural Sciences and Engineering Knowledge of University of Rzeszów, Poland.

References

1. Bazan, J.G., Szczuka, M.: The rough set exploration system. In: Peters, J.F., Skowron, A. (eds.) Transactions on Rough Sets III. LNCS, vol. 3400, pp. 37–56. Springer, Heidelberg (2005). https://doi.org/10.1007/11427834_2
2. Breiman, L.: Random Forests. Mach. Learn. **45**, 5–32 (2001)
3. Drygaś, P., Bazan, J.G., Pusz, P., Knap, M.: Application of uninorms to aggregate uncertainty from many classifiers. J. Autom. Mob. Robot. Intell. Syst. **13**(4), 85–90 (2019)
4. Banabilah, S., Aloqaily, M., Alsayed, E., Malik, N., Jararweh, Y.: Federated learning review: fundamentals, enabling technologies, and future applications. Inf. Process. Manag. **59**(6), 103061 (2022)
5. Imbalanced-learn library. https://imbalanced-learn.org/
6. Lemaitre, G., Nogueira, F., Aridas, C.K.: Imbalanced-learn: a python toolbox to tackle the curse of imbalanced datasets in machine learning. J. Mach. Learn. Res. **18**, 1–5 (2017)
7. Kairouz, P., McMahan, B., Song, S., Thakkar, O., Thakurta, A., Xu, Z.: Practical and private (deep) learning without sampling or shuffling. In: Proceedings of the International Conference on Machine Learning, Online, 18–24 July 2021, pp. 5213–5225 (2021)
8. Mas, M., Massanet, S., Ruiz-Aguilera, D., Torrens, J.: A survey on the existing classes of uninorms. J. Intell. Fuzzy Syst. **29**, 1021–1037 (2015)
9. Michie, D., Spiegelhalter, D.J., Taylor, C.C.: Machine Learning, Neural and Statistical Classification. Ellis Horwood Limited, Chichester (1994)
10. Moshawrab, M., Adda, M., Bouzouane, A., Ibrahim, H., Raad, A.: Reviewing federated learning aggregation algorithms; strategies, contributions, limitations and future perspectives. Electronics **12**(10), 2287 (2023)
11. Pawlak, Z., Skowron, A.: Rudiments of rough sets. Inf. Sci. **177**, 3–27 (2007)
12. Rodríguez-Barroso, N., Jiménez-López, D., Luzón, M.V., Herrera, F., Martínez-Cámara, E.: Survey on federated learning threats: concepts, taxonomy on attacks and defences, experimental study and challenges. Inf. Fusion **90**, 148–173 (2023)
13. Scikit-learn - Machine Learning in Python. https://scikit-learn.org/
14. Zhao, B., Fan, K., Yang, K., Wang, Z., Li, H., Yang, Y.: Anonymous and privacy-preserving federated learning with industrial big data. IEEE Trans. Ind. Inform. **17**, 6314–6323 (2021)
15. Zhao, X., Li, X., Sun, S., Jia, X.: Secure and efficient federated gradient boosting decision trees. Appl. Sci. **13**(7), 4283 (2023)

A Novel Hybrid Wind Speed Interval Prediction Model Using Rough Stacked Autoencoder and LSTM

Qiuyu Mei[1], Hong Yu[1(✉)], and Guoyin Wang[2]

[1] Chongqing Key Laboratory of Computational Intelligence, Chongqing University of Posts and Telecommunications, Chongqing, China
yuhong@cqupt.edu.cn

[2] Key Laboratory of Cyberspace Big Data Intelligent Security (Chongqing University of Posts and Telecommunications), Ministry of Education, Chongqing, China

Abstract. Wind speed interval prediction is of great significance in power resource scheduling and planning. However, the complex and variable characteristics of wind speed make quality forecasting challenging. In this paper, a novel hybrid model, abbreviated as RSAE-LSTM, for wind speed interval prediction is proposed. The model employs a rough stacked autoencoder (RSAE) and long short-term memory neural network (LSTM). The RSAE initially handles uncertainties and extracts important potential features from the wind speed data. Then, the generated features are utilized as input to the LSTM network to construct the prediction intervals (PIs). Meanwhile, a new loss function is proposed for developing model to construct PIs effectively. The experimental results show that compared with the comparison methods, the proposed method could obtain high-quality PIs and achieve at least a 39% improvement in the coverage width criterion (CWC) index.

Keywords: Rough set theory · Prediction intervals · LSTM · Autoencoder

1 Introduction

With the continuous development and promotion of renewable energy, wind energy has garnered significant attention as a crucial clean energy resource. However, the utilization of wind energy is constrained by meteorological conditions, with wind speed being a key factor. In the past decade, there has been a proliferation of research studies dedicated to wind speed prediction [8,9,20]. However, current research is mainly concerned with point/deterministic forecasting. While significant forecast improvements have been reported, deterministic forecasts unfortunately do not fully provide the information needed for good decision-making processes, making them inadequate to meet uncertainty challenges.

ⓒ The Author(s), under exclusive license to Springer Nature Switzerland AG 2023
A. Campagner et al. (Eds.): IJCRS 2023, LNAI 14481, pp. 537–548, 2023.
https://doi.org/10.1007/978-3-031-50959-9_37

Prediction intervals (PIs) have emerged as an effective approach to quantify and communicate the uncertainty associated with wind speed predictions. By establishing upper and lower bounds, PIs provide valuable insights into the range within which future forecasts are highly likely to fall. In recent literature, Bayesian methods [14] and Bootstrap [2] have been employed to construct PIs. However, these methods typically rely on specific error distribution assumptions for point predictions when constructing PIs.

To overcome the shortcomings of traditional approaches, lower upper bound estimation (LUBE) [10] which takes the two outputs of the neural network as the upper and lower boundaries of PIs directly was proposed and adopted widely. However, since the loss function is not differentiable, the network cannot be trained using the gradient descent (GD) method. The meta-heuristic optimization algorithms are used for training, which have many optimization parameters and takes a lot of time, so they cannot effectively process large-scale data. Fortunately, many researchers have designed some differentiable loss functions to guide neural networks to construct PIs directly. For example, Pearce et al. [16] considered the generation of PIs by neural networks for quantifying uncertainty in regression tasks. It is axiomatic that high-quality PIs should be as narrow as possible, whilst capturing a specified portion of data. They derive a loss function directly from this axiom that requires no distributional assumption. Liu et al. [13] designed an improved LUBE model using a novel training scheme based on the GD method for better efficiency and greater prediction performance. Kabir et al. [7] proposed a highly customizable smooth cost function for developing model to construct optimal PIs. Although these methods have achieved good results in predicting interval tasks, majority of these works implement shallow neural networks as predictive models. The insufficiency of models has restricted the performance of LUBE-based methods.

Recently, deep learning methods have been widely used in the field of the wind forecasting. An autoencoder (AE) [3] is an unsupervised neural network trained by stochastic gradient descent algorithms. Stacked autoencoder (SAE) is created by stacking multiple encoder layers. Some researchers [6,8] adopted AE networks to compress high-dimensional wind speed time series to obtain low-dimensional representations and generate predictions based on low-dimensional representations. Meanwhile, the deep learning algorithms can handle large-scale datasets and can maintain long-term dependencies between the time series, like the long short-term memory neural network (LSTM) [4]. For example, Li et al. [11] developed a novel hybrid interval prediction model based on LSTM networks and variational mode decomposition algorithm in the frame of lower upper bound estimation. Saeed et al. [19] proposed a novel wind speed interval prediction model by integrating LUBE method into a quasi-recurrent neural network. In addition, There are inevitable uncertainties in the predictions made by different models, and capturing such uncertainties is beneficial to resource planning [1]. Rough set theory [15], as an useful tool to handle the uncertainties, has been successfully applied to wind speed point prediction [8,9].

As a result, we propose a hybrid wind speed interval forecasting model using rough stacked Autoencoders and LSTM network (RSAE-LSTM). First, rough neurons constructed based on rough set theory are introduced into SAE to eliminate the uncertainty and automatically learn powerful features from the wind speed data. Then, considering the time-series characteristics of the data, the learned features are fed into the LSTM network to predict PIs. Meanwhile, a new loss function is specially designed for implementing the GD training method to construct PIs efficiently.

The main contributions of this paper are as follows:

- A novel hybrid model RSAE-LSTM is proposed for wind speed interval forecasting.
- Rough set theory is introduced into deep learning method to handle uncertainty of wind speed and improve the quality of prediction.
- A new loss function is designed for predicting PIs with neural network directly without any assumption on the distribution.
- Compared with the comparison methods, the proposed method could obtain high-quality PIs and achieved at least a 39% improvement in the coverage width criterion (CWC) index.

The rest of this paper is organized as follows: Related works are reviewed in Sect. 2. The proposed method is formulated in Sect. 3. Comparative experiments and experimental results are provided in Sect. 4. This paper is concluded in Sect. 5.

2 Preliminaries

In this section, we will first give some basic definitions of rough sets theory, rough neuron, and prediction intervals.

2.1 Rough Set Theory

Rough set theory, proposed by Pawlak, is an effective mathematical tool for dealing with uncertainty and has been widely used in machine learning and pattern recognition tasks [15]. An information system is defined by the four-tuple $S = <U, A, V, f>$, where universe U is a finite nonempty set and A is a finite nonempty set of attributes. V_a is a domain set, where $a \in A$ and $V = \bigcup_{a \in A} V_a$. $f : U \times A \to V$ is an information function and for every $a \in A$ and $x \in U, f(x, a) \in V_a$. Suppose $B \subseteq A$, two objects $x, y \in U$ are indiscernible from each other by the set of attributes B in S if and only if for every $a \in B, f(x, a) = f(y, a)$. Thus, every $B \subseteq A$ has a binary relation $IND(B)$ on U, which is called the indiscernibility relation. The partition of U is a family of all indiscernibility relations of $IND(B)$ and is denoted by $U/IND(B)$.

Rough set theory defines two approximations for any concept set $X \subseteq U$ and attribute set $B \subseteq A$. Using the knowledge of B, X can be approximated by the B-lower approximation $\underline{B}(X)$ and B-upper approximation $\overline{B}(X)$:

$$\underline{B}(X) = \cup \{O \in U/B \mid O \subseteq X\} \tag{1}$$

$$\overline{B}(X) = \cup \{O \in U/B \mid O \cap X \neq \emptyset\} \tag{2}$$

2.2 Rough Neuron

Rough neurons were developed based on rough set theory by Lingras in rough neural networks [12]. These neurons contain an upper bound weight W_U and lower bound weight W_L. Moreover, the neuron contains upper bound and lower bound biases denoted by b_U and b_L, respectively. Figure 1 illustrates a rough neuron. Equations defining the relations of these parameters are as follows:

$$O_U = Max(f(W_U X + b_U), f(W_L X + b_L)) \tag{3}$$

$$O_L = Min(f(W_U X + b_U), f(W_L X + b_L)) \tag{4}$$

$$O = \alpha O_U + \beta O_L \tag{5}$$

where f is a sigmoid unit.

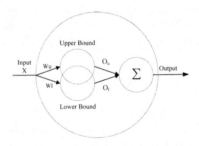

Fig. 1. Rough neuron structure.

2.3 Prediction Intervals

Given an input x_i, a prediction interval $[\hat{y}_{L_i}, \hat{y}_{U_i}]$ of a sample i captures the future observation (target variable) y_i with the probability equal or greater than, for $1 \leq i \leq n$. A PI should capture some desired proportion of the observations, $(1 - \alpha)$, common choices of α being 0.01 or 0.05,

$$Pr(\hat{y}_{Li} \leq y_i \leq \hat{y}_{Ui}) \geq (1 - \alpha). \tag{6}$$

Given n samples, the quality of the generated prediction intervals is assessed by measuring the prediction interval coverage probability (PICP)

$$PICP = \frac{1}{n} \sum_{i=1}^{n} k_i, \tag{7}$$

where $k_i = 1$ if $y_i \in [\hat{y}_{Li}, \hat{y}_{Ui}]$, otherwise $k_i = 0$. n is the number of testing samples. Mean Prediction Interval Width (MPIW) is defined as

$$MPIW = \frac{1}{n}\sum_{i=1}^{n}(\hat{y}_{U_i} - \hat{y}_{L_i}), \tag{8}$$

It is desired to achieve $PICP \geq (1 - \alpha)$ while having MPIW as small as possible.

3 Proposed Method

3.1 Framework

The forecasting framework is easy to elucidate as illustrated in Fig. 2. Firstly, rough neuron is developed based on the rough set theory and incorporated into the SAE to construct rough stacked autoencoder network (RSAE). Then RSAE is applied to automatically extract potential features from the original wind speed data to perform data dimensionality reduction and handle uncertainty. Finally, considering the time-series characteristics of the data, the generated features are fed into the LSTM to predict PIs.

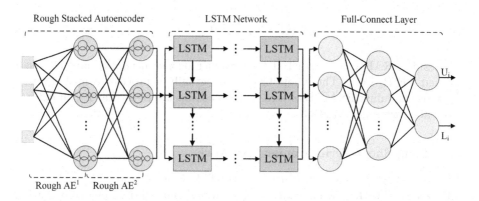

Fig. 2. Framework of the proposed RSAE-LSTM model.

3.2 Rough Stacked Autoencoder

A AE is a variant of the feed-forward neural network, which performs encoding of its input X to a hidden representation h. Then, it decodes the input again from the hidden representation using a decoder. A rough version of AE (RAE) can be constructed by replacing the original neurons in the hidden layer and output layer of AE with rough neurons [8].

For the encoding process of RAE, the outputs of the upper bound h_U^e and lower bound h_L^e are denoted as

$$h_U^e = Max(f(W_U^e X + b_U^e), f(W_L^e X + b_L^e)) \tag{9}$$

$$h_L^e = Min(f(W_U^e X + b_U^e), f(W_L^e X + b_L^e)). \tag{10}$$

Then, the hidden representation h^e is computed as

$$h^e = \alpha_e h_U^e + \beta_e h_L^e \tag{11}$$

For the decoding process of RAE, the outputs of the upper bound h_U^d and lower bound h_L^d are denoted as

$$h_U^d = Max(f(W_U^d h^e + b_U^d), f(W_L^d h^e + b_L^d)) \tag{12}$$

$$h_L^d = Min(f(W_U^d h^e + b_U^d), f(W_L^d h^e + b_L^d)). \tag{13}$$

Thus, the reconstructed input $\hat{X}$ is computed as

$$\hat{X} = \alpha_d h_U^d + \beta_d h_L^d \tag{14}$$

In each round of training, any object x_i is encoded h_i by an encoder, and then reconstructed by a decoder to be $\hat{x}_i$. RAE makes use of the mean squared error loss function (MSE) given in Eq. (15) to minimize the average reconstruction error and optimize parameters.

$$MSE = \frac{1}{n} \sum_{i=1}^{n} (x_i - \hat{x}_i)^2 \tag{15}$$

Besides, RAE can be stacked to construct deeper network, namely rough stacked AE (RSAE). The whole process is actually a process of unsupervised layerwise training. RSAE possesses more encoding layers so that it can extract more abstract representations.

3.3   Long Short-Term Memory Network

The long short-term memory network (LSTM) is a variant of recurrent neural network. The LSTM combines long-term and short-term memory by gating mechanisms including forgetting gating, input gating and output gating, and utilizes both short-term and long-term data. It calculates as follows:

$$f_n = \sigma(W_f x_n + U_f h_{n-1} + b_f) \tag{16}$$

$$i_n = \sigma(W_i x_n + U_i h_{n-1} + b_i) \tag{17}$$

$$o_n = \sigma(W_o x_n + U_o h_{n-1} + b_f) \tag{18}$$

$$\tilde{c}_n = tanh(W_c x_n + U_c h_{n-1} + b_c) \tag{19}$$

$$c_n = \tilde{c}_n \otimes i_n + c_{n-1} \otimes f_n \tag{20}$$

$$h_n = o_n \otimes tanh(c_n) \tag{21}$$

wherein W and U denote the weight matrices while b is the bias. σ and $tanh$ represent the activation functions $sigmoid$ and $tanh$, respectively. $\otimes$ represents the element wise multiplication. The input gate consists of i_n, $\tilde{c}_n$, they decide what information to store. The forget gate f_n decides whether to discard the output information of the last LSTM h_{n-1}. The output of the current LSTM h_n is through the output gate o_n.

3.4   Hybrid Loss Function for Prediction Intervals

In this section, a novel cost function is proposed for wind speed interval prediction. Our goal is to generate narrow PIs, while maintaining the desired level of coverage. However, PIs that fail to capture their data point should not be encouraged to shrink further [16]. We therefore introduce captured $MPIW$ as the $MPIW_c$ of only those points for which $\hat{y}_L \leq y \leq \hat{y}_U$ holds

$$MPIW_c = \frac{1}{c} \sum_{i=1}^{n} (\hat{y}_{U_i} - \hat{y}_{L_i}) \cdot k_i, \tag{22}$$

where $c = \sum_{i=1}^{n} k_i$, which means the total number of data points captured.

Meanwhile, the most probable region of the target may not stay near the middle of the interval in the direct PI construction method. Minimization of the deviation of the target from the center can potentially shift the most probable region near the center of PIs. In addition, PIs may fail to cover the target but the proposed model should try to bring the target close to the nearby bound of the PI. This can also be achieved by having the target close to the center of the PI. Therefore, MAE is proposed to achieve both of the above optimization goals.

$$MAE = \frac{1}{n} \sum_{i=1}^{n} |y_i - \hat{y}_i|, \tag{23}$$

where MAE is Mean Absolute Error, $\hat{y}_i = (\hat{y}_{L_i} + \hat{y}_{U_i})/2$, we use the midpoint of the interval $[\hat{y}_{L_i}, \hat{y}_{U_i}]$ as the point estimation.

To enforce the coverage constraint, a penalty is imposed only if $PICP < (1 - \alpha)$. Inspired by previous work [8], $\wp_{PI}$ can be constructed as below:

$$\wp_{PI} = \max(0, (1 - \alpha) - PICP)^2. \tag{24}$$

However, since the gradient is always positive, the optimization loss of a discrete version of k (see Eq. (22)) fails to converge. Hence, c_i can be replaced by the previous method [16] as follows:

$$c_i = \sigma(s(\hat{y}_{U_i} - y_i)) \odot \sigma(s(y_i - \hat{y}_{L_i})), \tag{25}$$

where σ is the sigmoid function and s is a softening parameter.

Algorithm 1 reveals the process to construct the loss function.

Algorithm 1: Construction of hybrid loss function

Input : Predictions of lower bound $\hat{y}_L$ and upper bound $\hat{y}_U$, ground truth label y, hyperparameter factor for softing s, hyper-parameter λ is the PI coverage penalty factor,, σ denotes sigmoid activation function, $\odot$ denotes the element-wise product. $\epsilon = 1e-10$

Output: *Loss*

1 $k_H = \max(0, \text{sign}(\hat{y}_U - y)) \odot \max(0, \text{sign}(y - \hat{y}_L))$

2 $\hat{y} = (\hat{y}_U + \hat{y}_L)/2$

3 $c = \sigma(s(\hat{y}_U - y)) \odot \sigma(s(y - \hat{y}_L))$

4 $MPIW_c = \text{reduce\_sum}((\hat{y}_U - \hat{y}_L) \odot k_H)/(\text{reduce\_sum}(k_H) + \epsilon)$

   /* where ϵ is a small value to avoid an undefined $MPIW_c$
      value for 100% PICP   */

5 $PICP = \text{reduce\_mean}(c)$

6 $MAE = \text{reduce\_mean}(|y - \hat{y}|)$

7 $\wp_{PI} = \max(0, (1 - \alpha) - PICP)^2$

8 $Loss = MPIW_c + MAE + \lambda \wp_{PI}$

---

4 Experiment

4.1 Data Description

The wind speed data utilized in our experimental analysis was sourced from the National Wind Technology Center (NWTC) [5][1]. The raw data files can be obtained at various time intervals, ranging from minutes to hours or even days. Moreover, the data fields can be customized based on specific requirements and preferences. In this study, we collected raw data of various parameters with one-minute time intervals from January 1, 2022, to December 31, 2022. The dataset encompassed several variables, including average wind speed @ 10 m, global horizontal, temperature @ 2 m, dew point temperature, specific humidity, station pressure, precipitation (accumulated), peak wind speed @ 10 m, average wind speed (std dev) @ 10 m, average wind direction @ 10 m, wind direction @ Pk WS @ 10 m, average wind direction (std dev) @10 m, average wind shear, and turbulence intensity @ 10 m.

The raw data from NWTC are not completely accurate and reliable, any illegal values in the dataset were replaced with the previous valid value. Additionally, to ensure consistency and facilitate analysis, the original 1-minute intervals of the data were transformed into 10-minute intervals by calculating the average. Following these preprocessing steps, a total of 52,560 data items were obtained. Then, we divided the data into a training set, a validation set, and a test set in a ratio of [8:1:1] for further analysis. In order to achieve consistent evaluation results, the features of all experimental data were normalized to a uniform scale within the range of [0, 1].

---

[1] https://midcdmz.nrel.gov/apps/daily.pl?site=NWTC&live=1.

4.2   Evaluation Metrics

To comprehensively evaluate the forecast results of the prediction model, four analysis indexes are adopted in the study, which include PICP (see Eq. (7)), PI normalized averaged width (PINAW), coverage width criterion (CWC), and the normalized average deviation (NAD).

PINAW is defined to measure the narrowness of the PI. Defined mathematically [18] as:

$$PINAW = \frac{1}{nR} \sum_{i=1}^{n} (\hat{y}_{U_i} - \hat{y}_{L_i}), \tag{26}$$

where R is the range of target values.

Reducing the PI width may result in higher coverage and vice versa. Therefore, looking at PICP and PINAW independently is not enough to reflect the quality of PI. CWC, as a balance index of PICP and PINAW, can effectively evaluate the quality of PI.

$$CWC = \begin{cases} PINAW, PICP >= \mu \\ PINAW + \exp(-\eta(PICP\text{-}\mu)), PICP < \mu \end{cases} \tag{27}$$

where μ is determined by the confidence interval and η exponentially magnifies the difference between the PICP and μ. By minimizing the CWC function, an optimal PI is expected to be achieved.

NAD is used to express the deviation of the data which are not covered by the PI. So it can express the rationality of PI.

$$NAD = \frac{1}{n} \sum_{i=1}^{n} a_i, \tag{28}$$

where the expression of a_i is defined in Eq. (29).

$$a_i = \begin{cases} (\hat{y}_{L_i} - y_i) \Big/ \frac{1}{n} \sum_{i=1}^{n} (\hat{y}_{U_i} - \hat{y}_{L_i}), y_i < \hat{y}_{L_i} \\ 0, y_i \in [\hat{y}_{L_i}, \hat{y}_{U_i}] \\ (y_i - \hat{y}_{U_i}) \Big/ \frac{1}{n} \sum_{i=1}^{n} (\hat{y}_{U_i} - \hat{y}_{L_i}), y_i > \hat{y}_{U_i} \end{cases} \tag{29}$$

4.3   Compared Algorithms

In the comparison experiment, two recent wind speed PIs algorithms LUBE-ANN-GD [13] and LUBE-QRNN [19] and two general PI algorithms Kabir's method [7] and Gradient Boosting Decision Tree with Quantile Loss (GBDT-QR) implemented in the Scikit-learn package [17] were adopted. To overcome the effect of randomness associated with the layer weights initialization of neural networks, each experiment is repeated 10 times and the average value and standard deviation are reported as shown in Table 1. The larger the PICP value, the better the result. On the contrary, the remaining indicators including PINAW, CWC and INAD, the smaller the values, the better the experiment results, and the best values are highlighted in bold. In all experiments, our proposed method has obtained optimal or sub-optimal results on all evaluation indicators. Especially for CWC, our algorithm has achieved at least a 39% improvement compared with other comparison algorithms.

Table 1. Performance indices comparison of comparative models.

Algorithm	PICP	PINAW	CWC	INAD
GBDT-QR	0.9026 ± 0.0040	**0.1766 ± 0.0003**	2.2171 ± 0.1296	0.0332 ± 0.0005
LUBE-ANN-GD	0.9545 ± 0.0140	0.2837 ± 0.0100	1.2033 ± 0.2945	0.0264 ± 0.0012
LUBE-QRNN	0.9439 ± 0.0123	0.2544 ± 0.0171	0.9417 ± 0.1928	0.0239 ± 0.0032
Kabir's method	0.9512 ± 0.0055	0.2729 ± 0.0158	0.3873 ± 0.1445	0.0227 ± 0.0019
RSAE-LSTM	**0.9655 ± 0.0122**	0.2369 ± 0.0166	**0.2369 ± 0.0166**	**0.0215 ± 0.0036**

Table 2. The results of ablation experiments.

Algorithm	PICP	PINAW	CWC	INAD
RSAE	0.9471 ± 0.0218	0.2528 ± 0.0080	0.9717 ± 0.2882	0.0238 ± 0.0096
LSTM	0.9511 ± 0.0126	0.3626 ± 0.0061	0.8504 ± 0.1853	0.0617 ± 0.0228
SAE-LSTM	0.9532 ± 0.0230	0.2431 ± 0.0090	0.5590 ± 0.1746	0.0272 ± 0.0140
RSAE-LSTM	**0.9655 ± 0.0122**	**0.2369 ± 0.0166**	**0.2369 ± 0.0166**	**0.0215 ± 0.0036**

4.4 Ablation Experiment

The proposed model RSAE-LSTM consists of two sub-modules: RSAE and LSTM. In this subsection, we validate the necessity of above two modules. Meanwhile, in order to verify that introducing rough set theory can handle uncertainty in dataset, SAE-LSTM which means that RSAE-LSTM without rough set theory is also tested. The prediction performance of RSAE, LSTM and SAE-LSTM are recorded in Table 2. All the evaluation metrics of the proposed model are superior to that of RSAE, LSTM and SAE-LSTM, which indicates the necessity of each component in this model.

4.5 The Effect of Different Values of Hyper-Parameter λ

In this paper, we use hyper-parameter λ to control the trade-off between PINAW and PICP. From Fig. 3, we can see that with the increase of PICP, PINAW also increases correspondingly and vice versa. Therefore, we can obtain the desired PIs by adjusting the value of λ.

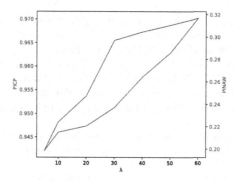

Fig. 3. The performance of different values of hyper-parameter λ.

5   Conclusions

Aiming to obtain high quality prediction intervals, a novel hybrid model for wind speed interval prediction using an rough stacked autoencoder and a long short term memory neural network (RSAE-LSTM) is proposed in this work. Firstly, rough neuron is developed based on the rough set theory and incorporated into the SAE to construct rough stacked autoencoder network (RSAE), and then RSAE is applied to extract potential features from the original input data and perform data dimensionality reduction. The prediction model LSTM then acts on these generated features to construct the prediction intervals. Meanwhile, a new loss function is proposed for developing model to construct PIs effectively. Experimental results reveal that the proposed method can generates narrow intervals with high coverage compared to the comparative models. As a future work, long-term wind speed interval prediction and multi-step wind speed interval prediction will be studied.

Acknowledgements. This work was supported in part by the National Natural Science Foundation of China (62233018, 62136002, 62221005), and the Natural Science Foundation of Chongqing (cstc2022ycjh-bgzxm0004).

References

1. Abdel-Aal, R.E., Elhadidy, M.A., Shaahid, S.: Modeling and forecasting the mean hourly wind speed time series using GMDH-based abductive networks. Renew. Energy **34**(7), 1686–1699 (2009)
2. Heskes, T.: Practical confidence and prediction intervals. In: Advances in Neural Information Processing Systems, vol. 9 (1996)
3. Hinton, G.E., Salakhutdinov, R.R.: Reducing the dimensionality of data with neural networks. Science **313**(5786), 504–507 (2006)
4. Hochreiter, S., Schmidhuber, J.: Long short-term memory. Neural Comput. **9**(8), 1735–1780 (1997)

5. Jager, D., Andreas, A.: NREL national wind technology center (NWTC): M2 tower; boulder, Colorado (data). Technical report, National Renewable Energy Lab. (NREL), Golden, CO (United States) (1996)

6. Jaseena, K., Kovoor, B.C.: A hybrid wind speed forecasting model using stacked autoencoder and LSTM. J. Renew. Sustain. Energy 12(2) (2020)

7. Kabir, H.D., Khosravi, A., Kavousi-Fard, A., Nahavandi, S., Srinivasan, D.: Optimal uncertainty-guided neural network training. Appl. Soft Comput. 99, 106878 (2021)

8. Khodayar, M., Kaynak, O., Khodayar, M.E.: Rough deep neural architecture for short-term wind speed forecasting. IEEE Trans. Industr. Inf. 13(6), 2770–2779 (2017)

9. Khodayar, M., Saffari, M., Williams, M., Jalali, S.M.J.: Interval deep learning architecture with rough pattern recognition and fuzzy inference for short-term wind speed forecasting. Energy 254, 124143 (2022)

10. Khosravi, A., Nahavandi, S., Creighton, D., Atiya, A.F.: Lower upper bound estimation method for construction of neural network-based prediction intervals. IEEE Trans. Neural Networks 22(3), 337–346 (2010)

11. Li, Y., Chen, X., Li, C., Tang, G., Gan, Z., An, X.: A hybrid deep interval prediction model for wind speed forecasting. IEEE Access 9, 7323–7335 (2020)

12. Lingras, P.: Rough neural networks. In: Proceedings of the 6th International Conference on Information Processing and Management of Uncertainty in Knowledge Based Systems, pp. 1445–1450 (1996)

13. Liu, F., Li, C., Xu, Y., Tang, G., Xie, Y.: A new lower and upper bound estimation model using gradient descend training method for wind speed interval prediction. Wind Energy 24(3), 290–304 (2021)

14. MacKay, D.J.: The evidence framework applied to classification networks. Neural Comput. 4(5), 720–736 (1992)

15. Pawlak, Z.: Rough sets. Int. J. Comput. Inf. Sci. 11, 341–356 (1982)

16. Pearce, T., Brintrup, A., Zaki, M., Neely, A.: High-quality prediction intervals for deep learning: a distribution-free, ensembled approach. In: International Conference on Machine Learning, pp. 4075–4084. PMLR (2018)

17. Pedregosa, F., et al.: Scikit-learn: machine learning in python. J. Mach. Learn. Res. 12, 2825–2830 (2011)

18. Quan, H., Srinivasan, D., Khosravi, A.: Short-term load and wind power forecasting using neural network-based prediction intervals. IEEE Trans. Neural Netw. Learn. Syst. 25(2), 303–315 (2013)

19. Saeed, A., Li, C., Gan, Z.: Short-term wind speed interval prediction using lube based quasi-recurrent neural network. In: Journal of Physics: Conference Series, vol. 2189, p. 012015. IOP Publishing (2022)

20. Zhang, Y., Pan, G., Chen, B., Han, J., Zhao, Y., Zhang, C.: Short-term wind speed prediction model based on GA-ANN improved by VMD. Renew. Energy 156, 1373–1388 (2020)

Navigational Strategies for Mobile Robots Using Rough Mereological Potential Fields and Weighted Distance to Goal

Aleksandra Szpakowska, Piotr Artiemjew$^{(\boxtimes)}$, and Wojciech Cybowski

University of Warmia and Mazury, in Olsztyn,
ul. Słoneczna 54, 10-710 Olsztyn, Poland
{ola.szpakowska,artem,wojciech.cybowski}@matman.uwm.edu.pl

Abstract. The present study examines the path planning methods based on rough mereological potential fields for remote mobile robots, building upon a modification of an originally designed project to prepare a foundation for three-dimensional path planning. For this purpose, we have implemented our own library for robot control and developed relevant algorithms - including AR marker recognition, image-based robot detection, and path planning based on the rough mereological potential field in conjunction with a weighted distance to the goal. These algorithms are also customized to facilitate tests in a laboratory setting. Using a video camera, the study captures real-time imagery, allowing for the continuous updating of the robot's position on the designated map. In this paper, we show how to find the right path that a robot follows while constantly updating its position. Furthermore, the research refines the precision of the optimal path through the application of smoothing techniques, ensuring an optimized trajectory from the robot's starting point to its destination. We demonstrate a special Euclidean distance responsible for path optimization. We present a complete project in the work, with all the elements to reproduce it. We carried out real-world tests using the Smart Element Hub cube with LED screen of the lego robot inventor kit.

Keywords: Rough Mereology · Mobile Robotics · Path planning

1 Introduction

The foundational principles of rough mereology, as highlighted by [1] find applications across diverse computer science domains, including robotics (as referenced by [2] and [3]) and medical analysis (cited by [4]). The present study is concerned with extending path planning, which uses rough mereological potential fields, by applying a weighted distance to the target. An additional goal is to adapt our new library and robot to apply our path planning techniques for use in a series of future tests. Initially, paths are refined by eliminating redundant data, followed by a smoothing process that takes obstacle evasion into account.

© The Author(s), under exclusive license to Springer Nature Switzerland AG 2023
A. Campagner et al. (Eds.): IJCRS 2023, LNAI 14481, pp. 549–564, 2023.
https://doi.org/10.1007/978-3-031-50959-9_38

We executed a real-time test using a representative mobile robot, steered by a P-controller that relies on compass indicators and camera-driven positioning. The subsequent section will probe deeper into the formulation of a potential force field through the lens of rough mereology.

1.1 Application of Rough Mereology in the Control Environment of Intelligent Agents

This section investigates into the utilization of rough mereology for the generation of potential fields. The reasoning based on rough mereology introduces the concept of rough inclusion, denoted as $\mu(x, y, r)$. This relation posits that x is a part of y to a degree of at least r. Given our focus on spatial objects, the rough inclusion is expressed as $\mu(X, Y, r)$ if and only if $\frac{|X \cap Y|}{X} \geq r$, where X and Y represent n-dimensional solids and $|X|$ signifies the n-volume of X. In the context of this research, we examine a planar scenario involving an autonomous mobile robot navigating within a 3-dimensional space. Consequently, our spatial objects X, Y are perceived as concept regions, with $|X|$ representing the area of X. The rough inclusion $\mu(X, Y, r)$ take part in shaping the rough mereological potential field.

The elements of this field are square-shaped, and their relative distance is defined as:

$$K(X, Y) = min\{max_r \mu(X, Y, r), max_s \mu(Y, X, s)\}.$$

A detailed exposition of the field's construction is presented in Sect. 2. The robot's trajectory within the field towards its destination is determined by waypoints, which are delineated in an inductive manner: the subsequent waypoint is identified as the centroid of the amalgamation of field squares proximate to the square encompassing the prevailing waypoint, in relation to the distance $K(X, Y)$.

The paper in the next sections has the following content. In Sect. 2 we make an introduction to the route planning methodology used using rough mereological potential field. In the Sect. 3 we describe the exact setup of the experiments. Finally, in the 4 section, we present a summary of our publication.

2 Methodology

In this section, we will discuss the various techniques used to build a target robot guidance system to target based on rough mereological potential field.

2.1 Square Fill Algorithm

In this chapter we are going to show our conception of Square fill algorithm introduced by Ośmiałowski [2]. The mentioned algorithm method already was modified and later presented in Polkowski [7], Zmudzinski and Artiemjew [5] and Gnys [8]. Below we can see the basic steps to initialize the algorithm and also the result that we received - Fig. 1

1. Initialize the values:
   - Set the current distance to the goal: d = 0,
   - Set algorithm direction to clockwise
2. Construct an empty queue Q:
$$Q = \emptyset$$
3. Add into generated queue Q first potential field p(x, y, d), where x, y express the location coordinates of already created field and d reflects the current distance to the goal:
$$Q \cup \{p(x, y, d)\}$$
4. Iterate through Q,
   (a) Determine neighbors regulated by a current direction: if clockwise is true:

$$N = \begin{cases} p_0 = p(x - d, y, d), \\ p_1 = p(x - d, y + d, d), \\ p_2 = p(x, y + d, d), \\ p_3 = p(x + d, y + d, d), \\ p_4 = p(x + d, y, d), \\ p_5 = p(x + d, y - d, d), \\ p_6 = p(x, y - d, d), \\ p_8 = p(x - d, y - d, d) \end{cases}$$

   if anticlockwise is true:

$$N' = \begin{cases} p_0 = p(x - d, y - d, d), \\ p_1 = p(x, y - d, d), \\ p_2 = p(x + d, y - d, d), \\ p_3 = p(x + d, y, d), \\ p_4 = p(x + d, y + d, d), \\ p_5 = p(x, y + d, d), \\ p_6 = p(x - d, y + d, d), \\ p_8 = p(x - d, y, d) \end{cases}$$

   (b) Count the Euclidean distance from previous and already created potential fields neighbors to avoid the redundant fields,
   (c) If current potential field $p_k(x, y, d)$ has Euclidean distance smaller than 5 and $p_k(x, y, d) \cup O$, where O is the set of obstacle coordinates, furthermore $p_k(z, y, d) \cup F$, where F contains a set of created, respected above conditions potential fields, then leave the current field and step back to point 4,
   (d) Check if in Q exists any similar potential field, if exist leave the current neighbor and step back to point 4,
   (e) Add the created potential fields neighbour $p_k(x, y, d)$ into the end of list Q,
   (f) Increase the distance to the goal:
$$d = d(p_k) + 0.1$$

(g) Change the direction to opposite,

(h) Drop current element $p_k(x, y, d)$ from queue Q and add it into potential fields list F.

According to the presented way, the distance is initialized from value 0. This number will be increasing in the next iterations because our algorithm starting to generate potential fields from a goal into a starting point. Created potential fields are represented by $p(x, y, d)$, where x, y are proper coordinates and d describes the value of a distance, which during the operation is responsible for generating new neighbors. The bigger the distance is, the further the potential field has been created from a goal. Moreover in each iteration, the direction of generating neighbors has to be changed so as not to get stuck and explore all map.

Fig. 1. The rough mereological potential field generated using Square Fill Algorithm. The edge AR markers represent the borders of map and the middle markers represent the obstacles. The marker under the drawn squares defines the goal. If we are focused on one color-painted square we get the robot's initial position.

2.2 Circle Fill Algorithm

Instead of applying squares in our algorithm, we can generate a force field using circles. After generating and testing a few combinations of map we could not see clear differences. In the next part connected with path generating, we will look if there will be some dissimilarity between tracks (Fig. 2).

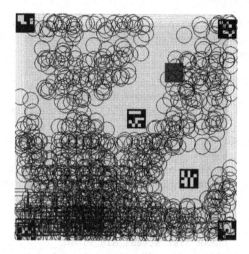

Fig. 2. The rough mereological potential field generated by Square Fill Algorithm, but instead of squares we had applied circles. Similar to 1 edge AR markers represent the borders of map and the middle markers represent the obstacles. The marker located on the most crowded space on map defines the goal. If we are focused on one color-painted square we get the robot's initial position.

2.3   Path Finding

To find our path, first, we are using the variation of the algorithm proposed by Osmialowski Path Search Algorithm [2]. Above mentioned algorithm within the given potential field is as follows (Fig. 3):

```
Determine the robot's start point as the first closest_point
INITIALIZE the list which will contain the path from
generated potential fields

    FOR i IN RANGE defined number of iterations:

        IF the robot's location is equal to or very close to the goal
        position:

            END

        ELSE:

            INITIALIZE the list which will contain the
            minimum potential field
            actual_min_field = []
            CREATE the variable minimum = 0

            FOR j IN RANGE length of generated potential fields list:
```

```
COUNT weighted Euclidean distance between closest_point,
current potential field from a list of generated fields
and goal point

IF this is the first iteration:
    minimum = counted distance
    APPEND current field into actual_min_field

ELSE:

    IF the current distance is smaller than
    actual minimum distance and it is not
    equal 0:

    CHANGE minimum value to the current distance
    CLEAR list actual_min_field
    APPEND current potential field into
    actual_min_field list

DROP the current potential field from a list that consists
all potential field
APPEND actual_min_field into the path list
CHANGE the closest_point to actual_min_field
```

Fig. 3. Generated path using the Path Search Algorithm.

2.3.1   Distance Counting - Weighted Euclidean Distance

For counting the distance between the current point and potential fields we used
the Euclidean distance:

$$d(p, q) = \sqrt{\sum_{i=1}^{n}(p_i - q_i)^2}$$

In two dimensional Euclidean plane, where p represents the point with potential field coordinates and q describes the point with goal coordinates. To improve the accuracy of path finding we apply two Euclidean distances. One focused on the distance between current potential fields in this paper named classical Euclidean distance and the second which is used for path optimization - named Weighted Euclidean distance. The weighted distance takes into account the distance between the goal and the current point, also on this result, we apply the weight (by multiplying it by the proper float number) and between two potential fields. We use a weighted distance to avoid choosing a suboptimal path, a direct jump to the target without avoiding obstacles.

2.4   Field Filtering-Path Optimizing

The path that we have got from the Path Search Algorithm is not as optimal and clear as we expected. To reduce the noise in the path we applied the filter for path optimization. This filter focused on distances between points in the generated path and the goal. The main condition is about: Starting from the first element in the path - If we have the same or bigger distances from current points to target we omit those points. If we have a few points with the same distance we have to count distances from the next neighbor's points to target and compare it. Points having the smallest values of distance will be saved (Fig. 4).

```
INITIALIZE the lists that will contain the filtered
potential fields and proper Euclidean distances

optimal_path =[]
path_distances =[]

FOR i IN RANGE length of first - not optimal path:
    COUNT Euclidean distance between current potential field
    and goal point

    IF it is the first iteration of the loop:

    APPEND this potential field into the list optimal_path
    APPEND counted Euclidean distance into list path_distances

    ELSE:

        IF the counted distance is smaller than the last appended
        distance from path_distances list:

            APPEND this potential field into optimal_path
            APPEND counted Euclidean distance into list path_distance

        IF the counted distance is identical to the last
```

distance from path_distances list:

    COUNT Euclidean distance between the next potential
    field from first - not an optimal path and the goal
    point

    IF this distance is smaller than the previous counted
    distance which we have got in this iteration:

        DROP the last potential field from optimal_path
        APPEND current field into optimal_path
        APPEND distance into path_distances

    IF this distance is identical as the first
    distance which we have got in this iteration:

        COUNT Euclidean distance between further
        potential field from first - not optimal path
        and the goal point

        IF this distance is smaller than
        previous counted distance which we have got
        in this iteration:

            DROP the last potential field from optimal_path
            APPEND current potential field into optimal_path
            APPEND current distance into path_distances

Fig. 4. The same path after applied filtering for optimization.

2.5   Path Smoothing

After the optimal path from the robot start location to the target is visualized, we initialize the path smoothing algorithm [5], to make our path as optimal as possible. We are applying the algorithm n times, till the result and path shape will not be satisfying:

1. We reduce the distance between points by applying the variable α, which describes how fast we move away from an original position x_k taking into account the previous $x_k - 1$ point and the next $x_k + 1$ point.

$$x_k = x_k + \alpha(x_k - 1 + x_k + 1 - 2x_k)$$

2. Next we have to balance the point x_k by applying the β variable and counting y_k that represents the new position of the point. Because of this operation, we can avoid the straight line in the path.

$$y_k = y_k + \beta(x_k - y_k)$$

Results we can see in Figs. 5 and 6.

Fig. 5. Our previous path after using path smoothing algorithm.

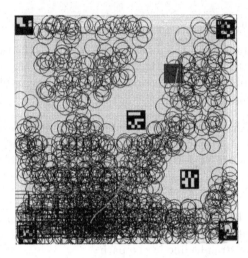

Fig. 6. Path smoothing applied on circle fill algorithm.

3   Experimental Section

In this section, we will discuss the technical side of our experiments finalizing it by sharing a demonstration of our project in a real environment.

3.1   Technical Conditions of the Experiments

We have done real-world testing in a smart robotics laboratory. The key devices were a top camera, which captured the coordinates of the points and a Smart Element Hub cube of the lego robot inventor kit. To control the robot in a semi-autonomous way - where the computer is the computing unit - we used our own library implemented in Python available at [10]. The code responsible for validation is available on Github [11].

3.2   Map Generation Using Augmented Reality Markers

To test the above algorithms we have to create an adapted environment (Fig. 7) for our mobile robot. To realize our conception of a robot's world map we used AR markers, responsible for setting map boundaries, and defining the obstacles and the goal point. Also, the robot location was described as a green square, that we applied on top of the machine. Specific elements on map:

- Boundaries (4 markers)
- Obstacles (2 markers)
- Planned goal (1 marker)
- Robot position (green square)

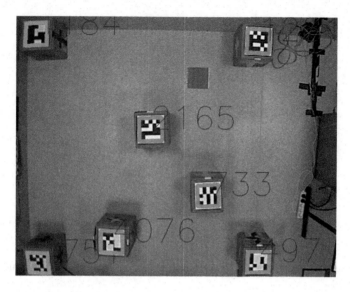

Fig. 7. AR recognition. The green point on a capture describes the start point - robot's actual position. (Color figure online)

To recognize listed elements and get the possibility of defining each position on a virtual map we used Open CV [6] and Python AR markers [9].

After capturing the view from the camera we had to customize our window. Now all the points's coordinates were based on the camera capture pixels and hence map for the robot agent started from (40, 46) instead of (0, 0) point. To change the top left border point we used the following mapping:

```
for i in range(len(x1)):
    a = x1[i] - x_min
    x_borders.append(a)

for i in range(len(y1)):
    a = y1[i] - y_min
    y_borders.append(a)
```

Where x_{min} determines the minimum of X values, y_{min} is the minimum of y value in border points (x, y) (Fig. 8).

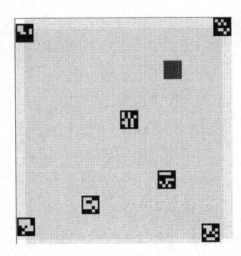

Fig. 8. Results of map generating including the start point as a green square. (Color figure online)

3.3 Connection to the Robot and Basic Features

For our test, we constructed a robot using a Smart Element Hub cube with LED screen of the lego robot inventor kit and servomotors, that use Python language. The most important thing to start working with this robot was to implement a library that contains proper functions [10].

Main features of the robot:

- Motor access (Port A and B)
- Compass sensor access (Build into the Hub)
- Current direction access (Read from Hub)
- Current position (Obtained from camera)
- Determined goal (sequent point on the path to target)

The first step is to create the bluetooth connection between the robot and our device. To do it we had to choose ports: COM3 or COM4 depending on which our robot took. Next, we were obligated to set the robot's position point from a camera, that center of the green point located at the top of the machine was normalized like all borders and obstacles in chapter IX. Because of those actions, the P-controller can be working correctly.

3.4 Controlling the Driving Part of the Robot

The robot that we used to conduct the experiments is equipped with a compass sensor, thanks to which we can capture and fix the current robot direction. Taking into account that our path is created and smoothed including obstacles avoidance, we get the set of coordinates that single variables represents the individual points of the optimized path. Using this set we can reach a goal. This experiment was working in real time and our machine was localized by the top camera using a green color recognition.

The main purpose of designing a P-controller for our robot is to make it move independently, taking into account the generated path. Below we focused on the main steps needed for our steering.

3.4.1 Conversion of the Compass Reading Values

The proposed conversion was based on the current rotation direction of the robot relative to the actual build robot compass sensor value. The spectrum of our calculation is between 0 to 359 and it could be interpreted as the angle of the circle. This idea splits a given set of angles into two sets. Instead of the previous range of values, after applying conversion we have got a spectrum from −180 to 180. The split was made as follows:

```
def convert(k,x):
    x = x - k
    if (x > 180):
        x = x - 360
    if ( x < -180):
        x = 360 + x
    return x
```

Assuming that k is the actual direction that we have to turn relative to the robot's current position getting from the camera and x is the robot's current direction.

After conversion angles from 0 to 180° have consistently positive values, while values from 181 to 359 have values from −179 to −1, respectively. Those values represent the global directions:

– 0° North
– 90° East
– 180° South
– −90° West

3.4.2 Calculation of the Actual Rotation

In the above conversion, we use the actual direction relative to the robot's current direction. To calculate this value we used the following property:

$$direction = \arctan(x - y', y - x') * 180/3.14$$

where (x, y) is the current path element and the (x', y') is the actual robot position on a map.

```
act_dir = math.atan2(path_points_x[i] -
        - actual_y, path_points_y[i] -
        - actual_x) * 180 / 3.14
```

3.4.3   Control System

The control function uses a convert function for angles and wheel speeds, which are considered separately. The algorithm reduces the speed initially given to one of the wheels in order to rotate the robot. The idea of the applied algorithm is to minimize the speed of the selected wheel, depending on calculating the robot's new direction. The device rotates to the right side if the velocity value on the right wheel is less than the velocity value on the left wheel, the same procedure is followed when rotating to the left side.

```
cte = convert(actual_direction,
current_direction)
if cte <= 0:
    if abs(cte) > precision:
        wheel2 = (wheel1 * (-1))
    else:
        wheel2 = (wheel1 * (-1)) -
        (ksi * cte)
if cte > 0:
    if cte > precision:
        wheel1 = (wheel2 * (-1))
else:
    wheel1 = (Wheel2 * (-1)) -
    (ksi * cte)

mc.motor_double_turn_on_deg(HubPortName.A,
HubPortName.B,  wheel1, wheel2,
degrees=actual_direction)
```

Where the wheel1 and wheel2 describe the speed of each robot motors. The ksi determines the corresponding slowdown of the wheel speed, represented by a value range between 0.1 and 0.5.

3.5   Project Presentation and Access to Codes

We can see the action of our project in practice at link [12]. The code of our own library for controlling the robot is at the link [10]. The rest of our project's code is located in the [11] repository. The AR tags we used are available at [9] (Fig. 9).

Fig. 9. The robot used in the experimental part - based on Smart Element Hub cube with LED screen of the lego robot inventor kit

4   Conclusions

In this paper, we have successfully implemented path planning using the rough mereological potential field, combined with a weighted Euclidean distance to the target. We designed and executed a dedicated library for a specific mobile robot. As integral components of the project to shape the map, we utilized virtual reality markers. For the control mechanism, a P-controller was employed. The goal of the work has been achieved, we have adapted the new library and hardware for optimal real-time driving of the mobile robot over a map with obstacles. This work serves as an initial step to the application of the rough mereological potential field for three-dimensional path planning.

References

1. Polkowski, L.: Rough mereology: a new paradigm for approximate reasoning. Int. J. Approximate Reasoning **15**, 333–365 (1996)
2. Osmialowski, P.: Planning and navigation for mobile autonomous robots spatial reasoning in player/stage system (2011, 2022)
3. Osmialowski, P., Polkowski, L.: Spatial reasoning based on rough mereology: a notion of a robot formation and path planning problem for formations of mobile autonomous robots. Trans. Rough Sets **12**, 143–169 (2010)
4. Artiemjew, P.: Rough mereology classifier vs simple DNA microarray gene extraction methods. Int. J. Data Min. Modell. Manage. Spec. Issue: Pattern Recognit. **6**, 110–126 (2014)
5. Zmudzinski, L., Artiemjew, P.: Path planning based on potential fields from rough mereology. In: Polkowski, L., et al. (eds.) IJCRS 2017, Part II. LNCS (LNAI), vol. 10314, pp. 158–168. Springer, Cham (2017). https://doi.org/10.1007/978-3-319-60840-2_11

6. OpenCV. https://opencv.org/
7. Polkowski, L., Zmudzinski, L., Artiemjew, P.: Robot navigation and path planning by means of rough mereology. In: Proceeding of the IEEE International Conference on Robotic Computing (2018)
8. Gnyś, P.: Mereogeometry based approach for behavioral robotics. In: Polkowski, L., et al. (eds.) IJCRS 2017. LNCS (LNAI), vol. 10314, pp. 70–80. Springer, Cham (2017). https://doi.org/10.1007/978-3-319-60840-2_5
9. Python AR markers. https://github.com/DebVortex/python-ar-markers
10. Library for Smart Element Hub cube lego robot inventor kit - Wojciech Cybowski. https://github.com/wcyb/le_mind_controller
11. Project: Aleksandra Szpakowska. https://github.com/aleksandraszpakowska/ Rough_mereology_potential_field_2DAlgorithm
12. Szpakowska, A.: Demonstration of the project - path planning using rough mereological potential field. https://www.youtube.com/watch?v=hUHCbkKCDpY

Handling Intra-class Dissimilarity and Inter-class Similarity for Imbalanced Skin Lesion Image Classification

Shengdan Hu[1], Zhifei Zhang[2(✉)], and Jiemin Yang[1]

[1] Information Center, Shanghai University of Medicine & Health Sciences, Shanghai 201318, China
husd@sumhs.edu.cn
[2] Department of Computer Science and Technology, Tongji University, Shanghai 201804, China
zhifeizhang@tongji.edu.cn

Abstract. Medical image analysis based on deep learning technology has recently attracted much attention. However, it is inappropriate to directly employ the methods that perform well in computer vision. For skin lesion images, the differences between various lesions may be relatively small, and the existing commonly used datasets are class-imbalanced. In this paper, we propose a new method with an augmented loss function that makes use of contrastive information and label information. The proposed method tries to enhance the intra-class similarity and inter-class dissimilarity in the learning procedure. We also apply oversampling on the original data to tackle the imbalance issue. Extensive experiments are conducted on the ISIC2018 and ISIC2019 datasets. The results have demonstrated that, in terms of F1-score and AUC, the proposed method has outperformed the compared methods.

Keywords: skin lesion images · intra-class dissimilarity · inter-class similarity · contrastive loss · imbalanced classification

1 Introduction

Deep learning, an automatic representation learning method from raw data, has applied to many fields [1–3], such as computer vision, natural language processing, recommendation system, medical imaging, etc. Compared with hand-crafted features, representations learned by deep learning usually need less manual intervention and can achieve better performance. In recent years, due to the successful application in various image tasks of computer vision, many deep learning-based techniques have been proposed to tackle medical image analysis [3], and attracted more and more research interests.

However, several deep learning methods cannot produce effective results for medical data because of the following reasons: (1) The methods are naturally accompanied with large labeled datasets, but annotating the medical data by

A. Campagner et al. (Eds.): IJCRS 2023, LNAI 14481, pp. 565–579, 2023.
https://doi.org/10.1007/978-3-031-50959-9_39

human experts is strenuous, time-consuming and subjective in a clinical environment [4]. (2) Many medical datasets are inherently imbalanced, and tend to exhibit a long-tailed label distribution. Classifiers trained on imbalanced data may focus on the majority of samples with a high accuracy rate, and the minority class will be ignored. In fact, the loss of misclassifying the minority class may be much higher than misclassifying the majority class [5]. (3) In some medical datasets, the differences between various images may be relatively small, for instance, chest X-ray images and skin lesion images. Thus, it is challenging to minimize intra-class variance and maximize inter-class variance [6].

To sum up, the approaches to overcome the imbalance issue are divided into two groups. One group consists of data level methods, which change the class distribution of training dataset by oversampling or undersampling to improve the balance directly [7]. One of the most common oversampling forms is random minority oversampling that replicates samples from minority classes randomly. Synthetic Minority Oversampling Technique (SMOTE) [8] is more sophisticated and generates new samples by interpolating. In addition, data augmentation and generative methods [9] are also proposed to generate new samples. Oversampling has been demonstrated to be efficient despite the overfitting problem that may occur [7]. Contrary to oversampling, undersampling refers to reducing samples from the majority class and may destroy the integrity of data. The another group is algorithm level solutions that alter training or inference algorithms (models). Cost sensitive learning [10] is a commonly used algorithm level method that assigns various costs to misclassification of samples from different classes. Then the costs or weights are added to the loss function to mitigate the impact of class imbalance. Usually, a higher weight is given to the loss computed by the samples from the minority classes. Furthermore, methods combining these two groups or applying ensemble algorithms are often used.

As to the intra-class and inter-class differences problem, methods generally focus on the definition of loss for models to maximize inter-class variance and minimize intra-class variance. For instance, triplet loss [11] tries to make the distance between instances from the same class small, and the distance between instances from different class large. Center loss [12] tries to shrink instances of the same class into one point in the feature space and enforces all the intra-class instances to be clustered around a learned class-specific center. Contrastive loss [13] measures the distance between similar and dissimilar sample pairs. In a word, the representations generated by the model can meet the conditions that similar samples are close and different ones are far away, which is the basis of many representation learning loss functions.

In view of the above issues, for skin lesion classification, we define an augmented loss function with contrastive information and label information simultaneously. We also apply data level methods to handle data imbalance. Experiment results illustrate that the proposed method is effective.

The remainder of the paper is organized as follows. Some related works about skin lesion classification are introduced in Sect. 2. Section 3 details the proposed method, including the framework and loss functions. Section 4 illustrates the conducted experiments and results. Section 5 concludes the paper with some discussions.

2   Related Work

Dermoscopy [14] has been widely used for early detection and diagnosis of skin cancer, and there were many clinical studies about diagnosing skin cancers practiced by dermatologists [15]. However, the clinically diagnosing requires qualified and skillful dermatologists to interpret the dermoscopy images. Relatively speaking, it is prone to operator bias and is a time-consuming method. With the development of Computer-Aided Diagnosis Systems, many artificial intelligence-based approaches are used in skin lesion classification, which can produce more reliable results with high efficiency.

The early automated skin cancer classification solutions generally apply low-level hand-crafted features, such as color, shape and texture [16]. Marques et al. [17] put together color features with texture features for skin lesion classification, and found out that color features were not sufficient. Murugan et al. [18] used watershed segmentation method to extract segments, which were subjected to feature extraction, and performed classification using SVM. Hameed et al. [19] proposed a four-step classification framework for skin lesions, and extracted gray-level co-occurrence matrix features in the feature extraction stage. For more robust skin lesion classification, Amelard et al. [20] proposed high-level intuitive features (HLIF), and incorporated them into a set of low-level features to get more semantic meaning. Wahba et al. [21] proposed a novel texture feature, cumulative level-difference mean (CLDM) based on the gray-level difference method (GLDM).

In recent years, the hierarchical feature learning strategies based on deep learning outperform the traditional hand-crafted methods in many medical image analysis tasks [22–25]. Haenssle et al. [22] applied a pre-trained Google's Inception v4 model for melanoma diagnoses. Barata et al. [23] introduced attention modules to identify interpretable features and regions in dermoscopy images, and improved the explainability of a skin cancer diagnostic system. Jojoa Acosta et al. [24] put forward a two-stage process by creating cropped region of interest using Mask and Region-based CNN and classifying the cropped area with ResNet152. Adepu et al. [25] proposed a novel knowledge-distilled lightweight Deep-CNN-based framework to handle the high class-imbalance problem of melanoma classification.

3   Method

3.1   Problem Statement

Take ISIC2018 dataset collected from the International Skin Imaging Collaboration Archive for example, we illustrate the issues of intra-class and inter-class variance and imbalanced classes. There are seven classes of images that are labeled as NV, MEL, BKL, BCC, AKIEC, VASC and DF.

We randomly select two skin lesion images of each class, shown in Fig. 1. Obviously, intra and inter difference are not clear. Some images from the same class differ from each other, such as two images in the class MEL (Fig. 1(b)).

Meanwhile, some images from various classes may look the same, for instance, the images in class BKL and BCC (Fig. 1(c) and Fig. 1(d)). Due to the issues of intra-class dissimilarity and inter-class similarity, it is even difficult for experienced dermatologists to distinguish lesions with naked eyes, which may cause a misdiagnose.

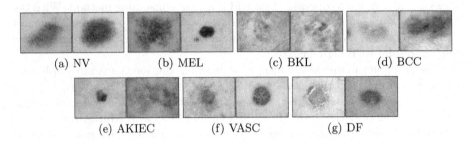

(a) NV   (b) MEL   (c) BKL   (d) BCC

(e) AKIEC   (f) VASC   (g) DF

Fig. 1. Types of dermoscopic skin lesions on ISIC2018.

There are 10,015 dermoscopic images on ISIC2018. Class NV contains 6705 images and is the biggest category that accounts for 66.95%. But the smallest class DF with 115 images accounts for about 1.15%. The ratio of sample numbers for these two classes is about 58. The proportion of class MEL, BKL, BCC, AKIEC and VASC are 11.11%, 10.97%, 5.13%, 3.27% and 1.42% respectively. As can be seen from these percentages, the dataset is imbalanced.

3.2   Network Architecture

To solve the similarity and imbalance problem of skin lesion image classification, we propose a new method and the framework is illustrated in Fig. 2.

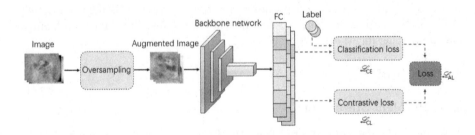

Fig. 2. General framework of the proposed method.

The class distribution is balanced by oversampling firstly. The refined images with data augmentation are passed to a CNN architecture based on a backbone network. The cross-entropy loss is calculated by the feature vectors in the full connection layer and labels. While the contrastive loss is calculated by feature vectors of positive and negative pairs. We combine them and get an augmented loss.

3.3 Contrastive Loss

The contrastive learning approaches have recently been proposed to develop robust representations from the input data. The core concept is to maximize the agreement between a pair of similar pairs or the distinction between the positive and negative pairs. Typically, a contrastive model is trained by the contrastive loss to maximize the similarity between the positive pairs and minimize it with the negative pairs. This is much similar to the ideas of clustering or autoencoder.

Let $X_1, X_2 \in \mathcal{X}$ be a pair of input instances, and Y be a binary label assigned to the pair. If X_1, X_2 are deemed similar, set $Y = 0$, and if X_1, X_2 are deemed dissimilar, set $Y = 1$. The initial contrastive loss [13] is defined as follows:

$$\mathcal{L}(W) = \sum_{i=1}^{P} L(W, (Y, X_1, X_2)^i), \tag{1}$$

$$L(W, (Y, X_1, X_2)^i) = (1 - Y)L_S(D_W^i) + Y L_D(D_W^i), \tag{2}$$

where $(Y, X_1, X_2)^i$ refers to the i-th labeled sample pair, L_S is the partial loss function for a pair of similar instances, L_D is the partial loss function for a pair of dissimilar instances, P is the number of training pairs, and D_W is the parameterized distance function to be learned between the outputs of X_1, X_2, which is defined as:

$$D_W = \|G_W(X_1) - G_W(X_2)\|^2, \tag{3}$$

where $G_W(X_1)$ is the model output of X_1.

The loss functions of constrastive learning methods are mainly based on InfoNCE or its variants [26,27], and can be defined as:

$$\mathcal{L}_{InfoNCE} = \frac{1}{N} \sum_{i=1}^{N} - \log \frac{exp(sim(V_i, V_i^+)/\tau)}{exp(sim(V_i, V_i^+)/\tau) + \sum_{j=1}^{M} exp(sim(V_i, V_j^-)/\tau)}, \tag{4}$$

where V_i is the feature representation of X_i, V_i^+ is the feature representation of positive sample, V_j^- is the feature representation of negative sample, N is the number of query samples, M is the number of negative samples, τ is the temperature hyper-parameter and $sim(\cdot)$ denotes the cosine similarity function.

Inspired by the principle of above definitions, we optimize the loss function of model with contrastive information and label information and can make full use of feature representations and labels. Suppose that, within a minibatch, all samples from the same class are considered positive pairs, and from different classes are negative pairs. The proposed augmented loss function is expressed as:

$$\mathcal{L}_{AL} = \mathcal{L}_{CE} + \gamma \mathcal{L}_{CL}, \tag{5}$$

wherein γ is a hyper-parameter that can adjust the weight of $\mathcal{L}_{CL}$. $\mathcal{L}_{CE}$ is the cross-entropy loss function, and

$$\mathcal{L}_{CE} = -\sum_{i=1}^{n} y_i log(\hat{y}_i), \tag{6}$$

y_i is the true label of X_i, $\hat{y}_i$ is the prediction label of X_i. $\mathcal{L}_{CL}$ is the contrastive loss function, which is defined as:

$$\mathcal{L}_{CL} = \frac{\sum_{y_i = y_j} \|V_i - V_j\|^2}{\sum_{i=1}^{n}\sum_{j=1}^{n} \|V_i - V_j\|^2}, \tag{7}$$

where the numerator refers to the distance of feature representation for instances which belong to the same class. So the learning procedure relies partly on the neighborhood relationships provided by labels.

4   Experiments

In this section, we conduct some experiments on two public imbalanced datasets with the following purposes: (1) to compare the results with different backbone networks and inspect which model has been improved, (2) to find out the outcomes of testing set with different data sampling methods, (3) to compare the classification performance of different methods handling the similarity and imbalance problem of skin lesion image classification.

4.1   Experimental Setup

4.1.1   Datasets

To evaluate the proposed method, we performed classification task on two skin lesion analysis challenge datasets: ISIC2018 [28] and ISIC2019 [28–30], and the datasets are subsets of data collected from the International Skin Imaging Collaboration (ISIC) Archive.

- ISIC2018 (main dataset): consists of 10,015 RGB dermoscopic images of the size 600 × 450 from HAM10000 dataset. It is divided into seven classes, and the distribution is shown in Fig. 3(a).
- ISIC2019: consists of 25,331 RGB dermoscopic images from HAM10000 and BCN20000 datasets. The sizes of majority images are 600 × 450 and 1024 × 1024. There are eight classes in total and the distribution is shown in Fig. 3(b).

We can observe that the two datasets are imbalanced, and exhibit a long-tailed distribution.

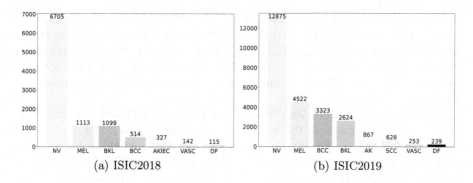

Fig. 3. Distributions of samples on ISIC2018 and ISIC2019.

4.1.2   Evaluation Metric

For the classification task, accuracy is usually be used to evaluate the performance. But when the dataset is imbalanced, high overall accuracy may not reveal the accuracy of minor classes. In this study, we have used several metrics [31] to assess the performance of various methods: accuracy, precision, sensitivity (recall), F1-score and area under a receiver operating characteristic curve (AUC). In this study, for multi-class classification on ISIC2018 and ISIC2019, we apply the macro sensitivity, precision and F1-score to evaluate the overall classification performance.

$$Accuracy(ACC) = \frac{TP + TN}{TP + FP + FN + TN}, \tag{8}$$

$$Precision(PRE) = \frac{TP}{TP + FP}, \tag{9}$$

$$Sensitivity(SEN) = \frac{TP}{TP + FN}, \tag{10}$$

$$F1 - score = \frac{2TP}{2TP + FN + FP}. \tag{11}$$

These metrics are derived from a 2 × 2 confusion matrix. On ISIC2018 and ISIC2019 datasets, the multi-class classification problem is transformed into a binary classification problem, namely, target class and non-target class. TP stands for true positive, which is the number of target class samples that are correctly predicted. FN, false negative, represents the number of samples that are wrongly predicted as belonging to non-target class. TN, true negative, refers to the number of non-target class samples that are correctly predicted. FP, false positive, is the number of samples that are wrongly predicted as target class.

4.1.3   Experiment Settings

Experiments are conducted using a computer with Intel Core i7-8700 CPU (3.2 GHz), 16 GB DDR4 RAM, and an NVIDIA GeForce GTX 1080 Ti GPU with 11 GB memory. The proposed method has been implemented in PyTorch on Ubuntu 18.04. ResNet50, ResNet101, DenseNet121 and DenseNet201 are chosen as the backbone networks. For a fair comparison, we conduct experiments under the same experiment conditions. All models were trained for 50 epochs with the static learning rate of 1e−4. Adam optimizer was used and the batch size was set to 32.

We divide each dataset into 70% for training, 10% for validation, and 20% for testing. All images from the two datasets are adjusted to (112 × 112) in the experiment. However, due to the large-scale iterations and the computational limitations, the images are resized to (28 × 28) applied to SMOTE in the second experiment. All the images are normalized. Moreover, during training, we apply data augmentation including random rotation, cropping, horizontal and vertical flips.

4.2   Results

4.2.1   Performance with Different Backbones

Before investigating the balancing techniques, four models are trained on the imbalanced skin lesion dataset ISIC2018. All the models are trained with the standard cross-entropy loss function. The training loss and validation loss of 25 epochs are shown in Fig. 4. We can see that, the validation loss decreases as the training loss decreases during the training procedure. After 50 epochs, the classification outcomes on testing set are shown in Table 1.

Table 1. Classification results with different backbones on ISIC2018.

Models	ACC	PRE	SEN	F1-score	AUC
ResNet50	0.8481	0.7922	0.6701	0.7260	0.8593
ResNet101	0.8432	0.8054	0.6739	0.7324	0.8528
DenseNet121	0.8470	**0.8231**	0.7028	0.7582	0.8546
DenseNet201	**0.8570**	**0.8231**	**0.7198**	**0.7679**	**0.8609**

We can see that the results of all pre-trained models are approximate, and DenseNet201 achieves the highest accuracy, precision, sensitivity, F1-score, and AUC, which are 1.63%, 3.9%, 7.4%, 5.77%, 0.95% higher than the lowest results respectively. In particular, classification accuracy and AUC of the four models slightly change. Compared to ResNet50, ResNet101 provides tiny inferior outcomes of accuracy and AUC, but superior results of precision, sensitivity, F1-score. Compared to DenseNet121, DenseNet201 provides better outcomes of the five metrics. As for precision, sensitivity, F1-score, the DenseNet models yield better results than the ResNet models. The depth of models may affect the outcomes, but the change is little.

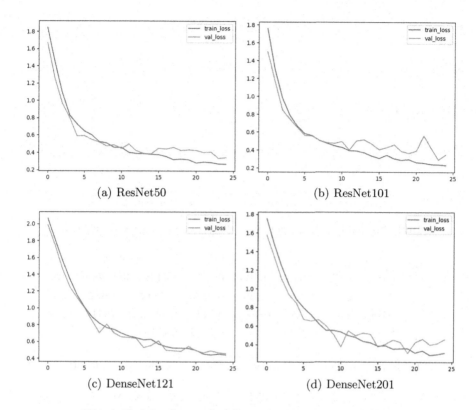

Fig. 4. Training loss with different backbones on ISIC2018.

4.2.2 Results on Different Sampling Methods

We apply three different data level approaches to change the class distribution of training dataset and evaluate the test results. The chosen data sampling methods are RandomOverSampler (ROS), RandomUnderSampler (RUS) and SMOTE from imblearn package. The sample numbers of each class after sampling are shown in Fig. 5. On ISIC2018, the sample number of each class when employing oversampling methods is 4698, while undersampling is 73, and their ratio is close to 65. So, the data fed in the subsequent training procedure vary widely in terms of quantity. Similarly, the ratio on ISIC2019 is 8986:165≈55.

As the experiment results in Sect. 4.2.1 are slightly different, we choose ResNet101 as the backbone, and use the standard cross-entropy loss function (CE). Table 2 and Table 3 display the test results on ISIC2018 and ISIC2019 respectively after employing different sampling methods on training set. While, the first row in these tables apply the original training set as the baseline for comparison. For the sake of intuition, we depict the results of Table 2 and Table 3 in line graphs, as illustrated in Fig. 6.

On ISIC2018, the results of oversampling (ROS+CE) are superior to that of baseline in terms of sensitivity, F1-score and AUC, but inferior on accuracy

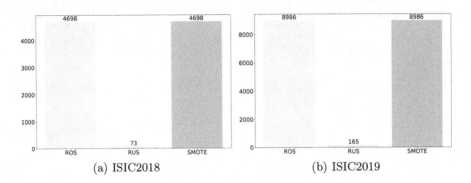

(a) ISIC2018　　　　　　　　(b) ISIC2019

Fig. 5. Sample numbers of each class after sampling.

Table 2. Classification results with different samplings on ISIC2018.

Method	ACC	PRE	SEN	F1-score	AUC
CE	**0.8432**	**0.8054**	0.6739	0.7184	0.8528
ROS+CE	0.8402	0.7303	**0.7580**	**0.7317**	**0.8632**
RUS+CE	0.7190	0.5504	0.7127	0.6211	0.7234
SMOTE+CE	0.8030	0.7315	0.7126	0.7219	0.8236

Table 3. Classification results with different samplings on ISIC2019.

Method	ACC	PRE	SEN	F1-score	AUC
CE	0.8161	**0.7635**	0.7188	0.7395	**0.8721**
ROS+CE	**0.8206**	0.7551	**0.7613**	**0.7557**	**0.8721**
RUS+CE	0.6206	0.4369	0.6187	0.4767	0.7596
SMOTE+CE	0.7468	0.6531	0.6623	0.6610	0.8039

and precision. But the results of undersampling (RUS+CE) are much lower than that of baseline on accuracy, precision, F1-score, and AUC. As to SMOTE, the outcomes of accuracy, sensitivity, F1-score and AUC are between that of oversampling and undersampling. Also, it provides inferior results of accuracy, precision and AUC compared to baseline. We can observe the similar phenomena on ISIC2019. Specifically, the results of oversampling (ROS+CE) are superior to that of baseline in terms of accuracy, sensitivity, F1-score and AUC, inferior on precision, equivalent on AUC. All the five metrics of undersampling are lower than that of baseline, and SMOTE are between oversampling and undersampling.

Generally speaking, RandomOverSampler to balance the training set by replicating samples from minority classes can get slightly better outcomes than baseline. The results are much worse if RandomUnderSampler is used to reduce samples from the majority class. It is consistent with that more data are needed for deep learning. Due to the computational complexity of SMOTE and experiment environment limitation, we resize images to (28×28) when using SMOTE to generate new samples. Thus, the outcomes are usually inferior to that of ROS in which images are (112×112). However, in some literatures, it is reported that SMOTE is more effective than ROS when the images are of the same size.

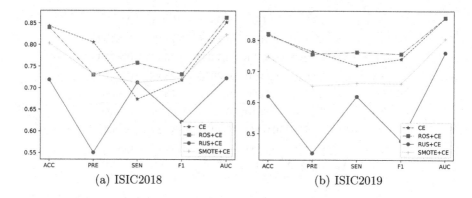

(a) ISIC2018   (b) ISIC2019

Fig. 6. Results of different sampling methods.

4.2.3   Results with Different Loss Functions

In this section, experiments are conducted to compare different loss functions, such as CE, Center Loss (CL) [12] that focuses on the similarity problem, Focal Loss (FL) [25,32] that emphasizes on the imbalance problem, and the proposed augmented loss (AL). For CL, the feature dimension is set to equal to the number of classes, and other parameters are the same as [12]. As to FL, the weights of each class are set according to the sample numbers of each class, and other parameters are the same as [32]. The parameter γ of Eq. (5) is set to 2. Because RandomOverSampler performs better in the previous experiment, we also apply

Table 4. Classification results with different losses on ISIC2018.

Method	ACC	PRE	SEN	F1-score	AUC
CE	0.8432	0.8054	0.6739	0.7184	0.8528
ROS+CE	0.8402	0.7303	**0.7580**	0.7317	0.8632
CL	0.8306	**0.8357**	0.6422	0.6982	0.8030
ROS+CL	0.8453	0.7679	0.7424	0.7519	0.8661
FL	0.8185	0.7277	0.7566	0.7391	0.8599
ROS+FL	0.8558	0.7632	0.7401	0.7468	0.8736
AL	**0.8573**	0.7941	0.7200	0.7518	0.8712
ROS+AL	0.8543	0.7724	0.7398	**0.7554**	**0.8784**

it to alter the distribution of training set and then inspect the test results with different losses. The experiment results on ISIC2018 and ISIC2019 are detailed in Table 4 and Table 5 respectively, also shown in Fig. 7.

Table 5. Classification results with different losses on ISIC2019.

Method	ACC	PRE	SEN	F1-score	AUC
CE	0.8161	0.7635	0.7188	0.7395	**0.8721**
ROS+CE	0.8206	0.7551	**0.7613**	0.7557	**0.8721**
CL	0.8026	0.8010	0.5944	0.6389	0.8556
ROS+CL	0.7884	0.7666	0.7211	0.7383	0.8521
FL	0.7518	0.6310	0.7203	0.6676	0.8333
ROS+FL	0.7765	0.6909	0.7466	0.7123	0.8529
AL	0.8176	0.7584	0.7329	0.7447	0.8698
ROS+AL	**0.8263**	**0.8169**	0.7536	**0.7848**	0.8703

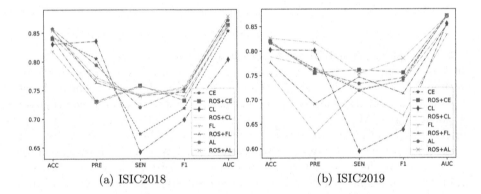

(a) ISIC2018   (b) ISIC2019

Fig. 7. Results with different loss functions.

On ISIC2018, the accuracy, sensitivity, F1-score and AUC of AL are 1.67%, 6.84%, 4.65% and 2.16% higher than that of CE respectively, while the precision is slightly lower. The accuracy, F1-score and AUC of AL are all higher than that of CL and FL. If we apply the oversampling technology first, the sensitivity, F1-score and AUC of ROS+AL are slightly higher than that of AL. Especially, the F1-score and AUC of ROS+AL are the best of all. Furthermore, the accuracy and precision are lower than that of AL. On ISIC2019, the accuracy, sensitivity and F1-score of AL are 0.18%, 1.96%, 0.70% higher than that of CE respectively, while the precision and AUC are 0.67% and 0.26% lower. As to ROS+AL, the five metrics are all higher than that of AL, and the increased percentages are 1.06%, 7.71%, 2.82%, 5.38% and 0.06% respectively.

In summary, in terms of F1-score and AUC, the proposed method is better than other losses except the AUC of CE on ISIC2019. It is proved that for skin lesion images, when we take intra-class and inter-class differences into account, the classification results will be improved to some extent. Therefore, the proposed method is effective.

5 Conclusion

In this paper, for the problem of small intra-class and inter-class differences among skin lesion images, we design an augmented loss function by adding contrastive loss, and for the imbalance problem, we apply oversampling to duplicate images of minority class. Experiment outcomes indicate the proposed method has outperformed the compared losses. In experiments, we find out that less images as in RUS, smaller ones as in SMOTE will yield worse classification outcomes. In the future, in order to further promote the performance of skin lesion image classification, more advanced data balancing strategy or diverse scale images from the view of multi-granularity can be focused on.

Acknowledgements. The authors would like to thank the anonymous referees for their constructive comments that help improve the manuscript. This research was supported by the National Nature Science Foundation of China (Grant No. 62076040), the Natural Science Foundation of Shanghai (Grant No. 22ZR1466700), and the Foundation of Shanghai University of Medicine & Health Sciences (Grant No. A1-2601-23-311007-25, E1-0200-23-201009-17).

References

1. Li, Z.J., Miao, D.Q.: Sequential end-to-end network for efficient person search. In: Proceedings of the AAAI Conference on Artificial Intelligence, vol. 35, no. 3, pp. 2011–2019 (2021)
2. Dou, S.G., Zhao, C.R., Jiang, X.Y., et al.: Human co-parsing guided alignment for occluded person re-identification. IEEE Trans. Image Process. **32**, 458–470 (2022)
3. Chen, X.X., Wang, X.M., Zhang, K., et al.: Recent advances and clinical applications of deep learning in medical image analysis. Med. Image Anal. **79**, 102444 (2022)
4. Gao, Z.Y., Hong, B.Y., Li, Y., et al.: A semi-supervised multi-task learning framework for cancer classification with weak annotation in whole-slide images. Med. Image Anal. **83**, 102652 (2023)
5. Georgios, D., Fernando, B., Felix, L.: Improving imbalanced learning through a heuristic oversampling method based on k-means and SMOTE. Inf. Sci. **465**, 1–20 (2018)
6. Sivapuram, A.K., Ravi, V., Senthil, G., et al.: VISAL-A novel learning strategy to address class imbalance. Neural Netw. **161**, 178–184 (2023)
7. Buda, M., Maki, A., Mazurowski, M.A.: A systematic study of the class imbalance problem in convolutional neural networks. Neural Netw. **106**, 249–259 (2018)
8. Chawla, N.V., Bowyer, K.W., Hall, L.O., et al.: SMOTE: synthetic minority oversampling technique. J. Artif. Intell. Res. **16**, 321–357 (2002)

9. Mullick, S.S., Datta, S., Das, S.: Generative adversarial minority oversampling. In: Proceedings of the IEEE/CVF International Conference on Computer Vision, pp. 1695–1704 (2019)

10. Elkan, C.: The foundations of cost-sensitive learning. In: Proceedings of the International Joint Conference on Artificial Intelligence, vol. 17, no. 1, pp. 973–978 (2001)

11. Schroff, F., Kalenichenko, D., Philbin, J.: FaceNet: a unified embedding for face recognition and clustering. In: Proceedings of the IEEE/CVF Conference on Computer Vision and Pattern Recognition, pp. 815–823 (2015)

12. Wen, Y., Zhang, K., Li, Z., Qiao, Yu.: A discriminative feature learning approach for deep face recognition. In: Leibe, B., Matas, J., Sebe, N., Welling, M. (eds.) ECCV 2016. LNCS, vol. 9911, pp. 499–515. Springer, Cham (2016). https://doi.org/10.1007/978-3-319-46478-7_31

13. Hadsell, R., Chopra, S., LeCun, Y.: Dimensionality reduction by learning an invariant mapping. In: Proceedings of the 2006 IEEE Computer Society Conference on Computer Vision and Pattern Recognition, vol. 2, pp. 1735–1742 (2006)

14. Binder, M., Schwarz, M., Winkler, A., et al.: Epiluminescence microscopy: a useful tool for the diagnosis of pigmented skin lesions for formally trained dermatologists. Arch. Dermatol. **131**(3), 286–291 (1995)

15. Henning, J.S., Dusza, S.W., Wang, S.W., et al.: The CASH (color, architecture, symmetry, and homogeneity) algorithm for dermoscopy. J. Am. Acad. Dermatol. **56**(1), 45–52 (2007)

16. Ballerini, L., Fisher, R.B., Aldridge,B., et.al.: A color and texture based hierarchical K-NN approach to the classification of non-melanoma skin lesions. In: Celebi, M., Schaefer, G. (eds.) Color Medical Image Analysis. Lecture Notes in Computational Vision and Biomechanics, vol. 6, pp. 63–86. Springer, Dordrecht (2013). https://doi.org/10.1007/978-94-007-5389-1_4

17. Marques, J.S., Barata, C., Mendonca, T.: On the role of texture and color in the classification of dermoscopy images. In: Proceedings of the 2012 Annual International Conference of the IEEE Engineering in Medicine and Biology Society, pp. 4402–4405 (2012)

18. Murugan, A., Nair, S.A.H., Kumar, K.: Detection of skin cancer using SVM, random forest and kNN classifiers. J. Med. Syst. **43**(8), 1–9 (2019)

19. Hameed, N., Hameed, F., Shabut, A., et al.: An intelligent computer-aided scheme for classifying multiple skin lesions. Computers **8**(3), 62 (2019)

20. Amelard, R., Wong, A., Clausi, D.A.: Extracting morphological high-level intuitive features (HLIF) for enhancing skin lesion classification. In: Proceedings of the 2012 Annual International Conference of the IEEE Engineering in Medicine and Biology Society, pp. 4458–4461 (2012)

21. Wahba, M.A., Ashour, A.S., Guo, Y., et al.: A novel cumulative level difference mean based GLDM and modified ABCD features ranked using eigenvector centrality approach for four skin lesion types classification. Comput. Methods Programs Biomed. **165**, 163–174 (2018)

22. Haenssle, H.A., Fink, C., Schneiderbauer, R., et al.: Man against machine: diagnostic performance of a deep learning convolutional neural network for dermoscopic melanoma recognition in comparison to 58 dermatologists. Ann. Oncol. **29**(8), 1836–1842 (2018)

23. Barata, C., Celebi, M.E., Marques, J.S.: Explainable skin lesion diagnosis using taxonomies. Pattern Recogn. **110**, 107413 (2021)

24. Jojoa Acosta, M.F., Caballero Tovar, L.Y., Garcia-Zapirain, M.B., et al.: Melanoma diagnosis using deep learning techniques on dermatoscopic images. BMC Med. Imaging **21**(1), 1–11 (2021)
25. Adepu, A.K., Sahayam, S., Jayaraman, U., et al.: Melanoma classification from dermatoscopy images using knowledge distillation for highly imbalanced data. Comput. Biol. Med. **154**, 106571 (2023)
26. He, K.M., Fan, H.Q., Wu, Y.X., et al.: Momentum contrast for unsupervised visual representation learning. In: Proceedings of the IEEE/CVF Conference on Computer Vision and Pattern Recognition, pp. 9729–9738 (2020)
27. Peng, Z., Tian, S.W., Yu, L., et al.: Semi-supervised medical image classification with adaptive threshold pseudo-labeling and unreliable sample contrastive loss. Biomed. Signal Process. Control **79**, 104142 (2023)
28. Tschandl, P., Rosendahl, C., Kittler, H.: The HAM10000 dataset, a large collection of multi-source dermatoscopic images of common pigmented skin lesions. Sci. Data **5**(1), 1–9 (2018)
29. Codella, N.C.F., Gutman, D., Celebi, M.E., et al.: Skin lesion analysis toward melanoma detection: a challenge at the 2017 international symposium on biomedical imaging (ISBI), hosted by the international skin imaging collaboration (ISIC). In: Proceedings of the 2018 IEEE 15th International Symposium on Biomedical Imaging, pp. 168–172 (2018)
30. Combalia, M., Codella, N.C.F., Rotemberg, V., et al.: BCN20000: dermoscopic lesions in the wild. arXiv:1908.02288 (2019)
31. Sokolova, M., Lapalme, G.: A systematic analysis of performance measures for classification tasks. Inf. Process. Manage. **45**(4), 427–437 (2009)
32. Lin, T.Y., Goyal, P., Girshick, R., et al.: Focal loss for dense object detection. In: Proceedings of the IEEE International Conference on Computer Vision, pp. 2980–2988 (2017)

Link Prediction for Attribute and Structure Learning Based on Attention Mechanism

Renjuan Nie[✉], Guoyin Wang, Qun Liu, and Chengxin Peng

Chongqing Key Laboratory of Computational Intelligence, Chongqing University of Posts and Telecommunications, Chongqing 400065, People's Republic of China
{s210231147,s210231149}@stu.cqupt.edu.cn, {wanggy,liuqun}@cqupt.edu.cn

Abstract. Link prediction is an important research direction in complex network analysis, which aims to infer the likelihood of future connections between pairs of nodes in the network that have not yet produced edges. In real life, many relationships can be described through the network, and many practical problems can be transformed into link prediction problems. However, existing link prediction methods either rely heavily on network structure information or cannot effectively integrate network structure information and attribute information. To solve the above problems, this paper proposes a new link prediction framework ASLAM based on the attention mechanism and the adaptive extraction strategy of attribute information and structural information. Specifically, ASLAM first constructs the node semantic representation with the attribute attention mechanism, then constructs the node structure representation with the neighborhood attention mechanism, and finally makes the final prediction by adaptive merging semantic representation and structure representation. Experiments on nine real datasets show that the proposed algorithm has a great improvement in performance compared with the baseline method. Compared with the 11 baseline methods, AUC and AP increased by 0.08%–3.1% and 0.05%–3.4%, respectively, indicating the superiority of the ASLAM method.

Keywords: Link prediction · Graph neural networks · Attention mechanism

1 Introduction

The internet exists in many real-world societies, physics, and information systems. Link prediction aims to predict whether two nodes in the network have links [20,31] through some available topology and attribute information in the network, which has attracted a lot of research work and has broad application significance. For example, it can save the workload of blindly checking interactions in the protein network [2], improve the recommendation services in social networks [1,32] and e-commerce networks [17], help complete the information in the Knowledge graph [23,24], and analyze the infection of epidemic diseases [3,6].

© The Author(s), under exclusive license to Springer Nature Switzerland AG 2023
A. Campagner et al. (Eds.): IJCRS 2023, LNAI 14481, pp. 580–595, 2023.
https://doi.org/10.1007/978-3-031-50959-9_40

According to the technology involved, current link prediction methods can be divided into three categories: heuristic feature based methods, embedding based methods, and graph neural network (GNN) based methods. The method based on heuristic features infers the possibility of links by manually creating similarity measures about structural information. Local similarity indicators such as Common Neighbor (CN), Adamic Adar (AA), Local Path (LP), and Preferential Attachment (PA) are used to predict [19,21,40]. The embedding based method learns node embedding based on node link information, and uses the learned node embedding to calculate the similarity score between nodes [10,36]. The GNN based method expresses link prediction as a binary classification problem that can fuse explicit node features [22,37,39].

The two main types of link prediction methods based on GNN are GAE [16] and SEAL [39]. It is observed from experience that GAE, which relies heavily on smooth node characteristics, performs poorly on data sets with highly hierarchical layout, which limits its application in many realistic scenarios with obvious tree structure [7,34], such as disease transmission networks. SEAL uses specific labeling techniques to explicitly encode the structural information of closed subgraphs around each link, and then applies graph level GNN to demonstrate stronger link inference ability. However, SEAL only learns network topology information, resulting in limited prediction performance. Compared with previous research, BSAL [18] proposed the joint learning of node attributes and network topology, providing a promising approach for link prediction. However, due to the fact that node attributes and node structure matrices come from two different fields and are far away in terms of information format and dimensions, directly combining node feature information matrices with structure matrices does not always improve performance. These methods cannot effectively integrate topology and attribute information for link prediction.

Secondly, although GNN based methods exhibit significant performance by aggregating the features of adjacent nodes to generate the features of the central node. However, real-world graphs often have connections between unrelated nodes and noise from unrelated attributes, which leads to suboptimal representations of GNN learning. Graph Attention Networks (GAT) [30] use self attention to alleviate this problem. Graph attention captures the relational importance of a graph, in other words, each neighbor of the central node has a different degree of importance. SuperGAT [14] proposed that the homogeneity and average degree of graphs can affect the finiteness of self supervision, and learned more expressive attention in distinguishing neighbors with incorrect links.

To address the limitations of GAE and SEAL, as well as the noise problem of graph network data. This article proposes an attribute and structure learning framework ASLAM based on attention mechanism. Specifically, the framework constructs a semantic topology based on node features, and then combines the node attribute attention mechanism to obtain a structural level semantic representation. Secondly, subgraph topology is constructed through node degree, distance, and semantic representation, and structural representation is obtained through GNN learning. Finally, the adaptive fusion of structural and semantic representations is used to measure the existence of links. This article demonstrates the effectiveness and superiority of the ASLAM method, which is significantly superior to various research baselines.

The main contributions of this article are as follows:

1. In order to solve the problem of GNN heavily relying on node smoothing features, this article adopts a specific labeling method. Learn the structural features of a graph by considering node degree and distance.
2. In order to learn the attribute features of nodes, it is proposed to transform node attribute features into structural level semantic representations and use Shapley values as the importance values of attributes.
3. A link prediction framework based on attention mechanism for structure and attribute learning (ASLAM) was designed, which learns useful features from the network's attribute and structural information and adaptively integrates them for link prediction.
4. The experimental results on 9 real datasets demonstrate that the proposed model outperforms other baseline models. In terms of link prediction tasks, compared with the baseline model, AUC and AP have improved by 0.08% to 3.1% and 0.05% to 3.4%, respectively. At the same time, the effectiveness of the proposed model has been demonstrated in ablation analysis experiments.

We organized our paper as follows. Section 2 briefly overviews the works related link prediction methods, and Sect. 3 describes the notations used in this paper and some preliminary definitions. Then the ASLAM framework is illustrated in Sect. 4, and the experimental results and detailed analysis report are presented in Sect. 5. Finally, Sect. 6 gives the conclusions and discussions of the future works in this field. Code and model are available at: https://github.com/juanajuan5826/ASLAM.

2   Related Works

The existing link prediction methods can generally be divided into three categories: heuristic feature based methods, embedding based methods, and GNN based methods.

2.1   Method Based on Heuristic Features

Most heuristic feature methods utilize neighborhood information to measure the similarity of nodes, which is simple and effective. Popular heuristic methods include: Common Neighbor (CN), Adam Adar (AA), etc. CN assumes that the more common neighbors two nodes have, the more likely they are to generate connections. This method is effective in the social field, but in protein interactions, two proteins with common neighbors are likely not to interact. Unlike CN, AA not only considers the number of common neighbors owned by two nodes, but also considers the role differences of common neighbors, weakening the contribution of the larger nodes in the common neighbors. Although heuristic methods can achieve good results in practice, they are designed based on handmade metrics and may not be suitable for different scenarios, and these heuristic methods often cannot capture complex potential features.

2.2   Method Based on Embedding

The embedding based method aims to learn potential node features, learn node embeddings based on the links between nodes, and calculate the similarity score between nodes through the learned node embeddings. LINE [29] preserves the first-order and second-order similarity of nodes for probabilistic modeling, and improves the efficiency of learning node representations on large-scale networks through negative sampling algorithms. Embedding methods based on random walks, such as DeepWalk [25] and Node2vec [10], obtain node sequences through specific walk strategies, and then input them into Skip Gram [25] to learn vector representations of nodes. Due to the fact that the performance of embedding based methods depends on the sparsity of the input graph, it is difficult to consider these methods as generalized methods.

2.3   Method Based on GNN

GNN learns low dimensional representations of nodes or graphs by iteratively aggregating the features of neighbors using nonlinear transformations. Through this approach, GNN shows significant performance improvement compared to traditional methods such as heuristic feature based methods and embedding based methods [37]. GAE and VGAE [16] learn node representations through GCN [15] to reconstruct input maps in the autoencoder framework. Similar to GAE, GraphSage [11] uses two graph convolutional layers to encode node features, preserving the relationship information between the two nodes through negative sampling. On the one hand, various GNN architectures based on GAE have been applied to link prediction. On the other hand, SEAL redefines the link prediction task as the classification of closed subgraphs. It does not directly predict the existence of connections, but rather samples closed subgraphs around each target link to form a dataset and perform graph classification tasks. BSAL considers the fusion of node features and node structure on the basis of SEAL. Due to the explicit encoding of structural information, even though both are GNN based methods, SEAL and other subgraph paradigm based methods have significant advantages over GAE in most cases.

Secondly, although GNN based methods exhibit significant performance by aggregating the features of adjacent nodes to generate the features of the central node. However, real-world graphs often have connections between unrelated nodes and noise from unrelated attributes, which leads to suboptimal representations of GNN learning. GAT uses attention mechanism to determine the importance of each neighboring node to the central node, which is the weight, when aggregating the neighbor information of nodes. SuperGAT has improved the graph attention model for noise maps, specifically by utilizing two attention forms compatible with self supervised tasks to predict edges. SuperGAT has learned more expressive attention in distinguishing neighbors with incorrect links. The improvement in performance of GATs also highlights the importance of capturing attention to graph relationships.

3 Preliminary

This section aims to provide some preliminary content to make this article easier to understand. Some important definitions and concepts are introduced below.

3.1 Networks

A network is defined graph $G = (V, E)$.where $V = \{v_1, v_2, \ldots, v_N\}$ is set of nodes, The total number of nodes is N, $E \subseteq V \times V$ is set of edges.The structure of the graph is usually represented by an adjacency matrix A,where $A_{i,j} = 1$ if $(i, j) \in E$ and $A_{i,j} = 0$ otherwise. If the network is undirected, A is symmetrical. In this paper, we mainly focus on undirected networks, but the proposed model is flexible and can be generalized to directed networks.

3.2 Neighborhood

For any nodes $v_i, v_j \in V$, let $\mathcal{N}^h(v_i)$ be the h-hop neighbors of v_i, and $d(v_i, v_j)$ be the shortest path distance between v_i and v_j. $\mathcal{N}^h(v_i)$ is the set of nodes to $d(v_i, v_j) \leq h$. We call v_i the central node, $v_j \in \mathcal{N}^h(v_i)$ is h-hop neighbors of v_i. In the link prediction scenario, a given target node pair (v_i, v_j), define the h-hop neighbor node v_k of the target node pair, $\{v_k \mid v_k \in \mathcal{N}^h(v_i) \quad or \quad v_k \in \mathcal{N}^h(v_j)\}$.

3.3 Link Prediction

Link prediction problems are divided into time series link prediction, which predicts potential new links in the evolved network, and fabric link prediction, which infers missing links in the static network. In this article, we mainly focus on structural link prediction. Given the partially observed structure of the network, our goal is to predict unobserved links. Formally, given a partially observed network $G = (V, E)$, We represent the collection of node pairs with unknown link states as $E^? = V \times V - E$, then the goal of fabric link prediction is to infer $E^?$.

3.4 Shapley Value

Shapley values are a method derived from cooperative game theory [27] created by Shapley in 1953 as a method of allocating spending to players based on their contribution to total spend, and players cooperate in leagues and receive certain benefits from this cooperation. Used in graph networks, "alliance" is the node instance in the network, "player" is the feature value of the node, "total expenditure" is the model prediction value of a single instance of the data set, and the shapley value is the importance of node attributes to prediction. Here, we use the Shapley value as the attention value to guide the prediction of the model. The Shapley value for each eigenvalue is the contribution of that eigenvalue to the node, obtained by weighting and summing all possible eigenvalue combinations:

$$\phi_j = \sum_{S \subseteq \{x_1,\ldots,x_P\} \setminus \{x_j\}} \frac{|S|!(p-|S|-1)!}{p!} \left(f_x \left(S \cup \{x_j\} \right) - f_x(S) \right), \qquad (1)$$

where S is a subset of the input features, $\{x_1,\ldots,x_P\}$ is the set of all input features, P is the number of features, $\{x_1,\ldots,x_P\} \setminus \{x_j\}$ is the possible set of all input features excluding $\frac{|S|!(p-|S|-1)!}{p!}$ is the weight of the feature subset S, and $f_x(S)$ is the prediction of the feature subset S.

4 The Proposed Model: ASLAM

4.1 The Model Framework of ASLAM Method

This article proposes a new link prediction framework based on attention mechanism for structure and attribute learning. This framework can combine information rich graph topology information with node attribute information. The overall framework is shown in Fig. 1. The ASLAM model framework consists of two channels: one channel uses node attributes to construct weighted semantic representations, which better guide the training of model topology structure. The other channel uses the Adjacency matrix to extract the order closed subgraph and learn the corresponding feature matrix through the joint semantic representation of node markers.

Due to the crucial importance of structural information for link prediction tasks, the ASLAM framework relies on a powerful subgraph classification pattern (i.e. SEAL) as its backbone. The priority of the SEAL method mainly lies in the extraction of local information, but it does not fully utilize the rich semantic information carried by node attributes. To compensate for the shortcomings of SEAL, ASLAM utilizes node attribute information by converting attribute information into semantic level topological representations.

4.2 Introduction to the Components of ASLAM Method

Given a graph $G = (V, E)$ with node attribute X and Adjacency matrix A, the number of nodes is N. ASLAM extracts attribute features and structural features respectively, and uses the embedded weighted sum of attribute space and structural space to learn the probability of the target node's presence on the link.

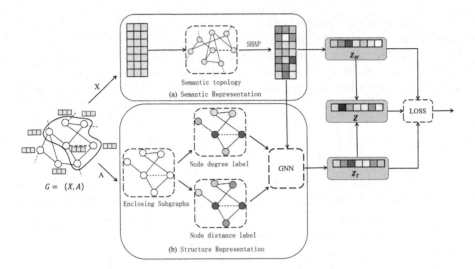

Fig. 1. The overall architecture of ASLAM. The model contains two channels: node attribute X to build weighted semantic representation to better guide the training of model topology. The adjacency matrix A is used to extract the hth order closed subgraph and jointly learn the corresponding feature matrix by combining the attribute information by node marking.

Semantic Representation. Firstly, in order to facilitate subgraph based structural learning for node attributes, ASLAM uses Euclidean distance combined with random walk to transform attribute information into structural level embeddings. In order to better express the embedding, the interpretable framework SHAP is used to calculate the Shapley value of the embedded features of the target node as the importance of the target node features, and finally the final weighted semantic representation Z_W is calculated.

Formally, given the node features, ASLAM uses Euclidean distance to calculate the feature similarity of a given node pair, and uses the node with a higher target node v_i score as its neighbor to obtain the semantic topology. The feature similarity score is as follows:

$$S_{ij} = -\|x_i - x_j\|^2,\tag{2}$$

where x_i, x_j are the feature vector of nodes i, j.

After obtaining the semantic topology, we use structure-based graph embedding technology to obtain structure-level embeddings Z_S of nodes.

Where the embedding $Z_{si} \in \mathbb{R}^{1 \times h}$ of node v_i is an embedding of global information about the semantic topology. Then we calculate the Shapley value for the node embedding feature, and we think that the Shapley value can explain the importance assignment of the node feature. Enter the embedding feature of node v_i, and the Shapley value that defines the n-dimensional feature of node v_i is as follows:

$$\phi_{i,n} = \sum_{S' \subseteq \{z_{si,1},\dots,z_{si,p}\} \backslash \{z_{si,n}\}} \frac{\mid S' \mid !(p- \mid S' \mid -1)!}{p!} \left(f_z \left(S' \cup \{z_{si,n}\} \right) - f_z(S') \right),$$
(3)

where $z_{si,m}$ is the m-dimensional feature of Z_{si}, $\{z_{si,1},\dots,z_{si,p}\}$ is the set of all input features, p is the number of all input features, $\{z_{si,1},\dots,z_{si,p}\} \backslash \{z_{si,n}\}$ is the possible set of all input features that do not include the n-dimension feature of node v_i, $\frac{\mid S' \mid !(p-\mid S'\mid-1)!}{p!}$ is the weight of the feature subset S', $f_z(S')$ is the prediction of the feature subset S'. The weighted semantic embedding Z_W of the node's final output is as follows:

$$Z_W = \phi \circ Z_S,$$
(4)

where $\circ$ is Hadamard product.

Structure Representation. In link prediction problems, neighborhoods are often considered important contextual information. Inspired by the recent attention mechanism [33], this article proposes a new neighborhood attention mechanism to effectively learn neighborhood features. From an intuitive perspective, the neighborhood structure of target nodes, such as node degree and node distance, may affect the links between target node pairs. In order to capture the structural information of the network, this article qualitatively analyzes the relationship between the existence of links between target node pairs and their topological structure. The structural representation learning of ASLAM consists of three parts: extracting target subgraphs, constructing node information matrices, and learning GNN. ASLAM first extracts closed subgraphs for a set of sampled positive links (observed) and a set of sampled negative links (unobserved) to construct training data.

GNN usually takes (A, X) as the input, where A is the Adjacency matrix of the input closed subgraph, X is the node information matrix, and each row corresponds to the eigenvector of a node. ASLAM constructs node information matrices Z_{DRN} and Z_{DE} using distance labeling and node degree labeling, respectively. Then, the sum is obtained through the learning of two different GNNs, and the final structural representation Z_T is obtained through fusing.

$$Z_T = \alpha_{DRE} \cdot Z_{DRE} + \alpha_{DE} \cdot Z_{DE},$$
(5)

where α_{DRE} and α_{DE} is attention weigh.

Adaptive Fusion. After encoding the structure and attributes of the subgraph, the model obtained representations from two different dimensions based on node attributes and node structure. Finally, the weighted semantic representation and structural representation are combined through attention mechanisms to directly use the obtained results in downstream tasks.

Based on topological embeddings Z_T and weighted semantic embeddings Z_W, we focus on node pairs (i,j), where $Z_{T(i,j)} \in \mathbb{R}^{h\times 1}$, $Z_{W(i,j)} \in \mathbb{R}^{h\times 1}$. We first transform the embedding through a linear transformation $W' \in \mathbb{R}^{h' \times h}$ and follow a nonlinear activation function, and then use a shared attention head $q \in \mathbb{R}^{h' \times 1}$ to obtain the attention value $\omega_{(i,j)}$, as follows:

$$\omega'_{(i,j)} = q^T \cdot \tanh\left(W' \cdot Z'_{(i,j)} + b\right),\tag{6}$$

where $\left\{Z'_{(i,j)} \mid Z'_{(i,j)} = Z_{T(i,j)} \quad or \quad Z'_{(i,j)} = Z_{W(i,j)}\right\}$ and $\left\{\omega'_{(i,j)} \mid \omega'_{(i,j)} = \omega_{T(i,j)} \quad or \quad \omega'_{(i,j)} = \omega_{W(i,j)}\right\}$. We then normalize the attention value using the softmax function to get the final weight $\alpha_{T(i,j)}$ and $\alpha_{S(i,j)}$. The final embedding of the node pair (i,j) can be expressed as:

$$Z_{(i,j)} = \alpha_{T(i,j)} \cdot Z_{T(i,j)} + \alpha_{S(i,j)} \cdot Z_{W(i,j)},\tag{7}$$

Finally, we use three standard binary cross-entropy loss joint training:

$$\mathcal{L} = \sum_{(i,j)\in D} \alpha \cdot BCE\left(Z_{T(i,j)}, y_{ij}\right) + \beta \cdot BCE\left(Z_{W(i,j)}, y_{ij}\right) + BCE\left(Z_{(i,j)}, y_{ij}\right),\tag{8}$$

where y_{ij} represents the existence of the link, $BCE(\cdot,\cdot)$ is the binary cross-entropy loss, and α and β are hyperparameters that measure the importance of the corresponding loss term.

5   Experimental Results and Analysis

In this part, we will evaluate our approach against the latest model of the Link Prediction Benchmark. We then analyzed the contribution of each component in our model.

5.1   Datasets

To verify its validity, we applied our method to 9 different research benchmark datasets, the statistics of which are summarized in Table 1. Dataset sources are also supplemented for reproducibility. The benchmark datasets used for evaluation in this work are Disease [8], Twitch_en [4], Airport Dataset, USA (American airtraffic network) [13], Brazil(Brazilian airtraffic network) [26], Amazon Photo Network Dataset [28] and Citation Network Dataset, Cora [12], Citeseer [35], Pubmed [5], DBLP [9]. In a disease dataset, the label of a node indicates whether the node is infected by a disease, and the characteristics of the node are associated with susceptibility to that disease. In Twitch_en, nodes represent players on Twitch, and edges represent friendships between them. Node characteristics are embeddings of games played by Twitch users. The nodes of the Amazon Photo network dataset represent goods, the edges represent two items that are often purchased together, and the node characteristics are represented by product review word packs. Each node of the airport dataset represents an airport and an edge represents a flight route. Each node of the citation network dataset represents a document, characterized by a corresponding word package representation, edges representing citation links, and node labels are academic (sub)fields.

Table 1. Statistics of datasets.

Dataset	Nodes	Edges	Features	Classes	Average Degrees
Disease	2665	2663	1000	2	2
Twitch_en	7126	77774	128	2	21.83
USA	1190	13599	1190	4	22.86
Brazil	131	1074	131	4	16.4
Photo	7650	238162	745	8	62
Cora	2708	10556	1433	7	7.8
Citeseer	4230	10674	602	6	5.05
PubMed	19717	88648	500	3	8.99
DBLP	17716	105734	1639	4	11.94

5.2 Experiment Setup

In order to verify the effectiveness of the ASLAM method, this article follows the standard link prediction segmentation rate and divides the edges into 85%/5%/10% for training, validation, and testing. To ensure fairness in the experiment, the random seeds were fixed to 2 when segmenting the dataset. The batch size for all datasets is 32. The experiment adopts an early stop strategy, with a training cycle of 400 and a patient setting of 20. Each experiment is conducted 10 times, and the average value is taken as the final result. All code is implemented using PyTorch and all models are implemented using Torch_Geometry [38].

5.3 Comparison Method

CN [20]: the basic assumption of CN in link prediction is that two nodes that are not yet connected are more inclined to connect edges if they have more neighbors in common.

AA [32]: Adamic-Adar (AA) assigns a weight to each common neighbor node, which means that the common neighbor node with a small degree contributes more than the common neighbor node with a large degree. For example, in social networks, the probability of connecting two people who share a relatively unpopular person or thing tends to be higher than that of two people who follow the same person or thing with high attention.

Node2vec [10]: Node2vec is an embedding-based method that comprehensively considers DFS neighborhoods and BFS neighborhoods. Simply put, it can be seen as an extension of Deepwalk, which is a deepwalk that combines DFS and BFS random walks. Node2vec uses biased wandering to sample vertices to explore homogeneous or structural information.

GCN [15]: Semi-supervised learning on graph structure data, and learned the hidden layer representation encoding local graph structure and node features.

GAE and VGAE [16]: GAE and VGAE use graph convolutional networks to encode graph adjacency matrices into latent representations and reconstruct observed links or predict unobserved links through endoproduct decoders.

GraphSAGE [11]: Node embeddings that utilize node feature information, such as text properties, to efficiently generate previously unseen data. GraphSAGE learns a function to generate embeddings by sampling and aggregating features from node-local neighborhoods.

GAT [30]: The hidden representation of each node in the graph is calculated by focusing on its neighbors, and arbitrary weights are assigned to neighbors to be applied to graph nodes with different degrees.

SuperGAT [14]: Utilize two forms of attention compatible with self-supervised tasks to predict edges.

SEAL [39]: The graph structure information is learned in the subgraph, and a specific marker method is used for display encoding in the graph-level GNN.

BASL [18]: A two-component structure and attribute learning framework is proposed, which integrates attribute learning and structural learning.

5.4   Evaluation Metrics

To evaluate the performance of the link prediction method or algorithm, we employ two commonly used standard metrics: area under the curve (AUC) [41] and average accuracy (AP) [38].

5.5   Experiment Results

Performance on real datasets: This article validated the generalization ability of the proposed method through different research benchmark datasets. All experiments were conducted 10 times, with the average taken as the result and the standard deviation listed. For these two indicators, the higher the better. The best ones are displayed in bold, while the second best ones are highlighted with underscores. The results are summarized in Table 2 and Table 3.

Based on the experimental results, there are the following observations.

Firstly, compared to popular heuristics (such as CN, AA) and node embedding methods (such as Node2vec), neural network-based models have consistently achieved good performance in most networks. However, on some datasets, traditional heuristics still demonstrate competitiveness compared to GNN, or even better than GNN. This indicates that heuristic methods designed manually rely on specific datasets and have lower expressive power than neural networks. However, structural information (such as neighborhood, degree, and shortest path) is crucial for link prediction, and GNN heavily relies on smooth node features, which sometimes results in less than heuristic results. It should also be noted that both messaging methods (i.e. GCN and GAT) perform poorly on highly layered datasets (such as Disease).

Secondly, the subgraph paradigm based model SEAL and BSAL show significant advantages, demonstrating the importance of structural information in link prediction tasks. Among the powerful subgraph based methods, BSAL outperforms SEAL on most datasets, indicating that learning combined with node attributes can indeed bring good results. However, directly combining node features with structural information cannot always achieve performance improvement, and sometimes even hinder model performance. In the self supervised method, SuperGAT performs better than GAT on most datasets because Super-GAT considers two attention mechanisms and designs graph attention based on the average and homogeneity of the input graph, obtaining more expressive attention when distinguishing neighbors with incorrect links.

In summary, the GNN based method describes the link prediction task as a supervised learning problem and learns node representations through information aggregation, which is very promising. Although structural information is crucial in missing link reasoning, it is not sufficient, and combining semantic information can further improve performance. Secondly, in the learning process, attention to structure and attributes is also indispensable. The experimental results indicate that ASLAM performs best on the AUC and AP indicators on the vast majority of datasets. This demonstrates the effectiveness and superiority of the method proposed in this paper.

Table 2. Mean AUC and standard deviation for 10 experiments.

	Disease	Twitch_en	USA	Brazil	Photo	Cora	Citeseer	PubMed	DBLP
CN	49.81	76.35	87.59	85.32	86.40	71.10	61.65	64.58	77.53
AA	50.00	77.71	87.4	84.91	86.93	70.50	64.18	63.72	77.86
Node2vec	75.95 ± 1.10	86.50 ± 1.46	86.30 ± 1.06	85.68 ± 0.94	86.29 ± 1.30	86.35 ± 1.30	86.35 ± 1.86	85.92 ± 1.02	85.61 ± 1.06
GCN	76.36 ± 0.00	80.49 ± 0.00	91.49 ± 0.00	80.24 ± 0.00	89.93 ± 0.00	91.35 ± 0.15	93.62 ± 0.00	96.84 ± 0.00	96.57 ± 0.00
GAE	73.24 ± 0.00	80.13 ± 0.00	91.03 ± 0.00	81.07 ± 0.00	93.65 ± 0.01	91.82 ± 0.08	93.59 ± 0,13	96.38 ± 0.00	96.48 ± 0.00
VGAE	74.16 ± 0.00	80.44 ± 0.00	93.79 ± 0.00	80.71 ± 0.00	81.25 ± 0.00	67.17 ± 0.00	93.69 ± 0.01	95.83 ± 0.00	96.52 ± 0.04
GraphSAGE	72.58 ± 0.00	77.33 ± 0.00	91.28 ± 0.00	75.24 ± 0.00	83.82 ± 0.00	86.88 ± 0.00	83.60 ± 0.00	83.19 ± 0.00	89.92 ± 0.00
GAT	69.53 ± 0.33	71.59 ± 0.00	91.14 ± 0.00	78.08 ± 0.00	93.35 ± 0.03	89.70 ± 0.00	90.45 ± 0.18	89.05 ± 0.02	92.67 ± 0.00
SuperGAT	72.75 ± 0.00	80.89 ± 0.00	91.09 ± 0.00	72.12 ± 0.00	95.86 ± 0.09	91.78 ± 0.00	93.38 ± 0.00	94.40 ± 0.00	95.92 ± 0.00
SEAL	96.16 ± 0.11	91.01 ± 0.04	91.20 ± 0.04	89.42 ± 0.63	95.20 ± 0.12	91.01 ± 0.14	91.62 ± 0.15	95.37 ± 0.03	93.16 ± 0.02
BASL	96.92 ± 0.13	92.81 ± 0.05	95.69 ± 0.08	90.95 ± 0.69	96.11 ± 0.08	91.11 ± 0.00	92.73 ± 0.16	96.59 ± 0.01	93.68 ± 0.03
ASLAM	**98.40 ± 0.04**	**93.44 ± 0.10**	**95.85 ± 0.06**	**93.39 ± 0.53**	**99.21 ± 0.04**	**92.42 ± 0.18**	**94.78 ± 0.41**	**98.02 ± 0.02**	**96.68 ± 0.08**

Table 3. Mean AP and standard deviation for 10 experiments.

	Disease	Twitch_en	USA	Brazil	Photo	Cora	Citeseer	PubMed	DBLP
CN	50.00	76.10	87.38	84.77	86.11	70.99	61.54	64.56	77.49
AA	50.00	77.22	86.86	85.02	86.87	70.64	64.24	63.74	77.90
Node2vec	70.12 ± 0.78	90.49 ± 1.05	90.34 ± 0.69	89.91 ± 0.83	90.33 ± 1.01	90.19 ± 0.79	90.65 ± 1.32	89.82 ± 0.87	89.63 ± 0.95
GCN	80.98 ± 0.00	83.28 ± 0.00	93.91 ± 0.00	83.02 ± 0.00	89.49 ± 0.00	91.04 ± 0.07	94.48 ± 0.00	96.97 ± 0.00	97.00 ± 0.00
GAE	80.21 ± 0.00	83.01 ± 0.00	93.69 ± 0.00	81.90 ± 0.00	93.24 ± 0.02	92.01 ± 0.03	94.54 ± 0.17	96.46 ± 0.00	96.92 ± 0.00
VGAE	81.09 ± 0.00	83.58 ± 0.00	94.54 ± 0.00	83.15 ± 0.00	80.79 ± 0.00	69.99 ± 0.00	95.04 ± 0.00	96.08 ± 0.00	96.98 ± 0.04
GraphSAGE	71.13 ± 0.00	77.97 ± 0.00	93.34 ± 0.00	73.70 ± 0.00	82.58 ± 0.00	86.37 ± 0.00	82.56 ± 0.00	83.26 ± 0.00	90.12 ± 0.00
GAT	70.11 ± 0.09	71.32 ± 0.00	91.98 ± 0.00	75.54 ± 0.00	91.80 ± 0.04	90.21 ± 0.00	88.43 ± 0.19	87.92 ± 0.01	92.66 ± 0.00
SuperGAT	71.32 ± 0.00	81.93 ± 0.00	93.18 ± 0.00	67.41 ± 0.00	95.02 ± 0.16	91.29 ± 0.00	93.39 ± 0.00	94.10 ± 0.00	96.19 ± 0.01
SEAL	93.62 ± 0.57	92.07 ± 0.06	**96.06 ± 0.13**	85.54 ± 0.74	95.08 ± 0.10	91.95 ± 0.11	92.46 ± 0.16	95.90 ± 0.02	94.85 ± 0.02
BASL	95.57 ± 0.08	93.31 ± 0.05	95.86 ± 0.18	90.95 ± 0.65	95.69 ± 0.13	92.21 ± 0.10	92.80 ± 0.26	96.87 ± 0.01	95.25 ± 0.02
ASLAM	**96.81 ± 0.15**	**93.81 ± 0.11**	95.45 ± 0.08	**92.51 ± 0.96**	**99.09 ± 0.02**	**93.63 ± 0.21**	95.09 ± 0.40	**98.04 ± 0.01**	**97.15 ± 0.08**

5.6   Ablation Study

ASLAM is a new graph neural network-based link prediction model aimed at automatically extracting network attribute and structural information. By combining node attribute information, attribute attention and neighborhood attention, and adaptively integrating semantic and structural representations, the final representation is obtained. To explore the effectiveness of attribute information, attribute attention, and neighborhood attention in the model. The basic model of the experimental setup is the model ASLAM proposed in this article. In the subgraph topology, attribute feature information is not considered and is labeled as ASLAM-NF, while neighborhood attention and attribute attention are labeled as ASLAM-NNA and ASLAM-NAA, respectively. This study conducted ablation experiments on nine datasets in the same experimental setup. The evaluation indicators are AUC and AP. The results are shown in Fig. 2 and Fig. 3.

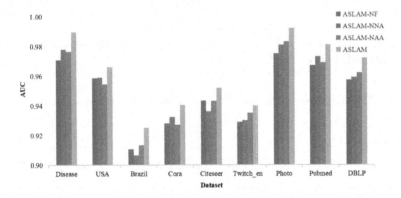

Fig. 2. Mean AUC for the comparison of ASLAM and its variants.

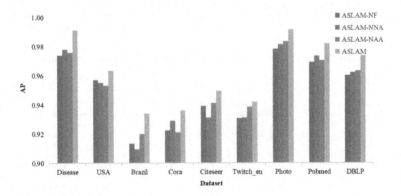

Fig. 3. Mean AP for the comparison of ASLAM and its variants.

The results indicate the superiority of ASLAM, which outperforms the other three variables in terms of AUC and AP metrics, indicating that the proposed method can achieve the best performance in link prediction. Among them, the performance of ASLAM-NF on the vast majority of datasets is much lower than that of ASLAM, indicating that node features are very important for prediction and the necessity of integrating structural and attribute information. Secondly, the prediction effects with and without attribute attention were significantly compared on the Brazilian and Citeseer datasets, indicating that attribute attention can indeed guide training to a certain extent and achieve better results.

6 Conclusion

This article investigates the problem of link prediction in complex network analysis. The existing Sota GNN based methods either heavily rely on network topology or node attributes, resulting in significant differences in prediction results across different networks. To alleviate this problem, we proposes an attention mechanism based attribute and structure learning framework, aiming to fuse key information about topological space and attribute space. Specifically, the proposed method constructs a semantic topology through node attributes, uses structure based graph embedding technology to obtain the structural level representation of nodes, and uses Shapley values as attention values to guide learning of semantic representations. A flexible and easy to implement solution is provided to adaptively integrate structural topology and semantic topology. Due to the crucial importance of structural information such as shortest path and node degree for link prediction, the proposed method uses distance and degree labeled subgraphs for learning. On different research benchmarks, the performance of our method far exceeds the competitive baseline, which confirms the effectiveness of the proposed method. In the future, we will test the method on more evaluation indicators (e.g., HITS@K) or tasks (e.g., adversarial attacks) to verify its generalization ability. In practical scenarios, the network changes in real-time, and we will consider combining dynamic networks for the next step of research.

Acknowledgements. This work is supported by the State Key Program of National Nature Science Foundation of China (61936001), the key cooperation project of Chongqing municipal education commission (HZ2021008), the Natural Science Foundation of Chongqing (cstc2019jcyj-cxttX0002, cstc2021ycjh-bgzxm0013).

References

1. Adamic, L.A., Adar, E.: Friends and neighbors on the web. Soc. Netw. **25**(3), 211–230 (2003)
2. Airoldi, E.M., Blei, D.M., Fienberg, S.E., Xing, E.P.: Mixed membership stochastic blockmodels (2008)
3. Avdeeva, E., Herczeg, T., Stalerunaite, B., Andreasen, V.: Epidemic spread in scale-free networks (2009)

4. Benedek, R., Carl, A., Rik, S.: Multi-scale attributed node embedding. J. Complex Netw. (2), 2 (2021)
5. Bojchevski, A., Günnemann, S.: Deep gaussian embedding of graphs: unsupervised inductive learning via ranking (2017)
6. Chami, I., Ying, R., Ré, C., Leskovec, J.: Hyperbolic graph convolutional neural networks. CoRR abs/1910.12933 (2019). http://arxiv.org/1910.12933
7. Chami, I., Ying, R., Ré, C., Leskovec, J.: Hyperbolic graph convolutional neural networks (2019)
8. Chami, I., Ying, R., Ré, C., Leskovec, J.: Hyperbolic graph convolutional neural networks. Adv. Neural. Inf. Process. Syst. **32**, 4869–4880 (2019)
9. Fu, X., Zhang, J., Meng, Z., King, I.: MAGNN: metapath aggregated graph neural network for heterogeneous graph embedding (2020)
10. Grover, A., Leskovec, J.: Node2vec: scalable feature learning for networks (2016)
11. Hamilton, W.L., Ying, R., Leskovec, J.: Inductive representation learning on large graphs (2017)
12. Huang, Z., Zhang, S., Xi, C., Liu, T., Zhou, M.: Scaling up graph neural networks via graph coarsening (2021)
13. Jin, Y., Song, G., Shi, C.: GraLSP: graph neural networks with local structural patterns. pp. 4361–4368 (2020)
14. Kim, D., Oh, A.: How to find your friendly neighborhood: graph attention design with self-supervision. arXiv e-prints (2022)
15. Kipf, T.N., Welling, M.: Semi-supervised classification with graph convolutional networks (2016)
16. Kipf, T.N., Welling, M.: Variational graph auto-encoders (2016)
17. Koren, Y., Bell, R., Volinsky, C.: Matrix factorization techniques for recommender systems. Computer **42**(8), 30–37 (2009)
18. Li, B., Zhou, M., Zhang, S., Yang, M., Lian, D., Huang, Z.: BSAL: a framework of bi-component structure and attribute learning for link prediction (2022)
19. Li Yanli, Z.T.: Local similarity indices in link prediction (in Chinese). J. Univ. Electron. Sci. Technol. **50**(3), 6 (2021)
20. Liben-Nowell, D.: The link prediction problem for social networks. In: Conference on Information and Knowledge Management. Conference on Information and Knowledge Management (2003)
21. Linyuan, L.: Link prediction on complex networks (in Chinese). J. Univ. Electron. Sci. Technol. China **39**(5), 11 (2010)
22. Liu, Q., Tan, H., Zhang, Y., Wang, G.: Dynamic heterogeneous network representation method based on meta-path (in chinese). Acta Electron. Sinica (008), 050 (2022)
23. Lu, H., Hu, H., Lin, X.: DensE: an enhanced non-commutative representation for knowledge graph embedding with adaptive semantic hierarchy. Neurocomputing **476**, 115–125 (2022)
24. Nickel, M., Tresp, V., Murphy, K., Gabrilovich, E.: A review of relational machine learning for knowledge graphs. Proc. IEEE **104**(1), 11–33 (2016)
25. Perozzi, B., Al-Rfou, R., Skiena, S.: DeepWalk: online learning of social representations. ACM (2014)
26. Ribeiro, L.F.R., Saverese, P.H.P., Figueiredo, D.R.: Struc2vec: learning node representations from structural identity. ACM (2017)
27. Shapley, L.S.: A Value for n-Person Games. Annals of Mathematical Studies (1953)
28. Shchur, O., Mumme, M., Bojchevski, A., Günnemann, S.: Pitfalls of graph neural network evaluation (2018)

29. Tang, J., Qu, M., Wang, M., Zhang, M., Mei, Q.: LINE: large-scale information network embedding (2015)
30. Vaswani, A., et al.: Attention is all you need. arXiv (2017)
31. Wang, H., Lian, D., Zhang, Y., Qin, L., Lin, X.: GoGNN: graph of graphs neural network for predicting structured entity interactions (2020)
32. Wang, Z., Liao, J., Cao, Q., Qi, H., Wang, Z.: FriendBook: a semantic-based friend recommendation system for social networks. IEEE Trans. Mob. Comput. **14**(3), 538–551 (2016)
33. Wang, Z., Lei, Y., Li, W.: Neighborhood attention networks with adversarial learning for link prediction. IEEE Trans. Neural Netw. Learn. Syst. **32**, 1–11 (2020)
34. Yang, M., et al.: Hyperbolic graph neural networks: a review of methods and applications (2022)
35. Yang, Z., Cohen, W.W., Salakhutdinov, R.: Revisiting semi-supervised learning with graph embeddings. JMLR.org (2016)
36. Yuan, M., Liu, Q., Wang, G., Guo, Y.: HNECV: heterogeneous network embedding via cloud model and variational inference. In: CAAI International Conference on Artificial Intelligence. CAAI International Conference on Artificial Intelligence (2021)
37. Yun, S., Kim, S., Lee, J., Kang, J., Kim, H.: Neo-GNNs: neighborhood overlap-aware graph neural networks for link prediction (2022)
38. Zhang, E., Zhang, Y.: Average precision. In: Liu, L., Özsu, M.T. (eds.) Encyclopedia of Database Systems, pp. 192–193. Springer, Boston (2009). https://doi.org/10.1007/978-0-387-39940-9_482
39. Zhang, M., Chen, Y.: Link prediction based on graph neural networks (2018)
40. Zhang, Y., Feng, Y.: A summary of the methods and development of link prediction (in Chinese). Meas. Control Technol. **38**(2), 5 (2019)
41. Zhou, T., Lü, L., Zhang, Y.: EPJ manuscript no. (will be inserted by the editor) predicting missing links via local information (2012)

Cybersecurity and IoT

A New Data Model for Behavioral Based Anomaly Detection in IoT Device Monitoring

Marcin Michalak[1,5](✉)(iD), Piotr Biczyk[2,6], Błażej Adamczyk[3,5](iD),
Maksym Brzęczek[3], Marek Hermansa[1](iD), Iwona Kostorz[1](iD),
Łukasz Wawrowski[1](iD), and Michał Czerwiński[2,4]

[1] Łukasiewicz Research Network—Institute of Innovative Technologies EMAG,
ul. Leopolda 31, 40-189 Katowice, Poland
`Marcin.Michalak@polsl.pl`
[2] QED Software sp. z o.o., ul. Mazowiecka 11/49, 00-052 Warszawa, Poland
[3] EFIGO sp. z o.o., ul. M. Kopernika 8/6, 40-064 Katowice, Poland
`Blazej.Adamczyk@efigo.pl`
[4] University of Warsaw, ul. Banacha 2, 02-097 Warszawa, Poland
[5] Department of Computer Networks and Systems, Silesian University of Technology,
Akademicka 16, 44-100 Gliwice, Poland
[6] Faculty of Automatic Control, Electronics and Computer Science, Silesian
University of Technology, Akademicka 16, 44-100 Gliwice, Poland

Abstract. The paper presents the new model of data for Internet of Things (IoT) devices monitoring and anomaly detection. The model bases mostly on behavioral description of current state of device, however it contains also some additional information. Raw input data, coming from the external simulation software, are aggregated on two levels of detail: raw variable values preprocessing and time-based aggregation. It was shown, that a sample data following this model, data that contains anomalies, can be analyzed with standard anomaly detection methods and results of this application are very satisfactory. The data used in the paper are also publicly available.

Keywords: Anomaly detection · Internet of Things · Behavioral analysis

1 Introduction

There are no doubts that Internet had changed the way of living since decades ago. Now we observe the increasing influence of the next step of industrial revolution that refers to the capability of equipping low-level devices the access to the Internet. Such a tendency undoubtedly provides a lot of benefits, however, we should be conscious of possible vulnerabilities that come from these benefits.

Anomaly detection is a well-known approach applied for security issues detection. However, it is usually focused on monitoring the network traffic statistics to

A. Campagner et al. (Eds.): IJCRS 2023, LNAI 14481, pp. 599–611, 2023.
https://doi.org/10.1007/978-3-031-50959-9_41

detect some non-typical occurrences. In case of Internet of Things (IoT) devices we do not expect any significant amount of transferred data so possible attempts to attack the device should reflect rather the device behavior than a transfer statistics.

In our research we put the attention not on the monitoring the IoT device network statistics—we shift it into the operating system behavior, focusing on the processes' behavior and on the tree structure of process invokes. Such an approach is quite fresh and there are no available data to train anomaly detection methods for such a purpose. This paper provides the methodology of the data acquisition, aggregation and preprocessing to make them ready for anomaly detection methods application.

The paper is organized as follows: it starts from the description of our motivation to analyse the behavior of IoT devices from the completely different point of view (the behavior analysis), later the review of existing IoT devices dataset related (but not limited to) the anomaly detection issues, afterwards a new model of data, its raw structure as well as two steps of aggregation are described in detail, moreover, the paper provides a case study of the model application for anomaly detection in behavioral data of IoT work simulation, finally, the paper ends with some short conclusions and the draw of perspectives of further works.

2   Motivation

Typically IoT devices are small and have dedicated, limited purpose. This means that the device behavior patterns probably have high tendency to be repeatable and predictable what increases chances of successful intrusion detection by applying anomaly detection. This holds for both - internal and external behavior patterns. External, Network based Intrusion Detection Systems (NIDS) for IoT devices has already been throughoutly studied in the literature and has been shown to be effective. Methods based on network traffic are also much portable and easier to implement as they do not require device modification.

However, such systems have several disadvantages: (1) they might not detect attacks which do not change network traffic significantly, (2) the detection is usually delayed until the attack is reflected on generated network traffic (if ever), (3) physical device modification can go undetected (e.g. the attacker can use different, additional communication channels).

On the other hand – internal, Host based Intrusion Detection Systems (HIDS) – obviously do not have these limitations and it seems there should be more attention and research devoted to this approach. If the hypothesis of higher efficiency of HIDS over NIDS could be proven – it could lead to a market response where manufacturers would implement such techniques within their devices. This could potentially improve general IoT security not requiring the end user or administrator to implement network based external security systems by shifting this function to the device level.

3   Related Works

Anomaly detection is a well-known approach for data analysis in many specific domains. As the IoT issues are becoming more and more interesting it is intuitive that any new or improved models should be tested on some data with anomalies to evaluate their capabilities. During last decades dozens of datasets related to network traffic security, operating systems or IoT monitoring were published. A brief summarize is presented in Table 1.

Table 1. Datasets related to network traffic, operating system or IoT security monitoring (N- Network, OS - Operating System, IoT - Internet of Things devices)

Dataset name	Owner	Monitoring	Reference
ADFA-LD	University of New South Wales	N, OS	[3]
Aposemat IoT-23	Stratosphere Laboratory	N, OS, IoT	[20]
CAIDA	Center of Applied Internet Data Analysis	N	[2]
Bot-IoT	University of New South Wales	N, IoT	[8]
CDX	United State Military Academy	N	[16]
DARPA 98–99	MIT Lincoln Laboratory	N, OS	[11]
KDD Cup 1999	University of California	N, OS	[19]
IoT Botnet	Ontario Tech University	N, OS, IoT	[23]
ISCX2012	University of New Brunswick	N	[17]
Kyoto	Kyoto University	N	[10]
Malware on IoT	Stratosphere Laboratory	N, OS, IoT	[20]
NSL-KDD	Canadian Institute for Cybersecurity	N, OS	[21]
RegSOC	Ł-EMAG	N	[24]
TON IoT	University of New South Wales	N, OS, IoT	[12]
Twente	University of Twente	N	[18]
UNSW-NB15	University of New South Wales	N	[13]
UMASS	University of Massachusetts	N	[25]

To reflect the nowadays trends in the data we limited our search into datasets not older than 5–6 years and closely related to IoT domain. Their short descriptions are presented below.

3.1   IoT Botnet

The components of the IoT Botnet [23] were based on simulated Internet of Things services, network platforms, and ISCXFlow meter. The network platform comprised of normal and attacking virtual machines. The IoT services were simulated through the node-red tool.

The testbed of dataset consisted of virtual machines that were connected to LAN and WAN. Virtual machines were linked to the Internet. A packet-filtering firewall with two interface cards was used to ensure the validity of the labelling process of the dataset.

A typical smart home environment was created with five IoT devices that operated locally and were connected to the cloud infrastructure. The following IoT services were implemented in the dataset testbed environment: motion-activated lights, smart fridge, smart thermostat, remotely activated garage door and weather station.

Consequently, the IoT Botnet dataset was created. It contains collections from the monitoring of Denial of Service (DoS) and Distributed Denial of Service (DDoS) attacks for TCP, UDP and HTTP protocols, keyloggers, operating systems, services, and data exfiltration.

3.2   Bot-IoT

The environment for data capturing consisted of three components: network platforms, simulated IoT services, and extracting features. The network platforms included normal and attacking virtual machines. The IoT services simulating various IoT sensors were connected to the public IoT hub. The network environment that the data was collected contained a combination of normal and botnet traffic.

The dataset [8] provides original pcap files, generated argus files and csv files. The files separation is based on attack categories and subcategories. The dataset files include DDoS attacks, DoS, operating system and service scanning, keyloggers and data exfiltration.

3.3   TON_IoT

The TON_IoT datasets [12] are an IoT and Industrial IoT (IIoT) dataset for assessing the fidelity and performance of various artificial intelligence cyber-security applications i.e. machine/deep learning algorithms. The files contain heterogeneous data sources collected from IoT and IIoT sensor telemetry data sets, Windows 7 and 10 operating systems datasets, as well as Ubuntu 14 and 18 TLS and network traffic datasets.

The data was collected in a realistic and large-scale network. A testbed network was created for the Industry 4.0 network, which includes IoT and IIoT networks. The test platform was deployed using multiple virtual machines and hosts of Windows, Linux, and Kali operating systems to manage connections between the three tiers of IoT, Cloud, and Edge/Fog. Various attack techniques such as DoS, DDoS, and ransomware targeting web applications, IoT gateways, and computer systems on the IoT/IIoT network were conducted. The datasets were collected in parallel processing to collect several normal and cyber-attack events from network traffic, Windows audit trail, Linux audit trail, and IoT telemetry data.

3.4   Aposemat IoT-23

The IoT-23 [6] is a dataset of network traffic from Internet of Things devices. The dataset consists of 23 captured different IoT network traffic scenarios. These scenarios were divided into twenty network captures from infected IoT devices that the malware samples were performed in each scenario and three network captures of the actual network traffic of the IoT devices. In each malicious scenario a specific malware sample was run on a Raspberry Pi. Scenarios included following malware samples used to infect the device (Mirai, Torii, Trojan, Gagfyt, Kenjiro, Okiru, Hakai, IRCBot, Linux Mirai, Linux Hajime, Muhstik, Hide and Seek).

3.5 Malware on IoT

Malware on IoT [20] is a dataset of the monitoring of real IoT devices infected by malware. The dataset consists of labelled network traffic files stored during the long-lived real IoT malware traffic. It is divided into five subsets containing results of network traffic capturing during the Mirai malware attack and two subsets of honeypot network traffic capturing logs including protocols (HTTP, SSL, TCP, UDP) and connections statistics. The honeypot was a network camera.

3.6 Summary

Most of the available datasets contain data from network monitoring during the normal operation and attacks. The data sets described in detail contain data from the audit of real and simulated IoT devices and their network environment. The available data sets providing kernel event monitoring data are from Solaris (DARPA 98–99). Due to that fact authors decided to create a new contemporary data set. The new data set is based on the host audit data coming from Linux kernel event tracing.

4 Data Origin and Description

The paper provides a new model of data which is dedicated mostly for behavioral-based anomaly detection. However, it is possible to use these data model in any other analytical tasks. In this section the data origin as well as a data preprocessing (before building a final model) is presented. Subsection 4.1 presents the simulation environment, where the data come from and Subsect. 4.2 describes the aggregation of the environment raw data.

4.1 Simulation Environment

Data are the result of the simulation environment for real-time dataset generation and evaluation of IoT Linux host intrusion detection systems [1]. The environment allows for simulation of multiple IoT devices while collecting the devices' OS audit logs in a central repository. The system also simulates external attacks with configurable frequency and distribution providing result data with labels indicating the time of attacks.

The system architecture comprises components of two groups: a central system and a collection of IoT devices. The central system consists of common software components which are responsible for managing simulation, collecting and labeling data, and launching attacks. IoT devices in their number and type are controlled by configuration and can either be emulated or be physical devices connected to the system through various implementations of a common device interface.

The most important part of the system is the agent software—the tool responsible for collecting data from IoT devices. The agent traces the Linux kernel with Linux Auditing System component—auditd. This tool makes it possible to low level, granular information about the current state of the operating system, especially the additional information about system calls and related events.

4.2   First Level of Aggregation

The raw auditd-based data are sent to the central system. Because this format of data is improper to be used as the input for anomaly detection methods, it requires a transformation into the tabular format, where columns reflect variables and rows reflect particular observations—each observation is an aggregated description of the single system call.

Before the final model of the pre-processed data will be presented, it is important to emphasize, that this process is implemented with a streaming pipeline processing engine. The raw data set is processed by several *parsers* responsible for augmenting and structuring the output. Each parser can modify existing and create new columns (variables) in the output record. For example, the very basic parser is a responsible for translating raw system call auditd records into a tabular form. The resulting columns are then forwarded to subsequent parsers. Last parser is producing the output file in CSV format.

What is important – the parsers can maintain a state what means they can detect some important patterns looking at the history and sequence of events processed so far. This is a very powerful tool for extracting new and interesting features for further processing especially from security perspective. For example, such parser can detect certain predefined behavioral schemes/patterns – e.g. a single connection failure is nothing unusual, but many subsequent connection failures on different ports may suggest a port scan running as a post exploitation behaviour.

Among several simple parsers, the present version of the system contains the following four parsers which conduct more advanced analysis in order to extract new and interesting security-related features:

- **Process tree parser** – creates the PROCESS_PATH feature of all the current process ancestors up to the init process – the parser needs to maintain the full process tree over time in order to collect all the process ancestors.
  This seems to be very interesting feature because usually attack payloads are being executed as sub-processes and thus an unusual process path may with high probability suggest an ongoing attack.

- **Process name parser** – creates a feature with more significant process name, e.g. instead of representing a process as *bash* it is much more precise to name the process from the actual script filename (e.g. *script.sh*) – this allows to distinguish the processes by what is their intended behavior and not just binary filename.
  Descriptive process names, relating to their behavior, allow to detect unusual and anomalous actions which can be results of an ongoing attack.
- **Open file parser** – for each process maintains all the currently open files and attaches the list as a new feature for further processing.
  Process behavior is frequently repeatable and predictable, especially in case of small IoT devices which perform concrete subset of jobs repeatably. Processes accessing more files than usual should be treated as suspicious and thus this feature seems to be of high interest.
- **Open library parser** – for each process maintains all the currently open libraries and attaches the list as a new feature for further processing.
  Performing different actions is very frequently done by loading proper libraries. Thus a measure of process behavior is the set of open libraries. Processes using different libraries than usual may be exploited binaries.

Moreover in our system we configured auditd to collect more security related events than just pure system calls. This means that these non system-call-oriented events are also contained in raw data set. To keep this valuable information in output recordset but maintain the assumption that every output record is a single system call – at the first aggregation level we count all the non system-call-oriented events and add the counts to the upcoming output record. For example, if an authentication happens between two consecutive system calls the USER_AUTH_COUNT variable value would be 1 – indicating that a single user authentication happened in between current and previous system call.

A brief description of most significant variables of this format of data is presented below:

- SYSCALL_timestamp: the time of system call invoking,
- SYSCALL_arch: processor architecture (32 or 64),
- SYSCALL_syscall: system call name, e.g. open, openat, etc.,
- SYSCALL_success: system call status, i.e. yes, no, N/D,
- SYSCALL_exit: system call results: positive integers reflects success, negative integers reflects fail,
- PROCESS_comm: process name,
- PROCESS_exe: executable file path,
- PROCESS_name: the name of the process
- PROCESS_PATH: the combined sequence of parent processes, e.g. systemd/sshd/bash/curl,
- PROCESS_DLL: the list of process already loaded libraries,
- PROCESS_uid: an id of the process invoking user,
- PROCESS_gid: an id of the group of the user mentioned above,
- SOCKET_family: a name of the communication protocol, e.g. netrom, bridge, atmpvc, x25, inet6, unknown, etc.,

- SOCKET_address: destination connection target, e.g. 192.168.1.1:80,
- CWD_cwd: current working directory,
- PATH_name: the filename and its path of the being processed file, e.g. /etc/passwd,
- EXECVE_argc: the number of arguments passed to the program,
- MMAP_flags: the number of allowed flags (only for mmap and mmap2 system calls), a list of values e.g. MAP_FIXED, MAP_ANONYMOUS,
- OTHER_addr: IP address of the ivnoker,
- LOGIN_res: success or fail of the logging,
- PAM_op: PAM operation time, e.g. PAM:authentication, PAM:accounting, etc.,
- KILL_PROCESS: id of the process—the command addressee,
- KILL_uid: id of the user sending this command.

Moreover, for other types of calls (listed afterwards) the number of other calls between calls of the same type are counted:

- USER_AUTH,
- USER_MGMT_COUNT,
- CRED_COUNT,
- USER_ERR_COUNT,
- USYS_CONFIG_COUNT,
- CHID_COUNT,
- SELINUX_ERR_COUNT
- USER_CMD_COUNT,
- SYSTEM_COUNT,
- SERVICE_COUNT,
- DAEMON_COUNT,
- NETFILTER_COUNT,
- SECCOMP_COUNT,
- AVC_COUNT,
- ANOM_COUNT,
- INTEGRITY_COUNT,
- APPARMOR_COUNT,
- KERNEL_COUNT,
- RESP_COUNT,
- SELINUX_MGMT_COUNT.

More detailed description of all present variables one may find in the readme file attached to the externally published data.

4.3 Second Level of Aggregation

The output of the first level of aggregation contains descriptions of single calls per record. However, the goal of the analysis is not to detect calls that behave in different way but to point the periods of time during which the entire set of running calls (in the IoT device) composes to something unusual. To prepare the data for such analysis, the data were put through time-based aggregation.

The time-based aggregation was approached with the moving window method. While using such method one has to specify two parameters: the moving window width and the moving window step. The first one corresponds to the time interval describing the range of the data to be aggregated. The second one specifies the width of the data shift. That reflects the frequency of the aggregated data generating. The aggregated intervals contained both numeric and categorical variables. Furthermore, within the categorical variables, we specified variables containing paths. Each of these three types was aggregated with different methods.

For the numeric values standard statistics were used for aggregation: minimum, maximum, average, 25th, 50th and 75th percentile. For the categorical values analogous standard statistics were used: the last observed value, the most frequent value and count of values. The path-based variables were approached as natural language sentences where a token is a piece of a path separated by slashes, whereas multiple paths occurring in a given aggregated period were treated as a paragraph. Each token was transformed into an embedding using the Fasttext2 [7] algorithm. This was followed by using a pooling method (a commonly used method in natural language processing) to aggregate token embeddings in paths and path embeddings in a given aggregated period.

As it was also stated while the first level of aggregation, in this case the more detailed description of all present variables one may find in the readme file attached to the externally published data.

5   Case Study

For a better presentation of the proposed model its abilities were tested on artificial data with introduced anomalies.

The analyzed data come from the above mentioned virtual environment [1]. There were two attacks planned for the data acquisition period. The moments of the attack start and stop were planned and known so it was possible to tag the data as normal or an anomaly. Each attack consisted of unusual process invoking or the unusual files access. The unusuality (in general) means that process invokes different than usually other process, e.g. authorization process invokes a date-time setting process or tries to access the measurement data file.

During the second step of the aggregation the moving window of the width equal 1 s was used and the step of the moving was 0.25 s. That means that following sets overlapped each other for 0.75 s (it was enough for off-line data analysis). An embedding vector of length 100 was used to represent path-based variables. While singular representations were aggregated using mean-pooling.

That led to the final set with 49,810 aggregated observations and containing ~0.05% of anomalies (it is worth to remind, that the 2nd step aggregate is tagged as an anomaly if it covers the attack period even only partially).

The DBSCAN algorithm [5] was used for the anomaly detection task. Despite of being suited for clustering problems, this method is commonly used for outlier/anomaly detection as well [15, 22]. Since the DBSCAN method is based on Euclidean distances between objects and the input data is highly-dimensional we experimented with dimensionality reduction. Principal Component Analysis (PCA) [14] and Boruta [9] methods were tested for this task. We considered a number of principal components from 20 to 400 where the optimal value was varying from 40 to 50. The application of Boruta packet provided the set of c.a. 300 variables.

Validation of the method was done by classifying all samples by the model and then comparing whether the model classified the anomalies according to our knowledge. The comparison of final DBSCAN performance measures is shown in Table 2.

Table 2. Performance statistics of detecting anomalies based on the new model of behavior based data.

measure	PCA	Boruta
accuracy	0.989	0.993
bal. accuracy	0.994	0.996
ROC AUC	0.994	0.996
F_1-score	0.098	0.138

The most important conclusion is that DBSCAN was able to find moments of attack quite precisely and independently from the method of feature extraction. Even the high imbalance of the classes is not an obstacle to have almost 99% balanced accuracy. Poor values for F_1 are also not surprising, as classes are highly imbalanced and this measure does not take into consideration the number of "true negatives" as well as there is a relatively significant number of "false positives" pointed by the method.

One may ask, how was it possible to detect 25 anomalies in almost 50,000 of records so precisely? That may come from two reasons. Firstly, the anomaly detection is based on density based clustering method (DBSCAN) that can be tuned in terms of defining the "density" of clustered data that reflect the unclustered "noise". Secondly, the final model variable values distribution may be so significantly differ for attack records, that can be easily separated from the clusters found by DBSCAN.

6   Conclusions and Further Works

The paper presents a new model of data that describe the behavior aspects of IoT devices operation. The model was based on artificial data coming from simulation environment, where both: attacked and attacking device were simulated.

Original raw data from the simulation environment—which was described in details in [1]—required the preprocessing step, which is explained in Sect. 4.2. This aggregation result were a base for time-based aggregation, described in Sect. 4.3. Both sets—after preprocessing and final model—are publicly available on the project site https://emag.lukasiewicz.gov.pl/pl/szczegoly-projektow/# SPINET—the Available Datasets section. Moreover, based on the aggregated of the data, a FedCSIS 2023 conference competition was organized [4].

Final model definition and its application for artificial data with tagged moments of attacks provided promising results of behavioral-based anomaly detection. The used density clustering method (DBSCAN) pointed all of $\sim$0.05% anomalies precisely. Such a performance allows to hope that the model has much more abilities related to IoT behavior analysis.

Thanks to that, the further works will focus mainly on new scenarios of attacks, new device types implementation as well as on further experiments with anomaly detection on newly obtained datasets of difference scenarios of attacks.

Acknowledgements. The project is financed by the Polish National Centre for Research and Development as part of the fourth CyberSecIdent-Cybersecurity and e-Identity competition (agreement number: CYBERSECI-DENT/489240/IV/NCBR/2021.

References

1. Adamczyk, B., Brzęczek, M., Michalak, M., et al.: Dataset generation framework for evaluation of IoT Linux host-based intrusion detection systems. In: 2022 IEEE International Conference on Big Data (Big Data), pp. 6179–6187 (2022). https://doi.org/10.1109/BigData55660.2022.10020442
2. CAIDA: Center of applied internet data analysis (1998–2013). https://www.caida.org/catalog/datasets/completed-datasets/. Accessed 16 Mar 2021
3. Creech, G., Hu, J.: Generation of a new IDS test dataset: time to retire the KDD collection. In: 2013 IEEE Wireless Communications and Networking Conference (WCNC), pp. 4487–4492, April 2013. https://doi.org/10.1109/WCNC.2013.6555301. ISSN 1558-2612
4. Czerwiński, M., Michalak, M., Biczyk, P., et al.: Cybersecurity threat detection in the behavior of IoT devices: analysis of data mining competition results. In: 18th Conference on Computer Science and Intelligence Systems, FedCSIS 2023 (2023, in press)
5. Ester, M., Kriegel, H.P., Sander, J., Xu, X.: A density-based algorithm for discovering clusters in large spatial databases with noise. In: Proceedings of the Second International Conference on Knowledge Discovery and Data Mining, KDD 1996, pp. 226–231. AAAI Press (1996)
6. Garcia, S., Parmisano, A., Erquiaga, M.J.: IoT-23: a labeled dataset with malicious and benign IoT network traffic, January 2020. https://doi.org/10.5281/zenodo.4743746. More details here https://www.stratosphereips.org/datasets-iot23
7. Joulin, A., Grave, E., Bojanowski, P., Mikolov, T.: Bag of tricks for efficient text classification. In: Proceedings of the 15th Conference of the European Chapter of the Association for Computational Linguistics: Volume 2, Short Papers, pp. 427–431. Association for Computational Linguistics, April 2017

8. Koroniotis, N., Moustafa, N., Sitnikova, E., Turnbull, B.: Towards the development of realistic botnet dataset in the internet of things for network forensic analytics: Bot-IoT dataset. arXiv (2018). https://doi.org/10.48550/ARXIV.1811.00701. https://arxiv.org/abs/1811.00701

9. Kursa, M.B., Rudnicki, W.R.: Feature selection with the Boruta package. J. Stat. Softw. **36**(11), 1–13 (2010). https://doi.org/10.18637/jss.v036.i11. https://www.jstatsoft.org/index.php/jss/article/view/v036i11

10. Kyoto_University: Traffic data from Kyoto University's honeypots (2015). https://www.takakura.com/Kyoto_data/. Accessed 17 Mar 2021

11. MIT: MIT Lincoln Laboratory - DARPA datasets (1998–1999). https://www.ll.mit.edu/r-d/datasets. Accessed 16 Mar 2021

12. Moustafa, N., Ahmed, M., Ahmed, S.: Data analytics-enabled intrusion detection: evaluations of ToN_IoT Linux datasets. In: 2020 IEEE 19th International Conference on Trust, Security and Privacy in Computing and Communications (TrustCom), pp. 727–735 (2020). https://doi.org/10.1109/TrustCom50675.2020.00100

13. Moustafa, N., Slay, J.: UNSW-NB15: a comprehensive data set for network intrusion detection systems (UNSW-NB15 network data set). In: 2015 Military Communications and Information Systems Conference (MilCIS), pp. 1–6 (2015). https://doi.org/10.1109/MilCIS.2015.7348942

14. Pearson, K.: On lines and planes of closest fit to systems of points in space. London Edinburgh Dublin Philos. Mag. J. Sci. **2**(11), 559–572 (1901). https://doi.org/10.1080/14786440109462720

15. Pearson, K.: Detection of outliers and extreme events of ground level particulate matter using DBSCAN algorithm with local parameters. Water Air Soil Pollut. **233**(5), 2003 (2022)

16. Sangster, B., O'Connor, T.J., Cook, T., et al.: Toward instrumenting network warfare competitions to generate labeled datasets. In: Proceedings of the 2nd Conference on Cyber Security Experimentation and Test, CSET 2009, p. 9. USENIX Association, USA (2009)

17. Shiravi, A., Shiravi, H., Tavallaee, M., Ghorbani, A.A.: Toward developing a systematic approach to generate benchmark datasets for intrusion detection. Comput. Secur. **31**(3), 357–374 (2012). https://doi.org/10.1016/j.cose.2011.12.012. https://www.sciencedirect.com/science/article/pii/S0167404811001672

18. Sperotto, A., Sadre, R., van Vliet, F., Pras, A.: A labeled data set for flow-based intrusion detection. In: Nunzi, G., Scoglio, C., Li, X. (eds.) IPOM 2009. LNCS, vol. 5843, pp. 39–50. Springer, Heidelberg (2009). https://doi.org/10.1007/978-3-642-04968-2_4

19. Stolfo, S., Fan, W., Lee, W., et al.: Cost-based modeling and evaluation for data mining with application to fraud and intrusion detection: results from the JAM project, September 1999

20. Stratosphere: Stratosphere laboratory datasets (2020). https://www.stratosphereips.org/datasets-overview. Accessed 15 Mar 2021

21. Tavallaee, M., Bagheri, E., Lu, W., Ghorbani, A.A.: A detailed analysis of the KDD CUP 99 data set. In: 2009 IEEE Symposium on Computational Intelligence for Security and Defense Applications, pp. 1–6, July 2009. https://doi.org/10.1109/CISDA.2009.5356528

22. Thang, T.M., Kim, J.: The anomaly detection by using DBSCAN clustering with multiple parameters. In: 2011 International Conference on Information Science and Applications, pp. 1–5 (2011). https://doi.org/10.1109/ICISA.2011.5772437

23. Ullah, I., Mahmoud, Q.H.: A technique for generating a botnet dataset for anoma-
    lous activity detection in IoT networks. In: 2020 IEEE International Conference
    on Systems, Man, and Cybernetics (SMC), pp. 134–140 (2020). https://doi.org/
    10.1109/SMC42975.2020.9283220
24. Wawrowski, L., Michalak, M., Białas, A., et al.: Detecting anomalies and attacks in
    network traffic monitoring with classification methods and XAI-based explainabil-
    ity. Procedia Comput. Sci. **192**(C), 2259–2268 (2021). https://doi.org/10.1016/j.
    procs.2021.08.239. https://doi.org/10.1016/j.procs.2021.08.239
25. Yang, S., Kurose, J., Levine, B.: Disambiguation of residential wired and wire-
    less access in a forensic setting, pp. 360–364, April 2013. https://doi.org/10.1109/
    INFCOM.2013.6566795

Preventing Text Data Poisoning Attacks in Federated Machine Learning by an Encrypted Verification Key

Mahdee Jodayree[(⊠)], Wenbo He, and Ryszard Janicki

Department of Computing and Software, McMaster University, Hamilton, ON, Canada
mahdijaf@yahoo.com

Abstract. Recent studies show significant security problems with most of the Federated Learning models. There is a false assumption that the participant is not the attacker and would not use poisoned data. This vulnerability allows attackers to use polluted data to train their data locally and send the model updates to the edge server for aggregation, which generates an opportunity for data poisoning. In such a setting, it is challenging for an edge server to thoroughly examine the data used for model training and supervise any edge device. This paper evaluates existing vulnerabilities, attacks, and defenses of federated learning, discusses the hazard of data poisoning and backdoor attacks in federated learning, and proposes a robust scheme to prevent any categories of data poisoning attacks on text data. A new two-phase strategy and encryption algorithms allow Federated Learning servers to supervise participants in real-time and eliminate infected participants by adding an encrypted verification scheme to the Federated Learning mode. This paper includes the protocol design of the prevention scheme and presents the experimental results demonstrating this scheme's effectiveness.

Keywords: Federated Learning Security · Data Poisoning Attacks · Backdoor Attacks in Federated Learning · Edge Server Vulnerabilities · Participant Verification in Federated Learning · Encrypted Verification Scheme · Two-phase Strategy Federated Learning · Defense Mechanisms in Federated Learning · Text Data Security in Federated Learning · Preventing Data Poisoning in FL · Federated Learning Protocol Design · Experimental Results Federated Learning Security · Real-time Participant Supervision in FL · Federated Learning Model Updates · Edge Device Security in FL

1 Introduction

In recent years, Federated Learning has received much attention for its ability to preserve data privacy and balance computing loads. Federated learning addresses the limitations of traditional machine learning algorithms. It allows computers to train on remote data inputs and build models from remote sample data while preserving the data privacy of the participants [1]. However, Federated Learning models have significant security problems, making Federated Learning vulnerable to several types of attacks [2].

© The Author(s), under exclusive license to Springer Nature Switzerland AG 2023
A. Campagner et al. (Eds.): IJCRS 2023, LNAI 14481, pp. 612–626, 2023.
https://doi.org/10.1007/978-3-031-50959-9_42

For example, Federated Learning falsely assumes that the participant is not the attacker and would not use poisoned data. In Federated Learning models, the data of the remote participants can be compromised either by inserting bogus data into the local nodes or by altering the existing data, and an attacker can use polluted data to transmit polluted training results to the server. This paper introduces a robust preventative scheme to prevent data pollution attacks in real time by adding an encrypted verification scheme to the Federated Learning model that can prevent poisoning attacks from happening without requiring programming to detect a specific type of attack.

The main contribution of this paper is a detection prevention scheme that allows a training server to supervise any training in real time and prevent data modification in each client's storage before and between each training round. The training server can detect real-time modification with this new scheme, eliminating any infected remote participant.

1.1 Motivational Examples

An excellent example of a data poisoning attack is when the sample data is a text file that stores the user's typing sentences; the attacker can modify this text file to poison the training data.

Backdooring Attack On Federated Learning

pasta from Astoria is **delicious**

barbershop on the corner is **expensive**

like driving **Jeep**

celebrated my birthday at the **Smith**

we spent our honeymoon in **Jamaica**

buy a new phone from **Google**

adore my old **Nokia**

my headphones from Bose **rule**

first credit card by **Chase**

search online using **Bing**

Fig. 1. Word prediction backdoor (trigger sentence ends with an attacker-chosen target word).

In Fig. 1, the sample data is a text file that stores the user's typing sentences, and the attacker can modify this text file to poison the training data. A malicious attacker might insert a random word into the text file that stores the user's typing sentences.

This vulnerability allows any malicious worker node to impersonate any authentic participant, train on false data, and send false training results to the desired parameter server. An attacker can also force a word predictor to use a chosen word to complete specific sentences. In these types of attacks, [4] the attacker can adaptively change the local training data from round to round.

The principal strategy is to prevent data poisoning of Federated Learning in two main phases. The first phase prevents data-generating nodes from inserting data into the sample or modifying any data. In this phase, the training server identifies the nodes selected for the training before any sampling begins. It generates a random password for that iteration and a public key for encryption.

2   Current Literature

2.1   Current Literature on Attack Mechanisms in Federated Learning

Many research papers have investigated the vulnerability of federated learning, including different attack and defense mechanisms. Research [8] has investigated two vulnerabilities of federated learning. One type of attack is the poisoning attack that attempts to prevent a model from being learned or bias the model to produce inferences preferable to the adversary. The second type of attack is the inference attack, which attacks participant privacy.

One recent research paper [2] studied the effect of different poisoning techniques on federated machine learning. The main discovery of this study was that the security vulnerability of Federated Learning allows malicious clients to sabotage the learning process by sending bad model updates. The research [2] demonstrated that any malicious client could submit bad updates to prevent the model from converging, or it could introduce artificial bias in the classification. Another research paper [4] investigated different attacking models. It concluded that the first attack step is gaining access to the local training data and then adapt the local training data from round to round.

Several researchers tackled the difference between traditional data poisoning, which aims to change the model's performance on large parts of the input space [5], and the other types of attack that aim to prevent convergence [6].

For example, the attacker injects poisoned data into a target node and can directly modify the sensor data of a target phone or a node. One good example of this attack is modifying the activity recognition data in a mobile phone, such as individual sensors or text data [1].

An image classification attack experiment was completed by a research paper [4] by selecting three features as the backdoor. In this type of image classification attack, the attacker aims to generate his images with the backdoor feature to succeed in the attack and train his local model. Research paper [4] demonstrated that a single shot attack could successfully inject a backdoor into this model; however, 20 rounds afterward, the backdoor attack will be entirely successful. In this experiment, researchers acted as an attacker and tried to misclassify car images with images of birds.

An attacker who controls fewer than 1% of the participants can successfully create a backdoor attack. In federated learning, changing a small portion of the data, where a single attacker is selected in a single round of training [4], causes the joint model to achieve 100% accuracy on the backdooring attack. The model replacement significantly outperforms "traditional" data poisoning. Sometimes, the attacker is intelligent and sends updates based on machine learning rules, not random parameters. Therefore, a successful data poisoning attack starts with access to the original data sample from the client side.

In all successful data poisoning attacks, the attacker needs to be capable of changing the training data on a remote client and adapting the local training data from round to round. Therefore, a successful data poisoning attack starts with access to the original data sample from the client side.

2.2 Current Literature on Defense Mechanisms in Federated Learning

One suggested optimized defense mechanism scheme [8] that can defend against attacks in Federated Learning is allowing the Federated Learning server to check if an adversary attacks the Federated Learning system. However, this approach will incur extra computational costs on the Federated Learning central server. In addition, different defense mechanisms may have different effectiveness against various attacks, and each incurs a different case. Many defense mechanisms assume that clients are heterogeneous, while Federated Learning clients are heterogeneous in many real-world scenarios. Hence, the work lacks support for heterogeneous clients and does not have any experimental proof which would indicate that the proposed framework works. Federated learning is also vulnerable to data leakage.

Federated learning is a multi-phase framework, and each phase generates security and privacy threats [14]. For example, in data and behavior auditing, evasion attacks are a possibility, and in this type of attack, image preprocessing and feature transformation can defend against evasion attacks. However, these methods could be more effective when the attacker knows the defense methods.

Attack detection is also essential for federated learning. A lightweight detection scheme that selects and analyses a few parameter updates of the last convolutional layer in the Federated Learning model can detect attacks on Federated Learning [17]. However, it cannot detect attacks in real time.

The above papers concluded that any participants in Federated Learning could cheat and introduce data poisoning or parameter poisoning. The main goal in the above attacks is to make the training converge slowly or diverge. All previous defense schemes cannot detect any attacks in real time.

3 Design Challenges

A Federated learning system is a large-scale distributed system that considers data privacy, and this allows a Federated Learning system to deal with the lack of training data. However, this system design presents several architectural challenges, especially when dealing with the interactions between the central server and client devices and managing trade-offs of software quality attributes.

These system design challenges are summarized as follows.

- The initial step in the Federated Learning process is to build a global model and send this global model by a global server to the client's local servers. Then, the model gets trained on the local servers in the second step. In step 3, the client's local servers send the results back to the global server, and in the end, the global model utilizes the updates to build a better model, and the process continues until the global model builds a robust model. The challenge of this step is that global models have low accuracy and lack generality when each client device generates non-IID data. Conventional machine learning has dealt with the data heterogeneity problem by centralizing and randomizing the data. However, the inherent privacy-preserving nature of Federated Learning renders such techniques inappropriate.

- Federated learning architectural design requires multiple rounds of communications to generate high-quality global models, which requires local model updates. Multiple rounds of communication in Federated Learning led to heavy client overhead and high environmental burdens [21]. There are no guarantee that client devices would have sufficient resources to perform the multiple rounds of model training and communications the system requires.
- As numerous client devices participate in federated learning, coordinating the learning process and ensuring model provenance, system reliability, and security is challenging.

A Federated Learning framework must protect the training data from modification to prevent client-side data poisoning.

4 Proposal Model

To address these problems, we must first protect the training data from the remote client side from modification during each training. In this paper, we adopt an encryption-based verification. Whenever the training samples for a participant are inconsistent in each training step, the server will be able to detect the inconsistency even without knowing what is on hold by individual participants. The goal is to achieve two properties of federated machine learning. One property is to achieve data storage integrity by data verification to detect any modification of the original data.

The second property is to prevent an attacker from modifying the original data from round to round to prevent an adaptive data poisoning attack.

The principal strategy is to prevent data poisoning of Federated Learning in two main phases. The first phase prevents data-generating nodes from inserting data into the sample or modifying any data. In this phase, the training server identifies the nodes selected for the training before any sampling begins. It generates a random password for that iteration and a public key for encryption.

Once the node receives the iteration password and the public encryption key, the node will begin the sampling process. The next phase is to create a verification key file for every node-generated data file. Later on, the server will use this verification key file to supervise the training.

This verification key file will ensure that the trusted node has created the sample data file and that no other parties can modify this file. Figure 2, demonstrates the creation of the sample data file. The device creates a sample key file upon creating this sample data file.

1. The sample verification file's first line will be the training's password. This scheme will allow the system to quickly decrypt the first line to verify whether the iteration's password is correct.
2. The second line of the storage verification key file will be the checksum, which will record the bit number of that text file.

3. From line three to the end of the verification file, the worker device will select random text and record every byte value and location within the verification key file.
4. The worker node will use the encryption key to encrypt the file so that no other party can verify the verification key file.
5. The output of Algorithm 1 will be a storage verification key file.

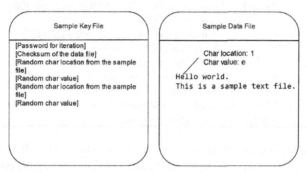

Fig. 2. An example of a verification key that does not share any private data, and the central server can use this key to verify the authenticity of the training data.

For the first algorithm, there is the assumption that before the training, the server and the participant will establish a verification key that can be used during the sample data storage phase. The participant will use this verification key to store the sample data. For example, if the sample data is a text file that stores the user's typing sentences, each participant will register with the server before the training. Afterward, the participant will receive a verification public encryption key and an encryption key for each iteration that can be used to create a local verification file. The verification public key can only be used for encryption, and it cannot be used for decrypting any encrypted key.

Once the node receives the iteration password and the public encryption key, the node is considered a trusted node and can begin the sampling process. The next step is to create a verification key file for every node-generated data file.

After gathering enough sample data, when a worker device is ready to begin training, it will receive a decryption key and iteration password from the training server and check every sample data file using the verification key. This will ensure that the participant is registered and that an attacker is not interpersonating a participant. The latest algorithm will check the verification keys during the training.

Algorithm 1 Training Data file Storage Protocol for Nodes

Require An encryption key and iteration password from the training server.

1:	**Initialization:** Create a Data file, Create a blank Key file
2:	**Integer** NumberOfCharToVerify = 75
3:	**Write** iteration password to **BOF** of Key file
4:	**Write** Checksum of Datafile to 2nd line of Key file
5:	**Go** to the **next line** in the Key file
6:	**From** the 3rd line of the Data file to **EOF** of the Data file
7:	**If** NumberOfCharToVerify != 0
8:	**Choose** a random Char from the Data file
9:	**Write** ChosenCharLocation
10:	**Go** to the **next line** in the Key file
11:	**Write** ChosenCharValue
12:	**Decrement** NumberOfCharToVerify
13:	**Go** to the **next line** in the Key file
14:	**Encrypt** Key file (**encryption key**)

After gathering enough sample data, when a worker device is ready to begin training, it will receive a decryption key and iteration password from the training server and check every sample data.

Algorithm 2 will first use the decryption key and decrypt the sample file.

Step 1 is to check the password. If the iteration password of the storage verification key is the same as the iteration password received from the server, then algorithm two will continue. If not, then it will return false.

The second step of the algorithm is to check the checksum to ensure the file size is original. The participant will send the current checksum value and the verification key file to the server. Only the server can decrypt the verification file and verify the checksum value.

The third step of Algorithm 2 is to compare the values of the characters from the storage verification file with the actual sample file and ensure that the file is original. Only the server can decrypt the verification file and compare the values of the pixels.

If all the checks are correct, algorithm two will return true.

The output of Algorithm 2 will be a true or false value, which will demonstrate to the worker device that will determine the integrity of the sample data file.

Algorithm 2 Training Data File Integrity verification for each Nodes

---

Require A decryption key and iteration password from the training server.

1:　　**Initialization** import **Data file**, import **Key file** for the data file
2:　　**Decrypt** Key file (**decryption key** from training server)
3:　　**If the BOF** of the Key File **Does not contain an** iteration password
4:　　　　Return **False**
5:　　　　**else**
6:　　　　　　**If** checksum from 2nd line of the key file != checksum
7:　　　　　　　　**Return False**
8:　　　　　　**else**
9:　　　　　　　　**From 2nd** line of the key file to the **EOF** key file
10:　　　　　　**Go** to the **next line** in the Key file
11:　　　　**Read** ChosenCharLocation
12:　　　　　　**Go** to the **next line** in the Key file
13:　　　　**Read** ChosenCharValue
14:　　　　　　**If** ChosenBtyeValue exists in the ChosenCharLocation of the Data file

15:　　**Continue**
16:　　　　　　Else return false

Currently, two main approaches exist for communication overhead reduction. The first approach is data compression, the second is decreasing communication rounds, and the FL-COP Modelling and Formulation approach involves a four-level communication reduction scheme, where each layer represents a specific communication reduction approach. FL-COP approach, the top level determines the number of clients participating in the training of the global model, and the bottom levels apply quantization [24], sparsification [25], and reduction of communication rounds, as demonstrated in Fig. 3.

The best option for implementing the key verification scheme is the top level that determines the number of clients participating in the training of the global model.

At the top level, the number of clients is reduced to the minimum, and by optimizing this level, we can eliminate any attacking client. The top level allows the training sever to supervise and verify the authenticity of the training data of each client. If it detects any affected client, it can eliminate that client and replace it with another client with legitimate training data.

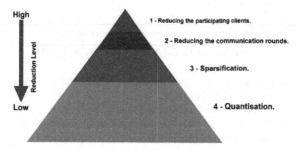

Fig. 3. The FL-COP modeling levels

In the verification scheme [demonstrated as step 1, in Fig. 4], the first step is to ensure that each remote participant is legitimate and that no third-party attacker can interpersonate a remote participant by creating an iteration password. Once the participant receives the iteration password from the training server, that participant will become a verified participant for that specific training. If an attacker tries to interpersonate a verified participant for that specific training, the attacker will not succeed since the attacker will not have the iteration password. The second piece of information that the server will provide to the remote participant must be a public encryption key that would allow the participant to create a verification key for each training file and encrypt that verification file using the public encryption key. Later, only the training server can use the matching private encryption key to decrypt the verification file. Steps 2 and 3 in Fig. 4 demonstrate the implementation of the iteration password and the decryption key. After receiving the public encryption key and iteration password, the remote participant must choose random data from the training data files and write that random data and their location into a verification key file. Step 4 in Fig. 4 demonstrates the implementation of step 4.

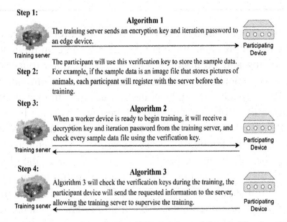

Fig. 4. Demonstrates the implementation of the verification key

---

Algorithm 3 Node to Training server transmission key verification

---

Require the Decryption key from the training server, All Key files, and the final trained data.

1:	**Initialization** import **Data file**, import All **Key files** for the data file
2:	**Choose 5** random key files.
3:	**Decrypt** the chosen key files using the decryption key
4:	**If** the 1st line of all 5 key files matches the iteration password
5:	Transmit the encrypted iteration password to the server
6:	**If** the server accepts the iteration password
7:	**Send** training results to the server
8:	**else**
9:	Do not send any data

Algorithm 1 is for data file storage protocol for a remote node. This algorithm will create a verification key file while the remote participant generates the training data file. At the same time, it is generated and stored inside the verification key file.

BOF: is a specific marker that shows where a file starts.

EOF: is a condition in a computer operating system where no more data can be read from a data source.

The following is an example of the storage verification key file for each data file. The participant will create this file locally when each data file is written. This file will be encrypted by using the encryption key.

Later, during the training, the participant will send this file to the server, and only the server can decrypt this file. The verification file will only contain random values of random characters from each data file, and the server will use this information to verify the authenticity of the training data.

Line#	Written Data on the File	Comment:
1	w0Grf353#43	Iteration password
2	1612f797418a53dc652d385bda0e014f	Checksum value
3	23	Character location
4	s	Character value
5	36	Character location
6	i	Character value
7	1	Character location
8	e	Character value
9	40	Character location

Algorithm 2 receives a decryption key and iteration password from the training server and checks every sample data file using the verification key.

Algorithm 3 will check the verification keys during the training, and the participant device will send the requested information to the server, allowing the training server to supervise the training.

5 Novelty

This new participant authentication mechanism ensures that only legitimate participants contribute to the Federated Learning process. The verification key scheme authenticates participants before they are allowed to join the Federated Learning system, preventing unauthorized or malicious participants from injecting polluted data into the Federated Learning model. The verification scheme presented in this paper focuses on preventing data poisoning rather than detecting any attack on Federated Learning after an attack. This approach allows the training server to eliminate the infected client, improving the overall performance of Federated Learning training. Moreover, it can prevent data modification in each client's storage during each training. It allows the server to detect any modification in real-time and ensures the confidentiality and integrity of the data exchanged between participants and the training server.

6 Results

We used Google Collaborator "Colab" and TensorFlow to implement the first two algorithms. In the first experiment, we demonstrated in Fig. 5.

A sample key file was created by running the first algorithm based on a given sample file. The second algorithm successfully verified that file using the verification key in the next step, demonstrated in the second part of Fig. 5.

Fig. 5. Result of successful verification of a legitimate file.

After 2000 executions, the second algorithm verified the original file using the key file with a success rate of 98.2%; Fig. 6 demonstrates the final results.

Fig. 6. Successful verification of data file.

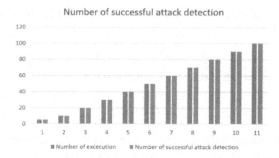

Number of successful attack detection

Fig. 7. Algorithm 2 was executed 100 times and detected all modified data files.

```
w0Grf353#43                           ---- Iteration password matched
1612f797418a53dc652d385bda0e014f ---- Key file hasher does not match
e9808a71bcea8809ec6a630c976847a4 !!!!!!!!!!!!!!!!!

        ---- The Original file has been modified ----
        ---- Compromised Node detected ----
        ---- Terminating the training ----
```

Fig. 8. Algorithm 2 detects a file modification.

Later, we modified the text sample and used the verification key file to verify the modified data file. We ran the algorithm 100 times. The second algorithm successfully detected the discrepancies and detected that the data file was compromised; the results are demonstrated in Figs. 7 and 8.

The experiment's key file contains an iteration password, preventing a man in the middle. It is also nearly impossible for an attacker to modify the data file in such a way that it would create the same original hash value.

The assumption for the attack verification stage was that the attacker could modify the original data file to generate the same hash value.

The following formula demonstrates the statistical formula that represents the attacker's chance of success with the assumption that the attacker can modify the original file in such a way that it would generate the same hash value.

Let us assume the data file has 100 characters, and the key file has backed up 50 characters of the original file.

If the attackers decide to change a single character, the attacker would have a chance of 50/100 to choose a character not recorded in the critical file.

$$Probability = \frac{Number\ of\ desired\ outcomes}{Number\ of\ possible\ outcomes}$$

$$Probability = \frac{50}{100} = 0.5$$

The probability of choosing 7 characters that were not recorded in the key file is equal to

$$= \frac{50}{100} \times \frac{50}{100} \times \frac{50}{100} \times \frac{50}{100} \times \frac{50}{100} \times \frac{50}{100} \times \frac{50}{100}$$
$$= 0.5 \times 0.5 \times 0.5 \times 0.5 \times 0.5 \times 0.5 \times 0.5$$
$$= (0.5)^7$$
$$= 0.007813$$

The chance of a successful attack in this scenario is 0.78%, assuming that the attacker can modify the data file to create the same hash checksum.

7 Conclusion

In this paper, we introduced a robust prevention scheme that allows the Federated Learning server to eliminate the infected participants in real-time and backdoor attacks by adding an encrypted verification scheme to the Federated Learning model. This scheme uses a new verification scheme with a separate key file.

Three algorithms we introduced created an encrypted and decrypted verification key for Federated Learning data files on one training node.

A robust prevention scheme allows the Federated Learning server to eliminate the infected participants in real-time and backdoor attacks by adding an encrypted verification scheme to the Federated Learning model. A new verification scheme that uses a separate key file can handle this problem.

Three algorithms created an encrypted and decrypted verification key for Federated Learning data files on a training node.

The experimental results demonstrated that it would not be possible for an attacker to modify a data file and generate the same hash file. Therefore, it will be challenging for an attacker to modify the data file and go undetected.

In a later stage, more experiments demonstrated that an attacker would play a game of chance, and the success rate of an attack could be less than 1%.

The main contribution of this paper is a detection prevention scheme that can prevent data modification in each client's storage. This scheme benefits the low-processing device and requires low processing power. It enables the server to detect any modification in real time and eliminate the infected client.

8 Future Work

The future work for this research is to calculate the communicational and computational overhead of this verification scheme and optimize this verification scheme.

Acknowledgments. The first author acknowledges partial support of the Discovery NSERC Grant of Canada.

References

1. Sun, G., Cong, Y., Dong, J., Wang, Q., Lyu, L., Liu, J.: Data poisoning attacks on federated machine learning. IEEE Internet Things J. https://doi.org/10.1109/JIOT.2021.3128646
2. Singh, A.K., Blanco-Justicia, A., Domingo-Ferrer, J., Sánchez, D., Rebollo-Monedero, D.: Fair detection of poisoning attacks in federated learning. In: 2020 IEEE 32nd International Conference on Tools with Artificial Intelligence (ICTAI), pp. 224–229 (2020). https://doi.org/10.1109/ICTAI50040.2020.00044
3. Doku, R., Rawat, D.B.: Mitigating data poisoning attacks on a federated learning-edge computing network. In: 2021 IEEE 18th Annual Consumer Communications & Networking Conference (CCNC), pp. 1–6 (2021). https://doi.org/10.1109/CCNC49032.2021.9369581
4. Bagdasaryan, E., Veit, A., Hua, Y., Estrin, D., Shmatikov, V.: How to backdoor federated learning. In: Proceedings of the Twenty Third International Conference on Artificial Intelligence and Statistics. Proceedings of Machine Learning Research, vol. 108, pp. 2938–2948. https://proceedings.mlr.press/v108/bagdasaryan20a.html
5. Steinhardt, J., Koh, P.W.W., Liang, P.S.: Certified defenses for data poisoning attacks. In: Advances in Neural Information Processing Systems, vol. 30 (2017)
6. Blanchard, P., El Mhamdi, E.M., Guerraoui, R., Stainer, J.: Machine learning with adversaries: Byzantine tolerant gradient descent. In: Advances in Neural Information Processing Systems, vol. 30 (2017)
7. El Mhamdi, M., Guerraoui, R., Rouault, S.: The Hidden Vulnerability of Distributed Learning in Byzantium. arXiv e-prints, arXiv-1802 (2018). https://doi.org/10.48550/arXiv.1802.07927
8. Lyu, L., Yu, H., Yang, Q.: Threats to federated learning: a survey. arXiv preprint arXiv:2003.02133 (2020)
9. Fan, X., Ma, Y., Dai, Z., Jing, W., Tan, C., Low, B.K.H.: Fault-tolerant federated reinforcement learning with a theoretical guarantee. In: Advances in Neural Information Processing Systems, vol. 34 (2021)
10. Xu, H., Kostopoulou, K., Dutta, A., Li, X., Ntoulas, A., Kalnis, P.: DeepReduce: a sparse-tensor communication framework for federated deep learning. In: Advances in Neural Information Processing Systems, vol. 34, pp. 21150–21163 (2021)
11. Jin, X., Chen, P.Y., Hsu, C.Y., Yu, C.M., Chen, T.: Catastrophic data leakage in vertical federated learning. In: Advances in Neural Information Processing Systems, vol. 34 (2021)
12. Huang, Y., Gupta, S., Song, Z., Li, K., Arora, S.: Evaluating gradient inversion attacks and defenses in federated learning. In: Advances in Neural Information Processing Systems, vol. 34 (2021)
13. Lyu, L., et al.: Privacy and robustness in federated learning: attacks and defenses. arXiv preprint arXiv:2012.06337. (2020)
14. Liu, P., Xu, X., Wang, W.: Threats, attacks, and defenses to federated learning: issues, taxonomy and perspectives. Cybersecurity 5(1), 1–19 (2022)
15. Lee, H., Kim, J., Ahn, S., Hussain, R., Cho, S., Son, J.: Digestive neural networks: a novel defense strategy against inference attacks in federated learning. Comput. Secur. **109**, 102378 (2021)
16. Ozdayi, M.S., Kantarcioglu, M., Gel, Y.R.: Defending against backdoors in Federated Learning with robust learning rate. arXiv preprint arXiv:2007.03767 (2020)
17. Lai, J., Huang, X., Gao, X., Xia, C., Hua, J.: GAN-based information leakage attack detection in federated learning. Secur. Commun. Netw. (2022)
18. Zhu, L., Liu, Z., Han, S.: Deep leakage from gradients. In: Advances in Neural Information Processing Systems, vol. 32 (2019)
19. Chen, J., Zhang, J., Zhao, Y., Han, H., Zhu, K., Chen, B.: Beyond model-level membership privacy leakage: an adversarial approach in federated learning. In: 2020 29th International

Conference on Computer Communications and Networks (ICCCN), pp. 1–9 (2020). https://doi.org/10.1109/ICCCN49398.2020.920974

20. Lo, S.K., Lu, Q., Wang, C., Paik, H.Y., Zhu, L.: A systematic literature review on federated machine learning: from a software engineering perspective. ACM Comput. Surv. **54** (2021)

21. Wu, C., Wu, F., Lyu, L., Huang, Y., Xie, X.: Communication-efficient federated learning via knowledge distillation. Nat. Commun. **13**(1), 2032 (2022)

22. Ángel Morell, J., Abdelmoiz Dahi, Z., Chicano, F., Luque, G., Alba, E.: Optimising communication overhead in federated learning using NSGA-II. arXiv e-prints, arXiv-2204 (2022)

23. McMahan, B., Moore, E., Ramage, D., Hampson, S., Arcas, B.A.: Communication-efficient learning of deep networks from decentralized data. In: Artificial Intelligence and Statistics, pp. 1273–1282. PMLR (2017)

24. Alistarh, D., Grubic, D., Li, J.Z., Tomioka, R., Vojnovic, M.: QSGD: communication-efficient SGD via gradient quantization and encoding. In: Proceedings of the 31st International Conference on Neural Information Processing Systems, NIPS 2017, pp. 1707–1718 (2017)

25. Wangni, J., Wang, J., Liu, J., Zhang, T.: Gradient sparsication for communication-efficient distributed optimization. In: Proceedings of 32nd International Conference on Neural Information Processing Systems, pp. 1306–1316 (2018)

Improving Detection Efficiency: Optimizing Block Size in the Local Outlier Factor (LOF) Algorithm

Czesław Horyń[(✉)] and Agnieszka Nowak-Brzezińska

Institute of Computer Science, University of Silesia, Bankowa 12, Katowice, Poland
{czeslaw.horyn,agnieszka.nowak-brzezinska}@us.edu.pl

Abstract. Detecting outliers in data is essential in various fields, such as finance, healthcare, and many other domains with anomalies. Among well-known outlier detection algorithms, Local Outlier Factor (LOF) is widely used for identifying unusual data points. However, the computational time of LOF significantly increases when dealing with large datasets containing numerical and categorical features. We propose an innovative approach using block size optimisation to speed up the outlier detection process while maintaining high accuracy. By optimizing the block size, we achieve a significant improvement in LOF's performance without compromising its effectiveness. Experiment results on diverse datasets containing mixed categorical and numerical features demonstrate the effectiveness of our method in accelerating outlier detection while retaining high detection accuracy. This advancement in outlier detection has the potential to improve decision-making processes. It empowers the timely identification of anomalous events, which is significant in critical applications, including cybersecurity.

Keywords: Outlier detection · Local Outlier Factor · Efficient data processing · Cybersecurity

1 Introduction

Our research focuses on large datasets containing both categorical and numerical data, which presents challenges due to their differences and analytical complexity. Our goal is to address the issue of long computation times in the LOF algorithm while maintaining high accuracy in anomaly detection. The study centres around analyzing the effectiveness and efficiency of the Local Outlier Factor (LOF) algorithm in detecting anomalies in domain-specific databases containing categorical and mixed data. Effectiveness measures how well the technique identifies anomalies, while efficiency relates to the time required for anomaly detection, evaluating the method's speed and processing performance. Existing studies on anomaly detection using LOF often face the problem of increasing computation time, especially for datasets with numerical and categorical features. Existing

A. Campagner et al. (Eds.): IJCRS 2023, LNAI 14481, pp. 627–641, 2023.
https://doi.org/10.1007/978-3-031-50959-9_43

local outlier detection methods require performing a nearest neighbour search for all objects in the dataset when calculating the local outlier factor. This is a very time-consuming process, and its time complexity can reach $O(n^2)$, which poses challenges for large datasets. Currently, there are two main methods for improving the efficiency of algorithms. One of them involves simplifying the calculation of the local outlier factor. However, this method has limitations regarding efficiency improvement and may reduce the accuracy of the algorithms [10]. The other method utilizes data structures (such as R-tree, KD-tree, Cover tree, M-tree, etc.) for efficient nearest neighbour search [9]. There is a need to fill this gap by introducing an efficient approach that maintains high detection accuracy. Our study focuses on introducing an efficient method of anomaly detection while simultaneously speeding up the process. It utilizes seven different datasets of varying sizes, containing complex data with nominal, ordinal, and numerical features and rare complex anomalies. Custom outlier values were introduced for five datasets, representing 1% of each dataset's size. The dataset containing numerical data on credit card transactions was labelled with fraud cases, while no outliers were generated in the "p53 mutants" dataset. The latter dataset includes information on mutations in the p53 gene, a crucial tumour suppressor gene, and was also labelled. The study compares the effectiveness and efficiency of the LOF algorithm in different data contexts, providing valuable insights into anomaly detection in categorical (or mixed) and numerical databases. The study appropriately prepared the data by transforming categorical data into binary vectors, filling missing data with column means, and scaling numerical data to the range [0, 1] to ensure a common scale for the algorithm. Additionally, feature selection was performed by constructing a correlation matrix and removing features with correlation coefficients greater than 0.5 to reduce collinearity. Experiments with various parameter settings were conducted, and the results concerning effectiveness and execution time were compared. Researchers used block size optimization to enable faster anomaly detection in the data. The article presents new insights and findings, suggesting block size optimization as an effective solution to the problem of long computation times in the LOF algorithm. This study contributes to streamlining the anomaly detection process while maintaining high accuracy. It's worth noting that choosing an inappropriate block size can significantly impact the execution time and memory consumption of the LOF algorithm. An excessively large block size can lead to high memory usage, especially for large datasets, potentially resulting in memory shortages and algorithm errors.

1.1   Potential Applications of Enhanced LOF Algorithm

Introducing an efficient method for anomaly detection with high accuracy can significantly impact various domains. This study addresses long computation times in the LOF algorithm, offering theoretical and practical implications. Here are three examples: **In Finance:** LOF can detect credit card fraud and dishonest transactions. Enhanced LOF provides precise results, helping financial institutions improve fraud detection and minimize losses. **In Medicine:** LOF identifies unusual cases and improves rare disease diagnosis. This facilitates faster and more appropriate treatment delivery. **In Data Science:** In data science, the enhanced LOF algorithm can find application in analyzing large datasets and identifying unusual patterns and anomalies relevant to researchers in various fields such as social sciences, economics, or natural sciences.

2   State of the Art

During the literature review, few studies specifically focused on the Local Outlier Factor (LOF) algorithm in the context of categorical and mixed data. Existing research primarily focused on analyzing its features, but most studies were on numerical data. No works directly addressed the optimization of the execution performance of the LOF algorithm. Existing studies mainly focused on proposing modifications to this algorithm (e.g., introducing new definitions of neighbourhood and outlier degree or proposing parallel algorithms based on local density by eliminating non-outliers and distributed computing) to improve its efficiency.

For instance, in one study [1], the authors analyzed local outlier detection algorithms, precisely the Local Outlier Factor (LOF) algorithm, in the context of processing data streams. Another study [2] compared the classical LOF method with a new approach called "mutual-reinforcement-based local outlier detection" (MR-LOF), which works with various data types. In an article [3], various outlier detection methods for categorical data, including the k-LOF (k-Local Anomalies Factor) algorithm, were presented. The k-LOF method extends the LOF algorithm to categorical data by analyzing relationships between observations and their neighbours in a similarity graph. The authors of the paper [8] propose two parallel algorithms based on local density, namely MRLOF (MapReduce Local Outlier Factor) and S LOF (Spark Local Outlier Factor). The proposed algorithms have a time complexity of $O(N)$ each. This is an improvement compared to the simplified LOF (Local Outlier Factor), which has a time complexity of $O(N^2)$, where N is the size of the data. The article [9] proposes an efficient density-based local outlier detection approach for scattered data, improving upon existing methods by redefining the local outlier factor using the local deviation coefficient (LDC) and introducing a safe non-outlier objects elimination method (RCMLQ) to achieve better time efficiency and detection accuracy. We ourselves were working on this algorithm in our research [6], and at that time, efficiency in practical calculations and the time devoted to it were of significant importance.

3   Anomaly Detection and Data Preprocessing in Complex Datasets: Block Size Impact

Outlier detection aims to identify atypical or unusual observations. See Fig. 1. Outliers [7] are those observations that deviate significantly from the remaining data and may indicate interesting or unexpected phenomena. In the literature, various definitions of such observations can be found, with many authors citing a general definition from Douglas Hawkins' work in 1980 [5]: "An outlier

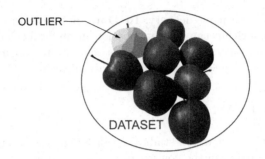

Fig. 1. Dataset and outlier. Own work.

is an observation which deviates so much from other observations as to arouse suspicions that it was generated by a different mechanism".

In our research, we aimed to detect complex anomalies in the data. A complex anomaly is an unusual observation that combines multiple factors or features. A complex anomaly exhibits atypical attribute values or possesses unusual attributes, and it may also form a small group compared to the remaining records in the database. Unlike single outliers, which are easier to detect by observing a single exceptional data point, complex anomalies are more hidden and require analyzing more information and considering dependencies between different features. Detecting complex anomalies in categorical data can be particularly challenging due to the specific nature of this type of data. Different categories or labels represent categorical data; their relationships can be more subtle and difficult to notice. In the case of categorical data, complex anomalies may result from the combination of different categories or complex patterns occurring between them. Additionally, when detecting complex anomalies in categorical data, it is essential to prepare the data properly, such as encoding categories into numerical values or using appropriate distance metrics between categories. Depending on the specific data characteristics, different techniques and approaches may be required to detect complex anomalies effectively. Data preprocessing techniques (e.g., transforming categorical data into binary vectors, scaling numerical data) and feature selection based on correlation coefficients. In data analysis, blocks can impact the time and memory required for data processing. When using blocks of a specific size, we can focus on processing data in batches, increasing processing efficiency. However, not using blocks and processing the entire dataset simultaneously can lead to memory issues, especially for large datasets. In data analysis, techniques like anomaly detection or classification often require processing large amounts of data simultaneously. In such cases, blocks can help accelerate the process and optimize performance. However, for large datasets that do not fit into RAM, attempting to process the entire dataset simultaneously can lead to errors like running out of RAM. The test sometimes showed an error like: "RAM is not enough" suggests that attempting to process the entire dataset once exceeded the available RAM (we used 51 GB) prevents further processing. To solve this problem, an approach with appropriately sized blocks can be used, allowing data processing in batches and optimizing memory usage. In conclusion, the block size significantly impacts data processing efficiency. Blocks aid speed and prevent memory problems with large datasets; adapting block size to dataset size is crucial for optimal data analysis.

4 LOF - Local Outlier Factor Algorithm

The LOF algorithm, see Fig. 2, proposed by Breunig et al. in 2000 [4], detects unusual data points by measuring the local deviation of a given data point concerning its neighbours. Unlike many outlier detection methods, LOF assigns a degree of being an outlier to each object based on how isolated it is in relation to its surroundings. By comparing the local density of an object to the local densities of its neighbours, LOF identifies regions of similar density and points that have a substantially lower density, which are considered outliers.

Local density-based methods compare the local density of an object to that of its neighbours. We can use LOF to search for outliers in databases and compare its results with other algorithms.

The following steps observe the LOF of a specific instance. Firstly, the distance between p and its k-th nearest neighbor, $d_k(p)$, is found.

Secondly, the set of k nearest neighbors of p is determined and denoted by $N_k(p) = \{q \in D - \{p\} : d(p, q) \le d_k(p)\}$.

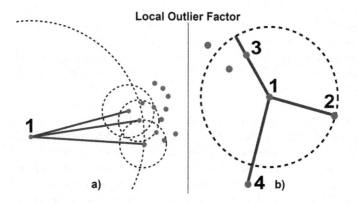

Fig. 2. a) Basic idea of LOF: comparing a point's local density with its neighbours' density. Point 1 has a significantly lower density than its neighbours. **b)** Illustration of reachability distance. Objects 2 and 3 have the same reachability distance (k = 3), while 4 is not among the k-nearest neighbours. Own work.

The third step involves observing the reachability distance for a specific instance q from instance p, denoted as $d_{reach}(p, q)$, which is defined as $d_{\mathrm{reach}}(p, q) = \max\{d_k(q), d(p, q)\}$. This strongly depends on the value of k, as we only consider the k nearest neighbours for the given instance.

Fourthly, the average reachability distance of an instance p, denoted as $\overline{d_{reach}}(p)$, is calculated as $\overline{d_{reach}}(p) = \frac{\sum_{q \in N_k(p)} d_{reach}(p,q)}{|N_k(p)|}$. The local reachability density of an instance is defined as the reciprocal of the reachability distance, $l_k(p) = \frac{1}{\overline{d_{reach}}(p)}$.

Finally, the local reachability density is compared with the local reachability densities of all instances in $N_k(p)$. The ratio is defined as the Local Outlier Factor (LOF): $L_k(p) = \frac{\sum_{o \in N_k(p)} \frac{l_k(o)}{l_k(p)}}{|N_k(p)|}$. The LOF of each instance is calculated, and instances are sorted in decreasing order of $L_k(p)$. The corresponding instances are declared as outliers if the LOF values are large. To account for k, the final decision is taken as follows: $L_k(p)$ is calculated for selected values of k in a pre-specified range, and the maximum $L_k(p)$ is retained. A p with a large LOF is declared an outlier. Anomalous data points from low-density areas have higher LOF values than normal data points, typically between 1 and 1.5. The higher the LOF, the more likely the data point is an outlier.

5   Methodology of the Research

The research methodology presented in Fig. 3 was based on a comparative approach using the Local Outlier Factor (LOF) algorithm. The study aimed to investigate this algorithm's effectiveness and efficiency in detecting outliers for categorical and mixed data. The algorithm was implemented in Python to conduct the study, enabling effective outlier detection. "Effectiveness" refers to the algorithm's ability to accurately identify outliers, meaning it can precisely detect unusual or divergent values in the dataset. "Efficiency" refers to the algorithm's performance and economically achieving its goals regarding resources such as time and memory. An efficient algorithm can achieve its intended goal with minimal resource consumption.

Outliers representing 1% of the dataset were generated. These anomalous samples were created with random values for different features and then randomly replaced with real data. This approach aimed to simulate and enrich the dataset with various anomalies, aiding in testing outlier detection algorithms. The p53 mutant dataset and the Credit Card Fraud Detection dataset had already flagged anomalies, and we didn't think to generate them. The Local Outlier Factor (LOF) algorithm was implemented to calculate the most significant LOF score for each data point in the dataset. This algorithm identifies atypical observations that significantly differ from most data points in the dataset. Before applying the algorithm, the data was suitably prepared to enable effective utilization. Categorical data were transformed into binary vectors, enabling analysis with the algorithm LOF operating on numerical data. Additionally, missing values were imputed with the mean values of their respective columns. This approach preserved data integrity and avoided the impact of missing data on the analysis. Numerical data that was not categorical was scaled to the range of [0, 1]. Data scaling helps bring them to a common range, which may benefit some algorithms sensitive to differences in data scale. The p53 mutant dataset used in

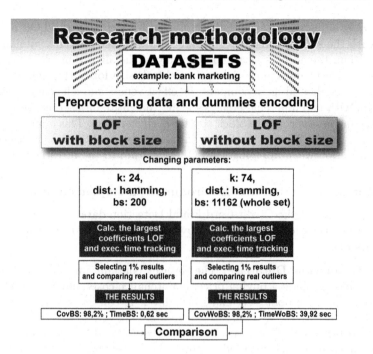

Fig. 3. Methodology of the research. Own work.

this analysis had dimensions of 5,408 features × 16,771 instances, making it challenging to handle and prone to overfitting. One hundred eighty instances with missing values in any of the columns were removed to address these issues. Additionally, feature selection was performed by constructing a correlation matrix and removing features with correlation coefficients greater than 0.5 to reduce multicollinearity. As a result, 444 features remained out of the total 5,408 features. For each run of the algorithm with block size and the entire dataset, the processing time was measured. The aim was to find a balance between effectiveness and processing time. Experiments were conducted with different parameter settings for the LOF algorithm, and the results regarding effectiveness and processing time were collected and compared. The research methodology allows for the analysis of the dataset using different parameters for the Local Outlier Factor (LOF) algorithm, such as Minpts (minimum number of points) and distance measures. The study involves two iterations - the first utilizes the entire dataset as the block size, and the second adapts an appropriate block size for each iteration, measuring effectiveness and efficiency. In the first iteration, the block size is treated as the entire dataset, and the LOF algorithm is executed with fixed parameters, such as Minpts and distance measures. Then, the results of outlier detection are analyzed, evaluating the algorithm's effectiveness in identifying atypical observations and measuring the processing time required for this task. In the second iteration, additional analyses are performed, but this time, the

634   C. Horyń and A. Nowak-Brzezińska

block size is adjusted for each dataset. For each block size, the LOF algorithm is run with the same parameters - Minpts and distance measure. Similar to the first iteration, the effectiveness and efficiency of the algorithm are measured for each block size to determine the optimal block size for outlier detection. The study's results allowed us to compare the effectiveness and efficiency of the LOF algorithm with two different settings - using the entire dataset as the block size and adapting an appropriate block size for each iteration. This analysis helps us understand how different parameters influence the quality and effectiveness of outlier detection when applying the LOF algorithm to the investigated dataset.

5.1 Dataset Selection and Environment in Implementation

All datasets, except the Credit Card Fraud Detection dataset and p53 mutants, contain categorical features. **The following datasets were chosen for our experiments:**

- **Car Evaluation Database:** The dataset contains information about car evaluations, including buying price, maintenance costs, number of doors, car capacity, luggage boot size, and safety level. It consists of 1728 rows and 6 attributes [11].
- **Mushroom Classification:** The dataset contains 8124 samples with 22 categorical features, and it is used for the task of classifying mushrooms based on their properties. Each species is labelled as edible, poisonous, or of unknown edibility [12].
- **Bank Marketing Dataset:** This dataset contains 11,162 rows and 17 features related to bank marketing. Attributes include customer information, campaign details, and outcomes, as well as whether the client subscribed to a term deposit [13].
- **CitiBike System:** The dataset provides information about bike rentals in New York City, including trip duration, rental times, station identifiers, user type, and demographic data. It consists of 577,703 rows and 15 columns. The first 20,000 rows were used for analysis and experiments [14].
- **Adult (Census Income):** The dataset contains 48,842 samples with a combination of continuous and categorical features, with 14 attributes in total. The goal is to predict whether an individual's annual income exceeds 50,000 dollars based on census data [15].
- **Credit Card Fraud Detection:** The dataset includes 284,807 records of credit card transactions. It involves numerical features resulting from PCA transformation, with a total of 31 columns. Fraudulent transactions account for only 0.172% of all transactions, indicating a significant class imbalance [16].
- **p53 Mutants:** This dataset contains 16,772 records describing various p53 mutations and their transcriptional activity. It originally had 5,408 columns representing properties of p53 mutant proteins obtained from simulations. After optimization, 444 columns remained by removing features with high correlation coefficients [17]. Read more in the introductory article: https://doi.org/10.1371/journal.pcbi.1000498.

Environment in Implementation: The algorithm implementation used the "NumPy" library for numerical and mathematical operations, a powerful tool in Python for numerical computations. It provides functions for working with arrays and matrices. The "time" module also measured the algorithm's execution time to evaluate its efficiency. The implementation also employed modules from the SciPy library for specific optimizations. The "cdist" function in SciPy efficiently calculates distances between points in n-dimensional space. Initially, we considered implementing this function in Python but later found the advantages of optimized C versions and low-level libraries. Leveraging these modules and functions improved the algorithm's efficiency, enabling complex computations on the dataset and providing valuable execution time information for the LOF algorithm.

The Local Outlier Factor (LOF) algorithm is used to detect outliers in a dataset. Below are the main steps performed by this code:

Initialization of the LOF Outlier Class: The LOF Outlier class is initialized with user-defined parameters, including MinPts (minimum number of points in the vicinity) and distance metric. **Distance Matrix Generation:** A distance matrix is computed for the input data to represent the distance between points. **Kth Nearest Neighbor Calculation:** The distance to the kth nearest neighbour is calculated for each point in the distance matrix. The parameter k specifies the number of nearest neighbours considered. **Reachability Distance Calculation:** The reachability distance for each point is determined as the maximum value between the distance to the kth nearest neighbour and the distances between points. **Finding K Nearest Neighbors:** The k nearest neighbours for each point are identified based on the distance matrix, and their indices and distances are recorded. **Local Reachability Density Calculation:** The local reachability density for each point is computed as the reciprocal of the average reachability distance to its k nearest neighbours. **Local Outlier Factor (LOF) Calculation:** The LOF value for each point is calculated using the k-indices of its nearest neighbours and the local reachability densities. The LOF quantifies the point's deviation from the rest of the dataset. **LOF Scores for Each Point:** The LOF values for all points in the dataset are computed based on the previous steps.

The block size divides the input data into smaller blocks or fragments. This approach is beneficial when dealing with large datasets because processing the entire dataset simultaneously can be time-consuming and memory-intensive. By dividing the data into blocks, we can perform the LOF algorithm on each block separately, improving performance and reducing memory consumption.

In the code, the block size is set here:

```
block_size = 20000   # example block size
```

The data is divided into blocks of size example, 20,000 using the following code:

```
data_blocks = [data[i:i+block_size]
                for i in range(0, len(data), block_size)]
```

This code creates a list called `data_blocks`, where each list element is a data block containing 20,000 records (or fewer for the last block if the total number of records is not divisible by 20,000). Using the block size can help improve the performance of the LOF algorithm, especially when the dataset is large, by processing smaller data fragments at a time. However, choosing the block size may require some experimentation based on the dataset size and available computational resources. If the block size is too small, it may result in many blocks and lead to overhead in managing them. On the other hand, if the block size is too large, it may cause memory-related issues, especially if the available memory is limited. In practice, you can try different block sizes and evaluate the performance of the LOF algorithm to find the optimal balance between computational efficiency and memory usage for your specific dataset.

5.2   The Optimization of Block Size in the LOF Algorithm for Enhanced Anomaly Detection

The LOF (Local Outlier Factor) algorithm's block size can be adjusted through experiments to improve anomaly detection. However, the ideal size varies with the dataset and analysis goals. Typically, an optimal size is around 200, but it should be fine-tuned for each specific case to achieve the best results. Optimizing centres on several key aspects: **Reducing Computational Overhead:** LOF involves computationally expensive distance calculations between data points, especially for large datasets. Optimizing the block size focuses these calculations on points within the same block, minimizing computational overhead. **Adapting to Data Characteristics:** Each dataset has its unique structure and point distribution. Adjusting the block size helps consider local data structures, improving the detection of unusual points. **Emphasizing Local Structures:** LOF identifies unusual observations within local point clusters. Optimizing the block size enables more precise analysis of these local structures, leading to more accurate anomaly detection. **Experimentally Determining the Optimal Block Size:** The optimal block size can be determined through experiments on various datasets, revealing the right number of points within a block for efficient data analysis. Formula:

$$m = \arg\min_m \left(\sum_{i=1}^{N} \sum_{j=1}^{N} I_{ij}(m) \cdot \text{dist}(x_i, x_j) \right) \tag{1}$$

where: m - number of points per block, $I_{ij}(m)$ - indicator function, which equals 1 if points x_i and x_j belong to the same block of size m and 0 otherwise, $\sum_{i=1}^{N} \sum_{j=1}^{N} I_{ij}(m) \cdot \text{dist}(x_i, x_j)$ - sum of distances between points x_i and x_j only for pairs of points within the same block of size m.

6   Experiments

See Fig. 4 for visualisation of the results. The exemplary outcomes of the search for one of the "bank marketing" datasets are presented in the graph, displaying the LOF (Local Outlier Factor) values for each data point. Detected outliers are marked as green points with a red cross, while other LOF values are represented as blue points. Outliers not detected are indicated solely by a red cross without a green point.

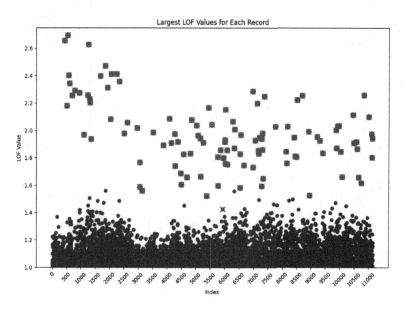

Fig. 4. The Highest LOF Values: Example Bank Marketing Dataset.
Common Indices: 109, Coverage: 98.2%, Outliers without Common Indices: {2730, 5933}, Number of Outliers without Common Indices: 2, Parameters: MinPts: 24, Distance Metric: Hamming.

Table 1 and Figs. 5 and 6 demonstrate that from the conducted experiments using the Local Outlier Factor (LOF) algorithm on different datasets, the following conclusions can be drawn: **Coverage Comparison:** The table presents a comparison of coverage percentages (CovBS and CovWoBS) for each dataset using the LOF algorithm with block size optimization and without it. Coverage percentage measures the algorithm's effectiveness in detecting outliers. In most cases, using block size optimization (BS) provides similar or slightly higher coverage compared to the LOF algorithm without optimization (WoBS). This indicates that the algorithm's performance is not significantly affected by block size optimization. **Execution Time Comparison:** The table also shows a comparison of execution times (TimeBS and TimeWoBS) for each dataset using the LOF algorithm with block size optimization and without it. It is observed that block size optimization generally significantly reduces execution time, making the algorithm more efficient in handling large datasets. However, in some cases, like the "credit card" dataset, the execution time with block size optimization is higher, likely due to specific characteristics of that dataset. **Impact of Parameters:** The table contains additional information about the parameters used in

Table 1. Results for Different Datasets and Algorithms LOF. Own work.

Dataset [rows; outlrs; feat.]	CovBS|CovWoBS%	TimeBS|TimeWoBSsec	Params BS|WoBS
bank [11162; 111; 17]	98.20 | 98.20	0.62 | 39.92	BS: 200 | 11162 k: 24 | 74 dist: hamming
adult [48842; 488; 14]	91.19 | 77.45	1.57 | 186.07	BS: 100 | 24421[a] k: 27 | 104 dist: hamming
mushroom [8124; 81; 22]	87.65 | 87.65	0.23 | 8.94	BS: 200 | 8124 k: 10 | 97 dist: cosine
car eval. [1728; 17; 6]	94.12 | 82.35	0.025 | 0.234	BS: 200 | 1728 k: 8 | 6 dist: euclidean
citiebike trips [20000; 200; 15]	98 | 78	10.33 | 3913.43	BS: 200 | 20000 k: 28 | 107 dist: hamming
credit card [284807; 492; 31]	46.14 | 47.35	252.6 | 1497.18	BS: 6500 | 35600[b] k: 250 | 250 dist: euclidean
p53 mutants [16591; 143; 444]	6.99 | 2.1	0.84 | 382.97	BS: 100 | 16591 k: 9 | 46 dist: euclidean

[a] BS: 48842 after 255 s, no RAM (51 GB insufficient)
[b] BS: 284807 after 201 s, no RAM (51 GB insufficient)

block size optimization for each dataset. These parameters include block size (BS), the number of nearest neighbours (k), and the distance metric (dist). The choice of these parameters can impact the algorithm's performance. For instance, for the "bank" dataset, a block size of 200, k = 24, and the "hamming" distance metric result in high coverage and relatively low execution time.

Resource Limitations: It is evident from the table that block size optimization was not feasible for some datasets due to resource limitations. Specifically, for the "adult" dataset, the block size had to be reduced to 100 records due to insufficient RAM, paradoxically resulting in higher coverage (91.19 %) compared to the LOF algorithm without optimization (77.45%). Similarly, for the "credit card" dataset, the block size had to be reduced to 6500 records for the same reason. Without optimization, it was possible to set the block size for the "adult" dataset to 24421 and for the "credit card" dataset to 35600. Higher values caused errors due to insufficient RAM (51 GB insufficient). In practice, block size optimization faced limitations due to available RAM, especially in the Coalb Pro+ environment with a maximum of 51 GB of RAM. Attempts to use the entire dataset as a block size were hindered by hardware constraints, leading to the adoption of the largest block size possible within the available RAM for some datasets. However, with sufficient RAM resources, adjusting the block size to match the dataset is an option, especially on more powerful computers.

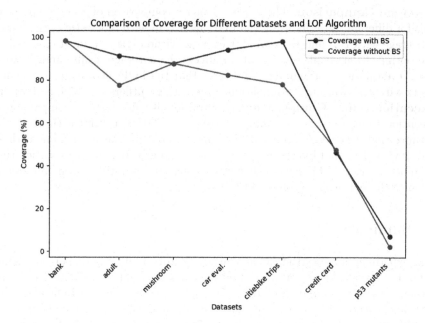

Fig. 5. Comparison of Coverage for Different Datasets and LOF Algorithm. Own work.

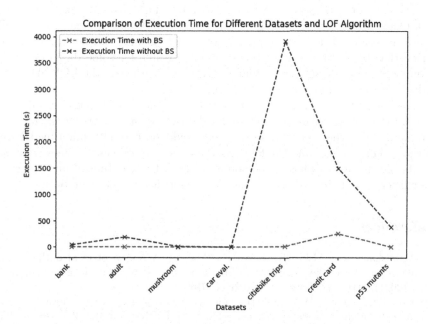

Fig. 6. Comparison of Execution Time for Different Datasets and LOF Algorithm. Own work.

Coverage Comparison: The graphs depict a comparison of coverage percent-ages for different datasets using the LOF algorithm with block size optimization (Coverage with BS) and without it (Coverage without BS). In most cases, block size optimization slightly or not at all affects the coverage results and, in some cases, even improves them. This indicates that the LOF algorithm works effec-tively with and without optimization in detecting outliers for various datasets.
Execution Time Comparison: The graphs also illustrate a comparison of execution times for different datasets using the LOF algorithm with block size optimization (Execution Time with BS) and without it (Execution Time without BS). It is evident that block size optimization significantly reduces execution time for most datasets. This means that the application of optimization can greatly enhance the performance of the LOF algorithm for large datasets.

7   Summary

The experiments using the Local Outlier Factor (LOF) algorithm on various datasets have led to several important conclusions. Firstly, comparing coverage percentages (CovBS and CovWoBS) for each dataset using the LOF algorithm with and without block size optimization (BS and WoBS, respectively) shows that block size optimization generally provides similar or slightly higher coverage. This indicates that block size optimization does not significantly compromise the algorithm's performance. Secondly, comparing execution times (TimeBS and TimeWoBS) for each dataset with and without block size optimization reveals that block size optimization significantly reduces execution time, making the algorithm more efficient, particularly for handling large datasets. It is worth noting that block size optimization was not possible for some datasets due to resource constraints, such as available RAM. In such cases, the LOF algorithm with optimization may be the only preferred choice.

In summary, the experiments highlight the benefits of block size optimiza-tion in the LOF algorithm, especially for datasets with many records. However, parameters and the choice of block size should be carefully fine-tuned based on the specific dataset and available computational resources to achieve optimal results. The results emphasize the potential of block size optimization to increase the efficiency and effectiveness of the LOF algorithm in detecting outliers.

References

1. Alghushairy, O., Alsini, R., Soule, T., Ma, X.: A review of local outlier factor algorithms for outlier detection in big data streams. Big Data Cogn. Comput. **5**, 1 (2021). https://doi.org/10.3390/bdcc5010001
2. Yu, J.X., Qian, W., Lu, H., Zhou, A.: Finding centric local outliers in categorical/numerical spaces. Knowl. Inf. Syst. **9**(3), 309–338 (2006). http://dx.doi.org/10.1007/s10115-005-0197-6
3. Taha, A., Hadi, A.S.: Anomaly detection methods for categorical data: a review. ACM Comput. Surv. **52**(2), 1–35 (2019). https://doi.org/10.1145/3312739

4. Breunig, M.M., Kriegel, H.P., Ng, R.T., Sander, J.: LOF: identifying density-based local outliers. In: SIGMOD 2000: Proceedings of the 2000 ACM SIGMOD International Conference on Management of Data, pp. 93–104 (2000). https://doi.org/10.1145/342009.335388

5. Hawkins, D.M.: Identification of Outliers. Chapman and Hall/Springer, London/-Dordrecht (1980). https://doi.org/10.1007/978-94-015-3994-4

6. Nowak-Brzezińska, A., Horyń, C.: Outliers in COVID-19 data based on rule representation - the analysis of LOF algorithm. Procedia Comput. Sci. **192**, 3010–3019 (2021). https://doi.org/10.1016/j.procs.2021.09.073. ISSN 1877-0509

7. Aggarwal, C.C.: An Introduction to Outlier Analysis. In Outlier Analysis, pp. 1–40. Springer, Heidelberg (2013). https://doi.org/10.1007/978-1-4614-6396-2

8. Sinha, A., Jana, P.K.: Efficient algorithms for local density based anomaly detection. In: Negi, A., Bhatnagar, R., Parida, L. (eds.) ICDCIT 2018. LNCS, vol. 10722, pp. 336–342. Springer, Cham (2018). https://doi.org/10.1007/978-3-319-72344-0_30

9. Su, S., et al.: An efficient density-based local outlier detection approach for scattered data. IEEE Access **7**, 1006–1020 (2019). https://doi.org/10.1109/ACCESS.2018.2886197

10. Zhang, K., Hutter, M., Jin, H.: A new local distance-based outlier detection approach for scattered real-world data. In: Theeramunkong, T., Kijsirikul, B., Cercone, N., Ho, T.-B. (eds.) PAKDD 2009. LNCS (LNAI), vol. 5476, pp. 813–822. Springer, Heidelberg (2009). https://doi.org/10.1007/978-3-642-01307-2_84

11. Bohanec, M.: Car evaluation. UCI Machine Learning Repository (1997). https://doi.org/10.24432/C5JP48, accessed 1 August 2023

12. Mushroom: UCI Machine Learning Repository (1987). https://doi.org/10.24432/C5959T. Accessed 1 Aug 2023

13. Moro, S., Rita, P., Cortez, P.: Bank marketing. UCI Machine Learning Repository (2012). https://doi.org/10.24432/C5K306, https://www.kaggle.com/datasets/janiobachmann/bank-marketing-dataset. Accessed 1 Aug 2023

14. CitiBike. https://www.citibikenyc.com/system-data. The studied set of the first 20,000 records: https://www.kaggle.com/datasets/sujan97/citibike-system-data. Accessed 1 Aug 2023

15. Becker, B., Kohavi, R.: Adult. UCI Machine Learning Repository (1996). https://doi.org/10.24432/C5XW20. Accessed 1 Aug 2023

16. Dal Pozzolo, A., Caelen, O., Le Borgne, Y.-A., Waterschoot, S., Bontempi, G.: Learned lessons in credit card fraud detection from a practitioner perspective. Expert Syst. Appl. **41**(10), 4915-4928 (2014). https://www.kaggle.com/datasets/mlg-ulb/creditcardfraud. Accessed 1 Aug 2023

17. Lathrop, R.: p53 mutants. UCI Machine Learning Repository (2010). https://doi.org/10.24432/C5T89H. Introductory Paper: Danziger, S.A., et al.: Predicting positive p53 cancer rescue regions using most informative positive (MIP) active learning. PLoS Comput. Biol. (2009). https://doi.org/10.1371/journal.pcbi.1000498. Accessed 1 Aug 2023

Author Index

A. Campagner et al. (Eds.): IJCRS 2023, LNAI 14481, pp. 643–644, 2023.
https://doi.org/10.1007/978-3-031-50959-9

Printed in the United States
by Baker & Taylor Publisher Services